Electrical

Level Four

Trainee Guide
2008 *NEC*® Revision

PEARSON

Prentice
Hall

Upper Saddle River, New Jersey
Columbus, Ohio

National Center for Construction Education and Research
President: Don Whyte
Director of Product Development: Daniele Stacey
Electrical Project Manager: Daniele Stacey
Production Manager: Tim Davis
Quality Assurance Coordinator: Debie Ness
Desktop Publishing Coordinator: James McKay
Editors: Rob Richardson, Matt Tischler, Brendan Coote

Writing and development services provided by Topaz Publications, Liverpool, NY
Lead Writer/Project Manager: Veronica Westfall
Desktop Publisher: Joanne Hart
Art Director: Megan Paye
Permissions Editors: Andrea LaBarge, Jackie Vidler
Writers: Tom Burke, Gerald Shannon, Nancy Brown, Charles Rogers

Pearson Education, Inc.
Product Manager: Lori Cowen
Product Development Editor: Janet Ryerson
Project Managers: Stephen C. Robb
Production Coordination: DeAnn Montoya, S4Carlisle Publishing Services
AV Project Manager: Janet Portisch
Senior Operations Supervisor: Pat Tonneman
Art Director: Diane Y. Ernsberger
Text Designer: Kristina D. Holmes
Cover Designer: Kristina D. Holmes
Cover Photo: Tim Davis
Director of Marketing: David Gesell
Executive Marketing Manager: Derril Trakalo
Senior Marketing Coordinator: Alicia Wozniak
Copyeditor: Sheryl Rose
Proofreader: Bret Workman

This book was set in Palatino and Helvetica by S4Carlisle Publishing Services and was printed and bound by Courier Kendallville, Inc. The cover was printed by Phoenix Color Corp.

10 9 8 7 6 5 4

Perfect bound	ISBN-13: 978-0-13-604475-8
	ISBN-10: 0-13-604475-1
Loose leaf	ISBN-13: 978-0-13-604476-5
	ISBN-10: 0-13-604476-X

Preface

Electricity powers the applications that make our daily lives more productive and efficient. Thirst for electricity has led to vast job opportunities in the electrical field. Electricians constitute one of the largest construction occupations in the United States, and they are among the highest-paid workers in the construction industry. According to the U.S. Bureau of Labor Statistics, job opportunities for electricians are expected to be very good as the demand for skilled craftspeople is projected to outpace the supply of trained electricians.

Electricians install electrical systems in structures. They install wiring and other electrical components, such as circuit breaker panels, switches, and light fixtures. Electricians follow blueprints, the *National Electrical Code*®, and state and local codes. They use specialized tools and testing equipment, such as ammeters, ohmmeters, and voltmeters. Electricians learn their trade through craft and apprenticeship programs. These programs provide classroom instruction and on-the-job training with experienced electricians.

We wish you success as you embark on your fourth year of training in the electrical craft and hope that you will continue your training beyond this textbook. There are more than 700,000 people employed in electrical work in the United States, and, as most of them can tell you, there are many opportunities awaiting those with the skills and desire to move forward in the construction industry.

NEW WITH *ELECTRICAL LEVEL FOUR*

NCCER and Pearson Education, Inc. are pleased to present the sixth edition of *Electrical Level Four*. This edition includes a new "Health Care Facilities" module that instructs trainees how to guard against power loss in hospitals where lives depend on consistent electrical service. Also, the upgraded "Motor Operation and Maintenance" module shows how proper maintenance can improve the life and service of electric motors.

The opening page of each of the 13 modules features an entry from the Solar Decathlon, an annual contest sponsored by the U.S. Department of Energy in which college and university teams compete to design, build, and operate the most attractive and energy-efficient solar-powered house. Check out these designs for a look at the future of energy-efficient housing.

We invite you to visit the NCCER website at **www.nccer.org** for the latest releases, training information, newsletter, and much more. You can also reference the Contren® product catalog online at that site. Your feedback is welcome. Email your comments to **curriculum@nccer.org** or send general comments and inquiries to **info@nccer.org**.

CONTREN® LEARNING SERIES

The National Center for Construction Education and Research (NCCER) is a not-for-profit 501(c)(3) education foundation established in 1996 by the world's largest and most progressive construction companies and national construction associations. It was founded to address the severe workforce shortage facing the industry and to develop a standardized training process and curricula. Today, NCCER is supported by hundreds of leading construction and maintenance companies, manufacturers, and national associations. The Contren® Learning Series was developed by NCCER in partnership with Pearson Education, Inc., the world's largest educational publisher.

Some features of NCCER's Contren® Learning Series are as follows:

- An industry-proven record of success
- Curricula developed by the industry for the industry
- National standardization providing portability of learned job skills and educational credits
- Compliance with the Office of Apprenticeship requirements for related classroom training *(CFR 29:29)*
- Well-illustrated, up-to-date, and practical information

NCCER also maintains a National Registry that provides transcripts, certificates, and wallet cards to individuals who have successfully completed modules of NCCER's Contren® Learning Series. *Training programs must be delivered by an NCCER Accredited Training Sponsor in order to receive these credentials.*

Special Features of This Book

In an effort to provide a comprehensive, user-friendly training resource, we have incorporated many different features for your use. Whether you are a visual or hands-on learner, this book will provide you with the proper tools to get started in the electrical industry.

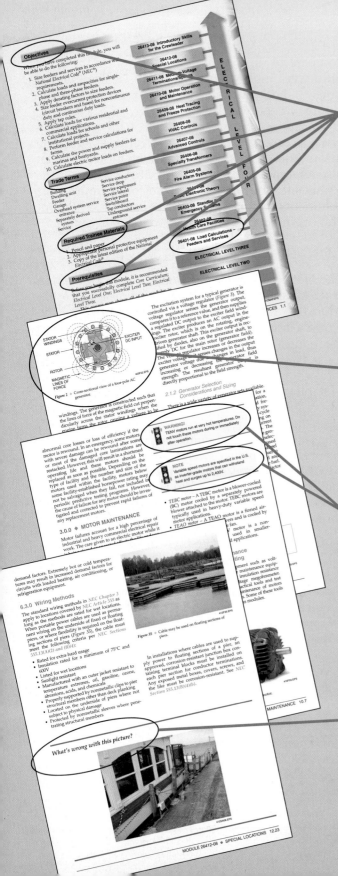

Introduction Page

This page is found at the beginning of each module and lists the Objectives, Trade Terms, Required Trainee Materials, Prerequisites, and Course Map for that module. The Objectives list the skills and knowledge you will need in order to complete the module successfully. The list of Trade Terms identifies important terms you will need to know by the end of the module. Required Trainee Materials list the materials and supplies needed for the module. The Prerequisites for the module are listed and illustrated in the Course Map. The Course Map also gives a visual overview of the entire course and a suggested learning sequence for you to follow.

Color Illustrations and Photographs

Full-color illustrations and photographs are used throughout each module to provide vivid detail. These figures highlight important concepts from the text and provide clarity for complex instructions. Each figure reference is denoted in the text in *italic type* for easy reference.

Notes, Cautions, and Warnings

Safety features are set off from the main text in highlighted boxes and are organized into three categories based on the potential danger of the issue being addressed. Notes simply provide additional information on the topic area. Cautions alert you of a danger that does not present potential injury but may cause damage to equipment. Warnings stress a potentially dangerous situation that may cause injury to you or a co-worker.

What's wrong with this picture?

What's wrong with this picture? features include photos of actual code violations for identification and encourage you to approach each installation with a critical eye.

Inside Track

Inside Track features provide a head start for those entering the electrical field by presenting technical tips and professional practices from master electricians in a variety of disciplines. Inside Tracks often include real-life scenarios similar to those you might encounter on the job site.

Think About It

Think About It features use "What if?" questions to help you apply theory to real-world experiences and put your ideas into action.

Trade Terms

Each module presents a list of Trade Terms that are discussed within the text and defined in the Glossary at the end of the module. These terms are denoted in the text with blue bold type upon their first occurrence. To make searches for key information easier, a comprehensive Glossary of Trade Terms from all modules is located at the back of this book.

Going Green

Going Green looks at ways to preserve the environment, save energy, and make good choices regarding the health of the planet. Through the introduction of new construction practices and products, you will see how the "greening of America" has already taken root.

Review Questions

Review Questions are provided to reinforce the knowledge you have gained. This makes them a useful tool for measuring what you have learned.

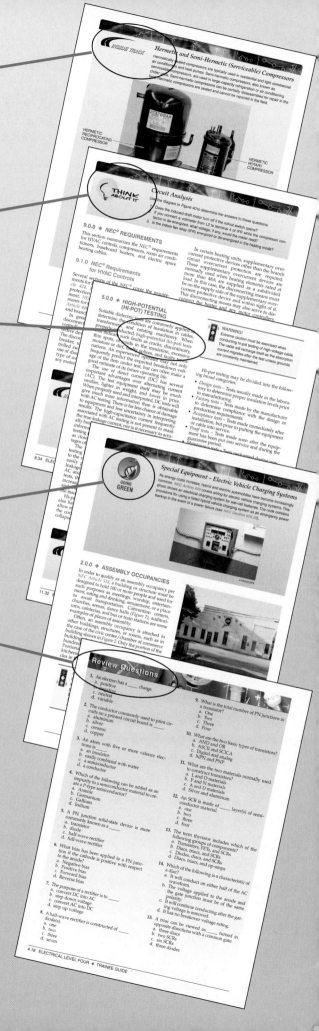

Contren® Curricula

NCCER's training programs comprise over 50 construction, maintenance, and pipeline areas and include skills assessments, safety training, and management education.

Boilermaking
Cabinetmaking
Carpentry
Concrete Finishing
Construction Craft Laborer
Construction Technology
Core Curriculum: Introductory Craft Skills
Drywall
Electrical
Electronic Systems Technician
Heating, Ventilating, and Air Conditioning
Heavy Equipment Operations
Highway/Heavy Construction
Hydroblasting
Industrial Maintenance Electrical and
    Instrumentation Technician
Industrial Maintenance Mechanic
Instrumentation
Insulating
Ironworking
Masonry
Millwright
Mobile Crane Operations
Painting
Painting, Industrial
Pipefitting
Pipelayer
Plumbing
Reinforcing Ironwork
Rigging
Scaffolding
Sheet Metal
Site Layout
Sprinkler Fitting
Welding

Pipeline

Control Center Operations, Liquid
Corrosion Control
Electrical and Instrumentation
Field Operations, Liquid
Field Operations, Gas
Maintenance
Mechanical

Safety

Field Safety
Safety Orientation
Safety Technology

Management

Introductory Skills for the Crew Leader
Project Management
Project Supervision

Spanish Translations

Andamios
Currículo Básico: Habilidades Introductorias del
    Oficio
Instalación de Rociadores Nivel Uno
Introducción a la Carpinteria
Orientación de Seguridad
Seguridad de Campo

Supplemental Titles

Applied Construction Math
Careers in Construction
Your Role in the Green Environment

Acknowledgments

This curriculum was revised as a result of the farsightedness and leadership of the following sponsors:

ABC of Iowa
ABC of New Mexico
ABC Pelican Chapter Southwest, Westlake, LA
Baker Electric
Beacon Electric Company
Cuyahoga Valley Career Center
M.C. Dean Inc.
Duck Creek Engineering
Hamilton Electric Construction Company
IMTI of New York and Connecticut

Lamphear Electric
Madison Comprehensive High School/Central Ohio ABC
Pumba Electric LLC
Putnam Career & Technical Center
Rust Constructors Inc.
TIC Industrial
Tri-City Electrical Contractors, Inc.
Trident Technical College
Vector Electric and Controls Inc.

This curriculum would not exist were it not for the dedication and unselfish energy of those volunteers who served on the Authoring Team. A sincere thanks is extended to the following:

John S. Autrey
Clarence "Ed" Cockrell
Scott Davis
Tim Dean
Gary Edgington
Tim Ely
Al Hamilton
William "Billy" Hussey
E. L. Jarrell
Dan Lamphear

Leonard R. "Skip" Layne
L. J. LeBlanc
David Lewis
Neil Matthes
Jim Mitchem
Christine Porter
Michael J. Powers
Wayne Stratton
Marcel Veronneau
Irene Ward

A final note: This book is the result of a collaborative effort involving the production, editorial, and development staff at Pearson Education, Inc., and the National Center for Construction Education and Research. Thanks to all of the dedicated people involved in the many stages of this project.

PARTNERING ASSOCIATIONS

ABC Texas Gulf Coast Chapter
American Fire Sprinkler Association
Associated Builders & Contractors, Inc.
Associated General Contractors of America
Association for Career and Technical Education
Association for Skilled & Technical Sciences
Carolinas AGC, Inc.
Carolinas Electrical Contractors Association
Center for the Improvement of Construction Management and Processes
Construction Industry Institute
Construction Users Roundtable
Design Build Institute of America
Electronic Systems Industry Consortium
Merit Contractors Association of Canada
Metal Building Manufacturers Association
NACE International

National Association of Minority Contractors
National Association of Women in Construction
National Insulation Association
National Ready Mixed Concrete Association
National Systems Contractors Association
National Technical Honor Society
National Utility Contractors Association
NAWIC Education Foundation
North American Crane Bureau
North American Technician Excellence
Painting & Decorating Contractors of America
Portland Cement Association
SkillsUSA
Steel Erectors Association of America
U.S. Army Corps of Engineers
University of Florida
Women Construction Owners & Executives, USA

Contents

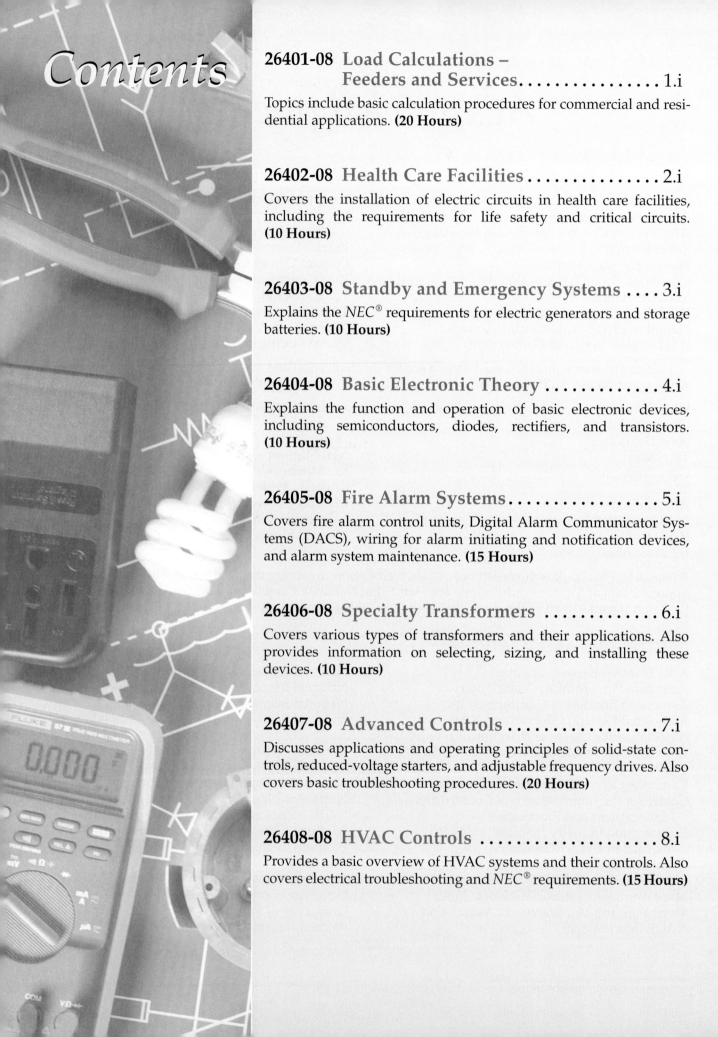

26401-08 Load Calculations –
Feeders and Services 1.i

Topics include basic calculation procedures for commercial and residential applications. **(20 Hours)**

26402-08 Health Care Facilities 2.i

Covers the installation of electric circuits in health care facilities, including the requirements for life safety and critical circuits. **(10 Hours)**

26403-08 Standby and Emergency Systems 3.i

Explains the *NEC®* requirements for electric generators and storage batteries. **(10 Hours)**

26404-08 Basic Electronic Theory 4.i

Explains the function and operation of basic electronic devices, including semiconductors, diodes, rectifiers, and transistors. **(10 Hours)**

26405-08 Fire Alarm Systems 5.i

Covers fire alarm control units, Digital Alarm Communicator Systems (DACS), wiring for alarm initiating and notification devices, and alarm system maintenance. **(15 Hours)**

26406-08 Specialty Transformers 6.i

Covers various types of transformers and their applications. Also provides information on selecting, sizing, and installing these devices. **(10 Hours)**

26407-08 Advanced Controls 7.i

Discusses applications and operating principles of solid-state controls, reduced-voltage starters, and adjustable frequency drives. Also covers basic troubleshooting procedures. **(20 Hours)**

26408-08 HVAC Controls 8.i

Provides a basic overview of HVAC systems and their controls. Also covers electrical troubleshooting and *NEC®* requirements. **(15 Hours)**

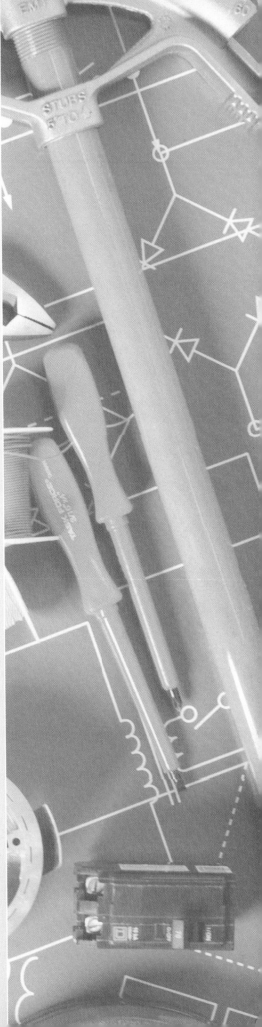

26409-08 Heat Tracing and Freeze Protection... 9.i

Covers various heat tracing systems along with their applications and installation requirements. **(10 Hours)**

26410-08 Motor Operation and Maintenance 10.i

Covers motor cleaning, testing, and preventive maintenance. Also describes basic troubleshooting procedures. **(10 Hours)**

26411-08 Medium-Voltage Terminations/Splices............... 11.i

Offers an overview of the *NEC®* and cable manufacturers' requirements for medium-voltage terminations and splices. **(10 Hours)**

26412-08 Special Locations.................. 12.i

Describes the *NEC®* requirements for selecting and installing equipment, enclosures, and devices in various special locations including places of assembly, theaters, carnivals, agricultural buildings, marinas, temporary installations, wired partitions, and swimming pools. **(20 Hours)**

26413-08 Introductory Skills for The Crew Leader............... 13.i

Teaches the basic leadership skills required to supervise personnel. Discusses principles of project planning, scheduling, estimating, and management, and presents several case studies for student participation. **(10 Hours)**

Glossary of Key Terms G.1

Figure Credits... FC.1

Index I.1

Load Calculations – Feeders and Services

Carnegie Mellon University – Plug and Play House

Much like adding a new "plug and play" device to a computer, the Carnegie Mellon home can be upgraded with smaller or larger rooms. All the rooms are arranged around the home's central core, which contains all the home's mechanical systems. Connections to mechanical supports are installed in the core and are easily accessible and adjustable.

26401-08

26401-08

Load Calculations – Feeders and Services

Topics to be presented in this module include:

1.0.0   Introduction 1.2

2.0.0   Basic Calculation Procedures 1.2

3.0.0   Load Calculations for a Minimum Size Service 1.13

4.0.0   Commercial Occupancy Calculations 1.24

5.0.0   Restaurants 1.31

6.0.0   Optional Calculation for New Restaurants 1.33

7.0.0   Services for Hotels and Motels 1.33

8.0.0   Optional Calculations for Schools 1.33

9.0.0   Shore Power Circuits for Marinas and Boatyards 1.34

10.0.0   Farm Load Calculations 1.34

11.0.0   Motors and Motor Circuits 1.36

Overview

Rules and regulations pertaining to branch circuit, feeder, and service calculations are covered in *National Electrical Code Article 220*. Load calculations must be performed and documented before service equipment, panelboards, circuit breakers, or conductors are installed because all of these electrical components depend on the calculated loads. Calculating the connected load is not as simple as listing the amperage ratings found on equipment labels. For instance, general lighting loads must be calculated based on the square footage of occupied space. General lighting loads include both general-purpose receptacles and lighting.

The general lighting load is added to the total load calculations for all other equipment. In residential occupancies, these loads include small appliances, HVAC equipment, cooking appliances, laundry room appliances, and any other individual load not included in the general lighting load calculation. Total loads for commercial and industrial facilities are calculated in a similar manner. However, demand factors may be applied to reduce the total nameplate loads in locations such as apartment complexes, commercial laundry facilities, and commercial kitchens because not all of the equipment will be used at the same time.

Objectives

When you have completed this module, you will be able to do the following:

1. Size feeders and services in accordance with *National Electrical Code®* (*NEC®*) requirements.
2. Calculate loads and ampacities for single-phase and three-phase feeders.
3. Apply derating factors to size feeders.
4. Size feeder overcurrent protection devices (circuit breakers and fuses) for noncontinuous duty and continuous duty loads.
5. Apply tap rules.
6. Calculate loads for various residential and commercial applications.
7. Calculate loads for schools and other institutional projects.
8. Perform feeder and service calculations for farms.
9. Calculate the power and supply feeders for marinas and boatyards.
10. Calculate electric motor loads on feeders.

Trade Terms

Building
Dwelling unit
Feeder
Garage
Overhead system service entrance
Separately derived system
Service
Service conductors
Service drop
Service equipment
Service lateral
Service point
Switchboard
Tap conductors
Underground service entrance

Required Trainee Materials

1. Pencil and paper
2. Appropriate personal protective equipment
3. Copy of the latest edition of the *National Electrical Code®*

Prerequisites

Before you begin this module, it is recommended that you successfully complete *Core Curriculum; Electrical Level One; Electrical Level Two; Electrical Level Three*.

This course map shows all of the modules in *Electrical Level Four*. The suggested training order begins at the bottom and proceeds up. Skill levels increase as you advance on the course map. The local Training Program Sponsor may adjust the training order.

ELECTRICAL LEVEL FOUR

26413-08 Introductory Skills for the Crewleader

26412-08 Special Locations

26411-08 Medium-Voltage Terminations/Splices

26410-08 Motor Operation and Maintenance

26409-08 Heat Tracing and Freeze Protection

26408-08 HVAC Controls

26407-08 Advanced Controls

26406-08 Specialty Transformers

26405-08 Fire Alarm Systems

26404-08 Basic Electronic Theory

26403-08 Standby and Emergency Systems

26402-08 Health Care Facilities

26401-08 Load Calculations – Feeders and Services

ELECTRICAL LEVEL THREE

ELECTRICAL LEVEL TWO

ELECTRICAL LEVEL ONE

CORE CURRICULUM: Introductory Craft Skills

401CMAP.EPS

1.0.0 ◆ INTRODUCTION

The purpose of **feeder** circuit load calculations is to determine the size of feeder circuit overcurrent protection devices and feeder circuit conductors using *NEC®* requirements. Once the feeder circuit load is accurately calculated, feeder circuit components may be sized to serve the load safely. Feeder circuits typically supply transformers, motor control centers, and subpanels. The load calculations and methods presented in this module follow the same pattern and use the same equations that were presented in the *Branch and Feeder Circuits* module in Level Three. Some of the information presented previously will be reviewed in this module; however, it is strongly suggested that you review *Branch and Feeder Circuits* prior to proceeding with this module.

NEC Article 215 covers feeder circuits. *NEC Section 215.2* states that feeders shall have an ampacity that is no less than that required to supply the load, as computed in *NEC Article 220, Parts III, IV, and V.* The minimum feeder-circuit conductor size, before the application of any adjustment or correction factors, shall have an allowable ampacity of not less than 100% of the noncontinuous load plus 125% of the continuous load.

2.0.0 ◆ BASIC CALCULATION PROCEDURES

The *NEC®* establishes minimum standards for electrical calculations. Since most electrical exams are based upon the minimum *NEC®* requirements, this module will focus on those requirements. Common practice in the electrical industry is to design and plan electrical systems to make allowance for future expansion. Providing conduit and raceways larger than the minimum required by the *NEC®* may reduce the labor required in pulling conductors, which may offset the higher cost of the larger conduit. **Services** are frequently sized larger than the minimum required by the *NEC®* so that future expansion of the electrical system will be possible.

It should be stressed once more that the journeyman and master electrician examinations are in most cases based upon calculations using *NEC®* minimum requirements, without an allowance for future expansion or ease of installation.

2.1.0 Load Calculations – Basic Considerations

Electrical calculations can be divided into three sections:

* Branch circuits
* Feeders
* Services

The total load connected to the branch circuits determines the load on feeders, and the total feeder load determines the load on the **service conductors** and equipment. When calculating loads, the logical sequence is to determine the load on branch circuits, calculate the feeder load, and then determine the service load. Branch circuit calculations were covered in your Level Three training, and you are strongly urged to review that material before proceeding because the information presented in this module is based upon the calculations developed using branch circuit calculations.

2.2.0 Conductor Adjustments

The allowable current-carrying capacity (ampacity) of conductors may require adjustment due to several conditions. After complying with *NEC Section 110.14(C)*, the conductors selected may have to be increased in size to accommodate the altered allowable ampacity for that conductor. Feeder conductor size is adjusted when:

* The load to be supplied is a continuous duty load per *NEC Section 215.2(A)(1)*.
* There are more than three current-carrying conductors in a raceway per *NEC Table 310.15(B)(2)(a)*.
* The ambient temperature that the conductors will pass through exceeds the temperature ratings for conductors listed in *NEC Table 310.16* correction factors.
* Although not mandatory, an increase in conductor size may be made when there is a voltage drop exceeding 3% for feeders or 5% for the combination of the feeder and branch circuit that was caused by excessive voltage drop per *NEC Section 215.2(A)(3), FPN No. 2.*

2.2.1 Continuous Duty, Raceway Fill, and Ambient Temperature

Feeder conductors are adjusted when the load or a part of the load connected to the feeder is a continuous duty load; that is, it operates without interruption for more than three hours consecutively. *NEC Table 310.16* lists allowable ampacities for insulated conductors rated at 0 to 2,000V. The ampacities listed apply when no more than three current-carrying conductors are installed in a raceway or are part of a cable assembly. The ampacities are based upon an ambient temperature of 30°C (86°F). If the number of current-carrying conductors in the conduit or cable exceeds three, the ampacity of the conductors must be adjusted using *NEC Table 310.15(B)(2)(a)*. If the ambient temperature exceeds 30°C (86°F), the ampacity of the conductors must be adjusted using the temperature correction factors listed at the bottom of *NEC Table 310.16*.

When derating conductors, it is permitted to use the allowable ampacity listed for the temperature rating of the conductor instead of the selection ampacity, which is restricted to the 60°C or 75°C column. For example, the minimum size of THHN copper permitted to supply a load of 125A is No. 1 AWG copper, based on the 75°C column of *NEC Table 310.16*. When derating this conductor, the ampacity used will be the 150A, as shown in the 90°C column.

NOTE

The selection ampacity for choosing a conductor to carry a load is taken from either the 60°C or 75°C column of *NEC Table 310.16* as required by *NEC Section 110.14(C)*. When determining the ability of that conductor to carry the load under adverse conditions, the allowable ampacity of the conductor is used based on its actual temperature rating.

The size of a conductor may also be increased when the length of the feeder conductor requires a larger circular mil area to ensure a low enough resistance to allow the needed current to flow.

Example 1:

What is the adjusted ampacity for each of six No. 4/0 THHN copper current-carrying conductors in a single conduit?

Per *NEC Table 310.16*, the ampacity of No. 4/0 THHN is 260A. Per *NEC Table 310.15(B)(2)(a)*, the ampacity must be adjusted by 80% (four through six current-carrying conductors).

$$260A \times 80\% = 208A \text{ for each conductor}$$

Example 2:

What is the maximum adjusted ampacity for each of nine No. 3/0 current-carrying THWN copper conductors in a single conduit?

Per *NEC Table 310.16*, the ampacity of No. 3/0 THWN is 200A. Per *NEC Table 310.15(B)(2)(a)*, the ampacity must be adjusted by 70% (seven through nine current-carrying conductors).

$$200A \times 70\% = 140A \text{ for each conductor}$$

Example 3:

A feeder is being installed to supply a load of 375A. The feeder must be routed through a room with an ambient temperature of 98°F. Will 500 kcmil THHN copper conductors be able to carry the load under these conditions?

Per *NEC Table 310.16*, 500 kcmil THHN conductors have an ampacity of 430A. Per the correction factors of *NEC Table 310.16*, the ampacity of a type THHN conductor must be multiplied by a factor of 0.91 to account for the ambient temperature.

$$430A \times 0.91 = 391.3A$$

The 500 kcmil THHN conductors will meet the requirements of the *NEC®*.

Example 4:

What is the minimum size THHN copper conductor required for a feeder supplying a continuous load of 85A and a noncontinuous load of 105A while in an ambient temperature of 34°C and with a total of four current-carrying conductors in the same raceway?

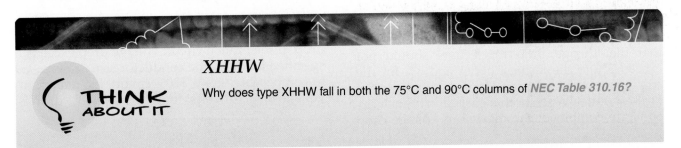

XHHW

Why does type XHHW fall in both the 75°C and 90°C columns of *NEC Table 310.16*?

THINK ABOUT IT

To select a feeder conductor, multiply the continuous load by 125%:

$$85A \times 125\% = 106.25A$$

Add the noncontinuous load:

$$106.25A + 105A = 211.25A$$

Per *NEC Table 310.16*, No. 4/0 THHN is selected to carry the load of 211.25A using the 75°C column.

The connected load before any adjustments are made is:

$$85A + 105A = 190A$$

Per *NEC Table 310.16*, the ampacity of a type THHN conductor when in an ambient temperature of 35°C must be multiplied by a factor of 0.96. The allowable ampacity of No. 4/0 THHN copper is 260A.

$$260A \times 0.96 = 249.6A$$

Per *NEC Table 310.15(B)(2)(a)* the allowable ampacity of a conductor run in a raceway with four to six current-carrying conductors must be reduced by 80%.

$$249.6A \times 80\% = 199.68A$$

The No. 4/0 THHN conductors at 199.68A will be able to carry the load of 190A under these adverse conditions. This conductor satisfies the requirements of *NEC Section 215.2(A)(1)*.

2.2.2 Voltage Drop

NEC Section 215.2(A)(3), FPN No. 2 recommends that the voltage drop for feeders should not exceed 3% to the farthest outlet and 5% for the combination of feeders and branch circuits to the farthest outlet. When both feeders and branch circuits are combined, the maximum voltage drop allowed for the branch circuits is usually 3%, then the voltage drop allowed for the feeder can be no more than 2%. There are two formulas used to calculate voltage drop for single-phase circuits with resistive loads (power factor = 1.0).

Formula 1:

$$VD = \frac{2 \times L \times R \times I}{1,000}$$

Formula 2:

$$VD = \frac{2 \times L \times K \times I}{CM}$$

Where:

L = one-way length in feet

R = conductor resistance
  (NEC Chapter 9, Table 8)

K = constant (12.9 for copper conductors or 21.2 for aluminum conductors)

I = load (in amps)

CM = conductor size in circular mils
  (NEC Chapter 9, Table 8)

VD = voltage drop (in volts)

To calculate the percentage of voltage drop, divide the voltage drop (VD) by the applied voltage (V) and multiply by 100:

$$\text{Percent voltage drop} = (VD \div V) \times 100$$

The result is divided by circuit voltage and then multiplied by 100 to determine the % voltage drop. If the branch circuit voltage drop exceeds a specified amount, the conductor size is increased to compensate. For a 120V circuit, the maximum voltage drop (in volts) for a permitted 3% drop is:

$$120V \times 3\% = 3.6V$$

The K Factor

To find the K factor, multiply the resistance per foot of the conductor (found in *NEC Chapter 9, Table 8*) by the circular mils of the conductor and divide by 1,000. For example, No. 6 AWG = 26,240 CM × 0.491 = 12,883.84 ÷ 1,000 = 12.88 (round to 12.9).

For a 240V, single-phase (Ø) feeder combined with branch circuits, the maximum voltage drop (in volts) would normally be:

$$240V \times 5\% = 12V$$

In commercial fixed load applications, if the feeder load is not concentrated at the end of the feeder but is spread out along the feeder due to multiple taps, the load center length of the feeder should be calculated and used in the voltage drop formulas. This is because the total current does not flow the complete length of the feeder. If the full length is used in computing the voltage drop, the drop determined will be greater than what would actually occur. The load center length of a feeder is that point in the feeder where, if the load were concentrated at that point, the voltage drop would be the same as the voltage drop to the farthest load in the actual feeder. To determine the load center length of a feeder with multiple taps, as shown in *Figure 1*, multiply each tap load by its actual physical routing distance from the supply end of the feeder. Add these products for all loads fed from the feeder and divide this sum by the sum of the individual loads. The resulting distance is the load center length (L) for the total load (I) of the feeder.

Example 1:

The length of a 240V, three-wire panel feeder is 145'. The total panel load consists of 180A of non-continuous duty loads. No. 3/0 THHN copper conductors would normally be selected to supply this load. If they are used, will the voltage drop for this panel feeder exceed 2%?

Use the first voltage drop formula to determine the voltage drop for the feeder. Look up the resistance for No. 3/0 AWG copper in *NEC Chapter 9, Table 8.* The DC resistance is 0.0766Ω/ft. Using the voltage drop formula:

$$2 \times 145' \times 0.0766\Omega/ft \times 180A \div 1,000 = 4V$$
$$4V \div 240V = 0.0167V \times 100 = 1.67\%$$

Since this percentage does not exceed 2%, the answer to the question is no, the voltage drop for this feeder using No. 3/0 AWG conductors will not exceed the recommended value. Note that in

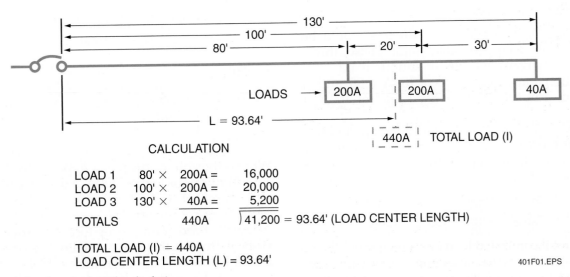

Figure 1 ◆ Load center length calculation.

this example, the second formula could have been used, if desired. Either formula will yield similar results.

Example 2:

What size THHN copper conductors would be required for a continuous feeder load of 180A, 208V, 1Ø for fixed utilization equipment? The total length of the feeder is 318'.

Since this is a continuous load, the load is multiplied by 125% to determine the feeder ampacity.

$$180A \times 125\% = 225A$$

This feeder will require a No. 4/0 THHN copper conductor to supply the continuous load. The connected load, however, is 180A, and this is the amount of current that will be used in our voltage drop calculations.

Use the second voltage drop formula as an example to determine the voltage drop for this feeder. Look up the area in CM for a No. 4/0 THHN copper conductor in *NEC Chapter 9, Table 8.* Using the voltage drop formula:

$$2 \times 318' \times 12.9 \times 180A \div 211,600 = 6.98V$$
$$6.98V \div 208V = 0.0336 \times 100 = 3.36\%$$

This percentage exceeds 3%; therefore, No. 4/0 AWG copper would not meet the *NEC®* recommendations for voltage drop. Using the area in CM for 250 kcmil (250,000 CM) conductors, the formula becomes:

$$2 \times 318' \times 12.9 \times 180A \div 250,000 = 5.91V$$
$$5.91V \div 208V = 0.0284 \times 100 = 2.84\%$$

No. 250 kcmil conductors would meet the *NEC®* voltage drop recommendations for this feeder.

Example 3:

What size THHN copper conductors would be required for a tapped 240V, single-phase, three-wire feeder consisting of continuous fixed 240V utilization equipment loads of 180A at 30' and at the end, a noncontinuous fixed load of 100A at 450'?

Since there is one continuous load tap and one noncontinuous load tap, the minimum ampacity required for the feeder conductors is computed as follows:

$$Tap\ 1 = 180A \times 125\% = 225A$$
$$Tap\ 2 = 100A \times 100\% = 100A$$

The total computed load of the feeder = 225A + 100A = 325A. The minimum conductor size required for this feeder would be 400 kcmil copper per *NEC Table 310.16.*

To determine the conductor size that will prevent excessive voltage drop, use the load center length calculation. Multiply the loads by the distance from the feeder source, and then divide the sum of the two products by the sum of the two loads as follows:

$$Tap\ 1 = 30' \times 180A = 5,400$$
$$Tap\ 2 = 450' \times 100A = 45,000$$

Sum the products:

$$5,400 + 45,000 = 50,400$$

Divide by the sum of the loads:

$$180A + 100A = 280A$$
$$50,400 \div 280 = 180'\ for\ the\ load\ center$$

Using the second single-phase voltage drop formula, substitute values using the load center length, 280A, and 400 kcmil (400,000 circular mils) to check the voltage drop:

$$2 \times 180' \times 12.9 \times 280A \div 400,000\ CM = 3.25V$$

The maximum recommended voltage drop is 240V × 3% or 7.2V; therefore, 400 kcmil conductors are more than adequate. Note that if the full length of the feeder (450') had been used in this example instead of the load center length, the voltage drop would be 8.1V and 500 kcmil conductors would have to be selected, which is excessive for this application.

Voltage drop for balanced three-phase circuits (with negligible reactance and a power factor of 1.0) can be calculated using the two voltage drop formulas by substituting $\sqrt{3}$ (1.732) for the value of 2 in the first two formulas, which results in the following formulas:

Formula 3:

$$VD = \frac{\sqrt{3} \times L \times R \times I}{1.000}$$

Formula 4:

$$VD = \frac{\sqrt{3} \times L \times K \times I}{CM}$$

Example 4:

What is the voltage drop of a 208V, 3Ø feeder with a noncontinuous load of 230A, a distance from the circuit breaker to the load of 315', and using No. 4/0 copper conductors?

Use the third voltage drop formula to determine voltage drop for this 3Ø feeder. Look up the resistance for No. 4/0 copper conductor in *NEC Chapter 9, Table 8.* Using the voltage drop formula:

$$\sqrt{3} \times 315' \times 0.0608 \times 230A \div 1,000 = 7.63V$$

The maximum voltage drop permitted for a 208V circuit is:

$$208V \times 3\% = 6.24V$$

Since, in this problem, 7.63V is more than the maximum allowable voltage drop, this circuit will not meet the minimum *NEC®* recommendation for voltage drop, and the conductor size should be increased to compensate for the voltage drop.

Example 5:

What size THHN copper conductors would be required for a panelboard used to supply continuous nonlinear loads of 100A for a 208V, three-phase, four-wire feeder circuit with a length of 150' from the source to the panelboard?

Since the feeder is supplying nonlinear loads, the neutral wire becomes a current-carrying conductor per *NEC Section 310.15(B)(4)(c),* and therefore, there are four current-carrying conductors. This will require the ampacity of the wire to be adjusted by 80% per *NEC Table 310.15(B)(2)(a):*

Actual current = rated current × derating %

Since the actual current is known, solve the equation for the rated current:

Actual current ÷ derating % = rated current

Substituting values yields:

$$100A \div 80\% = 125A$$

With the computed value of 125A, the minimum size for the THHN copper conductor per the 90°C column of *NEC Table 310.16* is No. 2. However, because this is a feeder conductor, it must also meet the requirements of *NEC Section 215.2(A)(1).* This requires a minimum ampacity of 125% for continuous loads before the application of any adjustment factors. Multiply the actual load of 100A by 125% to obtain 125A for the required conductor ampacity. Since the actual load is 100A, you are limited to ampacity values of 60°C per *NEC Section 110.14(C).* Refer to *NEC Table 310.16* and select No. 1/0 copper based on the ampacity requirements of 125A.

Apply the fourth formula, substituting values (using the actual connected load in amps) to check that the voltage drop does not exceed the recommended 3% of 208V, or 6.24V.

$$\sqrt{3} \times 150' \times 12.9 \times 100A \div 105,600 \text{ CM} = 3.17V$$

The No. 1/0 AWG conductors will be adequate for this application.

2.2.3 Temperature Limitations for Terminating Conductors

NEC Section 110.14(C) includes temperature limitations for terminating conductors. The ampacity rat-

ing of conductors must be selected and coordinated so that the ampacity rating of the conductor does not exceed the temperature rating of any component in the feeder circuit. If the feeder size is smaller than 100A, the 60°C rating for the feeder conductor must be used unless all feeder components are rated and listed for use at a higher temperature. If the feeder size is larger than 100A, the 75°C rating for the feeder conductor must be used unless all feeder components are rated and listed for use at a higher temperature. For example, if the termination rating on a 125A fuse or circuit breaker is 75°C, type THHN conductors may be used for the feeder; however, the 75°C rating for the conductor size must be used. Conductors with higher temperature ratings may be used for ampacity adjustment, correction, or both. As you work through this module, please keep in mind that, although type THHN conductors are often used in example problems, the limitations of *NEC Section 110.14(C)* still apply.

Example:

A 200A feeder is required to serve a new panelboard. The terminals in the panelboard are rated at 75°C, and the terminals in the fusible switch protecting the feeder are rated at 60°C. What is the minimum size THHN copper conductor required to supply the panelboard?

Per *NEC Table 310.16,* No. 2/0 THHN is rated at 195A and *NEC Section 240.4(B)* would permit connection to the 200A fuse. However, because the fuse terminals are rated at 60°C, the rating must be taken from the 60°C column. The minimum allowable size taken from the 60°C column for the 200A load is No. 4/0. The answer is No. 4/0 THHN copper conductor.

2.3.0 Calculating Feeder Ampacity

Feeder ampacity for single-phase circuits (with negligible reactance and a power factor of 1.0) is calculated by dividing VA by the circuit voltage:

$$I = \frac{VA}{V}$$

For example, the ampacity of a single-phase 240V load rated at 21,600VA is determined by dividing 21,600VA by 240V.

$$21,600VA \div 240V = 90A$$

Feeder ampacity for balanced three-phase circuits (wye or delta with negligible reactance and a power factor of 1.0) is calculated by dividing VA by the line-to-line circuit voltage times $\sqrt{3}$.

$$I = \frac{VA}{V \times \sqrt{3}}$$

For example, the ampacity of a 208V, three-phase load rated at 15,200VA is determined by dividing 15,200VA by (208V × √3).

$$15{,}200VA \div (208V \times \sqrt{3}) = 42.19A$$

The ampacity of a 144kVA three-phase load at 480V is 173.21A.

$$144{,}000VA \div (480V \times \sqrt{3}) = 173.21A$$

Example 1:

What is the ampacity of a single-phase resistive load with a nameplate rating of 42.5kW, 240V?

Multiply 42.5kW by 1,000 to determine VA (watts = volt-amperes) and then divide the result by 240V.

$$42.5kW \times 1{,}000 = 42{,}500VA \div 240V = 177.08A$$

The ampacity of this load is 177.08A.

Example 2:

What is the ampacity of a three-phase panel feeder with a calculated load of 135kVA, 208V?

Multiply 135kVA by 1,000 and divide the result by 208V × √3.

$$135kVA \times 1{,}000 = 135{,}000kVA \div (208V \times \sqrt{3})$$
$$= 374.72A$$

Example 3:

What size THWN copper conductors are required to supply the primary of a 75kVA, 3Ø transformer? The primary line-to-line voltage is 480V.

Multiply 75kVA by 1,000 to determine total VA and divide the result by 480V × √3 to determine the transformer primary current. Using *NEC Table 310.16*, determine the conductor size:

$$75kVA \times 1{,}000 = 75{,}000VA \div (480V \times \sqrt{3})$$
$$= 90.21A$$

Per *NEC Table 310.16*, No. 3 THWN copper (rated at 100A) is required if both the transformer and the panel are marked for 75°C terminations. If not, then No. 1 AWG copper would be required.

Example 4:

What is the secondary current for a 1,500kVA three-phase transformer with a voltage rating of 480V/277Y?

Multiply 1,500kVA by 1,000 to determine total VA and divide the result by 480V × √3 to determine the secondary current.

$$1{,}500kVA \times 1{,}000 = 1{,}500{,}000VA \div (480V \times \sqrt{3})$$
$$= 1{,}804.22A$$

The secondary current for this transformer = 1,804.22A.

2.4.0 Tap Rules

When the overcurrent protection for a conductor is located at the load end instead of at the supply side of a conductor, the *NEC®* generally considers that to be a **tap conductor**. This means that almost all secondary sides of transformers are classified as taps, since there is no overcurrent device within the transformer itself. The most frequently used tap rules are the 10' tap rule and the 25' tap rule *(NEC Section 240.21)*.

2.4.1 The 10' Tap Rule

Generally, smaller conductors may be tapped from larger conductors, providing that the total length of the tap conductor does not exceed 10'. However, there are very specific *NEC®* requirements for each tap rule, including the 10' tap rule *[NEC Section 240.21(B)(1)]*. It is important to note that, when applying the 10' tap rule, the total length of the tap conductors (from their source of supply to their termination point in an overcurrent device) cannot exceed 10'. The *NEC®* requirements for the 10' tap rule state that, in order for this rule to be applied, all of the following conditions must be met:

- The total length of the tap conductors must not exceed 10'.
- The tap conductors must be rated at no less than the calculated load(s) to be served or the device to which the tap conductors are connected.
- The tap conductors cannot extend beyond the **switchboard**, panel, or control devices which they supply.
- The tap conductors must be enclosed in a raceway.
- The rating of the line side overcurrent device supplying the tap conductors shall not exceed 1,000% of the rating of the tap conductors.

2.4.2 The 25' Tap Rule

NEC Section 240.21(B)(2) lists specific conditions for the application of the 25' tap rule. It is important to note that, when applying the 25' tap rule, the total length of the tap conductors (from their source of supply to their termination point in an overcurrent device) cannot exceed 25'. The *NEC®* requirements for the 25' tap rule state that, in order for this rule to be applied, all of the following conditions must be met:

- The total length of the tap conductors must not exceed 25'.
- The tap conductors must be rated at no less than one-third the capacity of the overcurrent device protecting the feeder.

Transformers

Transformer secondary conductors are considered taps and as such are limited in length to no more than 10' or 25' (depending on the application) when located inside buildings. See *NEC Section 240.21(C)*.

- The tap conductors must terminate in a single circuit breaker or a single set of fuses that will limit the load to the conductor rating.
- The tap conductors must be enclosed in a raceway or otherwise suitably protected from physical damage.

2.4.3 Taps Supplying a Transformer– Primary and Secondary Not Over 25'

NEC Section 240.21(B)(3) lists specific conditions for the application of the transformer 25' tap rule. It is important to note that, when applying the 25' tap rule, the total length of the tap conductors including one primary conductor and one secondary conductor (from their source of supply to their termination point in an overcurrent device) cannot exceed 25'. The requirements of this rule apply only when the primary conductors that supply a transformer are not protected by an overcurrent device at their rated ampacity. The *NEC®* requirements for this transformer 25' tap rule state that, in order for this rule to be applied, all of the following conditions must be met:

- The primary conductors must have an ampacity that is at least one-third the capacity of the overcurrent device protecting the primary conductors.
- Use the ratio of the primary to the secondary voltage multiplied by one-third the size of the overcurrent device protecting the feeder conductors.
- The total length of the primary and secondary conductors must not exceed 25'.
- Both primary and secondary conductors must be protected from physical damage by being enclosed in an approved raceway or other approved means.
- The secondary conductors must terminate in a single circuit breaker or set of fuses that limit the load per *NEC Section 310.15.*

2.4.4 The 100' Tap Rule

Use of the 100' tap rule is limited to installations in high-bay manufacturing **buildings** with wall heights over 35'. *NEC Section 240.21(B)(4)* lists specific conditions for the application of the 100' tap rule. This tap rule may be used in buildings where only qualified persons will service and maintain the systems. The length of this tap is limited to 25' horizontally, and the total length of the tap cannot exceed 100'. In addition, the *NEC®* requires that the following conditions be met:

- The ampacity of the tap conductors must be at least one-third the ampacity of the feeder overcurrent protection device.
- The tap conductors must terminate in a single circuit breaker or set of fuses that will limit the load to the conductor ratings.
- The tap conductors must be protected from physical damage.
- The tap conductors must not be spliced.
- The tap conductors must be a minimum size of No. 6 copper or No. 4 aluminum.
- The tap conductors must not penetrate walls, floors, or ceilings.
- The tap to the feeder must be made at least 30' above the floor.

2.4.5 Outside Taps of Unlimited Length

NEC Section 240.21(B)(5) states that where the conductors are located outside of a building, except at the point of load termination, there is no limit to the length of the tap conductors provided that all of the following conditions are met:

- The conductors are suitably protected from physical damage in an approved manner.
- The conductors terminate in a single circuit breaker or set of fuses that limit the load to the ampacity of the conductors. Additional loads may be supplied from the breaker or set of fuses.

- The overcurrent device for the conductors is an integral part of the disconnecting means or is located immediately adjacent to the disconnecting means.
- The disconnecting means is installed at a readily accessible location complying with one of the following:
  - Outside of a building or structure
  - Inside, nearest the point of entrance of the conductors
  - Where installed in accordance with *NEC Section 230.6,* nearest the point of entrance of the conductors

2.4.6 Transformer Secondary Conductors

NEC Section 240.21(C) provides the installation and length requirements for the secondary conductors of transformers. Although there are many similarities in these tap rules, there are a few differences. The 10' tap rule for a transformer secondary requires that the ampacity of the secondary conductors be at least 10% of the rating of the overcurrent device on the primary side multiplied by the primary to secondary transformer voltage ratio. The outside taps of unlimited length for transformers are similar to the tap rules previously discussed. To install secondary conductors longer than 10' inside buildings, the requirements in *NEC Section 240.21(C)(3)* or *NEC Section 240.21(C)(6)* must be met.

NEC Section 240.21(C)(3) will permit secondary conductors from a transformer in lengths up to 25' in industrial installations only, when the following conditions are met:

- Conditions of maintenance and supervision ensure that only qualified persons service the system.
- The ampacity of the secondary conductors is not less than the secondary current rating of the transformer, and the sum of the ratings of the overcurrent devices does not exceed the ampacity of the secondary conductors.
- All overcurrent devices are grouped.
- The secondary conductors are protected from physical damage by approved means.

NEC Section 240.21(C)(6) will permit secondary conductors from a transformer in lengths up to 25' when these conditions are met:

- Use the ratio of the primary to the secondary voltage multiplied by one-third the size of the overcurrent device protecting the feeder conductors.

- The secondary conductors terminate in a single circuit breaker or set of fuses that limit the load to not more than the conductor ampacity that is permitted by *NEC Section 310.15.*
- The conductors are protected from physical damage by approved means.

Example 1:

Using the 10' tap rule, what size THWN copper conductors are required for a 40hp, 460V, three-phase motor load tapped from a 400A feeder when all terminations are marked as suitable for 75°C connections?

According to *NEC Table 430.250,* the FLA for that motor is 52A. All motor loads are calculated at 125%, so 52A × 1.25 = 65A.

Using *NEC Table 310.16,* determine the THWN copper conductor size needed to supply 65A. No. 6 THWN copper is rated for 65A. Assuming all conditions are met for application of the 10' tap rule, No. 6 THWN copper conductors are required.

Example 2:

What size THWN copper conductors are required for a 25' tap supplying a load of 100A from a feeder protected by a 600A overcurrent device?

To apply the 25' tap rule, the tap conductors must be rated at least one-third the capacity of the feeder overcurrent device.

$$600A \div 3 = 200A$$

The conductors must be rated for at least 200A, even though the supplied load is substantially lower. Using *NEC Table 310.16,* determine the THWN copper conductor size. No. 3/0 THWN copper is rated for 200A.

2.5.0 Applying Demand Factors

The total load placed upon certain feeders may be reduced by applying demand factors. The *NEC®* permits the reduction of these loads by using the demand factors listed in the following *NEC®* tables.

- *NEC Table 220.42, Lighting Load Demand Factors*
- *NEC Table 220.44, Demand Factors for Nondwelling Receptacle Loads*
- *NEC Table 220.54, Demand Factors for Household Electric Clothes Dryers*
- *NEC Table 220.55, Demand Loads for Household Cooking Appliances*
- *NEC Table 220.56, Demand Factors for Kitchen Equipment – Other Than Dwelling Unit(s)*
- *NEC Section 220.82, Optional Calculation – Dwelling Unit*

- *NEC Section 220.83, Optional Calculation for Additional Loads in Existing Dwelling Unit*
- *NEC Table 220.84, Optional Calculations – Demand Factors for Three or More Multi-Family Dwelling Units*
- *NEC Table 220.86, Optional Method – Demand Factors for Feeders and Service-Entrance Conductors for Schools*
- *NEC Section 220.87, Optional Method for Determining Existing Loads*
- *NEC Table 220.88, Optional Method – Demand Factors for Service-Entrance and Feeder Conductors for New Restaurants*
- *NEC Table 220.102, Method for Computing Farm Loads for Other Than Dwelling Unit(s)*
- *NEC Table 220.103, Method for Computing Total Farm Load*

Calculations utilizing these demand factors will be applied throughout this module.

Example:

The total general-purpose receptacle load in an office building is 26,800VA. Calculate the demand load.

Per *NEC Table 220.44*, the first 10kVA is calculated at 100% and the remaining load is calculated at 50%. Subtract 10,000VA from the receptacle load in this example, and multiply the result by 50%. Add the result to the 10,000VA to determine the demand receptacle load.

$$26,800VA - 10,000VA = 16,800VA$$

$$16,800VA \times 50\% = 8,400VA$$

$$8,400VA + 10,000VA = 18,400VA$$

The total general-purpose receptacle load is calculated to be 18,400VA.

2.5.1 Demand Factors for Neutral Conductors

The neutral conductor of electrical systems generally carries only the current imbalance of the phase conductors. For example, in a single-phase feeder circuit with one phase conductor carrying 50A and the other carrying 40A, the neutral conductor would carry 10A. Since the neutral in many cases will not be required to carry as much current as the phase conductors, the *NEC*® allows us to apply a demand factor. Refer to *NEC Section 220.61* for specific information. Also, *NEC Section 310.15(B)(4)(a)* allows the exclusion of a neutral conductor carrying only unbalanced current when determining any adjustment factor for three or more conductors in a conduit.

NOTE

In certain circumstances such as electrical discharge lighting, data processing equipment, and other similar equipment, you cannot apply a demand factor on the neutral conductors because these types of equipment produce harmonic currents that increase the heating effect in the neutral conductor. Also see *NEC Section 220.61(C)(2)*.

2.5.2 Service Conductor Considerations

You should be aware that the **overhead system service entrance, underground service entrance,** and primary feeder conductor sizing determined in the examples in the remainder of this module are based only on the calculated demand load for the premises served.

Except in a few instances, derating factors for temperature, number of conductors, or voltage drop due to conductor length have not been considered. In actual situations, these factors would have to be included when sizing the conductors.

Depending on the **service point** of the utility serving a particular location and, in some cases, the type of premises involved, sizing of a **service drop** or **service lateral** may also be required, and the same adjustment factors will apply to their conductors as well.

2.6.0 Lighting Loads

NEC Section 220.12 provides specific information for determining lighting loads for occupancies. *NEC Table 220.12, General Lighting Load by Occupancy,* provides unit load per square foot according to the type of occupancy. For example, a **dwelling unit** has a unit lighting load of 3VA per square foot, and a storage warehouse has a unit lighting load of ¼VA per square foot.

To determine the lighting load, the square footage area of the occupancy is calculated using the outside dimensions of the occupancy area. For dwellings, this area does not include open porches, **garages,** or unused or unfinished spaces not adaptable for future use.

2.7.0 Basic Steps for Load Calculations

Figure 2 is a block diagram that illustrates the basic steps in performing load calculations. This chart may be useful as a guide when calculating the loads of various types of electrical installations.

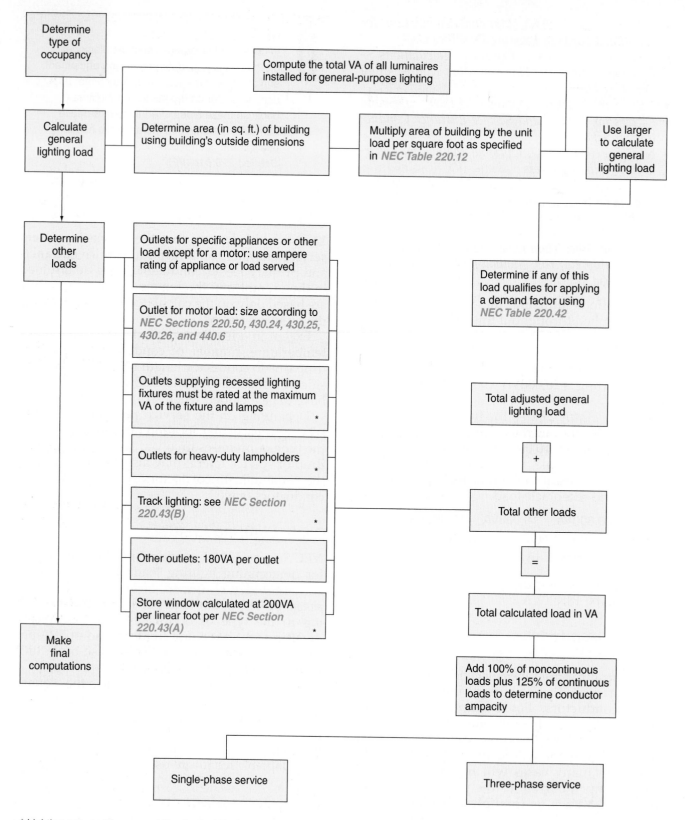

* Lighting not used for general illumination. Typical applications are task or display accent lighting.

Figure 2 ◆ Basic steps in load calculation.

401F02.EPS

3.0.0 ◆ LOAD CALCULATIONS FOR A MINIMUM SIZE SERVICE

The steps below describe the method of determining the service load for a given type of occupancy. These are basic steps, and specific examples will be provided in later sections of this module.

Step 1 Compute the floor area of the building using the outside dimensions of the building. The area arrived at is used to determine the general lighting load for the particular occupancy. *NEC Table 220.12* allows different general lighting loads based upon the use of the area. For example, an office in a warehouse building would require a multiplication factor of 3½VA per square foot, while another calculation of the general lighting load for a warehouse floor area requires a multiplication factor of ¼VA per square foot. If the use of the area is in question, the authority having jurisdiction (*NEC Section 90.4*) will make the determination of the use of the building as it relates to the *NEC®*.

Step 2 Multiply the floor area determined in Step 1 by the unit load per square foot (volt-amperes), and calculate the general lighting load per *NEC Table 220.12*. Compare this value to the total installed lighting used for general illumination, and use the larger of the two loads.

Step 3 Calculate the lighting loads used for other than general illumination, such as for task lighting, exterior lighting, or accent lighting. Refer to the applicable sections of the *NEC®*:

- Light fixtures, *NEC Section 220.14(D)*
- Heavy-duty lampholders, *NEC Section 220.14(E)*
- Sign or outline lighting, *NEC Section 220.14(F)*
- Show windows, *NEC Section 220.14(G) and 220.43(A)*
- Track lighting, *NEC Section 220.43(B)*

Step 4 Calculate the receptacle loads. Refer to the applicable sections of the *NEC®*:

- Specific-purpose receptacles, *NEC Section 220.14(A)*
- Fixed multi-outlet assemblies, *NEC Section 220.14(H)*
- General-purpose receptacles, *NEC Section 220.14(I)*

Step 5 Apply applicable demand factors. For specific applications, refer to the following sections of the *NEC®*:

- General-purpose receptacles, *NEC Section 220.14(I)*
- Cranes, *NEC Article 610*
- Elevators, *NEC Article 620*
- Welders, *NEC Article 630*
- X-ray equipment, *NEC Article 660*
- Fire pumps, *NEC Article 695* (and other articles as needed for specific equipment or locations)

Step 6 Calculate all the motor loads in accordance with *NEC Section 220.50.*

- Use *NEC Section 430.17* to determine the largest motor in the calculation.
- Add together 125% of the FLA of the largest motor plus 100% of the FLA for all other motors.

Step 7 Calculate both the heating and air conditioning loads, and use only the largest of the two in the total load calculation (*NEC Section 220.60*).

Step 8 Add all of the loads. The resultant load in volt-amps is divided by the system voltage for single-phase service or by the system voltage times the square root of three for three-phase services.

Step 9 Add 100% of noncontinuous loads plus 125% of continuous loads to determine conductor ampacity.

3.1.0 Minimum Service Ratings

The total calculated load of an occupancy is determined by adding all the individual loads after any demand factors have been applied. Generally, the total calculated load determines the minimum size service for a particular occupancy (*NEC Section 230.42*).

3.1.1 Rural Pump House Sample Calculation

Figure 3 shows the floor plan of a type of small rural pump house that may be used on some farms to supply water for livestock some distance from the farmhouse or other buildings containing electricity. Therefore, a separate service must be supplied for this pump house.

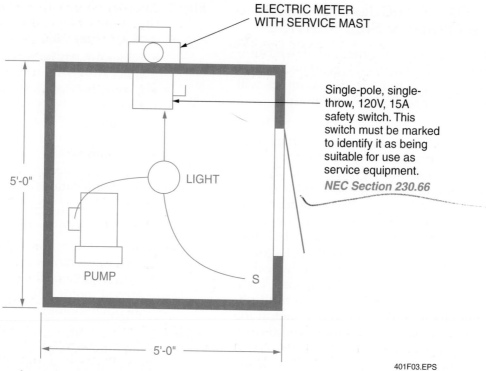

ELECTRIC METER
WITH SERVICE MAST

Single-pole, single-throw, 120V, 15A safety switch. This switch must be marked to identify it as being suitable for use as service equipment.
NEC Section 230.66

LIGHT

5'-0"

PUMP

S

5'-0"

401F03.EPS

Figure 3 ◆ Floor plan of a small pump house.

The total loads for this facility consist of the following:

- Shallow well pump with a ⅓hp, 115V single-phase motor
- One wall switch-controlled lighting fixture containing one 60W lamp

The load and service size are calculated as follows:

Step 1 NEC Table 220.12 does not list a small pump house. Therefore, use *NEC Section 220.14(A)* for the calculation.

Step 2 *NEC Section 430.6(A)(1)* requires that where the current rating of a motor is used to determine the ampacity of conductors, the values given in *NEC Tables 430.247 through 430.250,* including notes, shall be used instead of the actual current rating marked on the motor nameplate. From *NEC Table 430.248,* we find that the listed ampacity for a ⅓hp single-phase 115VAC motor is 7.2A.

Also, according to *NEC Section 430.22(A),* a single motor shall have an ampacity of no less than 125% of the motor full-load current rating.

Therefore, the motor load is:

$$7.2A \times 1.25 = 9.0A$$
$$9.0A \times 120V = 1,080VA$$

This, added to the 60W lamp load (60VA), gives a total connected load of 1,140VA.

Step 3 Determine the size of the service-entrance conductors.

$$1,140VA \div 120V = 9.5A$$

Step 4 Check *NEC Section 230.42(B)* for the minimum size of the service-entrance conductors. Check *NEC Section 230.79(A)* for the minimum rating of the service disconnecting means. The service-entrance conductors must have an ampacity that is not less than the load served and not less than the minimum required disconnecting means. The disconnect in this case would be 15A.

Since the total connected load for the pump house is less than 10A and the single branch circuit feeding the pump and light fixture need only be No. 14 AWG copper (15A), a No. 14 AWG copper or No. 12 aluminum conductor will qualify for the service-entrance conductors.

NEC Section 230.23 gives the minimum size of service drop conductors, and *NEC Section 230.31* gives the minimum size of service lateral conductors. In addition to being able to serve the load, service drops and service laterals must have a minimum mechanical strength.

> **NOTE**
> Many inspection jurisdictions and utility companies now require a fault current study prior to issuance of a permit. The fault current study may result in the use of a conductor size that is larger than the minimum *NEC®* requirement.

3.1.2 Roadside Vegetable Stand

Another practical application of *NEC Sections 230.42 and 230.79* is shown in *Figure 4*. This is a floor plan of a typical roadside vegetable stand. Again, *NEC Table 220.12* does not list this facility. Therefore, use *NEC Section 220.14* for the calculation.

Step 1   Determine the lighting load. Two fluorescent fixtures are used to illuminate the 9 × 12 prime area. Since each fixture contains two 40W fluorescent lamps, and the ballast is rated at 0.83A (which translates to approximately 100VA), the total connected load for each fixture is 100VA, or a total of 200VA for both fixtures.

Step 2   Determine the remaining loads. The only other electrical outlets in the stand consist of two receptacles: one furnishes power to

a refrigerator with a nameplate full-load rating of 12.2A (total VA = 120V × 12.2A = 1,464VA); the other furnishes power for an electric cash register rated at 300VA and an electronic calculator rated at 200VA (total VA for other loads is 300VA + 200VA = 500VA).

Step 3   Determine the total connected load.

Fluorescent fixtures =	200VA
Receptacle for refrigerator =	1,464VA
Receptacle for other loads =	500VA
Total calculated load =	2,164VA

Step 4   Identify and total the continuous loads. In this example, the continuous loads are the fluorescent fixtures and the refrigerator.

$$200VA + 1,464VA = 1,664VA$$

Step 5   Determine the size and rating of the service-entrance conductors. Add 100% of the noncontinuous loads plus 125% of the continuous loads.

$$500VA + (1,664VA \times 1.25) = 2,580VA \div 240 = 10.75A$$

Step 6   Check *NEC Section 230.42(B)* for the minimum size of the service-entrance conductors. Check *NEC Section 230.79(B)* for the minimum rating of the service disconnecting means. No. 10 AWG copper or No. 8 aluminum is the minimum size allowed by the *NEC®* for the service-entrance conductors on this project. The disconnecting means in this instance would be 30A.

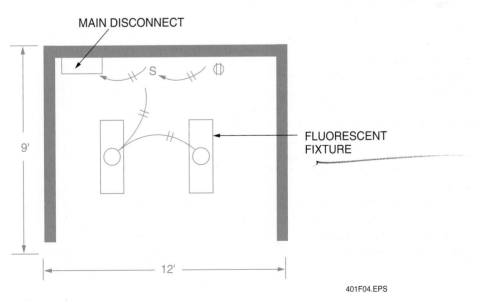

401F04.EPS

Figure 4 ◆ Roadside vegetable stand.

See *NEC Section 230.23* for the minimum size of service drop conductors or *NEC Section 230.31* for the minimum size of service lateral conductors when they are not under the jurisdiction of the serving utility.

3.1.3 Single-Family and Multi-Family Dwellings

A floor plan of a single-family dwelling is shown in *Figure 5*. This building is constructed on a concrete slab with no basement or crawlspace. There is an unfinished attic above the living area and an open carport just outside the kitchen entrance. Appliances include a 12kW electric range (12,000W = 12,000VA) and a 4.5kW water heater (4,500W = 4,500VA). There is also a washer and a 5.5kW dryer (5,500W = 5,500VA) in the utility room. Gas heaters are installed in each room with no electrical requirements. What size service entrance should be provided for this residence if no other information is specified?

Figure 6 shows a typical form used for service load calculations. A blank form is included in the *Appendix* for future use. The steps necessary to perform these calculations are as follows:

Step 1 Compute the area of the occupancy *[NEC Section 220.12]*. The general lighting load (including general receptacle load) is determined by finding the area of the occupancy using the outside dimensions of the structure. Remember that this does not include open porches, garages, or unused or unfinished spaces that are not adaptable for future use. For example, an unfinished basement would be included in this area measurement because it might later be converted into living space. The area in square feet is used to calculate the general lighting load in Step 2.

Using the floor plan and an architect's scale, measure the longest width of the building using the outside dimensions. In this case, it is 33'. The longest length of the building is 48'. The product of these two measurements is:

$$33' \times 48' = 1,584 \text{ sq. ft.}$$

The area of the carport within this total area is:

$$12' \times 19.5' = 234 \text{ sq. ft.}$$

The net living area is then:

$$1,584 - 234 = 1,350 \text{ sq. ft.}$$

Step 2 Determine the general lighting load *[NEC Sections 220.10 and 220.12]*. Multiply the resulting area found in Step 1 by the unit load per square foot (volt-amperes) for dwelling units. In this example, it is 3VA per square foot. The lighting load is now:

$$1,350 \text{ sq. ft.} \times 3\text{VA/sq. ft.} = 4,050\text{VA}$$

Step 3 Multiply the number of small appliance branch circuits *[NEC Section 210.11(C)(1)]* by 1,500VA (minimum of two required).

The *NEC®* requires at least two 120V, 20A small appliance branch circuits to be installed in each kitchen, dining area, breakfast nook, and similar areas where toasters, coffee makers, and other small appliances will be used. *NEC Section 220.52* also requires small appliance

Figure 5 ◆ Single-family dwelling.

401F05.EPS

General Lighting Load								Phase		Neutral	
Square footage of the dwelling	[1]	1,350	× 3VA =	[2]	4,050						
Kitchen small appliance circuits	[3]	2	× 1500 =	[4]	3,000						
Laundry branch circuit	[5]	1	× 1500 =	[6]	1,500						
Subtotal of general lighting loads per *NEC Section 220.52*				[7]	8,550						
Subtract 1st 3000VA per *NEC Table 220.42*				[8]	3,000	× 100% =	[9]	3,000			
Remaining VA times 35% per *NEC Table 220.42*				[10]	5,550	× 35% =	[11]	1,943			
Total demand for general lighting loads =							[12]	4,943	[13]	4,943	

Fixed Appliance Loads (nameplate or NEC FLA of motors) per *NEC Section 220.53*							
Hot water tank, 4.5kVA, 240V	[14]	4,500					
	[15]						
	[16]						
	[17]						
	[18]						
	[19]						
Subtotal of fixed appliances [20]	4,500						
If 3 or less fixed appliances take @ 100% =	[21]	4,500	[22]	0			
If 4 or more fixed appliances take @ 75% =	[23]		[24]				

Other Loads per *NEC Section 220.14*					
Electric range per *NEC Table 220.55* (neutral @ 70 % per *NEC Section 220.61*)	[25]	8,000	[26]	5,600	
Electric dryer per *NEC Table 220.54* (neutral @ 70% per *NEC Section 220.61*)	[27]	5,500	[28]	3,850	
Electric heat per *NEC Section 220.51* Air conditioning per *NEC Section 220.14(A)*	Omit smaller load per *NEC Section 220.60*	[29]		[30]	
Largest Motor = 0	× 25% per *NEC Section 430.24* =	[31]	0	[32]	0

Total VA Demand =	[33]	22,943	[34]	14,393		
VA/240V = Amps =	[35]	96	[36]	60		
Service OCD and Minimum Size Grounding Electrode Conductor =	[37]	100	[38]	8 AWG		
AWG per *NEC Tables 310.15(B)(6)* and *310.16* **for neutral**	[39]	4 AWG	[40]	6 AWG		

401F06.EPS

Figure 6 ◆ Completed calculation form.

branch circuits to be rated at 1,500VA for each kitchen area served. Since the sample residence has only one kitchen area, the load for these circuits would be:

$$2 \times 1,500VA = 3,000VA$$

Step 4 Multiply the laundry receptacle *[NEC Section 210.11(C)(2)]* circuit by 1,500VA *[NEC Section 220.52]*.

An additional 20A branch circuit must be provided for the exclusive use of each laundry area. This circuit must be an individual branch circuit and must not supply any other load except for the laundry receptacles.

$$1 \times 1,500VA = 1,500VA$$

Step 5 Calculate the lighting load demand. In *NEC Table 220.42*, note that the demand factor is 100% for the first 3,000VA, 35% for the next 117,000VA, and 25% for the

remainder. Since all residential electrical outlets are never used at the same time, the *NEC*® allows the use of demand factors.

First 3,000VA at 100% =	3,000VA
The remainder (8,550 − 3,000) = 5,550 at 35% =	1,942.5VA
Net general lighting and small appliance load =	4,942.5VA

Step 6 Calculate the loads for other appliances *(NEC Section 220.53)*, such as a water heater, disposal, dishwasher, furnace blower motor, etc.

Water heater calculated at 100% = 4,500VA

Step 7 Calculate the range load and other cooking appliances *(NEC Section 220.55 and Table 220.55)*.

Electric range (using demand factor) = 8,000VA

Step 8 Calculate the load for the clothes dryer *(NEC Section 220.54 and Table 220.54).* Note that this load must be either 5,000VA or the nameplate rating of the dryer, whichever is larger. If a gas dryer is used, it will be part of the 1,000VA demand in Step 4.

Clothes dryer (nameplate rating) = 5,500VA

NOTE

Four or more other appliances may have a demand factor of 75% applied.

Step 9 Calculate the larger of the heating or air conditioning loads *(NEC Section 220.60).* (Negligible in this example.)

NOTE

If a gas forced air furnace is used, the blower motor will have been included in Step 6. If some form of electric heat is used (baseboard, heat pump, or furnace), this load will be compared to the air conditioning load and the larger load then added to the calculations *(NEC Section 220.60).*

Step 10 Calculate 25% of the largest motor load *(NEC Section 430.24).* (None in this example.)

Step 11 Total all the loads.

General lighting and appliance load =	4,942.5VA
Electric range (using demand factor) =	8,000.0VA
Clothes dryer =	5,500.0VA
Water heater =	4,500.0VA
Total load =	22,942.5VA

Step 12 Convert the volt-amps to amperes. The total load amperage is derived by dividing the total volt-amps of the load by the circuit voltage. In this example, the system voltage is 240V.

$$22,942.5VA \div 240V = 95.6A$$

Step 13 Determine the required service size *[NEC Section 230.42(B)].*

Typical electric service for residential use is 120/240V, three-wire, single-phase.

Services are sized in amperes after the total load in volt-amperes is determined.

For this ampere load, the minimum 100A service would be required *[NEC Section 230.79(C)].*

3.2.0 Sizing Neutral Conductors

The neutral conductor in a three-wire, single-phase service carries only the unbalanced load between the two hot legs. Since there are several 240V loads in the example residence, these 240V loads will be balanced and therefore reduce the load on the service neutral conductor. Consequently, in most cases, the service neutral does not have to be as large as the ungrounded (hot) conductors.

In this example, the water heater does not have to be included in the neutral conductor calculation since it is strictly 240V with no 120V loads. The clothes dryer and electric range, however, have 120V lights that will unbalance the current between phases. *NEC Section 220.61* allows a demand factor of 70% for these two appliances. Using this information, the neutral conductor may be sized accordingly:

General lighting and appliance load =	4,942.5VA
Electric range (8,000VA × 0.70) =	5,600VA
Clothes dryer (5,500VA × 0.70) =	3,850VA
Total =	14,392.5VA

To find the total line-to-line amperes, divide the total volt-amperes by the voltage between lines.

$$14,392.5VA \div 240V = 59.97A \text{ or } 60A$$

The service-entrance conductors have now been calculated and must be rated at 100A with a neutral conductor rated for at least 60A.

In *NEC Table 310.15(B)(6),* special consideration is given to 120/240V, single-phase residential services. This table shows that the NEC® allows a No. 4 AWG copper or No. 2 AWG aluminum or copper-clad aluminum for a 100A service.

When sizing the grounded conductor for services, all of the provisions stated in *NEC Sections 215.2, 220.61, and 230.42* must be met, along with other applicable sections.

3.2.1 Single-Family Dwelling Standard Calculation

To calculate the service size for the example dwelling, first determine the square footage. This dwelling consists of 1,624 square feet of living area. The following loads must be accounted for:

Motor Sizes

THINK ABOUT IT

Which is the largest motor, a 1hp, 240V single-phase motor or a ¾hp, 120V single-phase motor?

- 12kW range
- 5.5kW dryer
- 1,250VA dishwasher
- ¾hp, 120V disposal
- 1hp, 240V pump
- ⅛hp, 120V blower for gas furnace

Step 1 Calculate the general lighting and small appliance loads.

General lighting: 1,624 sq. ft. × 3VA =	4,872VA
Small appliance load: 1,500VA × 2 =	3,000VA
Laundry load =	1,500VA
Total general and small appliance loads =	9,372VA

Step 2 Apply demand factors for general lighting and small appliances *(NEC Table 220.42)*.

First 3,000VA at 100% =	3,000VA
9,372VA − 3,000VA = 6,372 at 35% =	2,230VA
Total general and small appliance loads =	5,230VA

Step 3 Range load *(NEC Table 220.55, Column C)* = 8,000VA

Step 4 Dryer load *(NEC Table 220.54)* = 5,500VA

Step 5 Calculate fixed appliance loads.

Dishwasher =	1,250VA
Disposal *(NEC Table 430.248)*: 13.8A × 120V =	1,656VA
Pump *(NEC Table 430.248)*: 8A × 240V =	1,920VA
Blower *(NEC Table 430.248)*: 7.2A × 120V =	864VA
Total fixed appliance loads =	5,690VA

Step 6 Since four or more fixed appliances exist, apply the demand factor for fixed appliance loads *(NEC Section 220.53)*.

Total fixed appliance loads: 5,690VA × 75% =	4,268VA

Step 7 Use *NEC Section 430.17* to determine the largest motor.

25% of the largest motor load: 1,920VA × 25% =	480VA

Step 8 Combine the computed loads.

Total load: 5,230VA + 8,000VA + 5,500VA + 4,268VA + 480VA = 23,478VA

The minimum size service required for this dwelling is then determined by dividing the total calculated VA by the line-to-line service voltage. The line-to-line voltage is 240V, single phase.

$$23,478VA \div 240V = 97.83A$$

NEC Section 230.79(C) requires a minimum service size of 100A for a single-family dwelling.

3.2.2 Standard Calculation for Single-Family Dwelling Using Electric Heat

This section provides an example load calculation for a single-family residence using electric heat. In this example, the dwelling consists of 1,775 square feet of living area. The home has the following loads:

- 8.75kW range
- 5.5kW dryer
- 1,000VA dishwasher
- ¾hp, 120V disposal
- ½hp, 120V attic fan
- ⅛hp, 120V blower for electric furnace
- 20kW electric furnace
- Central air conditioning unit with a nameplate rating of 25A at 240V

Step 1 Calculate the general lighting and small appliance loads.

General lighting:
1,775 sq. ft. × 3VA =   5,325VA

Small appliance load:
1500VA × 2 =   3,000VA

Laundry load =   1,500VA

Total general and small
appliance loads =   9,825VA

Step 2 Apply demand factors for general lighting and small appliances *(NEC Table 220.42)*.

First 3,000VA at 100% =   3,000VA

9,825VA − 3,000VA = 6,825
at 35% =   2,389VA

Total general and small
appliance loads =   5,389VA

Step 3 Range load *(NEC Table 220.55, Column B)*.

8,750VA × 80% =   7,000VA

Step 4 Dryer load *(NEC Table 220.54)*.

5,500VA × 100% =   5,500VA

Step 5 Calculate fixed appliance loads.

Dishwasher =   1,000VA

Disposal *(NEC Table 430.248)*:
13.8A × 120V =   1,656VA

Furnace blower *(NEC Table 430.248)*:
7.2A × 120V =   864VA

Attic fan *(NEC Table 430.248)*:
9.8A × 120V =   1,176VA

Total fixed appliance loads =   4,696VA

Step 6 Since four or more fixed appliances exist, apply the demand factor for fixed appliance loads *(NEC Section 220.53)*.

Total fixed appliance loads:

4,696VA × 75% =   3,522VA

Step 7 Calculate the larger of the heating and cooling loads (use *NEC Section 220.60* for noncoincidental loads to omit the smaller load).

Heating load =   20,000VA

Cooling load: 25A × 240V =   6,000VA

Larger of the two loads =   20,000VA

Step 8 Use *NEC Section 430.17* to determine the largest motor.

25% of the largest motor load:
1,656VA × 25% =   414VA

Step 9 Combine the computed loads.

Total load: 5,389VA + 7,000VA + 5,500VA
+ 3,522VA + 20,000VA + 414VA
= 41,825VA

The minimum size service for this dwelling is determined by dividing the total calculated VA by the line-to-line service voltage. The line-to-line voltage is 240V, single phase.

41,825VA ÷ 240V = 174.27A

The next larger standard size overcurrent device is 175A. Although a 175A device would satisfy the *NEC*® requirements, it is more likely that a 200A service would be installed.

3.2.3 Optional Calculation for a Single-Family Dwelling

The *NEC*® provides an alternative method of computing the load for a dwelling unit in *NEC Section 220.82*. Using this method will almost always result in a calculation that is lower than the standard method when electric heat is used.

For example, suppose you have a single-family dwelling with 1,624 square feet of living area and the following loads:

INSIDE TRACK

Optional Calculations

When using the optional calculations, *NEC Section 220.82(B)* permits the use of motor nameplate values rather than the values listed in *NEC Tables 430.248 through 430.250*.

- 12kW range
- 5.5kW dryer
- 1,250VA dishwasher
- ½hp, 120V disposal at 1,176VA
- ¾hp, 120V attic fan at 1,656VA
- 14.4kW electric furnace
- ⅛hp, 120V (864VA) blower motor for electric furnace
- Air conditioning unit with a nameplate rating of 6,000VA

Step 1 Calculate the general lighting, small appliance, and fixed appliance loads. Use full nameplate values for each.

General lighting: 1,624 sq. ft. × 3VA =	4,872VA
Small kitchen appliance load: 1,500VA × 2 =	3,000VA
Laundry load =	1,500VA
Range =	12,000VA
Dryer =	5,500VA
Dishwasher =	1,250VA
Disposal =	1,176VA
Attic fan =	1,656VA
Blower =	864VA
Subtotal load =	31,818VA

Step 2 Apply the demand factor in *NEC Section 220.82(B)*.

First 10kVA at 100% =	10,000VA
Remaining load at 40%: 21,818VA × 40% =	8,727VA
Subtotal for general loads =	18,727VA

Step 3 Compute the demands permitted for electric heating and air conditioning per *NEC Section 220.82(C)*. Use the largest of the computed demand loads.

Air conditioner at 100% = 6,000VA × 100% =	6,000VA
Central electric heating at 65%: 14,400 × 65% =	9,360VA

Step 4 Add together the demand for general loads and the largest of the heating or air conditioning load.

Subtotal general loads =	18,727VA
Heating load (larger than the air conditioning load) =	9,360VA
Total load =	28,087VA

The minimum size service required for this dwelling is determined by dividing the total calculated VA by the line-to-line service voltage. The line-to-line voltage is 240V, 1Ø.

$$28,087\text{VA} \div 240\text{V} = 117\text{A}$$

The minimum service required for this dwelling would be 125A.

3.3.0 Multi-Family Calculations

The service load for multi-family dwellings is not simply the sum of the individual dwelling unit loads because of demand factors that may be applied when either the standard calculation or the optional calculation is used to compute the service load.

When the standard calculation is used to compute the service load, the total lighting, small appliance, and laundry loads as well as the total load from all electric ranges and clothes dryers are subject to the application of demand factors. Additional demand factors may be applied to the portion of the neutral load contributed by electric ranges and the portion of the total neutral load greater than 200A.

When the optional calculation is used, the total connected load is subject to the application of a demand factor that varies according to the number of individual units in the dwelling.

A summary of the calculation methods for designing wiring systems in multi-family dwellings and the applicable *NEC®* references are shown in *Figure 7*. The selection of a calculation method for computing the service load is not affected by the method used to design the feeders to the individual dwelling units.

The rules for computing the service load are also used for computing a main feeder load when the wiring system consists of a service that supplies main feeders that, in turn, supply a number of sub-feeders to individual dwelling units.

3.3.1 Multi-Family Dwelling Sample Calculation

A 30-unit apartment building consists of 18 one-bedroom units of 650 sq. ft. each, 6 two-bedroom units of 775 sq. ft. each, and 6 three-bedroom units of 950 sq. ft. each. The kitchen in each unit contains the following loads:

- 7.5kW electric range
- 1,250VA dishwasher
- ⅛hp, 120V garbage disposal at 864VA
- Air conditioning unit rated at 3,600VA

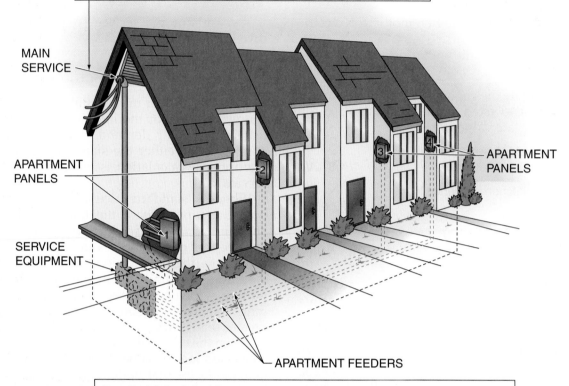

Standard Calculation: Feeder sized per *NEC Article 220, Part III*
Optional Calculation: Per *NEC Section 220.84*, if each dwelling unit has:
    1. Single feeder
    2. Electric cooking equipment
    3. Electric space heating or air conditioning, or both

MAIN
SERVICE

APARTMENT
PANELS

APARTMENT
PANELS

SERVICE
EQUIPMENT

APARTMENT FEEDERS

Standard Calculation: Feeder sized per *NEC Article 220, Part III*
Optional Calculation: Use *NEC Section 220.84* if each dwelling unit has single 3-wire, 120/240V or 208Y/120V feeders with ampacity of 100A or greater.

401F07.EPS

Figure 7 ◆ *NEC*® requirements for multi-family dwellings.

The entire building is heated by a boiler that uses three electric pump motors: a 2hp squirrel-cage motor and two ½hp squirrel-cage induction motors, all 240V single-phase, two-wire. There is a community laundry room in the building with four washers with a nameplate rating of 12.5A at 120V (1,500VA) and four clothes dryers, with the latter rated at 4,500VA each. The house lighting consists of 40 incandescent fixtures rated at 60W each. There are three general-purpose receptacle outlets for servicing of the boilers and laundry room. The building is furnished with a 120/240V, single-phase, three-wire service. What size service is required?

Step 1   Calculate the general lighting and small appliance loads in each apartment.

General lighting: 650 sq. ft. × 3VA = 1,950VA × 18 units =	35,100VA
General lighting: 775 sq. ft. × 3VA = 2,325VA × 6 units =	13,950VA
General lighting: 950 sq. ft. × 3VA = 2,850VA × 6 units =	17,100VA
Small appliance circuits: 1,500VA × 2 × 30 units =	90,000VA
Total general lighting and small appliance load =	156,150VA

Step 2 Apply the demand factors using *NEC Table 220.42*.

First 3,000VA (out of 120,000VA) at 100% =	3,000VA
Next 117,000VA (out of 120,000VA) at 35% =	40,950VA
Remaining 36,150VA at 25% =	9,038VA
Net general lighting and small appliance load =	52,988VA

Step 3 Calculate the appliance loads using *NEC Section 220.53*.

Dishwashers: 30 × 1,250VA × 75% =	28,125VA
Disposals: 30 × 864VA × 75% =	19,440VA

Step 4 Calculate the range load using *NEC Table 220.55*.

Range load *(NEC Table 220.55, Column C)* =	45,000VA

NOTE

The range load could have been calculated using *NEC Table 220.55, Column B, Note 3*. However, the value obtained (54kVA) is higher, so Column C was used.

Step 5 Calculate the air conditioning load *[NEC Sections 220.60 and 220.14(A)]*.

Air conditioners: 30 × 3,600VA × 100% =	108,000VA

Step 6 Calculate the house load.

2hp boiler *(NEC Table 430.248)*: 12A × 240V =	2,880VA
Two ½hp boiler motors: 4.9A × 240V × 2 =	2,352VA
Laundry: 1,500VA × 4 =	6,000VA
Dryers *(NEC Table 220.54)*: 4,500 × 4 × 100% =	18,000VA
House lighting: 60W × 40 =	2,400VA
General-purpose receptacles: 180VA × 3 =	540VA
Total house load =	32,172VA

Step 7 Determine the largest motor load and multiply the FLA or VA by 25% *(NEC Section 430.24)*.

Boiler motor: 2,880VA × 25% =	720VA

Step 8 Determine the total calculated load.

Net general lighting and small appliance loads =	52,988VA
Dishwasher loads =	28,125VA
Disposal loads =	19,440VA
Range loads =	45,000VA
Air conditioning loads =	108,000VA
House loads =	32,172VA
25% of largest motor load =	720VA
Total load =	286,445VA
Add 25% of continuous loads (house lighting) =	600VA
Total adjusted load =	287,045VA

Step 9 Determine the total load in amperes in order to select service conductors, equipment, and overcurrent protective devices.

$$287,045VA \div 240V = 1,196A$$

Step 10 Select the service-entrance conductors, service equipment, and overcurrent protection.

NOTE

Since the exact calculated load is not a standard size, the rating of the service for this apartment building would be 1,200A; this is the next largest standard size and is permitted by the *NEC®*. See *NEC Sections 240.4 and 240.6*.

3.3.2 Multi-Family Dwelling Calculation Using Optional Method

NEC Section 220.84 permits an alternative method of computing the load for multi-family dwellings. The conditions that must be met to use this calculation are that each individual dwelling must be supplied by only one feeder, have electric cooking (or an assumed cooking load of 8kW), and have electric heating, air conditioning, or both. The demand factors of *NEC Table 220.84* are not permitted to be applied to the house loads.

For example, suppose you have a 30-unit apartment building consisting of 18 one-bedroom units of 650 sq. ft. each, six two-bedroom units of 775 sq. ft. each, and six three-bedroom units of 950 sq. ft. each. The kitchen in each unit contains one 7.5kW electric range, one 1,250VA dishwasher, and one ⅓hp, 120V garbage disposal at 864VA. Each unit has an air conditioning unit rated at 3,600VA.

The entire building is heated by a boiler that uses three electric pump motors: a 2hp squirrel-cage motor and two ½hp squirrel-cage induction motors, all 240V single-phase, two-wire. There is a community laundry room in the building with four washers and four clothes dryers, with the latter rated at 4,500VA each. The house lighting consists of 40 incandescent fixtures rated at 60W each. There are three general-purpose receptacle outlets for servicing of the boilers and laundry room. The building is to be furnished with a 120/240V, single-phase, three-wire service.

Step 1 Calculate the lighting and small appliance loads [NEC Sections 220.84(C)(1) and (2)].

General lighting: 650 sq. ft. × 3VA = 1,950 × 18 units =	35,100VA
General lighting: 775 sq. ft. × 3VA = 2,325 × 6 units =	13,950VA
General lighting: 950 sq. ft. × 3VA = 2,850 × 6 units =	17,100VA
Small appliance circuits: 1,500VA × 2 × 30 units =	90,000VA
Total general lighting and small appliance loads =	156,150VA

Step 2 Calculate the appliance and motor loads [NEC Sections 220.84(C)(3), (4), and (5)].

Dishwashers: 1,250VA × 30 units =	37,500VA
Disposals: 864VA × 30 units =	25,920VA
Ranges: 7,500VA × 30 units =	225,000VA
Air conditioners = 3,600VA × 30 units =	108,000VA
Total appliance and motor loads =	396,420VA

Step 3 Apartment unit loads:

156,150VA + 396,420VA =	552,570VA

Step 4 Apply the demand factor (NEC Table 220.84).

552,570VA × 33% =	182,348VA

Step 5 Calculate the total house load [NEC Section 220.84(B)].

2hp boiler (NEC Table 430.248): 12A × 240V =	2,880VA
Two ½hp boiler motors: 4.9A × 240V × 2 =	2,352VA
Laundry: 1,500VA × 4 =	6,000VA
Dryers (NEC Table 220.55): 4,500VA × 4 × 100% =	18,000VA
House lighting: 60W × 40 =	2,400VA
General-purpose receptacles: 180VA × 3 =	540VA
Total house load =	32,172VA

Step 6 Total load:

182,348VA + 32,172VA + 600VA (25% of continuous load) =	215,120VA

Using the optional method, the current would be:

$$215,120VA \div 240V = 896A$$

The next standard size of service equipment greater than 896A is 1,000A, so this apartment building would require a 1,000A service. Note that this is less than the 1,200A required when using the standard method of calculation for the same building.

4.0.0 ◆ COMMERCIAL OCCUPANCY CALCULATIONS

Calculating load requirements for commercial occupancies is based on specific NEC® requirements that relate to the loads present. The basic approach is to separate the loads into the following:

- Lighting
- Receptacles
- Motors
- Appliances
- Other special loads

In general, all loads for commercial occupancies should be considered continuous unless specific information is available to the contrary. Smaller commercial establishments normally use single-phase, three-wire services; larger projects almost always use a three-phase, four-wire service. Many installations have secondary feeders supplying panelboards, which, in turn, supply branch circuits operating at different voltages. This requires separate calculations for the feeder and branch circuit loads for each voltage. The rating of the main service is based on the total load with the load values transformed according to the various circuit voltages, if necessary.

Demand factors also apply to some commercial establishments. For example, the lighting loads in hospitals, hotels, motels, and warehouses are subject to the application of demand factors. In restaurants and similar establishments, the load of electric cooking equipment is subject to a demand factor if there are more than three cooking units. Optional calculation methods to determine feeder or service loads for schools and similar occupancies are also provided in the *NEC*®.

Special occupancies, such as mobile homes and recreational vehicles, require the feeder or service load to be calculated in accordance with specific *NEC*® requirements. The service for mobile home parks and recreational vehicle parks is also designed based on specific *NEC*® requirements that apply only to those locations. The feeder or service load for receptacles supplying shore power for boats in marinas and boatyards is also specified in the *NEC*®.

When transformers are not involved, a relatively simple calculation involving only one voltage results. If step-down transformers are used, the transformer itself must be protected by an overcurrent device that may also protect the circuit conductors in most cases.

Switches and panelboards used for the distribution of electricity within a commercial building are also subject to *NEC*® rules. These requirements could affect the number of feeders required when a large number of lighting or appliance circuits are needed. See *NEC Article 408*.

4.1.0 Commercial and Industrial Load Calculations

The type of occupancy determines the methods employed to calculate loads. Industrial and commercial buildings are two of these categories addressed in the *NEC*®. Buildings that fall into these categories are ones that manufacture, process, market, or distribute goods. Loads are calculated based on the type of occupancy and the electrical requirements of the installed equipment.

Standard calculations used to compute the load and determine the service equipment requirement are arranged differently from those for dwelling units. There are specific types of loads that must be considered when making load calculations of commercial and industrial occupancies.

Demand factors may be required for certain types of loads.

The following are *NEC*® references for feeder and service loads:

- *NEC Section 215.2, Minimum Rating and Size*
- *NEC Section 220.42, General Lighting*
- *NEC Section 220.43, Show Window and Track Lighting*
- *NEC Section 220.44, Receptacle Loads – Nondwelling Units*
- *NEC Section 220.50, Motors*
- *NEC Section 220.51, Fixed Electric Space Heating*
- *NEC Section 220.56, Kitchen Equipment – Other Than Dwelling*
- *NEC Section 220.60, Noncoincident Loads*
- *NEC Section 220.61, Feeder or Service Neutral Load*

Each load must be designated as either continuous or noncontinuous. Most commercial loads are generally considered to be continuous. One exception would be a storage warehouse that is occupied on an infrequent basis. Demand factors may be applied to various noncontinuous loads according to the respective section of the *NEC*®. For example, *NEC Section 220.44* permits both general-purpose receptacles at 180VA per outlet and fixed multi-outlet assemblies to be adjusted for the demand factors given in *NEC Tables 220.42 and 220.44*.

4.1.1 Lighting Loads

Within this category, there are seven different categories to take into account:

- *NEC Table 220.12, General Lighting Loads*
- *NEC Section 220.14(D), Lighting Fixtures*
- *NEC Section 220.14(E), Heavy-Duty Lampholders*
- *NEC Section 220.14(F), Sign and Outline Lighting*
- *NEC Section 220.18(B), Inductive Lighting Loads*
- *NEC Section 220.14(G), Show Windows*
- *NEC Section 220.43(B), Track Lighting*

These lighting loads must be designated as either continuous or noncontinuous. Some of these lighting loads may have demand factors applied. The lighting loads should be calculated before any other loads are determined.

4.1.2 General Lighting Loads

This is the main lighting used for general illumination within the building and is in addition to any lighting installed for accent, display, task, show windows, or signs. It is computed based upon the type of occupancy and volt-amps per square foot given in *NEC Table 220.12*. For types of occupancies not listed in *NEC Table 220.12,* the general lighting load should be calculated according to the *NEC®* sections listed previously.

The minimum load for general lighting must be compared to the actual installed lighting used within a building, and the larger of the two computations must be used. In most cases, due to energy codes, the installed lighting will generally be lower than that required by *NEC Table 220.12.*

Any of these loads that are continuous duty are multiplied by 125% to determine the overcurrent protection device and conductor size. All others are calculated at 100%. Noncontinuous loads may have a demand factor applied in accordance with *NEC Section 220.42 and Table 220.42.*

Example:

Suppose a warehouse building has 57,600 square feet of floor area.

Step 1 From *NEC Table 220.12:*

    57,600 sq. ft. × ¼ VA/sq. ft. =   14,400VA

Step 2 Apply demand factors per *NEC Table 220.42.*

First 12,500VA at 100% =	12,500VA
14,400VA − 12,500VA = 1,900 at 50%	950VA
Net lighting load =	13,450VA

4.1.3 Show Window Loads

NEC Section 220.43(A) requires that the lighting load for show window feeders be calculated at 200VA per linear foot of show window.

Example:

Suppose a show window is 55' in length. What is the feeder load?

    Load = 55' × 200VA per linear ft. = 11,000VA

NOTE

Any loads that are continuous duty are multiplied by 125% to determine the overcurrent protection device and conductor size. All others are calculated at 100%.

4.1.4 Track Lighting Loads

Track lighting loads are calculated by allowing 150VA for every two feet of track or fraction thereof. Refer to *NEC Section 220.43(B).*

Example 1:

Determine the lighting load for 140' of track lighting.

$$140' \div 2 = 70'$$
$$70' \times 150VA = 10,500VA$$

Example 2:

Determine the lighting load of 21' of track lighting.

$$21' \div 2 = 10.5' \text{ (round fractions up)} = 11'$$
$$11' \times 150VA = 1,650VA$$

NOTE

Any loads that are continuous duty are multiplied by 125% to determine the overcurrent protection device and conductor size. All others are calculated at 100%.

4.1.5 Outside Lighting Loads

Outside lighting loads are calculated by multiplying the rating in VA by the number of fixtures. Refer to *NEC Section 215.2(A).*

Example:

Determine the load for 30 outside lighting fixtures, each rated at 250VA, including the ballast. These lights are rated at 240V.

$$250VA \times 30 = 7,500VA$$

NOTE

Any loads that are continuous duty are multiplied by 125% to determine the overcurrent protection device and conductor size. All others are calculated at 100%.

4.1.6 Sign and Outline Lighting Loads

NEC Section 600.5(A) requires that occupancies with a ground floor entry accessible to pedestrians have a 20A outlet installed for each tenant space in an accessible location. This must be a 20A branch circuit that supplies no other load.

NEC Section 220.14(F) states that the load for the branch circuit installed for the supply of exterior signs or outline lighting must be computed at a minimum of 1,200VA. However, if the actual rating of the sign lighting is greater than 1,200VA, the actual load of the sign shall be used.

NOTE

Any loads that are continuous duty are multiplied by 125% to determine the overcurrent protection device and conductor size. All others are calculated at 100%.

4.2.0 Retail Stores with Show Windows

Figure 8 shows a small store building with a show window in front. Note that the storage area has four general-purpose duplex receptacles, while the retail area has 14 wall-mounted duplex receptacles and two floor-mounted receptacles for a

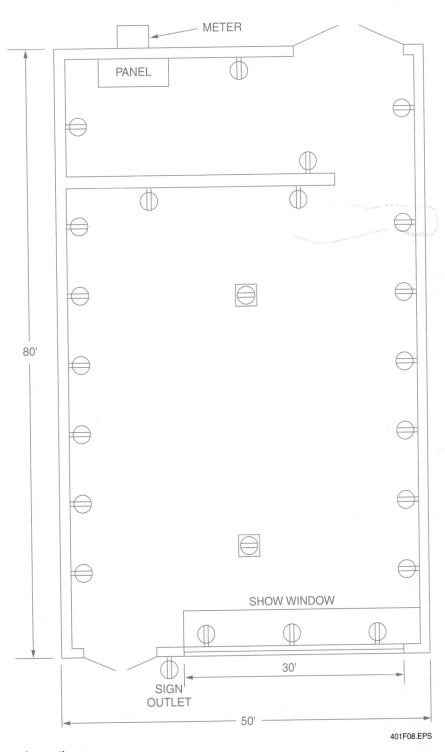

Figure 8 ◆ Floor plan of a retail store.

total of 16 in this area. These, combined with the storage area receptacles, bring the total to 20 general-purpose duplex receptacles that do not supply a continuous load. What are the conductor sizes for the service entrance if a 120/240V single-phase service will be used?

Step 1 Determine the total area of the building by multiplying length by width.

$$50' \times 80' = 4,000 \text{ sq. ft.}$$

Step 2 Calculate the lighting load using *NEC Table 220.12*; according to this table, the load is 3VA per square foot.

$$4,000 \text{ sq. ft.} \times 3\text{VA} = 12,000\text{VA}$$

Step 3 Determine the total volt-amperes for the 20 general-purpose duplex receptacles.

$$20 \times 180\text{VA} = 3,600\text{VA}$$

Step 4 Calculate the load for the 30' show window on the basis of 200VA per linear foot.

$$30' \times 200\text{VA} = 6,000\text{VA}$$

Step 5 Allow one 20A outlet for sign or outline lighting if the store is on the ground floor *[NEC Section 600.5(A)]*.

If the actual load of the sign is not known, a 1,200VA load is used in the calculation per *NEC Section 220.14(F)*.

Step 6 Calculate the total load in volt-amperes.

Noncontinuous load	
Receptacle load =	3,600VA
Continuous loads	
Lighting load =	12,000VA
Show window =	6,000VA
Sign =	1,200VA
Total continuous load	19,200VA
Multiply × *125%*	
19,200VA × 1.25 =	24,000VA
Add noncontinuous load	3,600VA
Total VA =	27,600VA

Step 7 Calculate the service size in amperes.

$$27,600\text{VA} \div 240\text{V} = 115\text{A}$$

Consequently, the service-entrance conductors must be rated for no less than 115A. The next highest standard rating of an overcurrent device is 125A. The conductors would be No. 1 AWG copper or No. 2/0 AWG aluminum.

4.3.0 Office Building

A 20,000 sq. ft. office building is served by a 480Y/277V, three-phase service. The building contains the following loads:

- 10,000VA, 208V, three-phase sign
- 100 duplex receptacles supplying continuous loads rated at 180VA each
- 30' long show window
- 12kVA, 208Y/120V, three-phase electric range
- 10kVA, 208Y/120V, three-phase electric oven
- 20kVA, 480V, three-phase water heater
- Seventy-five 150W, 120V incandescent outdoor lighting fixtures
- Two hundred 200VA input, 277V fluorescent lighting fixtures
- 7.5hp, 480V, three-phase motor for fan coil unit
- 40kVA, 480V, three-phase electric heating unit
- 60A, 480V, three-phase air conditioning unit

NOTE

The 200VA listing represents the input rating rather than the sum of loads of the installed lamps. For example, these fixtures could represent four-lamp, 40W luminaires with a 200VA rating.

The ratings of the service equipment, transformers, and feeders are to be determined, along with the required size of the service grounding conductor. Circuit breakers are used to protect each circuit, and THWN copper conductors are used throughout the electrical system.

A one-line diagram of the electrical system is shown in *Figure 9*. Note that the incoming three-phase, four-wire, 480Y/277V main service terminates into a main distribution panel containing six overcurrent protective devices. Because there are only six circuit breakers in this enclosure, no main circuit breaker or disconnect is required (*NEC Section 230.71*). Five of these circuit breakers protect feeders and branch circuits to 480/277V equipment, while the sixth circuit breaker protects the feeder to a 480/208Y/120V transformer. The secondary side of this transformer feeds a 208/120V lighting panel with all 120V loads balanced. Start at the loads connected to the 208Y/120V panel and perform the required calculations.

Step 1 Calculate the load for the 100 receptacles.

$$100 \times 180\text{VA} = 18,000\text{VA}$$

Apply demand factor from *NEC Table 220.44*:

First 10,000VA at 100% + remainder (8,000VA) at 50% = 14,000VA

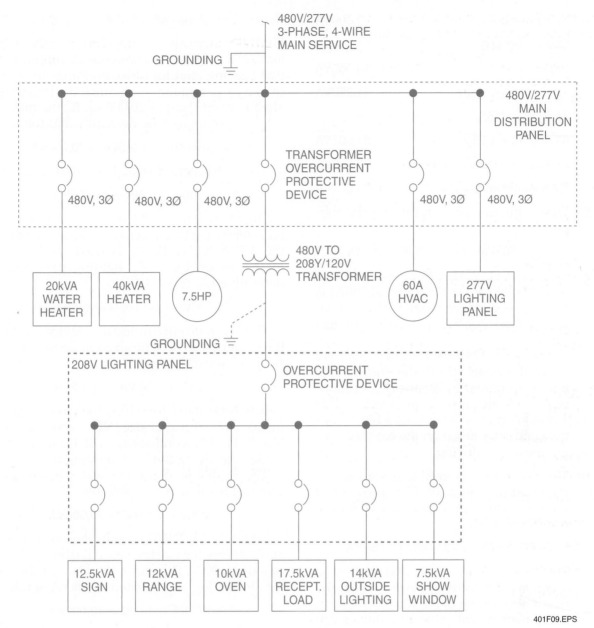

Figure 9 ◆ One-line diagram of an office building.

401F09.EPS

Step 2   Calculate the load for the show window using 200VA per linear foot.

$$200\text{VA} \times 30' = 6{,}000\text{VA}$$

Step 3   Calculate the load for the incandescent outside lighting.

$$75 \times 150\text{VA} = 11{,}250\text{VA}$$

Step 4   Calculate the load for the 10kVA sign.

$$10\text{kVA} \times 1{,}000 = 10{,}000\text{VA}$$

Step 5   Calculate the load for the 12kVA range *(NEC Section 220.56)*.

$$12\text{kVA} \times 1{,}000 = 12{,}000\text{VA}$$

Step 6   Calculate the load for the 10kVA oven *(NEC Section 220.56)*.

$$10\text{kVA} \times 1{,}000 = 10{,}000\text{VA}$$

Step 7   Determine the sum of the loads on the 208/120V lighting panel.

Noncontinuous loads

Range =	12,000VA
Oven =	10,000VA
Total noncontinuous loads =	22,000VA
Continuous loads	
Receptacles =	14,000VA

Show window =	6,000VA
Outside lighting =	11,250VA
Sign =	10,000VA
Total continuous loads =	41,250VA
Multiply × 125%	
41,250VA × 1.25 =	51,563VA
Add noncontinuous loads	22,000VA
Total feeder load =	73,563VA

Step 8 Determine the feeder rating for the subpanel.

$$73.563VA \div (\sqrt{3} \times 208V) = 204A$$

Step 9 Refer to *NEC Table 310.16* and find that 4/0 THWN conductor (rated at 230A) is the closest conductor size that will handle the load. A 225A breaker will protect this feeder.

The 208Y/120V feeder is a **separately derived system** from the transformer and is grounded by means of a grounding electrode conductor *(NEC Table 250.66)* that must be at least a No. 2 copper conductor based on the No. 4/0 copper feeder conductors.

Step 10 The transformer is sized to accommodate the computed load of 73,563VA. Select the overcurrent protective device for the transformer.

$$73.563VA \div (\sqrt{3} \times 480V) = 88.5A$$

Per *NEC Table 450.3(B)* and because a secondary protective device is used, the maximum setting of the transformer primary overcurrent protective device can be up to 250% or 2.5 × 88.5 = 221A; that is, it may not exceed this rating. The secondary protective device maximum setting is 125% of the 208V load:

$$73.563VA \div (\sqrt{3} \times 208V) = 204A$$

204A × 125% = 255A, next
highest standard overcurrent
protective device is 300A

4.3.1 Calculations for the Primary Feeder

For the feeder calculations, start with the lighting load. *NEC Table 220.12* requires a minimum general lighting load for office buildings to be based on 3.5VA per square foot. Since the building has already been sized at 20,000 sq. ft., the minimum VA for lighting may be determined as follows:

20,000 sq. ft. × 3.5VA = 70,000VA

The actual connected load, however, is as follows:

200 × 200VA = 40,000VA

Since this connected load is less than the *NEC®* requirement, it is neglected in the calculation, and the 3.5VA/sq. ft. figure is used. Therefore, the total load on the 277V lighting panel may be determined as follows:

$$70.000VA \div (\sqrt{3} \times 480V) = 84.2A$$

The overcurrent protective device protecting the feeder panel will be sized in accordance with *NEC Section 215.2(A)(1)*.

125% × 84.2A = 105.25A

It will be rated for 110A, the closest standard fuse or circuit breaker size *[NEC Section 240.6(A)]*. No. 2 THWN conductors will be used for the feeder supplying the 277V lighting panel.

The feeder load for the water heater in this example is calculated as follows:

20kVA × 1,000 = 20,000VA

NEC Section 220.51 permits the load on feeders and services for electric space heating to be computed at 100%. Therefore, the load on the feeder for the heating is computed at 40kVA as follows:

40kVA × 1,000 = 40,000VA

NEC Section 220.50 requires that motor loads be computed in accordance with *NEC Sections 430.24, 430.25, and 430.26*. Therefore, the FLA of this motor (11A) taken from *NEC Table 430.250* must be multiplied by 1.25 since this is the largest (only) motor.

$$11A \times (\sqrt{3} \times 480V) \times 1.25 = 11,431.5VA$$

Receptacles

When the number of general-purpose receptacles is unknown, use 1VA per sq. ft. for office buildings.

NEC Section 220.50 requires that hermetic refrigerant motor compressors be sized in accordance with *NEC Section 440.6.* Therefore, the load for the air conditioning unit is calculated as follows:

$$60A \times (\sqrt{3} \times 480V) = 49.883VA$$

4.3.2 Main Service Calculations

When performing the calculations for the main service, assume that all three-phase loads are balanced and may be computed in terms of volt-amperes or in terms of amperes. Calculation of the loads in amperes simplifies the selection of the main overcurrent protective device and service conductors. A summary of this calculation is shown below.

Type of Load	Computed Load	Neutral
208/120V system =	88.5A	0
277V lighting panel =	105.25A	105.25A
Water heater =	24A	0
Electric heater (neglected) =	0	0
7.5hp motor =	11A	0
Air conditioner =	60A	0
25% of largest motor =	15A	0
Service load =	303.75A	105.25A

5.0.0 ◆ RESTAURANTS

The combined load of three or more cooking appliances and other equipment for a commercial kitchen may be reduced in accordance with *NEC®*

demand factors. This provision would apply to restaurants, bakeries, and similar locations. For example, a small restaurant is supplied by a 120/208V, four-wire, three-phase service. The restaurant has the following loads:

- 1,000 sq. ft. area lighted by 120V lamps
- Ten duplex receptacles
- 20A, 208V, three-phase motor compressor
- 5hp, 208V, three-phase roof ventilation fan, running continuously, protected by an inverse time circuit breaker
- More than six units of kitchen equipment with a total connected load of 80kVA (all units are 208V, three-phase equipment)
- Two 20A sign circuits

The main service uses type THHN copper conductors and is calculated as shown in *Figure 10.* Lighting and receptacle loads contribute 35.8A to either phase A or C and 35.8A to the neutral. The 80kVA kitchen equipment load is subject to the application of a 65% demand factor (per *NEC Section 220.56*), which reduces it to a demand load of 52kVA. This load requires a minimum ampacity of 144.3A per phase at 208V. The load of the three-phase motors and 25% of the largest motor load bring the service load total to 246.8A for phase A or C and 186A for phase B.

If the phase conductors are three 250 kcmil THHN copper conductors, the grounding electrode conductor and the neutral conductor must each be at least a No. 2 copper conductor.

The fuses are selected in accordance with the *NEC®* rules for motor feeder protection. The ungrounded conductors, therefore, are protected at 300A each.

Service Loads	Line A, C	Neutral	Line B

Note: Unbalanced loads are calculated entirely on either phase A or C for a worst-case neutral (and conductor) load (*NEC Section 220.61*).

Service Loads	Line A, C	Neutral	Line B
A. 208/120V loads	20.8	20.8	0

$$\text{Lighting} = \frac{1.25 \times 2VA \text{ per sq. ft.} \times 1,000 \text{ sq. ft.}}{120V} = 20.8A$$

	Line A, C	Neutral	Line B
Receptacles $= \dfrac{180VA \times 10}{120V} = 15.0A$	15.0	15.0	0
B. Three-phase loads Kitchen equipment (6 or more units) =	144.3	0	144.3

$$\frac{80,000W \times .65}{\sqrt{3} \times 208V} = 144.3A$$

	Line A, C	Neutral	Line B
C. 20A three-phase motor compressor Breaker rating = 1.75 × 20A = 35A Use 35A (*NEC Table 430.52*)	20	0	20
D. Three-phase 5hp fan (16.7A) (*NEC Table 430.250*) Breaker rating = 2.5 × 16.7A = 41.75A Use 45A inverse-trip breaker (*NEC Table 430.52*)	16.7	0	16.7
E. 25% of largest motor load = .25 × 20A = 5A	5	0	5
F. Sign circuit $= \dfrac{1,200VA}{120V} = 10A$ each × 2 = 20A × 125% = 25A	25	25	0
	246.8A	60.8A	186A

Service Load
NEC Table 310.16
NEC Table 250.66
NEC Section 310.15(B)(4)(a)

NEC Section 230.90

NEC Section 240.6(A)

1. Conductors: Use No. 250 kcmil THHN copper at 75°C for ungrounded conductors; use No. 2 THHN copper conductor for neutral (neutral based on size of grounding electrode conductor).

2. Overcurrent protective device:
   Phases A and C = 45A (largest motor device) + 20.8A + 15.0A + 144.3A + 20A + 25A = 270.1A
   Use standard size 300A fuses.

3. Grounding electrode conductor required to be No. 2 copper.

401F10.EPS

Figure 10 ◆ Restaurant service specifications.

6.0.0 ◆ OPTIONAL CALCULATION FOR NEW RESTAURANTS

NEC Section 220.88 allows an optional method for the service calculation of a new (or completely rewired) restaurant in lieu of the standard method.

According to *NEC Table 220.88,* add all electrical loads, including heating and cooling loads, to compute the total connected load. Then select a single demand factor from the column that applies to the total computed load, and multiply the load by that factor to obtain the total demand load used for sizing the primary feeder or service-entrance conductors.

7.0.0 ◆ SERVICES FOR HOTELS AND MOTELS

The portion of the feeder or service load contributed by general lighting in hotels and motels without provisions for cooking by tenants is subject to the application of demand factors. In addition, the receptacle load in the guest rooms is included in the general lighting load at 2VA per square foot. However, the demand factor (load reduction) does not apply to any area where the entire lighting is likely to be used at one time, such as the dining room or a ballroom (see *NEC Table 220.42*).

Example:

Determine the 120/240V feeder load contributed by general lighting in a 100-unit motel. Each guest room is 240 sq. ft. in area. The general lighting load is:

2VA × 240 sq. ft. × 100 units = 48,000VA

The demand lighting load (per *NEC Table 220.42*) is:

First 20,000 at 50% =   10,000VA

48,000 − 20,000 = 28,000 at 40% =   11,200VA

Total =   21,200VA

This load would be added to any other loads on the feeder or service to compute the total capacity required.

8.0.0 ◆ OPTIONAL CALCULATIONS FOR SCHOOLS

NEC Section 220.86 provides an optional method for determining the feeder or service load of a school equipped with electric space heating, air conditioning, or both. This optional method applies to the building load, not to feeders within the building.

The optional method for schools basically involves determining the total connected load in volt-amperes, converting the load to volt-amperes per square foot, and applying the demand factors from *NEC Table 220.86*. If both air conditioning and electric space heating loads are present, only the larger of the loads is to be included in the calculation.

A school building has 200,000 square feet of floor area and a 480Y/277V service. The electrical loads are as follows:

- Interior lighting at 3VA per square foot = 600,000VA
- 300kVA power load
- 100kVA water heating load
- 100kVA cooking load
- 100kVA miscellaneous loads
- 200kVA air conditioning load
- 300kVA heating load

The service load in volt-amperes is to be determined by the optional calculation method for schools.

The combined connected load is 1,500,000VA. (This excludes the 200kVA air conditioning load since the heating load is larger.) Based on the 200,000 square feet of floor area, the load per square foot is:

1,500,000VA ÷ 200,000 sq. ft. = 7.5VA per sq. ft.

The demand factor for the portion of the load up to and including 3VA per square foot is 100%. The remaining 4.5VA per square foot in the example is added at a 75% demand factor for a total load of 1,275,000VA.

(200,000 × 3VA/sq. ft.) + (200,000 × 4.5kVA/sq. ft. × 0.75) = 1,275,000VA

To size the service-entrance conductors, *NEC Section 230.42* must be applied to this demand.

1,275,000VA × 1.25 = 1,593,750VA

Assuming a relatively short, rigid conduit overhead service where voltage drop will be negligible, calculate the size of THHW service-entrance conductors as follows for the 480Y/277V, four-wire, three-phase load:

Load current = 1,593,750VA ÷ (480V × $\sqrt{3}$)
= 1,917A

NEC Section 240.4(C) requires that the conductors have an ampacity that is not less than the rating of the overcurrent device when the device is rated more than 800A. The nearest standard rating of a nonadjustable trip overcurrent device needed for a 1,917A load is 2,000A. The conductors for this service must have an ampacity of at least 2,000A.

Because no standard size copper conductor can carry this load, the service will be accomplished by a number of parallel, four-wire runs in separate conduit between the service equipment and the service drop. The desired result is to establish the minimum number of runs with the minimum sized conductors to accommodate the load. To accomplish this, divide the load by a number of different runs and select a wire size from *NEC Table 310.16* that is very close to or exceeds the resulting current load.

Option 1 = 2,000A ÷ 4 runs = 500A = 900 kcmil conductor

Option 2 = 2,000A ÷ 5 runs = 400A = 600 kcmil conductor

Option 3 = 2,000A ÷ 6 runs = 333A = 400 kcmil conductor

Option 4 = 2,000A ÷ 7 runs = 286A = 350 kcmil conductor

Evaluate the options:

- *Option 1* – Discarded because the size of the conductors makes them difficult to install and also because it provides more capacity than needed.
- *Option 2* – Discarded because of the handling problem due to conductor size.
- *Option 3* – Selected because it meets the load capacity desired with the minimum number of runs.
- *Option 4* – Although this option meets the load requirements, it is discarded because Option 3 meets the requirement and has one less run.

The overhead service entrance is therefore minimally sized as six paralleled four-wire runs of 400 kcmil XHHW-2 copper conductor. [Type XHHW-2 was selected because it can be used in either wet or dry locations. See *NEC Table 310.13(A)*.] At this point, if the length of the run had been known, the voltage drop could be checked using the appropriate formula to see if the drop exceeded 2%. If so, the next larger standard wire could be selected.

Note that Option 4 with 350 kcmil conductors could be installed to allow for future needs. Also note that an alternate solution is to bring the higher line voltage to a small high-voltage substation inside or adjacent to the building, where it would be stepped down to an intermediate voltage and/or then stepped down to 480V and busbarred directly to switchboard equipment. This is a fairly common practice, especially in medium to large industrial establishments.

9.0.0 ◆ SHORE POWER CIRCUITS FOR MARINAS AND BOATYARDS

The wiring systems for marinas and boatyards are designed using the same *NEC®* rules as for other commercial occupancies, except for the application of several special rules dealing primarily with the design of circuits supplying power to boats (*NEC Article 555*).

The smallest sized locking and grounding-type receptacle that may be used to provide shore power for boats is 30A. Each single receptacle that supplies power to boats must be supplied by an individual branch circuit with a rating corresponding to the rating of the receptacle [*NEC Sections 555.19(A)(3) and 555.19(A)(4)*].

The feeder or service ampacity required to supply the receptacles depends on the number of receptacles and their rating, but demand factors may be applied that will reduce the load of five or more receptacles (*NEC Section 555.12*).

For example, a feeder supplying ten 30A shore power receptacles in a marina requires a minimum ampacity of:

$$10 \times 30A \times 120V = 36,000VA$$
$$36,000VA \times 0.8 = 28,800VA$$
$$28,000VA \div 240V = 120A$$

This is the minimum required by the *NEC®* unless individual watthour meters are provided for each shore power outlet.

10.0.0 ◆ FARM LOAD CALCULATIONS

NEC Article 220, Part V provides a separate method for computing farm loads (other than the dwelling). Tables of demand factors are provided for use in computing the feeder loads of individual buildings as well as the service load of the entire farm. See *NEC Sections 220.102 and 220.103*. See *NEC Article 547* for specific code requirements for agricultural buildings.

The demand factors may be applied to the 120/240V feeders for any building or load (other than the dwelling) that is supplied by two or more branch circuits. All loads that operate without diversity—that is, the entire load is on at one time—must be included in the calculation at 100% of the connected load. All other loads may be included at reduced demands. The load to be included at 100% demand, however, cannot be less than 125% of the largest motor and no less than the first 60A of the total load. In other words, if the nondiverse and largest motor load is less than 60A, a portion of the other loads will have to be included at 100% in order to reach the 60A minimum.

After the loads from individual buildings are computed, it may be possible to reduce the total farm load further by applying additional demand factors.

For example, a farm has a dwelling and two other buildings supplied by the same 120/240V service. The electrical loads are as follows:

- Dwelling = 100A load as computed by the calculation method for dwellings.
- Building No. 1 = 5kVA continuous lighting load operated by a single switch; 10hp, 240V motor; and 21kVA of other loads.
- Building No. 2 = 2kVA continuous load operated by a single switch and 15kVA of other loads.

Determine the individual building loads and the total farm load, as illustrated in *Figure 11*. The

Building No. 1 Feeder Load	240V Load	Neutral
Lighting (5kVA nondiverse load) = 5,000VA/240V	20.8	20.8
10hp motor = 1.25 × 50A	62.5	-0-
Total motor and nondiverse load	83.3	20.8
Other loads = 21,000VA/240V	87.5	87.5
Application of demand factors		
Motor and nondiverse loads @ 100%	83.3	20.8
Next 60A of other loads @ 50%	30.0	30.0
Remainder of other loads (87.5 − 60) @ 25%	6.9	6.9
Feeder Load	**120.2A**	**57.7A**
Building No. 2 Feeder Load		
Lighting (2kVA nondiverse load) = 2,000VA/240V	8.3	8.3
Other loads = 15,000VA/240V	62.5	62.5
Application of demand factors		
Nondiverse load @ 100%	8.3	8.3
Remainder of first 60 (60 − 8.3) @ 100%	51.7	51.7
Remainder of other loads (62.5 − 51.7) @ 50%	5.4	5.4
Total Farm Load	**65.4A**	**65.4A**
Application of demand factors		
Largest load (Building No. 1) @ 100%	120.2	57.7
Next largest load (Building No. 2) @ 75%	49.1	49.1
Farm load (less dwelling)	169.3	106.8
Farm dwelling load	100.0	100.0
Total Farm Load	**269.3A**	**206.8A**

401F11.EPS

Figure 11 ◆ Summary of farm calculations.

nondiverse load for building No. 1 consists of the 5kVA lighting load and the 10hp motor for a total of 83.3A. This value is included in the calculation at the 100% demand factor. Since the requirement for adding at least the first 60A of load at the 100% demand factor has been satisfied, the next 60A of the 87.5A from all other loads are added at a 50% demand factor and the remainder of 27.5A (87.5 – 60) is added at a 25% demand factor.

In the case of building No. 2, the nondiverse load is only 8.3A; therefore, 51.7A of other loads must be added at the 100% demand factor in order to meet the 60A minimum.

Using the method given for computing the total farm load, the service load is:

Largest load at 100% (Building No. 1)	= 120.2A
Second largest at 75% (Building No. 2)	= 49.1A
Total for both buildings	= 169.3A
Dwelling	= 100.0A
Total load in amperes	= 269.3A

The total service load of 269A requires the ungrounded service-entrance conductors to be at least 300 kcmil THHW copper conductors. The neutral load of the dwelling is assumed to be 100A, which brings the total farm neutral load to 207A.

11.0.0 ◆ MOTORS AND MOTOR CIRCUITS

Two or more motors may be connected to the same feeder circuit instead of routing an individual circuit to each motor. This saves money and is just as efficient if properly designed. *NEC Section 430.24* covers this type of installation. For two or more motors connected to the same feeder

conductors, the rule is to compute the sum of all motors plus 25% of the largest motor. Note the exceptions in *NEC Section 430.24* regarding motors with duty cycle ratings.

Care must be taken when designing circuits such as this to prevent several motors connected to the same circuit from starting at the same time. If not designed correctly, this would result in a severe voltage dip that could affect the operation of other equipment or prevent one or more of the motors from starting.

Example:

The following three-phase, 208V motors are supplied by the same feeder circuit:

- 20hp
- 5hp
- 10hp

Determine the total load and minimum size conductors to supply these motors.

Step 1 Determine the FLA of each motor. Refer to *NEC Table 430.250.*

$$20hp = 59.4A$$
$$5hp = 16.7A$$
$$10hp = 30.8A$$

Step 2 Take 125% of the largest motor FLA.

$$59.4A \times 125\% = 74.25$$

Step 3 Add all loads.

$$74.25A + 16.7A + 30.8A = 121.75A$$

Step 4 Find the minimum size THWN copper conductor from *NEC Table 310.16.*

A No. 1 AWG, type THWN copper conductor is rated at 130A.

1. A feeder circuit supplying a lighting and appliance panelboard is sized by adding _____% of the noncontinuous load and _____% of the continuous load.
   a. 80; 100
   b. 100; 100
   c. 100; 110
   d. 100; 125

2. What is the ampacity for each of eight No. 2 THWN copper current-carrying conductors installed in a single conduit?
   a. 57.5A
   b. 80.5A
   c. 91A
   d. 92A

3. What is the rating for a No. 250 kcmil THHN copper conductor in an ambient temperature of 101°F?
   a. 232.05A
   b. 263.9A
   c. 278.4A
   d. 290A

4. What is the rated ampacity per phase for six No. 3/0 THHN copper current-carrying conductors with two conductors per phase (parallel) and all conductors installed in the same conduit?
   a. 280A
   b. 315A
   c. 320A
   d. 360A

5. The voltage drop percentage for a panel feeder with a noncontinuous load of 180A, using 3/0 THWN copper conductors at 480V, 3Ø, and a length of 248' is _____.
   a. 1.23%
   b. 1.28%
   c. 1.42%
   d. 2.03%

6. What is the required size of THHN copper conductors for a feeder serving two 208Y/120V, four-wire, three-phase, 75°C branch circuit panelboards, one with a continuous load of 150A and the other with a noncontinuous load of 180A?
   a. 250 kcmil
   b. 400 kcmil
   c. 500 kcmil
   d. 600 kcmil

7. What is the voltage drop, in volts and in percent, for a 208V, 3Ø feeder serving a load of 200A at a distance of 290' using 4/0 THWN copper conductors?
   a. 3.53V; 1.7%
   b. 6.11V; 2.94%
   c. 7.07V; 3.39%
   d. 10.05V; 4.83%

8. For a THHN feeder rated at 225A, the _____ column from *NEC Table 310.16* must be used unless all components in the feeder circuit are rated for a higher temperature.
   a. 60°F
   b. 75°C
   c. 90°C
   d. 140°C

9. What size THWN copper conductors are required to supply a 120/208V, 3Ø panelboard with 20.5kVA of noncontinuous load and 25kVA of continuous load?
   a. 250 kcmil
   b. No. 1
   c. 1/0
   d. 4/0

10. What is the load, in amps, for a 1Ø, 240V feeder supplying a load calculated at 23,800VA?
    a. 57.25A
    b. 66.06A
    c. 99.17A
    d. 114.42A

11. What is the load, in amps, for a 3Ø, 208V feeder supplying a load calculated at 96.75kVA?
    a. 116.37A
    b. 268.55A
    c. 403.13A
    d. 465.14A

12. What is the load, in amps, for a 3Ø, 480V feeder supplying a load calculated at 112.5kVA?
    a. 135.32A
    b. 312.27A
    c. 468.75A
    d. 540.87A

13. Using the 10' tap rule, what size copper THWN conductor is required for a tap from a 400A, 480V, 3Ø feeder to serve a 150A load?
    a. No. 1
    b. No. 2
    c. 1/0
    d. 2/0

14. Using the 25' tap rule, what is the minimum size copper THWN conductor required for a tap from a 400A, 480V, 3Ø feeder?
    a. No. 1
    b. No. 2
    c. 1/0
    d. 2/0

15. What is the minimum general lighting load, in VA per square foot, required by the *NEC*® for a single-family dwelling?
    a. 1
    b. 2
    c. 3
    d. 3½

16. What is the minimum general lighting load, in VA per square foot, required by the *NEC*® for an office building?
    a. 2
    b. 3
    c. 3½
    d. 4½

17. What is the minimum general lighting load, in VA per square foot, required by the *NEC*® for a store?
    a. 2
    b. 2½
    c. 3
    d. 5

18. The lighting load for dwellings and other *NEC*®-listed occupancies is generally determined by _____.
    a. total connected load for the entire building
    b. volt-amperes per linear foot
    c. volt-amperes per square foot
    d. anticipated connected load for the building

19. When determining the building area for load calculations, building measurements are taken using the _____.
    a. outside dimensions of the building
    b. inside wall dimensions of the building
    c. width and length of each individual room
    d. property line perimeters

20. When both heating and cooling systems are used in a building, which of the following statements is true?
    a. The total load for both systems must be combined and then multiplied by a factor of 125% (1.25).
    b. The larger of the two loads is used in the calculation.
    c. The smaller of the two loads is used in the calculation.
    d. The total load for both systems must be multiplied by a factor of 150% (1.50).

21. When calculating the service size for a single-family dwelling using the standard calculation, what part of the total general lighting and small appliance loads must be calculated at 100%?
    a. First 3,000VA
    b. First 10,000VA
    c. First 15,000VA
    d. First 30,000VA

22. Using the standard calculation for a single-family dwelling, what is the demand factor for the general lighting and small appliance load that exceeds 120,000VA?
    a. 10%
    b. 15%
    c. 25%
    d. 30%

23. What is the minimum demand required for the portion of a feeder that supplies 30' of show windows?
    a. 2,700VA
    b. 3,000VA
    c. 5,400VA
    d. 6,000VA

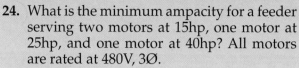

24. What is the minimum ampacity for a feeder serving two motors at 15hp, one motor at 25hp, and one motor at 40hp? All motors are rated at 480V, 3Ø.

   a. 100A
   b. 128A
   c. 141A
   d. 153A

25. When calculating the feeder load for a motor control center supplying several motors, which of the following statements is correct?

   a. Determine the sum of all motor loads.
   b. Determine the sum of all motor loads plus 30% of the largest motor.
   c. Determine the sum of all motor loads plus 25% of the largest motor.
   d. Determine the sum of all motor loads plus 125% of the largest motor.

Summary

Load calculations are necessary for determining sizes and ratings of conductors, equipment, and overcurrent protection required by the *NEC®* to be included in each electrical installation, including the service, feeders, and branch circuit loads. These calculations are necessary in every electrical installation, from the smallest roadside vegetable stand to the largest industrial establishment. Therefore, every electrician must know how to calculate services, feeders, and branch circuits for any given installation and must also know what *NEC®* requirements apply.

Notes

Trade Terms Introduced in This Module

Building: A structure that stands alone or that is cut off from adjoining structures by fire walls with all openings therein protected by approved fire doors.

Dwelling unit: One or more rooms for the use of one or more persons as a housekeeping unit with space for eating, living, and sleeping, as well as permanent provisions for cooking and sanitation. A one-family dwelling consists solely of one dwelling unit. A two-family dwelling consists solely of two dwelling units. A multi-family dwelling is a building containing three or more dwelling units.

Feeder: All circuit conductors between the service equipment or the source of a separately derived system and the final branch circuit overcurrent device.

Garage: A building or portion of a building in which one or more self-propelled vehicles carrying volatile, flammable liquid for fuel or power are kept for use, sale, storage, rental, repair, exhibition, or demonstration purposes, and all that portion of a building that is on or below the floor or floors in which such vehicles are kept and that is not separated by suitable cutoffs.

Overhead system service entrance: The service conductors between the terminals of the service equipment and a point usually outside the building, clear of building walls, where it is joined by a tap or splice to the service drop.

Separately derived system: A premises wiring system whose power is derived from a battery, a solar photovoltaic system, or from a generator, transformer, or converter windings, and that has no direct electrical connection, including a solidly-connected grounded circuit conductor, to supply conductors originating in another system.

Service: The conductors and equipment for delivering electric energy from the serving utility to the wiring system of the premises served.

Service conductors: The conductors from the service point to the service disconnecting means.

Service drop: The overhead service conductors from the last pole or other aerial support up to and including splices, if any, connecting to the service-entrance conductors to the building or other structure.

Service equipment: The necessary equipment, usually consisting of a circuit breaker or switch and fuses, and their accessories, located near the point of entrance of supply conductors to a building or other structure, or an otherwise defined area, and intended to constitute the main control and means of cutoff of the supply.

Service lateral: The underground service conductors between the street main, including any risers at a pole or other structure or from transformers, and the first point of connection to the service-entrance conductors in a terminal box or meter or other enclosure with adequate space, inside or outside the building wall. Where there is no terminal box, meter, or other enclosure with adequate space, the point of connection shall be considered to be the point of entrance of the service conductors into the building.

Service point: The point of connection between the facilities of the serving utility and the premises wiring.

Switchboard: A large single panel, frame, or assembly of panels on which are mounted, on the face, back, or both, switches, overcurrent and other protective devices, buses, and instruments. Switchboards are generally accessible from the rear as well as from the front and are not intended to be installed in cabinets.

Tap conductors: As defined in *NEC Article 240,* a tap conductor is a conductor, other than a service conductor, that has overcurrent protection ahead of its point of supply that exceeds the value permitted for similar conductors that are protected as described in *NEC Section 240.4.*

Underground service entrance: The service conductors between the terminals of the service equipment and the point of connection to the service lateral.

Load Calculation Form

General Lighting Load					Phase	Neutral
Square footage of the dwelling	[1]	× 3VA =	[2]			
Kitchen small appliance circuits	[3]	× 1500 =	[4]			
Laundry branch circuit	[5]	× 1500 =	[6]			
Subtotal of general lighting loads per *NEC Section 220.52*			[7]			
Subtract 1st 3000VA per *NEC Table 220.42*			[8]	× 100% =	[9]	
Remaining VA times 35% per *NEC Table 220.42*			[10]	× 35% =	[11]	
Total demand for general lighting loads =					[12]	[13]
Fixed Appliance Loads (nameplate or NEC FLA of motors) per *NEC Section 220.53*						
Hot water tank, 4.5kVA, 240V			[14]			
			[15]			
			[16]			
			[17]			
			[18]			
			[19]			
Subtotal of fixed appliances			[20]			
If 3 or less fixed appliances take @ 100% =					[21]	[22]
If 4 or more fixed appliances take @ 75% =					[23]	[24]
Other Loads per *NEC Section 220.14*						
Electric range per *NEC Table 220.55* (neutral @ 70 % per *NEC Section 220.61*)					[25]	[26]
Electric dryer per *NEC Table 220.54* (neutral @ 70% per *NEC Section 220.61*)					[27]	[28]
Electric heat per *NEC Section 220.51*						
Air conditioning per *NEC Section 220.14(A)*			*Omit smaller load per NEC Section 220.60*		[29]	[30]
Largest Motor =		× 25% per *NEC Section 430.24* =			[31]	[32]
Total VA Demand =					[33]	[34]
VA/240V = Amps =					[35]	[36]
Service OCD and Minimum Size Grounding Electrode Conductor =					[37]	[38]
AWG per *NEC Tables 310.15(B)(6)* and *310.16* for neutral					[39]	[40]

401A01.EPS

This module is intended to present thorough resources for task training. The following reference work is suggested for further study. This optional material is for continuing education rather than for task training.

National Electrical Code® Handbook, Latest Edition. Quincy, MA: National Fire Protection Association.

NCCER makes every effort to keep these textbooks up-to-date and free of technical errors. We appreciate your help in this process. If you have an idea for improving this textbook, or if you find an error, a typographical mistake, or an inaccuracy in NCCER's Contren® textbooks, please write us, using this form or a photocopy. Be sure to include the exact module number, page number, a detailed description, and the correction, if applicable. Your input will be brought to the attention of the Technical Review Committee. Thank you for your assistance.

Instructors – If you found that additional materials were necessary in order to teach this module effectively, please let us know so that we may include them in the Equipment/Materials list in the Annotated Instructor's Guide.

Write: Product Development and Revision
National Center for Construction Education and Research
3600 NW 43rd St., Bldg. G, Gainesville, FL 32606

Fax: 352-334-0932

E-mail: curriculum@nccer.org

Craft   Module Name

Copyright Date   Module Number   Page Number(s)

Description

(Optional) Correction

(Optional) Your Name and Address

Health Care Facilities

Cornell University – Sustainability

Cornell's organizational strategy is reflected in the unique construction of a "Light Canopy," which is adapted to their solar house. The Light Canopy's streamlined framework of steel trusses serves as a support for photovoltaic cells, evacuated tubes for solar water heating, and a series of vegetated screens that provide shade in the summer.

26402-08

26402-08
Health Care Facilities

Topics to be presented in this module include:

1.0.0	Introduction	.2.2
2.0.0	Essential Electrical System Types	.2.3
3.0.0	Electrical Distribution Systems	.2.7
4.0.0	Wiring and Devices	.2.11
5.0.0	Communication, Signaling, Data, and Fire Alarm Systems	.2.21
6.0.0	Isolated Power Systems	.2.22

Overview

Health care facilities must have adequate backup power should normal power be lost. In a health care facility, power distribution is divided into two categories: essential loads and nonessential loads. Essential loads consist of things that are needed for the safety of all occupants and for safe and effective patient care. Nonessential loads are things that are not necessary for the safety of building occupants. When the normal electrical source is lost, an alternate source must be able to supply the essential loads. In addition to backup power, health care facilities also require the use of special communication systems and wiring devices.

Objectives

When you have completed this module, you will be able to do the following:

1. List the types of electrical distribution systems used in the medical industry.
2. Describe the categories and branch portions of the distribution circuits.
3. List the items allowed in the life safety branch and critical branch.
4. Describe ground fault protection required to ensure a safe environment.
5. List the required wiring methods in a health care facility.
6. Explain the application of special wiring devices in critical care locations.
7. Describe the requirements for the installation of specialty equipment.
8. Describe the applications of isolated power systems.

Trade Terms

Equipotential ground plane
Inpatient
Outpatient
Throwover

Required Trainee Materials

1. Pencil and paper
2. Copy of the latest edition of the *National Electrical Code*®

Prerequisites

Before you begin this module, it is recommended that you successfully complete *Core Curriculum*; *Electrical Level One*; *Electrical Level Two*; *Electrical Level Three*; and *Electrical Level Four*, Module 26401-08.

This course map shows all of the modules in *Electrical Level Four*. The suggested training order begins at the bottom and proceeds up. Skill levels increase as you advance on the course map. The local Training Program Sponsor may adjust the training order.

26413-08 Introductory Skills for the Crewleader

26412-08 Special Locations

26411-08 Medium-Voltage Terminations/Splices

26410-08 Motor Operation and Maintenance

26409-08 Heat Tracing and Freeze Protection

26408-08 HVAC Controls

26407-08 Advanced Controls

26406-08 Specialty Transformers

26405-08 Fire Alarm Systems

26404-08 Basic Electronic Theory

26403-08 Standby and Emergency Systems

26402-08 Health Care Facilities

26401-08 Load Calculations – Feeders and Services

ELECTRICAL LEVEL FOUR

ELECTRICAL LEVEL THREE

ELECTRICAL LEVEL TWO

ELECTRICAL LEVEL ONE

CORE CURRICULUM: Introductory Craft Skills

402CMAP.EPS

1.0.0 ◆ INTRODUCTION

This module covers electrical requirements for health care facilities. It is based on *NFPA 70®, National Electrical Code® (NEC®)*, and *NFPA 99, Standard for Health Care Facilities*. However, state and local codes vary and may use different standards than those listed here. In addition, the Joint Commission (formerly the Joint Commission on Accreditation of Health Care Organizations or JCAHO) addresses emergency electrical power systems in their standards *EC.7.20* and *EC.7.40*, and addresses emergency procedures for utility system disruptions in standard *EC.7.10*. The standards for health care facilities are much more rigidly enforced than for other facilities. This is especially true for fireproofing wall or floor penetrations because patients cannot be evacuated quickly in case of fire. The Joint Commission inspects the health care facility at the time of occupancy, and then performs an extensive on-site inspection at least once every three years. Always check the most recent national, state, and local codes as well as Joint Commission requirements before beginning any installation.

Just like homes, health care facilities run on electrical power. Unlike homes, a loss of power at a hospital can cost lives. Just imagine the results of an unexpected power failure during a delicate surgery. The loss of electrical service to a health care facility can be devastating.

Earthquakes, hurricanes, floods, and tornados can interrupt electrical service. Fires, explosions, and electrical failures in the facility due to shorts or overloads can prevent the successful transfer of power. A well-planned electrical system is required to prevent or limit the internal disruption of power in these facilities. Loss of power can be corrected in seconds depending on the system employed. Systems should be designed to cope with the longest probable power outage. Safeguards are needed to ensure the proper operation and maintenance of electrical circuitry and mechanical components in vital areas.

The installation of electrical systems in health care facilities is governed by *NEC Article 517* and *NFPA 99*. (*OSHA Standard 29 CFR 1910.307* also applies, but this standard supports the *NEC®*.) Heath care facilities are defined as a place where medical, dental, or nursing care is provided. This includes hospitals, nursing homes, limited care facilities, ambulatory facilities, psychiatric hospitals, dental and medical clinics, and stand-alone day surgery centers. These types of facilities may be permanent construction, such as multilevel hospitals, or movable, such as mobile medical clinics.

> **NOTE**
>
> In addition to the *NEC®* and *NFPA 99*, some parts of a health care facility may fall under *IEEE Standard 241, Recommended Practice for Electrical Systems in Commercial Buildings*.

To understand the electrical code requirements for health care facilities, it is important to be clear about the definitions for these facilities. The *NEC®* definitions for various facilities are as follows:

- *Ambulatory health care occupancy* – A building or a portion thereof used to provide on an outpatient basis one or more of the following services or treatments that temporarily prevent patients from caring for themselves during an emergency:
  - Treatment or anesthesia that renders patients incapable of taking action for self-preservation without assistance
  - Emergency or urgent care for patients who due to the nature of their injury or illness are incapable of taking action for self-preservation without assistance
- *Hospital* – A building or a portion thereof that provides continuous (around-the-clock) inpatient services or treatment for medical, psychiatric, obstetrical, or surgical procedures.

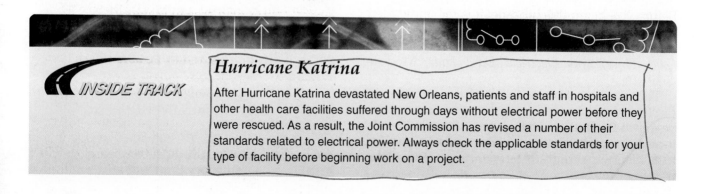

INSIDE TRACK

Hurricane Katrina

After Hurricane Katrina devastated New Orleans, patients and staff in hospitals and other health care facilities suffered through days without electrical power before they were rescued. As a result, the Joint Commission has revised a number of their standards related to electrical power. Always check the applicable standards for your type of facility before beginning work on a project.

- *Limited care facility* – A building or a portion thereof that provides around-the-clock care for patients who are incapable of self-preservation due to a physical or mental condition or debilitation due to an accident, illness, or chemical dependency.
- *Nursing home* – A building or a portion thereof that provides around-the-clock care for patients who might be unable to provide for their own needs and safety because of a physical or mental incapacity. *elderly*

NOTE

To qualify as any of the facilities defined here, they must be used simultaneously for four or more patients.

Within a health care facility, there are also various definitions for patient care areas. A patient care area is any portion of a health care facility where patients undergo examination or treatment. Normally, business offices, corridors, lounges, day rooms, dining rooms, and similar areas are not classified as patient care areas. Patient care areas include the following:

- *General care areas* – Patient bedrooms, examining rooms, treatment rooms, clinics, and similar areas where patients will come in contact with ordinary appliances including nurse call systems, electric beds, telephones, and entertainment equipment.
- *Critical care areas* – Special care units, intensive care units, coronary care units, angiography laboratories, cardiac catheterization laboratories, delivery rooms, operating rooms, and similar areas where patients are normally subjected to invasive procedures and are connected to line-operated electromedical equipment.
- *Wet procedure locations* – Any spaces within patient care areas where a procedure is performed and that are normally subject to wet conditions while patients are present. These include standing fluids on the floor or drenching of the work area that is intimate to a patient or staff. Routine housekeeping procedures or incidental liquid spills do not define a wet location.

2.0.0 ◆ ESSENTIAL ELECTRICAL SYSTEM TYPES

In a health care facility, power is divided into two categories: the nonessential loads and the essential electrical system (EES) (*Figure 1*). Nonessential loads are things that are not necessary

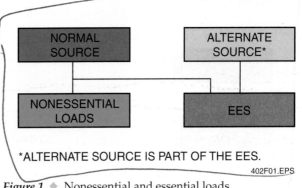

Figure 1 ◆ Nonessential and essential loads.

for the safety of building occupants and for safe and effective patient care, such as decorative lighting or televisions. The EES consists of systems or devices that are needed for the safety of all occupants, such as fire alarms, exit lighting, and hallway lighting, and for patient care, such as respirators, cardiac bypass equipment, and suction devices. When the normal electrical source is lost, the facility's alternate source must be able to supply the needs of the EES.

The EES is divided into the equipment system and the emergency system. The equipment system consists of those items that are needed to operate the facility but are not directly related to patient care, such as the facility's telephone system. The emergency system consists of those items that are essential to patient care. This system is further divided into the life safety branch and the critical branch. The life safety branch supplies power to those items needed to ensure the safety of building occupants as well as patients. It includes exit signs, stairway lighting, alarm system, and communication systems. The critical branch supplies power to areas that are directly related to patient care. These include pharmacies, emergency rooms, intensive care units, and so on. See *Figure 2*.

Typically, normal power for a system is supplied by electric companies, while an on-site power source (battery system or generators) is used as an alternate power source. Naturally, this depends on the specifics of the health care facility. However, when normal power is from on-site power generators, the alternate power source can be either the utility or other generators. In hospitals, when the normal power for a system is supplied by the utility, a generator is required as the alternate power source. Some facilities that do not admit patients on life support equipment may be able to use battery systems as their alternate power source.

In hospitals and similar facilities, the equipment system consists of electrical equipment and circuits for three-phase distribution including manual transfer or delayed automatic devices to feed essential equipment loads as described in *NFPA 99*. The emergency system includes distribution

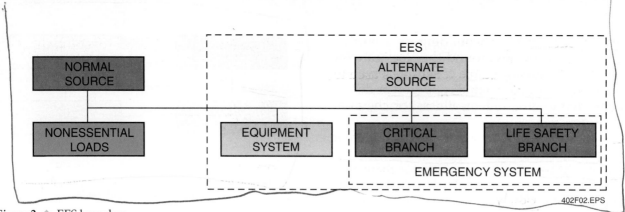

Figure 2 ◆ EES branches.

equipment and circuitry, and the automatic transfer devices needed to transfer from normal power to emergency power. To improve system reliability, each circuit must be installed separately from all other electrical circuits.

The *NEC®* and *NFPA 99* both require that the emergency system automatically restore power within 10 seconds of power interruption or loss. These standards also define the electrical loads to be served by the life safety branch and critical branch. In addition, *NEC Section 517.33(A)(9)* allows the installation of additional task illumination, receptacles, and selected power circuits needed for the effective operation of the hospital on the critical branch of the emergency system. This permits some flexibility in customizing the system to fit the needs of the hospital. *NEC Section 517.30(B)* requires the following:

- Essential electrical systems for hospitals are made up of two separate systems to supply a limited amount of lighting and power service that has been designated as essential for life safety and effective hospital operation during the time the normal electrical service is interrupted. These two systems are the emergency system and the equipment system.
- The emergency system is limited to circuits essential to life safety and critical patient care. These are designated the life safety branch and the critical branch.
- The equipment system must supply major electric equipment necessary for patient care and basic hospital operation.
- The number of transfer switches used to transfer power from the normal source to the alternate source is selected based on reliability, design, and load considerations. Each branch of the EES must be served by one or more transfer switches, as shown in *NEC FPN Figure 517.30, No. 1*. One transfer switch may serve one or more branches or systems in a facility with a maximum demand on the EES of 150kVA, as shown in *NEC FPN Figure 517.30, No. 2*.

Naturally, this is assuming that the transfer switch has sufficient capacity to serve the combined additional loads and that the alternate source of power is large enough to support the simultaneous transfer of both systems when normal power is lost.

The requirements for the EES are based on the care provided at the facility. There are three levels of EES: Type 1, Type 2, and Type 3. Type 1 is the most stringent. Type 3 is the least stringent.

2.1.0 Type 1 EES

Type 1 EES is required in any facility with a critical care unit or that admits patients requiring electrical life support equipment, regardless of the name of the facility. Type 1 must automatically switch to the backup system within 10 seconds of a power loss.

The life safety branch of a Type 1 EES supplies power for the following equipment, lighting, and receptacles:

- Hospital communication systems, where used for the dissemination of instructions during emergencies
- Task illumination battery chargers for emergency battery-powered lighting units and selected receptacles at the generator set location
- Illumination of exits, such as lighting required for corridors, passageways, stairways, landings at exit doors, and all approaches to exits

 CAUTION

Equipment to transfer patient corridor lighting from general illumination circuits to night illumination circuits is allowed as long as only one of two circuits can be selected, and both circuits cannot be off at the same time.

- Exit direction and exit signs
- Alarm and alerting systems, including fire alarms and alarms required for systems used to pipe nonflammable medical gases
- Elevator cab communication, control, lighting, and signal systems
- Automatically operated exit doors

CAUTION

No capacity or service other than those listed above may be connected to the life safety branch electrical system.

The critical branch of a Type 1 EES may consist of one or more branches that are designed to serve a limited number of receptacles and locations to reduce the load demand and the chances of a fault condition. Receptacles in general patient care corridors are allowed on the critical branch; however, they must be identified as part of the critical branch (either labeled or color-coded), in compliance with *NFPA 99*.

Task illumination and selected receptacles – The emergency system's critical branch supplies power for task illumination, selected receptacles, fixed equipment, and special power circuits feeding the following locations and functions related to patient care:

- Critical care areas that use anesthetizing gases – task illumination, selected receptacles, and fixed equipment
- The isolated power systems located in special environments
- Patient care areas – task illumination and selected receptacles in infant nurseries, medication preparation areas, pharmacy dispensing areas, selected acute nursing areas, psychiatric bed areas (omit receptacles), ward treatment rooms, and nurses' stations unless adequately lighted by corridor luminaires
- Additional specialized patient care task illumination and receptacles, where needed
- Nurse call systems
- Blood, bone, and tissue banks
- Telephone equipment rooms and closets
- Task illumination, selected receptacles, and selected power circuits for general care beds (at least one duplex receptacle per patient bedroom), angiographic labs, cardiac catheterization labs, coronary care units, hemodialysis rooms or areas, emergency room treatment areas (selected), human physiology labs, intensive care, and postoperative recovery rooms (selected)

Code Blue

Sometimes communication systems are used to call a code blue alarm. When a code blue is called, trained staff responds with a crash cart, which contains special equipment and medication that is used to revive a patient whose heart has stopped beating. Some hospitals use a wall-mounted alarm device such as the one shown here. It is included in the life safety branch of a Type 1 EES.

402SA01.EPS

- Additional task illumination, receptacles, and selected power circuits needed for effective hospital operation

NOTE

Single-phase fractional horsepower exhaust fan motors that are interlocked with three-phase motors on the equipment may be connected to the critical branch.

2.2.0 Type 2 EES

Many nursing homes and limited care facilities require Type 2 EES. These facilities do not have critical care and intensive care units, but they do provide patient care that is equivalent to hospital inpatient care; therefore, the branches of the emergency system in Type 2 EES are similar to Type 1.

NEC Section 517.35 and *NFPA 99* require an alternate source of power for any health care facilities where patients are treated in primarily the same manner as in hospitals, even though the facility may not be designated a hospital. Nursing homes and limited care facilities that are adjoining a hospital may use the essential electrical systems supplied by the hospital, but both facilities must have their own transfer devices.

Essential electrical systems for limited care facilities and nursing homes must have two separate branches that are able to supply the lighting and power required for the protection of life safety and the effective operation of the facility during the interruption of normal electrical service. These two branches are the life safety branch and the critical branch. The number of transfer switches used is based on reliability, design, and load considerations. Every branch of the EES must be served by one or more transfer switches as shown in *NEC FPN Figure 517.41, No. 1*. A single transfer switch will serve one or more branches or systems in a facility with a maximum demand on the EES of 150kVA, as shown in *NEC FPN Figure 517.41, No. 2*.

The life safety branch must be completely independent of all other equipment and wiring. It cannot enter the same cabinets, boxes, or raceways with other wiring except in transfer switches or in exit or emergency lighting fixtures (or their attached junction boxes), as long as the fixtures are supplied from two sources. The wiring of the critical branch can occupy the same cabinets, boxes, or raceways of other circuits that are not part of the life safety branch. See *NEC Section 517.41(D)*.

The life safety branch must be installed and connected to the alternate power source so that designated functions are automatically restored to operation within 10 seconds after the normal source interruption. The life safety branch supplies power for the following equipment, lighting, and receptacles:

- Hospital communication systems, where used for dissemination of instructions during emergencies
- Task illumination battery chargers for emergency battery-powered lighting units and selected receptacles at the generator set location

- Illumination of exits, such as lighting required for corridors, passageways, stairways, landings at exit doors, and all approaches to exits

CAUTION

Equipment to transfer patient corridor lighting from general illumination circuits to night illumination circuits is allowed as long as only one of two circuits can be selected, and both circuits cannot be off at the same time.

- Exit direction and exit signs
- Alarm and alerting systems, including fire alarms and alarms required for systems used for the piping of nonflammable medical gases
- Elevator cab communication, control, lighting, and signal systems
- Lighting in recreation and dining areas must be sufficient to illuminate exits.

CAUTION

No function other than those listed above may be connected to the life safety branch.

The critical branch must be installed and connected to the alternate power source so that the equipment listed in *NEC Section 517.43(A)* is automatically restored to operation at appropriate time-lag intervals following the restoration of the life safety branch to operation. Its arrangement must also provide for the additional connection of equipment listed in *NEC Section 517.43(B)* by either delayed automatic or manual operation.

The following equipment may be connected to the critical branch and arranged for delayed automatic connection to the alternate power source:

- Patient care areas – task illumination and selected receptacles in medication preparation areas, pharmacy dispensing areas, and nurses' stations unless adequately lighted by corridor luminaires
- Sump pumps and other equipment required to operate for the safety of major apparatus and associated control systems and alarms
- Smoke control and stair pressurization systems
- Kitchen hood supply and/or exhaust systems, if required to operate during a fire in or under the hood
- Supply, return, and exhaust ventilating systems for airborne infectious isolation rooms

The following equipment may be connected to the critical branch and arranged for either delayed automatic or manual connection to the alternate power source:

- Equipment to provide heating for patient rooms, except where:
  - The outside design temperature is higher than +20°F.
  - The outside design temperature is lower than +20°F but selected areas are heated for the comfort of all confined patients.
  - The facility is served by a dual source of normal power as described in *NEC Section 517.44(C), FPN*.
- In instances where disruption of power would result in elevators stopping between floors, throwover (transfer) switches must be provided to allow temporary operation of any elevator for the release of passengers. For elevator cab lighting, control, and signal system requirements, see *NEC Section 517.42(G)*.
- Additional illumination, receptacles, and equipment may be connected only to the critical branch.

2.3.0 Type 3 EES

Type 3 EES is used in some nursing homes and limited care facilities. It is the least stringent of the three EES types and is used only in facilities that admit the most self-sufficient patients. It must provide battery backup power for 90 minutes in the event of a primary power failure. Type 3 is permitted in freestanding buildings used as nursing homes and limited care facilities under the following conditions:

- The facility maintains admission and discharge policies that preclude the provision of care for any patient or resident who may need to be sustained by electrical life-support equipment.
- The facility offers no surgical treatment requiring general anesthesia.
- The facility provides automatic battery-operated power that is effective for at least 90 minutes and operates in accordance with *NEC Section 700.12*. The backup system must be capable of supplying lighting for exit lights, exit corridors, stairways, nursing stations, medical preparation areas, boiler rooms, communication areas, and all alarm systems.

Limited care facilities and nursing homes that furnish inpatient hospital care must comply with the requirements of *NEC Sections 517.30 through 517.35*. Regardless of the name given to the facility, the category of electrical system will depend on the type of patient care. Where care is evidently inpatient hospital care, a hospital-type electrical system must be installed.

Nursing homes and limited care facilities that are contiguous with or located on the same site as a hospital may have their essential electrical systems supplied by that of the hospital. Where a limited care facility or nursing home is in the same building with a hospital, the nursing home is not required to have its own EES if it gets its power from the hospital. This rule does not allow sharing of transfer devices.

3.0.0 ◆ ELECTRICAL DISTRIBUTION SYSTEMS

Providing a health care facility with reliable electrical power begins with careful system design followed by proper installation. To ensure continuous electrical service from the utility provider, hospitals and certain other health care facilities are placed on a spot network. This means the facility receives power from two or more distribution points from a provider rather than just one point as is typically done in a residential area. Ideally, these feeders are installed in parallel so that a fault in one feeder will not affect or produce faults in another. Each feeder must be sized to carry the full load of the system. This type of installation helps to ensure an adequate amount of power during high use periods.

3.1.0 Double-Ended System Arrangement

When transformation over 750kVA is needed, a double-ended substation is normally used. The substation shown in *Figure 3* uses a normally open tie circuit breaker interlocked with the mains circuit breakers so that not all three can be closed simultaneously. When a single mains feeder has loss, the tie circuit breaker may be closed, manually or automatically, adding additional load to the remaining feeder. The substation feeders and equipment can be medium-voltage equipment to reduce current rating requirements or low-voltage equipment. In either case, each of the utility feeders and circuit breakers used must be rated to carry the full load of the facility in the event that one of the mains sources is lost. Double-ended substations have additional benefits, such as lower fault current and the ability to differentiate between motor loads that require specific voltage regulation.

If the double-ended substation design uses a normally open, electrically operated tie circuit

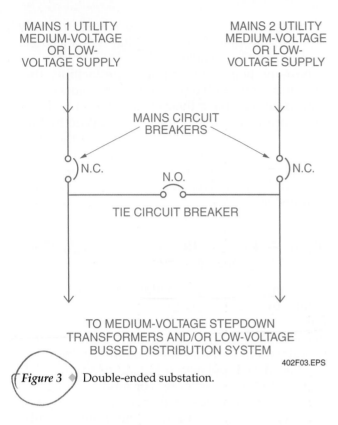

MAINS 1 UTILITY MEDIUM-VOLTAGE OR LOW-VOLTAGE SUPPLY

MAINS 2 UTILITY MEDIUM-VOLTAGE OR LOW-VOLTAGE SUPPLY

MAINS CIRCUIT BREAKERS

N.C.

N.O.

N.C.

TIE CIRCUIT BREAKER

TO MEDIUM-VOLTAGE STEPDOWN TRANSFORMERS AND/OR LOW-VOLTAGE BUSSED DISTRIBUTION SYSTEM

402F03.EPS

Figure 3 ◆ Double-ended substation.

breaker that automatically closes upon loss of either of the incoming mains feeders, then control and protective relaying must be added to prevent the bus tie circuit breaker from closing when a mains circuit breaker has tripped due to overload or short circuit conditions.

Utility companies may allow double-ended substations to operate with a closed tie and mains circuit breakers. Double-ended substations with mains and normally closed ties have distinct advantages such as better voltage regulation, transfer capabilities when power loss occurs, immediate switching when a power source is lost, and greater reliability of system operation. However, the disadvantages are greater fault current, greater cost, and a more complex design. The decision to use a closed tie is made by the designer of the system.

3.2.0 Alternate Power Source Arrangement

For a Type 1 or 2 EES, one or more diesel engine generators are normally used (*Figure 4*). In some cases where multiple generators are used, one is a designated backup generator. Generators may be housed internally or externally to the health facility. If housed in a nonheated environment where temperatures fall below about 30°F, the starting

batteries must be heated and cold-weather diesel engine starting equipment used in order to meet the 10-second startup requirement for restoration of power. *Figure 5* is a one-line diagram showing the arrangement of an alternate power source with transfer switching for an EES. As noted in the figure, delayed switching of the EES equipment system may be used if the alternate power source cannot tolerate the total EES load at startup.

3.3.0 Ground Fault Protection

Phase overcurrent device settings are primarily established by the load requirements. They are set to be insensitive to full-load and inrush current and to provide selectivity between downstream and upstream devices. However, phase overcurrent devices cannot distinguish between normal load current and a low-magnitude ground fault short circuit current of the same magnitude. For this reason, ground fault detection is used in addition to the phase overcurrent devices to provide protection.

NEC Sections 215.10 and 230.95 describe requirements for ground fault protection at the feeder. In health care facilities, additional ground fault protection may be required downstream. *NEC Section 517.17* requires two levels of ground fault protection in health care facilities to improve the selectivity between the feeder and building devices. The wiring configuration and estimated ground currents must be carefully analyzed before selecting appropriate ground fault protective devices.

402F04.EPS

Figure 4 ◆ Typical diesel engine generator.

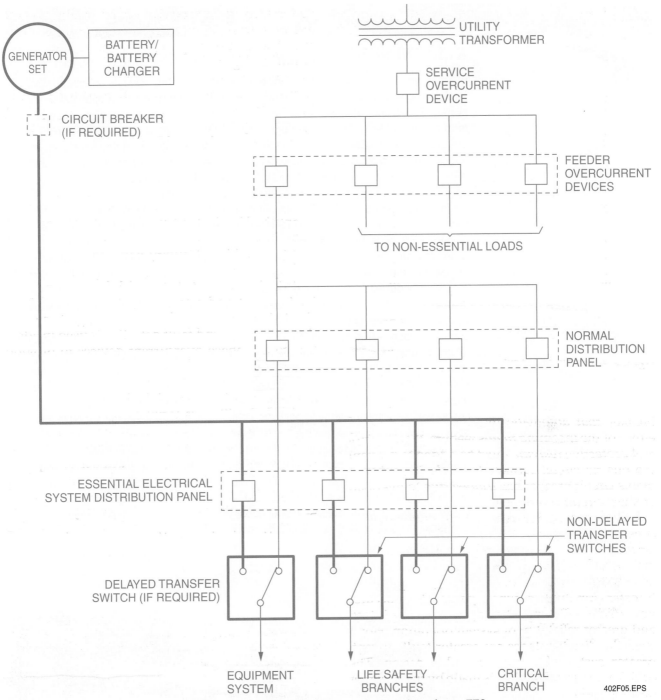

Figure 5 ◆ One-line diagram of an alternate power source with transfer switching for an EES.

3.4.0 Additional Distribution System Grounding and Bonding Requirements

Besides the usual grounding and bonding required by *NEC Article 250*, health care facilities require additional precautions for patient care areas. As specified in *NEC Section 517.14*, where branch circuit panelboards for normal and essential power serve the same patient care vicinity, they must be connected together with an insulated copper conductor no smaller than No. 10 AWG. In addition, where two or more panelboards that serve the same patient care vicinity are served from two different transfer switches, their equipment grounding terminals must be connected

Ride-Through Power Systems

In areas with numerous short-term power interruptions and brownouts or poor power quality, electrical design engineers may choose a flywheel-type ride-through power system. These units also provide a seamless transition to generator power for longer-term power problems. The ride-through module shown below is rated at 1,250kVA. They are available from 450kVA up to 1,250kVA and multiple modules can be paralleled. In addition, they are available as medium- or low-voltage units and with dual outputs that allow loadshedding of non-critical loads during the ride-through period. These units can operate in environments from 0°F to 140°F. Colder temperatures require only the addition of cold-weather diesel engine starting equipment. These modules are paralleled with and synchronized to three-phase power for the entire facility or across only the EES loads. Full-load ride-through time can be up to 15 seconds to accommodate short-term power problems plus the startup requirement of the diesel engine. Electric power for diesel startup is included as part of the full load. In the normal mode (utility power present), the generator becomes a synchronous condenser (a no-load, over-excited AC motor) that is connected to and maintains the speed of the outer rotor of an induction coupling. The outer rotor, via internal three-phase windings, causes a massive free-running inner rotor to rotate at a speed about three times faster than the outer rotor. The synchronous condenser function of the generator supplies reactive power to the load and functions along with a reactor, which decouples the load from the utility power as a dynamic active filter. During momentary power problems, the utility power input is switched off and the kinetic energy stored by the heavy rotor is converted to electrical energy to power the outer rotor and maintain the speed of the generator, which now supplies the power. If the utility power problem lasts more than a prescribed time, the diesel engine is started and the generator and induction coupling are powered through a mechanical override clutch on the engine side of the induction coupling that locks up when the engine reaches the required generator speed. During non-operation, the engine is pre-heated and pre-lubricated so that its startup time is only a few seconds. The automatic switchgear control and reactor for a module are normally remotely located in an electrical room.

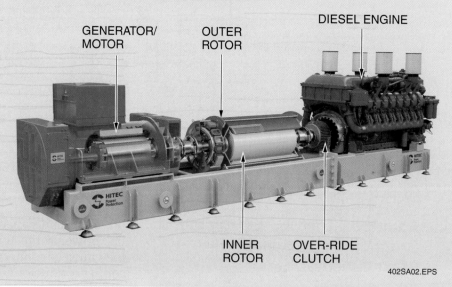

GENERATOR/ MOTOR   OUTER ROTOR   DIESEL ENGINE

INNER ROTOR   OVER-RIDE CLUTCH

402SA02.EPS

with an insulated conductor no smaller than No. 10 AWG.

NEC Section 517.19(D) specifies that where a grounded electrical distribution system is used, and feeder metal raceway or Type MC or MI cable that qualifies as an equipment grounding conductor per *NEC Section 250.118* is installed, the grounding of a panelboard or switchboard must be ensured by one of the following:

- A grounding bushing and continuous copper bonding jumper sized per *NEC Section 250.122*, with the jumper connected to the junction enclosure or the ground bus of the panel
- Connection of the feeder raceways or Type MC or MI cable to threaded hubs or bosses on the terminating enclosures
- Other approved devices such as bonding-type locknuts or bushings

4.0.0 ◆ WIRING AND DEVICES

Generally, hospitals have two sources of available power: normal and emergency. Devices on emergency power must be easy to identify. This reduces the time wasted in locating receptacles during a power outage. These devices may be identified with a distinctive color such as red, or by labeling. (When color-coding is used, it must be the same throughout the facility.) Color-coding is easier and less expensive than labeling, and it may prevent later confusion. For example, if painters remove the labeled cover plates, there is no longer a distinction between devices. In addition, since receptacles in critical care areas must have panelboard and circuit number labels, the device cover plates tend to become cluttered. *Figure 6* shows an example of color-coding used to identify various circuits in a patient headwall. Lighted emergency power receptacles are useful in patient areas that do not have emergency lighting as they make the receptacle easier to locate in the near-darkness of a power outage.

Devices in a hospital should be mounted so they are easy to use. Because a large number of hospital patients spend time in wheelchairs, particular attention should be given to these requirements. All receptacles should be mounted approximately 24" above the floor for the convenience of patients and staff.

4.1.0 Hospital-Grade Receptacles

Hospital-grade receptacles (*Figure 7*) are listed as suitable for use in health care facilities. Per *NEC Article 517*, health care facilities include (but are not limited to) hospitals, nursing homes, limited care facilities, clinics, medical and dental offices, and ambulatory care centers, whether permanent

402F06.EPS

Figure 6 ◆ Example of circuit color-coding for a headwall.

Figure 7 ◆ Hospital-grade receptacles.

402F07.EPS

or movable. Hospital-grade receptacles are often marked with a green dot on the outer face of the receptacle.

Hospital-grade receptacles meet nationally recognized testing laboratory criteria for superior electrical characteristics and mechanical strength. This is needed in a medical environment to ensure that equipment plugs fit tightly into the receptacles and the receptacles are durable.

4.1.1 Hospital-Grade Isolated Ground Receptacles

Isolated ground receptacles are used where separation of the device ground and the building ground is desired. They are normally used with digital electronic equipment, including computer cash registers, computer peripherals, and digital processing equipment. They prevent transient voltages on the ground system from causing malfunctions in digital circuits.

4.1.2 Hospital-Grade Tamper-Resistant Receptacles

Tamper-resistant receptacles prevent foreign objects from coming into contact with energized components inside the receptacle by only operating when both the hot and neutral blades of a plug are simultaneously inserted. These receptacles must be used throughout psychiatric and pediatric locations. When choosing tamper-resistant receptacles, be careful not to use the type that makes and breaks contacts where life support apparatus may be required.

4.2.0 General Care Areas

It is difficult to prevent a conductive path from a grounded object to a patient's body in a health care facility. The path may be created accidentally or through instruments directly connected to the patient that act as a source of electric current. This hazard worsens as more equipment is used near a patient, such as in surgical suites and intensive care units. Methods of reducing the likelihood of electric shock to a patient include the following:

- Increasing the resistance of the conductive circuit that includes the patient
- Insulation of exposed surfaces that may become energized
- Lowering the potential difference between exposed conductive surfaces in the patient's vicinity
- Using a combination of the above methods

NEC Section 517.18 requires that every patient bed location be supplied with at least two branch circuits, one from the normal system and the other from the emergency system. All branch circuits from the normal system must originate in the same panelboard. The *NEC*® lists three exceptions to this requirement:

- Branch circuits serving only special-purpose outlets or receptacles, such as portable X-ray outlets, do not have to be served from the same distribution panel or panels.
- Patient beds located in clinics; medical and dental offices; outpatient facilities; psychiatric, substance abuse, and rehabilitation hospitals; nursing homes; and limited care facilities meeting the requirements of *NEC Section 517.10(B)(2)* are exempt from this requirement.
- A general care patient bed location served from two separate transfer switches on the emergency system is not required to have circuits from the normal system.

In addition to two branch circuits, each patient bed location must have a minimum of four receptacles per *NEC Section 517.18(B)*. Receptacles may be of the single or duplex type or both. Some receptacles are supplied as part of a prefabricated patient headwall unit, as shown in *Figure 8*. All receptacles must be hospital-grade and grounded by means of an insulated copper equipment grounding conductor sized in accordance with *NEC Table 250.122*.

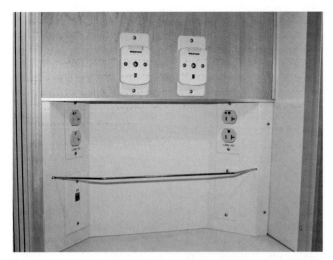

(A) RECEPTACLES ABOVE BED

(B) RECEPTACLES BEHIND BED

402F08.EPS

Figure 8 ◆ Patient bed location receptacles.

NOTE

This does not apply to psychiatric, substance abuse, and rehabilitation hospitals meeting the requirements of *NEC Section 517.10(B)(2)*. In addition, psychiatric security rooms are not required to have receptacle outlets.

NEC Section 517.18(C) requires that all receptacles meant to supply patient care areas of pediatric and psychiatric wards, rooms, or areas (bathrooms, playrooms, activity rooms, and patient care areas) must be tamper-resistant or use a tamper-resistant cover.

Patient Headwall Column

This patient headwall column supplies both an oxygen line and electrical power. This combination requires the use of a protective bollard to guard it against damage. When devices are combined in an assembly such as this one, a nationally recognized testing laboratory may be required to come in and label the entire assembly. Check with the authority having jurisdiction in your area.

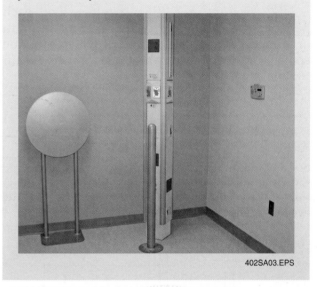

402SA03.EPS

NEC Section 517.19(C) requires that an equipment bonding jumper no smaller than No. 10 AWG be used to connect the grounding terminal of all grounding-type receptacles of a patient headwall to the patient grounding point.

4.3.0 Critical Care Areas

NEC Section 517.19(A) requires that every patient bed location in critical care areas be supplied by at least two branch circuits, one or more from the normal system and one or more from the emergency system so that patients will not be without electrical power when there is a fault. At least one branch circuit from the emergency system must supply an outlet only at that bed location. All branch circuits from the normal system must be from a single panelboard. Emergency system receptacles must be identified as such and labeled with the supply panelboard and circuit number.

> **NOTE**
>
> Branch circuits serving only special-purpose receptacles or equipment in critical care areas may be served by other panelboards; also, when the location is served by two individual transfer switches on the emergency system, it does not require a branch on the normal system. See *NEC Section 517.19(A), Exceptions.*

NEC Section 517.19(B) requires that each patient bed location be provided with a minimum of six receptacles, at least one of which must be connected to the normal system branch circuit required in *NEC Section 517.19(A)* or an emergency system branch circuit supplied by a different transfer switch than the other receptacles at the same location. They may be of the single or duplex types, or a combination. All receptacles must be listed as hospital-grade and connected to the reference grounding point by means of an insulated copper equipment grounding conductor.

A typical receptacle configuration in a critical care area is shown in *Figure 9*. The normal circuits must be supplied from the same panel (L-1). The emergency circuits may be supplied from different panels (EML-1 and EML-2). However, the emergency branch circuit to patient bed location A cannot supply emergency receptacles for patient bed location B.

The required power level for each bedside is typically 90VA for single-outlet receptacles and 180VA for duplex receptacles. The following recommendations are for bedside stations:

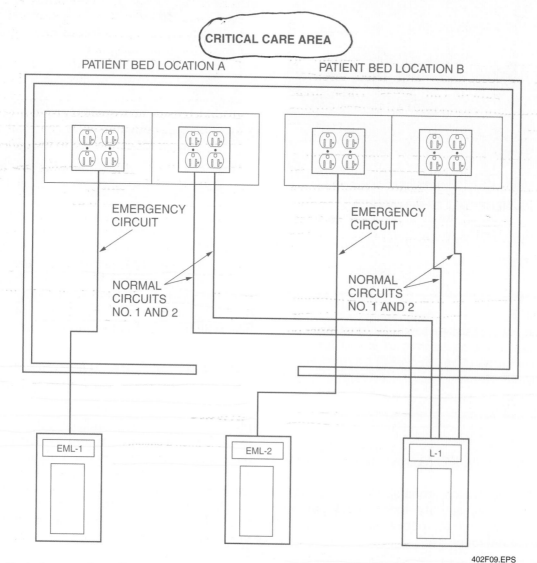

402F09.EPS

Figure 9 ◆ Typical receptacle configuration in a critical care area.

- Use only hospital-grade permissible power connectors.
- Use no less than two duplex outlets on each side of the patient's bed. The outlets must be protected by overcurrent protective devices in accordance with the *NEC*®.
- Furnish power outlets within a few feet of the junction boxes and equipment support locations.
- Allocate a grounding system in accordance with *NFPA 99* and the *NEC*®.
- Furnish an exposed terminal connected directly to the patient common reference bus for testing purposes. These terminals should be within 5' of each bed and each may serve multiple beds.
- Plan and design the system to keep power cable to a minimum length. This contributes both to physical and electrical safety.
- Ensure adequate ventilation in all areas where patient monitoring equipment is to be installed.

4.4.0 Grounding of Receptacles and Fixed Electric Equipment

NEC Sections 517.13(A) and (B) pertain to grounding of receptacles and fixed electrical equipment in patient care areas. These areas require redundant grounding. All branch circuits serving patient care areas must have an effective ground-fault current path by installation in a grounded metal raceway or a cable having a metallic armor or sheath assembly. The metal raceway, metallic armor, or sheath assembly must itself qualify as an equipment grounding conductor per *NEC Section 250.118*. Type MI and MC cable must have a health care facility rating and be clearly marked as such.

Per *NEC Section 517.13(B)*, the grounding terminals of receptacles and all noncurrent-carrying conductive surfaces of fixed electric equipment likely to become energized that are subject to personal contact and operating at over 100V must be connected to an insulated copper equipment grounding conductor. The grounding conductor must be sized in accordance with *NEC Table 250.122* and installed in a metal raceway with the branch circuit conductors supplying these receptacles or fixed electric equipment.

The insulated copper conductor is either installed with the branch circuit conductors or as part of listed cable with a metallic armor or sheath assembly. The reason for redundant grounding using an insulated ground wire within a grounded raceway or grounded sheathed cable is so that an effective ground-fault current path exists if an insulated grounding conductor or device fails. *NEC Section 517.13(B)* includes the following two exceptions:

- *Exception No. 1* – Metal faceplates may be connected to the equipment grounding conductor by means of metal mounting screws securing the faceplate to a grounded outlet box or grounded wiring device. This exception permits metal faceplates to be grounded without having to run a separate equipment grounding conductor to the faceplate.
- *Exception No. 2* – Light fixtures more than 7½' above the floor and switches located outside of the patient vicinity are not required to be grounded using an insulated grounding conductor. (These are both unlikely to contact the patient or any equipment connected to the patient.)

Metal-sheathed cable assemblies are not typically authorized for emergency circuits because *NEC Section 517.30(C)(3)* requires such wiring to be protected by installation in a nonflexible metal raceway or encased in concrete. Type MC cable is allowed where it is placed in listed prefabricated medical headwalls, fished into existing walls or ceilings where it is not likely to be damaged or casually accessed, located in listed office furniture, or where the equipment requires a flexible connection.

Electrical Risks to Patients

One of the most dangerous situations is when a patient has an external direct conductive path to the heart, such as a cardiac catheter. In this situation, the patient may be electrocuted at current levels so low that additional protection in the design of equipment, catheter insulation, and control of operational practices is necessary. Patients are at increased risk of electric shock when their body resistance is compromised either accidentally or by needed medical procedures. Situations such as catheter insertion or incontinence may make a patient much more susceptible to the effects of an electric current.

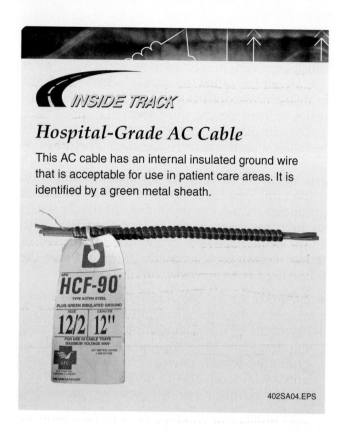

NOTE

Special grounding requirements are required in wet locations. See *NEC Section 517.20*. These locations are designated by the facility governing body. Patient bathrooms are *not* considered a wet location.

4.4.1 Equipotential Grounding

Equipotential grounding is one method of grounding used for sensitive equipment. An equipotential ground plane is a mass of conducting material that, when bonded together, provides a low impedance to current flow over a wide range of frequencies.

Equipotential ground planes shield adjacent sensitive circuits or equipment, contain electromagnetic (EM) noise fields between their source (cable, etc.) and the plane, increase filtering effectiveness of contained EM fields, and provide a low-impedance return path for radio frequency (RF) noise current.

Equipotential plane structures include metallic screens or sheet metal under floor tile, the supporting grids of raised access flooring (such as in computer rooms), conductive grids rooted in or attached to a concrete floor, and ceiling grids above sensitive equipment.

An equipotential reference plane can be used within a section of a single sensitive equipment enclosure, among various pieces of interconnected equipment, or over an entire facility. In all cases, the equipotential plane is bonded to both the local building ground and the grounding electrode conductor.

Within sensitive equipment cabinets, all significant components, signal return leads, backplanes, etc., must be connected via short (less than 5% to 10% of the wavelength of the highest frequency) conductors to the equipment chassis that forms the equipotential plane. All similar equipment-level equipotential planes should be coupled to the grounding electrode conductor as well as the room-level equipotential plane by way of multiple short conductors. The room-level equipotential plane must in turn be linked to one or more building-level equipotential plane(s) via multiple short conductors. This process is continued until the total sensitive electronic equipment system is interconnected to one large continuous equipotential plane. It is best to interlink conductors with single thin-wire cross sections to minimize their impedance at higher frequencies.

4.5.0 Inhalation Anesthetizing Locations

Anesthetizing locations are divided into hazardous (classified) locations, where flammable or nonflammable anesthetics may be interchangeably used, and other-than-hazardous (classified) locations, where only nonflammable anesthetics are used. Although flammable anesthetics are virtually obsolete in the United States, the *NEC*® and *NFPA 99* still cover them because they are in use in other countries that use these codes. In areas where flammable anesthetics are used, the entire location is considered a Class I, Division 1 location up to a level 5' above the floor. The area remaining up to the structural ceiling is considered to be above a hazardous (classified) location.

Any location or room in which flammable anesthetics or volatile flammable disinfecting agents are stored is considered a Class I, Division 1 location from floor to ceiling. Any inhalation anesthetizing location assigned for the exclusive use of · nonflammable anesthetizing agents is considered an other-than-hazardous (classified) location.

4.5.1 Wiring and Equipment within Hazardous Anesthetizing Locations

Per *NEC Section 517.61(A)*, wiring and equipment within hazardous (classified) anesthetizing locations must comply with the following:

- Except as permitted by *NEC Section 517.160*, each power circuit within, or partially within, a flammable anesthetizing location as referred to in *NEC Section 517.60* must be isolated from any distribution system by the use of an isolated power system.
- Isolated power system equipment must be listed for the purpose and the system must be designed and installed in accordance with *NEC Section 517.160*.
- In hazardous locations referred to in *NEC Section 517.60*, all fixed wiring and equipment, and all portable equipment, including lamps and other utilization equipment operating at more than 10V between conductors, must comply with *NEC Sections 501.1 through 501.25*, *NEC Sections 501.100 through 501.150*, and *NEC Sections 501.30(A) and (B)* for Class I, Division 1 locations. All such equipment must be specifically approved for the hazardous atmospheres involved.
- Where a box, fitting, or enclosure is partially, but not entirely, within a hazardous (classified) location, the entire box, fitting, or enclosure is considered within the hazardous (classified) location.
- Attachment plugs and receptacles in hazardous locations must be listed for use in Class I, Group C hazardous locations and have provision for the connection of a grounding conductor.
- Flexible cords used in hazardous locations for connection to portable utilization equipment, including lamps operating at more than 8V between conductors, must be approved for extra-hard usage in accordance with *NEC Table 400.4* and include an additional grounding conductor. Storage devices must be provided for flexible cords. These storage devices must not subject the cord to a bending radius of less than 3".

4.5.2 Wiring and Equipment Installed Above Hazardous Anesthetizing Locations

Per *NEC Section 517.61(B)*, wiring and equipment installed above hazardous (classified) anesthetizing locations must comply with the following:

- Wiring above a hazardous location referred to in *NEC Section 517.60* must be installed in rigid metal conduit, electrical metallic tubing, intermediate metal conduit, Type MI cable, or Type MC cable that employs a continuous gas/vaportight metal sheath.
- Equipment that may produce sparks, arcs, or particles of hot metal, such as lamps and lampholders for fixed lighting, switches, generators, motors, cutouts, or other equipment having sliding contacts, must be of the totally enclosed type or constructed in such a manner as to prevent the escape of sparks or hot metal particles. (This does not apply to wall-mounted receptacles installed above the hazardous location.)
- Surgical and other luminaires must conform to *NEC Section 501.130(B)*. See also *NEC Section 517.61(B)(3), Exceptions*, which state that the surface temperature limitations set forth for fixed lighting in *NEC Section 501.130(B)(1)* do not apply, and integral or pendant switches that are located above and cannot be lowered into the hazardous (classified) location(s) are not required to be explosionproof.
- Only listed seals may be used per *NEC Section 501.15*. *NEC Section 501.15(A)(4)* applies to both the horizontal and vertical boundaries of the defined hazardous (classified) locations.
- Attachment plugs and receptacles positioned over hazardous (classified) anesthetizing locations must be listed as hospital grade for services of prescribed voltage, frequency, rating, and number of conductors with provision for the connection of the grounding conductor. This requirement applies to attachment plugs and receptacles of the two-pole, three-wire grounding type for single-phase, 120V, nominal AC service.
- Plugs and receptacles rated at 250V for connection of 50A and 60A, AC medical equipment for use above hazardous (classified) locations must be arranged so that the 60A receptacle will accept either the 50A or the 60A plug. 50A receptacles must be designed so as not to accept a 60A attachment plug. The plugs must be of the two-pole, three-wire design with a third contact connecting to the insulated (green or green with yellow stripe) equipment grounding conductor of the electrical system.

4.5.3 Wiring in Other-Than-Hazardous Anesthetizing Locations

Per *NEC Section 517.61(C)*, wiring in other-than-hazardous (classified) anesthetizing locations must comply with the following:

- Wiring serving other-than-hazardous locations, as defined in *NEC Section 517.60*, must be installed in a metal raceway system or cable assembly. The metal raceway system, cable armor, or sheath assembly must qualify as an equipment grounding conductor in accordance with *NEC Section 250.118*. Type MC and MI cable must have an outer metal armor or sheath that is identified as an acceptable equipment grounding conductor. Per *NEC Section 517.61(C)(1), Exception*, pendant receptacles employing at least Type SJO or equivalent flexible cords suspended not less than 6' from the floor are not required to be installed in a metal raceway or cable assembly.

- Receptacles and attachment plugs installed and used in other-than-hazardous (classified) locations must be listed as hospital grade for services of prescribed voltage, frequency, rating, and number of conductors with provision for connection of the grounding conductor. This requirement applies to the two-pole, three-wire grounding type for single-phase 120V, 208V, or 240V nominal AC service.
- Plugs and receptacles rated at 250V for connection of 50A and 60A, AC medical equipment for use in other-than-hazardous (classified) locations must be arranged so that the 60A receptacle will accept either the 50A or the 60A plug. 50A receptacles must be designed so as not to accept a 60A attachment plug. The plugs must be of the two-pole, three-wire design with a third contact connecting to the insulated (green or green with yellow stripe) equipment grounding conductor of the electrical system.

4.5.4 Grounding in Anesthetizing Locations

In any anesthetizing location, all metal raceways and metal-sheathed cable, and all normally non-current-carrying conductive portions of fixed electric equipment must be connected to an equipment grounding conductor per *NEC Section 517.62*. Grounding in Class I locations must comply with *NEC Section 501.30*.

NOTE

Equipment operating at not more than 10V between conductors is not required to be grounded.

Grounding specifications and requirements for anesthetizing areas apply only to metal-sheathed cable, metal raceways, and electric equipment. It is not a requirement to ground tables, carts, or any other non-electrical items. However, in anesthetizing locations where flammables are present, portable carts and tables usually have an acceptable resistance to ground through conductive tires and wheels and conductive flooring to avoid the buildup of static electrical charges. See *NFPA 99* for information on static grounding.

4.5.5 Grounded Power Systems in Anesthetizing Locations

As a secondary backup, at least one battery-powered emergency lighting unit must be installed in every anesthetizing location per *NEC Section 517.63(A)*. Without this backup, failure of the emergency circuit feeder that supplies the operating room with power would likely place the room in complete darkness, which could have fatal results.

Per *NEC Section 517.63(B)*, branch circuits supplying only listed, stationary, diagnostic, and therapeutic equipment, permanently installed above the hazardous (classified) area or in other-than-hazardous (classified) locations may be fed from a normal grounded service (single- or three-phase system), provided that:

- Wiring for isolated and grounded circuits does not occupy the same cable or raceway.
- All conductive surfaces of the equipment are connected to an equipment grounding conductor.
- Equipment (except enclosed X-ray tubes and the leads to the tubes) is located at least 8' above the floor or outside the anesthetizing location, and switches for the grounded branch circuit are located outside the hazardous (classified) location. (This does not apply to other-than-hazardous locations.)

Per *NEC Section 517.63(C)*, fixed lighting branch circuits feeding only fixed lighting may be supplied by a normal grounded service, provided that:

- The fixtures are located at least 8' above the floor and switches are wall-mounted and located above hazardous (classified) locations. (This does not apply to other-than-hazardous locations.)
- All conductive surfaces of fixtures are connected to an equipment grounding conductor.
- Wiring for circuits supplying power to fixtures does not occupy the same raceway or cable as circuits supplying isolated power.

Wall-mounted remote control stations may be installed in any anesthetizing location for the operation of remote control switching operating at 24V or less per *NEC Section 517.63(D)*.

An isolated power supply and its grounded primary feeder may be located in an anesthetizing location, providing it is installed above a hazardous (classified) location or in an other-than-hazardous (classified) location and is listed for the purpose. See *NEC Section 517.63(E)*.

Except as permitted above, each power circuit within, or partially within, a flammable anesthetizing location as referred to in *NEC Section 517.60* must be isolated from any distribution system supplying other-than-anesthetizing locations.

4.6.0 Low-Voltage Equipment and Instruments

Low-voltage equipment that is typically in contact with patients or has exposed current-carrying elements must be moisture-resistant and operate on 10V or less, or be approved as intrinsically safe or double-insulated equipment. See *NEC Section 517.64*.

Electrical power to low-voltage equipment must be supplied from an individual portable isolation transformer (autotransformers cannot be used). This equipment must be connected to an isolated power circuit receptacle by means of an appropriate cord and attachment plug, a standard low-voltage isolation transformer installed in an other-than-hazardous (classified) location, individual dry-cell batteries, or common batteries made up of storage cells located in an other-than-hazardous (classified) location.

Isolating-type transformers for supplying low-voltage circuits must have an approved means of insulating the secondary circuit from the primary circuit. In addition, the case and core must be connected to an equipment grounding conductor.

Impedance and resistance devices are allowed to control low-voltage equipment. However, they cannot be used to limit the maximum available voltage supplied to the equipment.

Battery-supplied appliances must not be capable of being charged during operation unless the charging circuitry uses an integral isolating-type transformer. In addition, receptacles and attachment plugs placed on low-voltage circuits must be of a type that does not permit interchangeable connection with circuits of higher voltage.

WARNING!
Any circuit interruptions (even in circuits as low as 10V) caused by a switch or loose or defective connections anywhere in the circuit may produce a spark sufficient to ignite flammable anesthetic gases.

4.7.0 X-Ray Installations

The installation of X-ray equipment requires specialized training in order for the equipment to both operate as intended and to protect patients and personnel from unintended exposure to radiation. Any equipment for new X-ray installations and all used or reconditioned X-ray equipment transported to and reinstalled in a new location must be of an approved type. *NEC Article 517, Part V* covers X-ray equipment.

CAUTION
Nothing in this section shall be construed as an indication of safeguards against the useful beam or stray X-ray radiation. Radiation safety and performance requirements of several classes of X-ray equipment are regulated under Public Law 90-602 and are enforced by the Department of Health and Human Services. In addition, information on radiation protection is available from the National Council on Radiation Protection and Measurements (www.ncrppublications.org).

4.7.1 Connection to Supply Circuit

X-ray equipment, whether fixed or stationary, must be connected to the power supply in such a way that the wiring methods meet the requirements of the *NEC®*. Equipment supplied by a branch circuit rated at no more than 30A may be

Box Shielding

If an outlet or switch box must be installed in an X-ray area, lead shielding must be molded around the box. This ensures the integrity of the overall X-ray shield in the area. Some electricians peel the lead backing from scrap pieces of lead-shielded sheetrock and save it for shielding the boxes in the area.

supplied through a suitable attachment plug and hard-service cable or cord.

Individual branch circuits are not required for portable, mobile, and transportable medical X-ray equipment requiring a capacity of no more than 60A per *NEC Section 517.71(B)*. Circuits and equipment functioning on a supply circuit of over 600V must comply with the requirements of *NEC Article 490*.

4.7.2 Disconnecting Means

Per *NEC Section 517.72(A)*, a disconnecting means of adequate capacity for at least 50% of the input required for the momentary rating, or 100% of the input required for the long-term rating of the X-ray equipment (whichever is greater) must be provided in the supply circuit.

The disconnecting means must be in a location readily accessible from the X-ray control. A grounding-type attachment plug and receptacle of proper rating may serve as a disconnecting means for equipment supplied by a 120V branch circuit of 30A or less.

4.7.3 Rating of Supply Conductors and Overcurrent Protection

For diagnostic equipment, the amperage of supply branch circuit conductors and the current rating of overcurrent protective devices must be no less than 50% of the momentary rating or 100% of the long-term rating, whichever is greater. See *NEC Section 517.73(A)(1)*.

The amperage rating of supply feeders and the current rating of overcurrent protective devices feeding two or more branch circuits supplying X-ray units must be no less than 50% of the momentary demand rating of the largest unit, plus 25% of the momentary demand rating of the next largest unit, plus 10% of the momentary demand rating of each additional unit. Where concurrent bi-plane tests are undertaken with the X-ray units, the supply conductors and overcurrent protective devices must be 100% of the momentary demand rating of each X-ray unit.

The minimum conductor sizes for branch and feeder circuits are governed by voltage regulation requirements. In specific installations, manufacturers typically specify the minimum distribution transformer and conductor sizes required, the disconnect rating, and overcurrent protection requirements.

For therapeutic equipment, the amperage rating of conductors and the rating of overcurrent protective devices must not be less than 100% of the current rating of medical X-ray therapy equipment. The amperage rating of the branch circuit conductors and the ratings of disconnecting means and overcurrent protection for X-ray equipment are usually designated by the manufacturer for the specific installation.

4.7.4 Control Circuit Conductors

The number of control circuit conductors installed in a raceway must be determined in accordance with *NEC Section 300.17*. No. 18 AWG or No. 16 AWG fixture wires as specified in *NEC Section 725.49* and flexible cords are permitted for the control and operating circuits of X-ray and auxiliary equipment where protected by overcurrent devices of no more than 20A. See *NEC Section 517.74(B)*.

4.7.5 Transformers and Capacitors

Transformers and capacitors that are a part of X-ray equipment are not required to comply with *NEC Articles 450 and 460*. However, capacitors must be mounted within enclosures of insulating material or grounded metal. See *NEC Section 517.76*.

4.7.6 Installation of High Tension X-Ray Cable

Cable with a grounded shield connecting X-ray tubes and image intensifiers may be installed in cable trays or troughs along with X-ray equipment control and power supply conductors without the necessity of barriers to separate the wiring. See *NEC Section 517.77*.

4.7.7 Guarding and Grounding

All high-voltage parts, including X-ray tubes, must be mounted within grounded enclosures per *NEC Section 517.78(A)*. The link from high-voltage equipment to the X-ray tubes and other high-voltage devices must be made using high-voltage shielded cable. Air, oil, gas, or other suitable insulating material must be used to insulate the high voltage from the grounded enclosure.

All cable providing low voltages and connecting to oil-filled units that is not completely sealed (such as transformers, condensers, oil coolers, and high-voltage switches) must have insulation of the oil-resistant type. See *NEC Section 517.78(B)*.

5.0.0 ◆ COMMUNICATION, SIGNALING, DATA, AND FIRE ALARM SYSTEMS

Communication, signaling, data, and fire alarm systems, as well as systems operating on less than 120V in patient care areas must conform to the same insulation and isolation requirements as electrical distribution systems. Those systems not located in patient care areas must adhere to requirements in *NEC Articles 640, 725, 760, and 800*, as applicable. Nurse call systems (*Figure 10*) may use non-electrified patient control as an alternative isolation method.

Appliances with permanently installed signal cabling that are located in patient care areas and communicate with an appliance located in another area must use a transmission system that prevents grounding interconnection of the appliances. A common signal ground may be used

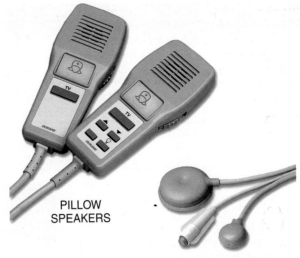

PILLOW SPEAKERS

CALL BUTTONS

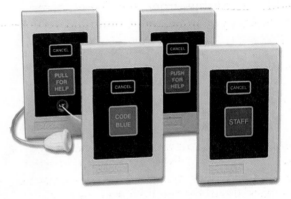

CALL STATIONS

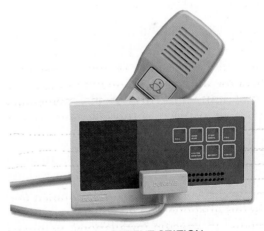

PATIENT STATION

402F10.EPS

Figure 10 ◆ Common signaling devices for nurse call systems.

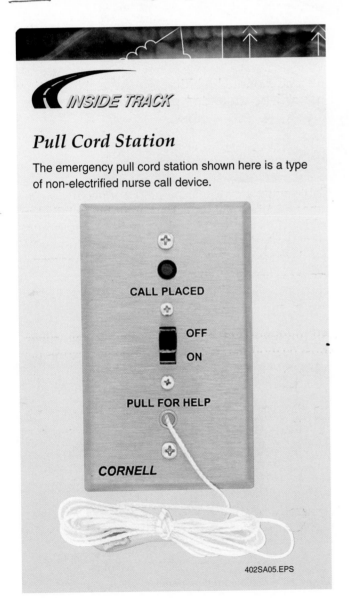

INSIDE TRACK

Pull Cord Station

The emergency pull cord station shown here is a type of non-electrified nurse call device.

CALL PLACED

OFF

ON

PULL FOR HELP

CORNELL

402SA05.EPS

Visibility

Many nurse call master stations have video touch screens that allow staff to quickly access information. This screen shows pending calls from patients and indicates that the responsible staff member for each call has been notified. The lower left portion of the screen shows that the staff members will be reminded if the call signal hasn't been turned off within a set time. At the bottom right of the screen, the staff locator system shows the location of selected staff members.

402SA06.EPS

between appliances in patient care areas as long as the appliances use the same ground reference point. See *NEC Sections 517.82(A) and (B)*.

6.0.0 ◆ ISOLATED POWER SYSTEMS

Isolated power systems are used to prevent power interruption due to a single line-to-ground fault and to isolate power from any electrical ground in certain critical areas. *Figure 11* shows an isolated power panel that may be installed in a corridor outside an operating room or other locations where power interruption cannot be tolerated. These panels are also available with various types of receptacles and up to seven ground jacks on the panel face for use in intensive care and coronary care units, cardiac catheterization labs, neonatal units, emergency rooms, critical care areas, and recovery areas.

402F11.EPS

Figure 11 ◆ Isolated power panel.

6.1.0 Installation of Isolated Power Systems

Per *NEC Section 517.160(A)*, each isolated power circuit must be controlled by a switch having a disconnecting pole in each isolated circuit conductor to simultaneously disconnect all power. Isolation is achieved using one or more isolation transformers, motor generator sets, or electrically isolated batteries.

Circuits feeding the primaries of isolation transformers must not exceed 600V between conductors and must be furnished with the proper overcurrent protection. The secondary voltage of isolation transformers must not exceed 600V between conductors of each circuit. All circuits fed from these secondaries must be ungrounded and have an approved overcurrent device of the proper rating in each conductor. If an electrostatic shield is present, it must be connected to the reference grounding point.

An isolated branch circuit supplying an anesthetizing location must not supply any other location except as noted in *NEC Sections 517.160(A)(4)(a) and (b)*.

Per *NEC Section 517.160(A)(5)*, the isolated circuit conductors must be identified as follows:

- Isolated Conductor No. 1 – Orange with a distinctive colored stripe other than white, green, or gray
- Isolated Conductor No. 2 – Brown with a distinctive colored stripe other than white, green, or gray
- Isolated Conductor No. 3 (for three-phase conductors) – Yellow with a distinctive colored stripe other than white, green, or gray

No wire-pulling compounds that increase the dielectric constant of a conductor may be used on the secondary conductors of the isolated power supply. Special lubricants are available for this application.

To meet impedance requirements, it may be desirable to limit the size of the isolation transformer to 10kVA and to use conductor insulation with low leakage. To reduce leakage from line to ground, steps should be taken to minimize the length of branch circuit conductors and use conductor insulation that meets the requirements of *NEC Section 517.160(A)(6), FPN No. 2*.

6.2.0 Line Isolation Monitors

Every isolated power system must be provided with an approved, continually functioning line isolation monitor that indicates possible leakage or fault current from either isolated conductor to ground, in addition to the usual control and protective devices. See *NEC Section 517.160(B)(1)*.

Grounding systems are the primary protection devices for patients. The ungrounded secondary of an isolation transformer reduces the maximum current in the grounding system in the event of a single fault between either of the isolated power conductors and ground. The line isolation monitor provides a warning when a single fault occurs, or when excessively low impedance to ground develops, which might expose the patient to an unsafe condition should an additional fault occur. Excessive current in the grounding conductors will not result from an initial fault. A hazard exists if a second fault occurs before the first fault is cleared.

Monitors must include a green signal lamp, visible to individuals, in the anesthetizing area, which remains illuminated when the system is sufficiently isolated from ground. An adjacent red signal lamp must illuminate with an audible warning siren when the total hazard current (consisting of possible capacitive and resistive leakage current) from either isolated conductor to ground reaches a threshold value of 5mA under normal line voltage conditions. The line isolation monitor should not alarm for any hazard fault current of 3.7mA or less or a total hazard current of less than 5mA. The audible siren may sound in another area if desired.

A line isolation monitor must have enough internal impedance so that when it is properly connected to the isolated system, the maximum internal current that will flow through the line isolation monitor when any point of the isolated system is grounded is 1mA.

An ammeter connected in series to verify the total hazard current of the system (the fault hazard current and monitor hazard current) must be mounted in a visible place on the line isolation monitor with the Alarm On zone at approximately the center of the scale. Locate the ammeter where it is visible to persons in the anesthetizing location.

Like all major electrical installations, health care facilities require extensive testing and performance verification. Typical testing forms are included in the *Appendix*.

1. Electrical work in health care facilities is governed by _____.
   a. NEC Section 511
   b. NEC Section 517
   c. NEC Section 550
   d. NEC Section 820

2. Which of the following provide around-the-clock services to patients?
   a. Hospitals, nursing homes, ambulatory health care facilities, and limited care facilities
   b. Nursing homes, ambulatory health care facilities, and limited care facilities
   c. Ambulatory health care facilities, hospitals, and limited care facilities
   d. Limited care facilities, hospitals, and nursing homes

3. The life safety branch of the essential electrical system includes _____.
   a. intensive care units
   b. patient care areas
   c. pharmacies
   d. exit signs

4. The maximum demand on one transfer switch in an essential electrical system is _____.
   a. 10kVA
   b. 50kVA
   c. 100kVA
   d. 150kVA

5. The maximum time allowed for transfer switches to engage an emergency power system is _____.
   a. 5 seconds
   b. 10 seconds
   c. 2 minutes
   d. 10 minutes

6. Many _____ must have Type 2 EES.
   a. doctors' offices
   b. nursing homes
   c. dialysis centers
   d. surgical centers

7. Throwover switches are often used to allow temporary operation of _____.
   a. elevators
   b. suction devices
   c. heating systems
   d. pharmacies

8. In a Type 2 EES, which of the following must have alternate power connected within 10 seconds of a power failure?
   a. Nurse's stations
   b. Medication preparation area
   c. Exit signs
   d. Heating equipment

9. Hospitals often receive power from a utility provider from two different distribution points.
   a. True
   b. False

10. Outlets that are on emergency power are often identified by _____.
    a. color
    b. stripes
    c. flags
    d. arrows

11. Hospital-grade tamper-resistant receptacles _____.
    a. provide power only when the hot and neutral blades of a plug are simultaneously inserted
    b. can be used when an isolated ground is needed
    c. are used in wet areas only
    d. are obsolete

12. Hospital-grade tamper-resistant receptacles must be used in _____.
    a. psychiatric locations
    b. critical care units
    c. wet areas
    d. laboratories

13. Each patient bed in the general patient care area must have a minimum of _____ receptacles.
    a. two
    b. four
    c. six
    d. eight

14. A Class I, Division 1 location extends upward to a level of _____ above the floor in an area where flammables are employed.

    a. 2 feet
    b. 5 feet
    c. 8 feet
    d. 5 meters

15. X-ray equipment supplied by a branch circuit rated at no more than _____ may use a suitable attachment plug and hard-service cable.

    a. 15A
    b. 30A
    c. 45A
    d. 60A

Summary

Unexpected loss of electrical power at a health care facility can be fatal. For this reason, it is very important that these electrical systems conform to the requirements of the *National Electrical Code®* and *NFPA 99*, as well as state and local codes. The first step in establishing which requirements apply to a facility is to determine the type of care provided at the facility. The type of care provided at the facility determines which parts of the *NEC®* apply.

Health care facilities must have adequate backup power should normal power be lost. In a health care facility, power is divided into two categories: essential loads and nonessential loads. When the normal electrical source is lost, the facility's alternate source must be able to supply the needs of the essential loads.

Notes

Trade Terms
Introduced in This Module

Equipotential ground plane: A mass of conducting material that, when bonded together, provides a low impedance to current flow over a wide range of frequencies.

Inpatient: A patient admitted into the hospital for an overnight stay.

Outpatient: A patient who receives services at a hospital but is not admitted for an overnight stay.

Throwover: A transfer switch used for supplying temporary power.

Typical Testing and Performance Verification Forms

BED TOWER EXPANSION
Part 3 – Bed Tower Expansion & Renovation
SECTION 16020
TESTS AND PERFORMANCE VERIFICATION

PART 1 - GENERAL

1.1   RELATED DOCUMENTS

General: Drawings and general provisions of the Contract, including General and Supplementary Conditions and Division 1 Specification sections, apply to work specified in this section.

1.2   DESCRIPTION

A.   Time: Perform verification work as required to show that the System is operating correctly in accordance with contract documents and manufacturer's literature. All verification shall be done after 3-day full operational period.

B.   Submission: Submit check out memos and completed testing results of all systems, cable, equipment, devices, etc., for acceptance prior to being energized or utilized.

1.3   QUALITY ASSURANCE

A.   Compliance: Testing shall comply to the following standards;

1.   NEMA
2.   ASTM
3.   NETA
4.   ANSI C2
5.   ICEA
6.   NFPA

1.4   QUALIFICATIONS OF TESTING FIRM

A.   Qualification: The testing firm shall be an independent testing organization which can function as an unbiased testing authority, professionally independent of the manufacturers, supplier, and installers of equipment or systems evaluated by the testing firm.

B.   Experience: The testing firm shall be regularly engaged in the testing of electrical equipment devices, installations, and systems.

C.   Accreditation: The testing firm shall meet OSHA criteria for accreditation of testing laboratories, Title 29, Part 1907, or be a Full Member company of the International Electrical Testing Association.

D.   Certification: The lead, on-site, technical person shall be currently certified by the International Electrical Testing Association (IETA) or National Institute for Certification in Engineering Technologies (NICET) in electrical power distribution system testing.

16020 - 1

E. Personnel: The testing firm shall utilize engineers and technicians who are regularly employed by the firm for testing services.

F. Proof of Qualifications: The testing firm shall submit proof of the above qualifications when requested.

G. Companies: Must use NETA-certified pre-qualified testing firms.

PART 2 - TEST

2.1 EQUIPMENT

A. Instruments: Supply all instruments required to read and record data. Calibration date shall be submitted on test reports. All instruments shall be certified per NETA standards.

B. Adjustments: Adjust system to operate at the required performance levels and within all tolerances as required by NETA Standards.

2.2 APPLICATIONS

A. Switchboards, Panelboards and Mechanical Equipment Feeders: After feeders are in place, but before being connected to devices and equipment, test for shorts, opens, and for intentional and unintentional grounds.

B. Ratings 250V or Less: Cables 250V or less in size #1/0 and larger shall be meggered using an industry approved "megger" with 500V internal generating voltage. Readings shall be recorded and submitted to the Engineer for acceptance prior to energizing same. Submit (5) copies of tabulated megger test values for all cables.

C. Cables 600 Volts or Less: Cables 600 volts or less in size #1/0 and larger shall be meggered using an industry approved "megger" with 1000V internal generating voltage. Readings shall be recorded and submitted to the Engineer for acceptance prior to energizing same. If values are less than recommended NETA values notify Engineer. Submit 5 copies of tabulated megger test values for all cables.

D. Ratings 600 Volts or Less: Cables 600 volts or less in size #1/0 AWG and larger shall be meggered using an industry approved "megger" with 500V internal generating voltage. Readings shall be recorded and submitted to the Engineer for acceptance prior to energizing same. Submit 5 copies of tabulated megger test values for all cables.

E. Ratings Above 600V: Cables above 600 volts in all sizes shall first be meggered, using an industry approved "megger" having 2500V internal generating voltage. When proper readings are obtained, the cables shall be "hi-potted" using (5) potentials and periods as recommended by NETA, cable manufacturer for the type and voltage class of cables installed.   Do not exceed (80%) of factory test voltage. Readings ("megger" and "hi-pot") shall be recorded and submitted to the Engineer, for acceptance prior to energizing it. Submit 5 copies of tabulated megger test values for all cables.

16020 - 2

402A01B.EPS

F.   Main circuit breakers and feeder circuit breakers 200 amps and greater shall be tested using primary injection testing as per NETA Specifications. Reports to include manufacturer's time current curve number and trip time. Submit five (5) copies to the Engineer at substantial completion.

G.   Transformers (75) KVA and larger. Perform Insulation resistance test and turns ratio test. Submit five (5) copies to Engineer at substantial completion.

2.3   MOTORS

A.   Procedure: Test run each motor, (25 HP) and larger. Tabulate and submit 5 copies of the Test Information at substantial completion for final inspection. Refer to form at the end of this Section.

B.   Provisions: With the system energized, line-to-line voltage and line current measurements shall be made at the motors under full load conditions. The condition shall be corrected when measured values deviate plus or minus 10% from the nameplate ratings.

C.   Insulation: Test the insulation resistance's of all motor windings to ground with an appropriate test instrument as recommended by the motor manufacturer, before applying line voltage to the motors. If these values are less than the manufacturer's recommended values, notify the contractor providing the motor for correction before initial start up.

D.   Power Factor: Check power factor of all motors (5 HP) and larger while driving its intended load, and at all operating speeds.

2.4   GROUNDS

A.   Electrode Ground: The resistance of electrodes (main service, generators, transformer, etc.) shall not exceed 10 ohms and shall be measured before equipment is placed in operation. Testing shall be performed on all grounding electrode installations. Testing of main ground shall be (3) point method in accordance with IEEE No. 81 Section 9.04 Standard.9. (2) point method for distribution equipment. Testing to be completed before service energized. Submit all ground test readings to the Engineer in tabulated format at substantial completion.

B.   Electrode Ground: The resistance of electrodes (main service, generators, transformer, etc.) shall not exceed 10 ohms and shall be measured by The Contractor before equipment is placed in operation. Testing shall be performed on all grounding electrode installations. Testing shall be 3 point method in accordance with IEEE Standard 81. Submit all ground test readings to the Engineer in tabulated format at substantial completion.

C.   Electrode Ground: The resistance of electrodes (main service, generators, transformer, etc.) shall not exceed 5 ohms and shall be measured by The Contractor before

16020 - 3

402A01C.EPS

equipment is placed in operation. Testing shall be performed on all grounding electrode installations. Testing shall be 2-point method in accordance with IEEE Standard 81. Submit all ground test readings to the Engineer in tabulated format at substantial completion.

2.5   EQUIPOTENTIAL GROUND

A.   Equipotential Ground: Test all metal conductive surfaces likely to become energized within patient care areas. Test all large conductive surfaces likely to become energized within a volume defined as 6 foot from the patient bed horizontally or 7 foot 6 inches vertically.

   1.   Large metal surfaces not likely to be energized, which do not require testing:
      a.   Window frames
      b.   Door frames
      c.   Floor drains
      d.   Moveable metal cabinets

   2.   Test Method:
      a.   Use impedance and voltage measurements
      b.   Utilize established ground bus or ground bar in panel serving area.
      c.   Measure voltage from reference point to conductive surfaces and all receptacle ground contacts.
      d.   Measure impedance between reference point and receptacle ground contacts.
      e.   Check for proper polarity.
      f.   Identify the reference ground for each room on the ground test report. Provide a blue dot label with a permanent adhesive backing located on the bottom center of the reference ground outlet cover.

   3.   Maximum Acceptable Values:
      a.   Voltage: 20 mV plus or minus 20 percent
      b.   Impedance: 0.1 ohm plus or minus 20 percent
      c.   Quiet ground impedance: 0.2 ohm plus or minus 20 percent

   4.   Equipment:
      a.   Millivolt meter with 1 Kohm impedance and proper frequency response, in accordance with NFPA 99.
      b.   Polarity tester

   5.   Ground Test Report. Complete ground test report included at the end of this specification section, and make available copies of such to engineer and inspecting authority at final inspection.

2.6   PRIMARY CABLE

A.   Utilize D.C. proof test: Test new cable from primary circuit utility interface to utilization equipment, after complete installation.

16020 - 4

402A01D.EPS

1. Defer test until all terminations have been completed. Disconnect permanently connected switches and other devices, except for connection to power grid.

2. Apply a D.C. voltage for 5 minutes from the appropriate test equipment to the conductor of the cable to be tested, and the low potential or grounded terminal connected to the cable shield.

3. Record leakage currents after 15, 20, 30, 45, and 60 seconds and at one minute intervals. Apply test voltage in accordance with IPCEA Standards.

B. Repair or Replacement: Repair or replace as directed any cables or terminals not meeting or exceeding minimum standards. Retest replaced cable or repaired splices.

2.7 DRY TYPE TRANSFORMERS

A. Required Factory Tests: Required factory tests shall be as follows;

1. Ratio
2. Polarity
3. Losses
    a. No load
    b. Full load
4. Resistance Measurements
5. Impedance
6. Temperature
7. Impulse Strength
8. Sound Level
9. Exciting Current
10. Low-frequency Dielectric Strength
11. ANSI Point and Curve

B. Submission: Submit test results with shop drawings.

2.8 EMERGENCY SYSTEM

A. General: Submit emergency system tests in accordance with NFPA 110. Refer to emergency section of the specification for additional information.

PART 3 - EXECUTION

3.1 SUBMITTALS

A. Equipotential Ground Test Report: Complete report form at the end of this specification.

B. Cable Test Report: Submit Cable Test Report in Triplicate.

C. Transformer Test Report: Indicate comparative data of ANSI and NEMA Standards. Indicate all characteristic values as specified herein. Certified copies of tests on electrically duplicate units are acceptable.

16020 - 5

402A01E.EPS

D. Check out Memos: Complete all information on forms at the end of this specification, project information, and certificate of completed demonstration memo. Submit data for examination and acceptance prior to final inspection request.

E. Tabulated Data: Submit data on 8-1/2 x 11 inch sheets with names of the personnel who performed the test.

F. Final: Submit accepted memos before a request for final inspection.

3.2 QUANTITIES

A. Quantity: Submit 5 copies of the check out memo on each major item of equipment. Insert accepted memos in each brochure with the performance verification information and submittal data.

END OF SECTION

16020 - 6

402A01F.EPS

FACILITY NAME:_____ PROJECT NAME:_____ AHCA LOG NO.

DATE: _____TESTED BY:

MAXIMUM TEST INTERVALS: NAME:
GENERAL CARE - 12 MOS.
CRITICAL CARE - 6 MOS.   COMPANY:
WET LOCATIONS - 12 MOS.

<u>GROUND TEST REPORT</u>

TYPE METER USED AND EXTERNAL NETWORK IF USED:

NOTE: MAXIMUM READINGS PERMITTED -   20 MV NEW CONSTRUCTION
                                   0.1 OHM NEW CONSTRUCTION

Room No.	AREA TYPE Description (C) = CRITICAL CARE (G) = GENERAL CARE	VOLTAGE MEASUREMENT			IMPEDANCE MEASUREMENT		REMARKS - IF VOLTAGE READINGS MORE THAN 20MV IN EXISTING CONST. NOTE TESTS & INVESTIGATION REQUIRED.
		NO. OF RECEPTS.	NO. OF OTHER	MAX. READING IN MILLIVOLTS	NO. OF RECEPTS.	MAX READING IN OHMS	

16020 - 7

402A01G.EPS

PROJECT NAME: _____

MOTOR TEST INFORMATION

Name of Checker: _____

Date Checked: _____

(a)	Name and identifying mark of motor	_____
(b)	Manufacturer	_____
(c)	Model Number	_____
(d)	Serial Number	_____
(e)	RPM	_____
(f)	Frame	_____
(g)	Code Letter	_____
(h)	Horsepower	_____
(i)	Nameplate Voltage and Phase	_____
(j)	Nameplate Amps	_____
(k)	Actual Voltage	_____
(l)	Actual Amps	_____
(m)	Starter Manufacturer	_____
(n)	Starter Size	_____
(o)	Heater Size, Catalog No. and Amp Rating	_____
(p)	Manufacturer of dual-element fuse	_____
(q)	Amp rating of fuse	_____
(r)	Power Factor at _____ Speed (For variable speed motors provide recording chart over operating range)	_____

16020 - 8

402A01H.EPS

TABULATED DATA

VOLTAGE AND AMPERAGE READINGS

SWITCHGEAR OR PANELBOARD

FULL LOAD AMPERAGE READINGS:

DATE

TIME

PHASE

      A

      B

      C

      N

FULL LOAD VOLTAGE READINGS:

DATE

TIME

PHASE   A TO N _____ A TO B

               B TO N _____ A TO C

               C TO N _____ B TO C

NO LOAD VOLTAGE READINGS

DATE

TIME

PHASE   A TO N _____ A TO B

               B TO N _____ A TO C

               C TO N _____ B TO C

_____ENGINEER'S REPRESENTATIVE

_____CONTRACTOR'S REPRESENTATIVE

16020 - 9

402A01I.EPS

Additional Resources

This module is intended to present thorough resources for task training. The following reference works are suggested for further study. These are optional materials for continuing education rather than for task training.

National Electrical Code® Handbook, Latest Edition. Quincy, MA: National Fire Protection Association.

Standard for Health Care Facilities (NFPA 99), Latest Edition. Quincy, MA: National Fire Protection Association.

NCCER makes every effort to keep these textbooks up-to-date and free of technical errors. We appreciate your help in this process. If you have an idea for improving this textbook, or if you find an error, a typographical mistake, or an inaccuracy in NCCER's Contren® textbooks, please write us, using this form or a photocopy. Be sure to include the exact module number, page number, a detailed description, and the correction, if applicable. Your input will be brought to the attention of the Technical Review Committee. Thank you for your assistance.

Instructors – If you found that additional materials were necessary in order to teach this module effectively, please let us know so that we may include them in the Equipment/Materials list in the Annotated Instructor's Guide.

Write: Product Development and Revision
National Center for Construction Education and Research
3600 NW 43rd St., Bldg. G, Gainesville, FL 32606

Fax: 352-334-0932

E-mail: curriculum@nccer.org

Craft _____ Module Name _____

Copyright Date _____ Module Number _____ Page Number(s) _____

Description _____

_____

_____

_____

(Optional) Correction _____

_____

_____

(Optional) Your Name and Address _____

_____

_____

Standby and Emergency Systems

University of Texas at Austin – The BLOOM House

This house is about life and its boundless possibilities; it's also about a budding solar way of life. In fact, the name symbolizes a home that "blooms" like a rose under the sun. The building's "skin" responds to the wind through shutters that allow for enormous flexibility in terms of light, heat, fresh air, and privacy.

26403-08

26403-08
Standby and Emergency Systems

Topics to be presented in this module include:

1.0.0   Introduction 3.2

2.0.0   Emergency and Standby Power System Components 3.2

3.0.0   Storage Batteries 3.11

4.0.0   Static Uninterruptible Power Supply 3.17

5.0.0   *NEC*® Requirements for Emergency Systems 3.21

6.0.0   Emergency System Circuits for Light and Power 3.23

Overview

When power is lost for short periods of time, we may experience mild to moderate levels of inconvenience based on our degrees of dependence. In some situations, it may be desirable to maintain uninterrupted power for the sake of convenience. Examples are large hotels, apartment complexes, and other facilities that house large numbers of people. However, in medical facilities and some industrial applications, there are situations in which uninterrupted power is essential. Standby or emergency electrical systems are usually installed to provide a backup source of electrical power.

Switching from commercial power to standby or emergency power is not always a simple task due to the risk of undesirable connections between the two systems. The initial activation of a standby system is typically controlled by sensing devices that continually monitor any changes in the characteristics of the connected commercial power. Although no switch between commercial power and standby power can be accomplished without some delay, critical switching can be completed in seconds in finely-designed switching control systems. Some standby systems share a portion of the load demand with the commercial power during normal periods. Thus, no interruption is noticed as the standby system assumes full responsibility of the power requirements during a power outage.

Objectives

When you have completed this module, you will be able to do the following:

1. Explain the basic differences between emergency systems, legally required standby systems, and optional standby systems.
2. Describe the operating principles of an engine-driven standby AC generator.
3. Describe the different types and characteristics of standby and emergency generators.
4. Recognize and describe the operating principles of both automatic and manual transfer switches.
5. Recognize the different types of storage batteries used in emergency and standby systems and explain how batteries charge and discharge.
6. For selected types of batteries, describe their characteristics, applications, maintenance, and testing.
7. Recognize double-conversion and single-conversion types of uninterruptible power supplies (UPSs) and describe how they operate.
8. Describe the *National Electrical Code®* (*NEC®*) requirements that pertain to the installation of standby and emergency power systems.

Trade Terms

Armature
Electrolyte
Exciter
Synchronous generator

Required Trainee Materials

1. Pencil and paper
2. Appropriate personal protective equipment
3. Copy of the latest edition of the *National Electrical Code®*

Prerequisites

Before you begin this module, it is recommended that you successfully complete *Core Curriculum; Electrical Level One; Electrical Level Two; Electrical Level Three; Electrical Level Four,* Modules 26401-08 and 26402-08.

This course map shows all of the modules in *Electrical Level Four.* The suggested training order begins at the bottom and proceeds up. Skill levels increase as you advance on the course map. The local Training Program Sponsor may adjust the training order.

26413-08 Introductory Skills for the Crewleader

26412-08 Special Locations

26411-08 Medium-Voltage Terminations/Splices

26410-08 Motor Operation and Maintenance

26409-08 Heat Tracing and Freeze Protection

26408-08 HVAC Controls

26407-08 Advanced Controls

26406-08 Specialty Transformers

26405-08 Fire Alarm Systems

26404-08 Basic Electronic Theory

26403-08 Standby and Emergency Systems

26402-08 Health Care Facilities

26401-08 Load Calculations – Feeders and Services

ELECTRICAL LEVEL FOUR

ELECTRICAL LEVEL THREE

ELECTRICAL LEVEL TWO

ELECTRICAL LEVEL ONE

CORE CURRICULUM: Introductory Craft Skills

403CMAP.EPS

1.0.0 ◆ INTRODUCTION

This module covers the principles of operation and maintenance of both standby and emergency power systems.

Emergency systems are defined by *NEC Article 700* as those systems legally required and classified as such by municipal, state, federal, and other codes, or by the government agency having jurisdiction. These systems are intended to automatically supply illumination and/or power to designated areas and equipment in the event of failure of the normal supply or in the event of an accident to elements of a system intended to supply, distribute, and control power and illumination for safety of human life.

Emergency systems are typically installed in places of assembly where artificial illumination is required for safe exiting and for panic control in buildings subject to occupancy by larger numbers of people, such as hotels, theaters, hospitals, sports arenas, health care facilities, and similar institutions. They may also be used to provide power for such functions as:

- Ventilation where essential to maintain life
- Fire detection and alarm system operation
- Elevator operation
- Fire pump operation
- Public safety communication system operation
- Industrial processes where current interruption would produce serious life safety or health hazards
- Similar functions

In the event of failure of the normal power supply, emergency lighting, emergency power, or both must be available within the time required for the application, but not to exceed 10 seconds.

Legally required standby systems are defined by *NEC Article 701* as those systems required and classified as such by municipal, state, federal, and other codes, or by the government agency having jurisdiction. These systems are intended to automatically supply power to selected loads (other than those classified as emergency systems) in the event of failure of the normal source.

Legally required standby systems are typically installed to serve loads such as heating and refrigeration systems, communication systems, ventilation and smoke removal systems, sewage disposal, lighting systems, and industrial processes that, when stopped during any interruption of the normal electrical supply, could create hazards or hamper rescue or firefighting operations.

The requirements for legally required standby systems are basically the same as for emergency systems; however, there are some differences. Upon loss of normal power, legally required systems are required to be able to supply standby power in 60 seconds or less, instead of 10 seconds or less as required for emergency systems.

Optional standby systems are defined by *NEC Article 702* as those systems intended to protect public or private facilities or property where life safety does not depend on the performance of the system. Optional standby systems are intended to supply site-generated power to selected loads either automatically or manually. Optional standby systems are typically installed to provide an alternate source of electric power for such facilities as industrial and commercial buildings, farms, and residences. These systems serve loads such as heating and refrigeration systems, data processing and communication systems, and industrial processes that, when stopped during any power outage, could cause discomfort, serious interruption of the process, or damage to the product or process.

2.0.0 ◆ EMERGENCY AND STANDBY POWER SYSTEM COMPONENTS

Figure 1 shows a one-line diagram of a typical electrical distribution system that incorporates a basic emergency power system (shown as heavy lines). The components that form an emergency or standby power system are basically the same. They include:

- An engine-driven AC generator set
- Automatic transfer switches

2.1.0 Engine-Driven Generator Sets

There are many designs and sizes of engine-driven **synchronous generator** sets used as power sources for emergency and standby systems.

2.1.1 Principles of Engine-Driven AC Generator Set Operation

An AC generator (alternator) converts rotating mechanical energy, typically supplied by an air or water-cooled diesel or gas engine, into electrical energy. The AC generator consists of a rotor (*Figure 2*) also called the **armature**, and stator, with the shaft of the rotor being driven by the engine. When the engine is running, it turns the rotor, which carries the generator field windings. These field windings are energized (or excited) by a DC source called the **exciter**, which is connected to the positive (+) and negative (−) ends of the field

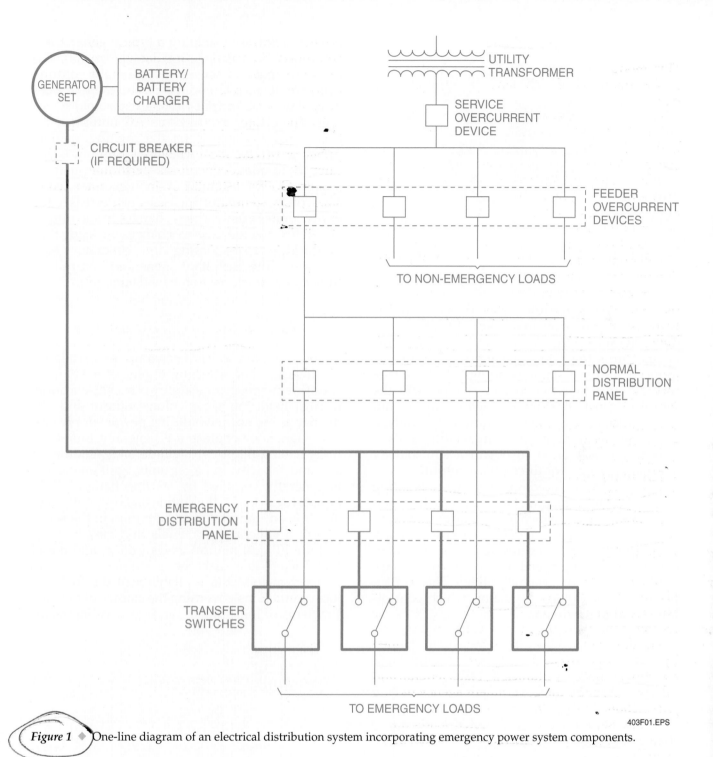

Figure 1 ◆ One-line diagram of an electrical distribution system incorporating emergency power system components.

403F01.EPS

Engine-Driven Generator Sets

Many facilities are adding engine-driven generator sets to assist them through brownout situations, rather than just for emergency purposes.

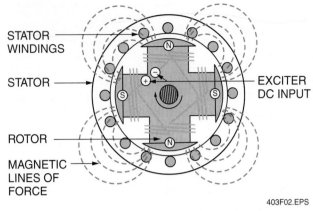

Figure 2 ◆ Cross-sectional view of a four-pole AC generator.

403F02.EPS

windings. The generator is constructed such that the lines of force of the magnetic field cut perpendicularly across the stator windings when the engine turns the rotor, causing a voltage to be induced into the stator windings. This voltage reverses each time the polarity changes. For the four-pole generator shown in *Figure 2*, this would occur twice during each revolution. Typically, a generator has four times as many winding slots as shown and is wound to obtain a sinusoidally alternating, single- or three-phase output.

The voltage induced in each stator winding depends on the strength of the magnetic field and the velocity with which the lines of force cut across the windings. Therefore, in order to vary the output voltage of a generator of a given diameter and operating speed, it is necessary to vary the strength of the magnetic field. This is controlled by the voltage regulator, which controls the output of the exciter.

The excitation system for a typical generator is controlled via a voltage regulator (*Figure 3*). The voltage regulator senses the generator output, compares it to a reference value, and then supplies a regulated DC output to the exciter field windings. The exciter produces an AC output in the exciter rotor, which is on the rotating, engine-driven generator shaft. This exciter output is rectified by diodes, also on the generator shaft, to supply DC for the main rotor (generator field). The voltage regulator increases or decreases the exciter voltage as it senses changes in the output generator voltage due to changes in load, thus increasing or decreasing the generator field strength. The resultant generator output is directly proportional to the field strength.

2.1.2 Generator Selection Considerations and Sizing

There is a wide variety of generator sets available. Obviously, when selecting a generator set for a specific emergency or standby power application, it must have the same voltage output and frequency as the normal building power supply system. Generator engines are generally four-cycle units having one to six cylinders, depending on size and capacity. In larger units, ignition current is generally supplied by starting batteries. The type of fuel used by the engine that drives the generator is an important consideration in the selection of a generator set. Fuels that may be used include LP gas, natural gas, gasoline, and diesel fuel. Factors that affect the selection of the fuel to be used include the availability of the fuel and local regulations governing the storage of the fuel. Problems of fuel storage can be a deciding factor

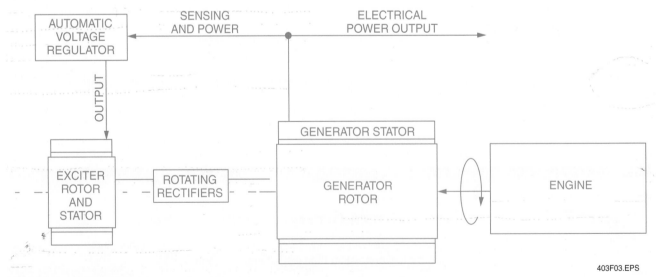

403F03.EPS

Figure 3 ◆ Self-exciter generator schematic.

Typical Engine-Generator Set

A large water-cooled engine-generator set is shown here. This unit is diesel-powered and is designed for long periods of operation. Critical applications may require one or more additional units as backup(s). These are commonly referred to as *N + 1*, *N + 2*, or *N + X* designs. Codes normally require that generators installed in habitable buildings be located inside a separate fire-rated room protected by a fire suppression system.

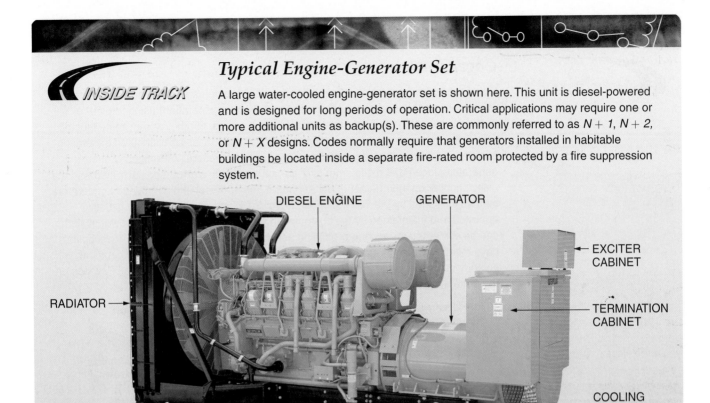

DIESEL ENGINE   GENERATOR

EXCITER CABINET

TERMINATION CABINET

RADIATOR

COOLING DUCT

CAT

403SA01.EPS

in the selection and types of generators used. If natural gas is used, its Btu heating content must be greater than 1,000 Btus per cubic foot. Diesel-powered generators are widely used because they require less maintenance and have a longer life.

Smaller generator sets are available with either air or liquid cooling systems. Larger units typically use liquid-cooled engines. Liquid-cooled engines used with generator sets operate in basically the same way as a liquid-cooled automobile engine. They are cooled by pumping a mixture of water and antifreeze through passages in the engine cylinder block and head(s) by means of an engine-driven pump. The engine, pump, and radiator form a closed-loop pressurized cooling system. The most common generator set arrangement has a mounted radiator and engine-driven fan to cool the coolant and ventilate the generator room. Alternative methods include a remotely located radiator, or a local or remotely located liquid-to-liquid heat exchanger used instead of a forced air-cooled radiator.

Depending on the generator size, compressed air and battery starting systems can be used to start generator engines. Compressed air starting systems can be used on larger generator sets. Battery starting systems are generally preferred for

most applications involving generator sets smaller than 2,000kW because they are considered to be as reliable as compressed air systems, but are much less expensive. Battery starter systems used with generator sets are usually 12V or 24V. The batteries must have enough capacity to provide the cranking motor current indicated on the generator set specification sheet. The batteries are normally either the lead-acid or nickel cadmium type. They must be designated for this use or may have to be approved by the authority having jurisdiction. For emergency power systems, a float-type battery charger, powered by the normal power source, must be provided to keep the batteries fully charged during standby without damage. As the battery approaches full charge, the charging current automatically tapers off to zero amperes or to a steady-state load on the battery. A high-output engine-driven alternator and automatic voltage regulator (part of the generator set) are provided to recharge the batteries during generator set operation. Batteries and battery charging systems are described in more detail later in this module.

The sizing of a generator set for a particular application is typically done using computer software developed specifically for that purpose or

Cogeneration

If an application requires an around-the-clock electric demand and a substantial heat requirement, combined heat and power generation (cogeneration) using turbine- or engine-generators may offer substantial energy savings. In other cases, some electric utilities/suppliers offer peak-sharing rate arrangements by using distributed generation facilities. Under distributed generation (electric cogeneration), intermittently operated generators, located at a customer site, are automatically started and paralleled with the utility supplier on their demand. Either natural gas or diesel engine-powered generators may be furnished by the utility/supplier and/or the customer, but they are usually operated and maintained by the customer. In many cases, idle standby-power systems can be converted to revenue-producing assets. With peak sharing, the utility/supplier pays monthly credits or grants special interruptible pricing that effectively caps the customer's energy price and provides profit from the sale of excess power. Payback on the generating equipment is often three years or less.

403SA02.EPS

procedures outlined in the product literature produced by most generator set manufacturers. As a minimum, the generator set must be sized to supply the maximum starting (surge) demands and steady-state running loads of the connected equipment. Once the starting and running loads have been determined, it is common to add a margin factor of up to 25% for future expansion or to select a generator set of the next largest standard rating. A large connected load that does not run during usual power outages, such as a fire pump, can serve as part of the margin factor. For fuel efficiency, the running load should stay within approximately 50% to 80% of the generator set kW rating.

In applications where the voltage and frequency dip performance specifications are stringent, it is often necessary to oversize a generator set. Applications with the following types of loads

and similar devices may also require that the generator be oversized:

- Static uninterruptible power supplies (UPSs)
- Variable frequency drives
- Battery charging rectifiers
- Medical diagnostic imaging equipment
- Fire pump(s)

2.1.3 Generator Set Maintenance and Service

The failure of an emergency or standby generator to start and run could lead to loss of life, personal injury, or property damage. For this reason, a comprehensive maintenance and service program carried out on a scheduled basis is absolutely necessary. This maintenance and service should be performed when scheduled and as recommended by the generator set manufacturer.

2.2.0 Transfer Switches

Transfer switches are a primary part of a building's emergency or standby power system. They are made in both automatic and nonautomatic/remote types.

2.2.1 Transfer Switch Operation

An automatic transfer switch automatically transfers electrical loads from a normal source (utility power) to an emergency source (generator set) when the normal source fails or its voltage is substantially reduced. The transfer switch automatically retransfers the load back to the normal power source when it is restored.

 WARNING!
Always follow lockout procedures even if the circuit is de-energized.

The typical automatic transfer switch consists of two major component areas, an electrically operated double-throw transfer switch and related control panel circuits. *Figure 4* shows a basic application of an automatic transfer switch. As shown, voltage inputs from the normal power source and the generator set (when running) are applied to the voltage monitor and voltage/frequency monitor circuits, respectively, of the

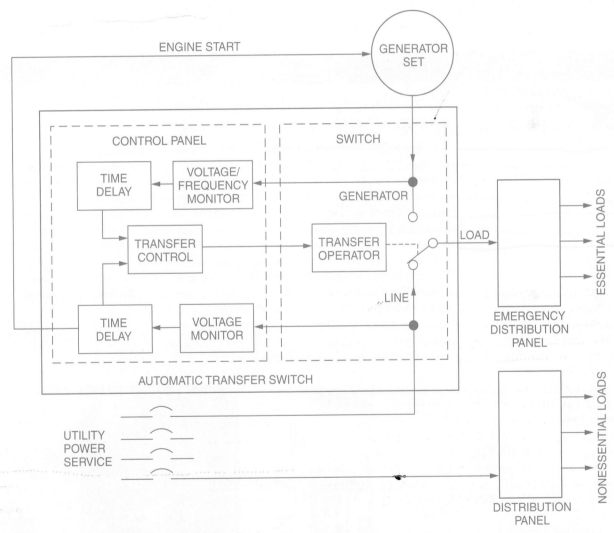

Figure 4 Simplified emergency/standby power distribution with automatic transfer switch.

403F04.EPS

control panel. The transfer switch operates to perform the following functions:

- It initiates starting the emergency generator engine via contacts in the transfer switch control panel when the utility voltage drops to an unacceptable level.
- It maintains connection of the load circuits to the normal power source during the starting period to provide use of any existing service on the normal source.
- It measures the output voltage and frequency of the voltage being generated by the generator set through the use of a voltage/frequency-sensitive monitor. Note that combined frequency and voltage monitoring protects against transferring loads to a generator set with an unacceptable output.
- It automatically switches the load circuits to the generator set when both its voltage and frequency are about normal.
- It provides a visual indication and auxiliary contact for remote indication when the generator set is supplying the loads.
- It automatically switches back to the utility power source when it is deemed to be stable and normal.

Time delays are provided in the transfer switch control circuits to program the operation of the switch. One such delay, typically adjustable between 0 and 15 seconds in duration, is provided to avoid unnecessary starting and transfer of the generator set power source for power interruptions of short duration. If the duration of the interruption exceeds the time delay, the transfer switch initiates generator set engine startup.

Once the load has been transferred to the generator set, another timer delays switching back (retransfer) to the normal source for an adjustable delay period (typically up to 30 minutes) until the source has had time to stabilize. This timer also

Microprocessor-Controlled Automatic Transfer Switches

INSIDE TRACK

Microprocessor-controlled automatic transfer switches are commonly available for voltages up to 600VAC and thousands of amperes of load-switching capability. The unit shown here can be programmed to start an engine-generator when the voltage drops to a predetermined level (usually 80% of nominal). When the emergency source reaches 90% of rated voltage and 95% of rated frequency, the unit switches and locks the load to the emergency source. When the outside line voltage is restored to a predetermined point as sensed by the unit (usually 90%), the unit switches and locks the load to the outside source. The microprocessor control allows programming the unit to switch when the outside line and emergency source are in phase and at or near zero phase difference. Optional programming allows momentary disconnection of a large motor load during switching, along with staged motor restarts, if necessary, after switching. If necessary, the unit also allows a delayed transition on transfer to ensure decay of rotating motor or transformer fields for UPS system sequencing. The unit can be programmed to activate load-shedding functions prior to switching.

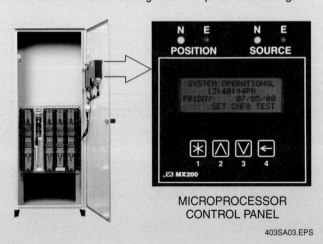

MICROPROCESSOR
CONTROL PANEL

403SA03.EPS

serves the function of allowing the generator set to operate under load for a preselected minimum time, thus assuring continued good performance of the engine and its starting system. An intermediate bypass of this transfer time delay is automatic should the generator set stop inadvertently during this delay.

A third time delay, usually adjustable from 0 to 10 minutes, is provided to allow for generator set engine cooldown. During cooldown, the generator set runs unloaded after the load has been switched back to the utility power. Note that running an unloaded engine for more than 10 minutes is not recommended since it can cause deterioration in engine performance.

The operation of a nonautomatic/remote transfer switch is initiated either by an operator at the transfer switch or by an external signal from a remote control. The nonautomatic/remote transfer switch does not include an automatic controller, but can be slaved to an automatic transfer switch with field wiring interconnections.

2.2.2 Transfer Switch Configurations and Sizing

Common transfer switch configurations are three-pole with a solid neutral for single and three-phase, three-wire or four-wire circuits, or four-pole with a switched neutral for use with separately derived three-phase, four-wire systems. Because of their location in the electrical distribution system, transfer switches carry load current continuously, whether fed by the normal or the alternate power source. By design, transfer switches must be highly reliable because failure to operate as intended could leave critical load equipment without power, even though power may be available from at least one source.

The sizing of transfer switches is based on the calculated continuous current, in a manner similar to sizing the circuit conductors. Use transfer equipment continuous current ratings that equal the calculated continuous current for the load equipment or the next higher rating. If future load expansion is a consideration, a transfer switch rating based on the maximum rating of the circuit overcurrent protective device may be selected. The available fault current and upstream overcurrent device may also affect sizing of the transfer equipment.

2.3.0 Automatic Sequential Paralleling Emergency/Standby System

Automatic sequential paralleling of two or more electric generating sets performing as a single power source offers important advantages to the user, such as increased power capability, greater versatility in installation and operation, and, most importantly, the greater reliability of a multiple-unit system. A typical automatic sequential paralleling emergency system for hospitals is illustrated in

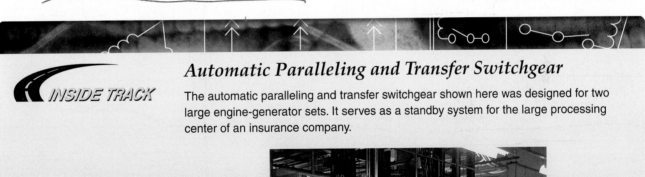

Automatic Paralleling and Transfer Switchgear

INSIDE TRACK

The automatic paralleling and transfer switchgear shown here was designed for two large engine-generator sets. It serves as a standby system for the large processing center of an insurance company.

403SA04.EPS

Figure 5. This system consists of engine-generator sets, paralleling switchboards, a totalizing board, and automatic transfer switches that monitor the total system and control the operation of each individual generating set.

If utility power is interrupted, an automatic transfer switch initiates starting of all three generator sets (A, B, and C) simultaneously. When the first of the three generating sets reaches operating voltage, a special emergency bus directs the power to the top-priority requirement areas, such as operating rooms and life safety and support systems.

With automatic sequential paralleling, each set generates power as soon as it starts, and any one can be the lead unit or the first on the bus. Speed in assuming the major or critical load is a major advantage of this system. With all sets in the system answering a request for power, there is a much higher probability of at least one of them assuming the emergency load sooner than a single-set system.

Each generating set has its own synchronizer and electronically governed load sensor designed to automatically disconnect, shed, or add loads on a programmed priority basis. In the event of an engine-generator failure, a reverse power relay in that set senses the reverse power condition, opens its circuit breaker, and disconnects that load from the bus. Noncritical equipment loads are usually shed first, assuring continuous power to the critical life safety and support loads.

There are other advantages offered by sequential paralleling systems, such as automatic operation and system expansion. If an addition is built or power needs increase, the system can be expanded. When a component is serviced, it can be done with only a limited loss of power—rather than total power loss, as is the case with a single-unit source. Furthermore, a sequentially paralleled system usually operates at higher overall efficiencies than individual generators.

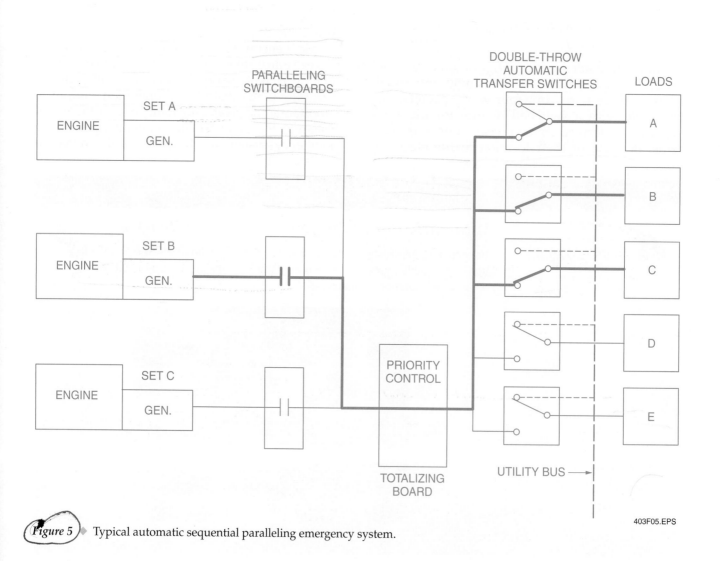

Figure 5 ◆ Typical automatic sequential paralleling emergency system.

403F05.EPS

3.0.0 ◆ STORAGE BATTERIES

Storage batteries are widely used in starting systems for generator sets, emergency lighting, and many other applications. The demand for reliable battery backup is a priority in any standby and/or emergency system. No matter what type of equipment the battery backs up, the cost of failure is unacceptable. It is therefore a top priority to apply, operate, and maintain batteries correctly.

NEC Section 701.11(A) specifies that a storage battery of suitable rating and capacity is required to supply and maintain at least 87½% of the system voltage for the total load of circuits supplying legally required standby power for a period of at least 1½ hours. Batteries, whether acid or alkali types, must be designed and constructed to meet the service requirements of emergency service and must be compatible with the charger for that particular installation. Some *NEC®* regulations governing storage batteries used in legally required standby systems are shown in *Figure 6*.

3.1.0 Lead-Acid Batteries

There are many types of lead-acid batteries. Regardless of the design, with all lead batteries the chemical reaction that takes place to store and produce electricity is the same (*Figure 7*). The lead-acid battery is a secondary or storage type of battery. In secondary batteries, electricity or power must be put into the battery to start the chemical reaction. When the battery is being charged, power is stored chemically in the electrodes and

Battery voltage is computed on the basis of 2.0V per cell for lead-acid types and 1.2V per cell for alkali types.

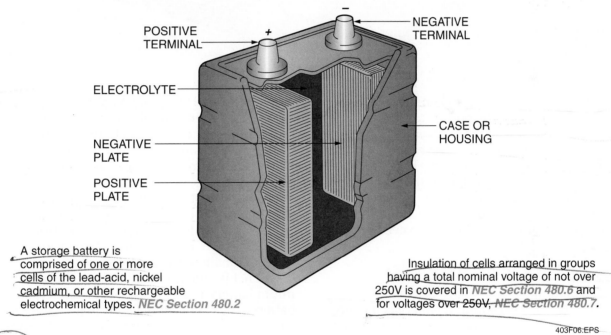

POSITIVE TERMINAL

NEGATIVE TERMINAL

ELECTROLYTE

CASE OR HOUSING

NEGATIVE PLATE

POSITIVE PLATE

A storage battery is comprised of one or more cells of the lead-acid, nickel cadmium, or other rechargeable electrochemical types. *NEC Section 480.2*

Insulation of cells arranged in groups having a total nominal voltage of not over 250V is covered in *NEC Section 480.6* and for voltages over 250V, *NEC Section 480.7*.

403F06.EPS

Figure 6 ◆ *NEC®* requirements governing storage batteries for legally required standby systems.

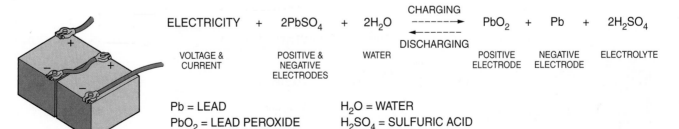

$$\text{ELECTRICITY} + 2PbSO_4 + 2H_2O \xrightarrow[\text{DISCHARGING}]{\text{CHARGING}} PbO_2 + Pb + 2H_2SO_4$$

| VOLTAGE & CURRENT | POSITIVE & NEGATIVE ELECTRODES | WATER | POSITIVE ELECTRODE | NEGATIVE ELECTRODE | ELECTROLYTE |

Pb = LEAD
PbO_2 = LEAD PEROXIDE
$PbSO_4$ = LEAD SULFATE

H_2O = WATER
H_2SO_4 = SULFURIC ACID

403F07.EPS

Figure 7 ◆ Chemical reaction in a lead-acid battery.

Lead-Acid Battery Construction

The internal construction of a typical standard lead-acid battery is shown here. Note the anti-flame propagation vent caps and the built-in lifting brackets.

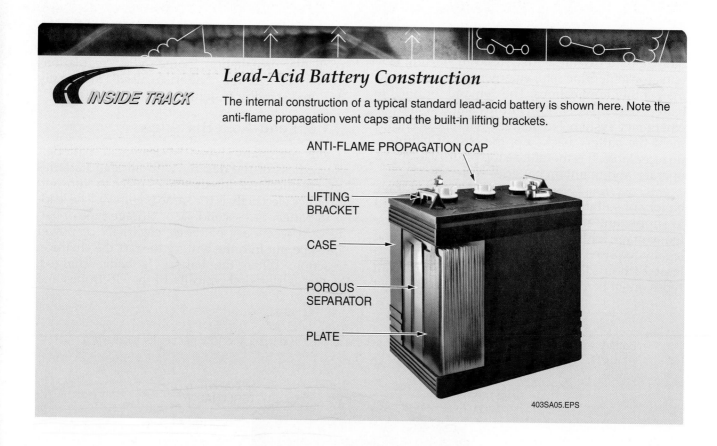

ANTI-FLAME PROPAGATION CAP

LIFTING BRACKET

CASE

POROUS SEPARATOR

PLATE

403SA05.EPS

electrolyte of the battery. When the battery is called upon to provide power to its load, the resulting chemical reaction is reversed, discharging the battery, and the flow of electrons in the external conductors or posts of the battery is from the negative electrode to the positive electrode. This constitutes the actual direction of the flow of electrons, as opposed to the conventional direction of current flow from positive to negative.

3.2.0 Nickel Cadmium Batteries

Nickel cadmium batteries have been widely used for years. The nickel cadmium battery is the most rugged of the battery technologies, mainly because of its durability in extreme-temperature environments. Its reliability has been demonstrated both in arctic and tropical regions. However, the initial cost of the nickel cadmium battery may be difficult to justify for a large number of applications. It is in the extremes of cold or heat, and where maintenance practices are not all that they might be, that the nickel cadmium battery is well worth the investment.

The performance of the nickel cadmium battery in extreme temperatures results from the chemical composition of its electrolyte. The electrolyte is an alkali solution of potassium hydroxide, which remains unaltered during the charge/discharge reaction. Like the lead-acid battery, the nickel cadmium battery is also a secondary or storage-type battery. The chemical reaction that takes place to store and produce electricity in a nickel cadmium battery is shown in *Figure 8*. Nickel cadmium batteries fall into three main categories: pocket plate, fiber plate, and sintered plate.

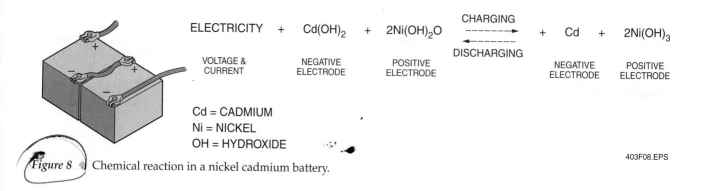

$$\text{ELECTRICITY} + Cd(OH)_2 + 2Ni(OH)_2O \underset{\text{DISCHARGING}}{\overset{\text{CHARGING}}{\rightleftharpoons}} + Cd + 2Ni(OH)_3$$

VOLTAGE & CURRENT   NEGATIVE ELECTRODE   POSITIVE ELECTRODE   NEGATIVE ELECTRODE   POSITIVE ELECTRODE

Cd = CADMIUM
Ni = NICKEL
OH = HYDROXIDE

Figure 8 ◆ Chemical reaction in a nickel cadmium battery.

403F08.EPS

Sealed Valve-Regulated Lead-Acid (VRLA) Batteries

High-amperage modular battery units are available that incorporate absorbed glass mat (AGM) battery technology with a 20-year design life when used in float service. The units can be connected in parallel and in series to obtain the desired voltage and amperage. The units use sealed valve-regulated lead-acid (VRLA) batteries that usually require no special venting under normal operating conditions. However, explosive gases may be vented during use if the batteries are overcharged, damaged, or subjected to a high-temperature environment.

403SA06.EPS

3.3.0 Battery Maintenance

Proper maintenance of storage batteries is extremely important in order to have a reliable emergency/standby system.

3.3.1 Battery Maintenance Overview

It is very important to note the differences in the charging or float voltage for each type of battery. Setting and maintaining proper float voltage is critical to battery life. Overcharging a lead-acid battery will destroy the positive plate, evident by a characteristic dark sediment at the bottom of the cell jar. Undercharging a lead-acid battery will destroy the negative plate, producing a light gray sediment and a crystal-like sparkle on the plates.

The nominal open circuit voltage of a lead-acid cell is 2.0V, and of a nickel cadmium cell is 1.3V, but the charging voltage is higher. An equalizing charge is even higher than a charging float voltage. An equalizing charge is applied to a battery when any individual cell voltage falls below a maintenance-specified minimum voltage. The equalizing overcharge tends to bring all cells back to float voltage when complete.

When a lead-acid battery is being discharged, the sulfuric acid of the battery is consumed and water is produced. During maintenance, this is observed and noted to determine the battery's state of charge. This is done by using a hydrometer to read the specific gravity of the sulfuric acid. The specific gravity of water is 1.0, whereas the specific gravity of sulfuric acid is near 1.15 when a battery is discharged and 1.28 when it is near full charge. Note that measuring the specific gravity on nickel cadmium batteries is not required because the electrolyte (potassium hydroxide) does not change during charge or discharge.

Adverse temperature environments affect the life and performance of lead-acid batteries. High temperatures will reduce the life of the lead-acid battery significantly. A general rule of thumb is that for every 15°F above 77°F, the life of the battery will be reduced by 50%. At low temperatures, the performance of discharge quickly falls off, and at extremely low temperatures, the battery can freeze once it goes into discharge. The freezing point of sulfuric acid is 31°F.

The nickel cadmium battery provides excellent durability in temperature extremes. Nickel cadmium batteries can operate down to −40°F without damage or the threat of freezing. They have some degradation in life expectancy at high temperatures, but not as bad as lead-acid batteries. The general rule of thumb for nickel cadmium batteries is that every 15°F above 77°F results in a 20% reduction in expected life.

Corrosion of the intercell connections in a bank of battery cells causes high resistance and even open circuits when the battery is being discharged under load. These problems have the potential to cause major system failures and battery fires. An intercell resistance test during maintenance using a digital low-resistance ohmmeter is used to detect these high-resistance connections. It is common practice to use a reading of greater than 20% deviation from the average reading to initiate corrective action. Retorque or replace the weak connections upon detection to prevent conduction path failures in the battery bank.

A battery bank can be effectively maintained with routine checks and inspections. Visually inspecting the bank's physical condition for damage or corrosion, as well as electrolyte level, along with keeping a monitored record of specific gravity, voltage, and temperature readings is a necessary part of any successful battery maintenance program. These readings will provide a measure of the state of charge of the battery. The state of charge, however, is not a complete indication of battery life or capacity. Battery capacity is the single most important criterion for replacing a battery.

The method of determining battery capacity is by performing a load test to measure the amount of power a fully charged battery is capable of delivering over a specific time period. Batteries are rated in amp-hours. A load test simply measures

Standard UPS Batteries

A UPS battery bank with vented standard lead-acid batteries is shown here. Codes normally require that these types of batteries and battery banks be located in a dedicated fire-rated room vented to the outdoor air and protected by a fire suppression system. Some codes may require a special explosion-resistant room with reinforced walls, ceilings, floors, and steel doors. Continuous outdoor air exchange for the room is monitored, and all battery charging must be shut down automatically if the ventilation fails. UPS units using sealed batteries do not normally require this type of facility.

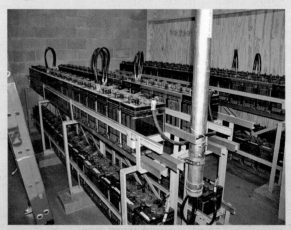

403SA07.EPS

the percent of time that the battery actually carries the load versus the time the battery is supposed to carry the load based on its amp-hour rating. The test is carried out down to a minimum operating voltage, typically 1.75V per cell. Any battery that exhibits less than 80% time capacity is considered failed and should be replaced. Various types of load boxes and monitors are used to perform load tests.

WARNING!
Batteries are hazardous and must be disposed of properly.

3.3.2 Battery Maintenance Guidelines

Much of the maintenance performed on batteries is done with the batteries in service. Because of this, the methods used should follow the protective procedures outlined in the appropriate standards and the battery manufacturer's instructions. In addition, these methods should preclude circuit interruption or arcing in the vicinity of the battery. One of the reasons for this is that hydrogen is liberated from all vented cell batteries while they are on float or equalize charge, as well as when they are nearing completion of a recharge. Valve-regulated lead-acid cells will also liberate hydrogen whenever their vents open. Therefore, adequate ventilation by natural or mechanical means must be available to minimize hydrogen accumulation. In addition, the battery area should be a mandated nonsmoking area.

NOTE
There are specific manufacturer procedures to be followed for receiving, handling, and installing battery systems and chargers. This includes receiving, inspection, storage, filling of lead-acid batteries, interconnection, and initial charge. Manufacturer services are often used to oversee these steps.

All work performed on a battery shall be done only with the proper and safe tools and with the appropriate protective equipment, including:

- Chemical-resistant goggles and face shield
- Acid-resistant or alkali-resistant gloves
- Chemical-resistant protective aprons and overshoes
- Portable or stationary water facilities for rinsing eyes and skin in case of contact with electrolyte

- Acid- or alkali-neutralizing solution
- Insulated tools
- Class C fire extinguisher or other type as recommended by the battery manufacturer
- Respirator if required

WARNING!
Some battery manufacturers do not recommend the use of CO_2 (Class C) fire extinguishers due to the possibility of thermal shock and possible cracking of the battery containers. Always refer to the manufacturer's instructions for safety and maintenance procedures.

Guidelines for maintenance inspections and tests are provided below. For more detailed information on battery maintenance scheduling, refer to *IEEE Standard 446*. Anyone performing maintenance should be familiar with the appropriate standards and the battery manufacturer's instructions for the battery to be maintained. This is particularly true for valve-regulated lead-acid cells since some testing is recommended that is designed to determine if dry-out of the cell is occurring. Dry-out can be caused by a number of factors, such as high float voltage and high temperature.

As applicable to the types of batteries involved, the following visual inspections should be performed in accordance with the battery manufacturer's recommendations and/or the facility schedule. A typical maintenance interval for each task is shown.

Monthly checks:

- General appearance of the battery/rack area
- Dirt or electrolyte on jars, covers, etc.
- Charger voltage and current output
- Electrolyte level (vented lead-acid and vented nickel cadmium cells)
- Jar or cover for cracks or leaks
- Jar or post seals (valve-regulated lead-acid cells)
- Excessive jar or cover distortion (valve-regulated lead-acid cells)
- Evidence of corrosion

Annual checks:

- Flame arrester clogged (vented lead-acid and vented nickel cadmium cells)
- Detailed rack inspection
- Insulating covers on racks
- Seismic rack part and spacers
- Detailed cell inspection (vented lead-acid cells)
- Plates for cracks, sulfate, and hydration (vented lead-acid cells)

- Abnormal sediment accumulation (vented lead-acid cells)
- Jar and post seals (other than valve-regulated lead-acid cells)
- Excessive jar or cover distortion (other than valve-regulated lead-acid cells)
- Excessive gassing (vented lead-acid cells)
- Signs of vibration

As applicable to the types of batteries involved, the following measurements should be performed in accordance with the battery manufacturer's recommendations and/or the user facility schedule. A typical maintenance interval for each task is shown.

Monthly checks:

- Battery float voltage
- Pilot cell voltage (vented lead-acid cells)
- Pilot cell electrolyte temperature (vented lead-acid cells)
- Pilot cell electrolyte specific gravity (vented lead-acid cells)
- Ventilation equipment adequacy

Quarterly checks:

- Individual cell voltage
- Individual cell electrolyte specific gravity (vented lead-acid cells)
- Pilot cell electrolyte temperature (vented nickel cadmium, vented lead-acid, and valve-regulated lead-acid cells)
- Intercell connection resistance (valve-regulated lead-acid cells)
- Cell impedance, conductance, and resistance (valve-regulated lead-acid cells)
- Ambient temperature (vented lead-acid and valve-regulated lead-acid cells)

Annual checks:

- AC ripple current and voltage (valve-regulated lead-acid cells)
- Intercell connection torque (vented nickel cadmium cells)
- Intercell connection resistance (vented lead-acid cells)
- Battery rack torque
- Ground connections

3.4.0 Battery and Battery Charger Operation

In addition to providing voltage to start generator sets as previously described, standby batteries are widely used in many other emergency power applications, particularly to power emergency lighting systems. Standby batteries used in these applications are normally operated in the float mode of operation where the battery, battery charger, and load are connected in parallel (*Figure 9*). The charging equipment should be of a size that provides all the power normally required by the loads plus enough additional power to keep the battery at full charge. Relatively large intermittent loads will draw power from the battery. This power is restored to the battery by the charger when the intermittent load ceases. If AC input power to the system is lost, the battery instantly carries the full load. If the battery and charger are properly matched to the load and to each other, there is no discernible voltage dip when the system goes to full battery operation.

When AC charging power is restored, a constant potential charger will deliver more current than would be necessary if the battery were fully charged. Some constant potential chargers will automatically increase their output voltage to the equalizer setting after power to the charger is restored. Similarly, a constant current charger may automatically increase its output to the high rate. The battery charger must be sized such that it can supply the load and restore the battery to full charge within an acceptable time. The

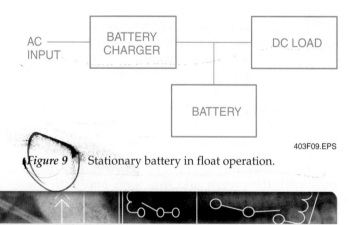

Figure 9 Stationary battery in float operation.

403F09.EPS

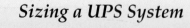

Sizing a UPS System

THINK ABOUT IT

What other factors besides the direct load should be taken into account when sizing a UPS system?

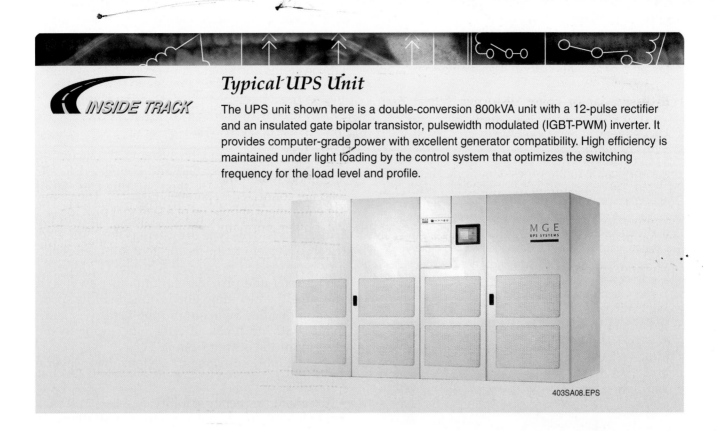
increased current delivered by a constant potential charger during this restoration period will decrease as the battery approaches full charge.

The battery charger is an important part of the emergency power system, and consideration must be given to redundant chargers on critical systems. The charger should be operated only with the battery connected to the DC bus. Without this connection, excessive ripple on the system could occur, which could affect the connected equipment, causing improper operation or failure of the equipment. If a system must be able to operate without the battery connected, then a device called a battery eliminator should be used. Also note that a battery charger's output must be derated for altitude when it is installed at locations above 3,300' and for temperature when the ambient temperature exceeds 122°F. The battery charger manufacturer can provide these derating factors.

4.0.0 ◆ STATIC UNINTERRUPTIBLE POWER SUPPLY

A static uninterruptible power supply (UPS) system is an electronically controlled, solid-state power control system designed as an alternate source to provide regulated AC power to critical loads in the event of a partial or total failure of the normal source of power, typically the power supplied by the electric utility. Today, the term *UPS* commonly refers to a system consisting of a dedicated stationary-type battery, rectifier/battery charger, static inverter, and accessories such as a transfer switch. Note that in the past, the term *UPS* was commonly used to describe a type of static inverter often used in large DC systems.

There is a wide diversity in UPS configurations that provide for voltage regulation, line conditioning, lightning protection, redundancy, electromagnetic interference (EMI) management, extended run time, load transfer to other units, and other such features required to protect the critical loads against failure of the normal AC source and against other power system disturbances. UPS systems are available in a number of designs and sizes, ranging from less than 100W to several megawatts. They may provide single-phase or three-phase power at frequencies of 50Hz, 60Hz, or 400Hz. In addition, static UPS systems may provide power of which the output purity may vary from a near-perfect sine wave with less than 1% total harmonic distortion to a power wave that is essentially a square wave.

Sealed VRLA Battery UPS Systems

The UPS system shown here has operating characteristics similar to the 800kVA unit illustrated earlier. The difference is that this particular unit is equipped with sealed VRLA batteries in a battery cabinet attached to the UPS cabinet. Sealed battery UPS systems can be located in most areas without special ventilation requirements.

SEALED VRLA BATTERIES

BATTERY CABINET   UPS

403SA09.EPS

There are two basic types of UPS systems used in a variety of configurations: double-conversion UPS systems and single-conversion UPS systems.

4.1.0 Double-Conversion UPS Systems

In double-conversion UPS systems, the incoming AC power is first rectified and converted to DC (*Figure 10*). The DC is then supplied as input power to a DC-to-AC converter (inverter). The inverter output is AC, which is used to power the critical loads. This type of system is the static electrical equivalent to the motor-generator set. A battery is connected in parallel with the DC input to the inverter and provides continuous power to it any time the incoming line is outside of its specification or fails. Switching to the battery is automatic, with no break in either the input to the inverter or the output from it.

The double-conversion system has several advantages:

- Excellent frequency stability
- High degree of isolation from variations in incoming line voltage and frequency
- Zero transfer time possible
- Quiet operation
- Can provide a sinusoidal output waveform with low distortion

In lower-power UPS applications (0.1kVA to 20kVA), the double-conversion UPS can have some of the following disadvantages:

- Lower efficiency
- Large DC power supply required (typically 1.5 times the full-load rating of the UPS)
- Poor noise isolation from line to load
- Possible shortened service life due to greater heat dissipation

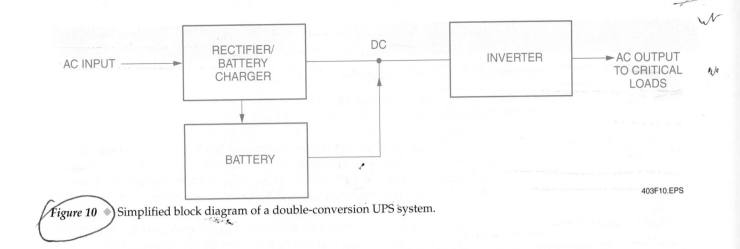

Figure 10 ◆ Simplified block diagram of a double-conversion UPS system.

403F10.EPS

4.2.0 Single-Conversion UPS Systems

In the single-conversion system (*Figure 11*), the incoming line to the UPS is not rectified to produce DC power to provide input to the inverter. The normal AC power is supplied directly to the critical loads through a series inductor or transformer. The normal AC power also supplies a small battery charger used to maintain the UPS batteries in a fully charged condition. Thus, the battery is only used when the inverter requires the battery's output to supplement or replace the normal power source.

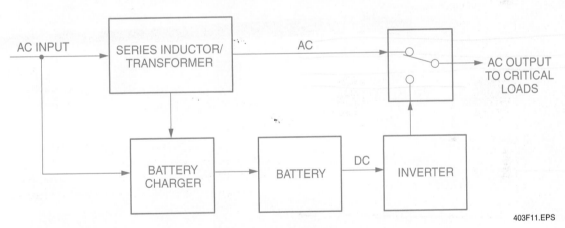

403F11.EPS

Figure 11 ◆ Simplified block diagram of a single-conversion UPS system.

Stored-Energy Flywheel UPS Systems

Battery-free, double-conversion UPS systems use a type of motor-generator and flywheel technology to provide 15 seconds of power at 100% rated load during power outages. These types of systems are sometimes referred to as ride-through UPS systems. In most cases, 15 seconds is sufficient to start and stabilize emergency engine-generators, including backup generators, and bring them online for the load, as well as to reestablish the UPS capability. Within the 15-second time frame, there is enough time to wait a few seconds to avoid unnecessary engine-generator starts resulting from momentary outages. The startup time for an engine-generator is normally five to six seconds with another five or six seconds required to stabilize the output. If the primary engine-generators start, any backup units will be shut down after a predetermined delay. The engine-generator starting voltage and current are normally furnished directly from the UPS so that engine starting batteries can be eliminated. This type of UPS system is highly reliable and economical because it is free of all batteries that deteriorate over time and because it is a simple rotating electrical device that requires little maintenance. The motor-generator also provides power conditioning by isolating the facility power from the utility power, thus eliminating power line transients and noise. The unit shown here is a 600kVA UPS with two paralleled motor-generator cabinets. The integrated motor-generator/flywheel devices in the cabinets are very compact, vertically mounted devices that occupy the lower third of each cabinet. The unit furnishes regulated output power with harmonics cancellation. It is microprocessor-controlled and can be scheduled for automatic, non-interfering periodic operational tests and engine-generator operational tests.

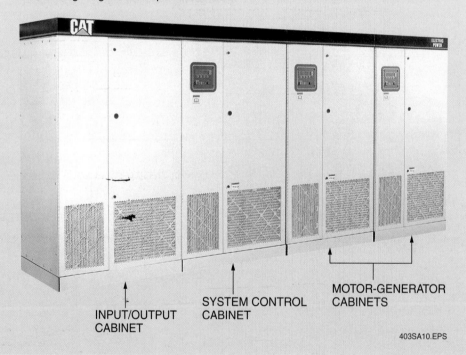

INPUT/OUTPUT CABINET

SYSTEM CONTROL CABINET

MOTOR-GENERATOR CABINETS

403SA10.EPS

5.0.0 ◆ NEC® REQUIREMENTS FOR EMERGENCY SYSTEMS

Several areas of the NEC® deal with emergency standby systems, along with the means of generating or producing electricity for such systems. In general, NEC Article 700 deals with emergency systems. However, there are other NEC® requirements that are directly related to the sources of power used for emergency systems. Branch circuits for emergency power are covered to some extent in NEC Chapters 2 and 3, transfer switches are dealt with in NEC Section 700.6, and storage batteries used for emergency electrical systems are covered in NEC Section 700.12(A) and NEC Article 480. Consequently, several areas of the NEC® will have to be referred to when designing, installing, or maintaining these systems.

5.1.0 Legally Required Standby Systems

NEC Article 701 covers installation requirements of legally required standby electrical systems. The article's intent is to provide a system that is electrically safe and covers circuits and equipment intended to supply, distribute, and control electricity to required facilities for illumination or power (or both) when the normal electrical supply is interrupted.

This article covers only those systems that are permanently installed in their entirety, including the power source.

In general, legally required standby systems are typically installed to serve loads such as heating and refrigeration systems, communication systems, ventilation and smoke removal systems, sewage disposal, lighting systems, and industrial processes that, when stopped during any interruption of the normal electrical supply, could create hazards or hamper rescue or fire-fighting operations.

A legally required standby system must have adequate capacity and rating for the supply of all equipment intended to be operated at one time. Furthermore, it must be suitable for the maximum available fault current at its terminals.

The alternate power source is permitted to supply legally required standby and optional standby system loads when automatic selective load pickup and load shedding are provided as needed

to ensure adequate power to the legally required standby circuits.

Transfer equipment, including automatic transfer switches, must be identified for standby use and approved by the authority having jurisdiction. Transfer equipment must also be designed and installed to prevent the inadvertent interconnection of normal and alternate sources of supply in any operation of the transfer equipment. Consequently, transfer equipment, including transfer switches, must operate in such a manner that all ungrounded conductors of one source of supply are completely disconnected before any ungrounded conductors of the second source are connected.

A means may be provided to bypass and isolate the transfer switch equipment. However, where bypass isolation switches are used, inadvertent parallel operation must be avoided. See Figure 12.

Where possible, audible and visual signal devices are usually provided to indicate:

- Derangement of the standby source
- A load on the standby source
- A malfunctioning battery charger

A sign must be placed at the service entrance indicating the type and location of on-site legally required standby power sources. Furthermore, where the grounded conductor that is connected to the emergency source is connected to a grounding electrode conductor at a location remote from the emergency source, there must be a sign at the grounding location to identify all emergency and normal sources connected at that location.

5.2.0 Sources of Power

Current must be supplied to legally required standby systems so that in the event of a normal power failure, legally required standby power will be available within the time required for the application, but not to exceed 60 seconds.

5.2.1 Storage Batteries

As described earlier in this module, NEC Section 701.11(A) specifies that a storage battery of suitable rating and capacity is required to supply and maintain at least 87½% of system voltage for the total load of the circuits supplying legally required standby power for a period of at least 1½ hours.

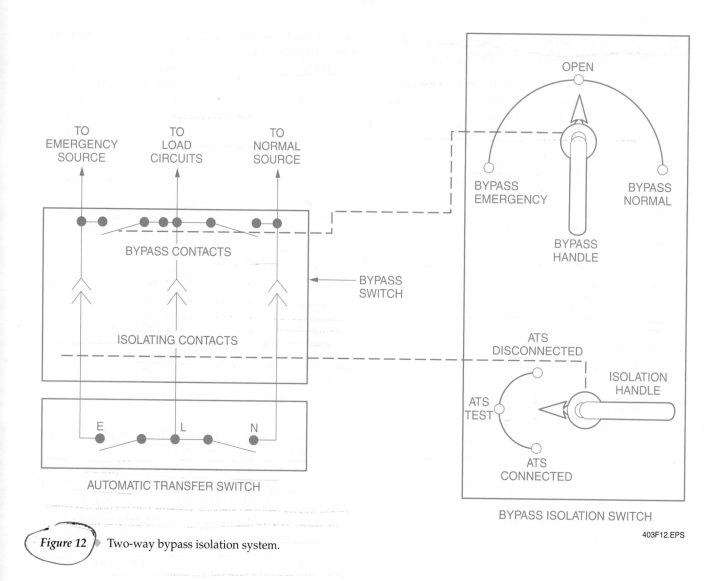

Figure 12 ◆ Two-way bypass isolation system.

5.2.2 Generator Sets

Generator sets suitable for use as a power supply for legally required standby systems were covered earlier in this module. However, a few additional notes are in order. In general, the prime mover (usually an engine) must be acceptable to the authority having jurisdiction and sized in accordance with *NEC Section 701.6*—that is, it must have adequate capacity and rating for the supply of all equipment intended to be operated at one time.

A means must be provided for automatic starting when the normal service fails. Furthermore, the system must be controlled so that the standby power is automatically transferred to all required electrical circuits. A short-time feature permitting a 15-minute setting must be provided to avoid retransfer in case of short-time re-establishment of the normal service.

See *Figure 13* for a summary of other *NEC®* requirements governing legally required standby systems. Refer to the listed *NEC®* sections for greater detail. Also be aware that when other *NEC®* sections apply, they should be adhered to where applicable.

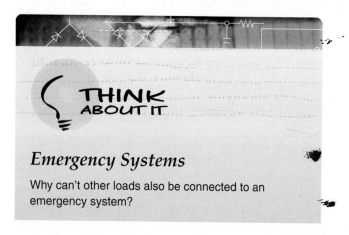

THINK ABOUT IT

Emergency Systems

Why can't other loads also be connected to an emergency system?

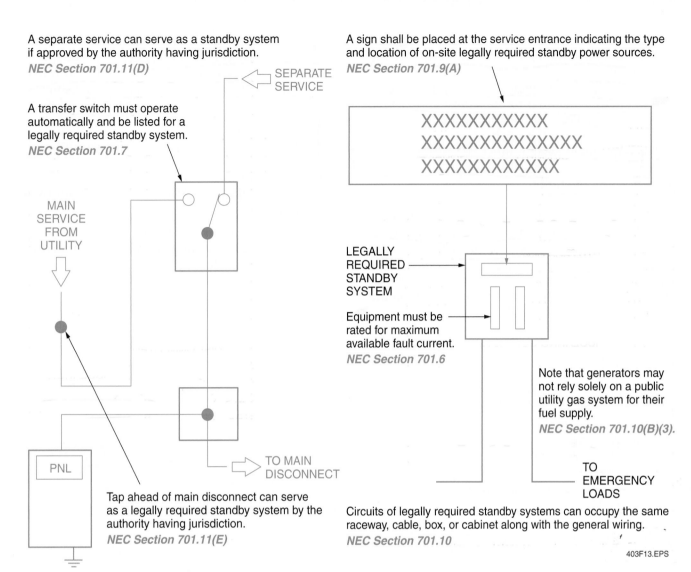

A separate service can serve as a standby system if approved by the authority having jurisdiction.
NEC Section 701.11(D)

A transfer switch must operate automatically and be listed for a legally required standby system.
NEC Section 701.7

SEPARATE SERVICE

MAIN SERVICE FROM UTILITY

PNL

Tap ahead of main disconnect can serve as a legally required standby system by the authority having jurisdiction.
NEC Section 701.11(E)

TO MAIN DISCONNECT

A sign shall be placed at the service entrance indicating the type and location of on-site legally required standby power sources.
NEC Section 701.9(A)

XXXXXXXXXX
XXXXXXXXXXXX
XXXXXXXXXX

LEGALLY REQUIRED STANDBY SYSTEM

Equipment must be rated for maximum available fault current.
NEC Section 701.6

Note that generators may not rely solely on a public utility gas system for their fuel supply.
NEC Section 701.10(B)(3).

TO EMERGENCY LOADS

Circuits of legally required standby systems can occupy the same raceway, cable, box, or cabinet along with the general wiring.
NEC Section 701.10

403F13.EPS

Figure 13 ◆ Summary of *NEC*® installation requirements governing legally required standby systems.

6.0.0 ◆ EMERGENCY SYSTEM CIRCUITS FOR LIGHT AND POWER

The loads on emergency circuits for light and power will vary with the building type. However, in almost all areas throughout the United States, all public buildings must have exit lights and emergency white lights (to enable occupants to see while exiting the building) connected to an emergency electric system. In general, the following items must be fed from an emergency circuit in public buildings:

- All exit lights
- Sufficient emergency white lights
- Fire alarm systems
- Life safety systems
- Sprinkler system water pumps

6.1.0 Health Care Facilities

NEC Article 517 requires an alternate power source consisting of one or more generator sets or battery systems (where permitted) to be installed in health care facilities. Each patient bed location must be supplied by at least two branch circuits, one or more from the emergency system and one or more circuits from the normal electrical system. Emergency system receptacles must be identified and must also indicate the panelboard and circuit number supplying them. All receptacles used in patient bed locations must be listed as hospital grade.

Essential electrical systems for hospitals must be comprised of two separate systems capable of supplying a limited amount of lighting and power service that is considered essential for life safety and effective hospital operation during the time the normal electrical service is interrupted. These two systems are the emergency system and the equipment system.

In general, the emergency system must be limited to circuits essential to life safety and critical patient care. These are designated the life safety branch and the critical branch.

The equipment system must supply the major electric equipment necessary for patient care and basic hospital operation.

The number of transfer switches to be used must be based upon reliability, design, and load considerations. Each branch of the essential electrical system must be served by one or more transfer switches, as shown in *Figure 14*. One transfer switch may serve one or more branches or systems in a facility with a maximum demand on the essential electrical system of 150kVA.

The life safety branch and critical branch of the emergency system must be kept entirely independent of all other wiring and equipment and must not enter the same raceways, boxes, or cabinets with each other or other wiring, except as follows:

- In transfer switches
- In exit or emergency lighting fixtures supplied from two sources
- In a common junction box attached to exit or emergency lighting fixtures supplied from two sources

Testing of the system is required upon installation and periodically thereafter. A written record must be kept of all testing and maintenance.

The wiring of the equipment system is permitted to occupy the same raceways, boxes, or cabinets of other circuits that are not part of the emergency system.

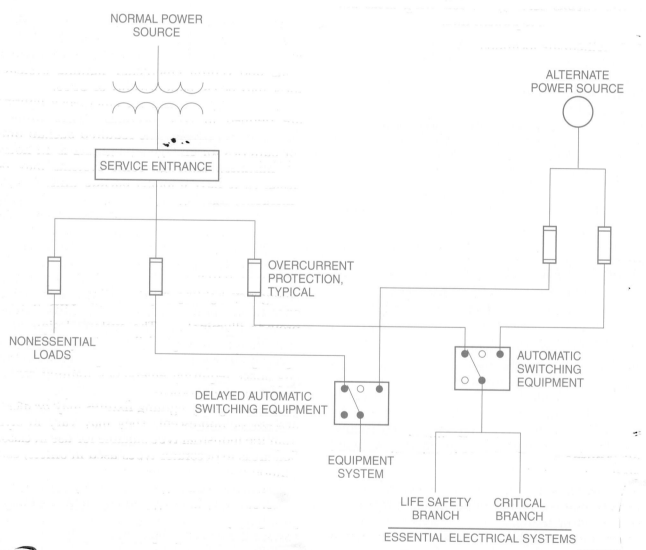

Figure 14 ◆ Essential electrical systems must be served by one or more transfer switches.

403F14.EPS

The essential electrical system must have adequate capacity to meet the demand for the operation of all functions and equipment to be served by each system and branch. Furthermore, the emergency system must be divided into two mandatory branches: the life safety branch and the critical branch, as described in *NEC Sections 517.32 and 517.33*.

The life safety branch of the emergency system must supply power for the following:

- Lighting required for corridors, passageways, stairways, and landings at all exit doors
- Exit signs
- Alarm and alerting systems
- Communication systems
- Generator set location
- Elevators

The critical branch of the emergency system must supply power for task illumination, fixed equipment, selected receptacles, and special power circuits serving the following areas and functions related to patient care:

- Anesthetizing locations
- Isolated power systems in special environments
- Patient care areas
- Nurse call system
- Blood, bone, and tissue banks
- Telephone equipment room and closets
- Task illumination, receptacles, and selected power circuits needed for effective hospital operation

NEC Section 700.9(A) also requires labels, signs, or some other permanent marking to be used on all enclosures containing emergency circuits to readily identify them as components of an emergency system. See *Figure 15*.

All boxes and enclosures for emergency circuits must be painted red, marked with red labels saying *EMERGENCY CIRCUITS*, or marked in some other manner clearly identifiable to electricians and maintenance personnel.

6.2.0 Battery-Powered Emergency Lighting

Emergency lighting systems are used to provide the required lighting to a facility for egress or other purposes such as security when the normal electrical supply is interrupted. The purpose for this lighting is to illuminate evacuation routes and exits so that the chance of personal injury or loss of life can be minimized in the event of a fire or other emergency that can cause the illumination

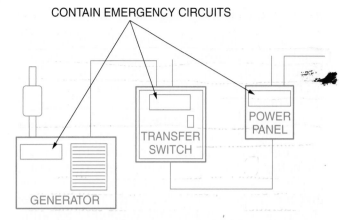

CONTAIN EMERGENCY CIRCUITS

GENERATOR TRANSFER SWITCH POWER PANEL

Labels, signs, or some other permanent marking must be used on all enclosures containing emergency circuits to readily identify them as components of an emergency system. *NEC Section 700.9(A)*

403F15.EPS

Figure 15 ◆ Summary of *NEC*® requirements for emergency circuit identification.

in an area to be lost or reduced to a level where evacuation may be difficult or impossible. Batteries are used in self-contained emergency lighting units and central emergency lighting systems; these may be DC batteries only or UPSs.

Specific requirements for emergency lighting are defined in *NEC Sections 700.16 through 700.22 and 701.11(G)*. The required backup time for batteries for emergency lighting is 1½ hours (90 minutes), although some systems may be designed to have a longer backup time. Also, if emergency lighting depends upon the changing of sources (for instance, from AC to DC), the change must occur automatically with no appreciable interruption (less than 10 seconds). In areas that are normally lighted solely by high-intensity discharge (HID) systems, any battery-powered emergency lighting must continue to function, once power is restored, until the HID fixtures provide illumination. The system design must also consider a single failure of any individual lighting element, such as a lamp failure, so that any space requiring emergency lighting cannot be left in total darkness.

The emergency lighting fixtures may be incandescent or fluorescent. They may vary in style from the industrial type suitable for use in classified areas to decorator types used in offices, convention centers, etc. Emergency lighting fixtures also include exit signs. The lamps illuminate stairs, stairway landings, aisles, corridors, ramps, passageways, exit doors, and any other areas requiring emergency lighting.

6.3.0 Emergency Lighting Units

Emergency lighting units, also called unit equipment, are the most popular type of emergency lighting equipment in use today (*Figure 16*). *NEC Section 701.11(G)* defines the requirements for unit equipment. The unit itself consists of a rechargeable battery, a battery charger, and one or more lamps, provisions for remote lamps, or both. The unit has the necessary controls to maintain the battery fully charged while the normal source of power is available and to automatically energize the lamps upon the loss of the normal power supply. Most units have the capability for automatic testing and self diagnostics. The controls will de-energize the lamps immediately or after a time delay (for HID lamps) upon restoration of the normal power supply. If the unit is used with a fluorescent fixture, it will also have a high-frequency inverter and controls intended for that type of fixture.

These units can be supplied with any of the lead-acid or pocket-plate nickel cadmium batteries described earlier in this module. In addition, the units may be supplied with chargers that have

an automatic temperature compensation of float/charge voltage. This feature is a must when a valve-regulated lead-acid battery is supplied with the unit. The selection of the battery type depends on the installation itself. For example, pocket-plate nickel cadmium batteries may be used in high-temperature areas, such as near an industrial boiler. On the other hand, valve-regulated lead-acid or sealed sintered-plate batteries may be used in an office area. All of these batteries require periodic maintenance and testing.

The batteries used in emergency lighting units are normally either 6VDC or 12VDC. For the lead-acid cells, the electrolyte specific gravity is normally 1.225 or greater, and the cell capacity is often expressed in ampere-hours at the 10-hour or 20-hour discharge rate of 1.75V per cell at 77°F.

6.4.0 Places of Assembly

NEC Article 518 covers places of assembly—that is, all buildings or portions of buildings or structures designed or intended for assembly of 100 or more persons. Such facilities include assembly halls, restaurants, dance halls, courtrooms, auditoriums, churches, and similar locations. The installation and control of emergency electrical systems in these facilities must comply with *NEC Article 700, Emergency Systems*. All equipment must be approved for use on emergency systems.

The authority having jurisdiction must conduct or witness a test on the complete system upon installation and periodically afterward. A written record must be kept of such tests and maintenance.

All boxes and enclosures (including transfer switches, generators, and power panels) for emergency circuits must be permanently marked so they can be readily identified as a component of an emergency circuit or system. These boxes and enclosures for emergency circuits must be painted red, marked with red labels saying *EMERGENCY CIRCUITS*, or marked in some other manner clearly identifiable to electricians and maintenance personnel.

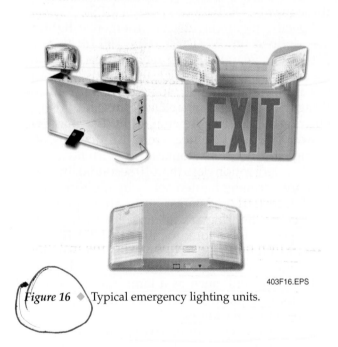

403F16.EPS

Figure 16 ◆ Typical emergency lighting units.

1. In the event of a normal power supply failure, standby power for an emergency system must be available within _____ of the failure.
   a. 10 seconds
   b. 30 seconds
   c. 60 seconds
   d. 5 minutes

2. Legally required standby systems are covered in _____.
   a. NEC Article 700
   b. NEC Article 701
   c. NEC Article 702
   d. NEC Article 705

3. The field windings of an AC generator are located on the _____ of the generator.
   a. rotor
   b. stator
   c. transfer switch
   d. exciter

4. A stator is _____.
   a. the rotating element of a motor or generator
   b. the stationary part of a motor or generator
   c. a device that supplies direct current
   d. not a part of a motor or generator

5. All of the following are parts of an AC generator *except* a(n) _____.
   a. stator
   b. rotor
   c. transfer switch
   d. exciter

6. When sizing a generator set once the starting and running loads have been determined, it is common to add a margin factor of up to _____ for future expansion.
   a. 10%
   b. 15%
   c. 25%
   d. 33%

7. In an automatic transfer switch, transfer to emergency/standby power is initiated by an external signal from a remote control.
   a. True
   b. False

8. The specific gravity of sulfuric acid when a battery is discharged is near _____.
   a. 1.10
   b. 1.15
   c. 1.20
   d. 1.28

9. How often should the intercell connections of a lead-acid battery bank be resistance tested with a digital low-resistance ohmmeter?
   a. Monthly
   b. Quarterly
   c. Annually
   d. Bi-annually

10. Each of the following is an advantage of a double-conversion UPS system *except* _____.
    a. zero transfer time is possible
    b. quiet operation
    c. excellent efficiency in all applications
    d. excellent frequency stability

11. In a single-conversion UPS system, the incoming AC utility power is first applied to the _____.
    a. battery charger
    b. inverter
    c. critical load
    d. battery charger and transformer

12. Which of the following covers the installation of emergency systems?
    a. NEC Article 518
    b. NEC Article 700
    c. NEC Section 250.34
    d. NEC Article 703

13. The maximum time allowed for a legally required standby generator system to be started and switched online in the event of a power failure is _____ seconds.
    a. 20
    b. 40
    c. 60
    d. 80

14. What is the minimum number of essential standby electrical systems that must be provided for in hospitals?
    a. One
    b. Two
    c. Three
    d. Four

15. All of the following contain both the life safety branch and critical branch wiring of the emergency system used in hospitals *except* _____.
    a. raceways
    b. transfer switches
    c. exit lighting fixtures supplied from two sources
    d. emergency lighting fixtures supplied from two sources

Summary

The provisions of *NEC Articles 700, 701, and 702* apply to the electrical design, installation, operation, and maintenance of emergency systems *(NEC Article 700)*, legally required standby systems *(NEC Article 701)*, and optional standby systems *(NEC Article 702)*. These systems consist of the circuits and equipment intended to supply, distribute, and control electricity for illumination or power (or both) to required facilities when the normal electrical supply or system is interrupted.

Notes

Trade Terms
Introduced in This Module

Armature: The assembly of windings and metal core laminations in which the output voltage is induced. It is the rotating part in a revolving field generator.

Electrolyte: A substance in which the conduction of electricity is accompanied by chemical action (for example, the paste that forms the conducting medium between the electrodes of a dry cell or storage cell).

Exciter: A device that supplies direct current (DC) to the field coils of a synchronous generator, producing the magnetic flux required for inducing output voltage in the stator coils.

Synchronous generator: An AC generator having a DC exciter. Synchronous generators are used as stand-alone generators for emergency and standby power systems and can also be paralleled with other synchronous generators and the utility system.

This module is intended to present thorough resources for task training. The following reference works are suggested for further study. These are optional materials for continuing education rather than for task training.

Liquid-Cooled Generator Sets Application Manual, Latest Edition. Minneapolis, MN: Cummins Onan.

National Electrical Code® Handbook, Latest Edition. Quincy, MA: National Fire Protection Association.

OT III Transfer Switches Application Manual, Latest Edition. Minneapolis, MN: Cummins Onan.

NCCER makes every effort to keep these textbooks up-to-date and free of technical errors. We appreciate your help in this process. If you have an idea for improving this textbook, or if you find an error, a typographical mistake, or an inaccuracy in NCCER's Contren® textbooks, please write us, using this form or a photocopy. Be sure to include the exact module number, page number, a detailed description, and the correction, if applicable. Your input will be brought to the attention of the Technical Review Committee. Thank you for your assistance.

Instructors – If you found that additional materials were necessary in order to teach this module effectively, please let us know so that we may include them in the Equipment/Materials list in the Annotated Instructor's Guide.

Write: Product Development and Revision
   National Center for Construction Education and Research
   3600 NW 43rd St., Bldg. G, Gainesville, FL 32606

Fax: 352-334-0932

E-mail: curriculum@nccer.org

Craft   Module Name

_____

Copyright Date   Module Number   Page Number(s)

_____

Description

_____

_____

_____

_____

(Optional) Correction

_____

_____

_____

(Optional) Your Name and Address

_____

_____

_____

Basic Electronic Theory

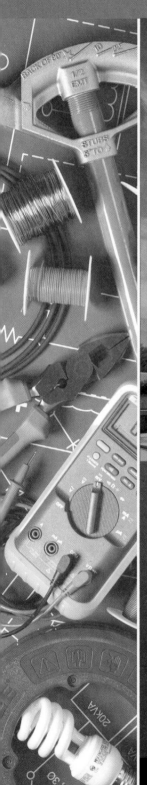

Lawrence Technological University – Blend of Engineering and Architecture

Michigan's Lawrence Technological University drew on locally sourced, sustainable materials, such as decking material made of a composite of rice hulls and polymer. As well, the small house is driven by a high level of technology, which required the close collaboration of engineering and architecture students.

26404-08

26404-08
Basic Electronic Theory

Topics to be presented in this module include:

1.0.0	Introduction	4.2
2.0.0	Electricity under Magnification	4.2
3.0.0	Semiconductor Fundamentals	4.3
4.0.0	Diodes	4.5
5.0.0	Light-Emitting Diodes	4.11
6.0.0	Transistors	4.12
7.0.0	Silicon-Controlled Rectifiers	4.14
8.0.0	Diacs	4.16
9.0.0	Triacs	4.16

Overview

Electronic components play an important part in regulating and controlling power characteristics, such as current, voltage, and frequency. As an example, the speed of some motors is now controlled by regulating the current or frequency delivered to the motor. These types of systems are referred to as DC or AC drive controls. Diodes have been used for many years as rectifiers to convert alternating current into direct current. Other applications of electronic devices in electrical systems include sequential switching and regulating devices used in the distribution of power at commercial power plants and large industrial facilities. Many metering systems for utility usage measurements now incorporate electronic devices, including global positioning systems (GPS), allowing meters to be read from remote locations.

It is essential for electricians to understand the basic theory associated with common electronic devices, such as diodes, transistors, and other semiconductor devices, in order to systematically troubleshoot electrical systems and equipment that incorporate these devices. It is merely a matter of time before most electromechanical control devices are replaced by electronic devices incorporating various types of solid-state, semiconductor components.

Objectives

When you have completed this module, you will be able to do the following:

1. Identify electronic system components.
2. Describe the electrical characteristics of solid-state devices.
3. Describe the basic materials that make up solid-state devices.
4. Describe and identify the various types of transistors and explain how they operate.
5. Interpret electronic schematic diagrams.
6. Describe and connect diodes.
7. Describe and connect light-emitting diodes (LEDs).
8. Describe how to connect silicon-controlled rectifiers (SCRs).
9. Identify the leads of various solid-state devices.

Trade Terms

Diac
Field-effect transistor (FET)
Forward bias
Junction field-effect transistor (JFET)
N-type material
P-type material
Rectifier
Reverse bias
Semiconductor
Silicon-controlled rectifier (SCR)
Solid-state device
Triac

Required Trainee Materials

1. Pencil and paper
2. Appropriate personal protective equipment
3. Copy of the latest edition of the *National Electrical Code*®

Prerequisites

Before you begin this module, it is recommended that you successfully complete *Core Curriculum; Electrical Level One; Electrical Level Two; Electrical Level Three; Electrical Level Four,* Modules 26401-08 through 26403-08.

This course map shows all of the modules in *Electrical Level Four.* The suggested training order begins at the bottom and proceeds up. Skill levels increase as you advance on the course map. The local Training Program Sponsor may adjust the training order.

ELECTRICAL LEVEL FOUR

26413-08 Introductory Skills for the Crewleader

26412-08 Special Locations

26411-08 Medium-Voltage Terminations/Splices

26410-08 Motor Operation and Maintenance

26409-08 Heat Tracing and Freeze Protection

26408-08 HVAC Controls

26407-08 Advanced Controls

26406-08 Specialty Transformers

26405-08 Fire Alarm Systems

26404-08 Basic Electronic Theory

26403-08 Standby and Emergency Systems

26402-08 Health Care Facilities

26401-08 Load Calculations – Feeders and Services

ELECTRICAL LEVEL THREE

ELECTRICAL LEVEL TWO

ELECTRICAL LEVEL ONE

CORE CURRICULUM: Introductory Craft Skills

404CMAP.EPS

1.0.0 ◆ INTRODUCTION

Many controls and related components used in electrical systems are operated by solid-state devices. In fact, many types of motor controls that were once operated either magnetically or mechanically have been replaced with solid-state electronic devices. Therefore, electricians in every branch of the industry should have a basic knowledge of electronics in order to properly install, maintain, and troubleshoot electronic control devices.

Electronics is a science that deals with the behavior and effect of electron flow in specific substances such as semiconductors, which will be discussed shortly. Electronics can be distinguished from electricity by thinking in terms of the voltages and currents used. In the electrical circuits you have been studying, you have dealt primarily with AC voltages that are measured in hundreds of volts and currents that are measured in tens of amps.

In electronic circuits, low-level DC voltages such as 5V, 10V, 15V, and 24V are common. Current is measured in milliamps (mA) and even microamps (μA).

2.0.0 ◆ ELECTRICITY UNDER MAGNIFICATION

If it were possible to view the flow of electrons through a high-powered microscope, at first glance, you might think you were studying astronomy rather than electricity. The atom consists of a central nucleus composed of protons and neutrons, surrounded by orbiting electrons, as shown in *Figure 1*. The nucleus is relatively large when compared with the orbiting electrons, just as the sun seems large when compared with its orbiting planets.

In an atom, the orbiting electrons are held in place by the electric force between the negative electrons and positive protons together with the centrifugal force that results from the motion of the electrons rotating around the nucleus. It is similar to how the Earth's gravity keeps its satellite (the moon) from drifting off into space. The law of charges states that opposite charges attract and like charges repel. The positively charged protons in the nucleus, therefore, attract the negatively charged electrons. If this force of attraction were the only one in effect, the electrons would be pulled closer and closer to the nucleus and eventually be absorbed into the nucleus. However, this force of attraction is balanced by the centrifugal force that results from the motion of the electrons around the nucleus (*Figure 2*). The

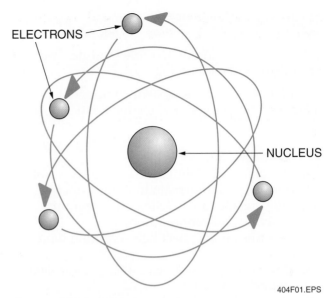

Figure 1 ◆ Structure of an atom.

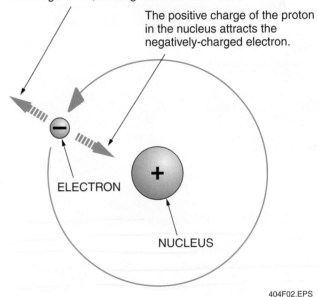

Figure 2 ◆ Electron in orbit around the nucleus.

law of centrifugal force states that a spinning object will pull away from its center point. The faster an object spins, the greater the centrifugal force becomes.

Since the protons and orbital electrons of an atom are equal in number and equal and opposite in charge, they neutralize each other electrically. Thus, each atom is normally electrically neutral—that is, it exhibits neither a positive nor a negative charge. However, under certain conditions, an atom can become unbalanced by losing or gaining electrons. If an atom loses a negatively charged electron, the atom will exhibit a positive charge and is then referred to as a positive ion. Similarly, an atom that gains an additional negatively charged electron becomes negatively charged itself and is then called a negative ion. In either case, an unbalanced condition is created in the atom, causing the formerly neutral atom to become charged. When one atom is charged and there is an unlike charge in another nearby atom, electrons can flow between the two. As you know, this flow of electrons is an electrical current.

3.0.0 ◆ SEMICONDUCTOR FUNDAMENTALS

Semiconductors are the basis for what is known as solid-state electronics. Solid-state electronics is, in turn, the basis for all modern microminiature electronics such as the tiny integrated circuit and microprocessor chips used in computers.

The ability to control the amount of conductivity in semiconductors makes them ideal for use in integrated circuits. In order to understand how semiconductors work, it is first necessary to review the principles of conductors and insulators.

3.1.0 Conductors

Conductors are so called because they readily carry electrical current. Good electrical conductors are usually also good heat conductors.

In each atom, there is a specific number of electrons that can be contained in each orbit or shell. The outer shell of an atom is the valence shell and the electrons contained in the valence shell are known as valence electrons. If the valence shell is not full, these electrons can be easily knocked out of their orbits and become free electrons.

Conductors are materials that have only one or two valence electrons in their atoms, as shown in *Figure 3*. An atom that has only one valence electron makes the best conductor because the electron is loosely held in orbit and is easily released to create current flow.

Gold and silver are excellent conductors, but they are too expensive to use on a large scale. However, in special applications requiring high conductivity, contacts may be plated with gold or silver. You would be most likely to find such conductors in precision devices where small currents are common and a high degree of accuracy is essential.

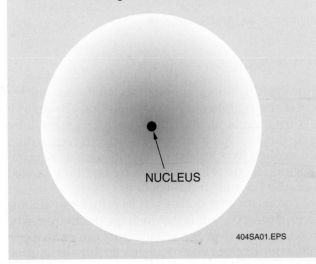

The Number of Electrons in Each Classical Shell of an Atom

The first (innermost) shell of an atom has a maximum of two electrons, and the second shell has a maximum of eight electrons. Unless it is the outermost shell, the third shell can have up to 18 electrons and the fourth and fifth shells can have up to 32 electrons. The maximum number of electrons in the outermost shell cannot exceed eight.

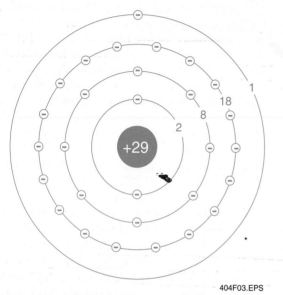

404F03.EPS

Figure 3 ◆ Atom of copper (a conductor).

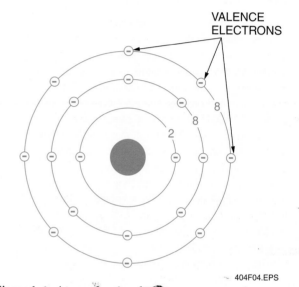

404F04.EPS

Figure 4 ◆ Atom of an insulator.

Copper is the most widely used conductor because it has excellent conductivity, while being much less expensive than precious metals such as gold and silver. Copper is used as the conductor in most types of wire and provides the current path on printed circuit boards. Aluminum is also used as a conductor, but it is not as good as copper.

3.2.0 Insulators

As you already know, insulators are materials that resist (and sometimes totally prevent) the passage of electrical current. Rubber, glass, and some plastics are common insulators. The atoms of insulating materials are characterized by having more than four valence electrons in their atomic structures. *Figure 4* shows the structure of an insulator atom. Note that it has eight valence electrons; this is the maximum number of electrons for the outer shell of an atom. Therefore, this atom has no free electrons and will not easily pass electric current.

3.3.0 Semiconductors

Semiconductors are materials that are neither good conductors nor good insulators. The materials used as semiconductors, such as germanium and silicon, have fewer free electrons in their outer shell than an insulator, but more than a conductor. Silicon (*Figure 5*) is more commonly used because it withstands heat better.

The factor that makes semiconductors valuable in electronic circuits is that their conductivity can be readily controlled. Semiconductors can be made to have positive or negative characteristics by adding certain impurities in a process known as doping.

When a substance with five valence electrons (e.g., arsenic or antimony) is added to the semiconductor material, the semiconductor material will no longer be electrically neutral. Instead, it will take on a positive charge. This is known as a P-type material.

When substances like indium or gallium, which have three valence electrons, are added to the semiconductor material, the material takes on a negative charge and is known as an N-type material.

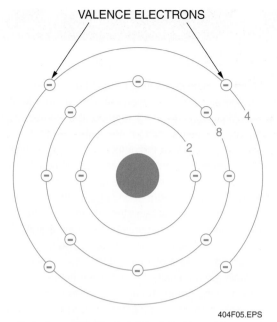

Figure 5 ◆ Atom of silicon (a semiconductor).

4.0.0 ◆ DIODES

A diode is made by joining a piece of P-type material with a piece of N-type material, as shown in Figure 6. The contacting surface is called the PN junction.

Diodes allow current to flow in one direction, but not in the other. This unidirectional current capability is the distinguishing feature of the diode. The activity occurring at the PN junction of the materials is responsible for the unidirectional characteristic of the diode.

Current does not normally flow across a PN junction. However, if a voltage of the correct polarity is applied, current will flow. This is known as **forward bias** and is shown in Figure 7. Note that the positive side of the voltage source is connected to the P material (the anode) and the negative side to the N material (the cathode). Also shown in Figure 7 is a PN junction with **reverse bias**. Note that the polarity of the applied voltage is opposite that of the forward biased junction. No current will flow in the reverse biased arrangement unless the reverse bias voltage exceeds the breakover voltage of the junction. This approach is used to create solid-state switching devices.

Modern systems rely heavily on electronic controls, which use low-level DC voltages. Some systems also use special controls powered by DC motors when very precise control is required. The electricity furnished by the power company is AC; it must be converted to DC to be suitable for most electronic circuits. The process of converting AC to DC is known as rectification and is accomplished using **rectifier** diodes.

As shown in Figure 7, a diode conducts current only when the voltage at its anode is positive with respect to the voltage at its cathode (forward bias). When the voltage at the anode is negative with respect to the cathode (reverse bias), current will not flow unless the voltage is so high that it overwhelms the diode. Most circuits using diodes are designed so that the diode will not conduct current unless the anode is positive with respect to the cathode.

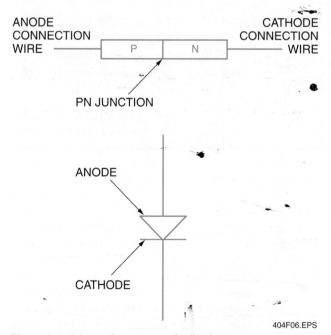

Figure 6 ◆ Material structure of a diode.

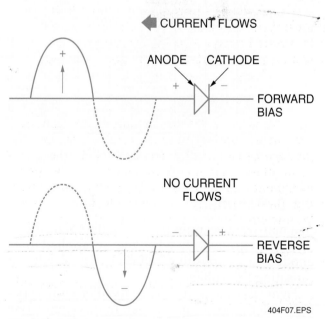

Figure 7 ◆ Forward and reverse bias.

The Periodic Table

Some versions of the periodic table of elements, such as the portion of one table shown here, contain a vertical list of numbers beside each element. The list represents the numbers of electrons in each of the classical shells of the atom for the element. The total of the electrons for an element equals the atomic number above the element symbol. The period number for each row corresponds to the number of classical shells that exist for the elements in the row. The group numbers across the top with an A or B are an older method of notation for each column of elements. The group number A elements are the representative elements that have a corresponding number of electrons in the outer shell with a corresponding valence even though an inner shell may be incomplete (except He). These representative elements become more stable from left to right as the number of electrons in the outer shell increases. The group number B elements are the transition elements. These elements sometimes have incomplete inner shells as well as incomplete outer shells. For these elements, the common valence is generally represented by the group number. Be aware that in quantum mechanics theory, the classical main shells shown in the table are made up of various energy-level sub-shells and the electrons that are allocated to the main shell are assigned to the various sub-shells.

PERIODIC TABLE OF THE ELEMENTS

Representative Elements

| Group | 1 — I A / 1A | 2 — II A / 2A |
| Period | | |

Transition Elements

Representative Elements

| 13 — III A / 3A | 14 — IV A / 4A | 15 — V A / 5A | 16 — VI A / 6A | 17 — VII A / 7A | 18 — VIII A / 8A |

Period	1A	2A	3B	4B	5B	6B	7B		8B		1B	2B	3A	4A	5A	6A	7A	8A
1	1 H 1.0080																	2 He 4.00260
2	3 Li 6.941	4 Be 9.01218											5 B 10.81	6 C 12.0111	7 N 14.0067	8 O 15.9994	9 F 18.9984	10 Ne 20.179
3	11 Na 22.9898	12 Mg 24.305	3 Sc...	4	5	6	7	8	9	10	11	12	13 Al 26.9815	14 Si 28.086	15 P 30.9738	16 S 32.06	17 Cl 35.453	18 Ar 39.948
4	19 K 39.102	20 Ca 40.08	21 Sc 44.956	22 Ti 47.90	23 V 50.941	24 Cr 51.996	25 Mn 54.9380	26 Fe 55.847	27 Co 58.9332	28 Ni 58.71	29 Cu 63.54	30 Zn 65.37	31 Ga 69.72	32 Ge 72.59	33 As 74.9216	34 Se 78.96	35 Br 79.904	36 Kr 83.80
5	37 Rb 85.4678	38 Sr 87.62	39 Y 88.906	40 Zr 91.22	41 Nb 92.906	42 Mo 95.94	43 Tc 99	44 Ru 101.07	45 Rh 102.906	46 Pd 106.4	47 Ag 107.870	48 Cd 112.40	49 In 114.82	50 Sn 118.69	51 Sb 121.75	52 Te 127.60	53 I 126.9045	54 Xe 131.30
6	55 Cs 132.91	56 Ba 137.34	57-71 Rare Earths	72 Hf 178.49	73 Ta 180.95	74 W 183.8	75 Re 186.2	76 Os 190.2	77 Ir 192.2	78 Pt 195.09	79 Au 196.967	80 Hg 200.59	81 Tl 204.37	82 Pb 207.2	83 Bi 208.9806	84 Pa 210	85 At 210	86 Rn 222
7	87 Fr 223	88 Ra 226.03	89-103 Radioactive Rare Earths	104 Rf 261	105 Ha 262	106 Sg 263	107 Bh 264	108 Hs 265	109 Mt 266	110 Uun 271.15	111 Uuu 272.15	112 Uub 277	113 Uut ?	114 Uuq 285	115 Uup ?	116 Uub 289	117 Uus ?	118 Uuo ?

404SA02.EPS

4.1.0 Rectifiers

In a half-wave rectifier (*Figure 8*), the single diode conducts current only when the AC applied to its anode is on its positive half-cycle. The result is a pulsating DC voltage.

Figure 9 shows a full-wave rectifier with a special center-tapped transformer. In this circuit, one of the diodes conducts on each half-cycle of the AC input. This produces a smoother pulsating DC voltage. Filter capacitors can be used to eliminate most of the ripple.

A bridge rectifier (*Figure 10*) contains four diodes, two of which conduct on each half-cycle. The bridge rectifier provides a smoother DC output and is the type most commonly used in electronic circuits. An advantage of the bridge rectifier

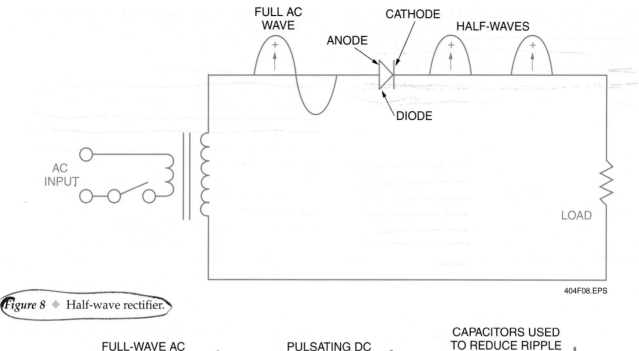

Figure 8 ◆ Half-wave rectifier.

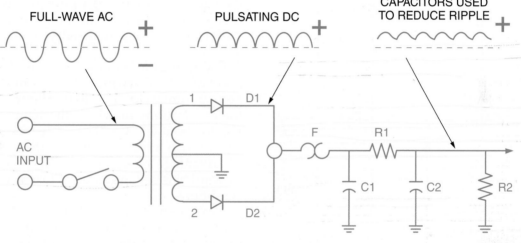

Figure 9 ◆ Full-wave rectifier.

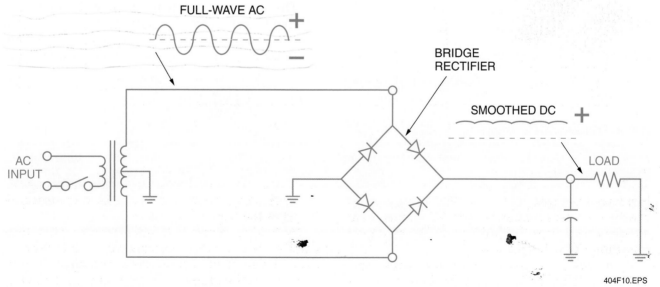

Figure 10 ◆ Bridge rectifier.

is that it does not need a center-tapped transformer. A filter and voltage regulator added to the output of the rectifier provide the precise, stable DC voltage needed for electronic devices.

In a three-phase power system, the three-phase rectifier shown in *Figure 11* is used. It contains six diodes to produce high-efficiency DC power.

4.2.0 Diode Identification

Most manufacturers' catalogs listing solid-state devices have hundreds of semiconductor diodes listed, along with the specifications of each. *Table 1* shows the specifications for two sample diodes. Note that the diodes are designated 1N34A and 1N58A. Manufacturers have agreed

upon certain standard designations for diodes having the same characteristics—the same as for wire sizes and types of insulation.

In *Table 1*, the peak inverse voltage (PIV) is the reverse bias at which avalanche breakover occurs. The ambient temperature rating is the range of temperatures over which the diode will operate and still maintain its basic characteristics. Forward current values are given for both the average current (that current at which the diode is usually operated) and the peak current (that current which, if exceeded, will damage the diode). The only difference between these two diodes is in the peak inverse voltage. Therefore, the 1N34A could be substituted for the 1N58A in applications involving signals of less than 60V peak-to-peak.

In general, there are two basic types of diodes: the silicon diode and the germanium diode. In most cases, silicon diodes have higher PIV and current ratings and wider temperature ranges than germanium diodes. Consequently, the silicon diode will be the type most often encountered in motor and HVAC solid-state controls or power supplies.

PIV ratings for silicon can be in the neighborhood of 1,000V, whereas the maximum value for germanium is closer to 400V. Silicon can be used for applications in which the temperature may rise to about 200°C (400°F), whereas germanium has a much lower maximum rating (100°C). The disadvantage of silicon, however, is the higher forward bias voltage required to reach operational conditions.

Diodes are used for many functions in solid-state devices. The most common use of silicon diodes is in constructing rectifiers. Germanium diodes are used mainly in signal and control circuits.

There are several ways in which diodes are marked to indicate the cathode and anode. See *Figure 12*. Note that in some cases diodes are marked with the schematic symbol, or there may

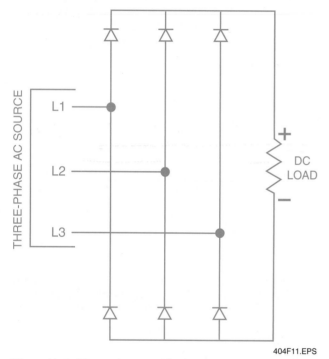

Figure 11 ◆ Three-phase rectifier.

404F11.EPS

Table 1	Diode Characteristics				
Type	Peak Inverse Voltage (PIV)	Ambient Temperature Range (°C)	Forward Peak (mA)	Current Average (mA)	Capacitance (µF)
1N34A	60	−50 to +75	150	50	150
1N58A	100	−50 to +75	150	50	150

Silicon Power Rectifier Diodes

Silicon power rectifier diodes are used in high-current DC applications including uninterruptible power supplies (UPSs), welders, and battery chargers, such as those used for electric forklifts. The stud-mounted silicon power rectifier diode shown here is rated for 300A and a PIV of 600V. The cathode is the threaded section, and it is screwed into or secured with a nut to an insulated heat sink that is at the DC potential. The flexible anode lead is connected to the AC power source. These diodes are also made with the anode as the threaded portion and the cathode as the flexible lead.

404SA03.EPS

be a band at one end to indicate the cathode. Other types of diodes use the shape of the diode housing to indicate the cathode end (i.e., the cathode end is either beveled or enlarged to ensure proper identification). When in doubt, the polarity of a diode may be checked with an ohmmeter, as shown in *Figure 13*. A forward bias will show a low resistance; a reverse bias will show a high resistance.

 CAUTION

The battery voltage of certain ohmmeters may exceed the PIV of the diode being tested. This may result in destruction of the diode.

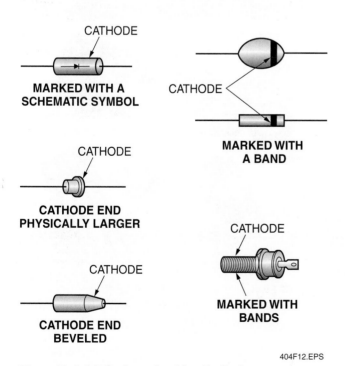

MARKED WITH A SCHEMATIC SYMBOL

CATHODE END PHYSICALLY LARGER

CATHODE END BEVELED

MARKED WITH A BAND

MARKED WITH BANDS

404F12.EPS

Figure 12 ◆ Methods used to identify diodes.

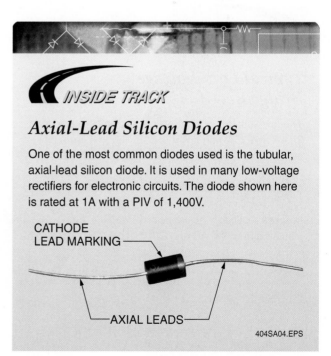

Axial-Lead Silicon Diodes

One of the most common diodes used is the tubular, axial-lead silicon diode. It is used in many low-voltage rectifiers for electronic circuits. The diode shown here is rated at 1A with a PIV of 1,400V.

CATHODE LEAD MARKING

AXIAL LEADS

404SA04.EPS

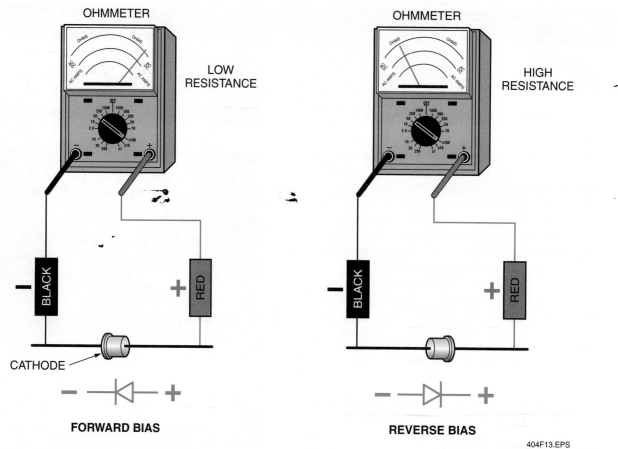

Figure 13 ◆ Testing a diode with an ohmmeter.

404F13.EPS

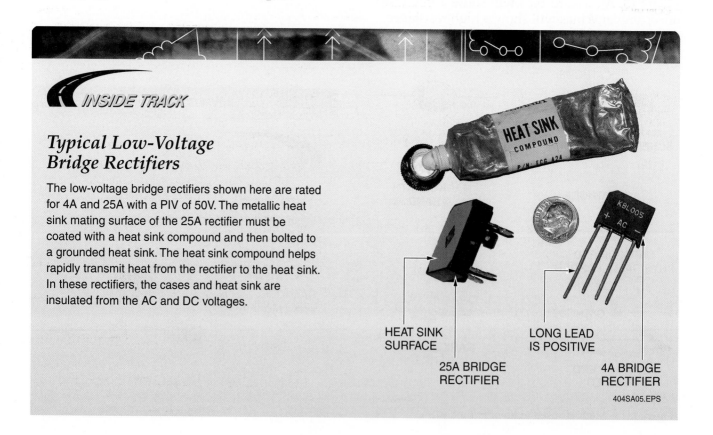

INSIDE TRACK

Typical Low-Voltage Bridge Rectifiers

The low-voltage bridge rectifiers shown here are rated for 4A and 25A with a PIV of 50V. The metallic heat sink mating surface of the 25A rectifier must be coated with a heat sink compound and then bolted to a grounded heat sink. The heat sink compound helps rapidly transmit heat from the rectifier to the heat sink. In these rectifiers, the cases and heat sink are insulated from the AC and DC voltages.

HEAT SINK COMPOUND P/N ECG A24

HEAT SINK SURFACE

LONG LEAD IS POSITIVE

25A BRIDGE RECTIFIER

4A BRIDGE RECTIFIER

KBL005

404SA05.EPS

5.0.0 ◆ LIGHT-EMITTING DIODES

A light-emitting diode (LED) is, as the name implies, a diode that will give off visible light when it is energized. All energized forward bias diodes give off some energy in the form of photons. In some types of diodes, the number of photons of light energy emitted is sufficient to create a visible light source.

The process of giving off light by applying an electrical source of energy is called electroluminescence. *Figure 14* shows that the conducting surface connected to the P-type material is much smaller to permit the emergence of the maximum number of photons of light energy in an LED. Note also in *Figure 15* that the symbol for an LED is similar to a conventional diode except that an arrow is pointing away from the diode.

Another solid-state device is activated by light and is known as a photo diode. The schematic symbol for the photo diode is exactly like a standard LED except that the arrow is reversed, as shown in *Figure 15*. A photo diode must have light in order to operate. It acts similarly to a conventional switch—that is, light turns the circuit on, and the absence of light opens the circuit.

When used in a circuit, an LED is generally operated at about 20mA or less. For example, if an LED is to be connected to a 9VDC circuit, a current-limiting resistor must be connected in

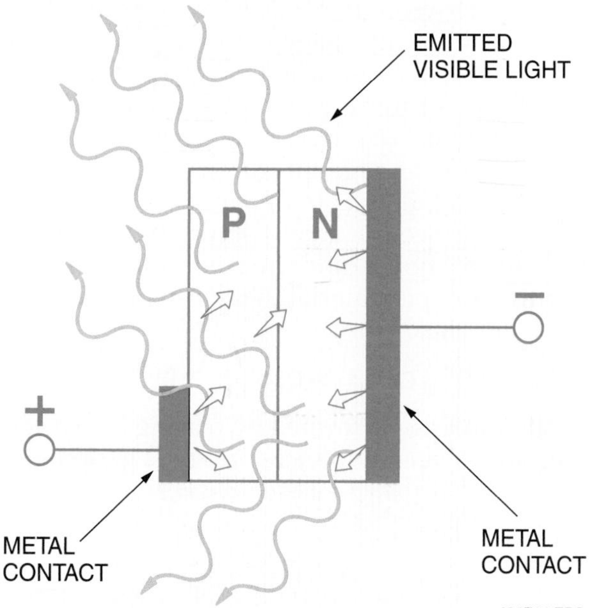

EMITTED VISIBLE LIGHT

P N

METAL CONTACT

METAL CONTACT

404F14.EPS

Figure 14 ◆ Process of electroluminescence in an LED.

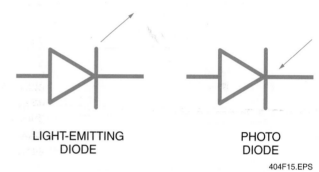

LIGHT-EMITTING DIODE

PHOTO DIODE

404F15.EPS

Figure 15 ◆ Schematic symbols for an LED and photo diode.

INSIDE TRACK

LED Lead Identification

Like conventional diodes, LEDs have identification methods. Most manufacturers use a flat surface in the LED case near one of the leads. The lead closest to this flat surface is the cathode. In other cases, the short lead is designated as the cathode lead. The LED shown here uses both methods.

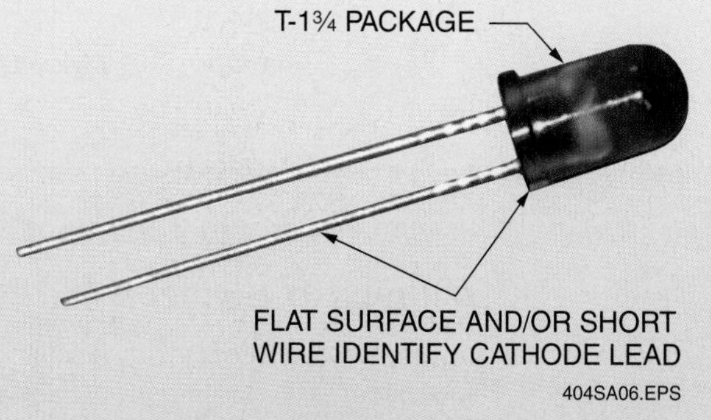

T-1¾ PACKAGE

FLAT SURFACE AND/OR SHORT WIRE IDENTIFY CATHODE LEAD

404SA06.EPS

series with the LED. Ohm's law may be used to calculate the required resistance as follows:

$$R = \frac{E}{I}$$

Where:

R = resistance (ohms)
E = voltage or emf (volts)
I = current (amperes)

$$R = \frac{9VDC}{0.020A}$$
$$R = 450\Omega$$

Therefore, a 450Ω resistor or the closest standard size without going under 450Ω should be used to limit the current flow through the LED.

LEDs are used as pilot lights on electronic equipment and as numerical displays. Many programmable HVAC controls use LEDs to indicate when a process is in operation. LEDs are also used in the opto-isolation circuit of solid-state relays for both motor controls and HVAC control systems.

6.0.0 ◆ TRANSISTORS

A transistor is made by joining three layers of semiconductor material, as shown in *Figure 16*. Regardless of the type of solid-state device being used, it is made by the joining of P-type and N-type materials to form either an NPN or PNP transistor. Each type has two PN junctions.

Control voltages are applied to the center layer, which is known as the base. These voltages control when and how much the transistor conducts. You will not encounter many transistors as discrete components. In today's self-contained electronic integrated circuit devices and microprocessors, there may be thousands of transistors.

They are so tiny that they can only be seen with a high-powered microscope.

Years ago, when computers were first invented, they required a roomful of electronic equipment. With the evolution of microminiature circuits, it became possible to perform the same work with a single tiny microprocessor chip containing thousands of microscopic electronic devices such as amplifiers and switches.

6.1.0 NPN Transistors

By sandwiching a very thin piece of P-type germanium between two slices of N-type germanium, an NPN transistor is formed. A transistor made in this way is called a junction transistor. The symbol for this type of transistor showing the three elements (emitter, base, and collector) is illustrated in *Figure 17*.

The current flow is in the opposite direction of the arrow. Since the arrow points away from the base, current flows from the emitter to the base or from the N-type material to the P-type material.

6.2.0 PNP Transistors

A PNP transistor is formed by placing N-type germanium between two slices of P-type germanium.

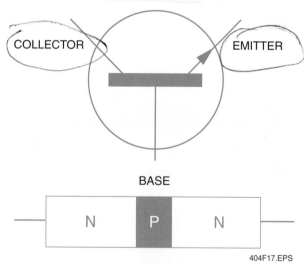

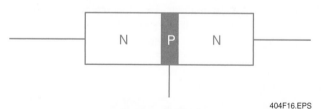

404F16.EPS

Figure 16 ◆ Material arrangement in an NPN transistor.

404F17.EPS

Figure 17 ◆ NPN transistor characteristics and schematic symbol.

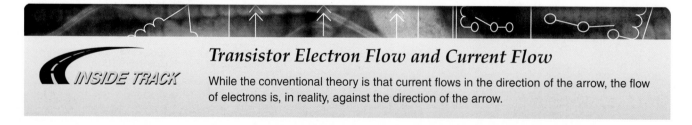

Transistor Electron Flow and Current Flow

INSIDE TRACK

While the conventional theory is that current flows in the direction of the arrow, the flow of electrons is, in reality, against the direction of the arrow.

The schematic symbol for the PNP transistor is almost identical to that of the NPN transistor. The only difference is the direction of the emitter arrow. In the NPN transistor, it points away from the base and in the PNP transistor, it points toward the base.

Electron flow in a PNP transistor is from the N-type germanium to the P-type germanium, and since the arrow in *Figure 18* points toward the base, the electron flow is in the opposite direction of the arrow—from base to emitter. This is the reverse of the NPN transistor discussed previously.

Note also the point-contact transistor shown in *Figure 18*. Junction and point-contact transistors are almost identical in operation. The main difference is in the method of assembly.

6.3.0 Identifying Transistor Leads

See *Figure 19*. Transistors are manufactured in a variety of configurations. Those with studs and heat sinks, as shown in *Figures 19(A)* and *19(B)*, are high-power devices. Those with a small can (top hat) or plastic body, as shown in *Figure 19(C)*, are low- to medium-power devices.

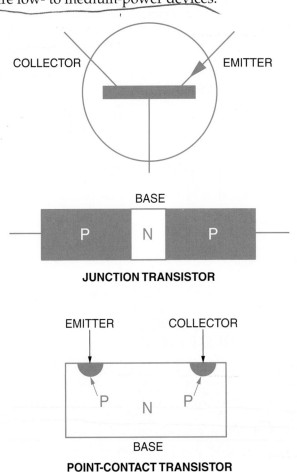

Figure 18 ◆ PNP transistor characteristics and schematic symbol.

404F18.EPS

Whenever possible, transistor casings will have some marking to indicate which leads are connected to the emitter, collector, or base of the transistor. A few of the methods commonly used are indicated in *Figure 20*.

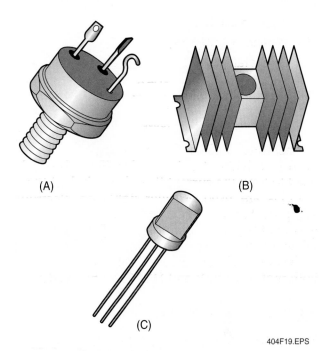

Figure 19 ◆ Various transistors.

404F19.EPS

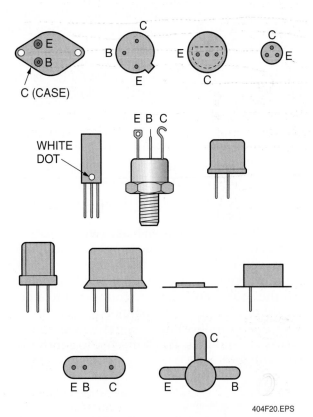

Figure 20 ◆ Lead identification of transistors.

404F20.EPS

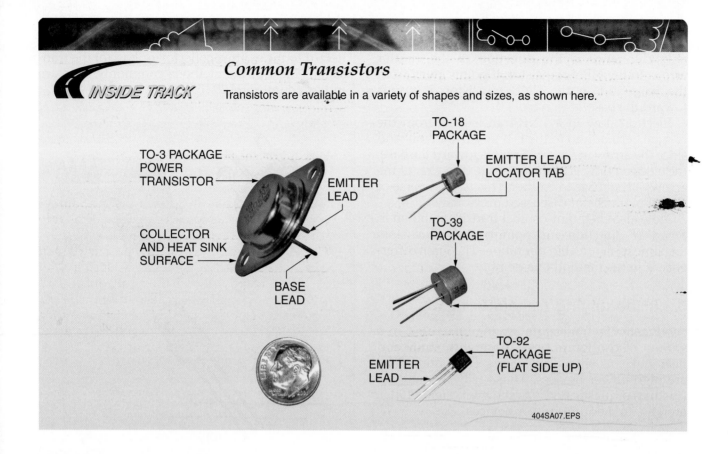
6.4.0 Field-Effect Transistors

Field-effect transistors (FETs) control the flow of current with an electric field. There are two basic types of FETs: the junction field-effect transistor (JFET) and the metal-oxide semiconductor field-effect transistor (MOSFET). A MOSFET is also referred to as an insulated gate field-effect transistor (IGFET).

6.4.1 Junction Field-Effect Transistors

There are two types of JFETs: an N-channel and a P-channel. *Figure 21* shows the schematic symbols for these two types and also denotes the terms for each JFET lead.

The difference between these two devices is in the polarity of the voltage to which the transistor is connected. In an N-channel device, the drain is connected to the more positive voltage, and the source and gate are connected to a more negative voltage. A P-channel device has the drain connected to a more negative voltage, and the source and gate are connected to a more positive voltage.

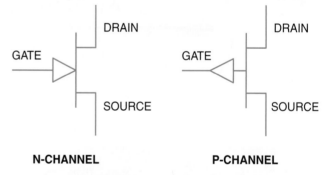

404F21.EPS

Figure 21 ◆ Junction field-effect transistor symbols.

7.0.0 ◆ SILICON-CONTROLLED RECTIFIERS

Silicon-controlled rectifiers (SCRs), along with diacs and triacs, belong to a class of semiconductors known as thyristors. One common characteristic of all thyristors is that they act as an open circuit until a triggering current is applied to their gate. Once that happens, the thyristor acts as a low-resistance current path from anode to cathode. It will continue to conduct, even if the gate signal is removed, until either the current is reduced below a certain level or the thyristor is turned off. These devices are used for a variety of purposes, including:

- Controlling the amount of AC power applied to a load
- Emergency lighting circuits
- Lamp dimmers
- Motor speed controls
- Ignition systems

The SCR is similar to a diode except that it has three terminals. Like a common diode, current will only flow through the SCR in one direction. However, in addition to needing the correct voltage polarity at the anode and cathode, the SCR also requires a gate voltage of the same polarity as the voltage applied to the anode. Once fired, the SCR will remain on until the cathode-to-anode current falls below a value known as the holding current. Once the SCR is off, another positive gate voltage must be applied before it will start conducting again.

The SCR is made from four adjoining layers of semiconductor material in a PNPN arrangement, as shown in *Figure 22*. The SCR symbol is also shown. Note that it is the same as the diode symbol except for the addition of a gating lead.

There is also a light-activated version of the SCR known as an LASCR. Its symbol is the same as that of the regular SCR, with the addition of two diagonal arrows representing light, similar to the photo diode previously covered.

The ability of an SCR to turn on at different points in the conducting cycle can be used to vary the amount of power delivered to a load. This type of variable control is called phase control. With such control, the speed of an electric motor, the brilliance of a lamp, or the output of an electric resistance heating unit can be controlled.

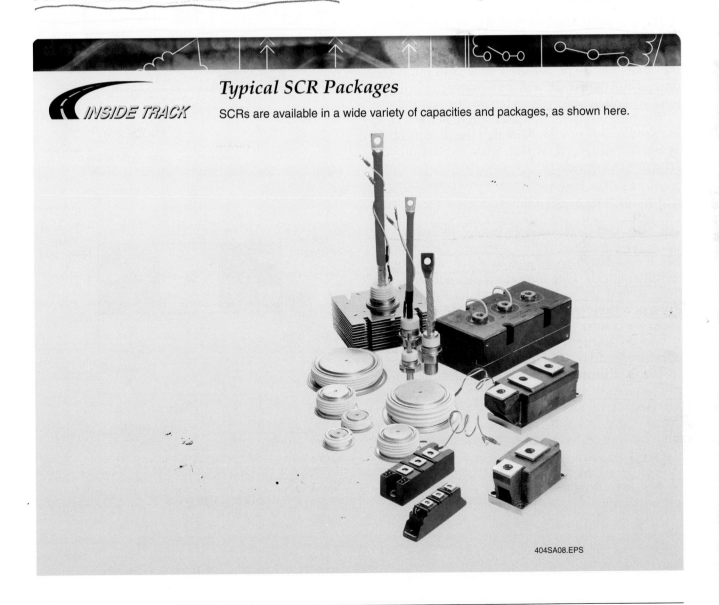

INSIDE TRACK

Typical SCR Packages

SCRs are available in a wide variety of capacities and packages, as shown here.

404SA08.EPS

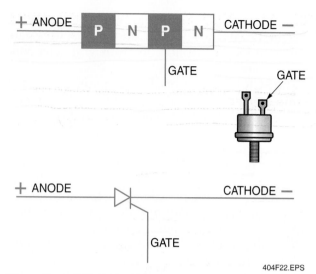

Figure 22 ◆ SCR characteristics and symbol.

404F22.EPS

8.0.0 ◆ DIACS

A diac can be thought of as an AC switch. Because it is bidirectional, current will flow through it on either half of the AC waveform. One major use of the diac is as a control for a triac. Although the diac is not gated, it will not conduct until the applied voltage exceeds its breakover voltage.

The basic construction and symbols for a diac are shown in *Figure 23*. Note that the diac has two symbols, either of which may be found on schematic diagrams.

9.0.0 ◆ TRIACS

The triac can be viewed as two SCRs turned in opposite directions with a common gate (*Figure 24*). It could also be viewed as a diac with a gate terminal added. One important distinction, however, is that the voltage applied across the triac does not have to exceed a breakover voltage in order for conduction to begin.

Like SCRs, triacs are used in phase control applications to control the average power applied to loads. Examples include light dimmers and photocell light switches.

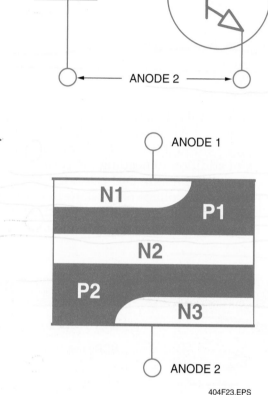

Figure 23 ◆ Basic diac construction and symbols.

404F23.EPS

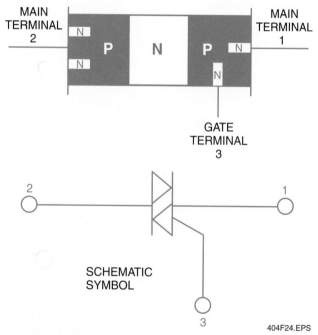

Figure 24 ◆ Basic triac construction and symbol.

404F24.EPS

Typical Triac

A typical triac and its associated heat sink are shown here. This particular version has no visual method of determining lead identification other than the emitter lead locator dot. The manufacturer's literature supplied with the device must be used to identify the lead numbers.

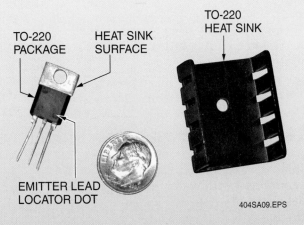

TO-220 PACKAGE

HEAT SINK SURFACE

TO-220 HEAT SINK

EMITTER LEAD LOCATOR DOT

404SA09.EPS

GOING GREEN

Solarvoltaic Systems

This photovoltaic array is part of a large power system that will eventually provide 15 megawatts of power. Constructed at Nellis Air Force Base in Nevada, this solar farm covers 140 acres of land and consists of about 70,000 solar panels. Because the panels track the sun throughout the day, they produce about 30 percent more power than fixed systems.

404SA10.EPS

1. An electron has a _____ charge.
   a. positive
   b. negative
   c. neutral
   d. variable

2. The conductor commonly used to print circuits on a printed circuit board is _____.
   a. aluminum
   b. silver
   c. ceramic
   d. copper

3. An atom with five or more valence electrons is _____.
   a. an insulator
   b. easily combined with water
   c. a semiconductor
   d. a conductor

4. Which of the following can be added as an impurity to a semiconductor material to create a P-type semiconductor?
   a. Arsenic
   b. Germanium
   c. Gallium
   d. Indium

5. A PN junction solid-state device is more commonly known as a _____.
   a. transistor
   b. diode
   c. half-wave rectifier
   d. full-wave rectifier

6. What bias has been applied to a PN junction if the cathode is positive with respect to the anode?
   a. Negative bias
   b. Positive bias
   c. Forward bias
   d. Reverse bias

7. The purpose of a rectifier is to _____.
   a. convert DC into AC
   b. step down voltage
   c. convert AC into DC
   d. step up voltage

8. A half-wave rectifier is constructed of _____ diode(s).
   a. one
   b. two
   c. three
   d. seven

9. What is the total number of PN junctions in a transistor?
   a. One
   b. Two
   c. Three
   d. Four

10. What are the two basic types of transistors?
    a. AND and OR
    b. ASCII and SCICA
    c. Digital and analog
    d. NPN and PNP

11. What are the two materials normally used to construct transistors?
    a. L and O materials
    b. P and N materials
    c. A and D materials
    d. Silver and aluminum

12. An SCR is made of _____ layer(s) of semiconductor material.
    a. one
    b. two
    c. three
    d. four

13. The term thyristor includes which of the following groups of components?
    a. Transistors, FETs, and SCRs
    b. Diacs, triacs, and SCRs
    c. Diodes, diacs, and SCRs
    d. Diacs, triacs, and op-amps

14. Which of the following is a characteristic of a diac?
    a. It will conduct on either half of the AC waveform.
    b. The voltage applied to the anode and the gate junction must be of the same polarity.
    c. It will continue conducting after the gating voltage is removed.
    d. It has no breakover voltage rating.

15. A triac can be viewed as _____ turned in opposite directions with a common gate.
    a. three diacs
    b. two SCRs
    c. six SCRs
    d. three diodes

Summary

Solid-state control devices are rapidly replacing electromechanical controls in all areas of the electrical industry. Common solid-state devices include:

- Rectifiers
- LEDs
- Transistors

- SCRs
- Diacs
- Triacs

With the ever-increasing number of solid-state devices in use, all electrical personnel should become familiar with the types of solid-state devices available and their applications.

Notes

1-2   conductor
3-4   semi-conductor
4 or more   insulator

Trade Terms
Introduced in This Module

Diac: A three-layer diode designed for use as a trigger in AC power control circuits, such as those using triacs.

Field-effect transistor (FET): A transistor that controls the flow of current through it with an electric field.

Forward bias: Forward bias exists when voltage is applied to a solid-state device in such a way as to allow the device to conduct easily.

Junction field-effect transistor (JFET): A field-effect transistor formed by combining layers of semiconductor material.

N-type material: A material created by doping a region of a crystal with atoms from an element that has more electrons in its outer shell than the crystal.

P-type material: A material created when a crystal is doped with atoms from an element that has fewer electrons in its outer shell than the natural crystal. This combination creates empty spaces in the crystalline structure. The missing electrons in the crystal structure are called *holes* and are represented as positive charges.

Rectifier: A device or circuit used to change AC voltage into DC voltage.

Reverse bias: A condition that exists when voltage is applied to a device in such a way that it causes the device to act as an insulator.

Semiconductor: A material that is neither a good insulator nor a good conductor. Such materials contain four valence electrons and are used in the production of solid-state devices.

Silicon-controlled rectifier (SCR): A device that is used mainly to convert AC voltage into DC voltage. To do so, however, the gate of the SCR must be triggered before the device will conduct current.

Solid-state device: An electronic component constructed from semiconductor material. Such devices have all but replaced the now-obsolete vacuum tube in electronic circuits.

Triac: A bidirectional triode thyristor that functions as an electrically controlled switch for AC loads.

This module is intended to present thorough resources for task training. The following reference works are suggested for further study. These are optional materials for continuing education rather than for task training.

National Electrical Code® Handbook, Latest Edition. Quincy, MA: National Fire Protection Association.

Solid-State Fundamentals for Electricians, Gary Rockis. Homewood, IL: American Technical Publishers, 1993.

CONTREN® LEARNING SERIES – USER UPDATE

NCCER makes every effort to keep these textbooks up-to-date and free of technical errors. We appreciate your help in this process. If you have an idea for improving this textbook, or if you find an error, a typographical mistake, or an inaccuracy in NCCER's Contren® textbooks, please write us, using this form or a photocopy. Be sure to include the exact module number, page number, a detailed description, and the correction, if applicable. Your input will be brought to the attention of the Technical Review Committee. Thank you for your assistance.

Instructors – If you found that additional materials were necessary in order to teach this module effectively, please let us know so that we may include them in the Equipment/Materials list in the Annotated Instructor's Guide.

Write: Product Development and Revision
National Center for Construction Education and Research
3600 NW 43rd St., Bldg. G, Gainesville, FL 32606

Fax: 352-334-0932

E-mail: curriculum@nccer.org

Craft

Module Name

Copyright Date

Module Number

Page Number(s)

Description

(Optional) Correction

(Optional) Your Name and Address

Fire Alarm Systems

Georgia Institute of Technology – Open to the Sun

The Georgia Tech house is all about sunlight, using it to transform and open up its living space. It was constructed using translucent walls, made of two sheets of polycarbonate that enclose an aerogel filler. Aerogel, sometimes referred to as "solid smoke," is one of the lightest solids known. It is an excellent insulator and is translucent, allowing filtered light to enter the home.

26405-08

26405-08
Fire Alarm Systems

Topics to be presented in this module include:

1.0.0	Introduction	5.2
2.0.0	Codes and Standards	5.2
3.0.0	Fire Alarm Systems Overview	5.4
4.0.0	Fire Alarm System Equipment	5.7
5.0.0	Fire Alarm Initiating Devices	5.7
6.0.0	Control Panels	5.23
7.0.0	FACP Primary and Secondary Power	5.26
8.0.0	Notification Appliances	5.26
9.0.0	Communications and Monitoring	5.31
10.0.0	General Installation Guidelines	5.34
11.0.0	Total Premises Fire Alarm System Installation Guidelines	5.41
12.0.0	Fire Alarm–Related Systems and Installation Guidelines	5.55
13.0.0	Troubleshooting	5.63

Overview

Installation and maintenance of fire alarm systems can be considered a specialty trade or branch of electrical work. In addition to basic electrical installation skills, fire alarm system technicians must have a solid understanding of electronic theory, as many fire alarm devices incorporate electronic components.

More than one article in the *National Electrical Code®* applies either directly or indirectly to the installation of fire alarm system components. *NEC Chapter Six* covers the installation of fire pumps, while an article in *NEC Chapter Seven* is devoted to the installation of fire alarm systems. These chapters provide rules and regulations promoting safe and properly operating fire alarm systems.

Note: *NFPA 70®*, *National Electrical Code®*, and *NEC®* are registered trademarks of the National Fire Protection Association, Inc., Quincy, MA 02269. All *National Electrical Code®* and *NEC®* references in this module refer to the 2008 edition of the *National Electrical Code®*.

Objectives

When you have completed this module, you will be able to do the following:

1. Define the unique terminology associated with fire alarm systems.
2. Describe the relationship between fire alarm systems and life safety.
3. Explain the role that various codes and standards play in both commercial and residential fire alarm applications.
4. Describe the characteristics and functions of various fire alarm system components.
5. Identify the different types of circuitry that connect fire alarm system components.
6. Describe the theory behind conventional, addressable, and analog fire alarm systems and explain how these systems function.

Trade Terms

The trade terms are too numerous to list. They appear in boldface (blue) type throughout the text and can be found in the Glossary at the end of this module.

Required Trainee Materials

1. Pencil and paper
2. Appropriate personal protective equipment
3. Copy of the latest edition of the *National Electrical Code®*

Prerequisites

Before you begin this module, it is recommended that you successfully complete *Core Curriculum; Electrical Level One; Electrical Level Two; Electrical Level Three; Electrical Level Four,* Modules 26401-08 through 26404-08.

This course map shows all of the modules in *Electrical Level Four.* The suggested training order begins at the bottom and proceeds up. Skill levels increase as you advance on the course map. The local Training Program Sponsor may adjust the training order.

26413-08 Introductory Skills for the Crewleader

26412-08 Special Locations

26411-08 Medium-Voltage Terminations/Splices

26410-08 Motor Operation and Maintenance

26409-08 Heat Tracing and Freeze Protection

26408-08 HVAC Controls

26407-08 Advanced Controls

26406-08 Specialty Transformers

26405-08 Fire Alarm Systems

26404-08 Basic Electronic Theory

26403-08 Standby and Emergency Systems

26402-08 Health Care Facilities

26401-08 Load Calculations – Feeders and Services

ELECTRICAL LEVEL FOUR

ELECTRICAL LEVEL THREE

ELECTRICAL LEVEL TWO

ELECTRICAL LEVEL ONE

CORE CURRICULUM: Introductory Craft Skills

405CMAP.EPS

1.0.0 ◆ INTRODUCTION

Fire alarm systems can make the difference between life and death. In a fire emergency, a quick and accurate response by a fire alarm system can reduce the possibility of deaths and injuries dramatically. However, practical field experience is necessary to master this trade. Remember to focus on doing the job properly at all times. Failure to do so could have tragic consequences.

Always seek out and abide by any National Fire Protection Association (NFPA), federal, state, and local codes that apply to each and every fire alarm system installation.

Different types of buildings and applications will have different fire safety goals. Therefore, each building will have a different fire alarm system to meet those specific goals. Most automatic fire alarm systems are configured to provide fire protection for life safety, property protection, or mission protection.

Life safety fire protection is concerned with protecting and preserving human life. In a fire alarm system, life safety can be defined as providing warning of a fire situation. This warning occurs early enough to allow notification of building occupants, ensuring that there is sufficient time for their safe evacuation.

Many times, fire alarm systems are taken for granted. People go about their everyday business with little or no thought of the behind-the-scenes safety devices that monitor for abnormal conditions. After all, no one expects a fire to happen. When it does, every component of the life safety fire alarm system must be working properly in order to minimize the threat to life.

2.0.0 ◆ CODES AND STANDARDS

Fire alarm system equipment and installation are regulated and controlled by various national, state, and local codes. Industry standards have also been developed as a means of establishing a common level of competency. These codes and standards are set by a variety of different associations, agencies, and laboratories that consist of fire alarm professionals across the country. Depending on the application and installation, you will need to follow the standards established by one or more of these organizations.

In the United States, local and state jurisdictions will select a model code and supporting documents that detail applicable standards. If necessary, they will amend the codes and standards as they see fit. Some amendments may be substantial changes to the code, and in certain situations, the jurisdictions will author their own sets of codes. Once a jurisdiction adopts a set of codes and standards, it becomes an enforceable legal document within that jurisdiction.

Installing and Servicing Fire Alarm Systems

In certain jurisdictions, the person who installs and/or services fire alarm systems must be a licensed fire alarm specialist.

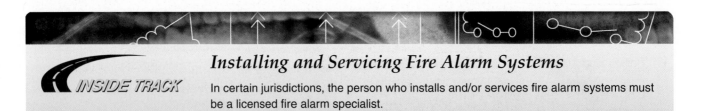

Property and Mission Protection

The design goal of most fire alarm systems is life safety. However, some systems are designed with a secondary purpose of protecting either property or the activities (mission) within a building. The goal of both property and mission protection is the early detection of a fire so that firefighting efforts can begin while the fire is still small and manageable. Fire alarm systems with a secondary goal of property protection are commonly used in museums, libraries, storage facilities, and historic buildings in order to minimize damage to the buildings or their contents. Systems with a secondary goal of mission protection are commonly used where it is essential to avoid business interruptions, such as in hospitals, financial businesses, security control rooms, and telecommunication centers.

The following list details some of the many national organizations responsible for setting the industry standards:

- *Underwriters Laboratories (UL)* – Establishes standards for fire equipment and systems.
- *The National Fire Protection Association (NFPA)* – Establishes standards for fire systems. The association publishes the *National Fire Alarm Code®* (NFPA 72®), the *Life Safety Code®* (NFPA 101®), and the *Uniform Fire Code™* (NFPA 1). These codes are the primary reference documents used by fire alarm system professionals.
- *Factory Mutual (FM)* – Establishes standards for fire systems.
- *National Electrical Manufacturers' Association (NEMA)* – Establishes standards for equipment.
- *The Federal Bank Protection Act* – Establishes fire alarm equipment and system standards for banks.
- *The Defense Intelligence Agency (DIAM-50-3)* – Establishes standards for military and intelligence installations.
- *The National Institute for Certification in Engineering Technologies (NICET)* – Establishes standardized testing of fire alarm designers and installers.

2.1.0 The National Fire Protection Association

The National Fire Protection Association (NFPA) is responsible for setting the national standards and codes for the fire alarm industry. Consisting of fire alarm representatives from all areas of business and fire protection services, the NFPA responds to the ever-changing needs of society by using a democratic process to form consensus standards that are acceptable to all members. The NFPA reviews equipment and system performance criteria and input from experienced industry professionals to set an acceptable level of protection for both life and property.

2.1.1 NFPA Codes

The following four widely adopted national codes are specified by the NFPA:

- *National Electrical Code®* (NFPA 70®) – The *National Electrical Code®* (NEC®) covers all of the necessary requirements for all electrical work performed in a building. The Fire Protective Signaling Systems portion of the code *(NEC Article 760)* details the specific requirements for

NFPA Codes

The three NFPA code books required for installing alarm systems are shown here. Nearly every requirement of *NFPA 101®, Life Safety Code®,* has resulted from the analysis of past fires in which human lives have been lost. The three code books shown are the current editions upon publication of this module. Code books are revised and changed periodically, typically every three or four years. For this reason, you should always make sure that you are using the most current edition of any code book.

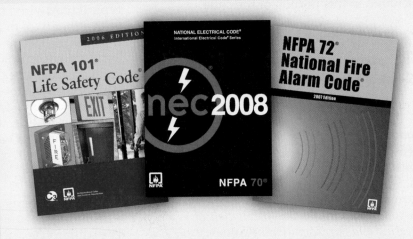

405SA01.EPS

wiring and equipment installation for fire protection signaling systems. Specifications include installation methods, connection types, **circuit** identification, and wire types (including gauges and insulation). The *NEC*® places restrictions on the number and types of circuit combinations that can be installed in the same enclosure.

* *National Fire Alarm Code*® *(NFPA 72*®*)* – The recommended requirements for installation of fire alarm systems and equipment in residential and commercial facilities are covered in this code. Included are requirements for installation of **initiating devices** (sensors) and notification appliances (visual or audible). Inspection, testing, and **maintenance** requirements for fire alarm systems and equipment are also covered.

* *Life Safety Code*® *(NFPA 101*®*)* – This document is focused on the preservation and protection of human life, as opposed to property. Life safety requirements are detailed for both new construction and existing structures. Specifically, necessary protection for unique building features and construction are detailed. In addition, chapters are organized to explain when, where, and for what applications fire alarm systems are required, the necessary means of initiation and occupant notification, and the means by which to notify the fire department. This code also details any equipment exceptions to these requirements.

* *Uniform Fire Code*™ *(NFPA 1)* – This code was established to help fire authorities continually develop safeguards against fire hazards. A chapter of this code is dedicated to fire protection systems. Information and requirements for testing, operation, installation, and periodic preventive maintenance of fire alarm systems are included in this portion of the code.

2.1.2 NFPA Standards

The NFPA also publishes specific standards that are used by fire alarm system professionals. These include:

* *NFPA 75, Standard for Protection of Electronic Computer/Data Processing Equipment*
* *NFPA 80, Standard for Fire Doors and Other Opening Protectives*
* *NFPA 90A, Standard for the Installation of Air Conditioning and Ventilating Systems*
* *NFPA 90B, Standard for the Installation of Warm Air Heating and Air Conditioning Systems*
* *NFPA 92A, Standard for Smoke-Control Systems Utilizing Barriers and Pressure Differences*

3.0.0 ◆ FIRE ALARM SYSTEMS OVERVIEW

Fire alarm systems are primarily designed to detect and warn of abnormal conditions, alert the appropriate authorities, and operate the necessary facility safety devices to minimize fire danger. Through a variety of manual and automatic system devices, a fire alarm system links the sensing of a fire condition with people inside and outside the building. The system communicates to fire professionals that action needs to be taken.

Although specific codes and standards must always be followed when designing and installing a fire alarm system, many different types of systems that employ different types of technology can be used. Most of these types of fire alarm systems are categorized by the means of communication between the detectors and the fire alarm control panel (FACP). Three major types of fire alarm systems are:

* Conventional hardwired
* Multiplex
* Addressable intelligent

3.1.0 Conventional Hardwired Systems

A hardwired system using conventional initiation devices (heat and **smoke detectors** and pull stations) and **notification devices** (**appliances**) (bells, **horns**, or lights) is the simplest of all fire alarm systems. Large buildings or areas that are being protected are usually divided into **zones** to identify the specific area where a fire is detected. A conventional hardwired system is limited to zone detection only, with no means of identifying the specific detector that initiated the alarm. A typical hardwired system might look like *Figure 1* with either two- or four-wire initiating or notification device circuits. In two-wire circuits, power for the devices is superimposed on the alarm circuits. In four-wire circuits, the operating power is supplied to the devices separately from the **signal** or alarm circuits. In either two- or four-wire systems, **end-of-line (EOL) devices** are used by the FACP to monitor circuit integrity.

3.2.0 Multiplex Systems

Multiplex systems are similar to hardwired systems in that they rely on zones for fire detection. The difference, however, is that **multiplexing** allows multiple signals from several sources to be sent and received over a single communication

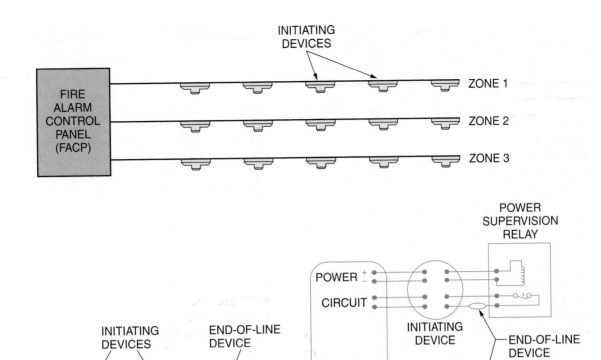

INITIATING DEVICES

FIRE ALARM CONTROL PANEL (FACP)

ZONE 1

ZONE 2

ZONE 3

INITIATING DEVICES

END-OF-LINE DEVICE

FIRE ALARM CONTROL PANEL

TWO-WIRE INITIATING CIRCUITS

POWER SUPERVISION RELAY

POWER

CIRCUIT

INITIATING DEVICE

END-OF-LINE DEVICE

BELL

FIRE ALARM CONTROL PANEL

NOTIFICATION APPLIANCE

FOUR-WIRE INITIATING CIRCUITS

405F01.EPS

Figure 1 ◆ Typical conventional hardwired system.

line. Each signal can be uniquely identified. This results in reduced control equipment, less wiring infrastructure, and a distributed power supply. *Figure 2* shows a simplified example of a multiplex system.

3.3.0 Addressable and Analog Addressable Systems

Two different versions of addressable intelligent systems are available. They are addressable and analog addressable systems. Analog addressable

systems are more sophisticated than addressable systems.

3.3.1 Addressable Systems

Addressable systems use advanced technology and detection equipment for discrete identification of alarm signals at the detector level. An addressable system can pinpoint an alarm location to the precise physical location of the initiating detector. The basic idea of an addressable system is to provide identification or control of individual initiation, control, or notification

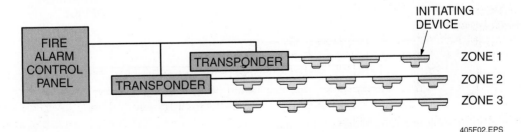

INITIATING DEVICE

FIRE ALARM CONTROL PANEL

TRANSPONDER

TRANSPONDER

ZONE 1

ZONE 2

ZONE 3

405F02.EPS

Figure 2 ◆ Typical multiplex system.

devices on a common circuit. Each component on the **signaling line circuit (SLC)** has an identification number or address. The addresses are usually assigned using switches or other similar means. *Figure 3* is a simplified representation of an addressable or analog addressable fire alarm system.

The fire alarm control panel (FACP) constantly polls each device using a signaling line circuit (SLC). The response from the device being polled verifies that the wiring **path (pathway)** is intact (wiring supervision) and that the device is in place and operational. Most addressable systems use at least three states to describe the status of the device: normal, trouble, and alarm. Smoke detection devices make the decision internally regarding their alarm state just like conventional smoke detectors. Output devices, like relays, are also checked for their presence and in some cases for their output status. Notification output modules also supervise the wiring to the horns, strobes, and other devices, as well as the availability of the power needed to run the devices in case of an alarm. When the FACP polls each device, it also compares the information from the device to the system program. For example, if the program indicates device 12 should be a contact **transmitter** but the device reports that it is a relay, a problem exists that must be corrected. Addressable fire alarm systems have been made with two, three, and four conductors. Generally, systems with more conductors can handle more **addressable devices.** Some systems may also contain multiple SLCs. These are comparable to multiple zones in a conventional hardwired system.

3.3.2 Analog Addressable Systems

Analog systems take the addressable system capabilities much further and change the way the information is processed. When a device is polled, it returns much more information than a device in a standard addressable system. For example, instead of a smoke detector transmitting that it is in alarm status, the device actually transmits the level of smoke or contamination present to the fire alarm control panel. The control panel then compares the information to the levels detected in previous polls. A slow change in levels (over days, weeks, or months) indicates that a device is dirty or malfunctioning. A rapid change, however, indicates a fire condition. Most systems have the capability to compensate for the dirt buildup in the detectors. The system will adjust the detector sensitivity to the desired range. Once the dirt buildup exceeds the compensation range, the system reports a trouble condition. The system can also administer self-checks on the detectors to test their ability to respond to smoke. If the airflow around a device is too great to allow proper detection, some systems will generate a trouble report.

The information in some systems is transmitted and received in a totally digital format. Others transmit the polling information digitally but receive the responses in an analog current-level format.

The panel, not the device, performs the actual determination of the alarm state. In many systems, the light-emitting diode (LED) on the detector is turned on by the panel and not by the detector. This ability to make decisions at the panel also allows the detector sensitivity to be adjusted at the panel. For instance, an increase in the ambient temperature can cause a smoke detector to become more sensitive, and the alarm level sensitivity at the panel can be adjusted to compensate. Sensitivities can even be adjusted based on the time of day or day of the week. Other detection devices can also be programmed to adjust their own sensitivity.

The ability of an analog addressable system to process more information than the three elementary alarm states found in simpler systems allows the analog addressable system to provide pre-alarm signals and other information. In many devices, five or more different signals can be received.

Most analog addressable systems operate on a two-conductor circuit. Most systems limit the number of devices to about one hundred. Because

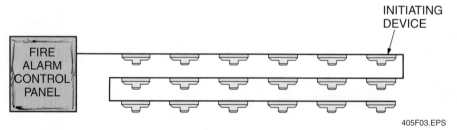

Figure 3 ◆ Typical addressable or analog addressable system.

405F03.EPS

of the high data rates on these signaling line circuits, capacitance also limits the conductor lengths. Always follow the manufacturer's installation instructions to ensure proper operation of the system.

3.3.3 Signaling Line Circuits (SLCs)

In addressable or analog addressable systems, there are two basic wiring or circuit types (classes). Class B is the most common and has six different styles. Class A, with four styles, provides additional reliability but is not normally required and is generally more expensive to install. Various codes will address what circuit types are required. The performance requirements for each of the styles of SLC circuits are detailed in *NFPA 72®*.

- **Class B circuits** – A Class B signaling line circuit for an addressable system essentially requires that two conductors reach each device on the circuit by any means as long as the wire type and physical installation rules are followed. It does not require wiring to pass in and out of each device in a series arrangement. With a circuit capable of 100 devices, it is permissible to go in 100 different directions from the panel. Supervision occurs because the fire alarm control panel polls and receives information from each device. The route taken is not important. A break in the wiring will result in the loss of communication with one or more devices.
- **Class A circuits** – A Class A signaling line circuit for an addressable system requires that the conductors loop into and out of each device. At the last device, the signaling line circuit is returned to the control panel by a different route. The control panel normally communicates with the devices via the outbound circuit but has the ability to communicate to the back side of a break through the return circuit. The panel detects the fact that a complete loop no longer exists and shifts the panel into Class A mode. All devices remain in operation. Some systems even have the ability to identify which conductor has broken and how many devices are on each side of the break.
- *Hybrid circuits* – A hybrid system may consist of a Class A main trunk with Class B spur circuits in each area. A good example would be a Class A circuit leaving the control panel, entering a junction box on each floor of a multistory building, and then returning to the control panel by a different route. The signaling line circuits on the floors are wired as Class B from the junction box. This provides good system reliability and keeps costs down.

4.0.0 ◆ FIRE ALARM SYSTEM EQUIPMENT

The equipment used in fire alarm systems is generally held to higher standards than typical electrical equipment. The main components of a fire alarm system include:

- Alarm initiating devices
- Control panels
- Primary (main) and secondary (standby) power supplies
- Notification appliances
- Communications and monitoring

5.0.0 ◆ FIRE ALARM INITIATING DEVICES

Fire alarm systems use initiating devices to report a fire and provide supervisory or trouble reports. Some of the initiating devices used to trigger a fire alarm are designed to sense the signs of fire automatically (automatic sensors). Some report level-of-fire conditions only. Others rely on people to see the signs of fire and then activate a manual device. Automatic sensors (detectors) are available that sense smoke, heat, and flame. Manual initiating devices are usually some form of pull station.

Figure 4 is a graphic representation of the application of various detectors for each stage of a fire. However, detection may not occur at any specific point in time within each stage, nor will the indicated sensors always provide detection in the stages represented. For example, alcohol fires can produce flame followed by heat while producing little or no visible particles.

Besides triggering an alarm, some fire alarm systems use the input from one or more detectors or pull stations to trigger fire suppression equipment such as a carbon dioxide (CO_2), dry chemical, or water deluge system.

It is mandatory that devices used for fire detection be **listed** for the purpose for which they will be used. Common UL listings for fire detection devices are:

- *UL 38* – Manually actuated signaling boxes
- *UL 217* – Single- and multiple-station smoke detectors
- *UL 268* – System-type smoke detectors
- *UL 268A* – Duct smoke detectors
- *UL 521* – **Heat detectors**

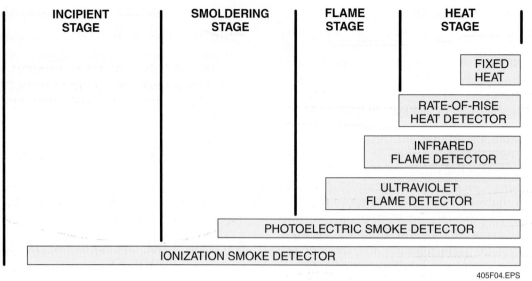

INCIPIENT STAGE	SMOLDERING STAGE	FLAME STAGE	HEAT STAGE

FIXED HEAT

RATE-OF-RISE HEAT DETECTOR

INFRARED FLAME DETECTOR

ULTRAVIOLET FLAME DETECTOR

PHOTOELECTRIC SMOKE DETECTOR

IONIZATION SMOKE DETECTOR

405F04.EPS

Figure 4 ◆ Detection versus stages of fires.

5.1.0 Conventional versus Addressable Commercial Detectors

Figure 5 shows some typical commercial detectors. Commercial detectors differ from stand-alone residential detectors in that a number of commercial detectors are usually wired in parallel and connected to a fire alarm control panel. In a typical conventional commercial fire alarm installation, a number of separate zones with multiple automatic sensors are used to partition the installation into fire zones.

As described previously, newer commercial fire systems use addressable automatic detectors, pull stations, and notification appliances to supply a coded identification signal. The systems also provide individual supervisory or **trouble signals** to the FACP, as well as any fire alarm signal, when periodically polled by the FACP. This provides the fire control system with specific device location information to pinpoint the fire, along with detector, pull station, or notification appliance status information. In many new commercial fire systems, analog addressable detectors (non-automatic) are used. Instead of sending a fire alarm signal when polled, these devices communicate information about the fire condition (level of smoke, temperature, etc.) in addition to their identification, supervisory, or trouble signals. In the case of these detectors, the control panel's internal programming analyzes the fire condition data from one or more sensors, using recent and historical information from the sensors, to determine if a fire alarm should be issued.

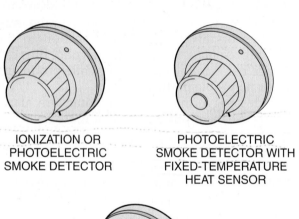

IONIZATION OR PHOTOELECTRIC SMOKE DETECTOR

PHOTOELECTRIC SMOKE DETECTOR WITH FIXED-TEMPERATURE HEAT SENSOR

RATE-OF-RISE HEAT DETECTOR WITH FIXED-TEMPERATURE HEAT SENSOR

405F05.EPS

Figure 5 ◆ Typical commercial automatic sensors.

NOTE

The descriptions of operation discussed in the sections covering automatic detectors, pull stations, and notification appliances are primarily for conventional devices.

5.2.0 Automatic Detectors

Automatic detectors can be divided into the following types:

- *Line detectors* – Detection is continuous along the entire length of the detector in these detection devices. Typical examples may include certain older pneumatic rate-of-rise tubing detectors, **projected beam smoke detectors**, and heat-sensitive cable.
- *Spot detectors* – These devices have a detecting element that is concentrated at a particular location. Typical examples include bimetallic detectors, fusible alloy detectors, certain **rate-of-rise detectors**, certain smoke detectors, and thermoelectric detectors.
- *Air sampling detectors* – These consist of piping or tubing distributed from the detector unit to the area(s) to be protected. An air pump draws air from the protected area back to the detector through air sampling ports and piping or tubing. At the detector, the air is analyzed for fire products.
- *Addressable or analog addressable detectors* – As mentioned previously, these detectors provide alarms and individual point identification, along with supervisory/trouble information, to the control panel. In certain analog addressable detectors, adjustable sensitivity of the alarm signal can be provided. In other analog addressable detectors, an alarm signal is not generated. Instead, only a level of detection signal is fed to the control panel. Alarm sensitivity can be adjusted, and the level of detection can be analyzed from the panel (based on historical data) to reduce the likelihood of false alarms in construction areas or areas of high humidity on a temporary or permanent basis. Because all detectors are continually polled, a T-tap splice is permitted with some signaling line circuit styles. T-taps are not permitted with all styles.

5.3.0 Heat Detectors

There are two major types of heat detectors in general use. One is a **rate-of-rise detector** that senses a 15° per minute increase in room/area temperature. The other is one of several versions of fixed-temperature detectors that activate if the room/area exceeds the rating of the sensor. Generally, rate-of-rise sensors are combined with fixed-temperature sensors in a combination heat detector. Heat detectors are generally used in areas where property protection is the only concern or where smoke detectors would be inappropriate.

5.3.1 Fixed-Temperature Heat Detectors

Fixed-temperature heat detectors activate when the temperature exceeds a preset level. Detectors are made to activate at different levels. The most commonly used temperature settings are 135°F, 190°F, and 200°F. The three types of fixed-temperature heat detectors are:

- *Fusible link* – The fusible link detector (*Figure 6*) consists of a plastic base containing a switch mechanism, wiring terminals, and a three-disc heat collector. Two sections of the heat collector are soldered together with an alloy that will cause the lower disc to drop away when the rated temperature is reached. This moves a plunger that shorts across the wiring contacts, causing a constant alarm signal. After the detector is activated, a new heat collector must be installed to **reset** the detector to an operating condition.
- *Quick metal* – The operation of the quick metal detector (*Figure 7*) is very simple. When the surrounding air reaches the prescribed temperature (usually 135°F or 190°F), the quick metal begins to soften and give way. This allows spring pressure within the device to push the top portion of the thermal element out of the way, causing the alarm contacts to close. After the detector has been activated, either the detector or the heat collector must be replaced to reset the detector to an operating condition.
- *Bimetallic* – In a bimetallic detector (*Figure 8*), two metals with different rates of thermal expansion are bonded together. Heat causes the two metals to expand at different rates, which causes the bonded strip to bend. This action closes a normally open circuit, which signals an alarm. Detectors with bimetallic elements automatically self-restore when the temperature returns to normal.

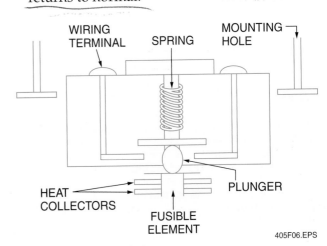

Figure 6 ◆ Fusible link detector.

Addressable and Analog Addressable Smoke Detectors

This is a typical addressable smoke detector. Addressable detectors send the fire alarm control panel (FACP) alarm status, detector location, and supervisory/trouble information. An analog addressable smoke detector performs the same functions; however, it has the additional capability of being able to report to the FACP information about the level of smoke that it is detecting.

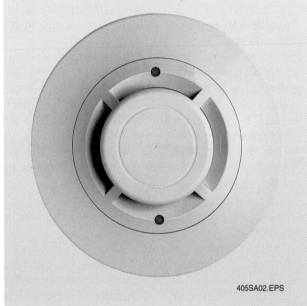

405SA02.EPS

5.3.2 Combination Heat Detectors

Combination heat detectors contain two types of heat detectors in a single housing. A fixed-temperature detector reacts to a preset temperature, and a rate-of-rise detector reacts to a rapid change in temperature even if the temperature reached does not exceed the preset level.

- *Rate-of-rise with fusible link detector* – (See *Figure 9*.) Rate-of-rise operation occurs when air in the chamber (1) expands more rapidly than it can escape from the vent (2). The increasing pressure moves the diaphragm (3) and causes the alarm contacts (4 and 5) to close, which results in an alarm signal. This portion of the detector automatically resets when the temperature stabilizes. A fixed-temperature trip occurs when the heat causes the fusible alloy (6) to melt, which releases the spring (7). The spring depresses the diaphragm, which closes the alarm contacts (4 and 5). If the fusible alloy is melted, the center section of the detector must be replaced.
- *Rate-of-rise with bimetallic detector* – (See *Figure 10*.) Rate-of-rise operation occurs when the air in the air chamber (1) expands more rapidly than it can escape from the valve (2). The increasing pressure moves the diaphragm (3) and causes the alarm contact (4) to close. A fixed-temperature trip occurs when the bimetallic element (5) is heated, which causes it to bend and force the spring-loaded contact (6) to mate with the fixed contact (7).

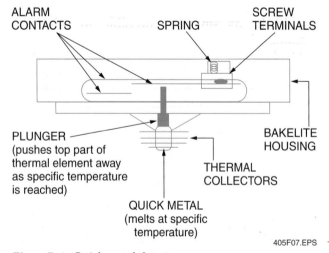

405F07.EPS

Figure 7 ◆ Quick metal detector.

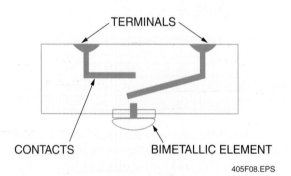

405F08.EPS

Figure 8 ◆ Bimetallic detector.

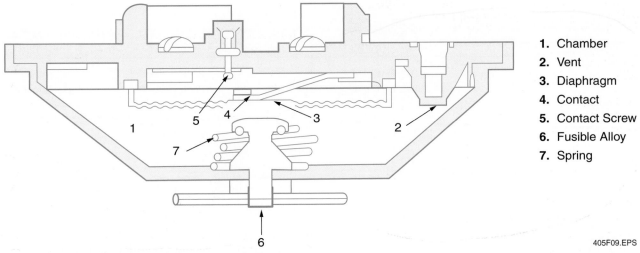

1. Chamber
2. Vent
3. Diaphragm
4. Contact
5. Contact Screw
6. Fusible Alloy
7. Spring

405F09.EPS

Figure 9 ◆ Rate-of-rise with fusible link detector.

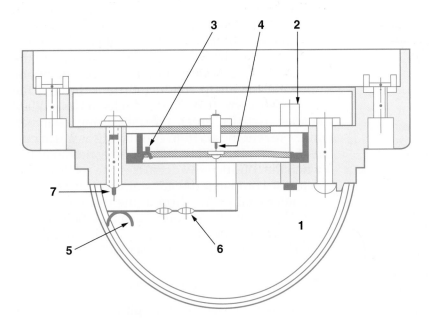

1. Air Chamber
2. Breather Valve
3. Diaphragm Assembly
4. Rate-of-Rise Contact
5. Bimetal Fixed-Temperature Element
6. Contact Spring
7. Fixed-Temperature Contact

405F10.EPS

Figure 10 ◆ Rate-of-rise with bimetallic detector.

5.3.3 Heat Detector Ratings

As with all automatic detectors, heat sensors are also rated for a listed **spacing** as well as a temperature rating (*Table 1*). The listed spacing is typically 50' × 50'. The applicable rules and formulas for proper spacing (from the standards) are then applied to the listed spacing. Caution is advised because it is difficult or impossible to distinguish the difference in the listed spacing for two different types of heat sensors based on their appearance.

The maximum **ceiling** temperature must be 20° or more below the detector rated temperature. The difference between the rated temperature and the maximum ambient temperature for the space should be as small as possible to minimize response time.

Heat sensors are not considered life safety devices. They should be used in areas that are unoccupied or in areas that are environmentally unsuitable for the application of smoke detectors. To every extent possible, heat sensors should be limited to use as property protection devices.

5.4.0 Smoke Detectors

There are two basic types of smoke detectors: photoelectric detectors (*Figure 11*), which sense the presence of smoke, and ionization detectors (*Figure 12*), which sense the presence of

Table 1 Heat Detector Temperature Ratings

Temperature Classification	Temperature Rating Range	Maximum Ceiling Temperature	Color Code
Low	100°F–134°F	80°F	No color
Ordinary	135°F–174°F	100°F	No color
Intermediate	175°F–249°F	150°F	White
High	250°F–324°F	225°F	Blue
Extra high	325°F–399°F	300°F	Red
Very extra high	400°F–499°F	375°F	Green
Ultra high	500°F–575°F	475°F	Orange

405F11.EPS

Figure 11 ◆ Typical photoelectric smoke detector.

405F12.EPS

Figure 12 ◆ Typical ionization smoke detector.

combustible gases. For residential use, both types are available combined in one detector housing for maximum coverage. For commercial use, either type is also available combined with a heat detector. As mentioned previously, the newer commercial fire systems use smoke detectors that are analog addressable units, which do not signal an alarm. Instead, they return a signal that represents the level of detection to the FACP for further analysis.

5.4.1 Ionization Detectors

An ionization detector (*Figure 13*) uses the change in the electrical conductivity of air to detect smoke. An alarm is indicated when the amount of smoke in the detector rises above a certain level. The detector has a very small amount of radioactive material in the sensing chambers. As shown in *Figure 14A*, the radioactive material ionizes the air in the measuring and reference chambers, allowing the air to conduct current through the space between two charged electrodes. When smoke particles enter the measuring chamber, they prevent the air from conducting as much current (*Figure 14B*). The detector activates an alarm when the conductivity decreases to a set level. The detector compares the current drop in the main chamber against the drop in the reference chamber. This allows it to avoid alarming when the current drops due to surges, radio frequency interference (RFI), or other factors.

5.4.2 Photoelectric Smoke Detectors

There are two basic types of **photoelectric smoke detectors**: light-scattering detectors and beam detectors. Light-scattering detectors use the reflective properties of smoke to detect the smoke. The **light scattering** principle is used for the most common single-housing detectors. Beam detectors rely upon smoke to block enough light to cause an alarm. Early smoke detectors operated on the **obscuration** principle. Today, photoelectric smoke detectors are used primarily to sense smoke in large open areas with high ceilings. In some cases, mirrors are used to direct the beam in a desired path; however, the mirrors reduce the overall range of the detector. Most commercial photoelectric smoke detectors require an external power source.

Light-scattering detectors are usually spot detectors that contain a light source and a photosensitive device arranged so that light rays do not normally fall on the device (*Figure 15*). When smoke particles enter the light path, the light hits

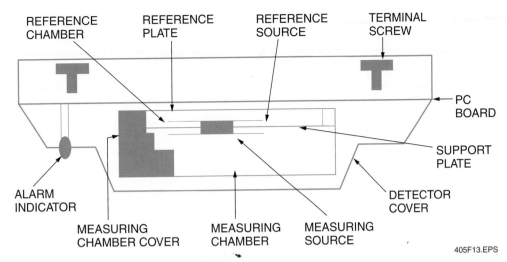

Figure 13 ◆ Ionization detector.

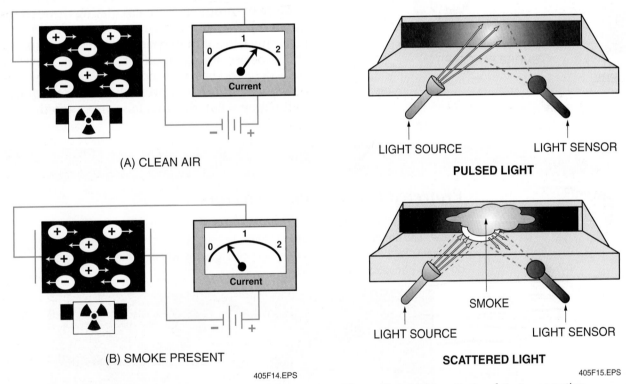

(A) CLEAN AIR

(B) SMOKE PRESENT

405F14.EPS

Figure 14 ◆ Ionization action.

PULSED LIGHT

LIGHT SOURCE   LIGHT SENSOR

SMOKE

LIGHT SOURCE   LIGHT SENSOR

SCATTERED LIGHT

405F15.EPS

Figure 15 ◆ Light-scattering detector operation.

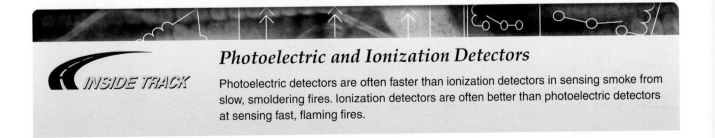

Photoelectric and Ionization Detectors

Photoelectric detectors are often faster than ionization detectors in sensing smoke from slow, smoldering fires. Ionization detectors are often better than photoelectric detectors at sensing fast, flaming fires.

the particles and scatters, hitting the photosensor, which signals the alarm.

Projected beam smoke detectors that operate on the light obscuration principle (*Figure 16*) consist of a light source, a light beam focusing device, and a photosensitive device. Smoke obscures or blocks part of the light beam, which reduces the amount of light that reaches the photosensor, signaling an alarm. The prime use of the light obscuration principle is with projected beam-type smoke detectors that are employed in the protection of large, open high-bay warehouses, high-ceiling churches, and other large areas. A version of a beam detector was used in older style duct detectors.

After a number of fires where smoke spread through building duct systems and contributed to numerous deaths, building codes began to require the installation of duct detectors. Two versions of duct detectors are shown in *Figure 17*. The projected beam detector is an older style of duct detector that may be encountered in existing installations.

Duct detectors enable a system to control the spread of smoke within a building by turning off the HVAC system, operating exhaust fans, closing doors, or pressurizing smoke compartments in the event of a fire. This prevents smoke, fumes, and fire by-products from circulating through the ductwork. Because the air in the duct is either at rest or moving at high speed, the detector must be able to sense smoke in either situation. A typical duct detector has a listed airflow range within which it will function properly. It is not required to function when the duct fans are stopped.

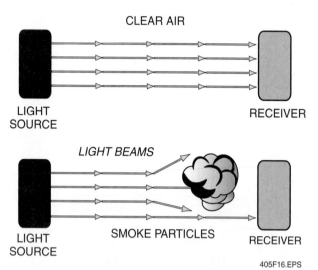

CLEAR AIR

LIGHT SOURCE

RECEIVER

LIGHT BEAMS

LIGHT SOURCE

SMOKE PARTICLES

RECEIVER

405F16.EPS

Figure 16 ◆ Light obscuration principle.

Projected Beam Smoke Detectors

The transmitter and receiver for a projected beam smoke detector are shown here. These units are designed for use in atriums, ballrooms, churches, warehouses, museums, factories, and other large, high-ceiling areas where conventional smoke detectors cannot be easily installed.

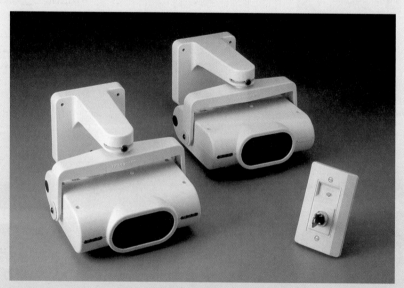

405SA03.EPS

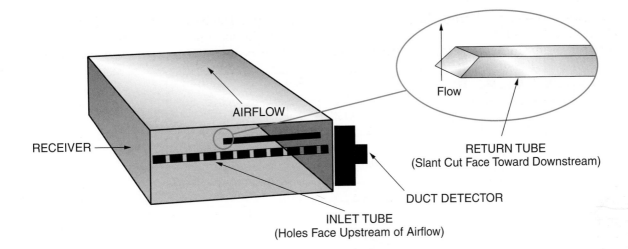

SAMPLING TUBE STYLE

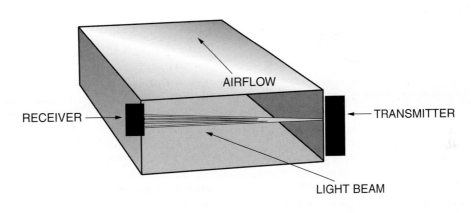

PROJECTED LIGHT BEAM STYLE

405F17.EPS

Figure 17 ◆ Typical duct detector installations.

At least one duct detector is available with a small blower for use in very low airflow applications. Duct detectors must not be used as a substitute for open area protection because smoke may not be drawn from open areas of the building when the HVAC system is shut down. Duct detectors can be mounted inside the duct with an access panel or outside the duct with sampling tubes protruding into the duct.

As mentioned previously, the primary function of a duct detector is to turn off the HVAC system to prevent smoke from being circulated. However, *NFPA 90A* and *NFPA 90B* require that duct detectors be tied into a general fire alarm system if the building contains one. If no separate fire alarm system exists, then remote audio/visual indicators, triggered by the duct detectors, must be provided in normally occupied areas of the building. Duct detectors that perform functions other than the shutdown of HVAC equipment must be supplied with backup power.

Cloud chamber smoke detectors (*Figure 18*) use sampling tubes to draw air from several areas (zones). The air is passed through several chambers where humidity is added and pressure is reduced with a vacuum pump. Reducing the pressure causes water droplets to form around the sub-micron smoke particles and allows them to become visible. A light beam is passed through the droplets and measured by a photoelectric detector. As the number of droplets increases, the light reaching the detector is reduced, which initiates an alarm signal.

5.5.0 Other Types of Detectors

This section describes special-application detectors. These include rate compensation detectors, semiconductor line-type heat detectors, fusible line-type heat detectors, ultraviolet flame detectors, and infrared flame detectors.

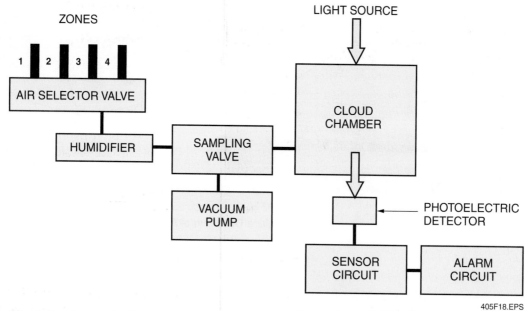

Figure 18 ◆ Cloud chamber smoke detector.

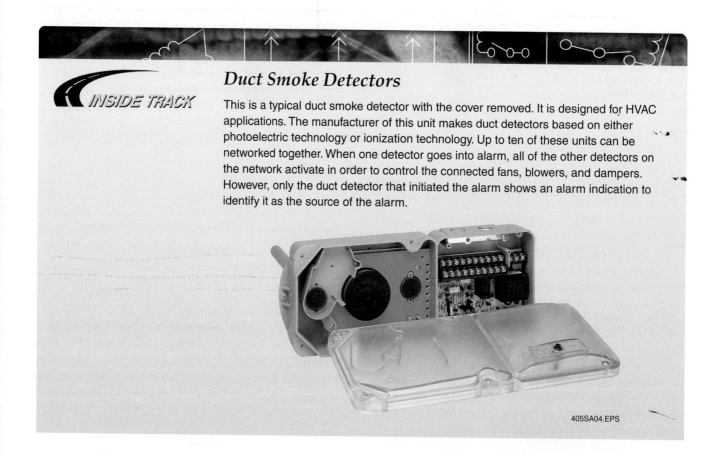

5.5.1 Rate Compensation Detectors

A rate compensation detector (*Figure 19*) is a device that responds when the temperature of the surrounding air reaches a predetermined level, regardless of the rate of temperature rise. The detector also responds if the temperature rises quickly over a short period. This type of detector has a tubular casing of metal that expands lengthwise as it is heated. A contact closes as it reaches a certain stage of elongation. These detectors are self-restoring. Rate compensation detectors are more complex than either fixed or rate-of-rise detectors. They combine the principles of both to

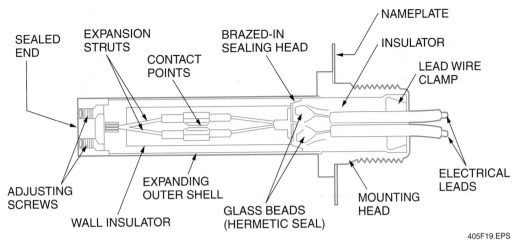

Figure 19 ◆ Rate compensation detector.

compensate for thermal lag. When the air temperature is rising rapidly, the unit is designed to respond almost exactly at the point when the air temperature reaches the unit's rated temperature. It does not lag while it absorbs the heat and rises to that temperature. Because of the precision associated with their operation, rate compensation detectors are well suited for use in areas where thermal lag must be minimized.

5.5.2 Semiconductor Line-Type Heat Detectors

A restorable semiconductor line-type detector (*Figure 20*) uses a semiconductor material and a stainless steel capillary tube. The capillary tube contains a coaxial center conductor separated from the tube wall by a temperature-sensitive semiconductor thermistor material. Under normal conditions, a small current (below the alarm threshold) flows. As the temperature rises, the resistance of the semiconductor thermistor decreases. This allows more current to flow and initiates the alarm. When the temperature falls, the current flow decreases below the alarm threshold level, which stops the alarm. The wire is connected to special controls or modules that establish the alarm threshold level and sense the current flow. Some of these control devices can pinpoint the location in the length of the wire where the temperature change occurs. Line-type heat detectors are commonly used in cable trays, conveyors, electrical switchgear, warehouse rack storage, mines, pipelines, hangars, and other similar applications.

5.5.3 Fusible Line-Type Heat Detector

A non-restorable fusible line-type heat detector (*Figure 21*) uses a pair of steel wires in a normally open circuit. The conductors are held apart by heat-sensitive insulation. The wires, under tension, are enclosed in a braided sheath to make a single cable. When the temperature limit is reached, the insulation melts, the two wires contact, and an alarm is initiated. The melted and

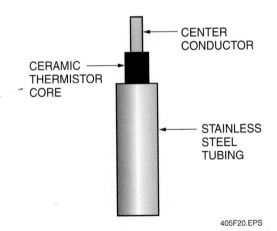

Figure 20 ◆ Restorable semiconductor line-type heat detector.

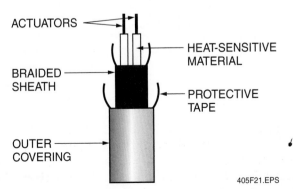

Figure 21 ◆ Non-restorable fusible line-type heat detector.

fused section of the cable must be replaced following an alarm to restore the system. The wire is available with different melting temperatures for the insulation. The temperature rating should be approximately 20°F above the ambient temperature. Special controls or modules are available that can be connected to the wires to pinpoint the fire location where the wires are shorted together.

5.5.4 Ultraviolet Flame Detectors

An ultraviolet (UV) flame detector (*Figure 22*) uses a solid-state sensing element of silicon carbide, aluminum nitrate, or a gas-filled tube. The UV radiation of a flame causes gas in the element or tube to ionize and become conductive. When sufficient current flow is detected, an alarm is initiated.

5.5.5 Infrared Flame Detectors

An infrared (IR) flame detector (*Figure 23*) consists of a filter and lens system that screens out unwanted radiant-energy wavelengths and focuses the incoming energy on light-sensitive components. These flame detectors can respond to the total IR content of the flame alone or to a combination of IR with flame flicker of a specific frequency. They are used indoors and have filtering systems or solar sensing circuits to minimize unwanted alarms from sunlight.

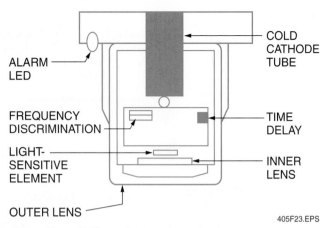

Figure 23 ◆ IR flame detector (top view).

5.6.0 Manual (Pull Station) Fire Detection Devices

When required by code, manual pull stations are required to be distributed throughout a commercial monitored area so that they are unobstructed, readily accessible, and in the normal path of exit from the area. Examples of pull stations are shown in *Figure 24*.

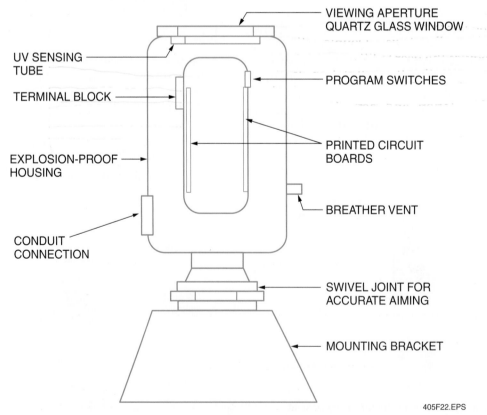

Figure 22 ◆ UV flame detector (top view).

SINGLE-ACTION
PULL STATION

GLASS-BREAK COVER
FOR PULL STATION

DOUBLE-ACTION
PULL STATION

405F24.EPS

Figure 24 ◆ Typical pull stations.

- *Single-action pull stations* – A single action or motion operates these devices. They are activated by pulling a handle that closes one or more sets of contacts and generates the alarm.
- *Glass-break pull stations* – In these devices, a glass rod, plate, or special element must be broken to activate the alarm. This is accomplished using a handle or hammer that is an integral part of the station. When the alarm is activated, one or more sets of contacts are closed and an alarm is actuated. Usually, the plate, rod, or element must be replaced to return the unit to service, although some stations will operate without the rod or plate.
- *Double-action pull stations* – Double-action pull stations require the user to lift a cover or open a door before operating the pull station. Two discretely independent actions are required to operate the station and activate the alarm. Using a stopper-type cover that allows the alarm to be tripped after a cover is lifted may turn a single-action pull station into a double-action pull station. According to Underwriters Laboratories, the stopper-type device is listed as an accessory to manual stations and is permissible for use as a double-action pull station device. With certain types of double-action stations, there have been instances when people have confused the sound of a tamper alarm that sounds when the cover is lifted with the sound

of fire alarm activation. The sounding of this tamper alarm sometimes causes them to fail to activate the pull station.

- *Key-operated pull stations* – Applications for key-operated pull stations (*Figure 25*) are restricted. Key-operated stations are permitted in certain occupancies where facility staff members may be in the immediate area and where use by other occupants of the area is not desirable. Typical situations would include certain detention and correctional facilities and some health care facilities, particularly those that provide mental health treatment.

405F25.EPS

Figure 25 ◆ Key-operated pull station.

Glass-Break Pull Station

INSIDE TRACK

When installing a glass-break pull station, consideration must be given to where the glass will go when broken and the danger it may pose to children and the disabled.

Manual Fire Alarm Station Reset

This manual fire alarm is operated by pulling on the pull cover. This engages a latching mechanism that prevents the pull cover from being returned to the closed position. The only way the station cover can be reset to the closed position is by using the appropriate reset key.

405SA05.EPS

5.7.0 Auto-Mechanical Fire Detection Equipment

This section describes various types of auto-mechanical fire detection equipment. It covers wet and dry sprinkler systems and water flow alarms.

5.7.1 Wet Sprinkler Systems

A wet sprinkler system (*Figure 26*) consists of a permanently piped water system under pressure, using heat-actuated sprinklers. When a fire occurs, the sprinkler heads exposed to high heat open and discharge water individually to attempt to control or extinguish the fire. They are designed to automatically detect and control a fire and protect a structure. Once a sprinkler head is activated, some type of water flow sensor signals a fire alarm. When activated, a sprinkler system may cause water damage. Wet systems should not be used in spaces subject to freezing.

WARNING!

If the air pressure drops and the system fills with water, the system must be drained. Consult a qualified specialist to restore the system to normal operation.

5.7.2 Dry Sprinkler Systems

A dry sprinkler system (*Figure 27*) consists of heat-operated sprinklers that are attached to a piping system containing air under pressure. Normally, air pressure in the pipes holds a water valve closed, which keeps water out of the piping system. When heat activates a sprinkler head, the open sprinkler head causes the air pressure to be released. This allows the water valve to open, and water flows through the pipes and out to the activated sprinkler head. Once a sprinkler head is activated and water starts to flow, a water flow

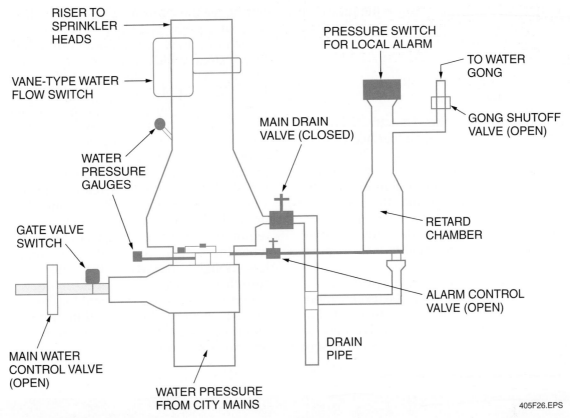

Figure 26 ◆ Wet sprinkler system.

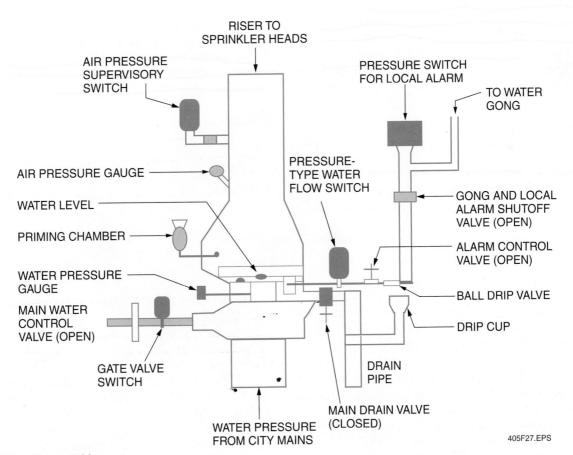

Figure 27 ◆ Dry sprinkler system.

sensor signals a fire alarm. Because the pipes are dry until a fire occurs, these systems may be used in spaces subject to freezing.

5.7.3 Wet or Dry Sprinkler System Water Flow Alarms

When a building sprinkler head is activated by the heat generated by a fire, the sprinkler head allows water to flow. As shown in *Figures 26* and *27*, pressure-type or vane-type water flow switches are installed in the sprinkler system along with local alarm devices. The water flow switches detect the movement of water in the system. Activation of these switches by the movement of water causes an initiation signal to be sent to an FACP that signals a fire alarm.

In addition to devices that signal sprinkler system activation, other devices may be used to monitor the status of the system using an FACP. For instance, the position of a control valve may be monitored so that a **supervisory signal** is sent whenever the control valve is turned to shut off the water to the sprinkler system. If this valve is turned off, no water can flow through the sprinkler system, which means the system is inactive. In some systems, water pressure from the municipal water supply may not be strong enough to push enough water to all parts of a building. In these cases, a fire pump is usually required. When the fire pump runs, a supervisory signal is sent indicating that the fire pump is activated. If the pump runs to maintain system pressure and does not shut down within a reasonable time, a site visit may be required. When water is scarce or unavailable, an on-site water tank may be required. Supervisory signals may be generated when the temperature of the water drops to a level low enough to freeze or if the water level or pressure drops below a safe level. In a dry system, a supervisory signal is generated if the air pressure drops below a usable level. If fire pump power is monitored, lack of power or power phase reversal will cause a trouble signal.

Water Flow Detectors

This water flow detector is typical of those used with wet-pipe sprinkler systems. Water flow through the associated pipe deflects the detector's vane. This activates the internal switch contacts, initiating an alarm or auxiliary indication. Water flow in the pipe can be caused by the opening of one or more sprinkler heads because of a fire, the opening of a test valve, or a leaking or ruptured pipe.

INSIDE TRACK

405SA06.EPS

6.0.0 ◆ CONTROL PANELS

Today, there are many different companies manufacturing hundreds of different control panels, and more are developed each year. *Figure 28* shows a typical intelligent, addressable control panel. The control panel shown has a voice command center added for use in high-rise buildings. It provides for automatic evacuation messages, firefighter paging, and two-way communication to a central station through a telephone network. Although many control panels may have unique features, all control panels perform some specific basic functions. As shown in *Figure 29*, they detect problems through the sensor devices connected to them, and they sound alerts or report these problems to a central location.

A control panel can allow the user to reprogram the system and, in some cases, to activate or deactivate the system or zones. The user may also change the sensitivity of detectors in certain zones of the system. On panels that can be used for both intrusion and fire alarms, controls are provided to allow the user to enter or leave the monitored area without setting off the system. In addition, controls allow some portions of the system (fire, panic, holdup, etc.) to remain armed 24 hours a day, 365 days a year. Control panels also provide the alarm user, responding authority, or inspector with a way to silence bells or control other system features.

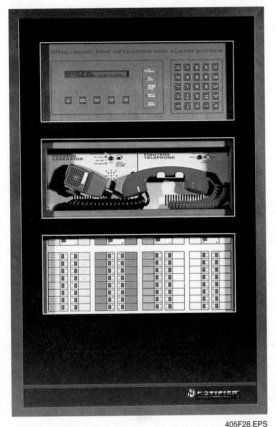

405F28.EPS

Figure 28 ◆ Typical intelligent, addressable control panel.

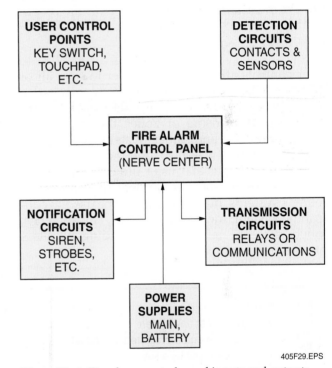

405F29.EPS

Figure 29 ◆ Fire alarm control panel inputs and outputs.

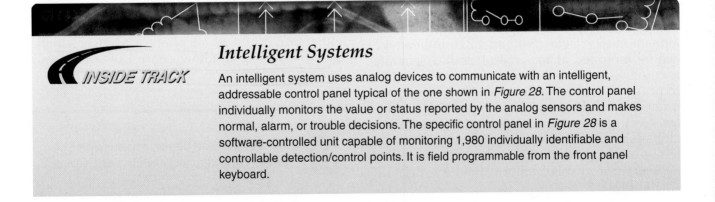

Intelligent Systems

An intelligent system uses analog devices to communicate with an intelligent, addressable control panel typical of the one shown in *Figure 28*. The control panel individually monitors the value or status reported by the analog sensors and makes normal, alarm, or trouble decisions. The specific control panel in *Figure 28* is a software-controlled unit capable of monitoring 1,980 individually identifiable and controllable detection/control points. It is field programmable from the front panel keyboard.

Control panels are usually equipped with LED, alphanumeric, or graphic displays to communicate information such as active zones or detectors to an alarm user or monitor. These types of displays are usually included on the FACP, but can also be located remotely if required.

The control panel organizes all the components into a working and functional system. The control panel is often referred to as the brain of the system. The control panel coordinates the actions the system takes in response to messages it gets from initiating devices or, in some cases, notification devices that are connected to it. Depending on the inputs from the devices, it may activate the notification devices and may also transmit data to a remote location via transmission circuits. In most cases, the control panel also conditions the power it receives from the building power system or from a backup battery so that it can be used by the fire alarm system.

6.1.0 User Control Points

User controls allow the alarm user to turn all or portions of the system on or off. User controls also allow the alarm user to monitor the system status, including the sensors or zones that are active, trouble reports, and other system parameters. They also allow the user to reset indicators of system events such as alarms or indications of trouble with phone lines, equipment, or circuits. In today's systems, many devices can be used to control the system, including keypads, key switches, touch screens, telephones (including wireless phones), and computers.

> **NOTE**
> UL requires that any system controls that affect system operation be protected behind a locked door or locked out by a key.

6.1.1 Keypads

Two general types of keypads are in use today: alphanumeric (*Figure 30*) and LED. Either type can be mounted at the control or at a separate location. Both allow the entry of a numerical code to program various system functions. An alphanumeric keypad combines a keypad that is similar to a pushbutton telephone dial with an alphanumeric display that is capable of showing letters and numbers. LED keypads use a similar keypad, but display information by lighting small LEDs.

405F30.EPS

Figure 30 ◆ Alphanumeric keypad and display.

6.1.2 Touch Screens

Touch screens allow the user to access multiple customized displays in graphic or menu formats. This enables rapid and easy interaction with the system.

6.1.3 Telephone/Computer Control

Since most fire alarm systems are connected to telephones for notification purposes, and telephones are often in locations where user control devices have traditionally been located, some control panel manufacturers have incorporated ways for the telephone to function as a user interface. When telephones or cell phones are used as an interface for the system, feedback on system status and events is given audibly over the phone. Because the system is connected to the telephone network, a system with this feature can be monitored and controlled from anywhere a land-line or cell phone call can be made. Most new systems can also be connected to a computer either locally or via a telephone modem or a high-speed data line to allow control of events and receipt of information.

> **NOTE**
> The NFPA does not permit remote resetting of fire alarm systems that have signaled an alarm.

6.2.0 FACP Initiating Circuits

An initiating circuit monitors the various types of initiating devices. A typical initiating circuit can monitor for three states: normal, trouble, and alarm. Initiating circuits can be used for monitoring fire devices, non-fire devices (supervisory), or devices for watch patrol or security. The signals from the initiating devices can be separately indicated at the premises and remotely monitored at a central station.

6.2.1 Initiating Circuit Zones

The term zone as it relates to modern fire alarm systems can have several definitions. Building codes restrict the location and size of a zone to enable emergency response personnel to quickly locate the source of an alarm. The traditional use of a zone in a conventional system was that each initiating circuit was a zone. Today, a single device or multiple initiating circuits or devices may occupy a zone. Generally, a zone may not cover more than one floor, exceed a certain number of square feet, or exceed a certain number of feet in length or width. In addition, some codes require different types of initiating devices to be on different initiating circuits within a physical area or zone. With modern addressable systems, each device is like a zone because each is displayed at the panel. It is common in addressable systems to group devices of several types into a zone. This is usually accomplished by programming the panel rather than by hardwiring the devices. Ultimately, the purpose of a zone is to provide the system monitor with information as to the location of a system alarm or problem. Local codes provide the specific guidelines.

6.2.2 Alarm Verification

To reduce false alarms, most FACPs allow for a delay in the activation of notification devices upon receiving an alarm signal from an initiating device. In some conventional systems using a positive alarm sequence feature, alarm delay is usually adjustable for a period up to three minutes, but must not exceed three minutes. This allows supervising or monitoring personnel to investigate the alarm. If a second detector activates during the investigation period, a fire alarm is immediately sounded. In other conventional systems, verification can be of the reset-and-resample type: The panel resets the detector and waits 10 to 20 seconds for the sensor to retransmit the alarm. If there is an actual fire, the sensor should detect it both times, and an alarm is then activated. In addressable systems, alarm verification can be of the wait-and-check type: The panel notes the initiation signal and then waits for about 10 to 20 seconds to see if the initiation signal remains constant before activating an alarm.

Another verification method for large areas with multiple sensors is to wait until two or more sensors are activated before an alarm is initiated. In this method, known as cross-zoning, a single detector activation may set a one minute or less pre-alarm condition warning signal at a manned monitoring location. If the pre-alarm signal is not cancelled within the allowed time, or if a second detector activates, a fire alarm is sounded. The pre-alarm signals are not required by the NFPA. In cross-zoning, two detectors must occupy each space regardless of the size of the space, and each detector may cover only half of the normal detection area. Cross-zoning along with other methods of alarm verification cannot be used on the same devices. Some building codes require alarm verification or some equivalent on all smoke detection devices.

Verification is commonly used in hotels, motels, hospitals, and other institutions with large numbers of detectors. It may also be used in monitored fire alarm systems for households to reduce false alarms.

6.2.3 FACP Labeling

Effective system design and installation requires that zones (or sensors) be labeled in a way that makes sense to all who will use or respond to the system. In some cases, this may require two sets of labels: one for the alarm user and another for the police and fire authorities. A central station operator should be aware of both sets of labels, since he or she will be talking to both the police and the alarm user. Labeling for the alarm user is done using names familiar to the user (Johnny's room, kitchen, master bedroom, and so on). Labeling for the police and fire authorities is done from the perspective of looking at the building from the outside (first floor east, basement rear, second floor west, and so on).

6.3.0 Types of FACP Alarm Outputs

Most panels provide one or more of the following types of outputs:

- *Relay or dry contacts* – Relay contacts or other dry contacts are electrically isolated from the circuit controlling them, which provides some protection from external spikes and surges. Additional protection is sometimes needed. Always check the contact ratings to determine how much voltage and current the contacts can handle. Check with the manufacturer to determine if this provides adequate protection.
- *Built-in siren drivers* – Built-in siren drivers use a transistor amplifier circuit to drive a siren speaker. Impedance must be maintained within the manufacturer's specifications. Voltage drop caused by long wire runs of small conductors can greatly reduce the output level of the siren.
- *Voltage outputs* – Voltage outputs are not isolated from the control, and some protection may be required. Spikes and surges such as those generated by solenoid bells or the back EMF generated when a relay is de-energized can be a problem.
- *Open collector outputs* – Open collector outputs are outputs directly from a transistor and have limited current output. They are often used to drive a very low-current device or relay. Filtering may be required. Great caution should be used not to overload these outputs, as the circuit board normally has to be returned to the manufacturer if an overload of even a very short duration occurs.

6.4.0 FACP Listings

It is mandatory that fire alarm control panels be listed for the purpose for which they will be used. Common UL listings for FACPs are *UL 864* (fire alarm control panels) and *UL 985* (household fire warning system units).

In combination fire/burglary panels, fire circuits must be on or active at all times, even if the burglary control is disarmed or turned off. Some codes prohibit using combination intrusion and fire alarm control panels in commercial applications.

A fire alarm control panel listed for household use by the UL may never be installed in any commercial application unless it also has the appropriate commercial listing, has been granted equivalency by the **authority having jurisdiction (AHJ)** in writing, or is specifically permitted by a document that supersedes the reference standards. The NFPA defines a household as a single- or two-family residential unit, and this term applies to systems wholly within the confines of the unit. Except for monitoring, no initiation from, or notification to, locations outside the residence are permitted. Although an apartment building or condominium is considered commercial, a household system may be installed within an individual living unit. However, any devices that are outside the confines of the individual living unit, such as manual pull stations, must be connected to an FACP listed as commercial.

7.0.0 ◆ FACP PRIMARY AND SECONDARY POWER

Primary power for a fire alarm system and the FACP is normally a source of power provided by a utility company. However, primary power for systems and panels can also be supplied from emergency uninterruptible backup primary power systems. To avoid service interruption, a fire alarm system may be connected on the line side of the electrical main service disconnect switch. The circuit must be protected with a circuit breaker or fuse no larger than 20 amperes (A).

Secondary power for a fire alarm system can be provided by a battery or a battery with an **approved** backup generator as defined by the *NEC®*. Secondary power must be immediately supplied to the fire alarm system in the event of a primary power failure. As defined by *NFPA 72®*, standby time for operation of the system on secondary power must be no less than 24 hours for central, local, proprietary, voice communications, and household systems and no less than 60 hours for auxiliary and remote systems. The battery standby time may be reduced if a properly configured battery-backup generator is available. During an alarm condition, the secondary power must operate the system under load for a minimum of 4, 5, or 15 minutes after expiration of the standby period, depending on the type of system. Any batteries used for secondary power must be able to be recharged within 48 hours.

8.0.0 ◆ NOTIFICATION APPLIANCES

Notification to building occupants of the existence of a fire is the most important life safety function of a fire alarm system. There are two primary types of notification: audible and visible. Concerns resulting from the **Americans with Disabilities Act (ADA)** have prompted the introduction of olfactory (sense of smell) and tactile (sense of touch) types of notification as well.

8.1.0 Visual Notification Devices

Strobes are high-intensity lights that flash when activated. They can be separate devices or mounted on or near the audible device. Strobe lights are more effective in residential, industrial, and office areas where they don't compete with other bright objects. Because ADA requirements dictate clear or white xenon (or equivalent) strobe lights, most interior visual notification devices are furnished with clear strobe lights (*Figure 31*). Strobe devices can be wall-mounted or ceiling-mounted and are usually combined with an audible notification device. When more than two strobes are visible from any location, the strobes must be synchronized to avoid random flashing, which can be disorienting and

CEILING-MOUNTED   **WALL-MOUNTED**

405F31.EPS

Figure 31 ◆ Typical ceiling-mounted and wall-mounted strobe devices.

may actually cause seizures in certain individuals. Some strobes are brighter than others to accommodate different applications. These units may be rated in candelas.

8.2.0 Audible Notification Devices

Audible alarm devices are noise-making devices, such as sirens, bells, or horns, that are used as part of a local alarm system to indicate an alarm condition. In some cases, low-current **chimes** or buzzers are also used. Audible signals used for fire alarms in a facility that contains other sound-producing devices must produce a unique sound pattern so that a fire alarm can be recognized. If more than one audible signal is used in a facility, they must be synchronized to maintain any sound pattern. *NFPA 72*® specifies that a standard signal known as Temporal three is required by most fire alarm systems including household systems. This signal has three ½-second tones, a pause, and then a repeat of the pattern until the alarm is manually reset. Only a fire alarm may use this signal.

Some bells (*Figure 32*) use an electrically-vibrated clapper to repeatedly strike a gong. These types of solenoid-operated bells can also produce electrical **noise**, which will interfere with controls unless proper filters are used. Solenoid-operated bells draw relatively high current, typically 750 to 1,500 milliamps (mA). Another type of bell is a motor-driven bell. In this type of bell, a motor drives the clapper and produces louder sounds than a solenoid-operated bell. Electrical interference is eliminated and less current is required to operate the bell.

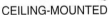

Strobe Lights

Strobe lights may draw a relatively high current compared to other types of notification devices. For this reason, they may require a separate circuit.

Audible Alarms

Audible alarms must have an intensity and frequency that attracts the attention of those with partial hearing loss. Such alarms shall produce a sound that exceeds the prevailing sound level in the space by 15 audible decibels (dBA) or exceeds the maximum sound level with a duration of 60 seconds by 5dBA, whichever is louder. Sound levels shall not exceed 120dBA.

Some jurisdictions prohibit the use of bell notification devices in schools as they can be confused with classroom bells.

Figure 32 ◆ Typical bell with a strobe light.

Self-contained sirens are combinations of speakers and sound equipment. They produce siren sounds and can be used for voice announcements. Self-contained siren packaging saves installation time. If more than one siren is required, they must be synchronized.

A horn (*Figure 33*) usually consists of a continuously vibrating membrane or a piezoelectric element. In a supervised notification circuit, all devices are polarized, allowing current to flow in one direction only. A buzzer uses less power and consists of a continuously vibrating membrane like a horn. Chimes are electronic devices, like self-contained sirens, and have a very low current draw.

8.3.0 Voice Evacuation Systems

With a voice evacuation system (*Figure 34*), building occupants can be given instructions in the event of an emergency. A voice evacuation system can be used in conjunction with an FACP or as a stand-alone unit with a built-in power supply and battery charger. Voice announcements (*Figure 35*) can be made to inform occupants what the problem is or how to evacuate. In many cases, the voice announcements are prerecorded and selected as required by the system. The announcements can also be made from a microphone (*Figure 36*) located at the FACP or at a remote panel. *Figure 36* shows several types of speakers used for voice evacuation systems. Temporal three signaling is not used with zoned voice evacuation systems.

8.4.0 Signal Considerations

Closed doors may drop the audible decibel (dBA) sound levels of alarms below those required to wake children or hearing-impaired adults. Air

Figure 33 ◆ Typical horn with a strobe light.

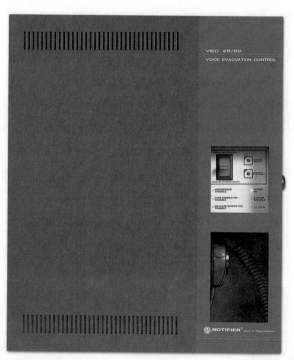

Figure 34 ◆ Voice evacuation system.

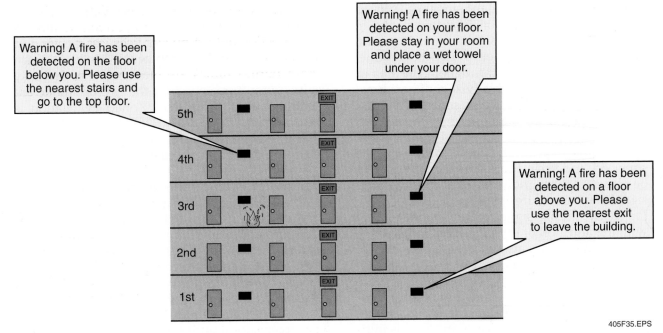

Figure 35 ◆ Typical voice evacuation messages.

Figure 36 ◆ Typical speakers.

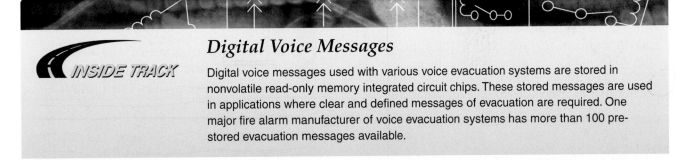

Digital Voice Messages

Digital voice messages used with various voice evacuation systems are stored in nonvolatile read-only memory integrated circuit chips. These stored messages are used in applications where clear and defined messages of evacuation are required. One major fire alarm manufacturer of voice evacuation systems has more than 100 pre-stored evacuation messages available.

conditioners, room humidifiers, and other equipment may cause noise levels to increase. Also, sirens, horns, and bells may fail or be disabled by the fire. Multiple fire sounders provide redundancy and, if properly placed, will provide ample decibel levels to wake all sleeping occupants.

The following is a description of the types of code-authorized signals supplied to various notification devices from a typical fire control panel:

General signal – General signals operate throughout the entire building. Evacuation signals require a distinctive signal. The code requires that a Temporal three signal pattern be used that is in accordance with *ANSI Standard S3.41* (and *ISO 8201*), *NFPA 101®*, and *NFPA 72®*. Temporal three signals consist of three short ½-second tones with ½-second pauses between the tones. This is followed by a 1-second silent period, and then the process is repeated.

• *Attendant signal* – Attendant signaling is used where assisted evacuation is required due to such factors as age, disability, and restraint. Such systems are commonly used in conjunction with coded chimes throughout the area or a coded voice message to advise staff personnel of the location of the alarm source. A coded message might be announced such as "Mr. Green, please report to the third floor nurses' station." No **general alarm** needs to be sounded, and no indicating appliances need to be activated throughout the area. The attendant signal feature requires the approval of the AHJ.

• *Pre-signal* – A pre-signal is an attendant signal with the addition of human action to activate a general signal. It is also used when the control delays the general alarm by more than one minute. The pre-signal feature requires the approval of the AHJ. A signal to remote locations must activate upon an initial alarm signal.

• *Positive alarm sequence* – When an alarm is initiated, the general alarm is not activated for 15 seconds. If a staff person acknowledges the alarm within the 15-second window, the general alarm is delayed for three minutes so that the staff can investigate. Failure to manually acknowledge the initial attendant signal will automatically cause a general alarm. Failure to abort the acknowledged signal within 180 seconds will also automatically cause a general alarm. If any second selected automatic detector is activated during the delay window, the system will immediately cause a general alarm. An activation of a manual station will automatically cause a general alarm. This is an excellent technique for the prevention of unwanted general alarms; however, trained personnel are an integral part of these systems. The positive alarm sequence feature requires the approval of the AHJ.

• *Voice evacuation* – Voice evacuation may be either live or prerecorded and automatically or manually initiated. Voice evacuation systems are a permitted form of general alarm and are required in certain occupancies, as specified in the applicable chapter of *NFPA 101*®. Voice evacuation systems are often zoned so that only the floors threatened by the fire or smoke are immediately evacuated. Zoned evacuation is used where total evacuation is physically impractical. Temporal three signaling is not used with zoned evacuation. Voice evacuation systems must maintain their ability to communicate even when one or more zones are disabled due to fire damage. Buildings with voice evacuation usually include a two-way communication system (fireman's phone) for emergency communications. High-rise buildings are required to have a voice evacuation system.

• *Alarm sound levels* – Where a general alarm is required throughout the premises, audible signals must be clearly heard above maximum ambient noise under normal occupancy conditions. Public area audible alarms should be 75dBA at 10' and a maximum of 120dBA at minimum hearing distance. Public area audible alarms must be at least 15dBA above ambient sound level, or a maximum sound level that lasts over 60 seconds measured at 5' above finished floor (AFF) level. Audible alarms to alert persons responsible for implementing emergency plans (guards, monitors, supervisory personnel, or others) must be between 45dBA and 120dBA at minimum hearing distance. If an average sound level greater than 105dBA exists, the use of a visible signal is required. Typical average ambient sound levels from *NFPA 72*® are given in *Table 2*.

NOTE

The data given in *Table 2* are to be used as a guide. Actual conditions may vary.

Table 2	Typical Average Ambient Sound Levels		
Area	**Sound Level (dBA)**	**Area**	**Sound Level (dBA)**
Mechanical rooms	85	Educational occupancies	45
Industrial occupancies	80	Underground structures	40
Busy urban thoroughfares	70	Windowless structures	40
Urban thoroughfares	55	Mercantile occupancies	40
Institutional occupancies	50	Places of assembly	40
Vehicles and vessels	50	Residential occupancies	35
Business occupancies	45	Storage	30

Closed doors may drop the dB levels below those required to wake children or the hearing-impaired. Air conditioners, room humidifiers, and similar equipment may cause the noise levels to increase (55dBA typical). Sirens, horns, bells, and other indicating devices (appliances) may fail or be disabled by a fire. To reduce the effect of these disabled devices, multiple fire sounders should be considered. This may also help in providing ample dB levels to wake sleeping occupants throughout the structure (70dBA at the pillow is commonly accepted as sufficient to wake a sleeping person).

All residential alarm sounding devices must have a minimum dB rating of 85dBA at 10'. An exception to this rule is when more than one sounding device exists in the same room. A sounding device in the room must still be 85dBA at 10', but all additional sounding devices in that room may have a rating as low as 75dBA at 10'. Sound intensity doubles with each 3dB gain and is reduced by one-half with each 3dB loss. Doubling the distance to the sound source will cause a 6dB loss. Some other loss considerations are given in *Table 3*.

- *Coded versus non-coded signals* – A coded signal is a signal that is pulsed in a prescribed code for each round of transmission. For example, four pulses would indicate an alarm on the fourth floor. Temporal three is not a coded signal and is only intended to be a distinct, general fire alarm signal. A non-coded signal is a signal that is energized continuously by the control. It may pulse, but the pulsing will not be designed to indicate any code or message.
- *Visual appliance signals* – Notification signals for occupants to evacuate must be by audible and visible signals in accordance with *NFPA 72*® and *CABO/ANSI A117.1*. However, there may or may not be exceptions to this rule. Under the existing *NFPA 101*® building chapter, only audible signals are required in premises where:

- No hearing-impaired occupant is ever present under normal operation
- In hotels and apartments where special rooms are made available to the hearing-impaired
- Where the AHJ approves alternatives to visual signals (ADA codes may or may not allow these exceptions)

9.0.0 ◆ COMMUNICATIONS AND MONITORING

Communications is a means of sending information to personnel who are too far away to directly see or hear a fire alarm system's notification devices. It is the transmission and reception of information from one location, point, person, or piece of equipment to another. Understanding the information that a fire alarm system communicates makes it easier to determine what is happening at the alarm site. Knowing how that information gets from the alarm site to the monitoring site is helpful if a problem occurs somewhere in between.

9.1.0 Monitoring Options

There are several options for monitoring the signals of an alarm system. They include:

- *Central station* – A location, normally run by private individuals or companies, where operators monitor receiving equipment for incoming fire alarm system signals. The central station may be a part of the same company that sold and installed the fire alarm system. It is also common for the installing company to contract with another company to do the monitoring on its behalf.
- *Proprietary* – A facility similar to a central station except that the notification devices are located in a constantly staffed room maintained by the property owner for internal safety operations. The personnel may respond to alarms, alert local fire departments when alarms are activated, or both.
- *Certified central station* – Monitoring facilities that are constructed and operated according to a standard and are inspected by a listing agency to verify compliance. Several organizations, including the UL, publish criteria and list those central stations that conform to those criteria.

9.2.0 Digital Communicators

Digital communicators use standard telephone lines or wireless telephone service to send and

Table 3	Typical Sound Loss at 1,000Hz
Area	**Loss (dBA)**
Stud wall	41
Open doorway	4
Typical interior door	11
Typical fire-rated door	20
Typical gasketed door	24

receive data. Costs are low using this method because existing voice lines may be used, eliminating the need to purchase additional communication lines. Standard voice-grade telephone lines are also easier to repair than special fire alarm communication lines.

Digital communicators are connected to a standard, voice-grade telephone line through a special connecting device called the RJ31-X (*Figure 37*). The RJ31-X is a modular telephone jack into which a cord from the digital communicator is plugged. The RJ31-X separates the telephone company's equipment from the fire alarm system equipment and is approved by the Federal Communications Commission (FCC).

Although using standard telephone lines has several advantages, problems may arise if the customer and the fire alarm system both need the phone at the same time. In the event of an alarm, a technique known as line seizure gives the fire alarm system priority. The digital communicator is connected to the phones and can control or seize the line whenever it needs to send a signal. If the customer is using the telephone when the alarm system needs to send a signal, the digital communicator will disconnect the customer until the alarm signal has been sent. Once the signal is sent, the customer's phones are reconnected.

The typical sequence that occurs when the digital communicator for an alarm system is activated is shown in *Figure 38*. When an alarm is to be sent, the digital communicator energizes the seizure relay. The activated relay disconnects the

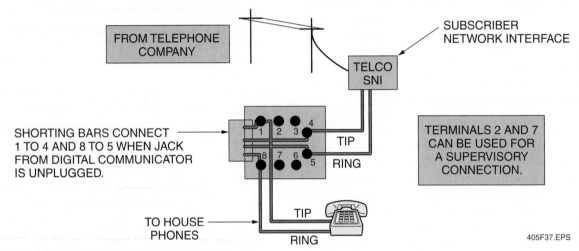

Figure 37 ◆ RJ31-X connection device.

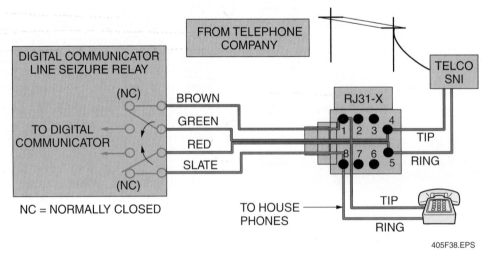

Figure 38 ◆ Line seizure.

house phones and connects the communicator to the Telco line. After the communicator detects a signal called the kiss-off tone, it de-energizes the seizure relay. This disconnects the communicator and reconnects the house phones.

In combination fire and security systems, the fire alarm supersedes the security alert. The RJ31-X will send the fire alarm signal before sending the security alert signal.

The RJ31-X is a modular connection and, like a standard telephone cord, it can unplug easily. If the cord remains unplugged from the jack, the digital communicator will be disconnected from the telephone line until the cord is reconnected. This creates a problem because the digital communicator cannot reach the digital receiver without the telephone line. This reduces the effectiveness of the alarm system, even though local notification devices are activated.

The NFPA uses the following terms to refer to digital communications:

- Digital Alarm Communicator Receiver (DACR) – This is a system component that will accept and display signals from Digital Alarm Communicator Transmitters (DACT) sent over the Public Switched Telephone Network.
- Digital Alarm Communicator System (DACS) – This is a system in which signals are transmitted from a DACT (located in the secured area) through the public switched telephone network to a DACR.
- Digital Alarm Communicator Transmitter (DACT) – This is a device that sends signals over the public switched telephone network to a DACR.

The following conditions have been established by NFPA 72® with regard to digital communicators:

- They can be used as a remote supervising station fire alarm system when acceptable to the AHJ.
- Only loop start and not ground start lines can be used.
- The communicator must have line seizure capability.
- A failure-to-communicate signal must be shown if ten attempts are made without getting through.
- They must connect to two separate phone lines at the protected premises. Exception: The secondary line may be a radio system (this does not apply to household systems).
- Failure of either phone line must be annunciated at the premises within four minutes of the failure.
- If long distance telephone service, including Wide Area Telephone Service (WATS) is used, the second telephone number shall be provided by a different long distance provider, where available.
- Each communicator shall initiate a test call to the central station at least once every 24 hours (this does not apply to household systems).

9.3.0 Cellular Backup

Some fire alarm systems that rely on phone lines for communications have a cellular backup system that utilizes wireless phone technology to restore communications in the event of a disruption in normal telephone line service (*Figure 39*).

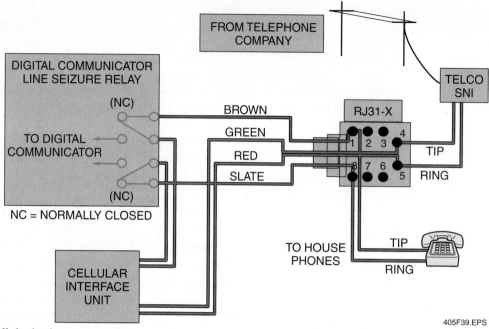

Figure 39 ◆ Cellular backup system.

405F39.EPS

10.0.0 ◆ GENERAL INSTALLATION GUIDELINES

This section contains general installation information applicable to all types of fire alarm systems. For specific information, always refer to the manufacturer's instructions, the building drawings, and all applicable local and national codes.

10.1.0 General Wiring Requirements

NEC Article 760 specifies the wiring methods and special cables required for fire protective signaling systems. The following special cable types are used in protective signaling systems:

- Power-limited fire alarm (FPL) cable
- Power-limited fire alarm riser (FPLR) cable
- Power-limited fire alarm plenum (FPLP) cable
- Nonpower-limited fire alarm (NPLF) circuit cable
- Nonpower-limited fire alarm riser (NPLFR) circuit riser cable
- Nonpower-limited fire alarm plenum (NPLFP) circuit cable

In addition to *NEC Article 760,* the following *NEC®* articles cover other items of concern for fire alarm system installations.

- *NEC Sections 110.11 and 300.6(A),(B), and (C), Corrosive, Damp, or Wet Locations*
- *NEC Section 300.21, Spread of Fire or Products of Combustion*

- *NEC Section 300.22, Ducts, Plenums, and Other Air Handling Spaces*
- *NEC Articles 500 through 516 and 517, Part IV, Locations Classified as Hazardous*
- *NEC Article 695, Fire Pumps*
- *NEC Article 725, Remote-Control and Signaling Circuits (Building Control Circuits)*
- *NEC Article 770, Fiber Optics*
- *NEC Article 800, Communications Circuits*
- *NEC Article 810, Radio and Television Equipment*

In addition to the *NEC®*, some AHJs may specify requirements that modify or add to the *NEC®*. It is essential that a person or firm engaged in fire alarm work be thoroughly familiar with the *NEC®* requirements, as well as any local requirements for fire alarm systems.

10.2.0 Workmanship

Fire alarm circuits must be installed in a neat and workmanlike manner. Cables must be supported by the building structure in such a manner that the cables will not be damaged by normal building use. One way to determine accepted industry practice is to refer to nationally recognized standards such as *Commercial Building Telecommunications Wiring Standard, ANSI/EIA/TIA 568; Commercial Building Standard for Telecommunications Pathways and Spaces, ANSI/EIA/TIA 569;* and *Residential and Light Commercial Telecommunications Wiring Standard, ANSI/EIA/TIA 570.*

10.3.0 Access to Equipment

Access to equipment must not be blocked by an accumulation of wires and cables that prevents removal of panels, including suspended ceiling panels.

10.4.0 Fire Alarm Circuit Identification

Fire alarm circuits must be identified at the control and at all junctions as fire alarm circuits. Junction boxes must be clearly marked as fire junction boxes to prevent confusion with commercial light and power. AHJs differ on what constitutes clear marking. Check the requirements before starting work. The following are examples of some AHJ-acceptable markings:

- Red painted cover
- The words Fire Alarm on the cover
- Red painted box
- The word Fire on the cover
- Red stripe on the cover

10.5.0 Power-Limited Circuits in Raceways

Power-limited fire circuits must not be run in the same cable, raceway, or conduit as high-voltage circuits. Examples of high-voltage circuits are electric light, power, and nonpower-limited fire (NPLF) circuits. When NPLF cables must be run in the same junction box, they must be run in accordance with the *NEC*®, including maintaining a ¼" spacing from Class 1, power, and lighting circuits.

10.6.0 Mounting of Detectors

Observe the following precautions when mounting detectors:

- Circuit conductors are not supports and must not be used to support the detector.
- Plastic masonry anchors should not be used for mounting detectors to gypsum drywall, plaster, or drop ceilings.
- Toggle or winged expansion anchors should be the minimum used for gypsum drywall.
- The best choice for mounting is an electrical box. Most equipment is designed to be fastened to a standard electrical box.

All fire alarm devices should be mounted to the appropriate electrical box as specified by the manufacturer. If not mounted to an electrical box, fire alarm devices must be mounted by other means as specified by the manufacturer. The use of plastic masonry anchors does not mean that a plastic anchor designed for use in drywall cannot be used in drywall. Masonry plastic anchors do not open as far on the end as drywall plastic anchors. Always read and follow the manufacturer's instructions; doing so is necessary to meet the requirements of the UL listing.

10.7.0 Outdoor Wiring

Fire alarm circuits extending beyond one building are governed by the *NEC*®. The *NEC*® sets the standards on the size of cable and methods of fastening required for cabling. Some manufacturers prohibit any aerial wiring. The *NEC*® also specifies clearance requirements for cable from the ground. Overhead spans of open conductors and open multi-conductor cables of not over 600V, nominal, must be at least 10' (3.05 m) above finished grade, sidewalks, or from any platform or projection from which they might be reached, where the supply conductors are limited to 150V to ground and accessible to pedestrians only. Additional requirements apply for areas with vehicle traffic. In addition, the *NEC*® states that fire alarm wiring can be attached to the building, but must be kept not less than 3' (914 mm) from the sides, top, bottom, and front of windows that are designed to be opened; doors; porches; balconies; ladders; stairs; fire escapes; and similar locations. However, conductors running above the top level of a window can be less than the 3' (914 mm) requirement.

10.8.0 Fire Seals

Electrical equipment and cables must not be installed in a way that might help the spread of fire. The integrity of all fire-rated walls, floors, partitions, and ceilings must be maintained. An approved sealant or sealing device must be used to fill all penetrations. Any wall that extends from the floor to the roof or from floor-to-floor should be considered a firewall. In addition, raceways and cables that go from one room to another through a fire barrier must be sealed.

10.9.0 Wiring in Air Handling Spaces

Wiring in air handling spaces requires the use of approved wiring methods, including:

- Special plenum-rated cable
- Flexible metal tubing (Greenfield)
- Electrical metallic tubing (EMT)
- Intermediate metallic conduit (IMC)
- Rigid metallic conduit (hard wall or Schedule 80 conduit)

Firestopping Materials

There are a wide variety of firestopping materials on the market. Shown here are just a few examples. Firestopping and fireproofing are not the same thing. Firestopping is intended to prevent the spread of fire and smoke from room to room through openings in walls and floors. Fireproofing is a thermal barrier that causes a fire to burn more slowly and retards the spread of fire.

FIRE SEALANT

CABLE PROTECTION SPRAY

INTUMESCENT PUTTY

INTUMESCENT PIPE SLEEVE

405SA07.EP

Standard cable tie straps are not permissible in plenums and other air handling spaces. Ties must be plenum rated. Bare solid copper wire used in short sections as tie wraps may be permitted by most AHJs. Fire alarm equipment is permitted to be installed in ducts and plenums only to sense the air. All splices and equipment must be contained in approved fire-resistant and low-smoke-producing boxes.

10.10.0 Wiring in Hazardous Locations

The *NEC®* includes requirements for wiring in hazardous locations. Some areas that are considered hazardous are listed below:

- *NEC Article 511, Commercial Garages, Repair and Storage*
- *NEC Article 513, Aircraft Hangars*
- *NEC Article 514, Gasoline Dispensing and Service Stations*
- *NEC Article 515, Bulk Storage Plants*
- *NEC Article 516, Spray Application, Dipping, and Coating*
- *NEC Article 517, Healthcare Facilities*
- *NEC Article 518, Places of Assembly*
- *NEC Article 520, Theaters and Similar Locations*
- *NEC Article 545, Manufactured Buildings*
- *NEC Article 547, Agricultural Buildings*

10.11.0 Remote Control Signaling Circuits

Building control circuits (*Figure 40*) are normally governed by *NEC Article 725*. However, circuit wiring that is both powered and controlled by the fire alarm system is governed by *NEC Article 760*. A common residential problem occurs when using a system-type fire alarm that is a combination burglar and fire alarm. Many believe that the keypad is only a burglar alarm device. If the keypad is also used to control the fire alarm, it is a fire alarm device, and the cable used to connect it to the control panel must comply with *NEC Article 760*. Also, motion detectors that are controlled and powered by the combination fire alarm and burglar alarm power must be wired in accordance with *NEC Article 760* (fire-rated cable).

10.12.0 Cables Running Floor to Floor

Riser cable is required when wiring runs from floor to floor. The cable must be labeled as passing a test to prevent fire from spreading from floor to floor. An example of riser cable is FPLR. This requirement does not apply to one- and two-family residential dwellings.

10.13.0 Cables Running in Raceways

All cables in a raceway must have insulation rated for the highest voltage used in the raceway. Power-limited wiring may be installed in raceways or conduit, exposed on the surface of a ceiling or wall, or fished in concealed spaces. Cable splices or terminations must be made in listed fittings, boxes, enclosures, fire alarm devices, or utilization equipment. All wiring must enter boxes through approved fittings and be protected against physical damage.

10.14.0 Cable Spacing

Power-limited fire alarm circuit conductors must be separated at least 2" (50.8 mm) from any electric light, power, Class 1, or nonpower-limited fire alarm circuit conductors. This is to prevent damage to the power-limited fire alarm circuits from induced currents caused by the electric light, power, Class 1, or nonpower-limited fire alarm circuits.

10.15.0 Elevator Shafts

Wiring in elevator shafts must directly relate to the elevator and be installed in rigid metallic conduit, rigid nonmetallic conduit, EMT, IMC, or up to 6' of flexible conduit.

10.16.0 Terminal Wiring Methods

The wiring for circuits using EOL terminations must be done so that removing the device causes a trouble signal (*Figure 41*).

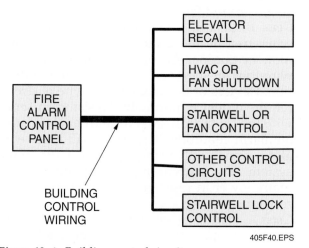

```
ELEVATOR
RECALL

HVAC OR
FAN SHUTDOWN

FIRE
ALARM          STAIRWELL OR
CONTROL        FAN CONTROL
PANEL

OTHER CONTROL
CIRCUITS

BUILDING
CONTROL        STAIRWELL LOCK
WIRING         CONTROL
```

405F40.EPS

Figure 40 ◆ Building control circuits.

Network Command Center

This PC-based network command center is used to display event information from local or wide area network devices in a text or graphic format. When a device initiates an alarm, the appropriate graphic floor plan is displayed along with operator instructions. Other capabilities of this command center include event history tracking and fire panel programming/control.

405SA08.EPS

10.17.0 Conventional Initiation Device Circuits

There are three styles of Class B and two styles of Class A conventional initiation device circuits listed in *NFPA 72®*. Various local codes address what circuit types are required. *NFPA 72®* describes how they are to operate. A brief explanation of some of the circuits follows:

- *Class B, Style A* – Fire alarm control panels (FACPs) using this style are no longer made in the United States, but a few systems remain in operation. Single (wire) opens and **ground faults** are the only types of trouble that can be indicated, and the system will not receive an alarm in a ground fault condition. An alarm is initiated with a wire-to-wire short.
- *Class B, Style B* – (See *Figure 42.*) The FACP is required to receive an alarm from any device up to a break with a single open. An alarm is initi-

ated with a wire-to-wire short. A trouble signal is generated for a circuit ground or open using an end-of-line (EOL) device that is usually a resistor. An alarm can also be received with a single ground fault on the system.

- *Class B, Style C* – (See *Figure 43.*) This style, while used in the United States, is more common in Europe. An open circuit, ground, or wire-to-wire short will cause a trouble indication. Devices or detectors in this type of circuit require a device (normally a resistor) in series with the contacts in order for the panel to detect an alarm condition. The panel will receive an alarm signal with a single ground fault on the system. The current-limiting resistor is normally lower in resistance than the end-of-line resistor.
- *Class A, Style D* – (See *Figure 44.*) In this type of circuit, an open or ground will cause a trouble signal. Shorting across the initiation loop will

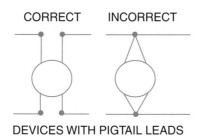

CORRECT   INCORRECT

DEVICES WITH PIGTAIL LEADS

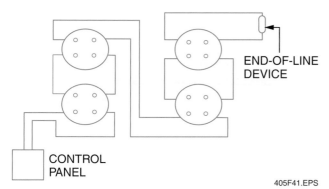

END-OF-LINE
DEVICE

CONTROL
PANEL

405F41.EPS

Figure 41 ◆ Correct wiring for devices with EOL terminations.

cause an alarm. Activation of any initiation device will result in an alarm, even when a single break or open exists anywhere in the circuit, because of the back loop circuit. The loop is returned to a special condition circuit, so there is no end of line and, therefore, no EOL device.

- *Class A, Style E –* (See *Figure 45.*) This style is an enhanced version of Style D. An open circuit, ground, or wire-to-wire short is a trouble condition. All devices require another device (normally a resistor) in series with the contacts to generate an alarm. Activation of any of the initiating devices will result in an alarm even if the initiation circuit has a single break or the system has a single ground.

The style of circuit can affect the following:

- The maximum quantity of each type of device permitted on each circuit
- The maximum quantity of circuits allowed for a fire alarm control panel/communicator
- The maximum quantity of buildings allowed for a signaling line circuit (SLC)
- The maximum quantity of signaling circuits and buildings allowed for a monitoring station

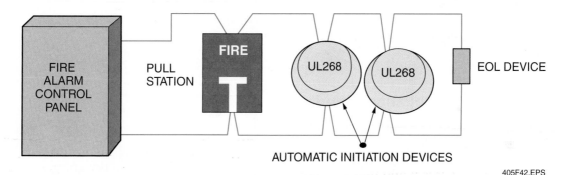

FIRE
ALARM
CONTROL
PANEL

PULL
STATION

FIRE

UL268   UL268

EOL DEVICE

AUTOMATIC INITIATION DEVICES

405F42.EPS

Figure 42 ◆ Typical Class B, Style B initiation circuit.

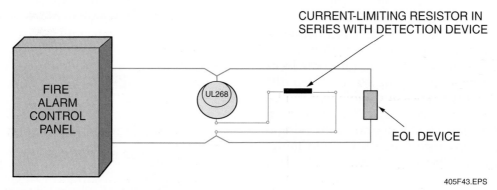

CURRENT-LIMITING RESISTOR IN
SERIES WITH DETECTION DEVICE

FIRE
ALARM
CONTROL
PANEL

UL268

EOL DEVICE

405F43.EPS

Figure 43 ◆ Typical Class B, Style C initiation circuit.

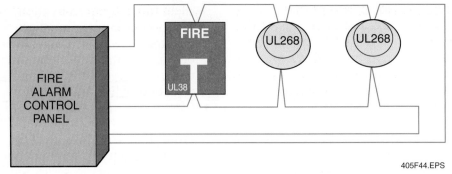

Figure 44 ◆ Typical Class A, Style D initiation circuit.

405F44.EPS

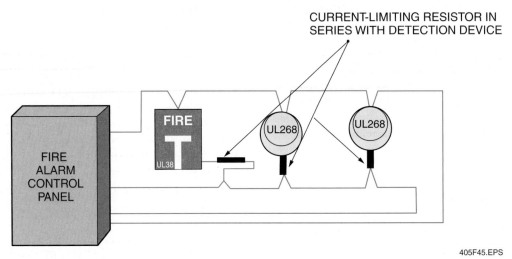

CURRENT-LIMITING RESISTOR IN
SERIES WITH DETECTION DEVICE

405F45.EPS

Figure 45 ◆ Typical Class A, Style E initiation circuit.

10.18.0 Notification Appliance Circuits

The following classes and styles of circuits are used for notification circuits (*Figure 46*):

- *Class B, Style W* – Class B, Style W is a two-wire circuit with an end-of-line device. Devices will operate up to the location of a **fault.** A ground may disable the circuit.
- *Class B, Style X* – Class B, Style X is a four-wire circuit. It has alarm capability with a single open, but not during a ground fault.
- *Class B, Style Y* – Class B, Style Y is a two-wire circuit with an end-of-line device. Devices on this style of circuit will operate up to the location of a fault. Ground faults are indicated differently than other circuit troubles.
- *Class A, Style Z* – Class A, Style Z is a four-wire circuit. All devices should operate with a single ground or open on the circuit. Ground faults are indicated differently than other circuit troubles.

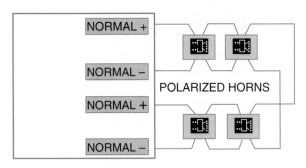

CLASS B, STYLE X OR CLASS A, STYLE Z

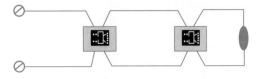

NOTIFICATION APPLIANCES

CLASS B, STYLES W AND Y

405F46.EPS

Figure 46 ◆ Typical notification appliance circuits.

Style X is similar to Style Z, except that during a ground fault, Style X will not operate and the panel will not be able to tell what type of trouble exists. Style W is similar to Style Y, except that during a ground fault, Style W will not operate and the panel will not be able to tell what type of trouble exists. Only Style Z operates all devices with a single open or a single ground fault. Only Style Z is a Class A notification appliance circuit (NAC).

All styles of circuits will indicate a trouble alarm at the premises with a single open and/or a single ground fault. Styles Y and Z have alarm capability with a single ground fault. Styles X and Z have alarm capability with a single open. In Class B circuits, the panel monitors whether or not the wire is intact by using an EOL device (wire supervision). Electrically, the EOL device must be at the end of the indicating circuit; however, Style X is an exception. Examples of EOL devices are resistors to limit current, diodes for polarity, and capacitors for filtering.

10.19.0 Primary Power Requirements

If more power is needed than can be supplied by one 20A circuit, additional circuits may be used, but none may exceed 20A. To help prevent system damage or false alarms caused by electrical surges or spikes, surge protection devices should be installed in the primary power circuits unless the fire alarm equipment has self-contained surge protection. In addition, some jurisdictions may require breaker locks and/or other means of identifying circuit breakers for FACPs.

10.20.0 Secondary Power Requirements

An approved generator supply or backup batteries must be used to supply the secondary power of a fire alarm system. The secondary power system must, upon loss of primary power, immediately keep the fire alarm functioning for at least as long as indicated in *Table 4*.

11.0.0 ◆ TOTAL PREMISES FIRE ALARM SYSTEM INSTALLATION GUIDELINES

This section covers the requirements for the proper installation, testing, and certification of fire alarm systems and related systems for totally protected premises.

11.1.0 Manual Fire Alarm Box (Pull Station) Installation

The following guidelines apply to the installation of a manual fire alarm box (pull station):

• Manual pull stations must be *UL 38*-listed or the equivalent. Multi-purpose keypads cannot be used as fire alarm manual pull stations unless *UL 38*-listed for that purpose.

• Manual pull stations must be installed in the natural path of escape, near each required exit from an area, in occupancies that require manual initiation. Ideally, they should be located near the doorknob edge of the exit door. In any case, they must be no more than 5' from the exit (*Figure 47*). In most cases, manual pull stations are installed with the actuators at the heights of 42" to 54" to conform to local ADA requirements. Most new pull stations are supplied with Grade II Braille on them for the visually impaired.

• The force required to operate manual pull stations must be no more than 5 foot-pounds.

• A manual pull station must be within 200' of horizontal travel on the same floor from any part of the building (*Figure 48*). If the distance is exceeded, additional pull stations must be installed on the floor.

Table 4	Secondary Power Duration Requirements	
NFPA Standard	**Maximum Normal Load**	**Maximum Alarm Load**
Central station	24 hours	See local system
Local system	24 hours	5 minutes
Auxiliary systems	60 hours	5 minutes
Remote stations	60 hours	5 minutes
Proprietary systems	24 hours	5 minutes
Household system	24 hours	4 minutes
Emergency voice alarm communications systems	24 hours	15 minutes maximum load 2 hours emergency operation

Wireless Smoke Detection System

The components that form a wireless smoke detection system are shown here. This type of system can be used in situations where the building design or installation costs make hard wiring of the detectors impractical or too expensive. The heart of this system is the translation unit, called a gateway. It can communicate with up to four remote receiver units. The receivers can monitor radio frequency signals from up to 80 wireless smoke detectors. Each receiver unit transmits the status of the related wireless detectors via communication wiring to the translator. The translator then communicates this status to an intelligent FACP via a signaling line circuit loop. (Note that some jurisdictions do not permit the use of these systems.)

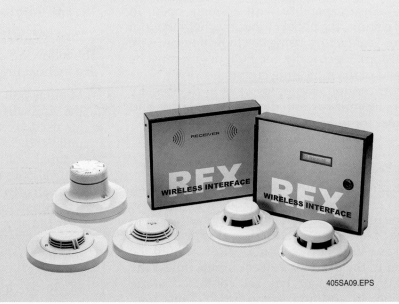

405SA09.EPS

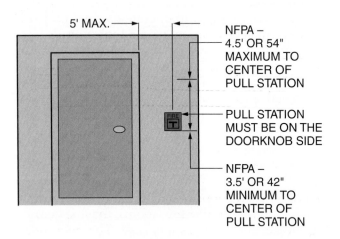

5' MAX.

NFPA –
4.5' OR 54"
MAXIMUM TO
CENTER OF
PULL STATION

PULL STATION
MUST BE ON THE
DOORKNOB SIDE

NFPA –
3.5' OR 42"
MINIMUM TO
CENTER OF
PULL STATION

405F47.EPS

Figure 47 ◆ Pull station location and mounting height.

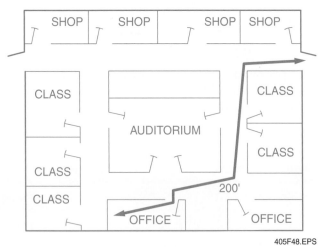

405F48.EPS

Figure 48 ◆ Maximum horizontal pull station distance from an exit.

11.2.0 Flame Detector Installation

When installing UV or IR flame detectors, the manufacturer's instructions and *NFPA 72®* should be consulted. The following should be observed when installing these detectors:

- *UV flame detectors*
  - Response is based on the distance from the fire, angle of view, and fire size.
  - While some units have a 180° field of view, sensitivity drops substantially with angles of more than 45° to 50°. Normally, the field of view is limited to less than 90° (*Figure 49*).
  - If used outdoors, make sure the unit is listed for outdoor use.
  - A UV detector is considered solar blind, but in order to prevent false alarms, it should never be aimed near or directly at any path that the sun can take.
  - Never aim the units into areas where electric arc welding or cutting may be performed.
- *IR flame detectors*
  - Response varies depending on the angle of view. At 45°, the sensitivity drops to 60% of the 0° sensitivity (*Figure 50*).

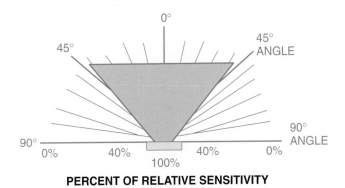

PERCENT OF RELATIVE SENSITIVITY

405F49.EPS

Figure 49 ◆ UV detector response.

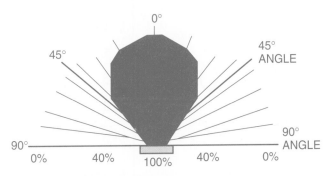

PERCENT OF RELATIVE SENSITIVITY

405F50.EPS

Figure 50 ◆ IR detector response.

- IR detectors cannot be used to detect alcohol, liquefied natural gas, hydrogen, or magnesium fires.
- IR detectors work best in low light level installations. High light levels desensitize the units. Discriminating units can tolerate up to ten footcandles of ambient light. Nondiscriminating units can tolerate up to two footcandles of ambient light.
- Never use the units outdoors.

11.3.0 Smoke Chamber Definition, Smoke Spread Phenomena, and Stratification Phenomena

The following sections cover smoke chamber definition, smoke spread phenomena, and **stratification** phenomena.

11.3.1 Smoke Chamber Definition

Before automatic smoke detectors can be installed, the area of coverage known as the smoke chamber must be defined. The smoke chamber is the continuous, smoke-resistant, perimeter boundary of a room, space, or area to be protected by one or more automatic smoke detectors between the upper surface of the floor and the lower surface of the ceiling. For the purposes of determining the area to be protected by smoke detectors, the smoke chamber is not the same as a smoke-tight compartment. It should be noted that some rooms that may have a raised floor and a false ceiling actually have three smoke chambers: the chamber beneath a raised floor, the chamber between the raised floor and the visible ceiling above, and the chamber between the room's visible ceiling and the floor above (or the lower portion of the roof). Few cases require detection in all these areas. However, some computer equipment rooms with a great deal of electrical power and communications cable under the raised floor may require detection in the chamber under the floor in addition to the chamber above the raised floor.

The simplest example of a smoke chamber would be a room with the door closed. If no intervening closed door exists between this and adjoining (communicating) rooms, the line denoting the barrier becomes less clear. The determining factor would be the depth of the wall section above the open archway or doorway, as follows:

- An archway or doorway that extends more than 18" down from the ceiling greatly delays smoke travel and is considered a boundary, just like a regular wall from floor to ceiling. A doorway that extends more than 4", but less than

18", must be considered a smoke barrier, and spacing of detectors must be reduced to ⅔ of the detectors' listed spacing distance on either side of the opening (*Figure 51*).

- Open grids above doors or walls that allow free flow of air and smoke are not considered barriers. To be considered an open grid, the opening must meet all of the requirements defined and demonstrated in *Figure 52*.
- Smoke doors, which are kept open by hold-and-release devices and meet all of the standards applicable thereto, may also be considered boundaries of a smoke chamber.
- If the space between the top of a low wall and the ceiling is less than 18", the wall is treated as if it extends to the ceiling and is a barrier. Smoke will still be able to travel to the other side of the wall, but it may be substantially delayed. This will delay notification. In this case, detector spacing must be reduced to ⅔ of the listed spacing on either side of the wall. If the space between the top of a low wall and the ceiling is 18" or more, the wall is not considered to substantially affect the smoke travel.

11.3.2 Smoke Spread and Stratification Phenomena

In a fire, smoke and heat rise in a plume toward the ceiling because they are lighter than the more dense, surrounding cooler air. In an area with a relatively low, flat, smooth ceiling, the smoke and heat quickly spread across the entire ceiling, triggering smoke or heat detectors. When the ceiling is irregular, smoke and heated air will tend to collect, perhaps stratifying near a peak or collecting in the bays of a beamed or joist ceiling.

In the case of a beamed or joist ceiling, the smoke and heated air will fill the nearest bays and begin to overflow to adjacent bays (*Figure 53*). The process continues until smoke and heat reach either a smoke or heat detector in sufficient quantity to cause activation. Because each bay must fill before it overflows, a substantial amount of time may pass before a detector placed at its maximum listed spacing activates. To reduce this time, the spacing for the detectors is reduced.

When smoke must rise a long distance, it tends to cool off and become denser. As its density becomes equal to that of the air around it, it stratifies, or stops rising (*Figure 54*). As a fire grows, additional heat is added, and the smoke will eventually rise to the ceiling. However, a great deal of time may be lost, the fire will be much larger, and a large quantity of toxic gases will be present at the floor level. Due to this delay, alternating detectors are lowered at least 3' (*Figure 55*). The science of stratification is very complex and requires a fire protection engineer's evaluation to determine if stratification is a factor and, if so, to determine how much to lower the detectors to compensate for this condition.

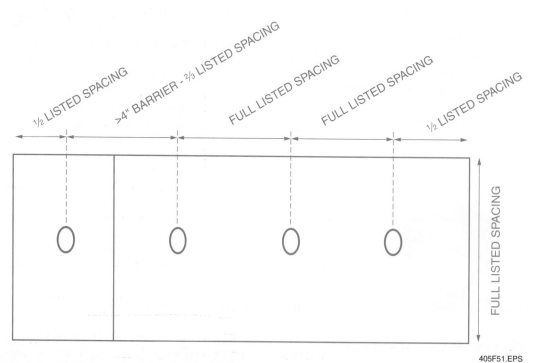

Figure 51 ◆ Reduced spacing required for a barrier.

405F51.EPS

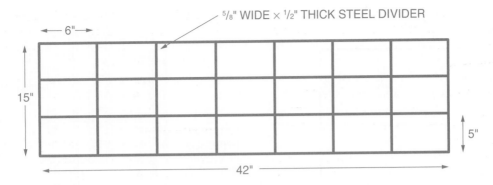

GRID OPENINGS MUST BE AT LEAST ¼" IN THE LEAST DIMENSION.
OPENINGS ARE 5³/₈" (6" MINUS ⁵/₈") × 4³/₈" (5" MINUS ⁵/₈").
REQUIREMENT MET.

THE THICKNESS OF THE MATERIAL DOES NOT EXCEED THE LEAST DIMENSION.
OPENINGS ARE 5³/₈" (6" MINUS ⁵/₈") × 4³/₈" (5" MINUS ⁵/₈"). THICKNESS IS ½".
REQUIREMENT MET.

THE OPENINGS CONSTITUTE AT LEAST 70% OF THE AREA OF THE PERFORATED MATERIAL.
OPENINGS ARE 5³/₈" (6" MINUS ⁵/₈") × 4³/₈" (5" MINUS ⁵/₈")
5³/₈" × 4³/₈" = 23½ sq. in.

21 OPENINGS × 23½ sq. in.   =   TOTAL OPENING = 493½ sq. in.
15" × 42" = 630 sq. in.   630 sq. in. × 0.70 = 441 sq. in.

REQUIREMENT MET.

405F52.EPS

Figure 52 ◆ Detailed grid definition.

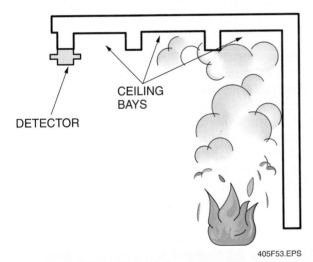

405F53.EPS

Figure 53 ◆ Smoke spread across a beamed or joist ceiling.

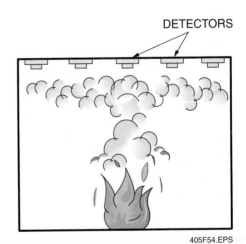

405F54.EPS

Figure 54 ◆ Smoke stratification.

Some conditions that cause stratification include:

- Uninsulated roofs that are heated by the sun, creating a heated air thermal block
- Roofs that are cooled by low outside temperatures, cooling the gases before they reach a detector

- HVAC systems that produce a hot layer of ceiling air
- Ambient air that is at the same temperature as the fire gases and smoke

There are no clear, set rules regarding which environments will and will not be susceptible to stratification. The factors and variables involved

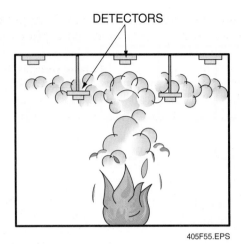

Figure 55 ◆ Smoke stratification countermeasure.

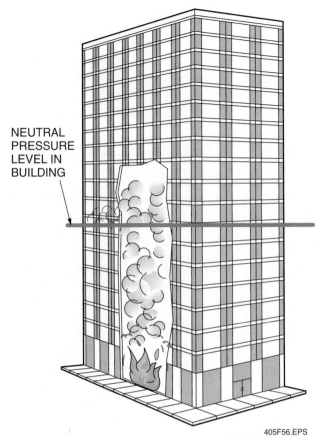

Figure 56 ◆ Stack effect in a high-rise building.

are beyond the scope of this course. For further reading on the subject, contact the NFPA reference department for a bibliography. Note that only alternating detectors are suspended for stratification. Some fires can result in stratification caused by superheated air and gases reaching the ceiling prior to the smoke, causing a barrier that holds the smoke below the detectors. Smoke detectors with integral heat detectors should be effective in such cases, without the need to suspend alternating detectors below the ceiling.

Stratification can also occur in high-rise buildings because of the stack effect principle. This occurs when smoke rises until it reaches a neutral pressure level, where it begins to stratify (*Figure 56*). Stack effect can move smoke from the source to distant parts of the building. Penetrations left unsealed can be a major contributor in such smoke migration. Stack effect factors include:

• Building height
• Air tightness
• Air leakage between floors
• Interior/exterior temperature differential
• Vertical openings
• Wind force and direction

Some factors that influence stack effect are variable, such as the weather conditions and which interior and exterior doors might be open. For this reason, the neutral pressure level may not be the same in a given building at different times. One item is very clear concerning stratification: Penetrations made in smoke barriers, particularly those in vertical openings such as shafts, must be sealed.

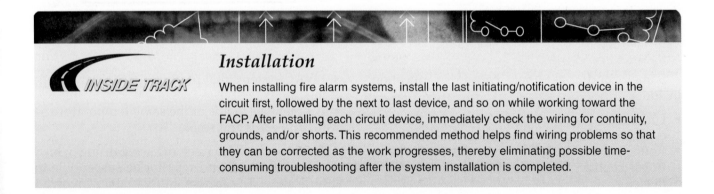

Installation

When installing fire alarm systems, install the last initiating/notification device in the circuit first, followed by the next to last device, and so on while working toward the FACP. After installing each circuit device, immediately check the wiring for continuity, grounds, and/or shorts. This recommended method helps find wiring problems so that they can be corrected as the work progresses, thereby eliminating possible time-consuming troubleshooting after the system installation is completed.

11.4.0 General Precautions for Detector Installation

The following are general precautions for detector installation:

- *Recessed mounting* – Detectors must not be recess-mounted unless specially listed for recessed mounting.
- *Air diffusers* – Air movement can have a number of undesirable effects on detectors. The introduction of air from outside the fire area can dilute the smoke, which delays activation. The air movement may create a barrier, which delays or prevents smoke from reaching the detector chamber. Smoke detectors are listed for a specific range of air velocity. This means that they are expected to function with air (and smoke) moving through the detection chamber at any speed within the listed range. Detectors may not function properly if air movement is above or below the listed velocity. Air movement can be measured with an instrument called a velocimeter. Unless the airflow exceeds the listed velocity, spot-type smoke detectors must be a minimum of 3' from air diffusers.
- *Problem locations and sources of false alarms* – Detectors, especially smoke detectors, can register false alarms as a result of the following:
  - *Electrical interference:* Keep detector locations away from fluorescent lights and radio transmitters, including cellular phones. Electrical noise or radio frequencies radiated from these devices can cause false alarms.
  - *Heating equipment:* High temperatures, dust accumulation, improper exhaust, and incomplete combustion from heating equipment are problems for detectors.
  - *Engine exhaust:* Exhaust from engine-powered forklifts, vehicles, and generators is a potential problem for ionization-type smoke detectors.
  - *Solvent and chemical fumes:* Cleaning solvents and adhesives are a problem for ionization detectors.
  - *Other gases and fumes:* Fumes from machining operations, paint spraying, industrial or battery gases, curing ovens/dryers, cooking equipment, sawing/drilling/grinding operations, and welding/cutting operations cause problems for smoke detectors.
  - *High temperatures:* Avoid very hot or cold environments for smoke detector locations. Temperatures below 32°F (0°C) can cause false alarms, and temperatures above 120°F (48.8°C) can prevent proper smoke

detector operation. Extreme temperatures affect beam, ionization, and photoelectric detectors.
  - *Dampness or humidity:* Smoke detectors must be located in areas where the humidity is less than 93%. In ionization detectors, dampness and high humidity can cause tiny water droplets to condense inside the sensing chamber, making it overly sensitive and causing false alarms. In photoelectric detectors, humidity can cause light refraction and loss of current flow, either of which can lead to false alarms. Common sources of moisture to avoid include slop sinks, steam tables, showers, water spray operations, humidifiers, and live steam sources.
  - *Lightning:* Nearby lightning can cause electrical damage to a fire alarm system. It may also cause electrical noise or spikes to be induced in the alarm system wiring or detectors, resulting in false alarms. Surge arrestors installed in the system's primary power supply to protect the system, in conjunction with alarm verification, can reduce the chance of system damage and false alarms.
  - *Dusty or dirty environments:* Dust and dirt can accumulate on a smoke detector's sensing chamber, making it overly sensitive. Avoid areas where fumigants, fog, dust, or mist-producing materials are consistently used.
  - *Outdoor locations:* Dust, air currents, and humidity typically affect outdoor structures, including sheds, barns, stables, and other open structures. This makes outdoor structures unsuitable for smoke detectors.
  - *Insect-infested areas:* Insects in a smoke detector can cause a false alarm. Good bug screens on a detector can prevent most adult insects from entering the detector. However, newly hatched insects may still be able to enter. An insecticide strip next to the detector may help solve the problem, but it may also cause false alarms because of fumes. Check with the manufacturer for the use of an approved strip. Ionization detectors are less prone to false alarms from insects.
  - *Construction:* Smoke detectors must not be installed until after construction cleanup unless required by the AHJ for protection during construction. Detectors that are installed prior to construction cleanup must be cleaned or replaced. Prior to 1993, the code permitted covering smoke detectors to protect them from dirt, dust, or paint mist. However, the covers did not work very well and resulted in clogged, oversensitive, and damaged detectors.

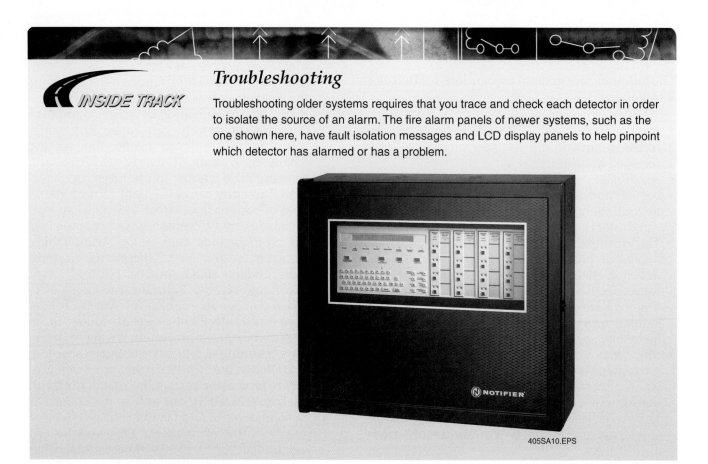

Troubleshooting

INSIDE TRACK

Troubleshooting older systems requires that you trace and check each detector in order to isolate the source of an alarm. The fire alarm panels of newer systems, such as the one shown here, have fault isolation messages and LCD display panels to help pinpoint which detector has alarmed or has a problem.

405SA10.EPS

11.5.0 Spot Detector Installations on Flat, Smooth Ceilings

Flat, smooth ceilings are defined as ceilings that have a slope equal to or less than 1.5" per foot (1' to 8') and do not have open joists or beams.

11.5.1 Conventional Installation Method

The following applies to conventional spot detector installation on flat, smooth ceilings:

- When a smoke detector manufacturer's specifications do not specify a particular spacing, a 30' spacing guide may be applied.
- The distance between heat detectors must not exceed their listed spacing.
- There must be detectors located within one-half the listed spacing measured at right angles from all sidewalls.

NOTE

This section assumes detectors listed for 30' spacing mounted on a smooth and flat ceiling of less than 10' in height, where 15' is one-half the listed spacing. Spacing is reduced as indicated in later sections when the **ceiling height** exceeds 10', or the ceiling is not smooth and flat as defined by the code.

In the following example, the detector locations for a simple room that is 30' × 60' long will be determined. The locations are determined by the intersection of columns and rows marked on a sketch of the ceiling:

- The first column is located by a line that is parallel to the end wall and not more than one-half the listed spacing from the end wall (*Figure 57*).
- The first row is located by a line parallel to the sidewall that is not more than one-half the listed spacing from that sidewall (*Figure 58*).
- The first detector is located at the intersection of the row and column lines.

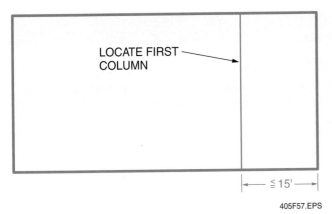

Figure 57 ◆ Locating the first column.

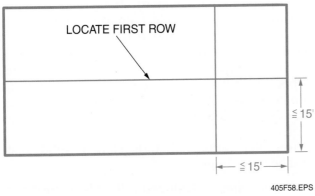

Figure 58 ◆ Locating the first row.

- The second column is located by a line that is parallel to the opposite end wall and not more than one-half the listed spacing from the end wall.
- The second detector is located at the intersection of the row and second column lines, provided that the distance between the first and second detectors does not exceed the listed spacing (*Figure 59*). The only time that the full listed spacing of any detector is used is when measuring from one detector to the next.

11.6.0 Photoelectric Beam Smoke Detector Installations on Flat, Smooth Ceilings

Two configurations of beam detectors are used for open area installations. Beam smoke detectors can be installed as straight-line devices or as angled-beam devices that employ mirrors. When used as angled-beam devices, the beam length is reduced for the number of mirrors used, as specified in *NFPA 72*®. When installing either type of beam smoke detector, always follow the manufacturer's instructions.

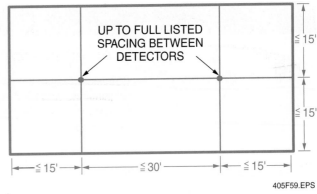

Figure 59 ◆ First and second detector locations.

11.6.1 Straight-Line Beam Detector Installations

Photoelectric beam smoke detectors have basically two listings: one for width of coverage and one for the minimum/maximum length of coverage (beam length), as shown in *Figure 60*. The beam length is listed by the manufacturer and, in most cases, so is the width coverage spacing (S). When a manufacturer does not list the width coverage, the 60' guideline specified in *NFPA 72*® must be used as the width coverage. The distance of the beam from the ceiling should normally be between 4" and 12". However, NFPA allows a greater distance to compensate for stratification. Always make sure the beam does not cross any expansion joint or other point of slippage that could eventually cause misalignment of the beam. In large areas, parallel beam detectors may be installed, separated by no more than their listed spacing (S).

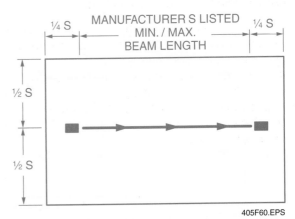

Figure 60 ◆ Maximum straight-line single-beam smoke detector coverage (ceiling view).

Beam-Type Detectors

Objects other than smoke can interfere with beam-type detectors. For example, balloons released by children in malls have been known to activate the fire protection system.

11.6.2 Angled-Beam Installations Using Mirrors

The use of mirrors to direct the light beam from the transmitter in a path other than a single straight line is permitted if the manufacturer's directions are followed and the mirror used is listed for the model of the beam detector being installed. The use of a mirror will require a reduction in the listed beam length of the detector. When using a mirror, the beam length is the distance between the transmitter and receiver (measured from the transmitter to the mirror, plus the distance from the mirror to the receiver). This distance is reduced to 66% of the manufacturer's listed beam length for a single mirror, and to 44% of the manufacturer's listed beam length for two mirrors. Maximum coverage for a two-mirror beam detector is shown in *Figure 61*. Width spacing (S) is defined as previously stated for single-beam and double-beam installations. The planner should also allow for additional installation and service time when mirrors are used. As with beam detectors without mirrors, make sure that the light beam does not cross a building expansion joint.

(A + B + C) MAY NOT EXCEED ⁴⁄₉ (44%) OF THE MANUFACTURER'S LISTED BEAM LENGTH

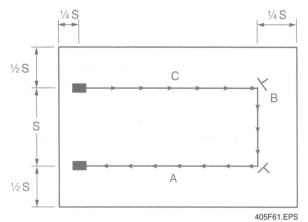

405F61.EPS

Figure 61 ◆ Coverage for a two-mirror beam detector installation.

11.7.0 Spot Detector Installations on Irregular Ceilings

Any ceiling that is not flat and smooth is considered irregular. For heat detectors, ceilings of any description that are over 10' above floor level are considered irregular. Heat detectors located on any ceiling over 10' above finished floor (AFF) level must have their spacing reduced below their listed spacing in accordance with *NFPA 72®*. This is because hot air is diluted by the surrounding cooler air as it rises, which reduces its temperature. Smoke detector spacing is not adjusted for high ceilings.

Irregular ceilings include sloped, solid joist, and beam ceilings:

- *Sloped ceilings* – A sloped ceiling is defined as a ceiling having a slope of more than 1.5" per foot (1' to 8'). Any smooth ceiling with a slope equal to or less than 1.5" per foot is considered a flat ceiling. Sloped ceilings are usually shed or peaked types.
  - *Shed:* A shed ceiling is defined as having the high point at one side with the slope extending toward the opposite side.
  - *Peaked:* A peaked ceiling is defined as sloping in two directions from its highest point. Peaked ceilings include domed or curved ceilings.
- *Solid joist or beam ceilings* – Solid joist or beam ceilings are defined as being spaced less than 3' center-to-center with a solid member extending down from the ceiling for a specified distance. For heat detectors, the solid member must extend more than 4" down from the ceiling. For smoke detectors, the solid member must extend down more than 8".

11.7.1 Shed Ceiling Detector Installation

Shed ceilings having a rise greater than 1.5"/1' run (1'/8') must have the first row of detectors (heat or smoke) located on the ceiling within 3' of the high side of the ceiling (measured horizontally from the sidewall).

Heat detectors must have their spacing reduced in areas with high ceilings (over 10'). Smoke detector spacing is not adjusted due to the ceiling height.

For a roof slope of less than 30°, determined as shown in *Figure 62*, all heat detector spacing must be reduced based on the height at the peak. For a roof slope of greater than 30°, the average ceiling height must be used for all heat detectors other than those located in the peak. The average ceiling height can be determined by adding the high side-wall and low sidewall heights together and dividing by two. In the case of slopes greater than 30°, the spacing to the second row is measured from the first row and not the sidewall for shed ceilings. See *NFPA 72*® for heat detector spacing reduction based on peak or average ceiling heights.

Once you have determined that the ceiling you are working with is a shed ceiling, use the following guidelines for determining detector placement:

- Place the first row of detectors within 3' of the high sidewall (measured horizontally).
- Use the listed spacing (adjusted for the height of the heat detectors on ceilings over 10') for each additional row of detectors (measured from the detector location, not the sidewall).
- For heat detectors on ceilings over 10' high, adjust the spacing per *NFPA 72*®.
- If the slope is less than 30°, all heat detector spacing is based on the peak height. If the slope is more than 30°, peak heat detectors are based on the peak height, and all other detectors are based on the average height.
- Additional columns of detectors that run at right angles to the slope of the ceiling do not need to be within 3' of the end walls.
- Smoke detector spacing is not adjusted for ceiling height.

To calculate the average ceiling height, use one of the following methods:

Method 1:

Step 1 Subtract the height of the low ceiling from the height of the high ceiling.

Step 2 Divide the above result by 2.

Step 3 Subtract the above result from the height of the high ceiling.

Method 2:

Step 1 Add the height of the low ceiling to the height of the high ceiling.

Step 2 Divide the above result by 2.

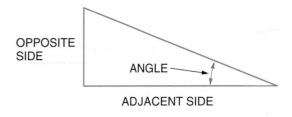

DETERMINING IF A SLOPE IS GREATER THAN 30°

OPPOSITE SIDE

ANGLE

ADJACENT SIDE

If the opposite side divided by the adjacent side is > 0.5774, the angle is > 30°.

The tangent of the smallest angle of a right triangle equals the opposite side divided by the adjacent side.

The sum of all the angles of a triangle equals 180°.

The tangent of a 30° angle is 0.5774.

405F62.EPS

Figure 62 ◆ Determining degree of slope.

11.7.2 Peaked Ceiling Detector Installation

A ceiling must be sloped as defined by the code in order to be considered a peaked ceiling. It must slope in more than one direction from its highest point. Because domed and curved ceilings do not have clean 90° lines to clearly show that the ceiling slopes in more than one direction, they should be viewed from an imaginary vertical center line. From this perspective, it can be seen if the ceiling slopes in more than one direction.

- The first row of detectors (heat or smoke) on a peaked ceiling should be located within 3' of the peak, measured horizontally. The detectors may be located alternately on either side of the peak, if desired.
- Regardless of where in the 3' space on either side of the peak the detector is located, the measurement for the next row of heat detectors is taken from the peak (measured horizontally, not from the heat detector located near the peak). This is different from the way detector location is determined for shed ceilings.
- Heat detectors must have their spacing reduced in areas with ceilings over 10'. Smoke detector spacing is not adjusted due to the ceiling height.
- As with sloped ceilings, use of the peak height for just the row of heat detectors at the peak, or for all the ceiling heat detectors, depends on whether or not the ceiling is sloped more than 30°. Use the peak height to adjust heat detector spacing of all the room's heat detectors when the slope of the ceiling is less than 30°. When the slope is greater than 30°, use the peak height for only the heat detectors located within 3' of the peak. Use the average height for all other heat detectors in the room.

Smoke detectors should be located on a peaked ceiling such that:

- A row of detectors is within 3' of the peak on either side or alternated from side to side.
- The next row of detectors is within the listed spacing of the peak measured horizontally. Do not measure from the detectors within 3' of the peak. It does not matter where within 3' of the peak the first row is; the next row on each side of the peak is installed within the listed spacing of the detector from the peak.
- Additional rows of detectors are installed using the full listed spacing of the detector.
- The sidewall must be within one-half the listed spacing of the last row of detectors.
- Columns of detectors are installed in the depth dimension of the above diagrams using the full listed spacing (the same as on a smooth, flat ceiling).
- There is no need to reduce smoke detector spacing for ceilings over 10'.

11.7.3 Solid Joist Ceiling Detector Installation

Solid joists are defined for heat detectors as being solid members that are spaced less than 3' center-to-center and that extend down from the ceiling more than 4". Solid joists are defined for smoke detectors as being solid members that are spaced less than 3' center-to-center and that extend down from the ceiling more than 8".

Heat detector spacing at right angles to the solid joist is reduced by 50%. In the direction running parallel to the joists, standard spacing principles are applied. If the ceiling height exceeds 10', spacing is adjusted for the high ceiling in addition to the solid joist as defined in *NFPA 72®*.

Smoke detector spacing at right angles to the solid joists is reduced by one-half the listed spacing for joists 1' or less in depth and ceilings 12' or less in height. In the direction running parallel to the joists, standard spacing principles are applied. If the ceiling height is over 12' or the depth of the joist exceeds 1', **spot-type detectors** must be located on the ceiling in every pocket. Additional reductions for sloped joist ceilings also apply as defined in *NFPA 72®*.

11.7.4 Beamed Ceiling Detector Installation

Beamed ceilings consist of solid structural or solid nonstructural members projecting down from the **ceiling surface** more than 4" and spaced more than 3' apart center-to-center as defined by *NFPA 72®*.

- *Heat detector installation:*
  - If the beams project more than 4" below the ceiling, the spacing of spot-type heat detectors at right angles to the direction of beam travel must not be more than two-thirds the smooth ceiling spacing.
  - If the beams project more than 12" below the ceiling and are more than 8' on center, each bay formed by the beams must be treated as a separate area.
  - Reductions of heat detector spacing in accordance with *NFPA 72®* are also required for ceilings over 10' in height.
- *Smoke detector installation:*
  - For smoke detectors, if beams are 4" to 12" in depth with an AFF level of 12' or less, the spacing of spot-type detectors in the direction perpendicular to the beams must be reduced 50%, and the detectors may be mounted on the bottoms of the beams.
  - If beams are greater than 12" in depth with an AFF level of more than 12', each beam bay must contain its own detectors. There are additional rules for sloped beam ceilings as defined in *NFPA 72®*.
  - The spacing of projected light beam detectors that run perpendicular to the ceiling beams need not be reduced. However, if the projected light beams are run parallel to the ceiling beams, the spacing must be reduced per *NFPA 72®*.

11.8.0 Notification Appliance Installation

There are several types of notification appliances:

- Audible devices such as bells, horns, chimes, speakers, sirens, and mini-sounders
- Visual devices such as strobe lights
- Tactile (sense of touch) devices such as bed shakers and special sprinkling devices
- Olfactory (sense of smell) devices, which are also accepted by the code

The appropriate devices for the occupancy must be determined. All devices must be listed for the purposes for which they are used. For example, *UL 1480* speakers are used for fire protective signaling systems, and *UL 1638* visual signaling appliances are used for private mode emergency and general utility signaling. While *NFPA 72®* recognizes tactile and olfactory devices, it does not specify installation requirements. If an occupant requires one of these types of notification, use equipment listed for the purpose and follow the manufacturer's instructions. As always, consult the local AHJ.

11.8.1 Notification Device Installation

The following guidelines apply to the installation of notification devices:

- Ensure that notification devices are wired using the applicable circuit style (Class A or B).
- In Class B circuits, the panel supervises the wire using an end-of-line device. Electrically, the EOL device must be at the end of the indicating circuit. Examples of EOL devices are resistors to limit current, diodes for polarity, and capacitors for filtering.
- It is extremely important that any polarized devices be installed correctly. The polarization of the leads or terminals of these devices are marked on the device or are noted in the manufacturer's installation data. If the leads of a polarized notification device are reversed, the panel will not detect the problem, and the device will not activate. Moreover, the device will act as an EOL device, preventing the panel from detecting breaks between the device and the actual EOL device. Because it is very easy to accidentally wire a device backwards, testing every device is extremely important. The general alarm should be activated and every notification device checked for proper operation.
- Sidewall-mounted audible notification devices must be mounted at least 90" AFF level or at least 6" below the finished ceiling. Ceiling-mounting and recessed appliances are permitted.
- **Visible notification appliances** must be mounted at the minimum heights of 80" to 96" AFF (*NFPA 72®*) or, for ADA requirements, either 80" AFF or 6" below the ceiling. In any case, the device must be within 16' of the pillow in a sleeping area. Combination audible/visible appliances must follow the requirements for visible appliances. Non-coded visible appliances should be installed in all areas where required by *NFPA 101®* or by the local AHJ. Consult ADA codes for the required illumination levels in sleeping areas.
- *Visual notification appliance spacing in corridors* – *Table 5* provides spacing requirements for corridors less than 20' wide. For corridors and rooms greater than 20', refer to *Tables 6* and *7*. In corridor applications, visible appliances must be rated at not less than 15 candelas (cd). Per *NFPA 72®*, visual appliances must be located no more than 15' from the end of the corridor with a separation of no more than 100' between appliances. Where there is an interruption of the concentrated viewing path, such as a fire door or elevation change, the area is to be considered as a separate corridor.
- *Visual notification appliance spacing in other applications* – The light source color for visual appliances must be clear or nominal white and must not exceed 1,000cd (*NFPA 72®*). In addition, special considerations apply when more than one visual appliance is installed in a room or corridor. *NFPA 72®* specifies that the separation between appliances must not exceed 100'. Visible notification appliances must be installed in accordance with *Tables 6* and *7*, using one of the following:
  - A single visible notification appliance
  - Two visible notification appliances located on opposite walls
  - More than two appliances for rooms 80' × 80' or larger (must be spaced a minimum of 55' from each other)

NOTE

More than two visual notification appliances within the field of view are required to flash in synchronization.

Table 5 Visual Notification Devices Required for Corridors not Exceeding 20'

Corridor Length (in ft.)	Minimum Number of 15cd Appliances Required
0–30	1
31–130	2
131–230	3
231–330	4
331–430	5
431–530	6

UL 1971 Listing

A *UL 1971* listing for a visual notification appliance indicates that it meets the ADA hearing-impaired requirements.

Table 6 Room Spacing for Wall-Mounted Visual Notification Appliances

Maximum Room Size (in ft.)	Minimum Required Light Output in Candelas (cd)		
	One Light per Room	Two Lights per Room*	Four Lights per Room**
20 × 20	15	Not allowable	Not allowable
30 × 30	30	15	Not allowable
40 × 40	60	30	Not allowable
50 × 50	95	60	Not allowable
60 × 60	135	95	Not allowable
70 × 70	185	95	Not allowable
80 × 80	240	135	60
90 × 90	305	185	95
100 × 100	375	240	95
110 × 110	455	240	135
120 × 120	540	305	135
130 × 130	635	375	185

*Locate on opposite walls
**One light per wall

Table 7 Room Spacing for Ceiling-Mounted Visual Notification Appliances

Maximum Room Size (in ft.)	Maximum Ceiling Height (in ft.)*	Minimum Required Light Output for One Light (cd)**
20 × 20	10	15
30 × 30	10	30
40 × 40	10	60
50 × 50	10	95
20 × 20	20	30
30 × 30	20	45
40 × 40	20	80
50 × 50	20	95
20 × 20	30	55
30 × 30	30	75
40 × 40	30	115
50 × 50	30	150

*Where ceiling heights exceed 30', visible signaling appliances must be suspended at or below 30' or wall mounted in accordance with *NFPA 72®*.
**This table is based on locating the visible signaling appliance at the center of the room. Where it is not located at the center of the room, the effective intensity (cd) must be determined by doubling the distance from the appliance to the farthest wall to obtain the maximum room size.

11.9.0 Fire Alarm Control Panel Installation Guidelines

The guidelines for installing a fire alarm control panel are as follows:

• When not located in an area that is continuously occupied, all fire alarm control equipment must be protected by a smoke detector, as shown in *Figure 63*. If the smoke detector is not designed to work properly in that environment, a heat detector must be used. It is not necessary to protect the entire space or room.

• Detector spacing must be adjusted if the ceiling over the control equipment is irregular in one or more respects. This is considered protection against a specific hazard under *NFPA 72®* and does not require the entire chamber (room) containing the control equipment to be protected under the 0.7 rule.

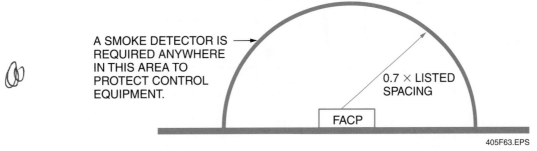

A smoke detector protecting control equipment must meet *NFPA 72* spacing and placement standards, but the entire space (room) containing the FACP need not be protected.

A SMOKE DETECTOR IS REQUIRED ANYWHERE IN THIS AREA TO PROTECT CONTROL EQUIPMENT.

0.7 × LISTED SPACING

FACP

405F63.EPS

Figure 63 ◆ Protection of an FACP.

- A means of silencing audible notification appliances from an FACP must be protected against unauthorized use. Most FACPs are located inside locked metal cabinets as shipped from the manufacturer. If the silencing switch is key-actuated or is locked within the cabinet, this provision should be considered satisfied. If the silencing means is within a room that is restricted to authorized use only, no additional measures should be required. The NFPA codes do not clearly define nor specify unauthorized or authorized use.
- FACP connections to the primary light and power circuits must be on a dedicated branch circuit with overcurrent protection rated at 20A or less. Any connections to the primary power circuit on the premises connected after the distribution panel (circuit breaker box) must be directly related to the fire alarm system. No other use is permitted. This requirement does not necessitate a direct tap into the power circuit ahead of the distribution panel, although connecting ahead of the main disconnect is acceptable with listed service equipment.
- The power disconnection means (circuit breaker) must be clearly identified as a fire alarm circuit control.

12.0.0 ◆ FIRE ALARM–RELATED SYSTEMS AND INSTALLATION GUIDELINES

This section discusses various fire alarm–related systems, as well as the installation guidelines for each system.

12.1.0 Ancillary Control Relay Installation Guidelines

Ancillary functions, commonly called auxiliary functions, include such controls as elevator capture (recall), elevator shaft pressurization, HVAC system shutdown, stairwell pressurization, smoke management systems, emergency lighting, door unlocking, door hold-open device control, and building music system shutoff. For example, sound systems are commonly powered down by the fire alarm system so that the evacuation signal may be heard. In the normal state (*Figure 64*), an energized relay completes the power circuit to the device being controlled.

NOTE

The circuit from the relay to the background music system is a remote control signaling circuit (see *NEC Article 725*). If this circuit is both powered by and controlled by the fire alarm, the circuit is a fire alarm circuit (see *NEC Article 760*).

12.2.0 Duct Smoke Detectors

Duct smoke detectors are not simply conventional detectors applied to HVAC systems. Conventional smoke detectors are listed for open area protection (OAP) under *UL 268*. Duct smoke detectors are normally listed for a slightly higher velocity of air movement and are tested under *UL 268A*. The primary function of a duct smoke detector is to turn off the HVAC system. This prevents the system from spreading smoke rapidly throughout the building and stops the system from providing a forced supply of oxygen to the fire. Duct smoke detectors are not intended primarily for early warning and notification. Relevant fire alarm-related provisions pertaining to HVAC systems can be found in *NFPA 90A* and *B*.

NFPA 90A covers when and where it will apply, which includes each of the following types of buildings:

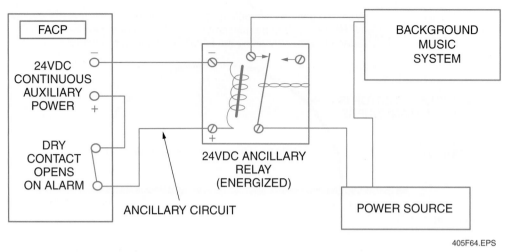

Figure 64 ◆ Music system control in normal state.

- HVAC systems serving spaces over 25,000 cubic feet in volume
- Buildings of Type III, IV, or V construction that are over three stories in height (see *NFPA 220*)
- Buildings that are not covered by other applicable standards
- Buildings that serve occupants or processes not covered by other applicable standards

NFPA 90B covers when and where it will apply, which includes:

- HVAC systems that service one- or two-family dwellings
- HVAC systems that service spaces not exceeding 25,000 cubic feet in volume

NOTE

No duct smoke detector requirements are found under *NFPA Standard 90B.*

12.2.1 Duct Detector Location

When determining the location of duct detectors prior to installation, use the following guidelines:

- In HVAC units over 2,000 cubic feet per minute (cfm), duct detectors must be installed on the supply side.
- Duct detectors must be located downstream of any air filters and upstream of any branch connection in the air supply.
- Duct detectors should be located upstream of any in-duct heating element.
- Duct detectors must be installed at each story prior to the connection to a common return and prior to any recirculation of fresh air inlet in the

air return of systems over 15,000 cfm serving more than one story.

- Return air system smoke detectors are not required when the entire space served by the HVAC system is protected by a system of automatic smoke detectors and when the HVAC is shut down upon activation of any of the smoke detectors.

12.2.2 Conversion Approximations

The approximations given in *Table 8* are useful when the protected-premises personnel do not know the cfm rating of an air-handling unit, but do know either the tonnage or British thermal unit (Btu) rating. Additionally, the cfm rating may not always appear on air handling unit (AHU) nameplates or in building specifications.

12.2.3 More Than One AHU Serving an Area

When more than one air handling unit (AHU) is used to supply air to a common space, and the return air is drawn from this common space, the total capacity of all units must be used in determining the size of the HVAC system (*Figure 65*).

Table 8	Conversions	
Capacity Rating	**CFM**
1 ton	400
12,000 Btus	400
5 tons	2,000
60,000 Btus	2,000
37.5 tons	15,000
450,000 Btus	15,000

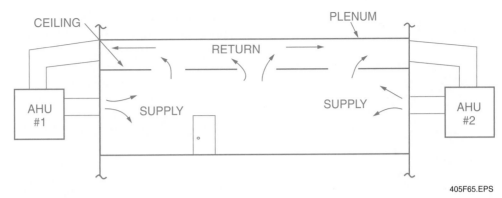

405F65.EPS

Figure 65 ◆ Non-ducted multiple AHU system.

This formal interpretation makes clear the fact that interconnected air handling units should be viewed as a system, as opposed to treating each AHU individually. When multiple AHUs serve a common space, the physical location of each AHU, relative to others interconnected to the same space, is irrelevant to the application of the formal interpretation.

The common space being served by multiple air handling units need not be contiguous (connected) for the formal interpretation to apply. In *Figure 66*, AHUs #3 and #4 serve common space and must be added together for consideration (1,100 + 1,100 = 2,200). Because 2,200 is greater than 2,000, duct smoke detectors are required on both. The same is true for AHU #1 and #2.

Duct smoke detectors may not be used as substitutes where open area detectors are required. This is because the HVAC unit may not be running when a fire occurs. Even if the fan is always on, the HVAC is not a listed fire alarm device.

Duct smoke detectors must automatically stop their respective fans upon detecting smoke. It is

also acceptable for the fire alarm control panel to stop the fan(s) upon activation of the duct detector. However, fans that are part of an engineered smoke control or management system are an exception and are not always shut down in all cases when smoke is detected.

12.2.4 Duct Smoke Detector Installation and Connections

When a fire alarm system is installed in a building, all duct smoke detectors in that building must be connected to the fire alarm system as either initiating devices or as supervisory devices. The code does not require the installation of a building fire alarm system. It does require that the duct smoke detectors be connected to the building fire alarm, if one exists. Duct smoke detectors, properly listed and installed, will accomplish their intended function when connected as initiating or supervisory devices.

When the building is not equipped with a fire alarm system, visual and audible alarm and trouble signal indicators (*Figure 67*) must be installed

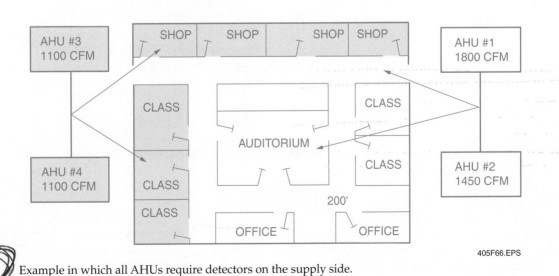

405F66.EPS

Figure 66 ◆ Example in which all AHUs require detectors on the supply side.

Figure 67 ◆ Typical remote duct indicator.

405F67.EPS

in a normally occupied area. Smoke detectors whose sole function is stopping fans do not require standby power.

AHUs often come with factory-installed duct detectors. If they are installed in a building with a fire alarm system, they must be connected to the fire alarm system, and the detector may have to have standby power. Consult with the AHJ concerning how many duct detectors can be placed in a zone. If it cannot be determined which detector activated after the power failure, the alarm indicator on the detector may be considered an additional function by the local AHJ.

Duct detectors for fresh air or return air ducts should be located six to ten duct widths from any openings, deflectors, sharp bends, or branch connections. This is necessary to obtain a representative air sample and to reduce the effects of stratification and dead air space.

INSIDE TRACK

Remote Duct Smoke Indicator

This duct smoke indicator combines a horn with a key-activated test and reset function. Green, yellow, and red LEDs provide a visual indication of system power, a trouble condition, and an alarm, respectively. This unit is also equipped with an optional strobe and smoke lens that provides an enhanced visual indication of an alarm.

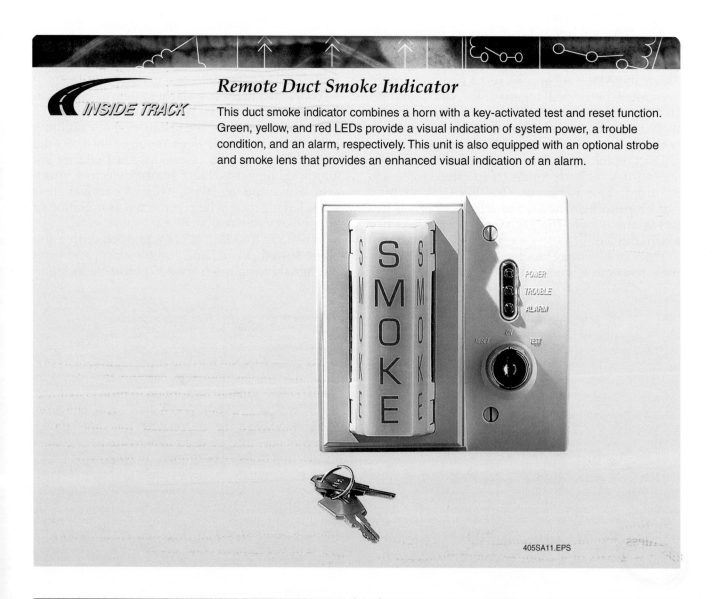

405SA11.EPS

12.3.0 Elevator Recall

Elevator recall is intended to route an elevator car to a non-fire floor, open its doors, and then put the car out of normal service until reset (Phase 1 recall). It is usually assumed that the recall floor will be the level of primary exit discharge or a grade level exit. In the event that the primary level of exit discharge is the floor where the fire has been detected, the system should route the car to a predetermined alternate floor, open its doors, and go out of normal service (Phase 2 recall). The operation of the elevators must be in accordance with *ANSI A17.1*.

Phase 1 recall (*Figure 68*) is the method of recalling the elevator to the floor that provides the highest probability of safe evacuation (as determined by the AHJ) in the event of a fire emergency.

Phase 2 recall (*Figure 69*) is the method of recalling the elevator to the floor that provides the next highest probability of safe evacuation (as determined by the AHJ) in the event of a fire emergency. Phase 2 recall is typically activated by the smoke detector in the lobby where the elevator would normally report during a Phase 1 recall.

Hoistway and elevator machine room smoke detectors must report an alarm condition to a building fire alarm control panel as well as an indicator light in the elevator car itself. However, notification devices may not be required to be activated if the control panel is in a constantly attended location. In facilities without a building fire alarm system, these smoke detectors must be connected to a dedicated fire alarm system control unit that must be designated as an Elevator Recall Control and Supervisory Panel.

Elevator recall must be initiated only by elevator lobby, hoistway, and machine room smoke detectors. Activation of manual pull stations, heat detectors, duct detectors, and any smoke detector not mentioned previously must not initiate elevator recall. In many systems, it would be inappropriate for the fire alarm control panel to initiate elevator recall. An exception is any system where the control function is selectable or programmable and the configuration limits the recall function to the specified detector activation only.

Caution should be used when using two-wire smoke detectors to recall elevators. Each elevator lobby, elevator hoistway, and elevator machine room smoke detector must be capable of initiating elevator recall when all other devices on the same initiating device circuit (IDC) have been manually or automatically placed in the alarm condition.

Unless the area encompassing the elevator lobby is a part of a chamber being protected by

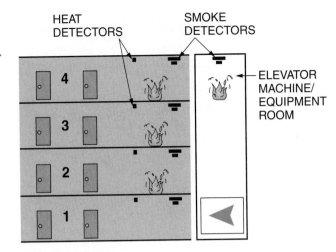

PHASE 1 RECALL:
Activation of any of these detectors sends all cars to the designated level.

405F68.EPS

Figure 68 ◆ Phase 1 recall.

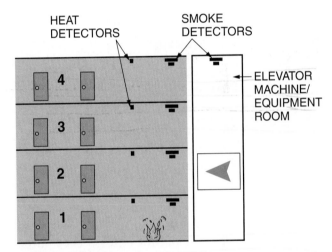

PHASE 2 RECALL:
Activation of the designated level detector sends all cars to the alternate floor.

405F69.EPS

Figure 69 ◆ Phase 2 recall.

smoke detectors, such as the continuation of a corridor, the smoke detectors serving the elevator lobby (*Figure 70*) may be applied to protect against a specific hazard per *NFPA 72®*. The elevator lobby, hoistway, or associated machine room detectors may also be used to activate emergency control functions as permitted by *NFPA 101®*.

Where ambient conditions prohibit installation of automatic smoke detection, other appropriate automatic fire detection must be permitted. For example, a heat detector may protect an elevator door that opens to a high-dust or low-temperature area. Always refer to local codes.

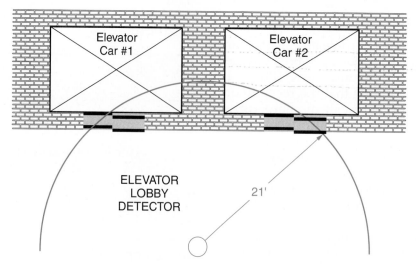

Assumes a 30' listed detector on a smooth and flat ceiling.
Target protection should be the extreme edges of the elevator door opening.

405F70.EPS

Figure 70 ◆ Elevator lobby detector.

12.4.0 Special Door Locking Arrangements

No lock, padlock, hasp, bar, chain, or other device intended to prevent free egress may be used at any door on which panic hardware or fire exit hardware is required (*NFPA 101®*). Note that these requirements apply only to exiting from the building. There is no restriction on locking doors against unrestricted entry from the exterior.

12.4.1 Stair Enclosure Doors

Upon activation of the fire alarm, stair enclosure doors must unlock to permit egress. They must also unlock to permit reentry into the building floors. This provision applies to buildings four stories and taller. The purpose of this requirement is to provide for an escape from fire and smoke entering a lower level of the stair enclosure and blocking safe egress from the stair enclosure. See *NFPA 101®* for additional details and exceptions.

12.5.0 Suppression System Supervision

The following section discusses both dry and wet chemical extinguishing systems.

12.5.1 Dry Chemical Extinguishing Systems

Dry chemical extinguishing systems must be connected to the building fire alarm system if a fire alarm system exists in the structure. The extinguishing system must be connected as an initiating device. The standard (*NFPA 17, Dry Chemical Extinguishing Systems*) does not require the installation of a building fire alarm system if one was not required elsewhere. Also see *NFPA 12* (*Carbon Dioxide Extinguishing Systems*), *NFPA 12A* (*Halon 1301 Fire Extinguishing Systems*), and *NFPA 2001* (*Clean Agent Fire Extinguishing Systems*).

12.5.2 Wet Chemical Extinguishing Systems

Wet chemical extinguishing systems must be connected to the building fire alarm system if a fire alarm system exists in the structure. As with dry chemical extinguishing systems, the standard (*NFPA 17A, Wet Chemical Extinguishing Systems*) does not require the installation of a building fire alarm system. Also, see *NFPA 16* (*Foam-Water Sprinkler and Foam-Water Spray Systems*).

12.6.0 Supervision of Suppression Systems

Each of the occupancy chapters of *NFPA 101®* will specify the extinguishing requirements for that occupancy. The code will specify either an approved automatic sprinkler or an approved supervised automatic sprinkler system if a sprinkler system is required.

12.6.1 Supervised Automatic Sprinkler

Where a supervised automatic sprinkler is required, various sections of *NFPA 101®* are applicable. These code sections begin with the phrase, "Where required by another section of this code." Use of the word supervised in the extinguishing requirement section of each occupancy chapter is the method used to implement these two provisions. Sprinkler systems do not automatically require electronic supervision. Supervised automatic sprinkler systems require a distinct supervisory signal. This signal indicates a condition that would impair the proper operation of the sprinkler system, at a location constantly attended by qualified personnel or at an approved remote monitoring facility. Water flow alarms from a supervised automatic sprinkler system must be transmitted to an approved monitoring station. When the supervised automatic sprinkler supervisory signal terminates on the protected premises in areas that are constantly attended by trained personnel, the supervisory signal is not required to be transmitted to a monitoring facility. In such cases, only alarm and trouble signals need to be transmitted.

The following are some of the sprinkler elements that are required to be supervised where applicable:

- Water supply control valves
- Fire pump power (including phase monitoring)
- Fire pump running
- Water tank levels
- Water tank temperatures
- Tank pressures
- Air pressure of dry-pipe systems

A sprinkler flow alarm must be initiated within 90 seconds of the flow of water equal to or greater than the flow from the sprinkler head, or from the smallest orifice (opening) size in the system. In actual field verification activities, the 90 seconds is measured from the time water begins to flow into the inspector's test drain, and *not* from the time the inspector's test valve is opened. The smallest orifice is the size of the opening at the smallest sprinkler head.

12.6.2 Sprinkler Systems and Manual Pull Stations

Sprinkler systems that initiate a fire alarm system by a water flow switch must include at least one manual station that is located where required by the AHJ. Manual stations required elsewhere in the codes or standards can be considered to meet this requirement. Some occupancy chapters of *NFPA 101*® will allow the sprinkler flow switch to substitute for manual stations at all the required exits. If such an option is utilized, this provision requires that at least one manual station be installed where acceptable to the AHJ.

Fire Pumps

A fire pump system is necessary when the available water supply is not adequate, in pressure or volume, to supply the fire suppression water needs of a sprinkler or standpipe system. The fire pump system is comprised of several components, including the pump, a driver (typically electric or diesel), and a controller. A typical fire pump system is shown here.

405SA12.EPS

12.6.3 Outside Screw and Yoke Control Valves and Tamper Switches

Each shutoff valve, also called an outside screw and yoke (OS&Y) control valve, must be supervised by a tamper switch. A distinctive signal must sound when the valve is moved from a fully open (off-normal) position (within two revolutions of the hand wheel, or when the stem has moved one-fifth from its normal open position). A common verification practice is to mark the 12:00 position on the valve in its normal (open) position, then rotate the hand wheel twice and stop at the 12:00 position on the second pass. The supervisory signal must be initiated by the time the wheel reaches the end of the second pass.

Water flow and supervisory devices, in addition to their circuits, must be installed such that no unauthorized person may tamper with them, open them, remove them, or disconnect them without initiating a signal. Publicly accessible junction boxes must have tamper-resistant screws or tamper-alarm switches. Most water flow switch and valve tamper switch housings come from the manufacturer with tamper-resistant screws (hex or allen head), and they may be configured to signal when the housing cover is removed. If the device, circuit, or junction box requiring protection is in an area that is not accessible to unauthorized personnel, no additional protective measures should be required. Simply sealing, locking, or removing the handle from a valve is not sufficient to meet the supervision requirement.

12.6.4 Tamper Switches versus Initiating Circuits

Water flow devices that are alarm-initiating devices cannot be connected on the same initiating circuit as valve supervisory devices. This is commonly done in violation of the code by connecting the valve tamper switch in series with the initiating circuit's EOL device, resulting in a trouble signal when activated. This method of wiring does not provide for a distinctive visual or audible signal. This statement is true for most conventional systems. It should be noted that at least one known addressable system has a listed module capable of distinctly separating the two types of signals. These devices must be wired on the same circuit to meet the standard.

> **NOTE**
>
> Addressable devices are not connected to initiating circuits. They are connected to signaling line circuits.

Supervisory Switches

This supervisory switch is typical of those used with sprinkler systems to provide a tamper indication of valve movement. The switch is mounted on a valve with the actuator arm normally resting against the movable target indicator assembly. If the normal position of the valve is altered, the valve stem moves, forcing the actuator arm to operate the switch. The switch activates between the first and second revolutions of the control valve wheel.

405SA13.EPS

12.6.5 Supervisory versus Trouble Signals

A supervisory signal must be visually and audibly distinctive from both alarm signals and trouble signals. It must be possible to tell the difference between a fire alarm signal, a valve being off-normal (closed), and an open (broken wire) in the circuit.

12.6.6 Suppression Systems in High-Rise Buildings

Where a high-rise building is protected throughout by an approved, supervised automatic sprinkler system, valve supervision and water flow devices must be provided on each floor

(*NFPA 101*®). In such buildings, a fire command center (central control system) is also required (*NFPA 101*®).

13.0.0 ◆ TROUBLESHOOTING

The troubleshooting approach to any fire alarm system is basically the same. Regardless of the situation or equipment, some basic steps can be followed to isolate problems:

Step 1 Know the equipment. For easy reference, keep specification sheets and instructions for commonly used and serviced equipment. Become familiar with the features of the equipment.

Step 2 Determine the symptoms. Try to make the system perform or fail to perform as it did when the problem was discovered.

Step 3 List possible causes. Write down everything that could possibly have caused the problem.

Step 4 Check the system systematically. Plan activities so that problem areas are not overlooked, in order to eliminate wasted time.

Step 5 Correct the problem. Once the problem has been located, repair it. If it is a component that cannot be repaired easily, replace the component.

Step 6 Test the system. After the initial problem has been corrected, thoroughly check all the functions and features of the system to make sure other problems are not present that were masked by the initial problem.

13.1.0 Alarm System Troubleshooting Guidelines

Figure 71 is a system troubleshooting chart, *Figure 72* is an alarm output troubleshooting chart, and *Figure 73* is an auxiliary power troubleshooting chart.

The following guidelines provide information for resolving potential problems for specific conditions:

- *Sensors* – Always check sensor power, connections, environment, and settings. Lack of detection can be caused by a loose connection, obstacles in the area of the sensor, or a faulty unit. If unwanted alarms occur, recheck the installation for changing environmental factors. If none are present, cover or seal the sensor to confirm that the alarm originated with the sensor. If alarms still occur, wiring problems, power problems or electromagnetic interference (EMI) could be the cause. If the alarm stops when the sensor is covered or sealed, the environment monitored by the sensor is the source of the problem. Replacing the unit should be the last resort after performing the above checks.
- *Open circuit problems* – Opens will cause trouble signals and account for the largest percentage of faults. Major causes of open circuits are loose connections, wire breaks (ripped or cut), staple cuts, bad splices, cold-solder joints, wire fatigue due to flexing, and defective wire.
- *Short circuit or ground fault problems* – Shorts or ground faults on a circuit will cause alarms or trouble signals. Events occurring past a short will not be seen by the system. Some of the common causes of shorts or ground faults are staple cuts, sharp edge cuts, improper splices, moisture, and cold-flow of wiring insulation.

13.2.0 Addressable System Troubleshooting Guidelines

There are a wide variety of designs, equipment, and configurations of addressable fire alarm systems. For this reason, some manufacturers of addressable systems provide troubleshooting data in their service literature, but other manufacturers do not. Most of the general guidelines for troubleshooting discussed earlier also apply to troubleshooting addressable systems. In the event general troubleshooting methods fail to find the problem and no troubleshooting information is provided by the manufacturer in the service literature for the equipment, the best thing to do is to contact the manufacturer and ask for technical assistance. This will prevent a needless loss of time.

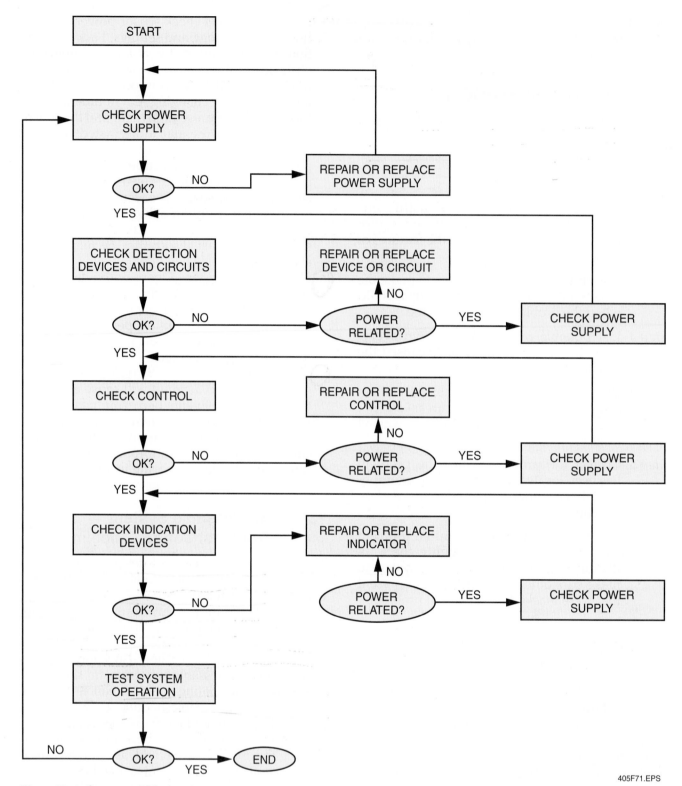

Figure 71 ◆ System troubleshooting.

405F71.EPS

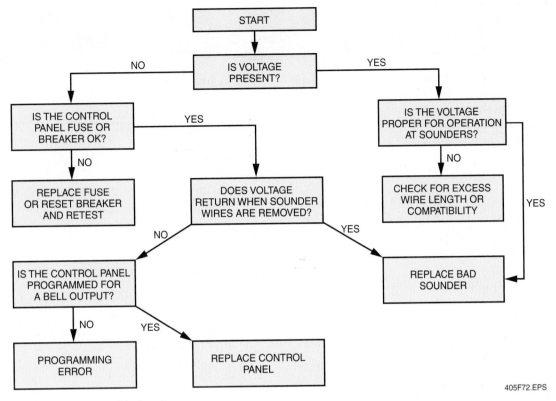

Figure 72 ◆ Alarm output troubleshooting.

405F72.EPS

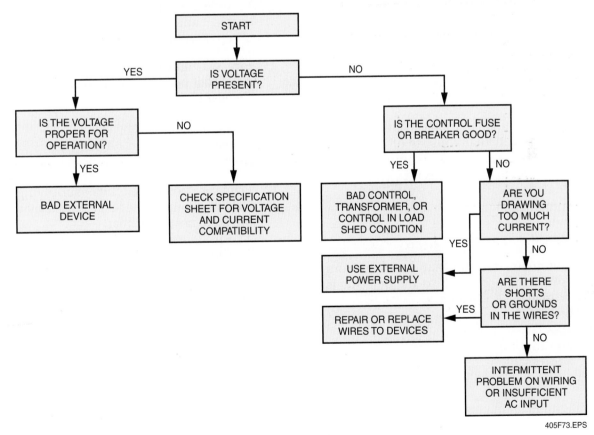

Figure 73 ◆ Auxiliary power troubleshooting.

405F73.EPS

1. The specific requirements for wiring and wiring equipment installation for fire protective signaling systems are covered in _____.
   a. NFPA 70®
   b. NFPA 72®
   c. NFPA 101®
   d. NFPA 1

2. A code that was established to help fire authorities continually develop safeguards against fire hazards is _____.
   a. NFPA 1
   b. NFPA 72®
   c. NFPA 70®
   d. NFPA 101®

3. A defined area within the boundaries of a fire alarm system is known as a(n) _____.
   a. section
   b. signal destination
   c. alarm unit area
   d. zone

4. A fire alarm system with zones that allow multiple signals from several sources to be sent and received over a single communication line is known as a(n) _____ system.
   a. addressable
   b. conventional hardwired
   c. zoned
   d. multiplex

5. An SLC circuit that consists of a Class A main trunk with a Class B spur circuit is known as an SLC _____ circuit.
   a. dual
   b. combined
   c. twin
   d. hybrid

6. An automatic detector that draws air from the protected area back to the detector is called a(n) _____ detector.
   a. line
   b. spot
   c. air sampling
   d. addressable

7. A heat detector that is *not* a fixed temperature device is a _____ detector.
   a. fusible link
   b. rate-of-rise
   c. quick metal
   d. bimetallic

8. Older type duct detectors used the _____ principle.
   a. ionization
   b. light scattering
   c. light obscuration
   d. smoke rate compensation

9. A _____ pull station is restricted for use in special applications.
   a. single-action
   b. glass-break
   c. double-action
   d. key-operated

10. Addressable systems commonly use a verification method called _____.
    a. positive alarm sequence
    b. wait and check
    c. reset and resample
    d. cross-zoning

11. Fire warning system units for residential use must be listed in accordance with *UL Standard* _____.
    a. 780
    b. 864
    c. 985
    d. 995

12. Primary power circuits supplying fire alarm systems must be protected by circuit breakers not larger than _____.
    a. 15A
    b. 20A
    c. 25A
    d. 30A

13. The *NFPA 72®*–specified Temporal three sound pattern for a fire alarm signal is _____ until the alarm is reset.
    a. three short (½ sec.), three long (1 sec.), and three short, a pause, then a repeat
    b. three long (1 sec.), a pause, then a repeat
    c. three short (½ sec.), a pause, then a repeat
    d. three short (½ sec.), three long (1 sec.), three short, three long, a pause, then a repeat

14. Public area audible notification devices must have a minimum dB rating of _____ at 10'.
    a. 65
    b. 75
    c. 85
    d. 95

15. The use of an attendant signal requires the approval of the AHJ.
    a. true
    b. false

16. Commercial fire alarm cables that run from floor to floor must be rated at least as _____ cable.
    a. general-purpose
    b. water-resistant
    c. riser
    d. plenum

17. Power-limited fire alarm circuit conductors must be separated from nonpower-limited fire alarm circuit conductors by at least _____.
    a. 2"
    b. 6"
    c. 12"
    d. 24"

18. For smoke chamber definition purposes, an archway that extends down 18" or more from the ceiling is considered the same as a(n) _____.
    a. barrier
    b. boundary
    c. open grid
    d. beam pocket

19. The primary function of a commercial duct smoke detector is to _____.
    a. sound an alarm for duct smoke
    b. shut down an associated HVAC system
    c. detect external open-area fires
    d. disable a music system for voice evacuation purposes

20. In a supervised automatic sprinkler system, a visual or audible supervisory signal must be _____.
    a. different from an alarm signal
    b. the same as an alarm signal
    c. different from a trouble signal
    d. different from both the alarm and trouble signals

Summary

A fire alarm system is a combination of components designed to detect and report fires, primarily for life safety purposes. A properly installed and fully functioning commercial or residential fire alarm system can save lives. This module covered the basics of fire alarm systems, components, and installation. Additionally, a basic introduction to some fire alarm system design criteria was provided. A fire alarm system will not function properly if it isn't designed, installed, and maintained according to specifications. No two system installations will be exactly alike. To help regulate the different applications that exist, codes and standards have been established that dictate the specifics of design and installation. The electrician must be familiar with requirements of these codes and standards as well as the basic design, components, installation, testing, and maintenance of typical fire alarm systems.

Notes

Trade Terms
Introduced in This Module

Addressable device: A fire alarm system component with discrete identification that can have its status individually identified or that is used to individually control other functions.

Air sampling detector: A detector consisting of piping or tubing distribution from the detector unit to the area or areas to be protected. An air pump draws air from the protected area back to the detector through the air sampling ports and piping or tubing. At the detector, the air is analyzed for fire products.

Alarm: In fire systems, a warning of fire danger.

Alarm signal: A signal indicating an emergency requiring immediate action, such as an alarm for fire from a manual station, water flow device, or automatic fire alarm system.

Alarm verification: A feature of a fire control panel that allows for a delay in the activation of alarms upon receiving an initiating signal from one of its circuits. Alarm verification must not be longer than three minutes, but can be adjustable from 0 to 3 minutes to allow supervising personnel to check the alarm. Alarm verification is commonly used in hotels, motels, hospitals, and institutions with large numbers of smoke detectors.

Americans with Disabilities Act (ADA): An act of Congress intended to ensure civil rights for physically challenged people.

Approved: Acceptable to the authority having jurisdiction.

Audible signal: An audible signal is the sound made by one or more audible indicating appliances, such as bells, chimes, horns, or speakers, in response to the operation of an initiating device.

Authority having jurisdiction (AHJ): The authority having jurisdiction is the organization, office, or individual responsible for approving equipment, installations, or procedures in a particular locality.

NOTE

The NFPA does not approve, inspect, or certify any installations, procedures, equipment, or materials, nor does it approve or evaluate testing laboratories. In determining the acceptability of installations, procedures, equipment, or materials, the authority having jurisdiction may base acceptance on compliance with NFPA, or other appropriate standards. In the absence of such standards, said authority may require evidence of proper installation, procedure, or use. The authority having jurisdiction may also refer to the listings or labeling practices of an organization concerned with product evaluations that is in a position to determine compliance with appropriate standards for the current production of listed items.

Automatic fire alarm system: A system in which all or some of the circuits are actuated by automatic devices, such as fire detectors, smoke detectors, heat detectors, and flame detectors.

CABO: Council of American Building Officials.

Ceiling: The upper surface of a space, regardless of height. Areas with a suspended ceiling would have two ceilings: one visible from the floor, and one above the suspended ceiling.

Ceiling height: The height from the continuous floor of a room to the continuous ceiling of a room or space.

Ceiling surface: Ceiling surfaces referred to in conjunction with the locations of initiating devices are as follows:

- *Beam construction* – Ceilings having solid structural or solid nonstructural members projecting down from the ceiling surface more than 4" (100mm) and spaced more than 3" (0.9m) center to center.
- *Girders* – Girders support beams or joists and run at right angles to the beams or joists. When the tops of girders are within 4" (100mm) of the ceiling, they are a factor in determining the number of detectors and are to be considered as beams. When the top of the girder is more than 4" (100mm) from the ceiling, it is not a factor in detector location.

Certification: A systematic program using randomly selected follow-up inspections of the certified system installed under the program, which allows the listing organization to verify that a fire alarm system complies with all the requirements of the *NFPA 72*® code. A system installed under such a program is identified by the issuance of a certificate and is designated as a certificated system.

Chimes: A single-stroke or vibrating audible signal appliance that has a xylophone-type striking bar.

Circuit: The conductors or radio channel as well as the associated equipment used to perform a definite function in connection with an alarm system.

Class A circuit: Class A refers to an arrangement of supervised initiating devices, signaling line circuits, or indicating appliance circuits (IAC) that prevents a single open or ground on the installation wiring of these circuits from causing loss of the system's intended function. It is also commonly known as a four-wire circuit.

Class B circuit: Class B refers to an arrangement of initiating devices, signaling lines, or indicating appliance circuits that does not prevent a single open or ground on the installation wiring of these circuits from causing loss of the system's intended function. It is commonly known as a two-wire circuit.

Coded signal: A signal pulsed in a prescribed code for each round of transmission. A minimum of three rounds and a minimum of three impulses are required for an alarm signal. A coded signal is usually used in a manually operated device on which the act of pulling a lever causes the transmission of not less than three rounds of coded alarm signals. These devices are similar to the non-coded type, except that instead of a manually operated switch, a mechanism to rotate a code wheel is utilized. Rotation of the code wheel, in turn, causes an electrical circuit to be alternately opened and closed, or closed and opened, thus sounding a coded alarm that identifies the location of the box. The code wheel is cut for the individual code to be transmitted by the device and can operate by clockwork or an electric motor. Clockwork transmitters can be prewound or can be wound by the pulling of the alarm lever. Usually, the box is designed to repeat its code four times before automatically coming to rest. Prewound transmitters must sound a trouble signal when they require rewinding. Solid-state electronic coding devices are also used in conjunction with the fire alarm control panel to produce coded sounding of the system's audible signaling appliances.

Control unit: A device with the control circuits necessary to furnish power to a fire alarm system, receive signals from alarm initiating devices (and transmit them to audible alarm indicating appliances and accessory equipment), and electrically supervise the system installation wiring and primary (main) power. The control unit can be contained in one or more cabinets in adjacent or remote locations.

Digital Alarm Communicator Receiver (DACR): A system component that will accept and display signals from digital alarm communicator transmitters (DACTs) sent over public switched telephone networks.

Digital Alarm Communicator System (DACS): A system in which signals are transmitted from a digital alarm communicator transmitter (DACT) located at the protected premises through the public switched telephone network to a digital alarm communicator receiver (DACR).

Digital Alarm Communicator Transmitter (DACT): A system component at the protected premises to which initiating devices or groups of devices are connected. The DACT will seize the connected telephone line, dial a preselected number to connect to a DACR, and transmit signals indicating a status change of the initiating device.

End-of-line (EOL) device: A device used to terminate a supervised circuit. An EOL is normally a resistor or a diode placed at the end of a two-wire circuit to maintain supervision.

Fault: An open, ground, or short condition on any line(s) extending from a control unit, which could prevent normal operation.

Fire: A chemical reaction between oxygen and a combustible material where rapid oxidation may cause the release of heat, light, flame, and smoke.

Flame detector: A device that detects the infrared, ultraviolet, or visible radiation produced by a fire. Some devices are also capable of detecting the flicker rate (frequency) of the flame.

General alarm: A term usually applied to the simultaneous operation of all the audible alarm signals on a system, to indicate the need for evacuation of a building.

Ground fault: A condition in which the resistance between a conductor and ground reaches an unacceptably low level.

Heat detector: A device that detects abnormally high temperature or rate-of-temperature rise.

Horn: An audible signal appliance in which energy produces a sound by imparting motion to a flexible component that vibrates at some nominal frequency.

Indicating device: Any audible or visible signal employed to indicate a fire, supervisory, or trouble condition. Examples of audible signal appliances are bells, horns, sirens, electronic horns, buzzers, and chimes. A visible indicator consists of an incandescent lamp, strobe lamp, mechanical target or flag, meter deflection, or the equivalent. Also called a notification device (appliance).

Initiating device: A manually or automatically operated device, the normal intended operation of which results in a fire alarm or supervisory signal indication from the control unit. Examples of alarm signal initiating devices are thermostats, manual boxes (stations), smoke detectors, and water flow devices. Examples of

supervisory signal initiating devices are water level indicators, sprinkler system valve-position switches, pressure supervisory switches, and water temperature switches.

Initiating device circuit (IDC): A circuit to which automatic or manual signal-initiating devices such as fire alarm manual boxes (pull stations), heat and smoke detectors, and water flow alarm devices are connected.

Labeled: Equipment or materials to which has been attached a label, symbol, or other identifying mark of an organization acceptable to the authority having jurisdiction, which is concerned with product evaluation and whose listing states either that the equipment or material meets appropriate standards or has been tested and found suitable for use in a specified manner.

Light scattering: The action of light being reflected or refracted off particles of combustion, for detection in a modern day photoelectric smoke detector. This is called the Tyndall effect.

Listed: Equipment or materials included in a list published by an organization acceptable to the authority having jurisdiction that is concerned with product evaluation and whose listing states either that the equipment or materials meets appropriate standards or has been tested and found suitable for use in a specified manner.

 NOTE

The means for identifying listed equipment may vary for each organization concerned with product evaluation [UL (Underwriters Laboratories), FM (Factory Mutual), IRI (Industrial Risk Insurers), and so on]. Some organizations do not recognize equipment listed unless it is also labeled. The authority having jurisdiction should use the system employed by the listed organization to identify a listed product.

Maintenance: Repair service, including periodic inspections and tests, required to keep the protective signaling system and its component parts in an operative condition at all times. This is used in conjunction with replacement of the system and its components when for any reason they become undependable or inoperative.

Multiplexing: A signaling method that uses wire path, cable carrier, radio, fiber optics, or a combination of these techniques, and characterized by the simultaneous or sequential (or both simultaneous and sequential) transmission and reception of multiple signals in a communication channel including means of positively identifying each signal.

National Fire Alarm Code®: This is the update of the NFPA standards book that contains the former *NFPA 71, NFPA 72®,* and *NFPA 74* standards, as well as the *NFPA 1221* standard. The NFAC was adopted and became effective May 1993.

National Fire Protection Association (NFPA): The NFPA administers the development and publishing of codes, standards, and other materials concerning all phases of fire safety.

Noise: A term used in electronics to cover all types of unwanted electrical signals. Noise signals originate from numerous sources, such as fluorescent lamps, walkie-talkies, amateur and CB radios, machines being switched on and off, and power surges. Today, equipment must tolerate increasing amounts of electrical interference, and the quality of equipment depends on how much noise it can ignore and withstand.

Non-coded signal: A signal from any indicating appliance that is continuously energized.

Notification device (appliance): See *indicating device (appliance).*

Obscuration: A reduction in the atmospheric transparency caused by smoke, usually expressed in percent per foot.

Path (pathway): Any conductor, optic fiber, radio carrier, or other means for transmitting fire alarm system information between two or more locations.

Photoelectric smoke detector: A detector employing the photoelectric principle of operation using either the obscuration effect or the light-scattering effect for detecting smoke in its chamber.

Positive alarm sequence: An automatic sequence that results in an alarm signal, even when manually delayed for investigation, unless the system is reset.

Power supply: A source of electrical operating power, including the circuits and terminations connecting it to the dependent system components.

Projected beam smoke detector: A type of photoelectric light-obscuration smoke detector in which the beam spans the protected area.

Protected premises: The physical location protected by a fire alarm system.

Protected premises (local) fire alarm system: A protected premises system that sounds an alarm at the protected premises as the result of the manual operation of a fire alarm box, or as a result of the operation of protection equipment or systems, such as water flowing in a sprinkler system, the discharge of carbon dioxide, the detection of smoke, or the detection of heat.

Public Switched Telephone Network: An assembly of communications facilities and central office equipment, operated jointly by authorized common carriers, that provides the general public with the ability to establish communications channels via discrete dialing codes.

Rate compensation detector: A device that responds when the temperature of the air surrounding the device reaches a predetermined level, regardless of the rate-of-temperature rise.

Rate-of-rise detector: A device that responds when the temperature rises at a rate exceeding a predetermined value.

Remote supervising station fire alarm system: A system installed in accordance with the applicable code to transmit alarm, supervisory, and trouble signals from one or more protected premises to a remote location where appropriate action is taken.

Reset: A control function that attempts to return a system or device to its normal, non-alarm state.

Signal: A status indication communicated by electrical or other means.

Signaling line circuits (SLCs): A circuit or path between any combination of circuit interfaces, control units, or transmitters over which multiple system input signals or output signals (or both input signals and output signals) are carried.

Smoke detector: A device that detects visible or invisible particles of combustion.

Spacing: A horizontally measured dimension related to the allowable coverage of fire detectors.

Spot-type detector: A device in which the detecting element is concentrated at a particular location. Typical examples are bimetallic detectors, fusible alloy detectors, certain pneumatic rate-of-rise detectors, certain smoke detectors, and thermoelectric detectors.

Stratification: The phenomenon in which the upward movement of smoke and gases ceases due to a loss of buoyancy.

Supervisory signal: A signal indicating the need for action in connection with the supervision of guard tours, the fire suppression systems or equipment, or the maintenance features of related systems.

System unit: The active subassemblies at the central station used for signal receiving, processing, display, or recording of status change signals. The failure of one of these subassemblies causes the loss of a number of alarm signals by that unit.

Transmitter: A system component that provides an interface between the transmission channel and signaling line circuits, initiating device circuits, or control units.

Trouble signal: A signal initiated by the fire alarm system or device that indicates a fault in a monitored circuit or component.

Visible notification appliance: A notification appliance that alerts by the sense of sight.

Wavelength: The distance between peaks of a sinusoidal wave. All radiant energy can be described as a wave having a wavelength. Wavelength serves as the unit of measure for distinguishing between different parts of the spectrum. Wavelengths are measured in microns, nanometers, or angstroms.

Wide Area Telephone Service (WATS): Telephone company service that provides reduced costs for certain telephone call arrangements. In-WATS or 800-number service calls can be placed from anywhere in the continental United States to the called party at no cost to the calling party. Out-WATS is a service whereby, for a flat-rate charge, dependent on the total duration of all such calls, a subscriber can make an unlimited number of calls within a prescribed area from a particular telephone terminal without the registration of individual call charges.

Zone: A defined area within the protected premises. A zone can define an area from which a signal can be received, an area to which a signal can be sent, or an area in which a form of control can be executed.

This module is intended to present thorough resources for task training. The following reference works are suggested for further study. These are optional materials for continuing education rather than for task training.

Certified Alarm Technician Level 1, Latest Edition. Silver Spring, MD: National Burglar and Fire Alarm Association.

Practical Fire Alarm Course, Latest Edition. Silver Spring, MD: National Burglar and Fire Alarm Association.

Understanding Alarm Systems, Latest Edition. Silver Spring, MD: National Burglar and Fire Alarm Association.

NCCER makes every effort to keep these textbooks up-to-date and free of technical errors. We appreciate your help in this process. If you have an idea for improving this textbook, or if you find an error, a typographical mistake, or an inaccuracy in NCCER's Contren® textbooks, please write us, using this form or a photocopy. Be sure to include the exact module number, page number, a detailed description, and the correction, if applicable. Your input will be brought to the attention of the Technical Review Committee. Thank you for your assistance.

Instructors – If you found that additional materials were necessary in order to teach this module effectively, please let us know so that we may include them in the Equipment/Materials list in the Annotated Instructor's Guide.

Write: Product Development and Revision
National Center for Construction Education and Research
3600 NW 43rd St., Bldg. G, Gainesville, FL 32606

Fax: 352-334-0932

E-mail: curriculum@nccer.org

Craft

Module Name

Copyright Date

Module Number

Page Number(s)

Description

(Optional) Correction

(Optional) Your Name and Address

Specialty Transformers

Massachusetts Institute of Technology – Learning from the Past

Although this is MIT's first Decathlon entry, it is by no means its first solar home. MIT has built six solar homes in the past, going back to the 1930s. Studying the history of these homes helped inspire the new home's primary technological feature, a Trombe wall of translucent tiles that are used to passively convert sunlight into stored heat.

26406-08
Specialty Transformers

Topics to be presented in this module include:

1.0.0   Introduction 6.2
2.0.0   Specialty Transformers 6.7
3.0.0   Instrument Transformers 6.13
4.0.0   Sizing Buck-and-Boost Transformers 6.17
5.0.0   Harmonics 6.19

Overview

When a circuit is connected to the secondary winding of a transformer and an AC voltage is applied to the primary, induced current will flow through the secondary winding and any circuits connected to it. Specialty transformers apply the same basic fundamentals of transformer induction, but they use different coil arrangements and access points (taps) to obtain unique voltage levels at the transformer secondary.

The basic electrical frequency generated by power plants in the U.S. is sixty cycles per second (60Hz), which graphically appears as a near perfect sinusoidal waveform on a scope meter. Most equipment or electrical loads are considered linear loads, meaning that they operate over the entire sine wave generated by sixty cycles. However, certain loads, such as fluorescent lamps and high-performance electronic equipment used in computers, operate on sharp, irregular pulses drawn from the sixty-cycle frequency. These types of loads are referred to as non-linear loads. The resulting effect on the sixty-cycle sinusoidal waveform caused by non-linear loads is a distorted wave, where distortion occurs in multiples of the basic sixty-cycle pattern. These harmonic currents can flow back to the source transformer, causing overheating and other problems. In installations where harmonics are a problem, the ampacity of the supply transformers must be derated, or a specially designed transformer must be used to avoid overheating.

Note: *NFPA 70*®, *National Electrical Code*®, and *NEC*® are registered trademarks of the National Fire Protection Association, Inc., Quincy, MA 02269. All *National Electrical Code*® and *NEC*® references in this module refer to the 2008 edition of the *National Electrical Code*®.

Objectives

When you have completed this module, you will be able to do the following:

1. Identify three-phase transformer connections.
2. Identify specialty transformer applications.
3. Size and select buck-and-boost transformers.
4. Calculate and install overcurrent protection for specialty transformers.
5. Ground specialty transformers in accordance with *National Electrical Code®* (*NEC®*) requirements.
6. Calculate transformer derating to account for the effects of harmonics.

Trade Terms

Ampere turn
Autotransformer
Bank
Core loss
Eddy currents

Harmonic
Hysteresis
Impedance
Isolation transformer
Reactance

Required Trainee Materials

1. Pencil and paper
2. Appropriate personal protective equipment
3. Copy of the latest edition of the *National Electrical Code®*

Prerequisites

Before you begin this module, it is recommended that you successfully complete *Core Curriculum; Electrical Level One; Electrical Level Two; Electrical Level Three; Electrical Level Four*, Modules 26401-08 through 26405-08.

This course map shows all of the modules in *Electrical Level Four*. The suggested training order begins at the bottom and proceeds up. Skill levels increase as you advance on the course map. The local Training Program Sponsor may adjust the training order.

26413-08 Introductory Skills for the Crewleader

26412-08 Special Locations

26411-08 Medium-Voltage Terminations/Splices

26410-08 Motor Operation and Maintenance

26409-08 Heat Tracing and Freeze Protection

26408-08 HVAC Controls

26407-08 Advanced Controls

26406-08 Specialty Transformers

26405-08 Fire Alarm Systems

26404-08 Basic Electronic Theory

26403-08 Standby and Emergency Systems

26402-08 Health Care Facilities

26401-08 Load Calculations – Feeders and Services

ELECTRICAL LEVEL FOUR

ELECTRICAL LEVEL THREE

ELECTRICAL LEVEL TWO

ELECTRICAL LEVEL ONE

CORE CURRICULUM: Introductory Craft Skills

406CMAP.EPS

1.0.0 ◆ INTRODUCTION

When the AC voltage needed for an application is lower or higher than the voltage available from the source, a transformer is used. The essential parts of a transformer are the primary winding (which is connected to the source) and the secondary winding (which is connected to the load), both wound on an iron core. The two windings are not physically connected.

The alternating voltage in the primary winding induces an alternating voltage in the secondary winding. The ratio of the primary and secondary voltages is equal to the ratio of the number of turns in the primary and secondary windings.

Transformers may step up the voltage applied to the primary winding and have a higher voltage at the secondary terminals, or they may step down the voltage applied to the primary winding and have a lower voltage at the secondary terminals.

NOTE

Transformers are applied in AC systems only and would not work in DC systems because the induction of voltage depends on the rate of change in the current.

A transformer can be constructed as a single-phase or three-phase apparatus. A three-phase transformer, such as the one shown in *Figure 1*, has three primary and three secondary windings, which may be connected in either a delta (Δ) or wye (Y) configuration. Combinations such as Δ-Δ, Δ-Y, Y-Δ, and Y-Y are possible. The first letter indicates the connection of the primary winding, and the second letter indicates that of the secondary winding.

A **bank** of three single-phase transformers can serve the same purpose as one three-phase transformer. The connections between the three primary windings and the three secondary windings are again Δ or Y, and they are available in all combinations.

1.1.0 Types of Transformers

Large transformers used in transmission systems are called power transformers. They may step up the voltage produced in the generator to make it suitable for commercial transmission, or they may step down the voltage in a transmission line to make it practical for distribution. Power transformers are usually installed outdoors in generating stations and substations. The transformer shown in *Figure 2* is a liquid-filled power transformer with radiators that disperse the heat from the transformer.

An **autotransformer** is a transformer with only one winding, which serves as both a primary and a secondary. Autotransformers are economical, space saving, and especially practical if the difference between the primary and secondary voltages is relatively small.

Three-winding, single-phase transformers have two secondary windings so that they can deliver two different secondary voltages.

In many applications of electrical measuring instruments, the voltage or current in the circuit to be measured is too high for the instruments. In such situations, instrument transformers are used to ensure safe operation of the instruments. An instrument transformer is either a current transformer, as

406F02.EPS

Figure 2 ◆ Typical liquid-filled, three-phase power transformer.

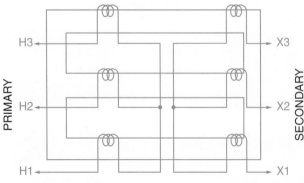

406F01.EPS

Figure 1 ◆ Typical three-phase transformer.

Transformer Management Systems

Some manufacturers provide monitoring and diagnostic systems for large liquid-filled power transformers used in substations or in large industrial applications. These systems are used to obtain warnings of impending faults and as a predictor for scheduling maintenance shutdowns. The systems use data from sensors incorporated in the transformers. The sensors monitor the following conditions:

- Dissolved gas-in-oil
- Top and bottom oil temperature
- Gas pressure
- Ambient temperature
- Oil level
- Winding temperature
- Load and meter values
- Bushing activity

This data is transmitted via a network to a computer and used with IEEE or IEC analytical modeling software offline to access the condition of the equipment and diagnose impending failures. This includes moisture-in-insulation modeling, predictive insulation aging rate, cumulative aging, partial discharge activity, overflux (overvoltage), fault event recording, harmonic monitoring, cooling system efficiency, and other trending information.

shown in *Figure 3*, or a potential (voltage) transformer, as shown in *Figure 4*. For example, if the current in a line is approximately 100A and the ammeter is rated at 5A, a bar-type current transformer with a current ratio of 100:5 can be connected in series with the line. The secondary winding is then connected to the ammeter. The current in the secondary is proportional to the current in the line. With the toroid current transformer, an insulated line passes through the center, and proportional current is induced in the toroid winding for application to the meter.

Similarly, the potential transformer reduces the voltage in a high-voltage line to the 120V for which the voltmeter is normally rated. The potential transformer is always connected across the line to be measured. Sometimes it is important to indicate the polarity of the current in both the instrument and the transformer. The polarity is the instantaneous direction of current at a specific moment. Wires with the same polarity usually contain a small black block or cross in electrical diagrams, as shown in *Figure 4*.

1.2.0 Internal Connections in Three-Phase Transformers

Various combinations of primary and secondary three-phase voltages are possible with the proper combination of delta and wye internal connections

BAR-TYPE CURRENT TRANSFORMER

TOROID (DONUT-TYPE)
CURRENT TRANSFORMERS

406F03.EPS

Figure 3 ◆ Two types of current transformers.

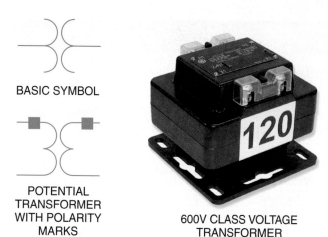

BASIC SYMBOL

POTENTIAL
TRANSFORMER
WITH POLARITY
MARKS

600V CLASS VOLTAGE
TRANSFORMER

406F04.EPS

Figure 4 ◆ Voltage (potential) transformer.

of three-phase transformers or with banks of single-phase transformers.

See *Figure 5*. In all examples, it is assumed that the primary line voltages between lines A, B, and C are all 1,000V and that the transformation ratio is 10:1. The primary connections are shown by the upper three windings, and the Δ or Y next to the winding indicates how the winding is connected.

In the Δ connections, as in *Figure 5(A)* and *Figure 5(D)*, the phase voltages of the primaries are the same as the line voltage or 1,000V. In the Y connections of the primaries, as in *Figure 5(B)* and *Figure 5(C)*, the phase voltages are 0.577 times the line voltages or 577V. The wye point (N) is indicated as a common point if the windings are wye connected.

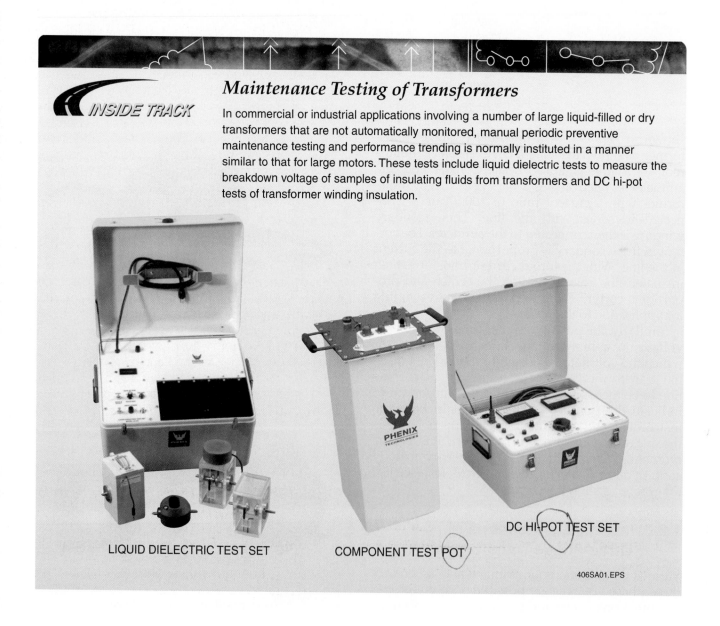

INSIDE TRACK

Maintenance Testing of Transformers

In commercial or industrial applications involving a number of large liquid-filled or dry transformers that are not automatically monitored, manual periodic preventive maintenance testing and performance trending is normally instituted in a manner similar to that for large motors. These tests include liquid dielectric tests to measure the breakdown voltage of samples of insulating fluids from transformers and DC hi-pot tests of transformer winding insulation.

LIQUID DIELECTRIC TEST SET

COMPONENT TEST POT

DC HI-POT TEST SET

406SA01.EPS

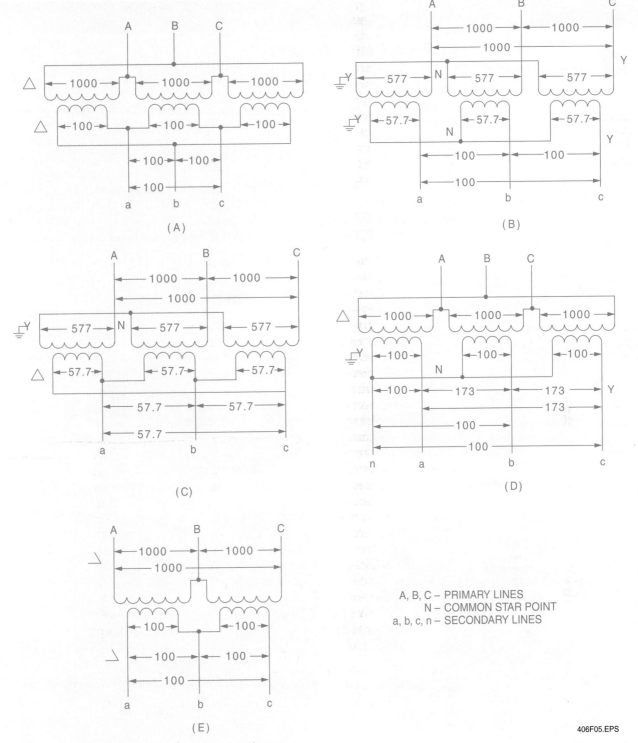

A, B, C – PRIMARY LINES
N – COMMON STAR POINT
a, b, c, n – SECONDARY LINES

406F05.EPS

Figure 5 ◆ Common power transformer connections.

The secondaries are shown in the lower row of windings, and their connections are also indicated by the symbol Δ or Y. Each secondary phase winding has only ¹⁄₁₀ of the turns used in the corresponding primary phase winding and therefore supplies a phase voltage that is ¹⁄₁₀ of the primary phase voltage. In *Figure 5(A)* and *Figure 5(D)*, the secondary phase voltages are 1,000 ÷ 10 = 100V, and in *Figure 5(B)* and *Figure 5(C)*, they are 577 ÷ 10 = 57.7V. The secondaries are delta connected in *Figure 5(A)* and *Figure 5(C)*, and their line voltages or the voltages between the secondary line wires (a, b, and c) are the same as the secondary phase windings. When the secondaries are wye

Cast-Coil Transformers

Some newer types of power transformers are made without using wire to wind the coils of the transformer. In the three-phase, step-down transformer shown here, strip foil technology is used to wind each of the primary coils. The primary coils are then placed in molds and encased with a mixture of epoxy resin and quartz powder under a high vacuum to remove any moisture. After heat curing, the coils are placed on the legs of the transformer core. Then, concentric secondary coils are wound using a sheet conductor. These secondary coils are encased in the same way as the primary coils. After curing, they are placed over the primary coils. Transformers made in this manner are capable of operating at temperatures ranging from −40°C to +180°C. They do not require liquid cooling or vaults and can be placed in NEMA Type 1 indoor or NEMA Type 3R outdoor enclosures. The epoxy-encased coils are highly resistant to caustic and humid environments.

CAST-COIL THREE-PHASE TRANSFORMER

STRIP-FOIL WINDING OF A PRIMARY COIL

SHEET-CONDUCTOR WINDING OF
A SECONDARY COIL

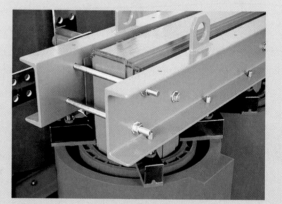

CONCENTRIC PRIMARY AND
SECONDARY CAST COILS

406SA02.EPS

connected, as in *Figure 5(B)* and *Figure 5(D)*, they have a common wye point (N) and the secondary line voltages are 1.732 times the phase voltage. In *Figure 5(B)*, the voltages between lines a, b, and c are $1.732 \times 57.7 = 100V$, and in *Figure 5(D)*, they are $1.732 \times 100 = 173V$. In addition, a fourth secondary wire (n) is brought out from the wye point (N) in *Figure 5(D)*, and the secondary voltage between any of the lines (a, b, or c and n) is equal to the secondary phase voltage or 100V.

As you become more experienced in reading electrical diagrams, you should be able to immediately recognize the differences between delta and wye connections in any three-phase system.

If the three windings have a common point, it is a Y connection, and the line voltages are 1.732 times higher than the individual phase voltages. If the three windings build a closed path, it is a delta connection, and the line voltages are the same as the phase voltages.

A connection using two single-phase transformers for a three-phase system is shown in *Figure 5(E)*. This is an open delta connection that provides a three-phase secondary with only two transformers.

2.0.0 ◆ SPECIALTY TRANSFORMERS

The principle of operation is the same for all transformers, but the forms, connections, and auxiliary devices differ widely. Among the many transformers designed for a specific purpose are single-phase transformers with two secondaries, single-phase transformers with three windings, autotransformers, constant-current transformers, series transformers, rectifier transformers, network transformers, and step-voltage regulators.

2.1.0 Transformers with Multiple Secondaries

One common type of transformer is a single-phase transformer with two secondaries (*Figure 6*). The first secondary has terminals X_1 and X_2 and is connected internally in series with the other

secondary, which has terminals X_3 and X_4. Terminals X_1 and X_4 are connected to outside leads, and terminals X_2 and X_3 are connected together, with the junction point connected to a third outside lead. Such transformers are commonly used as distribution transformers where three-wire service is needed from a two-wire, single-phase supply. The rated secondary voltage (120V in *Figure 6*) is obtained between either outside lead and the middle lead or neutral. In addition, double voltage (240V in *Figure 6*) is available between the two outside leads. This higher voltage is usually needed for HVAC equipment, electric ranges, and dryers, while lamps and small appliances are operated on 120V.

2.1.1 Three-Winding Transformers

Another widely used type of transformer is a three-winding, single-phase transformer. A third winding can be added to a transformer, and voltages will be induced in this winding proportional to the number of turns, the same as in the other windings. As a matter of fact, there is theoretically no limit to the number of windings that may be provided in a transformer. Practically, however, there is a limit because of the greater complexity, and transformers are seldom provided with more than four windings.

Three-winding transformers are used when transmission voltage is used to produce two secondary voltages. For example, a transformation may be desired from 230kV down to 69kV and 34.5kV to feed separate distribution networks.

Three-winding transformers are built in the same way as two-winding transformers, and they may be either the core or shell types. In the core construction, the additional winding is usually arranged so that it is concentric with the other two. In a shell transformer, the third winding is interleaved with the other two, which makes it more flexible than the core form.

2.2.0 Autotransformers

The usual transformer has two windings that are not physically connected. In an autotransformer, one of the windings is connected in series with the other, thereby forming the equivalent of a single winding, as shown in *Figure 7*. This illustration represents a step-up autotransformer, so called because the secondary voltage is higher than the voltage supplied to the primary.

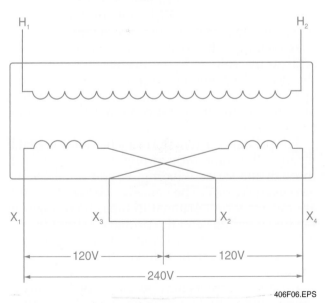

406F06.EPS

Figure 6 ◆ Single-phase transformer with two secondaries.

Multiple-Secondary Transformers

The two multiple-secondary transformers shown here are liquid-filled for cooling and insulation purposes and have internal, manually switched voltage taps. They are used for single-phase residential 120V/240V service.

PAD-MOUNTED TRANSFORMER

POLE-MOUNTED TRANSFORMER

406SA03.EPS

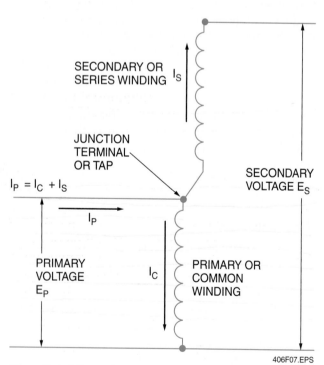

Figure 7 ◆ Wiring diagram of a typical autotransformer.

The primary voltage (E_P) is applied to the primary or common winding. The secondary or series winding is connected in series with the primary at the junction terminal. This point may be obtained by a tap, which will divide a single winding into a primary and a secondary.

A voltage induced in the secondary winding adds to the voltage in the primary winding, and the secondary voltage (E_s) is higher than the applied voltage. The ratio of transformation depends on the turns ratio, as in a two-winding transformer.

The primary current (I_p) branches into current I_c through the common winding and the current I_s through the series winding, as indicated by the arrows in *Figure 7*. The values of currents I_c and I_s are inversely proportional to the ratio of turns in the two windings and the primary current ($I_p = I_s + I_c$). Since the currents I_s and I_c oppose each other, the secondary current is lower than the primary current.

Autotransformers may be used economically to connect individual loads requiring voltages other than those available in the distribution system.

Three-Winding Control Transformers

INSIDE TRACK

Small three-winding transformers may be used in control systems. The one shown here has separate 24V and 26V secondary windings and a 120V primary winding.

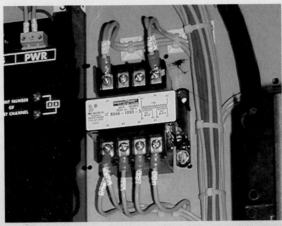

406SA04.EPS

Advantages of Autotransformers

INSIDE TRACK

Autotransformers are less costly than conventional two-winding transformers and have better voltage regulation and a better ratio of energy output to input (efficiency). They are used in motor starters so that a voltage lower than the line voltage may be applied during the starting period.

An example would be increasing the voltage on a branch circuit from a 120V/208V panel to 240V for more effective utilization of a motor or resistive load.

An autotransformer, however, cannot be used on a 240V or 480V, three-phase, three-wire delta system. A grounded neutral phase conductor must be available in accordance with *NEC Section 210.9*, which states that branch circuits shall not be supplied by autotransformers unless the system supplied has a grounded conductor that is electrically connected to a grounded conductor of the system supplying the autotransformer.

In general, the *NEC®* requires that separately derived alternating current systems be grounded. The secondary of an *isolation transformer* is a separately derived system. Therefore, it must be grounded in accordance with *NEC Section 250.30*. See *Figure 8*. In the case of an autotransformer, the grounded conductor of the supply is brought into the transformer to the common terminal and the ground is established to satisfy the *NEC®*.

2.3.0 Constant-Current Transformers

Constant-current transformers are used to supply series airport lighting and street lighting circuits in which the current must remain constant while the number of lamps in series varies because of burnout or bypass switching. This type of transformer has a stationary primary coil connected to a source of alternating voltage and a movable secondary coil connected to the lamp circuit. To allow it to move freely, the secondary coil is suspended from a shaft and rod attached to a rocker arm on which hinges are attached. The tendency of the secondary coil to move downward due to gravity is opposed by both the counterweight and the magnetic repulsion between the coils caused by the current in them.

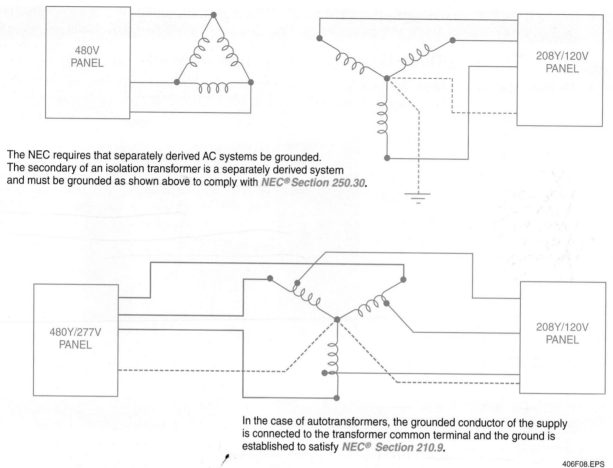

The NEC requires that separately derived AC systems be grounded.
The secondary of an isolation transformer is a separately derived system
and must be grounded as shown above to comply with *NEC® Section 250.30*.

In the case of autotransformers, the grounded conductor of the supply
is connected to the transformer common terminal and the ground is
established to satisfy *NEC® Section 210.9*.

406F08.EPS

Figure 8 ◆ *NEC®* grounding requirements for isolation and autotransformers.

The constant secondary current is usually between 4A and 7.5A, depending on the current rating of the lamps. In order to maintain a constant secondary current, the voltage of the secondary must vary directly with the number of lamps in series. If the resistance of the secondary is reduced by decreasing the number of lamps in series because of burnout or bypass switching, the current in the transformer will momentarily increase. This will increase the force of repulsion between the coils, which will move apart. The increase in distance between the coils increases the leakage **reactance** of the transformer and causes a greater voltage drop in the transformer, with a consequent decrease in the secondary voltage. The coil will continue to move until the current in the secondary winding reaches its original value, at which position the mechanical forces acting on the secondary coil are again balanced. Incandescent lamps used in series lighting circuits have a film cutout in the socket. When a lamp burns out, the full transformer secondary voltage appears between two spring contacts separated by a thin insulation film in the socket. The voltage

punctures the film, causing the series circuit to be re-established.

2.4.0 Control Transformers

Control transformers and their connections were covered in your Level Three training. However, due to the number of control transformers used in industrial establishments (both in new construction and in existing installations), and the importance of having a basic knowledge of how to size these transformers for any given application, a brief review is provided here.

In general, the selection of a proper control circuit transformer must be made from a determination of the maximum inrush VA and the maximum continuous VA to which it is subjected. This data can be determined as follows:

Step 1 Determine the inrush and sealed VA of all coils to be used.

Step 2 Determine the maximum sealed VA load on the transformer.

Step 3 Determine the maximum inrush VA load on the transformer at 100% of the secondary voltage. Add this value to any sealed VA present at the time inrush occurs.

Step 4 Calculate the power factor of the VA load obtained in Step 3. The actual coil power factor should be used. If this value is unknown, an inrush power factor of 35% may be assumed.

Step 5 Select a transformer with a continuous VA rating equal to or greater than the value obtained in Step 2 and whose maximum inrush VA from Step 3 at the calculated load power factor falls on or below the corresponding curve in *Figure 9*.

The regulation curves in *Figure 9* indicate the maximum permissible inrush loads (volt-amperes at 100% of the secondary voltage), which, if applied to the transformer secondary, will not cause the secondary voltage to drop below 85% of the rated voltage when the primary voltage has been reduced to 90% of the rated voltage.

2.5.0 Series Transformers

A power or distribution transformer is normally used with each of its primary terminals connected across the line. When a transformer is used in series with the main line, the term series transformer is applied.

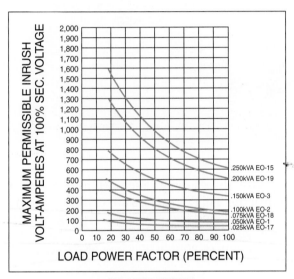

(A) 60/50 HERTZ

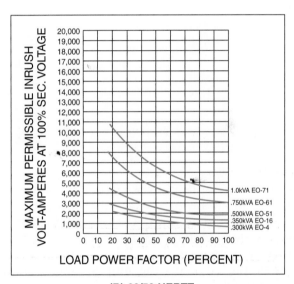

(B) 60/50 HERTZ

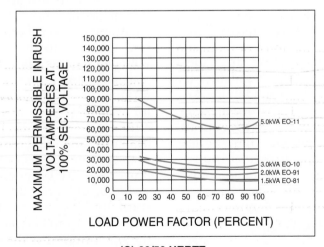

(C) 60/50 HERTZ

406F09.EPS

Figure 9 ◆ Regulation curves.

Control Power Transformers

The control power transformer shown here is rated for use in medium-voltage applications, including switchgear, with primary voltage ranges from 5kV to 34kV. Standard output voltages are 120V/240V for single-phase units and 208V/120V for three-phase units. Some versions can be mounted vertically to conserve space.

406SA05.EPS

A common application of a series transformer is its use in an airport system that has runway lamps connected in series. A small transformer is used with one or more lamp(s). The primary is connected in series with the line, and the secondary is connected across the lamp(s). The secondary winding is automatically short-circuited if a lamp burns out. The short circuit is obtained by a film cutout. When the circuit opens and the voltage across the secondary rises, it punctures the film, thereby short-circuiting the transformer secondary.

Series transformers are used in load tap changing circuits and with step-voltage regulators to reduce the operating voltage to ground when it is too high for the tap changes or to reduce the current in the tap changer contacts when the current exceeds the tap changer rating. These series transformers usually serve the purpose of an auxiliary transformer mounted in the main transformer tank.

2.6.0 Step-Voltage Regulators

Regulators of the step-voltage type are transformers provided with load tap changers. They are used to raise or lower the voltage of a circuit in response to a voltage-regulating relay or other control device. Regulators are usually designed to provide secondary voltages ranging from 10% below the supply voltage to 10% above it, or a total change of 20% in 32 steps of ⅝% each.

2.7.0 Other Specialty Transformers

Specialty transformers make up a large class of transformers and autotransformers used for changing a line voltage to some particular value best adapted to the load device. The primary voltage is generally 600V or less. Examples of specialty transformers are sign lighting transformers in which 120V is stepped down to 25V for low-voltage tungsten sign lamps; arc-lamp autotransformers in which 240V is stepped down to the voltage required for best operation of the arc; and transformers used to change 240V power to 120V for operating portable tools, fans, welders, and other devices. Also included in this specialty class are neon sign transformers that step 120V up to between 2,000V and 15,000V for the operation of neon signs. Many special stepdown transformers are used for small work, such as doorbells or other signaling systems, battery-charging rectifiers, and individual low-voltage lamps. Practically all specialty transformers are self-cooled and air-insulated. Sometimes the cases are filled with a special compound to prevent moisture absorption and to conduct heat to the enclosing structure.

3.0.0 ◆ INSTRUMENT TRANSFORMERS

Instrument transformers are so named because they are usually connected to an electrical instrument, such as an ammeter, voltmeter, wattmeter, or relay. As mentioned earlier, instrument transformers are of two types: current and potential (voltage).

The primary winding of a current transformer is connected in series in a line connecting the power source and the load, and it carries the full-load current. The turns ratio is designed to produce a rated current of 5A (or some other specified value) in the secondary winding when the rated current flows in the primary winding. The current transformer provides a small current suitable for the current coil of standard instruments and

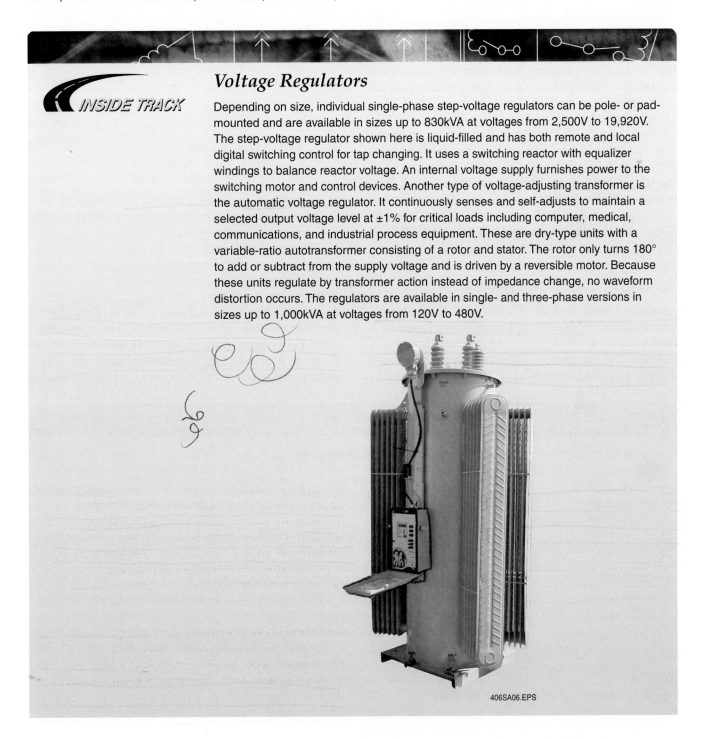

Voltage Regulators

Depending on size, individual single-phase step-voltage regulators can be pole- or pad-mounted and are available in sizes up to 830kVA at voltages from 2,500V to 19,920V. The step-voltage regulator shown here is liquid-filled and has both remote and local digital switching control for tap changing. It uses a switching reactor with equalizer windings to balance reactor voltage. An internal voltage supply furnishes power to the switching motor and control devices. Another type of voltage-adjusting transformer is the automatic voltage regulator. It continuously senses and self-adjusts to maintain a selected output voltage level at ±1% for critical loads including computer, medical, communications, and industrial process equipment. These are dry-type units with a variable-ratio autotransformer consisting of a rotor and stator. The rotor only turns 180° to add or subtract from the supply voltage and is driven by a reversible motor. Because these units regulate by transformer action instead of impedance change, no waveform distortion occurs. The regulators are available in single- and three-phase versions in sizes up to 1,000kVA at voltages from 120V to 480V.

406SA06.EPS

proportional to the load current. The low-voltage, low-current secondary winding providing the current is grounded for safety and economy in the secondary wiring and instruments.

A potential transformer is connected from one power line to another. Its secondary winding provides a low voltage, usually up to 120V, that is proportional to the line voltage. This low voltage is suitable for the voltage coil of standard instruments. The low-voltage secondary winding is grounded for the safety of the secondary wiring and the instrument, regardless of the power line voltage.

When connecting instrument transformers to wattmeters, watt-hour meters, power factor meters, etc., it is necessary to know the polarity of the leads. One primary lead and one secondary lead of the same polarity are clearly marked on all instrument transformers, usually by a white spot or white marker on the leads.

The direction of current in the two leads of the same polarity is such that, if it is toward the transformer in the marked primary lead, it is away from the transformer in the marked secondary lead. This is done to maintain the same phase sequence in the receiving device as is present on the power lines. In diagrams, the polarity mark is usually indicated in one of three ways, as shown in *Figure 10*.

3.1.0 Current Transformers

A current transformer is always a single-phase transformer. If current transformers are used in a three-phase system, one current transformer is inserted into each phase line between the power supply and the instrument. A current transformer

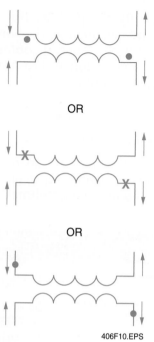

OR

OR

406F10.EPS

Figure 10 ◆ Polarity marks on transformers.

is considered a low kVA device. A connection of a current transformer into a single-phase, two-wire line is shown in *Figure 11*.

The primary winding of a current transformer must be connected in series with one of the main power lines; thus, the main line load current flows through the primary winding. The secondary winding at the current transformer is connected to a current-responsive instrument. The secondary circuit is grounded for safety. Instruments that use a current transformer may include ammeters, wattmeters, watt-hour meters, power factor

Instrument Transformers

INSIDE TRACK

Instrument transformers are classified according to the following factors:

- *Service* – Instrument transformers are designated for either metering or relaying service.
- *Burden* – The burden represents the size and characteristics of the load connected to the transformer secondary. For potential (voltage) transformers, the burden is expressed in VA at a specified power factor and voltage. For current transformers, the burden is expressed in total impedance at specified values of resistance and inductance. Generally, a transformer must perform as rated within the limits of its burden.
- *Accuracy* – For a metering transformer, an accuracy rating representing the amount of uncertainty (inaccuracy) is assigned for each rated burden of the transformer. For example, a transformer may have an accuracy rating of 0.3 at burden X and 0.6 at burden Y. Standard accuracy classes for instrument transformers are 0.3, 0.6, 1.2, and 2.4.

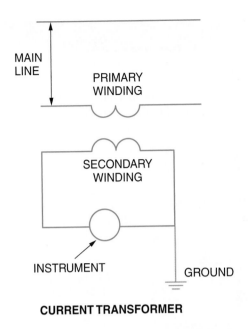

CURRENT TRANSFORMER

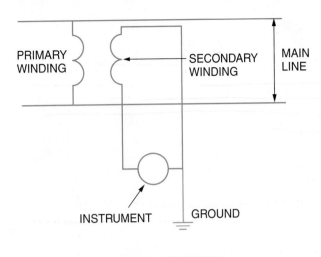

POTENTIAL TRANSFORMER

406F11.EPS

Figure 11 ◆ Connection of instrument transformers.

meters, some forms of relays, and trip coils of circuit breakers. One current transformer may feed through several devices in series.

Current transformers are built to standard IEEE burden ratings. Burden refers to the total load seen by the CT secondary. This load is the total impedance of conductors and instruments connected. For example, a CT with a B-1 designation is intended to work properly with no more than 1V impedance in the secondary circuit. This B-1 CT is rated to deliver 25VA at a secondary current of 5A. A CT with a B-8 designation is rated at 200VA at 5A. It is designed to work with an 8V impedance. When current flow is interrupted in the CT secondary, the secondary voltage can rise to dangerously high levels as the CT tries to deliver VA. Therefore, a shorting jumper should be placed across the secondary terminals when no device is connected.

 CAUTION
CTs shipped as part of equipment usually come with a shorting jumper installed. Be sure to remove the shorting jumper AFTER the connection to the relaying or metering circuit is verified.

3.1.1 Using Current Transformers

Before disconnecting an instrument, the secondary of the current transformer must be short-circuited. If the secondary circuit is opened while

the primary winding is carrying current, there will be no secondary ampere turns to balance the primary ampere turns. Therefore, the total primary current will become exciting current and magnetize the core to a high flux density, which will produce a high voltage across both the primary and secondary windings.

 WARNING!
The secondary circuit of a current transformer should never be opened while the primary is carrying current.

Because current transformers are designed for accuracy, the normal exciting current is only a small percentage of the full-load current. The voltage produced with the secondary open circuited is high enough to endanger the life of anyone coming in contact with the meters or leads. The high secondary voltage may also overstress the secondary insulation and cause a breakdown. Still other damage may be caused by operation with the secondary open circuited—the transformer core may become permanently magnetized, impairing the accuracy of the transformer. If this occurs, the core may be demagnetized by passing about 50% excess current through the primary, with the secondary connected to an adjustable high resistance that is gradually reduced to zero.

Three-Phase Current Transformers

A three-phase toroidal current transformer unit with the leads to a motor contactor passing through the center holes is shown here. The ground fault sensor is a similar type of transformer.

GROUND FAULT SENSOR

THREE-PHASE TOROIDAL CURRENT TRANSFORMER

406SA07.EPS

Single-Phase Current Transformers

This is an assembly of three wound-type single-phase current transformers serving a three-phase feeder.

406SA08.EPS

3.2.0 Potential Transformers

Potential or voltage transformers are single-phase transformers. If used on three-phase circuits, sets of two or three potential transformers are applied. The primary winding of a potential transformer is always connected across the main power lines.

A connection to a single-phase, two-wire circuit is shown in *Figure 11*. The primary of the potential transformer is connected across the main line, and the secondary is connected to an instrument. For safety, the secondary circuit is grounded.

The main circuit voltage exists across the primary winding. The secondary of the potential transformer is connected to one or more voltage-responsive devices, such as voltmeters, watt-meters, watt-hour meters, power factor meters, or some forms of relays or trip coils. The voltage across the secondary terminals of the potential transformer is always lower than the primary voltage and is rated at 120V. For example, if a potential transformer with a turns ratio of 12:1 is connected to a 1,380V main line, the voltage in the instrument connected across the secondary terminals will be about 120V.

3.2.1 Types of Potential Transformers

All potential transformers have a wound primary and a wound secondary. Mechanically, their construction is similar to that of wound current transformers. Their secondary thermal kVA rating seldom exceeds 1.5kVA.

Potential transformers are available in both oil-filled and dry types. In both types, the primary high-voltage leads are terminated at bushings and the housing contains the secondary low-voltage terminations. When supplied as part of a switch-gear assembly, they occupy dedicated spaces.

In certain small services, both current and potential transformers are packaged into a metering unit that is a single enclosure. This reduces the assembly time in field installations.

4.0.0 ◆ SIZING BUCK-AND-BOOST TRANSFORMERS

Manufacturers of buck-and-boost transformers normally offer easy-to-use selector charts that allow you to quickly select a buck-and-boost transformer for practically any application. These charts may be obtained from electrical equipment suppliers or ordered directly from the manufacturer—often at no charge. Instructions accompanying these charts will enable anyone familiar with transformers and electrical circuits to use them. An overview of the principles involved in using buck-and-boost transformers is given here.

When reviewing the selector charts, it may surprise you to discover that these transformers can handle loads that are much greater than their

Buck-and-Boost Transformers

A small buck-and-boost transformer is shown here. It is a single-phase compound-filled unit rated at 0.05kVA. However, depending on the percentage of voltage buck or boost required, it can be used in circuits with much higher loads.

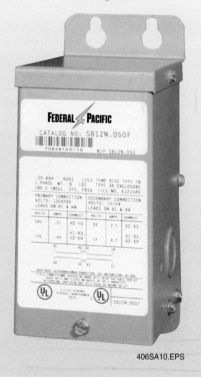

406SA10.EPS

nameplate ratings. For example, a typical 1kVA buck-and-boost transformer can easily handle an 11kVA load when the voltage boost is only 10%. In this case an isolation transformer will be incorporated as part of an autotransformer to form a buck-and-boost transformer.

Assume that you have a 1kVA (1,000VA) isolation transformer that is designed to transform 208V to 20.8V (see *Figure 12*). This results in a transformer winding ratio of 10:1. The primary current may be found using the following equation:

$$\text{Primary current} = \frac{1,000VA}{208V} = 4.8A$$

Because the transformation ratio is 10 to 1, the secondary amperes will be 48A (4.8A × 10 = 48A), or the amperage may be determined using the following equation:

$$\text{Secondary current} = \frac{1,000VA}{20.8V} = 48A$$

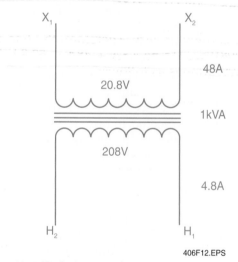

Figure 12 ◆ Isolation transformer.

406F12.EPS

Figure 13 shows how the windings are connected in series to form an autotransformer.

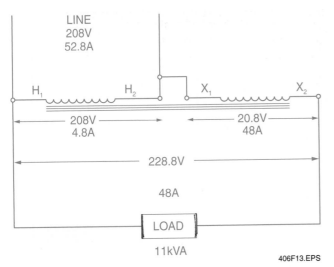

Figure 13 ◆ Boost transformer connection.

Because you started with 208V at the source and now add 20.8V to it, the load is now 208V + 20.8V = 228.8V. To find the kVA rating of the system at the load, use the following equation:

$$kVA = \frac{volts}{1,000} \times amps$$

$$= \frac{228.8V}{1,000} \times 48A$$

$$= 11kVA$$

Ten kVA is conducted from the source, and 1kVA is transformed from the source. The total kVA rating is 10kVA + 1kVA = 11kVA.

The H_1H_2 winding is rated at 4.8A, and the X_1X_2 winding is rated at 48A. Therefore, the line current at 208V would be 4.8A + 48A = 52.8A. The input kVA rating is as follows:

$$kVA = \frac{volts}{1,000} \times I$$

$$= \frac{208V}{1,000} \times 52.8A$$

$$= 11kVA$$

The diagrams shown in *Figures 12* and *13* are usually simplified even more in a line diagram, as shown in *Figure 14*. Actually, all three wiring diagrams indicate the same thing, and following the connections on any of these drawings will produce the same results at the load. *Figure 14* shows the calculations for a 240V source.

It should now be evident how a 1kVA buck-and-boost transformer, when connected in the circuit as described previously, can actually carry 11kVA in its secondary winding.

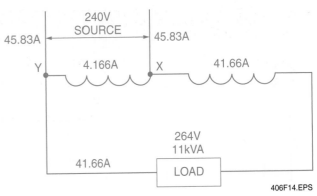

Figure 14 ◆ Typical line diagram of a transformer circuit.

5.0.0 ◆ HARMONICS

Harmonics are the byproducts of modern electronics. They are especially prevalent wherever there are large numbers of personal computers (PCs), adjustable-speed drives, and other types of equipment that draw current in short pulses.

This equipment is designed to draw current only during a controlled portion of the incoming voltage waveform. Although this dramatically improves efficiency, it causes harmonics in the load current. This results in overheated transformers and neutrals, as well as tripped circuit breakers.

The problem is evident when you look at a waveform. A normal 60-cycle power line voltage appears on the oscilloscope as a near sine wave, as shown in *Figure 15(A)*. When harmonics are present, the waveform is distorted, as shown in *Figure 15(B)* and *Figure 15(C)*. These waves are described as non-sinusoidal. The voltage and current waveforms are no longer simply related—hence the term nonlinear.

Finding the problem is relatively easy once you know what to look for and where to look. Harmonics are usually anything but subtle. This section will give you some basic pointers on how to find harmonics and some suggested ways to address the problem. However, in many cases, consultants must be called in to analyze the operation and design a plan for correcting the problem.

CAUTION

As part of a regular maintenance program, pay careful attention to overheating of the neutral conductors in distribution systems. Harmonics may cause deterioration of the insulation.

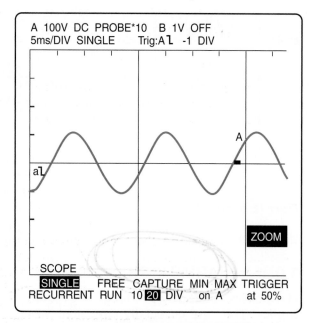

(A) NEAR SINE WAVE

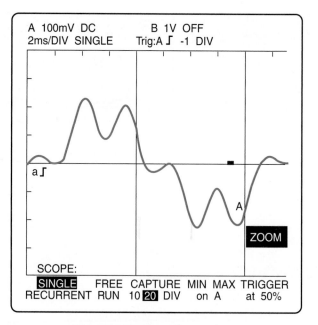

(B) DISTORTED WAVEFORM

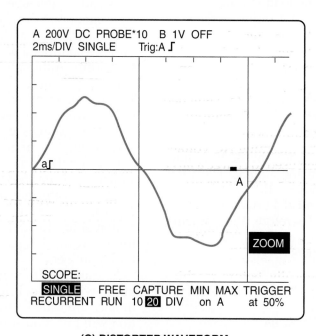

(C) DISTORTED WAVEFORM

406F15.EPS

Figure 15 ◆ Voltage waveforms.

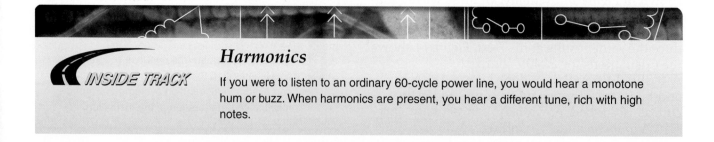

Harmonics

INSIDE TRACK

If you were to listen to an ordinary 60-cycle power line, you would hear a monotone hum or buzz. When harmonics are present, you hear a different tune, rich with high notes.

Nonlinear Loads

5.1.0 Defining the Problem

Harmonics are currents or voltages with frequencies that are integer multiples of the fundamental power frequency. For example, if the fundamental frequency is 60Hz, then the second harmonic is 120Hz, the third is 180Hz, and so on.

Harmonics are created by nonlinear loads that draw current in abrupt pulses rather than in a smooth sinusoidal manner. These pulses cause distorted current waveshapes, which in turn cause harmonic currents to flow back into other parts of the power system. This phenomenon is especially prevalent with equipment that contains diode/capacitor input or solid-state switched power supplies, such as personal computers, printers, and medical test equipment.

In a diode/capacitor, the incoming AC voltage is diode rectified and is then used to charge a large capacitor. After a few cycles, the capacitor is charged to the peak voltage of the sine wave (for example, 168V for a 120V line). The electronic equipment then draws current from this high DC voltage to power the rest of the circuit.

The equipment can draw the current down to a regulated lower limit. Typically, before reaching that limit, the capacitor is recharged to the peak in the next half cycle of the sine wave. This process is repeated over and over. The capacitor basically draws a pulse of current only during the peak of the wave. During the rest of the wave, when the voltage is below the capacitor residual, the capacitor draws no current.

5.1.1 Voltage Harmonics

The power line itself can be an indirect source of voltage harmonics. The harmonic current drawn by nonlinear loads acts in an Ohm's law relationship with the source impedance of the supplying transformer to produce the voltage harmonics.

The source impedance includes both the supplying transformer and the branch circuit components. For example, a 10A harmonic current being drawn for a source impedance of 0.1Ω will generate a harmonic voltage of 1.0V. Any loads sharing a transformer or branch circuit with a heavy harmonic load can be affected by the voltage harmonics generated.

Many types of devices are very susceptible to voltage harmonics. The performance of the diode/capacitor power supply is critically dependent on the magnitude of the peak voltage. Voltage harmonics can cause flat-topping of the voltage waveform, lowering the peak voltage. In severe cases, the computer may reset due to insufficient peak voltage.

In an industrial environment, the induction motor and power factor correction capacitors can also be seriously affected by voltage harmonics.

Power correction capacitors can form a resonant circuit with the inductive parts of a power distribution system. If the resonant frequency is near that of the harmonic voltage, the resultant harmonic current can increase substantially, overloading the capacitors and blowing the capacitor fuses. Fortunately, the capacitor failure detunes the circuit and the resonance disappears.

5.1.2 Classification of Harmonics

Each harmonic has a name, frequency, and sequence. The sequence refers to phasor rotation with respect to the fundamental (F); that is, in an induction motor, a positive sequence harmonic would generate a magnetic field that rotates in the same direction as the fundamental. A negative sequence harmonic would rotate in the reverse direction. The first nine harmonics, along with their effects, are shown in *Table 1*.

Table 1 Harmonic Rates and Effects

Name	F	2nd	3rd	4th	5th	6th	7th	8th	9th
Frequency	60	120	180	240	300	360	420	480	540
Sequence	+	−	0	+	−	0	+	−	0

Sequence	Rotation	Effects (skin effect, eddy currents, etc.)
Positive	Forward	Heating of conductors and circuit breakers
Negative	Reverse	Heating as above, plus motor problems
Zero	None	Heating, plus add-in neutral of three-phase, four-wire system

5.2.0 Office Buildings and Plants

Harmonics have a significant effect in office buildings and industrial establishments. Symptoms of harmonics usually show up in the power distribution equipment that supports the nonlinear loads. There are two basic types of nonlinear loads: single phase and three phase. Single-phase loads are prevalent in offices; three-phase loads are widespread in industrial plants.

Each component of the power distribution system manifests the effects of harmonics a little differently, but they are all subject to damage and inefficient performance.

5.2.1 Neutral Conductors

In a three-phase, four-wire system, the neutral conductor can be severely affected by nonlinear loads connected to the 120V branch circuits. Under normal conditions for a balanced linear load, the fundamental 60Hz portion of the phase currents will cancel in the neutral conductor.

In a four-wire system with single-phase, nonlinear loads, certain odd-numbered harmonics called triplens—odd multiples of the third harmonic: 3rd, 9th, 15th, etc.—do not cancel, but rather add together in the neutral conductor. In systems with many single-phase, nonlinear loads, the neutral current can actually exceed the phase current. The danger here is excessive overheating because there is no circuit breaker in the neutral conductor to limit the current as there are in the phase conductors.

Excessive current in the neutral conductor can also cause higher voltage drops between the neutral conductor and ground at the 120V outlet.

5.2.2 Circuit Breakers

Common thermal-magnetic circuit breakers use a bimetallic trip mechanism that responds to the heating effect of the circuit current. They are designed to respond to the true root-mean-square (rms) value of the current waveform and therefore will trip when they get too hot. This type of breaker has a better chance of protecting against harmonic current overloads.

A peak sensing electronic trip circuit breaker responds to the peak of the current waveform. As a result, it will not always respond properly to harmonic currents. Since the peak of the harmonic current is usually higher than normal, this type of circuit breaker may trip prematurely at a low current. If the peak is lower than normal, the breaker may fail to trip when it should.

5.2.3 Busbars and Connecting Lugs

Neutral busbars and connecting lugs are sized to carry the full value of the rated phase current. They can become overloaded when the neutral conductors are overloaded with the additional sum of the triplen harmonics.

5.2.4 Electrical Panels

Harmonics in electrical panels can be quite noisy. Panels that are designed to carry 60Hz current can become mechanically resonant to the magnetic fields generated by high-frequency harmonic currents. When this happens, the panel vibrates and emits a buzzing sound at the harmonic frequencies.

5.2.5 Telecommunications

Telecommunications cable is commonly run right next to power cables. To minimize the inductive interference from phase current, telecommunications cables are run closer to the neutral wire. Triplens in the neutral conductor commonly cause inductive interference that can be heard on a phone line. This is often the first indication of a harmonics problem and gives you a head start in detecting the problem before it causes major damage.

5.2.6 Transformers

Commercial buildings commonly have a 120V/208V transformer in a delta-wye configuration, as shown in *Figure 16*. These transformers commonly feed receptacles in a commercial building. Single-phase, nonlinear loads connected to the receptacles produce triplen harmonics that algebraically add up in the neutral. When this neutral current reaches the transformer, it is reflected into the delta primary winding, where it circulates and causes overheating and transformer failures.

Another transformer problem results from **core loss** and copper loss. Transformers are normally rated for a 60Hz phase current load only.

High-frequency harmonic currents cause increased core loss due to **eddy currents** and **hysteresis**, resulting in more heating than would occur at the same 60Hz current. These heating effects demand that transformers be derated for harmonic loads or replaced with specially designed transformers.

5.2.7 Generators

Standby generators are subject to the same types of overheating problems as transformers. Because they provide emergency backup for harmonics-producing loads such as data processing equipment, they are often even more vulnerable. In addition to overheating, certain types of harmonics produce distortion at the zero crossing of the current waveform, which causes interference and instability in the generator control circuits.

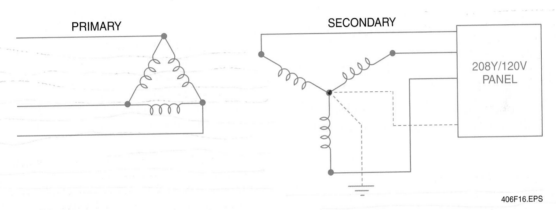

Figure 16 ◆ Three-phase, delta-wye transformer configuration.

K-Rated and Zig-Zag Transformers

Specially designed transformers that reduce or compensate for harmonic current include K-rated and zig-zag transformers. To prevent overheating, K-rated three-phase transformers for delta-to-wye connections are sized to handle 100% of the normal 60Hz load plus a specified nonlinear load defined by a K number. Manufacturers' specifications define the K numbers for various types of loads. If the various harmonics are known, a specific transformer can be custom built. The inherent phase shift of a K-rated transformer will cancel 5th and 7th harmonics but not the triplens. Because triplens are not cancelled, the neutral of the secondary is normally oversized at 200% of the maximum current rating of one of the phase connections. A zig-zag phase shift transformer cancels triplens and other harmonics at the load side of the transformer by using six windings in the secondary. An extra winding for each phase is connected in series with another phase to produce a phase shift. This type of transformer works well if the loads are balanced. A drawback is that zig-zag transformers use more material in the windings and are heavy. Other techniques such as active filters or ferroresonant transformers are also used for harmonic current suppression in certain applications.

5.3.0 Survey the Situation

A quick survey will help to determine whether or not you have a harmonics problem and where it is located. The survey procedure should include the following:

Step 1 Take a walking tour of the facility and look at the types of equipment in use. If there are a lot of personal computers and printers, adjustable-speed motors, solid-state heater controls, and certain types of fluorescent lighting, there is a good chance that harmonics are present.

Step 2 Locate the transformers feeding the non-linear loads and check for excess heating. Also, make sure that the cooling vents are unobstructed.

Step 3 Use a true rms meter to check transformer currents.
- Verify that the voltage ratings for the test equipment are adequate for the transformer being tested.
- Measure and record the transformer secondary currents in each phase and in the neutral (if used).
- Calculate the kVA delivered to the load, and compare it to the nameplate rating. If harmonic currents are present, the transformer can overheat, even if the kVA delivered is less than the nameplate rating.
- If the transformer secondary is a four-wire system, compare the measured neutral current to the value predicted from the imbalance in the phase currents. (The neutral current is the vector sum of the phase currents and is normally zero if the phase currents are balanced in both amplitude and phase.) If the neutral current is unexpectedly high, triple harmonics are likely, and the transformer may need to be derated.
- Measure the frequency of the neutral current. 180Hz would be a typical reading for a neutral current consisting of mostly third harmonics.

Step 4 Survey the subpanels that feed harmonic loads. Measure the current in each branch neutral, and compare the measured value to the rated capacity for the wire size used. Check the neutral busbar and feeder connections for heating or discoloration.

Step 5 Neutral overloading in receptacle branch circuits can sometimes be detected by measuring the neutral-to-ground voltage at the receptacle. Measure the voltage when the loads are on. A reading of 2V or less is normal. Higher voltages can indicate trouble, depending on the length of the run, quality of the connections, etc. Measure the frequency. 180Hz would suggest a strong presence of harmonics. 60Hz would suggest that the phases are out of balance. Pay special attention to under-carpet wiring and modular office panels with integrated wiring that use a neutral shared by three-phase conductors. Because the typical loads in these two areas are computer and office machines, they are often trouble spots for overloaded neutrals.

5.3.1 Meters

Having the proper equipment is crucial to diagnosing harmonics problems. The type of equipment used varies with the complexity of measurements required.

To determine whether you have a harmonics problem, you need to measure the true rms value and the instantaneous peak value of the waveshape. For this test, you need a true rms clamp-on multimeter or a handheld digital multimeter that makes true rms measurements and has a high-speed peak hold circuit.

The term true rms refers to the root-mean-square or equivalent heating value of a current or voltage waveshape. True distinguishes the measurement from those taken by average responding meters.

The vast majority of low-cost, portable clamp-on ammeters are average responding. These instruments give correct readings for pure sine waves only and will typically read low when confronted with a distorted current waveform. The result is a reading that can be up to 50% low.

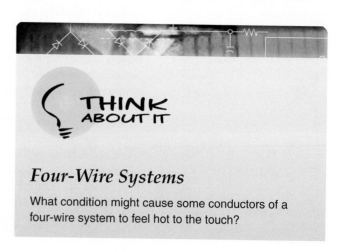

THINK ABOUT IT

Four-Wire Systems

What condition might cause some conductors of a four-wire system to feel hot to the touch?

True rms meters give correct readings for any waveshape within the instrument's crest factor and bandwidth specifications.

5.3.2 Crest Factor

The crest factor of a waveform is the ratio of the peak value to the rms value. For a sine wave, the crest factor is 1.414. A true rms meter will have a crest factor specification. This specification relates to the level of peaking that can be measured without errors.

A quality true rms handheld digital multimeter has a crest factor of 3.0 at full scale. This is more than adequate for most power distribution measurements. At half scale, the crest factor is double. For example, a meter may have a crest factor specification of 3.0 when measuring 400VAC and a crest factor of 6.0 when measuring 200VAC.

NOTE

Most true rms meters cannot be used for signals below 5% of scale because of the measurement noise problem. Use a lower range if it is available.

The crest factor can be easily calculated using a true rms meter with a peak function or a crest function. A crest factor other than 1.414 indicates the presence of harmonics. In typical single-phase cases, the greater the difference from 1.414, the greater the harmonics. For voltage harmonics, the typical crest factor is below 1.414 (i.e., a flat-top waveform). For single-phase current harmonics, the typical crest factor is well above 1.414. Three-phase current waveforms often exhibit the double-hump waveform; therefore, the crest factor comparison method should not be applied to three-phase load currents.

After you have determined that harmonics are present, you can make a more in-depth analysis of the situation using a harmonics analyzer.

5.4.0 Solving the Problem

The following are some suggested ways of addressing some typical harmonics problems. Before taking any measures, you should consult a power quality expert to analyze the problem and design a plan tailored to your specific situation.

Power Quality Analyzers

Portable power quality analyzers like the one shown here can be used to determine harmonics, as well as to make power measurements (kW, VA, and VAR), power factor and displaced power factor measurements, voltage and current readouts and waveforms, inrush current and duration recording, transient measurements, and sag and swell recording. Extensive power quality measurements can be made continuously using the software provided with most energy management systems.

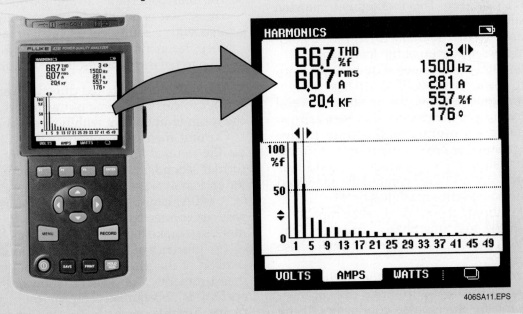

406SA11.EPS

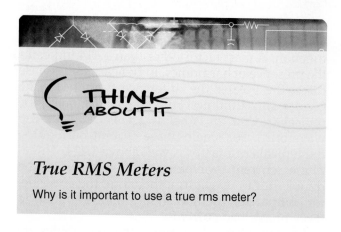

True RMS Meters

Why is it important to use a true rms meter?

5.4.1 Harmonics in Overloaded Neutrals

In a three-phase, four-wire wye system, the 60Hz portion of the neutral current can be minimized by balancing the loads in each phase. *NEC Section 220.61* prohibits reduced sizing of neutral conductors serving nonlinear loads. The triplen harmonics neutral current can be reduced by adding harmonic filters at the load. If neither of these solutions is practical, you can pull in extra neutrals (ideally one neutral for each phase) or you can install an oversized neutral to be shared by the three conductors.

In new construction, under-carpet wiring and modular office partition wiring should be specified with individual neutrals and possibly an isolated ground separate from the safety ground.

5.4.2 Derating Transformers

One way to protect a transformer from harmonics is to limit the amount of load placed on it. This is called derating the transformer. The most rigorous derating method is described in *ANSI/IEEE Standard C57.110-1986*. This method is somewhat impractical because it requires extensive loss data from the transformer manufacturer, plus a complete harmonics spectrum of the load current.

The Computer and Business Equipment Manufacturers' Association (CBEMA) has recommended a second method that involves several straightforward measurements that you can get using common test equipment. It appears to give reasonable results for 208V/120Y receptacle transformers that supply the low-frequency odd harmonics (3rd, 5th, and 7th) commonly generated by computers and office machines.

The test equipment you use must be capable of taking both the true rms phase current and the instantaneous peak phase current for each phase of the secondary.

To determine the derating factor for the transformer, take the peak and true rms current measurements for the three-phase conductors. If the phases are not balanced, average the three measurements and plug that value into the following equation:

Transformer harmonics derating factor (THDF) =

$$\frac{1.414 \times \text{true rms phase current}}{\text{instantaneous peak phase current}}$$

This equation generates a value between 0 and 1.0 (typically between 0.5 and 0.9). If the phase currents are purely sinusoidal (undistorted), the instantaneous current peaks at 1.414 times the true rms value and the derating factor is 1.0. If that is the case, no derating is required.

However, with harmonics present, the transformer rating is the product of the nameplate kVA rating times the THDF:

Derated kVA = THDF × nameplate kVA

For example, a modern office building dedicated primarily to computer software development contains a large number of PCs and other electronic office equipment. These electronic loads are fed by a 120V/208V transformer configured with a delta primary and a wye secondary. The PCs are fairly well distributed throughout the building, except for one large room that contains several machines. The PCs in this room, used exclusively for testing, are served by several branch circuits.

The transformer and main switchgear are located in a ground floor electrical room. An inspection of this room immediately reveals two symptoms of high harmonic currents:

- The transformer is generating a substantial amount of heat.
- The main panel emits an audible buzzing sound. The sound is not the chatter commonly associated with a faulty circuit breaker, but rather a deep, resonant buzz that indicates the mechanical parts of the panel itself are vibrating.

The ductwork installed directly over the transformer to carry off some of the excess heat keeps the room temperature within reasonable limits.

Current measurements (see *Table 2*) are taken on the neutral and on each phase of the transformer secondary using both a true rms multimeter and an average responding unit.

A 600A, clamp-on current transformer accessory is connected to each meter to allow the meters to make high current readings.

The presence of harmonics is obvious by a comparison of the phase current and neutral current measurements. As *Table 2* shows, the neutral current is substantially higher than any of the phase currents, even though the phase currents

Table 2 Current Measurements

Conductor Name	Multimeter (true rms)	Average Responding Multimeter	Instantaneous Peak Current
Phase 1	410A	328A	804A
Phase 2	445A	346A	892A
Phase 3	435A	355A	828A
Neutral	548A	537A	762A

are relatively well balanced. The average responding meter consistently shows readings that are approximately 20% low on all phases. Its neutral current readings are only 2% low.

The waveforms explain the discrepancy. The phase currents are badly distorted by large amounts of triplen harmonics, while the neutral current is not affected. The phase current readings listed in *Table 2* clearly demonstrate why true rms measurement capability is required to accurately determine the value of harmonic currents.

The next step is to calculate the transformer harmonic derating factor, or THDF, as explained previously.

The results indicate that, with the level of harmonics present, the transformer should be derated to 72.3% of its nameplate rating to prevent overheating. In this case, the transformer should be derated to 72.3% of its 225kVA rating, or 162.7kVA.

The actual load is calculated to be 151.3kVA. Although this figure is far less than the nameplate rating, the transformer is operating close to its derated capacity.

Next, a subpanel that supplies branch circuits for the 120V receptacles is also examined. The current in each neutral is measured and shown in *Table 3*.

When a marginal or overloaded conductor is identified, the associated phase currents and the neutral-to-ground voltage at the receptacle are also measured. When a check of neutral No. 6 reveals 15A in a conductor rated for 16A, the phase currents of the circuits (No. 25, No. 27, and No. 29) that share that neutral are also measured. See *Table 4*.

Note that each of the phase currents of these three branch circuits is substantially less than 15A and the same phase conductors have significant neutral-to-ground voltage drops.

In the branch circuits that have high neutral currents, the relationship between the neutral and the phase currents is similar to that of the transformer secondary. The neutral current is higher than any of the associated phase currents. The danger here is that the neutral conductors could become overloaded and not offer the warning signs of tripped circuit breakers.

The recommendations are:

- Refrain from adding additional loads to the receptacle transformer unless steps are taken to reduce the level of harmonics.
- Pull extra neutrals into the branch circuits that are heavily loaded.
- Monitor the load currents on a regular basis using true rms test equipment.

Table 3 Neutral Loads

Neutral Conductor Number	Current (Amps)
01	5.0
02	11.3
03	5.0
04	13.1
05	12.4
06	15.0
07	1.8
08	11.7
09	4.5
10	11.8
11	9.6
12	11.5
13	11.3
14	6.7
15	7.0
16	2.3
17	2.6

Table 4 Phase Currents and Neutral-to-Ground Voltages

Circuit Number	Phase Current	Neutral-to-Ground Voltage Drop at Receptacle
25	7.8A	3.75V
27	9.7A	4.00V
29	13.5A	8.05V

1. What is the name given to the large transformers used in transmission systems to step voltage up and down?
   a. Control transformers
   b. Power transformers
   c. Instrument transformers
   d. Potential transformers

2. The purpose of the two secondary windings in a three-winding, single-phase transformer is _____.
   a. so the transformer may be used on systems with different primary voltages
   b. to produce a non-magnetic field
   c. to produce a high-impedance field
   d. to produce two secondary voltages

3. How many windings are normally found in an autotransformer?
   a. One
   b. Two
   c. Three
   d. Four

4. What type of transformer is normally used with metering equipment when the current is too high for the metering equipment?
   a. Potential transformer
   b. Reactance transformer
   c. Current transformer
   d. Autotransformer

5. What type of transformer is normally used with metering equipment when the line voltage is too high for the metering equipment?
   a. Potential transformer
   b. Reactance transformer
   c. Current transformer
   d. Autotransformer

6. What is a transformer called when it is connected in series with the main line?
   a. Reactance transformer
   b. Series transformer
   c. High-voltage transformer
   d. Low-ampere transformer

7. A(n) _____ is always a single-phase transformer.
   a. power transformer
   b. reactance transformer
   c. current transformer
   d. autotransformer

8. The secondary thermal kVA rating of a conventional potential transformer seldom exceeds _____.
   a. 1kVA
   b. 1.5kVA
   c. 1,500kVA
   d. 15,000kVA

9. What type of transformer is normally able to handle a load that is much greater than its nameplate rating?
   a. Potential transformer
   b. Reactance transformer
   c. Current transformer
   d. Buck-and-boost transformer

10. The problem of harmonics is often first detected by the presence of inductive interference in _____ systems.
    a. lighting
    b. heating
    c. telecommunications
    d. emergency power

11. Potential transformers are available in _____ types.
    a. oil-filled and hydraulic
    b. oil-filled and dry
    c. dry and chemical
    d. dry and coiled

12. Regulators are usually designed to provide a total secondary voltage range of _____ of the supply voltage.
    a. 20%
    b. 35%
    c. 45%
    d. 55%

13. Each harmonic has a _____.
    a. name only
    b. name and frequency only
    c. name, frequency, and sequence
    d. name, frequency, and rotation

14. Neutral busbars and connecting lugs are sized to carry _____ of the rated phase current.
    a. 75%
    b. 80%
    c. 90%
    d. 100%

15. For a sine wave, the crest factor is _____.
    a. 1.000
    b. 1.414
    c. 2.212
    d. 2.414

Summary

Transformers are used in numerous applications to alter the output voltage to serve various purposes. This module covered types of specialty transformers, including:

- Power transformers
- Buck-and-boost transformers
- Current and potential transformers
- Transformers with multiple secondaries
- Autotransformers
- Constant-current transformers

- Control transformers
- Series transformers
- Rectifier transformers
- Step-voltage regulators

It also discussed the problem of harmonics and its associated symptoms and possible solutions. A complete understanding of transformer selection, application, and troubleshooting techniques is essential to the proper installation and servicing of electrical systems.

Notes

Trade Terms
Introduced in This Module

Ampere turn: The product of amperes times the number of turns in a coil.

Autotransformer: Any transformer in which the primary and secondary connections are made to a single winding. The application of an auto-transformer is a good choice where a 480Y/277V or 208Y/120V, three-phase, four-wire distribution system is used.

Bank: An installed grouping of a number of units of the same type of electrical equipment, such as a bank of transformers, a bank of capacitors, or a meter bank.

Core loss: The electric loss that occurs in the core of an armature or transformer due to conditions such as the presence of eddy currents or hysteresis.

Eddy currents: The circulating currents that are induced in conductive materials by varying magnetic fields; they are usually considered undesirable because they represent a loss of energy and produce excess heat.

Harmonic: An oscillation whose frequency is an integral multiple of the fundamental frequency.

Hysteresis: The time lag exhibited by a body in reacting to changes in the forces affecting it; hysteresis is an internal friction.

Impedance: The opposition to current flow in an AC circuit; impedance includes resistance (R), capacitive reactance (X_C), and inductive reactance (X_L). It is measured in ohms (Ω).

Isolation transformer: A transformer that has no electrical metallic connection between the primary and secondary windings.

Reactance: The imaginary part of impedance; also, the opposition to alternating current due to capacitance (X_C) and/or inductance (X_L).

This module is intended to present thorough resources for task training. The following reference work is suggested for further study. This optional material is for continuing education rather than for task training.

National Electrical Code® Handbook, Latest Edition. Quincy, MA: National Fire Protection Association.

CONTREN® LEARNING SERIES – USER UPDATE

NCCER makes every effort to keep these textbooks up-to-date and free of technical errors. We appreciate your help in this process. If you have an idea for improving this textbook, or if you find an error, a typographical mistake, or an inaccuracy in NCCER's Contren® textbooks, please write us, using this form or a photocopy. Be sure to include the exact module number, page number, a detailed description, and the correction, if applicable. Your input will be brought to the attention of the Technical Review Committee. Thank you for your assistance.

Instructors – If you found that additional materials were necessary in order to teach this module effectively, please let us know so that we may include them in the Equipment/Materials list in the Annotated Instructor's Guide.

Write: Product Development and Revision
National Center for Construction Education and Research
3600 NW 43rd St., Bldg. G, Gainesville, FL 32606

Fax: 352-334-0932

E-mail: curriculum@nccer.org

Craft _____   Module Name _____

Copyright Date _____   Module Number _____   Page Number(s) _____

Description _____

_____

_____

_____

_____

(Optional) Correction _____

_____

_____

(Optional) Your Name and Address _____

_____

_____

Advanced Controls

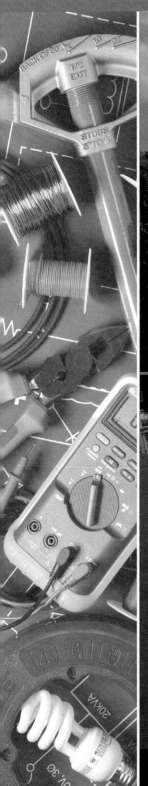

New York Institute of Technology – Open Minds, Open House

Students from the New York Institute of Technology named their dwelling "Open House" for two reasons. First, they are targeting beachfront homeowners to show them how an open solar design can complement shoreline properties. Second, the term "open house" is an expression of the team's ideal home: a home with influence extending beyond its physical walls.

26407-08

26407-08
Advanced Controls

Topics to be presented in this module include:

1.0.0	Introduction	7.2
2.0.0	Solid-State Relays	7.2
3.0.0	Solid-State Protective Relays	7.7
4.0.0	Timing Relays	7.9
5.0.0	Reduced-Voltage Starting Motor Control	7.16
6.0.0	Adjustable Frequency Drives	7.25
7.0.0	Motor Braking Methods	7.34
8.0.0	Precautions When Working with Solid-State Controls	7.37
9.0.0	Motor Control Maintenance	7.38
10.0.0	Motor Control Troubleshooting	7.39

Overview

At its most basic level, a motor control simply turns a motor on or off. With small motors, this can be accomplished with a simple toggle switch as long as the motor's current draw is small enough so that undesirable arcing does not occur between the switch contacts. In cases where the motor current draw is large enough to cause arcing between switching contacts, electromagnetic motor starters are normally used. In a motor starter, an electrically isolated magnetic control circuit opens and closes the motor's main line contacts.

With today's applications of increased motor horsepower ratings and a demand for finely controlled torque and motor speed, simple motor control methods are not sufficient to do the job. In order to achieve the desired control characteristics, advanced motor control systems are being designed and installed at a rapid pace. Electricians must understand conventional motor control and electronic motor control systems now being installed. Electricians must also maintain their training to keep up with technology in order to achieve success in motor control installation and service.

Objectives

When you have completed this module, you will be able to do the following:

1. Select and install solid-state relays for specific applications in motor control circuits.
2. Install non-programmable/programmable motor circuit protectors (solid-state overload relays) in accordance with the manufacturer's instructions.
3. Select and install electromechanical and solid-state timing relays for specific applications in motor control circuits.
4. Recognize the different types of reduced-voltage starting motor controllers and describe their operating principles.
5. Connect and program adjustable frequency drives to control a motor in accordance with the manufacturer's instructions.
6. Demonstrate and/or describe the special precautions used when handling and working with solid-state motor controls.
7. Recognize common types of motor braking and explain the operating principles of motor brakes.
8. Perform preventive maintenance and troubleshooting tasks in motor control circuits.

Trade Terms

Base speed
Closed circuit transition
Dashpot
Heat sink
Insulated gate bipolar transistor (IGBT)
Latch
Open circuit transition
Solid-state overload relay (SSOLR)
Solid-state relay (SSR)
Thermal resistance
Thyristor

Required Trainee Materials

1. Pencil and paper
2. Appropriate personal protective equipment
3. Copy of the latest edition of the *National Electrical Code*®

26413-08 Introductory Skills for the Crewleader

26412-08 Special Locations

26411-08 Medium-Voltage Terminations/Splices

26410-08 Motor Operation and Maintenance

26409-08 Heat Tracing and Freeze Protection

26408-08 HVAC Controls

26407-08 Advanced Controls

26406-08 Specialty Transformers

26405-08 Fire Alarm Systems

26404-08 Basic Electronic Theory

26403-08 Standby and Emergency Systems

26402-08 Health Care Facilities

26401-08 Load Calculations – Feeders and Services

ELECTRICAL LEVEL THREE

ELECTRICAL LEVEL TWO

ELECTRICAL LEVEL ONE

CORE CURRICULUM: Introductory Craft Skills

ELECTRICAL LEVEL FOUR

407CMAP.EPS

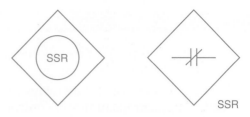

SCHEMATIC SYMBOLS

Before you begin this module, it is recommended that you successfully complete *Core Curriculum; Electrical Level One; Electrical Level Two; Electrical Level Three; Electrical Level Four,* Modules 26401-08 through 26406-08.

This course map shows all of the modules in *Electrical Level Four.* The suggested training order begins at the bottom and proceeds up. Skill levels increase as you advance on the course map. The local Training Program Sponsor may adjust the training order.

1.0.0 ◆ INTRODUCTION

This module is the third of three related modules that cover electrical motor controls and control circuits. It expands upon the material contained in the *Contactors and Relays* and *Motor Controls* modules studied earlier in your training. This module builds upon the knowledge you gained from the previous modules and provides new information with an emphasis on solid-state and other specialized motor controls and their applications. Maintenance and troubleshooting of motor control circuits and devices are also covered.

2.0.0 ◆ SOLID-STATE RELAYS

Solid-state relays (*Figure 1*) have many advantages over electromechanical relays. They have longer life, higher reliability, higher speed switching, and high resistance to shock and vibration. The absence of mechanical contacts eliminates contact bounce, arcing when the contacts open, and hazards from explosives and flammable gases. For this reason, solid-state relays (SSRs) are replacing electromechanical relays (EMRs) in many applications. The principles of operations of solid-state relays were briefly introduced in the *Motor Controls* module in Level Three of your training. They are covered in more detail here.

2.1.0 Solid-State Relay Operation

Solid-state relays can be used in many of the same control applications as EMRs and in many new applications involving electronic control circuit devices. An SSR is activated by the control circuit voltage applied to its input terminals. Depending on the SSR design, this input control voltage can be AC or DC. AC voltages typically range between 90VAC and 280VAC; DC voltages range between 3VDC and 32VDC. DC voltage inputs are normally applied from digitally controlled motor

Figure 1 ◆ Solid-state relay with heat sink.

control circuits. Application of the AC or DC input control voltage to the relay can be done by the closure of a mechanical switch or from the activation of a solid-state device, such as a transistor or integrated circuit.

Internally, an SSR consists of an input control circuit and an output circuit (*Figure 2*). The function of the input control circuit can be compared to that of the coil in an EMR in that it senses the input control signal and provides coupling between the input and output circuits. In most SSRs, the coupling of the switching command to the output circuit is done optically by the use of a light-emitting diode (LED) and photo detector.

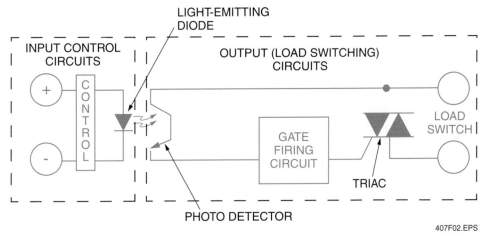

Figure 2 ◆ Block diagram of an optically isolated solid-state relay.

This type of SSR is called an optically isolated SSR. Since a light beam is used as the coupling medium, voltage spikes or electrical noise generated on the load side of the SSR are prevented from being coupled into the input control circuit side of the SSR. In SSRs that are not optically isolated, a small reed relay is commonly used to provide the coupling between the relay's input control circuit and the output switch circuit.

When the applied input control voltage level exceeds a predetermined threshold known as the pickup voltage, a light beam is generated by the LED and transmitted to the photo detector. This transmitted light is received by the photo detector and is used to trigger the output switch gate firing circuit, causing the output switch to be activated. The output switch will remain activated until the input control voltage level drops below the relay's minimum dropout voltage level, at which time the LED stops emitting light and the output switch is deactivated.

The output switching circuit electronically performs the same function as the mechanical contacts of an EMR. Depending on the design of the SSR and the type of load to be controlled, the output switching can be performed by power transistors, silicon-controlled rectifiers (SCRs), triacs, or thyristors. Power transistor and SCR switched outputs are typically used to control low-current and high-current DC loads, respectively. Triac switched outputs are typically used to control high-current AC loads, while thyristor switched outputs are used to control either low- or high-current AC loads, depending on their circuit configuration. The switching circuit can operate in one of four switching modes, depending on the SSR's design.

- *Instant on* – Turns on the load immediately at any point on the AC sine wave when the control voltage is present. It turns off the load when the control voltage is removed and the current in the load crosses zero. This mode of operation is the same as that which occurs with an EMR. Instant-on relays are commonly used to control contactors, magnetic starters, and similar loads.

- *Zero switching* – Turns on the load when the control voltage is applied and the voltage at the load crosses zero. The relay turns off the load when the control voltage is removed and the current in the load crosses zero. Zero switching relays are typically used to control heating elements, ovens, incandescent and tungsten lamps, and programmable controllers.

- *Peak switching* – Turns on the load when the control voltage is applied and the voltage at the load is at its peak amplitude. The relay turns off the load when the control voltage is removed and the current in the load crosses zero. Peak switching relays are typically used to control heavy inductive loads.

- *Analog switching* – In this type of relay, the amount of output voltage varies as a function of the input voltage. For any voltage within its input range (typically 0VDC to 5VDC), the output is a percentage of the available output voltage within its rated range. Analog switching relays are commonly used in closed-loop control applications.

2.2.0 Comparison of Electromechanical Relays to Solid-State Relays

Differences exist both in the terminology used to describe EMRs and SSRs and in their overall ability to perform certain functions. As an aid in developing a comparison between an EMR and an SSR, refer to *Tables 1* and *2*. *Table 1* shows the equivalent or comparable terminology between an EMR and an SSR; *Table 2* helps you to determine each

Table 1   Comparison Chart for EMR and SSR Technology

Comparable or Equivalent Terminology for Electromechanical and Solid-State Relays	
Electromechanical Relays (EMRs)	**Solid-State Relays (SSRs)**
1. *Coil voltage:* The minimum voltage necessary to energize or operate the relay. This value is also referred to as the pickup voltage.	1. *Control voltage:* The minimum voltage required to gate or activate the control circuit of the solid-state relay. Generally a maximum value is also specified.
2. *Coil current:* In conjunction with the coil voltage, the amount of current necessary to energize or operate the relay.	2. *Control current:* The minimum current required to turn on the solid-state control circuit. Generally a maximum value is also specified.
3. *Hold current:* The minimum current required to keep a relay energized or operating.	3. *See control current.*
4. *Dropout voltage:* The maximum voltage at which the relay is no longer energized.	4. *See control voltage.*
5. *Pull-in time:* The amount of time required to operate (open or close) the relay contacts after the coil voltage is applied.	5. *Turn-on time:* The elapsed time between the application of the control voltage and the application of the voltage to the load circuit.
6. *Dropout time:* The amount of time required for the relay contacts to return to their normal unoperated position after the coil voltage is removed.	6. *Turn-off time:* The elapsed time between the removal of the control voltage and the removal of the voltage from the load circuit.
7. *Contact voltage rating:* Maximum voltage rating that the contacts of a relay are capable of switching safely.	7. *Load voltage:* The maximum output voltage handling capability of a solid-state relay.
8. *Contact current rating:* Maximum current rating that the contacts of a relay are capable of switching safely.	8. *Load current:* The maximum output current handling capability of a solid-state relay.
9. *Surge current:* Maximum peak current that the contacts of a relay can withstand for short periods of time without damage.	9. *Surge current:* Maximum peak current that a solid-state relay can withstand for short periods of time without damage.
10. *Contact voltage drop:* Voltage drop across relay contacts when relay is operating (usually quite low).	10. *Switch-on voltage drop:* Voltage drop across a solid-state relay when operating.
11. *Insulation resistance:* Amount of resistance measured across relay contacts in open position.	11. *Switch-off resistance:* Amount of resistance measured across a solid-state relay when turned off.
12. *No equivalent.*	12. *Off state current leakage:* Amount of current leakage through a solid-state relay when turned off but still connected to the load voltage.
13. *No equivalent.*	13. *Zero current turn-off:* Turn-off at essentially the zero crossing of the load current that flows through an SSR. A thyristor will turn off only when the current falls below the minimum holding current. If input control is removed when the current is at a higher value, turn-off will be delayed until the next zero current crossing.
14. *No equivalent.*	14. *Zero voltage turn-on:* Initial turn-on occurs at a point near zero crossing of the AC line voltage. If input control is applied when the line voltage is at a higher value, initial turn-on will be delayed until the next zero crossing.

device's ability to perform certain functions. When making a choice between solid-state and electro-mechanical devices, you must compare the electrical and mechanical operating characteristics of each device with the application in which it is to be used.

The electronic nature of the SSR and its input circuit lends itself to use in digitally controlled logic circuits. Another advantage of the SSR over the EMR is its response time or ability to turn on and off very quickly. The reason for this is that the SSR may be turned on and off electronically much more rapidly than a relay can be electromagnetically pulled in and dropped out.

When comparing the voltage and current ratings of an EMR and an SSR in the load circuit, you are in effect comparing the maximum safe switching capability of a set of mechanical contacts to that of an electronic switching device such as an SCR or triac. The terminology is not equivalent,

Table 2 EMR and SSR Performance

Advantages and Disadvantages of Electromechanical and Solid-State Relays		
General Characteristics	**EMR**	**SSR**
1. Arcless switching of the load	−	+
2. Electronic (IC, etc.) compatibility for interfacing	−	+
3. Effects of temperature	+	−
4. Shock and vibration resistant	−	+
5. Immunity to improper functioning because of transients	+	−
6. Radio frequency switching	+	−
7. Zero voltage turn-on	−	+
8. Acoustic noise	−	+
9. Selection of multipole, multithrow switching capability	+	−
10. Contact bouncing	−	+
11. Ability to stand surge currents	+	−
12. Response time	−	+
13. Voltage drop in load circuit	+	−
14. AC & DC switching with same contacts	+	−
15. Zero current turn-off	−	+
16. Leakage current	+	−
17. Minimum current turn-on	+	−
18. Life expectancy	−	+
19. Initial cost	+	−
20. Real cost-lifetime	−	+

Note: Plus indicates advantages; minus indicates disadvantages.

but is somewhat comparable. Each device has certain limitations that will determine how much current and voltage it can safely handle. Since this varies from device to device and manufacturer to manufacturer, data sheets must be used to determine if a given device will safely switch a given load.

2.3.0 Two-Wire and Three-Wire SSR Control

Like an EMR, an SSR can be connected to achieve either two-wire or three-wire control. *Figure 3(A)* shows an SSR connected for two-wire control of an AC load. Pressing the START pushbutton in this circuit activates the SSR, which then energizes the load. The SSR remains activated only as long as the pushbutton remains pressed. Once it is released, the SSR will deactivate, turning off the load.

Figure 3(B) shows the SSR connected for three-wire control of an AC load. It also shows an SCR being used to provide memory for the START pushbutton after it has been pressed. Pressing the START pushbutton in this circuit activates the SSR, turning on the load. It also applies the control voltage to the gate input of an SCR, causing the SCR to energize and maintain (latch in) the on condition. This allows the DC control current to pass through the SSR input circuit and SCR even

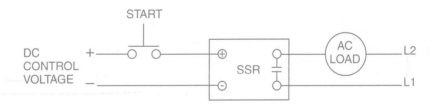

(A) TWO-WIRE CONTROL

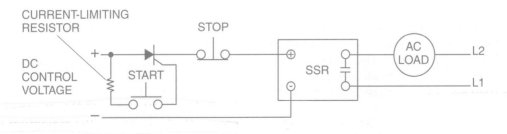

(B) THREE-WIRE CONTROL

407F03.EPS

Figure 3 ◆ Two-wire and three-wire SSR control.

after the START pushbutton is released. Resistor R1 is a current-limiting resistor for the SCR gate. The SSR will remain activated until the STOP pushbutton is pressed, interrupting the flow of current through the SCR.

2.4.0 Connecting SSRs to Achieve Multiple Outputs

SSRs are made with either single or multiple switched outputs. If an application requires that multiple SSR outputs be controlled by one control input signal, a multiple-output SSR would normally be used. However, the inputs of two or more single-output SSRs can be connected in series or in parallel to obtain the required number of switched outputs. *Figure 4* shows examples of three single-output SSRs controlled by a single switch connected both in parallel and in series to control the application of three-phase power to a load. When connecting SSRs in series, the DC control voltage divides equally across the individual SSRs. This reduces the voltage available to each one. For this reason, the supply voltage must be increased proportionally when multiple SSRs are connected in series. For the example shown, the control supply voltage for the three SSRs must be increased to at least three times the minimum operating voltage required for use with a single relay.

2.5.0 SSR Temperature Considerations

The operation of an SSR (or any solid-state device) can be affected by exposure to high temperatures. Manufacturers of SSRs specify the maximum temperature permitted for use with their relays. Typically this is 40°C (104°F). High temperatures at an SSR can result from high ambient temperatures in the surrounding area and in the enclosure in which the SSR is mounted. The power switching devices in an SSR itself also generate heat that contributes to its temperature. The larger the current passing through the relay, the more heat is produced.

Sometimes the temperature of SSRs and other devices is controlled by forced air cooling, but in most cases it is controlled by the use of a **heat sink** shown earlier in *Figure 1*. A heat sink is a metal base, commonly aluminum, used to dissipate the heat of the solid-state components mounted on it. In some SSRs, the package of the device often serves as its heat sink, but for an SSR used to handle higher power, a separate heat sink is often needed. A heat sink has a low **thermal resistance** as determined by the size of its surface area and

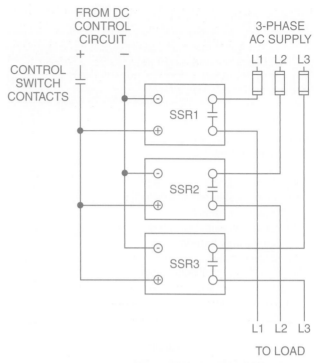

(A) THREE SSRs CONNECTED IN PARALLEL

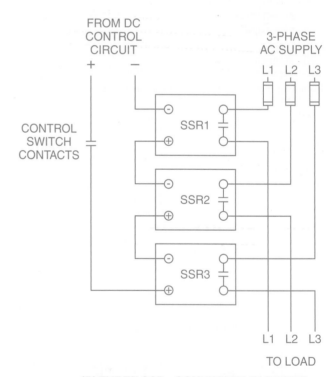

(B) THREE SSRs CONNECTED IN SERIES

407F04.EPS

Figure 4 ◆ Multiple SSRs connected in parallel and series to control a three-phase load.

the type of material from which it is made. Thermal resistance is the resistance of a material to the conduction of heat. The lower the thermal resistance number, the greater the ability to dissipate heat. The use of a heat sink enables the SSR to control higher current levels.

When selecting a heat sink for use with a particular SSR, follow the SSR manufacturer's recommendations. When installing a heat sink, follow these guidelines:

- When practical, use a heat sink with fins to provide the greatest heat dissipation per surface area.
- Make sure that the mounting surface between the heat sink and the SSR is flat and smooth.
- Use thermal grease or a pad between the SSR and the heat sink surfaces to eliminate any air gaps and enhance the thermal conductivity.
- Make sure all mounting hardware is securely fastened.

2.6.0 Solid-State Relay Overvoltage and Overcurrent Protection

An SSR's power transistors, SCRs, and triacs can be damaged by a shorted load. It is a good practice to install an overcurrent protection fuse to open the output circuit should the load current increase to a higher value than the nominal load current. This fuse should be an ultra-fast fuse designed for use with semiconductor devices.

To protect the power transistors, SCRs, triacs, etc. from overvoltages resulting from transients induced on the load lines, it is recommended that a peak voltage clamping device such as a varistor, zener diode, or a snubber network be installed in the output circuit.

3.0.0 ◆ SOLID-STATE PROTECTIVE RELAYS

Solid-state protective relays, commonly called solid-state overload relays (SSOLRs), protect the motor and control devices against overheating. Depending on the design, many also provide protection against phase loss, phase unbalance, phase reversal, and undervoltage. They do not protect against short circuits or ground faults. SSOLRs are used with solid-state contactors/controllers. Many are designed so that they also can be retrofitted into equipment where thermal overloads previously were used. Depending on the design, some SSOLRs are self-powered; others must be powered by a separate DC power supply. There are two types of SSOLRs: non-programmable and programmable.

3.1.0 Non-Programmable Solid-State Overload Relays

SSOLRs are made in many sizes and shapes and for stand-alone mounting, starter mounting, and DIN-rail mounting. *Figure 5* shows an example of one manufacturer's self-powered, stand-alone non-programmable SSOLR and a related control diagram. Control power is derived from the three-phase inputs. Regardless of manufacturer and internal circuitry differences, all SSOLRs operate in much the same way. As shown, the SSOLR senses the current drawn by each of the three motor leads via three built-in current transformer (CT) windings. Each CT winding consists of one, two, or three loop turns depending on the model of relay. The trip point for the SSOLR is determined by the setting of an adjustable trip current dial on the face of the unit. The dial setting is adjusted as directed in the manufacturer's instructions. The setting for the SSOLR shown is based on the motor's full load amperage (FLA) and service factor and the number of CT loop turns being used. For motors with a service factor of 1, the relay is adjusted for a setting equal to 0.9 times the motor FLA times 1, 2, or 3 (the number of CT loops). For motors with service factors of 1.15 to 1.25, the setting is equal to the motor FLA times 1, 2, or 3.

The SSOLR initiates a trip if the phase currents exceed 125% of the trip current dial setting. The time it takes to trip depends on the level of monitored current, the trip class (Class 10, Class 20, and so on), and the length of time since the last trip. A mechanically latched mechanism opens (unlatches) when a trip occurs. This action opens the normally closed (NC) overload trip contact, interrupting current flow in the motor control circuit. The relay's phase loss/phase unbalance circuitry will also initiate a trip if a current unbalance of 25% or greater exists or one of the three-phase currents is missing per *NEC Section 430.32*. The phase loss/phase unbalance circuitry remains operational at currents below the trip current to provide protection for lightly loaded motors. Once tripped, the SSOLR is reset by pressing the RESET pushbutton on the face of the unit. Most manufacturers also have a remote reset module that can be installed on their relays. Typically, this would be used with cranes, hoists, and in similar applications where the controls are mounted in a remote location, making them difficult to access.

An SSOLR can be an instantaneous-trip (instant removal with no time delay) device or inverse-time, Class 10, 20, 30, etc., device. Some have an adjustment to set the trip class. These models are usually shipped with trip Class 20 selected. Most

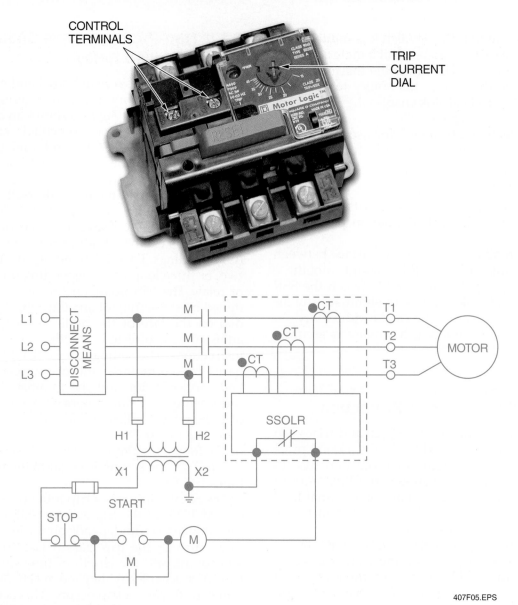

Figure 5 ◆ Solid-state overload relay.

407F05.EPS

SSOLRs are also equipped with LED indicators to show when power is applied and when a trip has occurred.

3.2.0 Programmable Solid-State Overload Relays

Programmable SSOLRs provide the same overload and phase loss/phase unbalance motor protection as non-programmable relays. Refer to *NEC Section 430.52(C)(5)*. In addition, most have an increased capability to detect several other types of motor-related faults. The programmable SSOLR is connected into the motor circuit in the same way as a non-programmable relay. However, most programmable SSOLRs require the use of an external power supply to power their solid-state circuitry. Like non-programmable SSOLRs, programmable models sense the motor current

via current transformer (CT) windings. For some models, these CT windings are internal to the relay. For others, they are mounted externally. External mounting typically occurs in motor circuit applications where high full-load currents (90A and above) are involved. Operation of the SSOLR is controlled by a microprocessor that can be programmed so that the relay can be used in a wide variety of applications. *Figure 6* shows an example of one manufacturer's programmable SSOLR.

Programming of an SSOLR is done using controls on the front panel of the unit. However, many SSOLR manufacturers also have software programs available that allow an SSOLR to be programmed using a personal computer (PC) connected to the relay via a communications network. Programming of SSOLRs is performed differently depending on the manufacturer and

Use of External Current Transformers

When using external current transformers, make sure that all the CT polarity marks face the same direction, and route all the positive terminal wires through the same side of the loop windows.

407F06.EPS

Figure 6 ◆ Typical programmable solid-state overload relay.

model of the relay. For this reason, it should be done in accordance with the manufacturer's instructions for the specific relay being used. The number and type of parameters that can be programmed differ from one SSOLR to another depending on their design. For the SSOLR shown, setting trip thresholds for the different circuit parameters is done by using the mode and display select switches and the display screen on the front of the unit. The types of functions and trip thresholds that can be programmed with most programmable SSOLRs include:

- *Overcurrent trip class* – Selects the trip class (Class 10, 20, 30, etc.). The specific trip class is determined by the motor and application. Class 20 is used for most NEMA-rated general purpose motors.
- *Effective turns ratio of CTs* – Used by the SSOLR to calculate the true current based on the turns ratio of the internal or externally mounted CTs.
- *Low- and high-voltage trip points* – Typically set to 10% of the motor nameplate rating.
- *Voltage unbalance trip point* – Sets the percentage of voltage unbalance allowed by the motor. Typically set for 5%.

- *Overcurrent trip point* – Typically set to between 110% and 120% of the motor FLA.
- *Undercurrent trip point* – Sets the level of acceptable undercurrent. Typically set for 80% of the motor FLA.
- *Restart delay times after a trip* – Controls the elapsed time between motor restarts after a trip. Different settings allow for different restart times, depending on the type of fault that has occurred.
- *Number of restarts* – If enabled, this determines the number of automatic restarts that can be initiated. After the programmed number of restarts have been attempted and failed, a manual reset must be performed.
- *Undercurrent trip delay* – Determines how long the SSOLR will allow an undercurrent condition to exist before it trips.
- *Ground fault current trip* – Determines when a Class II ground fault exists because degradation of the motor's insulation is allowing current leakage. The threshold is typically set for 10% to 20% of the motor FLA. This feature is for motor protection only; it is not used for ground fault protection of personnel.

Once tripped by an overload, the programmable SSOLR is reset locally by pressing the RESET/PROGRAM pushbutton on the face of the unit. If equipped with a remote reset module, it can be reset from a remote location.

4.0.0 ◆ TIMING RELAYS

Timing relays open or close electric circuits in order to perform selected operations according to a timed program. A typical application involves the use of timing relays to control the sequential energization or de-energization of two or more motors. An example of this process is described later in this section. Most timing relays have adjustable time cycles, allowing them to be used in more than one application. The focus of this section is on pneumatic, dashpot, and electronic (solid-state) timing relays. Before describing these specific types of devices, it is first necessary to introduce you to common terms used when describing timers and to the symbols used on schematic and ladder diagrams to identify timed contacts (*Figure 7*).

- *On-delay relay* – An on-delay relay provides the time delay after the relay is energized.
- *Off-delay relay* – An off-delay relay provides the time delay after the relay is de-energized.
- *Normally open, timed-closed (NOTC) contacts* – NOTC contacts (*Figure 7*) are on-delay, timed-closed contacts. The contacts are normally open. When the relay is energized, timing begins and the contacts close after the specified delay time has elapsed. The contacts remain closed until the relay is de-energized, at which time the contacts are immediately opened.
- *Normally closed, timed-open (NCTO) contacts* – NCTO contacts are on-delay, timed-open contacts. The contacts are normally closed. When

the relay is energized, timing begins and the contacts are opened after the specified delay time has elapsed. The contacts close immediately when the relay is de-energized.
- *Normally open, timed-open (NOTO) contacts* – NOTO contacts are off-delay, timed-open contacts. When the relay is energized, the normally open contacts close immediately and stay closed. When the relay is de-energized, timing begins and the contacts are opened after the specified delay time has elapsed.
- *Normally closed, timed-closed (NCTC) contacts* – NCTC contacts are off-delay, timed-closed contacts. When the relay is energized, the normally closed contacts open immediately and remain open. When the relay is de-energized, timing begins and the contacts are closed after the specified delay time has elapsed.

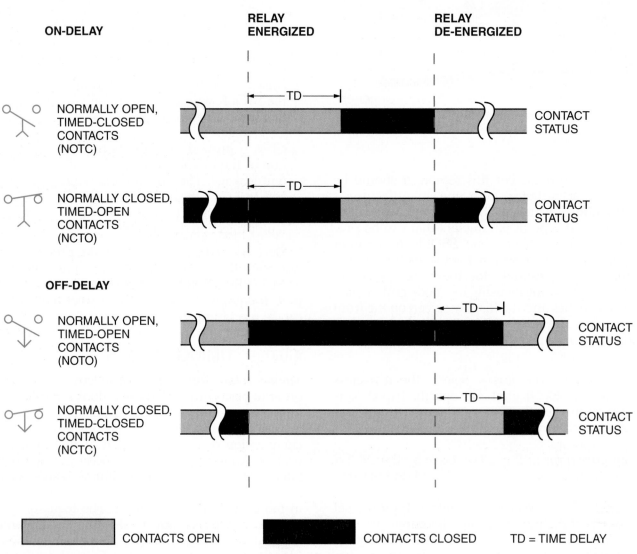

Figure 7 ◆ Timed contacts symbols and timing.

407F07.EPS

On-Delay and Off-Delay Relays

On-delay and off-delay relays are also commonly called *delay-on-make* and *delay-on-break* relays, respectively.

4.1.0 Pneumatic Timing Relays

Pneumatic timing relays are electromechanical relays that use air (pneumatic power) to retard the movement of the relay's moving contacts. The rate at which the air is allowed to pass through an internal restricted orifice determines the time delay. *Figure 8* shows a typical pneumatic timing relay that has one normally open (NO) and one normally closed (NC) contact. Most pneumatic timing relays operate in a similar manner. They have a synthetic rubber bellows (*Figure 9*) that is actuated by the plunger of an AC- or DC-operated solenoid. Many are field convertible to provide either an on-delay or an off-delay. The conversion from on-delay to off-delay and vice versa is done by removing the solenoid unit, rotating it 180°,

407F08.EPS

Figure 8 ◆ Pneumatic time delay relay.

and remounting it. An on-delay provides the time delay after the coil is energized. An off-delay provides the time delay after the coil is de-energized. Energizing and de-energizing of pneumatic relays can be controlled by pilot devices such as push-buttons, limit switches, or thermostatic relays.

When the on-delay relay solenoid is de-energized or the off-delay solenoid is energized, its plunger (*Figure 9*) causes the timing mechanism push rod to be forced into the down position. In this position, terminals 1 and 2 are the NO contact terminals and terminals 3 and 4 are the NC contact terminals. When the on-delay relay solenoid is energized or the off-delay solenoid is de-energized, the plunger moves away from the push rod, allowing the spring located inside the synthetic rubber bellows to push the plunger upward. As the plunger rises, it causes the over-center toggle mechanism to move the snap action toggle blade upward. This, in turn, picks up the push plate that carries the movable contacts, causing them to change position.

For the bellows to fully expand, air must enter the bellows through the air inlet. The speed at which the bellows can expand is determined by the setting of the needle valve. If it is nearly closed, an appreciable length of time will be required for air to pass through the air orifice, enabling the bellows to expand. The setting of the needle valve determines the time interval that must pass between operation of the solenoid plunger and the expanding of the bellows to operate the contact unit. Typically, the time delay can be adjusted from 0.2 to 60 seconds, with an accuracy of 10% of the setting. What this means is that the relay can provide a time delay as short as 0.2 second or as long as 60 seconds,

Repair of Pneumatic Timing Relays

Pneumatic timing relays made by most manufacturers can be disassembled to replace their contacts and other parts. Any such disassembly and reassembly should always be done in accordance with the manufacturer's instructions.

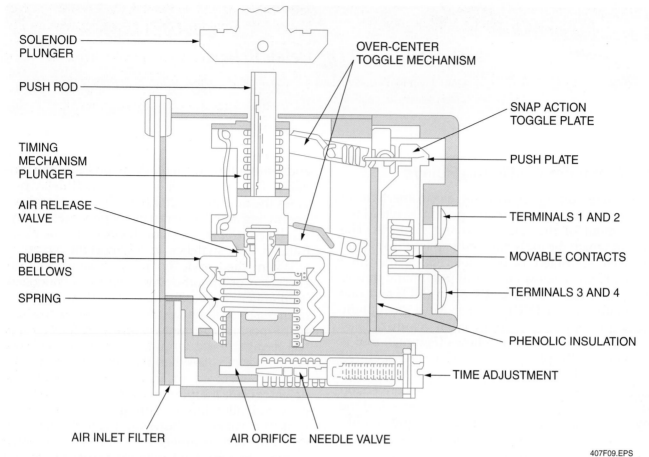

Figure 9 ◆ Cutaway view of pneumatic relay contact unit and timing mechanism.

407F09.EPS

but may vary as much as 10% from this value on a repeat basis.

When the on-delay relay solenoid is de-energized or the off-delay solenoid is energized, the push rod is again depressed by the solenoid plunger, forcing the timing mechanism plunger to its lower position. This causes the air in the bellows to be exhausted through the air release valve as the bellows contracts, causing the timer to be reset almost instantaneously, typically within 0.075 second.

When selecting a pneumatic timing relay, consult the manufacturer's specification sheets for a unit that is compatible with your application. Factors to consider are:

- Recommended application and mounting criteria
- Minimum and maximum time delay
- Reset time
- Repeat of accuracy (% of tolerance from the time delay setting)
- Operating coil voltage and frequency
- Operating temperature range
- Contact ratings

4.2.0 Dashpot Timing Relays

Dashpot timing relays operate essentially like pneumatic timing relays. In the case of the dashpot timing relay, the motion of the plunger of the solenoid is retarded by a fluid dashpot attached to the lower end of the solenoid plunger. The dashpot is filled with a silicone fluid, the viscosity of which is affected only slightly by changes in temperature. The delay time for closing the relay contacts is adjustable.

4.3.0 Solid-State Timing Relays

The solid-state timing relay derives its name from the fact that the time delay is provided by solid-state electronic devices enclosed within the relay. Solid-state timing relays are suitable for applications where high repeat accuracy is required and the time delay must frequently be changed. *Figure 10* shows a typical plug-in solid-state timing relay. This particular relay has the capability to provide one of several selectable fixed timing ranges as selected by the controls on the top of the relay.

Figure 10 ◆ Typical solid-state plug-in timing relay.

Figure 11 shows an example of connection diagrams and operational charts typical of those provided by most relay manufacturers for a timing relay. The relay connection diagrams in *Figure 11* show that the relay is made in 8-pin and 11-pin models. The operational charts for the 8-pin model show that the relay can be configured for one of two modes of operation: delay-on-make or interval. In the delay-on-make (on-delay) mode, the relay provides the timed delay (set time) after it is energized. In the interval mode, the relay contacts change on the application of power. At the end of the specified set time, the contacts return to the de-energized position. Reset occurs upon removal of the input power. The pin numbers and status of the relay's contacts (on or off) for each mode of operation are shown on the related operational chart.

The 8-pin timing relay is classified as a supply voltage timer relay. This means that the pilot device(s) used to control activation of the relay must be connected so that AC/DC control power is applied to the timer relay power input pins. For the relay shown, pins 2 and 7 are used. When using a supply voltage timer relay, remember that the pilot device(s) selected for use in the control circuit and the control circuit wiring size must be compatible with the voltage and current ratings of the timer relay being used.

The 11-pin timing relay is classified as a contact-controlled timer relay. This means that the pilot device(s) used to control activation of the relay are not connected in line with the timer relay power input (pins 2 and 10). Instead, they are connected to pins 5 and 6, the control signal switch. With a contact-controlled timing relay, the timer supplies the control voltage for the control circuit where the pilot device(s) are installed. This allows the control circuit pilot device(s) and wiring to operate at lower voltages.

The operational charts for the 11-pin relay show that this relay can also be configured for one of two modes: delay-on-break or single shot. These modes cannot be selected with the 8-pin relay. The delay-on-break (off-delay) mode provides the time delay (set time) after the relay receives an activation control signal applied at pins 5 and 6 from the external switch. In the single shot mode, the timing relay energizes when the external switch is closed. At the end of the set time, the relay will de-energize. Reset is accomplished by opening and closing the control switch. The pin numbers and the status of the relay's contacts for each mode of operation are shown in the related operational chart.

In addition to timing relays, many types of multi-function electronic timers are available, each with different physical features and operating characteristics. Electronic timer operation is typically based on microprocessor circuitry. This allows the timers to be programmed to select several special timing functions in addition to the basic timing functions described earlier for solid-state timing relays. Information about the operation and capabilities of an electronic

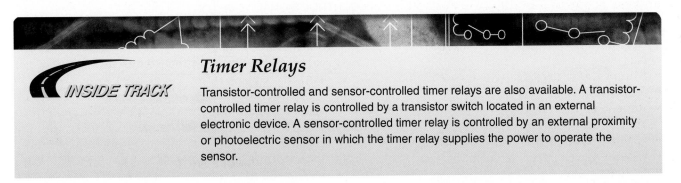

Timer Relays

Transistor-controlled and sensor-controlled timer relays are also available. A transistor-controlled timer relay is controlled by a transistor switch located in an external electronic device. A sensor-controlled timer relay is controlled by an external proximity or photoelectric sensor in which the timer relay supplies the power to operate the sensor.

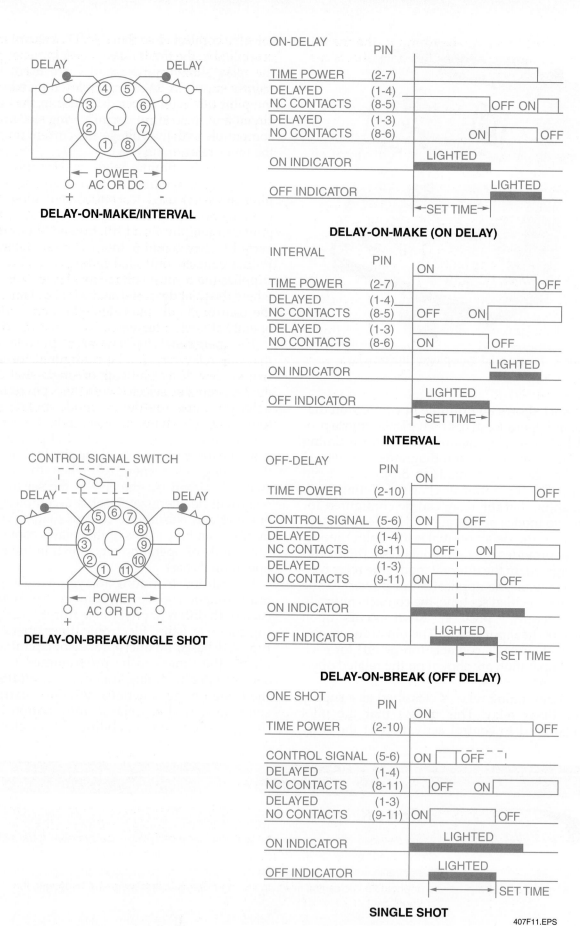

Figure 11 ◆ Plug-in solid-state relay connection diagrams and operation charts.

407F11.EPS

programmable timer can be found in the manufacturer's catalog or application data sheets for the specific timer being used.

4.4.0 Timing Relay Applications

There are hundreds of applications in which timing relays are used to provide timed control of equipment, operations, or processes. One common application is described here. It involves the use of timing relays to control the timed start of equipment containing three motors. This is necessary because the current surge to start all three motors at the same time would be too great for the system. To prevent this problem when the equipment is started, a delay of 20 seconds is allowed to elapse before the second motor, and then the third motor, can be started. *Figure 12* shows a START/ STOP pushbutton control used to operate three motor starters and two timing relays. The circuit is designed so that an overload on any motor will stop all motors.

When the START pushbutton is pressed, a circuit is completed from L1 through the STOP pushbutton, START pushbutton, the motor starter coil M1, and the closed overload contacts to L2. This energizes coil M1. When coil M1 energizes, motor No. 1 starts and auxiliary contacts M1 close. The closure of the M1 contacts provides memory for the START pushbutton when it is released. At the same time, a circuit is also completed through the coil of timing relay TR1 and the overload relays to L2, energizing the timer. After a 20-second interval, the normally open, timed close (NOTC) contacts of TR1 close. This completes a circuit from L1 through motor starter coil M2 and through parallel-connected timer relay coil TR2 to L2. When coil M2 energizes, motor No. 2 starts. Twenty seconds after the coil of timing relay TR2 is energized, its NOTC contacts close. This completes a circuit from L1 through motor starter coil M3 to L2, which causes motor No. 3 to start. As a result of the 20-second time delays, motor No. 2 starts 20 seconds after motor No. 1 and motor No. 3 starts 20 seconds after motor No. 2.

If the STOP pushbutton is pressed, the circuit to coils M1 and TR1 is broken. When motor starter M1 de-energizes, motor No. 1 stops and auxiliary contacts M1 open. TR1 is an on-delay relay; therefore, when coil TR1 is de-energized, the contacts

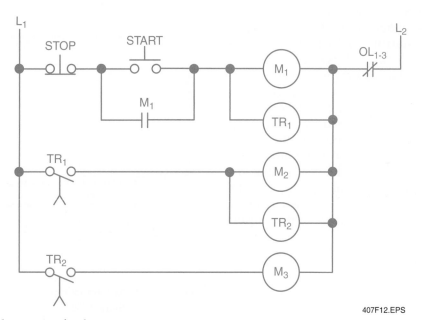

Figure 12 ◆ Timed delay starting for three motors.

407F12.EPS

Timing Relay Application

How would you rewire the circuit shown in *Figure 12* to absolutely ensure that motor M2 will not start until motor M1 has started?

of TR1 open immediately. When the TR1 contacts open, motor starter M2 and timing relay coil TR2 de-energize. This stops motor No. 2. Because timing relay TR2 is an on-delay relay, the TR2 contacts open immediately. This opens the circuit to motor starter M3, causing motor No. 3 to stop. The action described here for de-energizing the motors and timing relay coils is almost instantaneous. If one of the overload contacts opens while the circuit is energized, the effect is the same as pressing the STOP pushbutton. After the circuit stops, all contacts return to their normal positions.

5.0.0 ◆ REDUCED-VOLTAGE STARTING MOTOR CONTROL

Reduced-voltage starting reduces the amount of current an induction motor draws when starting. There are several reasons for using reduced-voltage starting. One common reason is to reduce the large amount of current drawn from power company lines by the across-the-line start of large motors. This occurs because the inrush current drawn by a motor can range from two to six times the motor's nameplate current rating. Such a sudden demand for large current can reflect back into the power lines and cause problems. Another common reason for using reduced-voltage starting is to control a motor's torque at startup so that the torque is applied to a load gradually. This is needed in applications where a high starting torque can cause damage to gears, belts, and chain drives being driven by the motor or where damage to a product can occur from a sudden forceful start. Reducing the voltage applied to a motor at startup reduces the current drawn by the motor. This reduces the amount of starting torque the motor can deliver because the starting torque is proportional to the current. Reduced-voltage starting is accomplished by using one of four types of reduced-voltage motor starters to control the motor:

- Autotransformer
- Part winding
- Wye-delta
- Solid-state

The application or the type of motor generally dictates the type of starter to use.

5.1.0 Autotransformer Reduced-Voltage Starting Motor Control

Autotransformer reduced-voltage starters are widely used because of their efficiency and flexibility. See *NEC Section 430.109(D)*. All power taken from the line, except transformer loss, is transmitted to the motor to accelerate the load. Several taps on the transformer allow for adjustment of the starting torque and inrush current to meet job requirements. Autotransformer reduced-voltage starters are typically used with hard-to-start loads such as reciprocating compressors, grinding mills, pumps, and similar devices. *Figure 13* shows a graph of the starting characteristics for an autotransformer reduced-voltage starting controller.

Depending on manufacturer and model, there are several designs for autotransformer reduced-voltage starting. Some use three-coil autotransformers; others use two-coil autotransformers. Some perform the transition from reduced voltage to full voltage using a **closed circuit transition**

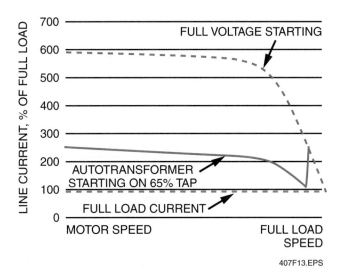

407F13.EPS

Figure 13 ◆ Autotransformer reduced-voltage starting characteristics.

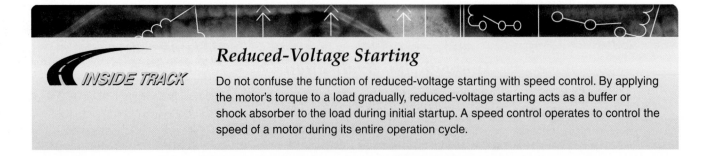

INSIDE TRACK

Reduced-Voltage Starting

Do not confuse the function of reduced-voltage starting with speed control. By applying the motor's torque to a load gradually, reduced-voltage starting acts as a buffer or shock absorber to the load during initial startup. A speed control operates to control the speed of a motor during its entire operation cycle.

method of motor connection. Others use an **open circuit transition** method of motor connection. In a closed circuit transition controller, the motor is never removed from the source of voltage while moving from one voltage level to another. In an open circuit transition controller, the motor being controlled may be temporarily disconnected from the line while moving from one voltage level to another. Controllers using the closed circuit transition method are more widely used because they cause the least amount of interference to the related electrical distribution system. This is because the motor is never disconnected from the line. Therefore, there is no interruption of line current that can cause a second inrush current during the transition period.

Figure 14 shows a schematic diagram for an induction motor being controlled by a closed circuit transition type of autotransformer reduced-voltage controller. As shown, it includes a three-coil autotransformer and two three-pole contactors (1S and 2S) used to connect the autotransformer for reduced-voltage starting. It also includes a three-pole run contactor (R) used to bypass the autotransformer and connect the motor directly across the line for full-voltage running. The size of these three-pole contactors is determined by the maximum horsepower of the motor and its voltage and frequency requirements. The timing of the voltage reduction cycle is controlled by a timing relay (TR) adjustable from 1.5 to 15 seconds.

Pressing the START pushbutton energizes timing relay TR. This causes its two sets of normally open instantaneous contacts to close. One set of closed instantaneous contacts provides memory for the timer; the other set completes an electrical path, via the normally closed contacts of contactor R and the timed open contacts (TRTO) of the timer relay, to the coil of contactor 1S. Completion of this path causes contactor 1S to energize. When contactor 1S energizes, its power contacts close, connecting the ends of the autotransformers together. It also causes the contactor 1S auxiliary contacts to close, thereby energizing contactor 2S. The normally closed auxiliary contacts of 1S in the control circuit for contactor R open. This provides an electrical interlock to prevent contactors R and 1S from being energized at the same time. When coil 2S energizes, its power contacts close and connect the motor through the autotransformer taps to the power line, starting the motor at reduced inrush current and starting torque. The normally open auxiliary contacts of 2S close to provide memory for contactor 2S.

After a predetermined time (1.5 to 15 seconds), timer relay TR times out and its normally closed, timed open (NCTO) contacts open, causing contactor 1S to de-energize. This opens its power contacts and returns its normally closed auxiliary

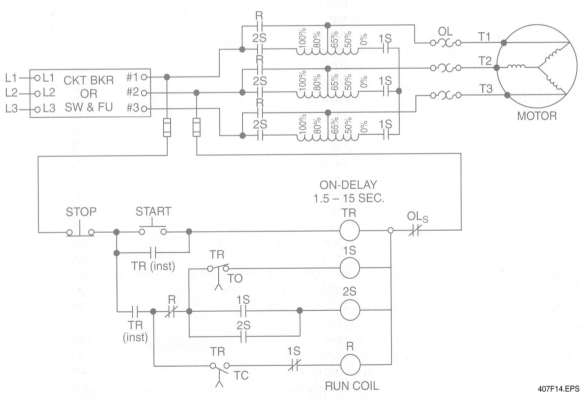

Figure 14 ◆ Typical autotransformer reduced-voltage starting circuit.

407F14.EPS

contacts to the closed position. Simultaneously, the normally open, timed closed (NOTC) contacts of timing relay TR close, causing contactor R to energize via the closed auxiliary contacts of contactor 1S. When contactor R energizes, its power contacts close, bypassing the autotransformer and connecting the motor to full line voltage. Its normally closed auxiliary contacts also open, causing contactor 2S to de-energize.

Note that during the transition from starting to full-line voltage, the motor was not disconnected from the circuit, indicating a closed circuit transition method of operation. As long as the motor is running in the full-voltage condition, timing relay TR and contactor R remain energized. Only an overload or pressing the STOP pushbutton stops the motor and resets the circuit.

5.2.0 Part-Winding, Reduced-Voltage Starting Motor Control

Part-winding, reduced-voltage starting is an older method of reduced-voltage starting in which an induction motor is started by first applying power to part of the motor's coil windings for starting, then after a short time delay applying power to the remaining coil windings for normal running. Part-winding starters are used with part-winding motors. A part-winding motor is one that has two sets of identical windings that are connected in parallel during full-voltage operation. Part-winding starters have typically been used with low starting torque loads such as fans, blowers, and motor-generator sets. These starters can be used with nine-lead, dual-voltage motors on the lower voltage and with special part-winding motors designed for any voltage. Most motors will produce a starting torque equal to between ½ and ⅔ of NEMA standard values with half of the winding energized and draw about ⅔ of normal line current inrush. *Figure 15* shows a graph of the typical starting characteristics for a part-winding, reduced-voltage starting mode of operation.

5.3.0 Wye-Delta, Reduced-Voltage Starting Motor Control

Wye-delta, reduced-voltage starting operates by first connecting the leads of the motor being controlled into a wye configuration for starting. Then, after a short time delay, the leads are reconfigured via switched contacts so that the motor is connected into a delta configuration for running. Both closed circuit transition and open circuit transition models are available. Wye-delta starters are used for controlling high-inertia loads with long acceleration times such as with

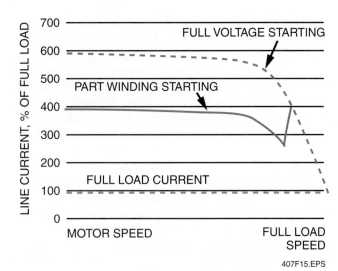

Figure 15 ◆ Part-winding, reduced-voltage starting characteristics.

centrifugal compressors, centrifuges, and similar loads. When six- or twelve-lead, delta-connected motors are started wye-connected, approximately 58% of the line voltage is applied to each winding. The motor develops 33% of its full-voltage starting torque and draws 33% of its normal locked-rotor current from the line. When the motor is accelerated, it is reconnected for normal delta operation. *Figure 16* shows a graph of the typical starting characteristics for a wye-delta, reduced-voltage starting mode of operation.

Figure 17 shows a schematic diagram for a wye-delta motor controlled by a closed circuit transition, wye-delta type of reduced-voltage controller. As shown, it includes a three-pole contactor 1S that is used to short motor leads T4, T5, and T6 during starting to connect the motor in the

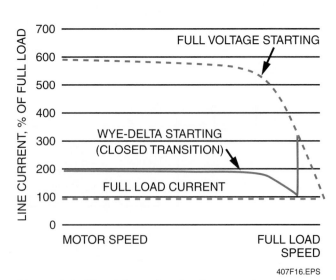

Figure 16 ◆ Wye-delta, reduced-voltage starting characteristics.

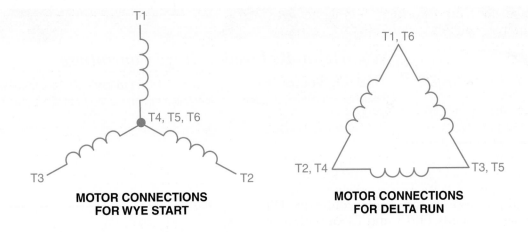

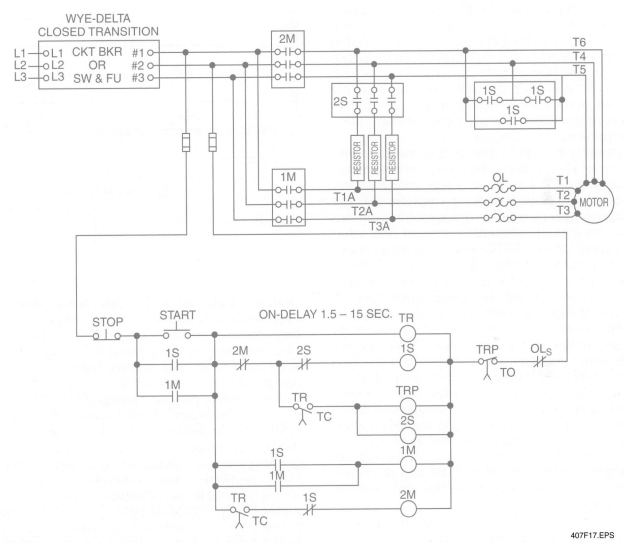

Figure 17 ◆ Typical wye-delta, reduced-voltage starting circuit.

407F17.EPS

INSIDE TRACK

Wye-Delta, Reduced-Voltage Controllers

When using a closed transition type of wye-delta, reduced-voltage controller, you must make sure that there is adequate ventilation to remove the heat dissipated by the resistors.

wye configuration. A three-pole contactor 1M energizes motor leads T1, T2, and T3 for both wye and delta connections. A three-pole contactor 2M energizes motor leads T4, T5, and T6 during running to connect the motor in the delta configuration. It also includes a three-pole contactor 2S that connects resistors in series with the motor windings during the start-to-run transition period. Note that contactor 2S and the resistor bank are unique to the closed circuit transition starter. An open circuit transition starter does not have these components.

Pressing the START pushbutton energizes timing relay TR and contactor 1S, whose power contacts connect the motor in a wye configuration. Interlocks 1S close, energizing contactor 1M. The 1M power contacts energize the motor windings in a wye configuration. After a preset time interval, timer contacts TRTC close, energizing contactor 2S and timing relay TRP. Interlock 2S opens, dropping out contactor 1S. The motor is now energized in series with the resistors. Interlock 1S closes, energizing contactor 2M, bypassing the resistors, and energizing the delta-connected motor at full voltage. Interlock 2M opens, de-energizing contactor 2S and timing relay TRP. Timing relay TRP opens the control circuit if the duty cycle for the transition resistors is exceeded. As long as the motor remains running in the full-voltage condition, contactors 1M and 2M and timing relay TR all remain energized. If an overload occurs or the STOP pushbutton is pressed, it de-energizes contactors 1M and 2M and the timer, removing the motor from the line and resetting the timer.

5.4.0 Solid-State, Reduced-Voltage Starting Motor Control

Solid-state, reduced-voltage starters perform the same function as electromechanical reduced-voltage starters but provide a smoother, stepless

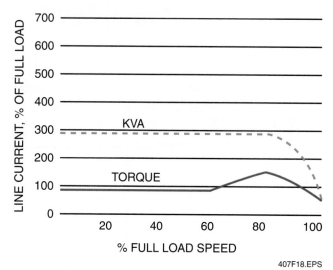

407F18.EPS

Figure 18 ◆ Solid-state, reduced-voltage starting characteristics.

start and acceleration. For this reason, they are often referred to as *soft-start* controllers. Because they provide controlled acceleration, they are ideal for use with many loads including conveyors, compressors, and pumps. Some models can also provide a soft stop where a sudden stop may cause system or product damage. *Figure 18* shows a graph of the typical starting characteristics for a solid-state (soft-start) reduced-voltage starting mode of operation.

There are many designs and models of solid-state, reduced-voltage controllers. *Figure 19* shows a simplified schematic diagram for a motor being controlled by a typical solid-state, reduced-voltage controller. As shown, the controller consists of an electronic circuit board and six silicon-controlled rectifiers (SCRs) connected back-to-back. The typical circuit board contains a microprocessor and related input/output (I/O) circuitry that functions both to control the operation of the SCRs and to protect the motor and

Solid-State, Reduced-Voltage Starters

Solid-state, reduced-voltage starters are available in a wide variety of sizes to suit various motor applications.

407SA01.EPS

starter circuits from damage. The current transformers sense the motor current and feed this information back to the control circuit for comparison and processing throughout the entire motor control process.

The use of six back-to-back SCRs provides control for the full cycle of the AC input voltage. Each single SCR controls one half cycle of the AC input voltage to the motor; the six together provide full-wave control of the motor input voltage. The amplitude of the voltage applied to the motor is determined by the precise point in the half cycle of the AC input waveform at which the SCRs are turned on. Turn-on of the SCRs is controlled by a gate provided by the circuit board and applied to the gates of the SCRs. By controlling the precise time that the gate is applied to the SCRs, the effective voltage delivered to the motor can be varied from zero to full voltage. This is done over time by sequentially allowing the SCRs to conduct for increasing portions of the half cycle. For operation at the lower voltage levels, the SCRs conduct over smaller portions of the half cycle; for higher voltage levels, they conduct over larger portions of the half cycle. During the time when the SCRs are

turned on, all the voltage is on the SCRs, and essentially all the voltage is on the motor. This means that the root-mean-square (rms) voltage applied to the motor is directly related to the amount of time per half cycle that the SCRs conduct. The SCRs are turned off when the negative half cycle of the input waveform is applied to their anodes.

The circuit shown is also equipped with a contactor (M) that is energized when the motor attains full voltage and speed at the end of the voltage ramp-up interval. This is common in some solid-state, reduced-voltage controller designs. The contacts of contactor M are connected in parallel with the SCRs. Contactor M is controlled by the closed contacts of the end-of-ramp relay K1 located on the circuit board inside the solid-state controller. When contactor M is energized at the end of the voltage ramp-up interval, its contacts close, causing the SCRs to be bypassed and the motor to be placed directly across the power line.

Settings for torque, time, and other parameters are made in the field to obtain switching of the SCRs as needed to achieve the desired motor performance. Typically, they are manually

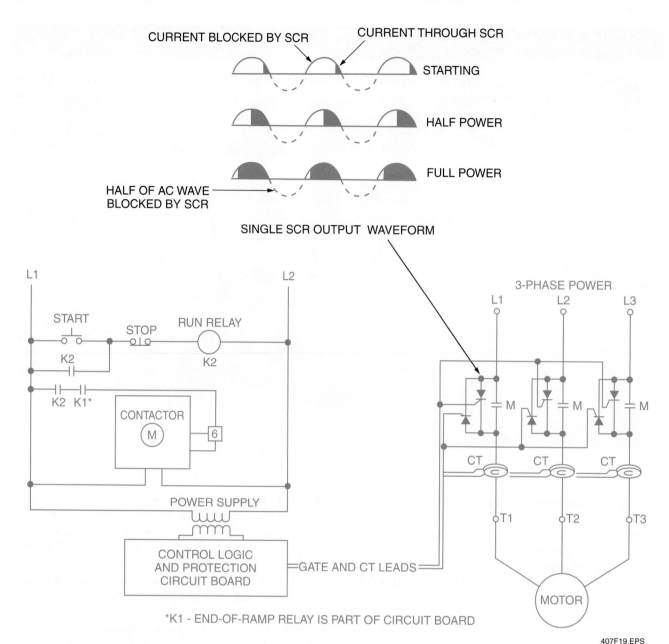

Figure 19 ◆ Simplified solid-state, reduced-voltage starting circuit.

407F19.EPS

programmed into the circuit board microprocessor via dipswitches and circuit board controls. Some starting characteristics that normally can be programmed include:

- *Ramp start* – Ramp start, shown in *Figure 20(A)*, is the most common form of soft start. It allows the initial torque value of the ramp to be set and then raises it to full voltage. The starting torque is typically adjustable from 0% to 85% of locked rotor torque; ramp time adjustment is from 0 to 180 seconds or more.

- *Kick start* – Kick start, shown in *Figure 20(B)*, provides an initial boost of current to the motor to help break free the rotor and start the motor. The starting torque is typically adjustable from 0% to 85% of locked rotor torque; the duration of the kick start interval is from 0 to 2 seconds.

- *Current limit* – Current limit, shown in *Figure 20(C)*, is used when it is necessary to limit the maximum starting current because of long starting times or to protect the motor. The starting torque is typically adjustable from 0% to

85% of the locked rotor current; the ramp time is typically from 0 to 180 seconds or more.

- *Soft stop* – Soft stop, shown in *Figure 20(D)*, is used when an extended coast-to-rest period is needed. It is often used with high-friction loads where a sudden stop can cause system or product damage such as in hydraulic pumps. The stop ramp time is typically adjustable from 0 to 60 seconds.

A solid-state, reduced-voltage controller typically also contains electronic circuits to protect the motor from overload and the starter from damage. Depending on the design, protection can be provided for shorted SCRs, over-temperature, current unbalance, undervoltage, phase loss, and/or phase reversal.

5.5.0 Selection of Reduced-Voltage Controllers

The following characteristics must be considered when selecting a reduced-voltage starter for use with a squirrel cage motor-driven load:

- The motor characteristics that will satisfy the starting requirements of the load
- The source of power and the effect of the motor starting current on the line voltage
- The load characteristics and the effect of the motor starting torque on the driven parts during acceleration
- The motor voltage
- The startup torque required
- The heater selection
- The enclosure type

Table 3 provides a brief summary of the types of reduced-voltage controllers covered in this module and their characteristics.

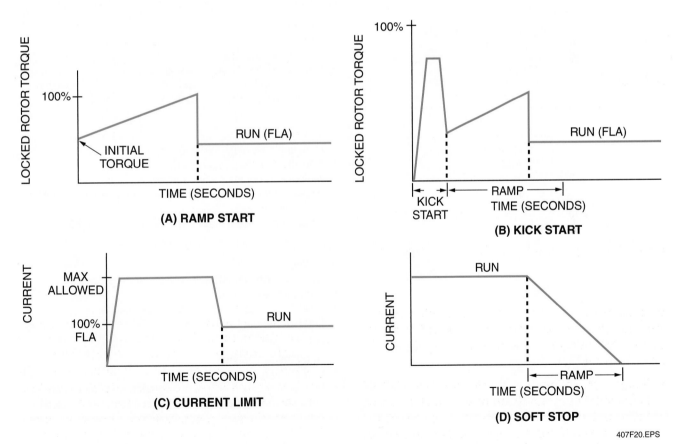

407F20.EPS

Figure 20 ◆ Starting characteristics of solid-state, reduced-voltage controllers.

Table 3 Comparison of Reduced-Voltage Controllers (1 of 2)

Type of Reduced-Voltage Controller	Starting Characteristics			Remarks
	Voltage at Motor (%)	Line Current (%)	Torque (%)	
Autotransformer	80	64	64	*Applications*–Blowers, pumps, compressors, conveyors
	65	42	42	
	50	25	25	
				Advantages • Provides maximum torque per ampere of line current • Starting characteristics easily adjusted • Different starting torques available through autotransformer taps • Suitable for relatively long starting periods • Motor current greater than line current during starting
				Disadvantages • Most complex of reduced-voltage controllers because proper sequencing of energization must be maintained • Large physical size • Low power factor • Expensive in lower hp ratings
Part-winding	100	65	48	*Applications*–Reciprocating compressors, pumps, blowers, fans
				Advantages • Least expensive • Small physical size • Suitable for low or high voltage • Full acceleration in one step for most standard induction or special part-winding motors when ⅔ winding connection is used
				Disadvantages • Unsuitable for high-inertia loads • Specific motor types required • Motor does not start when torque required by load exceeds that developed by motor when first half of motor is energized • Torque efficiency usually poor for high-speed motors
Wye-delta	100	33	33	*Applications*–Centrifugal compressors, centrifuges
				Advantages • Suitable for high-inertia, long-accelerating loads • High torque efficiency • Ideal for stringent inrush current restrictions • Ideal for frequent starts
				Disadvantages • Requires special motor • Low starting torque • Momentary inrush occurs during open transition when delta contactor is closed (open circuit transition models)

Table 3 Comparison of Reduced-Voltage Controllers (2 of 2)

| Type of Reduced-Voltage Controller | Starting Characteristics | | | Remarks |
	Voltage at Motor (%)	Line Current (%)	Torque (%)	
Solid-state	Adjustable	Adjustable	Adjustable	*Applications*–Machine tools, hoists, packaging equipment, conveyor systems *Advantages* • Voltage gradually applied during starting for a soft-start condition • Adjustable acceleration time • Adjustable kick start, current limit, soft stop *Disadvantages* • More expensive than electromechanical models • Requires specialized installation and maintenance • Electrical transients can damage solid-state components • Requires good ventilation

6.0.0 ◆ ADJUSTABLE FREQUENCY DRIVES

An adjustable frequency drive (AFD), also commonly called a variable frequency drive (VFD), converts three-phase 50Hz or 60Hz input power to an adjustable frequency and voltage output to control the speed of an AC motor. AFDs are discussed in *NEC Article 430, Part X*. The use of AFD-controlled systems provides energy savings by eliminating most of the losses associated with mechanical or electromechanical methods of adjusting speed used in the past. Many other benefits are derived from using an AFD. These include:

- *Controlled starting* – Provides for limited starting current, reduction of power line disturbance on starting, and low power demand on starting.
- *Controlled acceleration* – Provides for soft-start, adjustable acceleration based on time or load, and reduced motor size for pure inertial load acceleration.
- *Adjustable operating speed* – Allows process to be optimized or changed, provides energy savings, allows process start at reduced speed by programmable controller.
- *Adjustable torque limiting* – Current limit is available for quick and accurate torque control, protecting the machinery, product, or process from damage.
- *Controlled stopping* – Soft stop, timed stopping, and fast reversal with much less stress on AC motor than plug reverse.

6.1.0 Basic Adjustable Frequency Drive Operation

AFDs change the speed of AC squirrel cage induction motors to meet system requirements by converting three-phase 50Hz or 60Hz input power to an adjustable frequency and voltage. The speed of the motor is determined by the formula:

$$\text{Motor speed} = \frac{120 \times \text{frequency}}{\text{number of motor poles}}$$

For a specific motor, the number of poles is a constant since they are built into the motor. Examination of the formula shows that the speed of the motor is proportional to the applied frequency. As the frequency increases, the motor speed increases; as the frequency decreases, the motor speed decreases. To maintain a constant motor torque, AFDs are designed to automatically maintain a constant relationship between the voltage and frequency of the excitation output applied from the AFD to the motor terminals. This relationship is called the volts per hertz ratio (V/Hz). If this process is done properly, the speed of the motor can be controlled over a wide variation in shaft speed (0 rpm through twice nameplate) with the proper torque characteristics for the application.

AFDs are made in a wide variety of sizes and designs. Motor drive technology has changed greatly over the last decade. Today, digitally controlled, programmable, microprocessor-based, pulsewidth-modulated (PWM) variable frequency drives are commonly used. Their use provides a significant improvement in the elimination of dirty power problems (harmonics, EMI, etc.) generated by the analog SCR rectifier types of AFD units that were the state-of-the-art only a few years ago. For this reason, the remainder of this section will focus on the newer PWM type of AFD device.

Figure 21 shows a block diagram of a typical PWM AFD system. As shown, it consists of three basic parts: the operator control panel, the AFD controller, and an AC motor. The control panel provides an interface between the operator and the drive system. At the control panel, the operator can input commands for starting, stopping, and changing the speed and/or direction of the motor. He or she can also monitor various AFD and motor status signals. Typically, the control panel is removable and can be mounted externally and connected via a cable to the drive. Most AFDs have an input/output (I/O) interface board that allows communication with personal computers (PCs) or programmable logic controllers. The motor used with an AFD is typically a standard NEMA Design A or Design B, squirrel

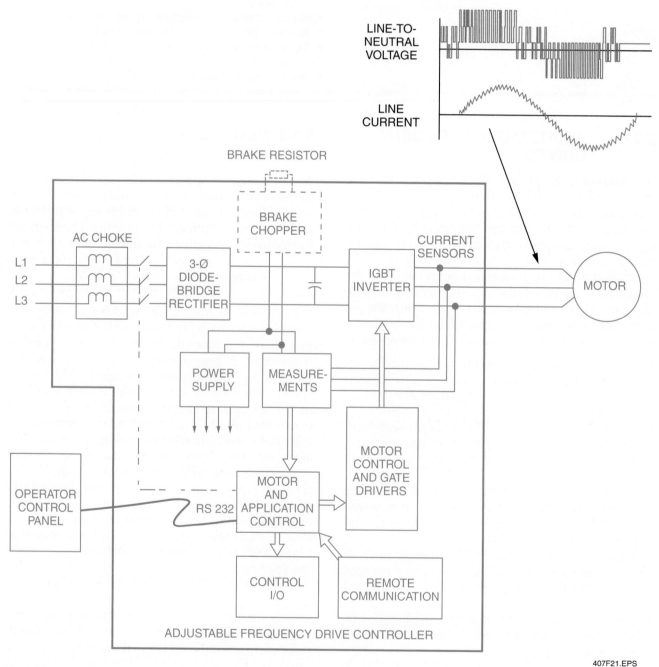

407F21.EPS

Figure 21 ◆ Block diagram of a typical adjustable frequency drive unit.

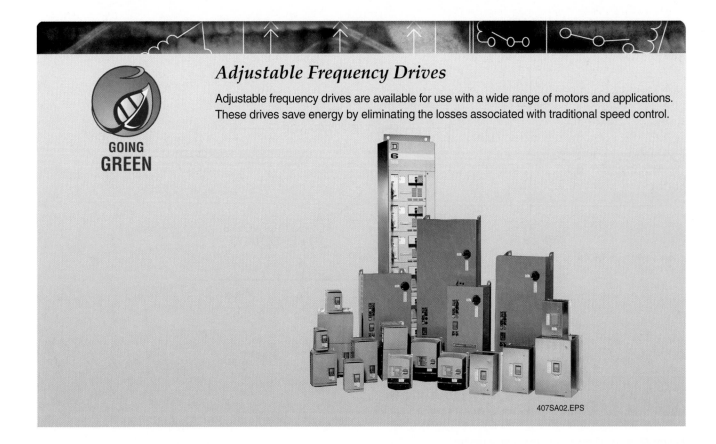

Adjustable Frequency Drives

Adjustable frequency drives are available for use with a wide range of motors and applications. These drives save energy by eliminating the losses associated with traditional speed control.

GOING GREEN

407SA02.EPS

cage induction motor rated for 240V or 480V, three-phase, 60Hz operation. In response to the AC power waveform applied from the AFD, the motor drives the load.

The AFD controller converts three-phase, 50Hz or 60Hz input power to an adjustable frequency and voltage output for controlling the speed of the AC motor. It typically receives 240V or 480V input power and provides adjustable frequency output power to the motor. The actual range of frequencies used is programmable and depends on the motor application. It also regulates the output voltage in proportion to the output frequency to provide a constant voltage-to-frequency ratio as required by the characteristics of the motor.

The three-phase AC-choke and the DC-link capacitor form an LC filter that, together with the three-phase, diode bridge-type full-wave rectifier, produces a regulated DC voltage for application to the inverter circuit. Note that the circuitry between the rectifier output and the inverter circuit input is sometimes referred to as the DC intermediate bus. The AC-choke smoothes any high-frequency (HF) disturbances from the utility to the drive and HF disturbances caused by the drive to the utility. It also improves the waveform of the input current to the drive.

Since the inverter section is powered by a fixed DC voltage, the maximum amplitude of its output waveform is fixed. The insulated gate bipolar transistor (IGBT) switches in the inverter circuit convert the DC voltage input into a three-phase, PWM, symmetrical AC voltage output for application to the AC motor. The frequency and amplitude of the waveform that is produced are determined by the duration and precise timing intervals of the gates applied to the IGBT switches from the gate driver circuit. The effective value of the output voltage applied to the motor is determined by the width of the zero voltage intervals in the output waveform. Use of IGBTs provides the high switching speed necessary for PWM inverter operation. They are capable of switching on and off several thousand times a second.

AFDs typically have one range of output frequencies over which the ratio of volts to hertz remains constant and another range of output frequencies over which the voltage remains constant while the frequency varies (*Figure 22*). The frequency at which the transition from one range to the other occurs is called the base frequency. Between the base frequency and the maximum frequency points, the rated output voltage remains constant. Below the base frequency point, the output voltage is less than the rated voltage and is based on the volts per hertz ratio (V/Hz). Depending on motor characteristics, these two ranges can correspond to constant torque (constant V/Hz) and constant power (constant voltage with an adjustable frequency).

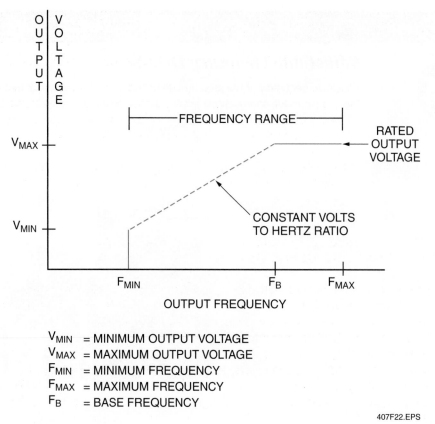

V_MIN = MINIMUM OUTPUT VOLTAGE
V_MAX = MAXIMUM OUTPUT VOLTAGE
F_MIN = MINIMUM FREQUENCY
F_MAX = MAXIMUM FREQUENCY
F_B = BASE FREQUENCY

407F22.EPS

Figure 22 ◆ AFD operating frequency range.

The motor and application control circuit consists of a programmable microprocessor and related circuits. The microprocessor controls the motor according to measured feedback signals and in accordance with parameter settings and commands. These commands may be applied from the control panel or from a remote device via the control input/output (I/O) circuits. The motor and application control circuit sends signals to and receives feedback signals from the motor control/gate driver circuits. The motor control/gate driver circuits calculate, then output switching position signals to the IGBT switches in the inverter.

When required, dynamic braking can be used with an AFD. This can be done by using a brake chopper circuit and external dynamic brake resistor that operate to absorb motor regenerative energy for stopping the load and to limit the energy flowing back to the drive. Dynamic braking is accomplished by continuing to excite the motor from the drive after the motor turns off. This causes a regenerative current at the drive's DC intermediate bus circuit. The dynamic brake resistor is then placed across the DC bus to dissipate the returned power. The brake resistors are switched into the circuit under the control of the brake chopper. Dynamic braking cannot occur in an induction AC drive unless the power circuit

and control circuit are operational. For this reason, a mechanical brake, actuated in case of a power failure, is often required to stop the motor. The resistor(s) used for dynamic braking must be capable of absorbing at least six times the stored rotational energy of the motor at **base speed**.

6.2.0 AFD Parameters That Can Be Programmed or Monitored

The parameters that can be programmed and/or monitored when using an AFD can vary widely based on the specific design and complexity of the AFD, the AFD manufacturer, and the application for which the AFD is to be used. Some of the parameters that can be programmed and/or monitored are briefly described here. Follow the instructions in the manufacturer's user manual for the specific AFD unit being used.

- *Minimum and maximum frequency* – This defines the frequency limits for the AFD operation.
- *Acceleration and deceleration time* – Acceleration time defines the time required for the output frequency to accelerate from the set minimum frequency to the set maximum frequency. Deceleration time defines the time required to decelerate from the set maximum frequency to the set minimum frequency.

PWM AFD Outputs to a Motor over Long Cable Runs

When a PWM AFD drive output is applied to a motor over long cable runs, the difference in the motor and cable impedance can cause voltage overshoots to be present at the motor terminals. These overshoots can greatly exceed 1,000V and have the potential to damage the motor's insulation. For small motors, this typically occurs with cable lengths that exceed 33', and for larger motors, with cable lengths that exceed 100'. For situations involving the use of long cable runs, install a reflective wave trap or similar filter at the motor terminals to prevent overshoots.

- *Current limit* – This determines the maximum motor current the AFD will provide on a short-term basis.
- *V/Hz ratio* – Typically there are two settings: linear and squared. The linear setting is used with applications requiring constant torque. In this mode, the voltage supplied to the motor changes linearly with the frequency from 0Hz to the nominal frequency of the motor. With the squared setting, the voltage to the motor changes following a squared curve from 0Hz to the nominal frequency of the motor. In this mode, the motor runs under-magnetized (below the normal frequency) and produces less torque and electromechanical noise. This mode is used in applications where the torque demand from the load is proportional to the square of the speed, such as with centrifugal fans and pumps.
- *V/Hz optimization* – This mode provides an automatic torque boost. The voltage to the motor changes automatically, which allows the motor to produce sufficient torque to start and run at low frequencies. The voltage increase depends on the motor type and horsepower. Automatic torque boost can be used in applications where starting torque due to starting friction is high, such as with a conveyor.
- *Nominal voltage of the motor* – This is the rated voltage from the nameplate of the motor.
- *Nominal frequency of the motor* – This is the frequency from the nameplate of the motor.
- *Nominal speed of the motor* – This is the speed from the nameplate of the motor.
- *Nominal current of the motor* – This is the current value from the nameplate of the motor.
- *Supply voltage* – This is set to a range compatible with the nominal voltage of the supply.
- *Basic frequency reference* – This selects the basis for the frequency reference such as a voltage reference supplied by a potentiometer or a current reference supplied by a transducer or other source.

6.3.0 Classifications and Nameplate Markings for AFDs

This section describes AFDs. It briefly lists their various classifications, service categories, and nameplate data.

6.3.1 Classifications

The following classifications are assigned by NEMA. They are used to describe the functional characteristics of drive converters, drives, and drive systems that produce an AC output.

- *Form FA converters* – Form FA converters have an AC input and an AC output. The frequency conversion is accomplished without an intermediate conversion to DC.
- *Form FB converters* – Form FB converters have a DC input and an AC output.
- *Form FC converters* – Form FC converters have an AC input and an AC output. The frequency conversion is accomplished with intermediate conversion to DC.

6.3.2 Nameplate Information

The information marked on the nameplate of an AFD unit reveals the operating characteristics of the unit. This information is useful when becoming familiar with an existing unit and when troubleshooting the unit. The following types of information are normally included on the nameplate of a drive unit:

- Manufacturer's name
- Equipment identification
- Input rating:
  - Voltage
  - Maximum continuous current, total rms amperes (including harmonics)
  - Frequency
  - Number of phases
  - Maximum allowable AC system symmetrical short-circuit current
  - Service category

- Output rating:
  - Maximum output voltage
  - Rated continuous current, fundamental rms amperes (excluding harmonics)
  - Overload capacity
  - Frequency range
  - Number of phases
  - Nominal hp or kW/kVA output (optional)
  - Phase rotation of output

6.4.0 Types of Adjustable Speed Loads

Because AFD systems are load dependent, a thorough understanding of the load characteristics is necessary when selecting an AFD for a particular application. There are three common types of adjustable speed loads: variable torque, constant torque, and constant horsepower.

6.4.1 Variable Torque Loads

A variable torque load requires a much lower torque at low speeds than it does at high speeds, as shown in *Figure 23(A)*. With this type of load,

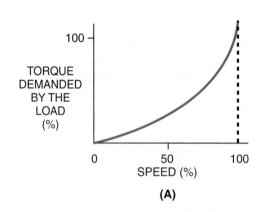

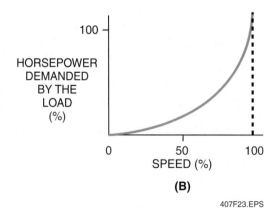

407F23.EPS

Figure 23 ◆ Variable torque load.

the horsepower varies approximately as the cube of the speed, as shown in *Figure 23(B)*, and the torque varies approximately as the square of the speed. This type of load is encountered in applications with high-inertia loads, such as machines that have flywheels, centrifugal fans, pumps, blowers, and punch presses.

6.4.2 Constant Torque Loads

Constant torque loads are friction-type loads where constant torque characteristics are needed to overcome the friction. A constant torque load requires the same amount of torque at both low and high speeds, as shown in *Figure 24(A)*. The torque remains constant throughout the speed range, and the horsepower increases or decreases in direct proportion to the speed, as shown in *Figure 24(B)*. This type of load is encountered in applications such as general machinery hoists, conveyors, printing presses, positive displacement pumps, and extruders, as well as for shock loads, overloads, or high-inertia loads.

6.4.3 Constant Horsepower Loads

A constant horsepower load requires constant horsepower over the speed range, as shown in *Figure 25(A)*. The load requires high torque at low speeds and lower torque at high speeds and, therefore, constant horsepower at any speed, as shown in *Figure 25(B)*. With a constant horsepower load, the speed and torque are inversely proportional to one another. As the speed increases, the torque decreases and vice versa. Constant horsepower loads are encountered in applications such as metal cutting tools operating over a wide speed range, mixers, center-driven winders, and some extruders.

6.5.0 AFD Selection Considerations

Selecting an AFD for a particular application requires an understanding of the capabilities of various drive units and the characteristics and/or problems encountered with different types of loads. For this reason, evaluating the needs of a drive system and selecting the components for use in the system is a task normally done by an electrical engineer, a drive manufacturer sales/application engineer, or other trained and qualified persons. *Figure 26* shows a typical checklist used by one manufacturer to gather the data needed to evaluate the requirements for an AFD system. The remainder of this section briefly describes some of the other factors that can affect the AFD selection process.

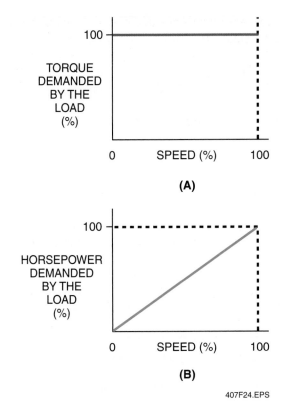

Figure 24 ◆ Constant torque load.

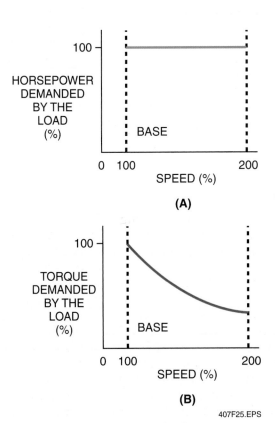

Figure 25 ◆ Constant horsepower load.

6.5.1 AFD System Motors

In most applications, the motors used with AFDs are standard NEMA Design A or Design B motors. The use of a motor other than a Design A or Design B usually requires the addition of special drive circuitry. Operating self-ventilated motors at reduced speeds may cause above-normal temperature rises. As the speed of a motor is reduced below its 60Hz base speed, motor cooling becomes less effective because of the reduced speed of the self-cooling fan. This limitation determines the minimum allowable motor speed for continuous torque operation. Derating or forced ventilation may be necessary to achieve the rated motor torque output at reduced speeds.

Using AFDs with older existing motors can cause an extra heating effect on the motor and subject the motor winding insulation to high-frequency induced voltage stress. The load capacity of older motors should be derated to at least the next lower Hp rating, particularly if the motor has a Class F temperature rise or a 1.0 service factor.

6.5.2 Adjustable Frequency Drives

The basic requirement of AFD sizing is to match the torque-versus-speed capability of the AFD to the torque-versus-speed requirement of the driven load.

If the load torque exceeds 150% for constant torque drives or 110% for variable torque drives during starting or intermittently while running the drive, oversizing of the drive may be required. The minimum and maximum motor speeds for the application will determine the drive's base speed.

The full-load current ratings of typical AFDs are matched to typical full-load, three-phase motor current ratings as listed in *NEC Table 430.250*. Generally, an AFD of a given horsepower rating will be adequate for a motor of the same rating, but the actual motor current required under operating conditions is the determining factor for AFD sizing. If the motor will be run at full load, the AFD output current rating must be equal to or greater than the motor nameplate current. If the motor is oversized to provide a wide speed range, the AFD should be sized to provide the current required by the motor at the maximum operating torque. Motor oversizing should generally be limited to an increase of one horsepower.

AFD APPLICATION CHECKLIST

Motor

New_____ Existing_____ Horsepower:_____ Base Speed:_____ Voltage:_____

FLA:_____ LRA:_____ NEMA Design:_____ Gearbox/Pulley Ratio:_____

Service Factor:_____

Load

Application:_____

Load Type: Constant Torque_____ Variable Torque_____ Constant Horsepower_____

Load inertia reflected to motor:_____

Required breakaway torque from motor:_____

Running load on motor:_____

Peak torques (above 100% running):_____

Shortest/longest required accel. time:_____ / _____ secs up to_____ Hz from zero speed

Shortest/longest required decel. time:_____ / _____ secs down to_____ Hz from max. speed

Operating speed range:_____ Hz to_____ Hz

Time for motor/load to coast to stop:_____ secs

AFD

Source of start/stop commands:_____

Source of speed adjustment:_____

Other operating requirements:_____

Will the motor ever be spinning when the AFD is started?_____

Is the load considered to be high inertia?_____

Is the load considered to be hard to start?_____

Distance from AFD to the motor:_____feet

Type of AFD (V/Hz, Flux Vector, Closed Loop Vector):_____

Options desired:_____

_____

_____

Other special requirements/conditions:_____

407F26A.EPS

Figure 26 ◆ AFD application checklist (sheet 1 of 2).

AFD APPLICATION CHECKLIST (Continued)

Power Supply

Supply Transformer: _____ kVA and _____ % Z or short circuit current at drive input: _____ amps

(If the drive does not include a built-in line reactor and the available feeder short circuit current is more than 100 times the drive FLA rating, a 1% line reactor or a drive isolation transformer is required.)

Total horsepower of all drives connected to supply transformer or feeder: _____hp

Is a drive transformer or line reactor desired? _____

Any harmonic requirements? _____ % Voltage THD: _____ % Current THD: _____ IEEE 519: _____

Total non-drive load connected to the same feeder as drive(s): _____amps

Service

Start-up Assistance: _____   Customer Training: _____

Preventive Maintenance: _____   Spare Parts: _____

Additional Issues

Will the AFD operate more than one motor? _____

Will the power supply source ever be switched with the AFD running? _____

Is starting or stopping time critical? _____

Are there any peak torques or impact loads? _____

Will user-supplied contactors be used on the input or output of the AFD? _____

Does the user or utility system have PF capacitors that are being switched? _____

Will the AFD be in a harsh environment or high altitude? _____

Does the utility system experience surges, spikes, or other fluctuations?_____

407F26B.EPS

Figure 26 ◆ AFD application checklist (sheet 2 of 2).

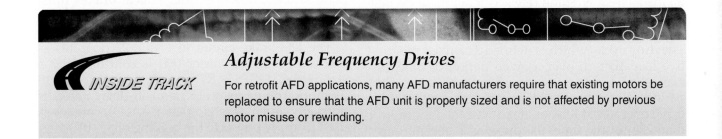

Adjustable Frequency Drives

INSIDE TRACK

For retrofit AFD applications, many AFD manufacturers require that existing motors be replaced to ensure that the AFD unit is properly sized and is not affected by previous motor misuse or rewinding.

AFD Environment

Heat rejection is a major concern with AFDs. The heat produced in the AFD cabinet can be substantial and can cause failure of the AFD SCRs and IGBTs if their operating temperature limits are exceeded. For this reason, operating an AFD at higher altitudes often requires that the unit be derated.

6.5.3 Environment

The environment in which the motor operates must be examined in order to properly select the equipment and enclosure. The ambient temperature, cooling air supply, and the presence of gas, moisture, and dust should all be considered when choosing a drive, its enclosures, and any protective features.

6.5.4 Torque Requirements

The starting, peak, and running torques should be considered when selecting a drive. Starting torque requirements can vary from a small percentage of the full load to a value several times full-load torque. The peak torque varies because of a change in load conditions or the mechanical nature of the machine. The motor torque available to the driven machine must be more than that required by the machine from start to full speed. The greater the excess torque, the more rapid the acceleration potential.

6.5.5 Duty Cycle

Selecting the proper drive depends on whether the load is steady, varies, follows a repetitive cycle of variation, or has pulsating torques. Certain applications may require continuous reversals, long acceleration times at high torque due to inertia loads, frequent high rate acceleration, or cyclic overloads, which may result in severe motor heating if not considered in the selection of the drive. The duty cycle, which is defined as a fixed repetitive load pattern over a given period of time, is expressed as the ratio of on-time to the cycle period. When the operating cycle is such that the drive operates at idle, or at reduced load for more than 25% of the time, the duty cycle becomes an important factor in selecting the proper drive.

7.0.0 ◆ MOTOR BRAKING METHODS

Braking provides a means of quickly stopping an AC motor when it is necessary to stop it faster than is normally possible by removing the power and letting the motor wind down. Braking can be accomplished in several ways. The method used depends on the application, available power, and circuit requirements. Some common methods of motor braking are:

- Dynamic braking (DC electric braking)
- Dynamic braking (AC drives)
- Electromechanical braking

7.1.0 Dynamic Braking (DC Electric Braking) of an AC Motor

One common method of braking an alternating current induction motor is by using a DC electric braking circuit. This is done by momentarily connecting a DC voltage to the stator winding (*Figure 27*). When DC is applied to the stator winding of an AC motor, the stator poles become electromagnets. Current is induced into the windings of the rotor as the rotor continues to spin through the magnetic field. This induced current produces a magnetic field around the rotor. The magnetic field of the rotor is attracted to the magnetic field produced in the stator. The attraction of these two magnetic fields produces a braking action in the motor. An advantage of using this method of braking is that motors can be stopped rapidly without having to use brake linings or drums. DC braking cannot be used to hold a suspended load, however. Mechanical brakes must be employed when a load must be suspended, as with a crane or hoist.

The DC brake circuit in *Figure 27* operates as follows. Magnetic starter (M) is controlled by a standard STOP/START pushbutton station with memory. Timing relay (TR) is an off-delay timer. Its normally open, timed-open (NOTO) contact is used to apply power to the braking relay (BR) after the STOP pushbutton is pressed. The off-delay of timing relay TR must be adjusted so that its contacts will remain closed until after the motor comes to a stop. At this time, the contacts open, disconnecting the DC power to the motor winding.

When the STOP pushbutton is pressed, start contactor M de-energizes, thus removing three-

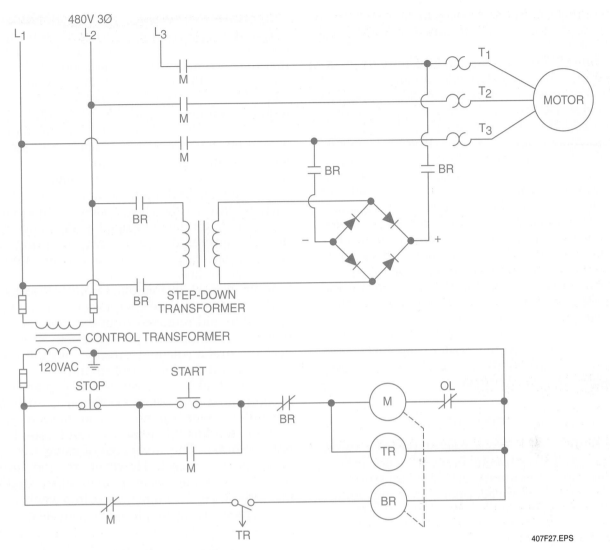

Figure 27 ◆ Circuit for DC braking of an AC motor.

phase input power from the motor. Also, braking relay BR is energized through the NC contact M and NOTO contact TR. This causes its NO contacts to close, energizing a DC supply consisting of a step-down transformer and full-wave bridge rectifier circuit. It also connects two of the motor leads (T1 and T3) to the output of the DC supply, initiating the motor braking action. A transformer with tapped windings is often used to adjust the amount of braking torque applied to the motor. Current-limiting resistors could be used for the same purpose. This allows for a low- or high-braking action depending on the application. The higher the applied DC voltage, the greater the braking force.

An electrical interlock is provided by the NC contacts of starter M and the NC contacts of braking relay BR. This interlock prevents the motor starter and braking relay from being energized at the same time. This is required because the AC

and DC power supplies must never be connected to the motor simultaneously. Total interlocking should always be used on electrical braking circuits. Total interlocking is the use of mechanical, electrical, and pushbutton interlocking.

7.2.0 Dynamic Braking (AC Drives)

Dynamic braking can be used with AC motors controlled by AFDs.

Torque will always act to cause the motor's rotor to run towards synchronous speed. If the synchronous speed is suddenly reduced, negative torque is developed in the motor. When this occurs, the motor acts like a generator by converting mechanical power from the shaft into electrical power that is returned to the AC drive unit. This is similar to driving a car downhill in a lower gear so that the car's engine acts as a brake. In order for an AC drive to operate without damage during this condition, a means must exist to deal with the electrical energy returned to the drive by the motor.

Figure 28 shows a simplified diagram of an AC drive unit that uses dynamic braking. Electrical energy returned by the motor can cause voltage in the drive DC bus to become excessively high when added to the existing supply voltage. This excess voltage can damage the drive components. To alleviate this problem, a braking resistor is connected across the DC bus through an IGBT switching transistor. The braking resistor is added and removed from the circuit by the switching action of the IGBT switch. Energy returned by the motor is impressed on the DC bus. When this energy reaches a predetermined level, the IGBT switch is turned on by the control logic. This places the braking resistor across the DC bus, causing the excess energy to be dissipated by the resistor, thus reducing the DC bus voltage. When the DC bus voltage is reduced to a safe level, the IGBT is turned off, removing the resistor from the DC bus. This process allows the motor to act as a brake, slowing the connected load.

7.3.0 Electromechanical Braking

Motor braking is also accomplished using solenoid-operated friction brakes. Friction brakes are used for applications such as printing presses, small cranes, overhead doors, hoisting equipment, and machine tool control.

Many friction brakes used with motors are similar to those used with vehicles. They normally consist of two shoe or pad-type friction surfaces that make contact with a wheel mounted on the motor shaft. Brake operation is controlled by a solenoid. The solenoid is energized when the motor is running, causing the brake shoes/pads to be retracted from the wheel. When the motor is turned off, the solenoid is de-energized, and the tension from a return spring in the brake mechanism brings the shoes/pads in contact with the wheel, causing braking to occur as a result of the friction between the shoes/pads and the wheel. The braking torque developed is directly proportional to the braking surface area and spring pressure. The spring pressure is adjustable on nearly all friction brakes.

See *Figure 29*. In this diagram, the brake solenoid energizes and de-energizes when voltage is applied to and removed from the motor, respectively. To prevent improper brake activation, connect the brake solenoid directly into the motor power circuit, not its control circuit.

Friction brakes are positive action, mechanical friction devices. Normal operation is such that when the power is removed from the motor, the brake is set. For this reason, it can be used as a holding brake. The advantage of using a friction brake is lower cost. However, friction brakes require more maintenance than other braking methods. Like the brakes used in a vehicle, the shoes/pads must be replaced periodically.

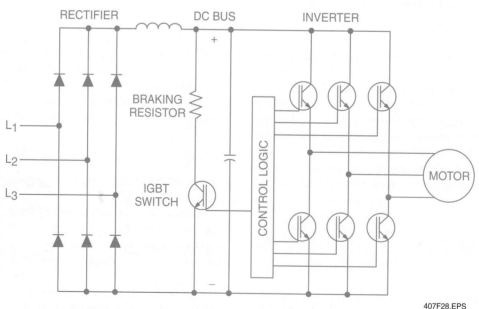

407F28.EPS

Figure 28 ◆ Simplified schematic of an AC drive using dynamic braking.

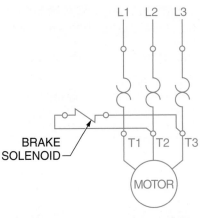

BRAKE
SOLENOID

T1 T2 T3

MOTOR

407F29.EPS

Figure 29 ◆ Friction brake solenoid connection.

8.0.0 ◆ PRECAUTIONS WHEN WORKING WITH SOLID-STATE CONTROLS

WARNING!

Solid-state equipment may contain devices that have potentially hazardous leakage current in the OFF state. Power should be turned off and the device disconnected from the power source before working on the circuit or load.

CAUTION

Some solid-state devices can be damaged by electrostatic charges. These devices should be handled in the manner specified by the manufacturer.

Solid-state equipment and/or components can be damaged if improperly handled or installed. The following guidelines should be followed when handling, installing, or otherwise working with solid-state equipment or components:

• Microprocessors and other integrated circuit chips are very sensitive to, and can be damaged by, static electricity from sources such as lightning or people. When handling circuit boards, avoid touching the components, printed circuit, and connector pins. Always ground yourself before touching a board. Disposable grounding wrist straps are sometimes supplied with boards containing electrostatic-sensitive components. If a wrist strap is not supplied with the equipment, a wrist strap grounding system should be used. It allows you to handle static-sensitive components without fear of electrostatic discharge. Store unused boards inside their special metallized shielded storage bags or a conductive tote box (*Figure 30*).

• Solid-state devices can be damaged by the application of reverse polarity or incorrect phase sequence. Input power and control signals must be applied with the polarity and phase sequence specified by the manufacturer.

• The main source of heat in many solid-state systems is the energy dissipated in the output power devices. Ensure that the device is operated within the maximum and minimum ambient operating temperatures specified by the manufacturer. Follow the manufacturer's recommendations pertaining to the selection of enclosures and ventilation requirements. For existing equipment, make sure that ventilation passages are kept clean and open.

• Moisture, corrosive gases, dust, and other contaminants can all have adverse effects on a system that is not adequately protected against atmospheric contaminants. If these contaminants are allowed to collect on printed circuit boards, bridging between the conductors may result in a malfunction of the circuit. This could lead to noisy, erratic control operation or a permanent malfunction. A thick coating of dust could also prevent adequate cooling of the board or heat sink, causing a malfunction.

• Excessive shock or vibration may cause damage to solid-state equipment. Follow all manufacturer's instructions for mounting.

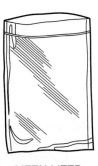
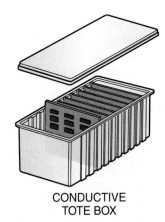

METALLIZED
SHIELDING BAG

CONDUCTIVE
TOTE BOX

407F30.EPS

Figure 30 ◆ ESD protection bag and tote box.

Ambient Temperature Surrounding Solid-State Devices

Do not overlook other sources of heat in or around an enclosure containing solid-state devices that might raise the ambient temperature to undesirable levels. Look for heat sources such as power supplies, transformers, radiated heat, adjacent furnaces, and sunlight.

• Electrical noise can affect the operation of solid-state controls. Sources of electrical noise include machines with large, fast-changing voltages or currents when they are energized or de-energized, such as motor starters, welding equipment, SCR-type adjustable speed devices, and other inductive devices. The following are basic steps to help minimize electrical noise:
  – Maintain sufficient physical separation between electrical noise sources and sensitive equipment to ensure that the noise will not cause malfunctions or unintended actuation of the control.
  – Maintain a physical separation between sensitive signal wires and electrical power and control conductors. This separation can be accomplished by using conduit, wiring trays, or methods recommended by the manufacturer.
  – Use twisted-pair wiring in critical signal circuits and noise producing circuits to minimize magnetic interference.
  – Use shielded wire to reduce the magnitude of the noise coupled into the low-level signal circuit by electrostatic or magnetic coupling.
  – Follow provisions of the *National Electrical Code®* with respect to grounding. Additional grounding precautions may be required to minimize electrical noise. These precautions generally deal with ground loop currents arising from multiple ground paths. The manufacturer's grounding recommendations must also be followed.

9.0.0 ◆ MOTOR CONTROL MAINTENANCE

A well-planned and well-executed preventive maintenance program is necessary to achieve satisfactory operation of electrical and electronic motor control equipment. Preventive maintenance keeps the equipment running with little or no downtime. Ideally, routine preventive maintenance for a specific piece of equipment is based on the manufacturer's recommendations, the severity of use, and the surrounding environment.

9.1.0 Preventive Maintenance Tasks

Before any preventive maintenance work begins that will shut down or otherwise interrupt equipment operation, always notify and/or coordinate your work with the building manager, shift leader, foreman, or other responsible person.

WARNING!

Before any preventive maintenance tasks are performed, de-energize, lock out, and tag all circuits and equipment in accordance with the prevailing lockout/tagout policy. If more than one incoming power source exists (such as when there is a separate control circuit) make sure to de-energize all such sources. Note that solid-state electronic equipment may contain devices that have potentially hazardous leakage current in the OFF state. Such equipment should be isolated from the power source by opening the related disconnect switch, rather than by simply turning off the device.

Preventive maintenance for motor control circuit components normally involves inspection and cleaning of the equipment. Control equipment should be kept clean and dry. Before opening the door or cover of a cabinet or enclosure, clean any foreign material, dirt, or debris from the outside top surfaces to avoid the risk of anything falling onto the equipment. After opening the cabinet or enclosure, inspect the equipment for the presence of any dust, dirt, moisture or evidence of moisture, or other contamination. If any is found, the cause must be determined and eliminated. This could indicate an incorrectly selected, deteriorated, or damaged enclosure; unsealed enclosure openings; internal condensation; condensate from unsealed conduit; or improper operating procedures such as operating the enclosure with the door open or the cover removed.

Clean the parts of the control equipment by vacuuming or wiping with a dry cloth or soft

Preventive Maintenance

INSIDE TRACK

For programmable overloads, reduced-voltage starters, adjustable frequency drives, and similar devices, refer to the user's manual/service manual supplied with the equipment for a description and schedule of specific preventive maintenance tasks that apply to the equipment.

brush. Use care not to damage delicate parts. Do not clean the equipment or components using compressed air because it may displace dirt, dust, or other debris into other parts or equipment, and the force may damage delicate components. Liquid cleaners, including spray cleaners, should not be used unless specified by the equipment manufacturer. This is because residues may cause damage or interfere with electrical or mechanical functions. If dust or dirt has accumulated on heat sinks and/or components that generate heat, it should be carefully removed because such accumulation can reduce these components' heat dissipation capability and lead to premature failures. Ventilation passages should be open and cleaned of any dust and dirt.

Check the mechanical integrity of the equipment. Physical damage or broken parts can usually be found quickly by inspection. A general inspection should be made to look for loose, broken, missing, or badly worn parts. Components and wiring should be inspected for signs of overheating. The movement of mechanical parts, such as the armature and contacts of contactors, disconnect switches, circuit breaker operator mechanisms, and mechanical interlocks should be checked for functional operation and freedom of motion. Any broken, deformed, or badly worn parts or assemblies should be replaced with manufacturer-recommended renewal parts. Any loose terminals or attaching hardware should be retightened securely to the torque specified by the equipment manufacturer.

The contacts of contactors should be checked for wear caused by arcing. During arcing, a small part of each contact melts, then vaporizes and is blown away. When the contacts are new, they are smooth and have a uniform silver color. As the device is used, the contacts become pitted, and the color may change to blue, brown, or black. These colors result from the normal formation of metal oxide on the contact surfaces and are not detrimental to contact life and performance. Contacts should be replaced under the following conditions:

- *Insufficient contact material* – When less than ¹⁄₆₄" remains, replace the contacts.
- *Irregular surface wear* – This type of wear is normal. However, if a corner of the contact material is worn away and a contact may mate with the opposing contact support member, the contacts should be replaced. This condition can result in contact welding.
- *Pitting* – Under normal wear, contact pitting should be uniform. This condition occurs during arcing, as described above. The contacts should be replaced if the pitting becomes excessive and little contact material remains.
- *Curling of contact surfaces* – This condition results from severe service that produces high contact temperatures and causes separation of the contact material from the contact support member.

Motor control mechanisms should not be lubricated unless recommended by the manufacturer. If lubrication is required, use only the recommended type and amount of lubricant. Remove any surplus lubricant to avoid the risk of establishing a tracking path across insulating surfaces, as well as the risk of excess lubricant migrating into areas that should not be lubricated.

10.0.0 ◆ MOTOR CONTROL TROUBLESHOOTING

Troubleshooting motor control systems requires a systematic approach. The process includes:

- Customer interface
- Physical examination of the system
- Basic system analysis with a clear understanding of the sequence of operation, interlocks, and limits
- Use of manufacturer's troubleshooting aids
- Troubleshooting the motor control circuit and components

10.1.0 Customer Interface

The process of troubleshooting motor control circuit problems should always begin with obtaining all the information possible about the equipment problem or malfunction. Talking to and asking questions of a customer with first-hand knowledge of the problem is always recommended. This can provide valuable information on equipment operation that can aid in the troubleshooting process. When interviewing the customer about a specific fault or problem, try to determine the following:

• The scope of the problem and when and how it began
• How often the problem occurs
• Whether the equipment has been worked on or tested recently
• Whether the equipment ever operated correctly
• Whether the problem occurred following a specific event, such as the addition of new electrical equipment, relocation of equipment, or new construction work

Before working on the motor control circuit equipment, always explain to the customer what course of action you intend to take and what system impact might be encountered, such as service interruptions or total system downtime. If the motor circuit with a problem is part of an emergency system (fire or backup), notify off-site monitoring agencies or authorities, such as the police or fire department, that you will be working on the system.

10.2.0 Physical Examination of the System

Problems can sometimes be identified by a visual inspection of the motor control circuit equipment, its wiring, and the adjacent area. For newly installed equipment, make sure that it has been wired correctly. For non-operating equipment, look for tripped circuit breakers, motor overloads, and signs of overheating. For operating equipment, check for odors of burning insulation, sounds of arcing, and signs that the unit is abnormally warm. For equipment with adjustable controls or mode selection switches, look for switches and/or controls that may be set to the wrong positions. For programmable equipment, check that the selected operating and parameter inputs are correct.

10.3.0 Basic System Analysis

A proper diagnosis of a problem requires that you know what the motor control circuit and application equipment being driven by the motor should be doing when they are operating properly. If you are not familiar with how a particular system or unit should operate, you must first study the manufacturer's service literature to familiarize yourself with the equipment modes and sequence of operation.

The second part of the diagnosis is to find out what symptoms are exhibited by the improperly operating system. This can be done by carefully listening to the customer's complaints and then analyzing the operation of the motor control circuit and application equipment. This usually means making electrical measurements at key points in the motor control circuit. Measured values can be compared with a set of previously recorded values or with values given in the manufacturer's service literature. When troubleshooting, avoid making assumptions about the cause of problems. Always secure as much information as possible before arriving at your diagnosis.

10.4.0 Use of Manufacturer's Troubleshooting Aids

Today's electronic equipment can be very complex, and no one person is an expert on the wide diversity of equipment being used to control motors. For this reason, a good troubleshooter should know how and where to locate information about the specific unit being serviced. This includes making phone calls to manufacturers' technical support centers, using manufacturers' website resources, and using manufacturers' user/service manuals. User/service manuals supplied with solid-state programmable drive units and similar types of complex equipment contain invaluable operating and troubleshooting information, such as the following:

• Principles of operation
• Functional description of the circuitry
• Illustrations showing the location of each control or indicator and its function
• Illustrations and text describing menu-driven control panel displays
• Step-by-step procedures for the proper use and maintenance of the equipment
• Safety considerations and precautions that should be observed when operating the equipment
• Schematic diagrams, wiring diagrams, mechanical layouts, and parts lists for the specific unit

Adjustable frequency drive units and similar microprocessor-controlled devices normally have built-in fault diagnostic circuits to aid in troubleshooting. When a fault occurs that interrupts the operation of the drive, these circuits give a visual indication that a fault trip has occurred. They also generate a fault code that can be selected for display on the unit control panel. When a fault occurs, look up the specific fault code displayed on the panel in a fault troubleshooting table in the user/service manual to help pinpoint the problem area (*Figure 31*).

10.5.0 Troubleshooting Motor Control Circuits and Components

A motor control circuit is a means of supplying power to and removing power from a motor. Most motor circuits consist of a combination of starting mechanisms, both automatic and manual. The simple motor control circuit shown in *Figure 32* is typical of most motor circuits. Troubleshooting electrical problems in motor control circuits may appear complex. However, it can be simplified if the electrical components are divided into smaller functional circuit areas based on the operations they perform. From your previous training, you know that motor control circuits can be divided into three functional circuit areas: power circuits, control circuits, and the load.

The power circuits, represented by the heavier lines shown in *Figure 32*, provide line power to the motor. Line voltages are usually 240VAC or 480VAC. As shown, the power circuits include the disconnecting means, the overcurrent protection devices (fuses or circuit breakers), the normally open contactor contacts, and the motor overload device. The load circuit is the AC motor.

The lighter-weight lines represent the control circuit used in a magnetic-type starter. The

FAULT CODES	FAULT	POSSIBLE CAUSES	CHECKING
F1	OVERCURRENT	FREQUENCY CONVERTER HAS MEASURED TOO HIGH A CURRENT IN THE MOTOR OUTPUT: – SUDDEN HEAVY LOAD INCREASE – SHORT CIRCUIT IN THE MOTOR CABLES – UNSUITABLE MOTOR	CHECK THE LOAD CHECK THE MOTOR SIZE CHECK THE CABLES
F2	OVERVOLTAGE	THE VOLTAGE OF THE INTERNAL DC-LINK OF THE FREQUENCY CONVERTER HAS EXCEEDED THE NOMINAL VOLTAGE BY 35%: – DECELERATION TIME IS TOO FAST – HIGH OVERVOLTAGE SPIKES AT UTILITY	ADJUST THE DECELERATION TIME
F3	GROUND FAULT	CURRENT MEASUREMENT HAS DETECTED THAT THE SUM OF THE MOTOR PHASE CURRENT IS NOT ZERO: – INSULATION FAILURE IN THE MOTOR OR THE CABLES	CHECK THE MOTOR CABLES
F4	INVERTER FAULT	FREQUENCY CONVERTER HAS DETECTED FAULTY OPERATION IN THE GATE DRIVERS OR IGBT BRIDGE: – INTERFERENCE FAULT – COMPONENT FAILURE	RESET THE FAULT AND RESTART AGAIN; IF THE FAULT OCCURS AGAIN, CONTACT YOUR DISTRIBUTOR
F5	CHARGING SWITCH	CHARGING SWITCH IS OPEN WHEN THE START COMMAND IS ACTIVE: – INTERFERENCE FAULT – COMPONENT FAILURE	RESET THE FAULT AND RESTART AGAIN; IF THE FAULT OCCURS AGAIN, CONTACT YOUR DISTRIBUTOR
F6	UNDERVOLTAGE	DC-BUS VOLTAGE HAS GONE BELOW 65% OF THE NOMINAL VOLTAGE: – MOST COMMON REASON IS FAILURE ~~THE UTILITY SUPPLY~~	IN CASE OF TEMPORARY SUPPLY VOLTAGE BREAK, RESET THE FAULT AND ~~START~~

407F31.EPS

Figure 31 ◆ Sample portion of a fault troubleshooting table.

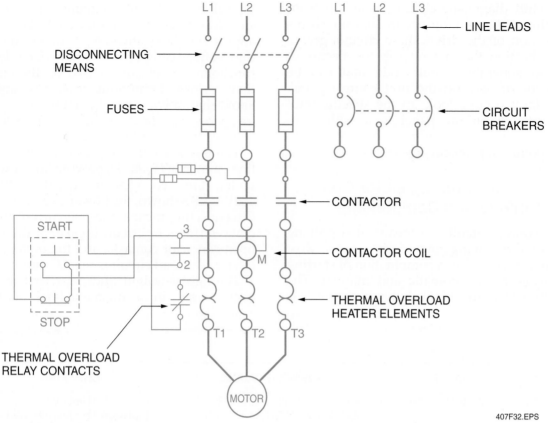

Figure 32 ◆ Basic motor control circuit.

407F32.EPS

control circuit is used to direct power to a magnetic contactor through the STOP/START station, thermal overload relay contacts, holding contacts, and the contactor coil. From your previous training, you also know that the control circuit can contain several other types of pilot devices used to control the motor. The control circuit can be connected directly to the line, as shown in *Figure 32*, or it can be isolated from the power circuit by a step-down control transformer. Lower voltages, such as 24VAC, 120VAC, and 240VAC, are commonly used with a control transformer. The voltage rating for the coil on the motor starter indicates which control voltage is used.

Isolation to the faulty circuit area (power, control, or load) is based on an analysis of equipment operation and a process of elimination. Troubleshooting can be done in several ways. When troubleshooting a motor circuit where the motor is not working or has a problem, the contactor or motor starter is usually checked first. This is because it is the point where the incoming power, load, and control circuit are connected. Basic voltage readings should be taken at a contactor or

motor starter to determine where the problem lies. The same basic procedure used to troubleshoot a motor starter works for contactors because a motor starter is a contactor with added overload protection.

The tightness of all terminals and busbar connections should be checked when troubleshooting control devices. Loose connections in the power circuit of contactors and motor starters cause overheating that can result in equipment malfunction or failure. Loose connections in the control circuit cause control malfunctions. Loose connections of grounding terminals can lead to electrical shock and cause electromagnetic interference.

Some general guidelines for isolating a fault in a motor control circuit follow. *Table 4* provides additional motor control troubleshooting information. Specific procedures for making electrical measurements and troubleshooting individual components in motor control circuits are given later in this section. Note that troubleshooting procedures for motors are covered in detail in Level Three, *Motor Maintenance One* and Level Four, *Motor Maintenance Two*.

Table 4   Motor Control Troubleshooting Chart

Malfunction	Possible Cause	Corrective Action
Constant chatter	Broken pole shader	Replace.
	Poor contact in control circuit	Improve contact or use holding circuit interlock (three-wire control).
	Low voltage	Correct voltage condition; check momentary voltage dip during starting.
Contactor welding or freezing	Abnormal inrush of current	Use larger contactor or check for grounds.
	Rapid jogging	Install larger device rated for jogging service.
	Insufficient contact pressure	Replace contact springs; check contact carrier for damage.
	Low voltage preventing magnet from sealing	Correct voltage condition; check momentary voltage dip during starting.
	Foreign matter preventing contacts from closing	Clean contacts with approved solvent.
	Short circuit	Remove fault and check to be sure fuse or breaker size is correct.
Short contact life or tip overheating	Filing or dressing	Do not file silver-faced contacts; rough spots or discoloration will not harm contacts.
	Interrupt excessively high	Install larger device or check currents for grounds, shorts, or excessive motor currents; use silver-faced contacts.
	Excessive jogging	Install larger device rated for jogging.
	Weak contact pressure	Adjust or replace contact springs.
	Dirt or foreign matter on contact surface	Clean contacts with approved solvent.
	Short circuit	Remove fault and check for proper fuse or breaker size.
	Loose connection	Clean and tighten.
	Sustained overload	Install larger device or check for excessive load current.
Coil overheating	Overvoltage or high ambient temperature	Check application and circuit.
	Incorrect coil	Check rating and if incorrect, replace with proper coil.
	Shorted turns caused by mechanical damage or corrosion	Replace coil.
	Undervoltage, failure of magnet to seal in	Correct system voltage.
	Dirt or rust on pole faces increasing air gap	Clean pole faces.
Overload relays tripping	Sustained overload	Check for grounds, shorts, or excessive currents.
	Loose connection on load wires	Clean and tighten.
	Incorrect heater	Replace relay with correct size heater unit.
Failure to trip causing motor burnout	Mechanical binding, dirt, corrosion, etc.	Clean or replace.
	Wrong heater or heaters omitted and jumper wires used	Check ratings; apply proper heater.
	Motor and relay at different temperatures	Adjust relay rating accordingly.
	Wrong calibration or improper calibration adjustment	Consult factory.
Magnetic and mechanical parts inoperative	Broken shading coil	Replace shading coil.
Noisy magnet humming	Magnet faces not mating	Replace magnet assembly; realign.
	Dirt or rust on magnet faces	Clean and realign.
	Low voltage	Check system voltage and voltage dips during starting.
Failure to pick up and seal	Low voltage	Check system voltage and voltage dips during starting.
	Coil open or shorted	Replace.
	Wrong coil	Check coil number.
	Mechanical obstruction	With power off, check for free movement of contact and armature assembly.
Failure to drop out	Gummy substance on pole faces	Clean with solvent.
	Voltage not removed	Check coil circuit.
	Worn or rusted parts causing binding	Replace parts.
	Residual magnetism due to lack of air gap in magnet path	Replace worn magnet parts.

General guidelines for isolating a fault in a motor control circuit are as follows. Be sure to consult the manufacturer's information for procedures specific to your application.

Step 1 Visually inspect the motor starter's contactor and overload relay for signs of heat damage, arcing, or wear. Repair or replace any motor starter components that show visual signs of such damage.

Step 2 If there are no visual signs of damage, reset the overload relay, then press the START pushbutton to energize the motor starter contactor. If the related motor starts, observe the motor starter for several minutes. If the overload relay trips, troubleshoot the motor and related wiring downstream from the motor starter for causes of an overload.

Step 3 If the motor does not start after resetting the overload, check the voltage applied to the line side of the contactor. If the voltage reading is 0V or is not within 10% of the motor's voltage rating, check the power wiring and components (fuses, circuit breaker, and disconnect switch) upstream of the motor starter.

Step 4 If voltage is applied to the line side of the contactor and it is at the correct level, press the START pushbutton to energize the contactor. If the contactor coil does not energize, troubleshoot the contactor coil to determine if it is open. If the coil is not open, troubleshoot the contactor coil control circuit components and wiring for a problem.

Step 5 If the contactor coil energizes, check the voltage at the load side of the contactor. If the voltage reading is acceptable (within 10% of the motor's voltage rating), the contacts are good. If there is no voltage reading or the reading is too low, turn the power off and replace the contactor contacts.

Step 6 If voltage is present at the load side of the contactor and it is at the correct level, check the output voltage from the overload relay. If the voltage reading is 0V, turn the power off and replace the overload relay. If the voltage reading is acceptable and the motor is not operating, troubleshoot the motor or power wiring downstream from the motor starter.

10.6.0 Electrical Troubleshooting Procedures Common to All Motor Control Circuits

Troubleshooting

When troubleshooting, observe the following list of guidelines:

- *Do* practice safe testing procedures.
- *Do* change only one component at a time.
- *Do* have a complete understanding of the machine's function.
- *Do* make sure that test equipment is operational and calibrated before using it for troubleshooting.
- *Do* use the correct test equipment with the proper rating for the circuit being tested.
- *Don't* change wiring connections in circuits that have worked in the past.
- *Don't* use jumpers around a device to confirm a hunch.

Electrical Troubleshooting

Most electrical troubleshooting in motor control circuits can be done using a multimeter (VOM/DMM) and an AC clamp-on ammeter. Some types of clamp-on instruments like the one shown here incorporate the functions of a clamp-on ammeter and multimeter into one instrument, allowing it to be used to measure current, voltage, and resistance.

407SA03.EPS

10.6.1 Input Voltage/Voltage Unbalance Measurements

All motors and related motor control equipment are designed to operate within a specific range of system voltages including a safety factor, typically 10%. This safety factor is added to compensate for temporary supply voltage fluctuations that might occur. Continuous operation of a motor outside the intended range of voltages can damage the motor. Insufficient operating voltage can cause overload/overheating and possible failure of motors and other devices.

Operating voltages applied to motors and related motor control equipment must be maintained within limits of the voltage value given on the motor nameplate. For single-voltage rated motors, the input supply voltage should be within ±10% of the motor's nameplate voltage. For example, a motor with a nameplate single voltage rating of 230V should have an input voltage that ranges between 207V and 253V (±10% of 230V).

For dual-voltage rated motors, the input supply voltage should be within ±10% of the motor's nameplate voltage. For example, a motor with a nameplate dual voltage rating of 208V/230V should have an input voltage that ranges between 187V (−10% of 208V) and 253V (+10% of 230V).

Voltage unbalance is very important when working with three-phase equipment. A small unbalance in phase-to-phase voltage can result in a much greater current unbalance. With a current unbalance, the heat generated in motor windings will be increased. Both current and heat can cause nuisance overload trips and may cause motor failure. For this reason, the voltage unbalance between any two legs of the voltage applied to a three-phase motor or system should not exceed 2%. If a voltage unbalance of more than 2% exists at the input to the equipment, correct the problem in the building or utility power distribution system before operating the equipment. *Figure 33* shows an example of how the

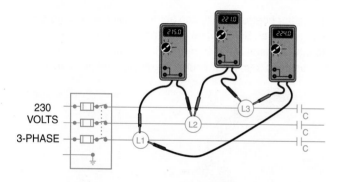

1)

PHASE	MEASURED READING
L1	215V
L2	221V
L3	224V

2) $\dfrac{\text{AVERAGE}}{\text{VOLTAGE}} = \dfrac{215 + 221 + 224}{3} = 220V$

3) INDIVIDUAL PHASE UNBALANCE FROM AVERAGE

L1 TO L2 = 220 − 215 = 5V
L2 TO L3 = 221 − 220 = 1V
L3 TO L1 = 224 − 220 = 4V

4) 5V = MAXIMUM UNBALANCE

5) % UNBALANCE = $\dfrac{\text{MAXIMUM UNBALANCE}}{\text{AVERAGE VOLTAGE}} \times 100$

% UNBALANCE = $\dfrac{5V}{220V} \times 100 = 2.27\%$ (OUT OF BALANCE)

> MAXIMUM VOLTAGE UNBALANCE BETWEEN ANY TWO LEGS MUST NOT EXCEED 2%.

(A) CALCULATING VOLTAGE UNBALANCE

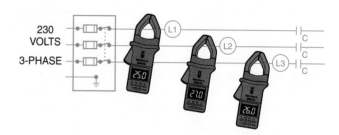

1)

PHASE	MEASURED READING
L1	25A
L2	27A
L3	26A

2) $\dfrac{\text{AVERAGE}}{\text{CURRENT}} = \dfrac{25 + 27 + 26}{3} = 26A$

3) INDIVIDUAL PHASE UNBALANCE FROM AVERAGE

L1 TO L2 = 25 − 27 = 2A
L2 TO L3 = 27 − 26 = 1A
L3 TO L1 = 26 − 25 = 1A

4) 2A = MAXIMUM UNBALANCE

5) % UNBALANCE = $\dfrac{\text{MAXIMUM UNBALANCE}}{\text{AVERAGE CURRENT}} \times 100$

% UNBALANCE = $\dfrac{2A}{26A} \times 100 = 7.7\%$ (IN BALANCE)

> MAXIMUM CURRENT UNBALANCE BETWEEN ANY TWO LEGS MUST NOT EXCEED 10%.

(B) CALCULATING CURRENT UNBALANCE

407F33.EPS

Figure 33 ◆ Three-phase input voltage checks.

amount of voltage unbalance is determined in a three-phase system.

Current unbalance in any one leg of a three-phase system should not exceed 10%. A current unbalance may occur without a voltage unbalance. For example, when an electrical terminal or contact becomes loose or corroded, it may cause a high resistance in the leg. Since current follows the path of least resistance, the current in the other two legs will increase, causing more heat to be generated in the devices supplied by those legs. Current unbalance in a three-phase system is determined in the same way as voltage unbalance. *Figure 33* shows an example.

10.6.2 Fuse Checks

Fuses or circuit breakers are normally the first components checked when a motor is totally inoperative. One way to test a fuse is by measuring continuity (*Figure 34*). To check the fuses, open the disconnect switch, then remove the fuses using an insulated fuse puller. Test the fuses for continuity using an analog or digital multimeter (VOM/DMM). If a short (zero ohms) exists across the fuse, it is usually good. If an open (infinite resistance) exists across the fuse, it is bad. A blown fuse is usually caused by some abnormal overload condition, such as a short circuit within the equipment or an overloaded motor. Replacing a blown fuse without locating and correcting the cause can result in damage to the motor or control equipment.

Fuses can also be tested with the circuit energized. Set the VOM/DMM to measure AC voltage on a range that is higher than the highest voltage expected. Place one VOM/DMM test lead on the line side of the L1 fuse. Touch the other test lead to the load side of the L2 fuse. If voltage is measured, the L2 fuse is good; if not, the fuse is bad. Repeat this procedure so that all fuses are measured with one test lead on the line side and the other test lead on the load side of a different fuse. This method tests one fuse at a time. If the measurement is performed with both test leads on the load side of the fuses, and the VOM/DMM shows no reading, you know that a fuse is blown, but not which one.

CAUTION

When checking fuses, you may get false readings if there is another component, such as a control transformer, that may provide a false indication that the fuse is good due to the component's feedback.

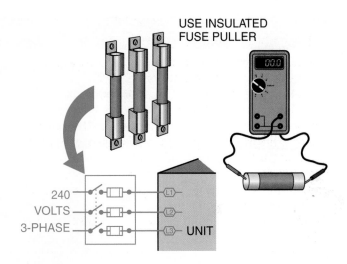

ZERO Ω READING = GOOD FUSE

MEASURABLE OR INFINITE RESISTANCE READING = BAD FUSE

CONTINUITY CHECK

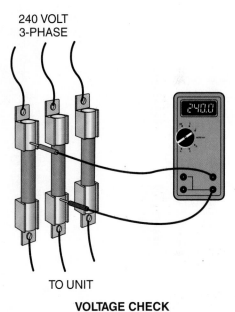

240 VOLT 3-PHASE

TO UNIT

VOLTAGE CHECK

407F34.EPS

Figure 34 ◆ Fuse checks.

10.6.3 Circuit Breaker Checks

Circuit breaker checks will also be necessary at times. To check a circuit breaker, set the circuit breaker to OFF. If required, remove any panel that covers the circuit breaker to expose the body of the breaker and the wires connected to its terminals. Set up a VOM/DMM to measure AC voltage on a range that is higher than the highest voltage expected. Measure the voltage

Power Monitoring

Power monitoring can be accomplished using either internal relays or external metering devices. Phase loss relays, such as those shown here, are used to monitor three-phase power and help prevent damage to motors caused by undervoltage, phase imbalance, phase loss, and phase reversal.

407SA04.EPS

Digital logging meters measure various values including voltage, current, resistance, frequency, and temperature. The logging meter shown here can be connected to a PC using special software for extended system monitoring and data storage.

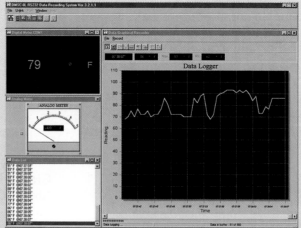

407SA05.EPS

applied to the circuit breaker input terminals (*Figure 35*):

- A to neutral or ground (single-pole breaker)
- A to B (two-pole breaker)
- A to B, B to C, and C to A (three-pole breaker)

Make sure that the breaker is closed by first setting it to the OFF position, then setting it to the ON position. Measure the voltage at the circuit breaker output terminals:

- A1 to neutral or ground (single-pole breaker)
- A1 to B1 (two-pole breaker)
- A1 to B1, B1 to C1, and C1 to A1 (three-pole breaker)

The measured input and output voltages should be the same. If the voltage is significantly lower than that measured at the input to the circuit breaker, visually inspect the circuit breaker for loose wires and terminals or signs of overheating.

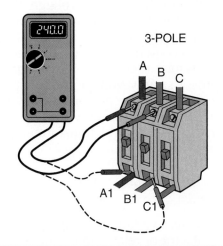

MEASURED INPUT AND OUTPUT VOLTAGES SHOULD BE THE SAME.

CIRCUIT BREAKER VOLTAGE CHECK

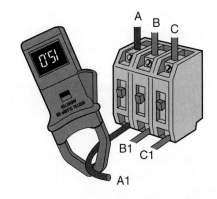

CIRCUIT BREAKER CURRENT CHECK

407F35.EPS

Figure 35 ◆ Circuit breaker checks.

If none are found, the circuit breaker should be replaced.

If the circuit breaker shows signs of overheating or trips when voltage is applied to the equipment, reset it, then check the current flow through the breaker using an AC clamp-on ammeter. Set up the AC clamp-on ammeter to measure AC current on a range that is higher than the highest current expected. Check the ampere rating marked on the breaker. It is usually stamped on the breaker lever or body. For one wire at a time, measure the current flow in the wires connected to the circuit breaker output terminals:

- A1 (single-pole breaker)
- A1, B1 (two-pole breaker)
- A1, B1, and C1 (three-pole breaker)

If the circuit breaker trips at a current below its rating or is not tripping at a higher current, the circuit breaker should be replaced. Be sure that the breaker is not being tripped because of high ambient temperature or other causes at the load or in the feeder.

10.6.4 Troubleshooting Control Circuits

Once the source of an electrical problem has been isolated to a control circuit and the control circuit voltage is known to be good, a series of voltage measurements can be made across the electrical devices in the control circuit to find the faulty device. As shown in *Figure 36*, the measurements can start from the line or control voltage side of the circuit and move toward the load device, typically a contactor coil or relay coil. Measurements are made until either no voltage is observed or until the voltage has been measured across all the devices in the circuit. Note that when there are many devices in the circuit being tested, the measurements can be made by starting at the midpoint in the circuit (divide-by-two method) and then working toward either the source of control voltage or the load device, depending on whether voltage was or was not measured at the midpoint. As a result of making the voltage measurements, one of the two situations described below should exist.

- *Zero Voltage* – At some point within the circuit, no voltage will be indicated on the VOM/DMM. This pinpoints an open set of switch or relay contacts between the last measurement point and the previous measurement point. *Figure 36(A)* shows an example of this situation, where the contacts of the low pressure switch are open, preventing the contactor coil (C) from energizing. If the open is caused by a set of contactor or relay contacts, you must find out if the related contactor or relay coil is not being

Troubleshooting Safety

Safe work practices dictate that troubleshooting should always be performed to the maximum extent possible in de-energized circuits or equipment. Typically, this is done by using a continuity tester or by making resistance measurements with a VOM/DMM. Troubleshooting in energized circuits or equipment should be performed only when the cause of a problem cannot be found by troubleshooting the de-energized circuit or equipment.

energized or is bad. *Figure 36(B)* shows an example of this situation, where the contacts (CR) in the control circuit are open, preventing the contactor (C) from energizing. These contacts close when the control relay coil (CR) is energized. Before assuming that the problem is caused by the open contacts (CR), you must troubleshoot the control circuit containing the related relay coil (CR) to find out if the coil is energized or de-energized. If it is de-energized, you must further troubleshoot its control circuit to find out why. For example, if the contacts of the thermostat COOL switch are open, the relay coil (CR) will not be energized.

- *Voltage Present at Coil* – If voltage is measured at the contactor coil, and the contactor is not energized, the contactor coil is most likely bad. Turn off the power to the circuit, then disconnect the coil from the circuit and test it as described later in this section. *Figure 36(C)* shows an example of a situation in which 24V is applied to the contactor coil (C) but it is not energized. In this case, the contactor coil is probably open.

10.6.5 Control Transformer Checks

Control transformers are checked by measuring the voltages across the secondary and primary windings (*Figure 37*). Typically, the secondary winding is measured first. The VOM/DMM should be set up to measure AC voltage on a range that is higher than the control voltage expected. If the voltage measured across the secondary winding is within 10% of the required voltage, the transformer is good. If no voltage is measured at the secondary winding, the voltage across the primary winding must be measured. Also check the secondary fuse (if provided) to see if it is blown.

If the voltage measured at the primary winding is within ±10% of the required voltage, the transformer is probably bad. This can be confirmed by performing a continuity check of the transformer primary and secondary windings. If no voltage or low voltage is measured across the primary winding, the power supply voltage to the equipment should be checked. If the power supply voltage is okay, troubleshoot the circuit wiring between the power supply and control transformer primary winding.

10.6.6 Contactor/Relay Coil Resistance Checks

When the correct voltage is being applied to the coil of a contactor or relay and it is not energized, the device is probably faulty. Once the coil of a contactor or relay has been identified as the probable cause of an electrical problem, it should be tested. The best way to test a coil is by measuring the resistance across the terminals of the coil. Before measuring resistance, make sure to electrically isolate the coil from the remainder of the circuit by removing all power to the circuit and

Measuring Coil Resistance

The actual resistance measured for a contactor or relay coil can vary widely depending on the device. Ideally, the exact resistance value for the device can be found in the manufacturer's service literature. Another way to judge whether the resistance reading is good is by comparing the resistance of the coil being tested with that of a similar coil that is known to be good.

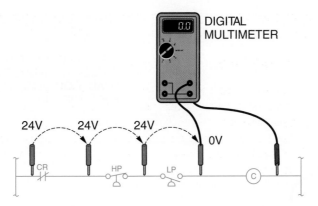

(A) OPEN LOW PRESSURE SWITCH CONTACTS

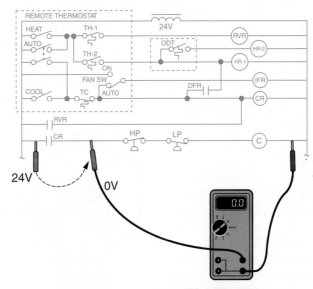

**(B) OPEN CR RELAY CONTACTS – CHECK
RELATED COIL CONTROL CIRCUIT**

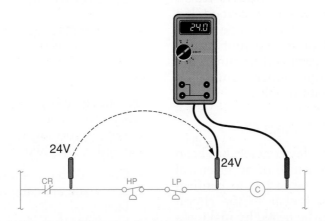

(C) LOAD DEVICE (CONTACTOR COIL) IS OPEN

407F36.EPS

Figure 36 ◆ Isolating to a faulty control circuit component.

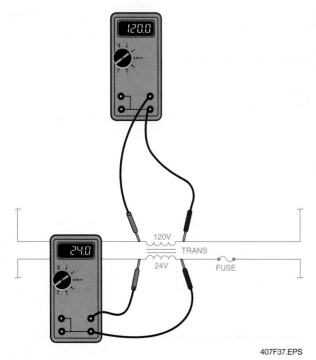

Figure 37 ◆ Control transformer checks.

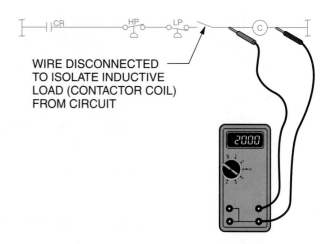

MEASURABLE RESISTANCE = GOOD LOAD
ZERO RESISTANCE = SHORTED LOAD
INFINITE RESISTANCE = OPEN LOAD

407F38.EPS

Figure 38 ◆ Coil resistance checks.

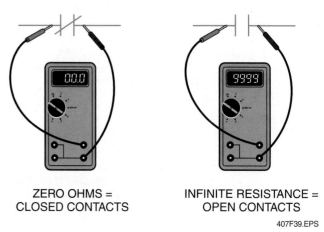

ZERO OHMS =
CLOSED CONTACTS

INFINITE RESISTANCE =
OPEN CONTACTS

407F39.EPS

Figure 39 ◆ Checking the continuity of contactor/relay contacts.

disconnecting at least one lead of the coil. This is important in order to get an accurate resistance reading. Otherwise, the meter may read the resistance of other components that are connected in parallel with the coil being measured. As shown in *Figure 38*, a reading of zero ohms indicates a shorted coil; a reading of infinite resistance indicates an open coil. In either case, replace the contactor or relay (or the coil, if replaceable). When the VOM/DMM indicates a measurable resistance, it usually indicates that the coil is good. If a low resistance is measured, connect one of the meter probes to ground or to the unit frame. Touch the other probe to each coil terminal. If a resistance is measured from either terminal to ground, replace the contactor or relay.

10.6.7 Relay/Contactor Contact Checks

Once a set of contactor or relay contacts has been identified as the probable cause of an electrical problem, the contacts can be tested to confirm their condition. With the power to the circuit turned off, the contacts can be tested by making a continuity measurement to determine whether the contacts are open or closed (*Figure 39*). If the contacts are open, the VOM/DMM indicates an infinite resistance reading. If the contacts are closed, the VOM/DMM indicates a short (zero ohms). When testing relay contacts with the power off, only the normally open or normally closed position of the contacts is being tested. It

may have no bearing on the status of the contacts when the system is powered up.

When working with contactor/relay contacts, remember the following:

* Contacts that open when the contactor/relay is energized are called normally closed (NC) contacts.
* Contacts that close when the contactor/relay is energized are called normally open (NO) contacts.

10.6.8 Troubleshooting Circuit Boards

Solid-state motor control devices such as adjustable frequency drives, reduced-voltage starters, and similar equipment contain printed

circuit boards. Once a board is installed and operating, it is unlikely to fail unless the failure is caused by some outside influence. Because of their complexity, many boards are difficult to troubleshoot and repair in the field. It often takes special automated test equipment at the manufacturer's plant to accomplish this task. For this reason, if a board fails, it is normally replaced with a new one. If the board is still under warranty, the manufacturer will generally replace the board at no cost. Do not try troubleshooting components on a board unless instructed to do so. Once a board has been worked on, the manufacturer has no way of knowing if the board was damaged through a manufacturing error or because you worked on it. In these cases, many manufacturers will not replace the board without payment.

Most boards are expensive, so you want to be absolutely sure that the board is bad before replacing it. A board should be replaced only when all the input signals to the board are known to be good, but the board fails to generate the proper output signal(s). Many boards contain a microprocessor control that controls all sequences of operation for the drive control system. To determine the condition (good or bad) of the board, it is necessary to check for the presence of the proper input signals. It is also necessary to verify that the board generates the proper output signal(s) in response to the input signal(s). Some boards have test points where measurements of key input and output signals can be made. The schematic or wiring diagram will identify the signal that should be available at each point.

To diagnose a board's condition correctly, you must understand the sequence of operation. If you are not familiar with the operation of the board used in the system, study the schematic and sequence of operation described in the manufacturer's service literature. Once you understand what the system and/or board should be doing, you must find out what the board is doing or what symptoms are present in the unit. To aid you in this analysis, many microprocessor-controlled boards have built-in diagnostic circuits that can run a check of motor functions and parameters and report their status. When troubleshooting, always use any available built-in diagnostic features and follow the manufacturer's related troubleshooting instructions.

Before replacing a board that appears to have failed, try to determine if an external cause might account for the problem. There are a variety of such sources:

- Electrical noise (EMI/RFI) from communications equipment, switching power supplies, and other sources will cause microprocessors to fail or operate erratically. The source of interference should be removed, or the board should be shielded.

- As explained earlier, microprocessors and other integrated circuit chips can be damaged by static electricity from sources such as lightning or people. When handling circuit boards, avoid touching the components, printed circuits, and connector pins. Always ground yourself before touching a board. Store unused boards properly.

- Voltage surges and excessive voltage can also cause board failures. Make sure that the applied voltage is within allowable limits and that the unit (and therefore the board) is properly grounded.

- Excessive heat can damage the board. The enclosure in which the board is installed should not be altered in any way that will restrict airflow to the board.

Many boards have dipswitch-selectable options or operating modes used to select site-specific parameters. When replacing a board with dipswitch-selectable options, make sure to set the dipswitch(es) to the correct position for your application. Determine the dipswitch position on the failed board, then set the dipswitch on the replacement board to the same position.

1. Which of the following statements is not true concerning solid-state relays?
   a. They have longer life than electro-mechanical relays.
   b. They have higher reliability than electromechanical relays.
   c. They have higher switching speeds than electromechanical relays.
   d. They can be operated at higher temperatures than electromechanical relays.

2. When connecting a solid-state relay for three-wire control of a load, a(n) _____ is typically used to provide memory for the START pushbutton.
   a. triac
   b. auxiliary solid-state relay contact
   c. SCR
   d. insulated gate bipolar transistor (IGBT)

3. The trip current setting for a solid-state protective relay (overcurrent relay) is based on the _____.
   a. number of loops on the current transformer
   b. motor's FLA, service factor, and number of current loops being used
   c. motor's FLA
   d. motor's service factor

4. A programmable threshold commonly called the _____ determines how long a solid-state overload relay will allow an undercurrent condition to exist before it trips.
   a. undercurrent trip delay
   b. undervoltage trip point
   c. ground fault current trip
   d. undercurrent trip point

5. A timing relay that provides the timed delay after the relay is energized is called a(n) _____.
   a. delay-on-break relay
   b. on-delay relay
   c. timed-closed relay
   d. off-delay relay

6. The contacts of timing relays TR1 and TR2 shown in *Figure 12* are _____ contacts.
   a. normally open, timed-closed (NOTC)
   b. normally closed, timed-open (NCTO)
   c. normally open, timed-open (NOTO)
   d. normally closed, timed-closed (NCTC)

7. For the 8-pin timing relay shown in *Figure 11*, the delayed normally closed contacts are _____.
   a. pins 1–3 and 8–6
   b. pins 1–4 and 8–5
   c. pins 2–7
   d. pins 1–3 and 8–5

8. Which method of reduced-voltage starting provides for adjustment of the motor starting torque and current using taps?
   a. Solid-state
   b. Wye-delta
   c. Resistive
   d. Autotransformer

9. The most common method of reduced-voltage starting used with solid-state reduced-voltage controllers is _____.
   a. kick start
   b. current limit
   c. ramp start
   d. quick start

10. All of the following statements are correct about adjustable frequency drives *except* _____.
    a. they have one range of output frequencies over which the voltage remains constant while the frequency varies
    b. they have two or more ranges of output frequencies over which the ratio of volts to hertz remains constant
    c. they have one range of output frequencies over which the ratio of volts to hertz remains constant
    d. the base frequency is the point at which the transition from one range of output frequencies to another occurs

11. The NEMA classification for the type of adjustable frequency drive described in this module is _____.
    a. FA
    b. FB
    c. FC
    d. FD

12. Which of the following best describes a motor's torque when the horsepower output of the motor varies directly with the speed of the motor?
    a. Constant torque
    b. Variable torque
    c. Constant horsepower
    d. Variable horsepower

13. In most applications, the motors used with AFDs are standard NEMA _____.
    a. Design A or Design C
    b. Design A or Design B
    c. Design B or Design C
    d. Design C or Design D

14. A method of braking where a DC voltage is applied to the stator winding of an AC motor is called _____ braking.
    a. electromechanical
    b. friction
    c. dynamic
    d. DC electric

15. A method of braking an AC motor that involves the use of braking resistors is called _____ braking.
    a. DC electric
    b. friction
    c. dynamic
    d. electromechanical

16. When some solid-state devices are in the OFF state, a hazard of electrical shock can exist from leakage current.
    a. True
    b. False

17. Each of the following should be a major consideration when initially determining the specific preventive maintenance tasks required and their schedule for performance for a motor control system *except* the _____.
    a. equipment manufacturer's recommendations
    b. availability of the equipment
    c. severity of equipment use
    d. equipment's surrounding environment

18. Liquid and spray cleaners are frequently recommended for cleaning the enclosure and components within a motor controller.
    a. True
    b. False

19. Worn or rusted parts in a contactor can cause failure to drop out.
    a. True
    b. False

20. Calculate the voltage unbalance for a motor control feeder with the following voltage readings: L1 to L2 = 442V, L2 to L3 = 456V, and L3 to L1 = 474V.
    a. 1.26%
    b. 2.56%
    c. 3.72%
    d. 4.15%

Summary

Solid-state relays have replaced electromechanical relays in many applications. Solid-state relays have several advantages, including longer life, higher reliability, higher speed switching, and high resistance to shock and vibration. The absence of mechanical contacts eliminates contact bounce, arcing when the contacts open, and hazards from explosives and flammable gases.

Solid-state overload relays operate to protect the motor and control devices against overheating. Depending on the design, many also provide protection against phase loss, phase unbalance, phase reversal, and undervoltage. They do not protect against short circuits or ground faults. Solid-state overload relays are used with solid-state contactors/controllers. Many are designed so that they can also be retrofitted into equipment where thermal overloads previously were used. Depending on the design, some solid-state overload relays are self-powered, and others must be powered by a separate DC power supply. Solid-state overload relays can be either non-programmable or programmable.

Timing relays operate to open or close electric circuits in order to perform selected operations according to a timed program. Most timing relays have adjustable time cycles, allowing them to be used in more than one application. Three commonly used types of timing relays are pneumatic, dashpot, and solid-state.

Four reduced-starting methods commonly used with motors are autotransformer starting, part-winding starting, wye-delta starting, and solid-state starting. Autotransformer starting uses a tapped transformer to limit inrush current. Part-winding starting uses a sequence of connections to connect the winding parts. Wye-delta starting uses a separate connection for starting and running the motor. Solid-state starting uses semiconductor electronic control devices.

An adjustable frequency drive, also commonly called a variable frequency drive, converts three-phase, 50 or 60Hz input power to an adjustable frequency and voltage output used to control the speed of an AC motor. The use of AFD-controlled systems provides for energy savings by eliminating most of the losses associated with mechanical or electromechanical methods of adjusting speed.

Motor braking can be accomplished in several ways. The method used depends on the application, available power, and circuit requirements. Three common methods of motor braking are DC electric braking, dynamic braking (AC drives), and electromechanical (friction) braking.

Preventive maintenance consists of inspecting, cleaning, and testing components in order to prevent equipment failures. Troubleshooting consists of making measurements and tests in order to diagnose the cause of a failure, then replacing the failed components.

Notes

Trade Terms
Introduced in This Module

Base speed: The rating given on a motor's nameplate where the motor will develop rated horsepower at rated load and voltage.

Closed circuit transition: A method of reduced-voltage starting where the motor being controlled is never removed from the source of voltage while moving from one voltage level to another.

Dashpot: A device that uses a gas or fluid to absorb energy from or retard the movement of the moving parts of a relay, circuit breaker, or other electrical or mechanical device.

Heat sink: A metal mounting base that dissipates the heat of solid-state components mounted on it by radiating the heat into the surrounding atmosphere.

Insulated gate bipolar transistor (IGBT): A type of transistor that has low losses and low gate drive requirements. This allows it to be operated at higher switching frequencies.

Latch: In a solid-state device, latching is the equivalent to the holding or memory circuit in relay logic. The current flowing through the relay maintains flow until the load is removed.

Open circuit transition: A method of reduced-voltage starting where the motor being controlled may be temporarily disconnected from the source of voltage while moving from one voltage level to another.

Solid-state overload relay (SSOLR): An electronic device that provides overload protection for electrical equipment.

Solid-state relay (SSR): A switching device that has no contacts and switches entirely by electronic means.

Thermal resistance: The resistance of a material or substance to the conductivity of heat.

Thyristor: A bi-stable semiconductor device that can be switched from the off to the on state or vice versa. All these devices require to turn on is a quick pulse of control current. They turn off only when the working current is interrupted elsewhere in the circuit. Note that SCRs and triacs are forms of thyristors.

This module is intended to present thorough resources for task training. The following reference works are suggested for further study. These are optional materials for continuing education rather than for task training.

Adjustable Frequency Drives, Application Guide, Latest Edition. Milwaukee, WI: Cutler-Hammer.

Consulting Application Guide, Distribution and Control, Latest Edition. Pittsburgh, PA: Cutler-Hammer.

Electrical Motor Controls, Gary Rockis and Glen A. Mazur. Homewood, IL: American Technical Publishers, Inc., 1997.

National Electrical Code® Handbook, Latest Edition. Quincy, MA: National Fire Protection Association.

NFPA 70B Recommended Practice for Electrical Equipment Maintenance. Quincy, MA: National Fire Protection Association, 1998.

NCCER makes every effort to keep these textbooks up-to-date and free of technical errors. We appreciate your help in this process. If you have an idea for improving this textbook, or if you find an error, a typographical mistake, or an inaccuracy in NCCER's Contren® textbooks, please write us, using this form or a photocopy. Be sure to include the exact module number, page number, a detailed description, and the correction, if applicable. Your input will be brought to the attention of the Technical Review Committee. Thank you for your assistance.

Instructors – If you found that additional materials were necessary in order to teach this module effectively, please let us know so that we may include them in the Equipment/Materials list in the Annotated Instructor's Guide.

Write: Product Development and Revision
National Center for Construction Education and Research
3600 NW 43rd St., Bldg. G, Gainesville, FL 32606

Fax: 352-334-0932

E-mail: curriculum@nccer.org

Craft _____ Module Name _____

Copyright Date _____ Module Number _____ Page Number(s) _____

Description _____

_____

_____

_____

_____

(Optional) Correction _____

_____

_____

(Optional) Your Name and Address _____

_____

_____

HVAC Controls

Penn State – Embracing the Possibilities

The Penn State students built their home with an "Energy Dashboard" which monitors and displays energy consumption and production to teach the inhabitants about how they are using their energy. A curtain wall system with photo-voltaic powered LED lighting glows in different colors depending on weather forecasts. Pennsylvania bluestone and reclaimed slate shingles provide thermal mass.

26408-08

26408-08
HVAC Controls

Topics to be presented in this module include:

1.0.0   Introduction 8.2

2.0.0   Heating 8.2

3.0.0   Ventilation 8.2

4.0.0   Air Conditioning 8.5

5.0.0   Thermostats 8.10

6.0.0   HVAC Control Systems 8.20

7.0.0   HVAC Digital Control Systems 8.26

8.0.0   Control Circuit Review 8.30

9.0.0   *NEC*® Requirements 8.34

10.0.0   Troubleshooting 8.44

Overview

Heating, ventilating, and air conditioning systems are equipped with unique control components, including a variety of sensing devices and specialized switches. Relays and contactors are used extensively in HVAC power distribution and control systems. The installation and servicing of HVAC systems is not typically done by general wiring electricians. This work is usually the responsibility of specialized contractors. However, an electrical contractor will be called upon to install the electrical service for the HVAC equipment and must, therefore, be familiar with the specialized requirements and control systems. Some companies that install and service HVAC systems also employ electricians.

Since many HVAC systems operate under very high-pressure gas conditions, special safety requirements including refrigerant handling and disposal requirements, in addition to those associated with electrical shock hazards, must be observed when working around HVAC equipment. Likewise, maintaining or troubleshooting combustion controls on furnace systems requires experienced personnel because of the hazards associated with faulty combustion or improper fuel-to-air mixtures. The *National Electrical Code*® devotes several articles to the installation of HVAC systems, including controls, compressors, room air conditioners, electric baseboard heating, and electric space heating.

Objectives

When you have completed this module, you will be able to do the following:

1. Identify the major mechanical components common to all HVAC systems.
2. Explain the function of a thermostat in an HVAC system.
3. Describe different types of thermostats and explain how they are used.
4. Demonstrate the correct installation and adjustment of a thermostat using proper siting and wiring techniques.
5. Explain the basic principles applicable to all control systems.
6. Identify the various types of electromechanical and electronic HVAC controls, and explain their function and operation.
7. State the *National Electrical Code*® (*NEC*®) requirements applicable to HVAC controls.

Trade Terms

Analog-to-digital converter
Automatic changeover thermostat
Bimetal
Compressor
Condenser
Cooling compensator
Deadband
Differential
Evaporator
Expansion device
Heat transfer
Invar®
Mechanical refrigeration
Refrigerant
Refrigeration cycle
Sub-base
Subcooling
Superheat
Thermostat

Required Trainee Materials

1. Pencil and paper
2. Appropriate personal protective equipment
3. Copy of the latest edition of the *National Electrical Code*®

Prerequisites

Before you begin this module, it is recommended that you successfully complete *Core Curriculum*; *Electrical Level One*; *Electrical Level Two*; *Electrical Level Three*; *Electrical Level Four*, Modules 26401-08 through 26407-08.

This course map shows all of the modules in *Electrical Level Four*. The suggested training order begins at the bottom and proceeds up. Skill levels increase as you advance on the course map. The local Training Program Sponsor may adjust the training order.

ELECTRICAL LEVEL FOUR

26413-08 Introductory Skills for the Crewleader

26412-08 Special Locations

26411-08 Medium-Voltage Terminations/Splices

26410-08 Motor Operation and Maintenance

26409-08 Heat Tracing and Freeze Protection

26408-08 HVAC Controls

26407-08 Advanced Controls

26406-08 Specialty Transformers

26405-08 Fire Alarm Systems

26404-08 Basic Electronic Theory

26403-08 Standby and Emergency Systems

26402-08 Health Care Facilities

26401-08 Load Calculations – Feeders and Services

ELECTRICAL LEVEL THREE

ELECTRICAL LEVEL TWO

ELECTRICAL LEVEL ONE

CORE CURRICULUM: Introductory Craft Skills

408CMAP.EPS

1.0.0 ◆ INTRODUCTION

Heating, ventilating, and air conditioning (HVAC) systems provide the means to control the temperature, humidity, and even the cleanliness of the air in our homes, schools, offices, and factories.

Electricians may be called upon to install the power distribution and control system wiring for this equipment and to troubleshoot the power distribution and control systems in the event of a malfunction. Many of the control devices used in heating and cooling systems are the same as, or similar to, devices you studied in earlier modules. What is unique is the method by which these devices are used to control HVAC equipment.

Before discussing control systems, this module briefly reviews the basic principles of heating and cooling systems and the equipment used in these systems.

2.0.0 ◆ HEATING

Early humans burned fuel as a source of heat. That hasn't changed; what's different between then and now is the way it's done. People no longer need to huddle around a wood fire to keep warm. Instead, a central heating source such as a furnace or boiler does the job using the heat transfer principle; that is, heat is created in one place and carried to another place by means of air or water.

For example, in a common household furnace, fuel oil, natural gas, or propane/butane gas is burned to create heat, which warms metal plates known as heat exchangers (*Figure 1*). Air from living spaces is circulated over the heat exchangers and returned to the living spaces as heated air. This type of system is known as a forced-air system and is the most common type of central heating system used in the United States.

Figure 2 shows the interior of a high-efficiency gas furnace. This design has two heat exchangers to extract the maximum amount of heat from the burning fuel. High-efficiency furnaces and boilers are often eligible for energy-saving cash incentives from local power companies.

Water is also used as a heat exchange medium. The water is heated in a boiler, then pumped through pipes to heat exchangers where the heat it contains is transferred to the surrounding air. The heat exchangers are usually baseboard heating elements located in the space to be heated. This type of system is known as a hydronic heating system and is more common in the Northeast and Midwest than in other parts of the country.

Natural gas and fuel oil are, by far, the most widely used heating fuels. Natural gas is currently the most popular fuel. Oil heat is more common in the Northeast than in other parts of the country. Propane/butane gas is used in many parts of the country in place of oil or natural gas. Oil and propane/butane fuels are primarily used in rural areas where natural gas pipelines are not available.

Electricity is also used as a heat source. In an electric heating system, electricity flows through coils of resistive wire, causing the coils to become hot. Air from the conditioned space is passed over the coils, and the heat from the coils is transferred to the air. Because electricity is so expensive, total electric heat is no longer common in cold climates. It is more likely to be used in warm climates where heat is seldom required. In some areas of the country, a main or supplementary heating system may be fueled by wood or coal. Due to increases in the cost of fossil fuels (such as oil and gas) over the last several decades, and the fact that wood is a relatively cheap and renewable resource, wood-burning stoves and furnaces have remained quite popular.

3.0.0 ◆ VENTILATION

Ventilation is the introduction of fresh air into a closed space in order to control air quality. Fresh air entering a building provides the oxygen we breathe. In addition to fresh air, we want clean air. The air in our homes, schools, and offices contains dust, pollen, and molds, as well as vapors and odors from a variety of sources. Relatively simple air circulation and filtration methods, including natural ventilation, are used to help keep the air in these environments clean and fresh. Among the common methods of improving the air in residential applications are the addition of humidifiers and electronic air cleaners to air handling systems such as forced-air furnaces (*Figure 3*). Many industrial environments, on the other hand, require special ventilation and air management systems. Such systems are needed to eliminate noxious or toxic particles and fumes that may be created by the processes and materials used at the facility.

The U.S. government has strict regulations governing indoor air quality (IAQ) in industrial environments and the release of toxic materials to the outside air. Where noxious or toxic fumes may be present, the indoor air must be constantly replaced with fresh air. Fans and other ventilating devices are normally used for this purpose. Special filtering devices may also be required; these not only protect the health of building occupants, but also prevent the release of toxic materials to the outside air.

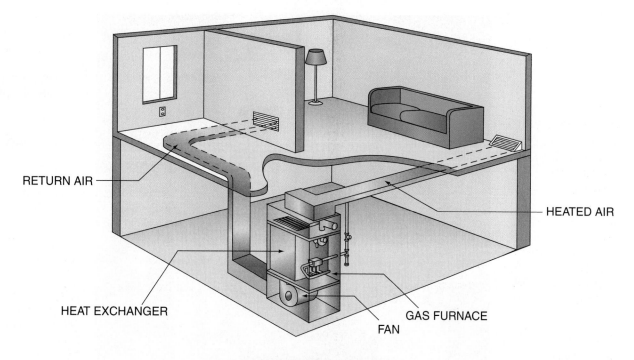

BASEMENT INSTALLATION

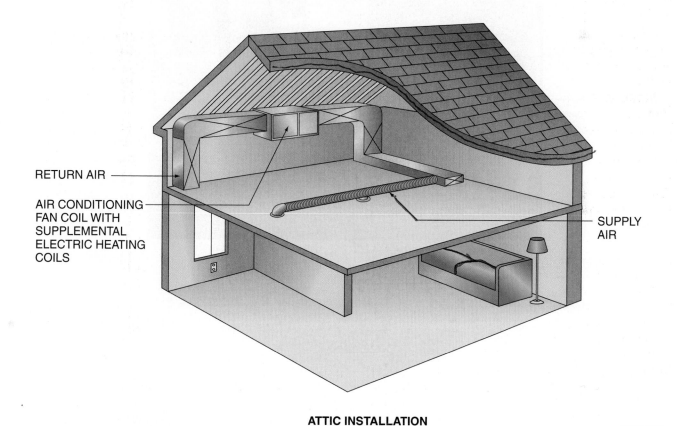

ATTIC INSTALLATION

Figure 1 ◆ Forced-air heating.

408F01.EPS

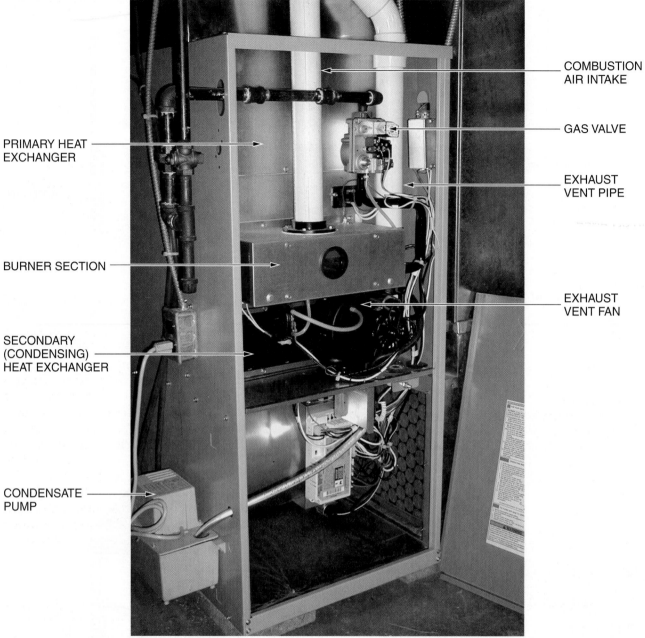

COMBUSTION AIR INTAKE

GAS VALVE

EXHAUST VENT PIPE

PRIMARY HEAT EXCHANGER

BURNER SECTION

EXHAUST VENT FAN

SECONDARY (CONDENSING) HEAT EXCHANGER

CONDENSATE PUMP

408F02.EPS

Figure 2 ◆ A high-efficiency condensing furnace.

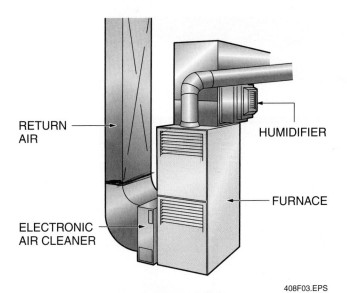

RETURN AIR

HUMIDIFIER

FURNACE

ELECTRONIC AIR CLEANER

408F03.EPS

Figure 3 ◆ Humidifier and electronic air cleaner in a residential heating system.

4.0.0 ◆ AIR CONDITIONING

All cooling systems contain the same basic components. These components function together in a process known as the refrigeration cycle. Understanding the basic refrigeration cycle will help you better understand how the various HVAC controls interact to keep the system operating properly. There are many types of systems used to provide cooling for personal comfort, food preservation, and industrial processes. Each of these systems uses a mechanical refrigeration system to produce the cooling. The operation of the mechanical refrigeration system is the same for all vapor compression systems. Varying from system to system are the type of refrigerant, the size and style of the components, and the installed locations of the basic components and the lines. We will review the basic refrigeration cycle before covering HVAC controls in more detail.

 WARNING!
It is unlawful to release refrigerants into the atmosphere.

Simply stated, the refrigeration cycle relies on the ability of chemical refrigerants to absorb heat. If a cold refrigerant flows through a warm space, it will absorb heat from the space. Having given up heat to the refrigerant, the space becomes cooler. The colder the refrigerant, the more heat it will absorb, and the cooler the space will become. If the hot refrigerant then flows to a cooler location, the outdoors for example, the refrigerant will

INSIDE TRACK

The Mechanical Refrigeration Cycle

Many people think that an air conditioner adds cool air to an indoor space. In reality, the basic principle of air conditioning and the mechanical refrigeration cycle is that heat is extracted from the indoor air and transferred to another location (the outdoors) by the refrigerant that flows through the system.

give up the heat it absorbed from the indoors and become cool again.

A mechanical refrigeration system (*Figure 4*) is a sealed system operating under high pressure. The relationship between temperature and pressure is critical to mechanical refrigeration. As you study the process, you will learn that the same refrigerant can be very cold at one point in the system (the evaporator input) and very hot at another (the condenser input). These two points are often only inches apart. This is possible because of pressure changes caused by the compressor and expansion device. In addition to the circulation of refrigerant, air must also circulate. Fans at the condenser and evaporator move air across the condenser and evaporator coils.

4.1.0 System Components

There are many types of systems used to provide cooling for personal comfort, food preservation, and industrial processes. Each of these uses a mechanical refrigeration system. Its components are:

- *Evaporator* – A heat exchanger where the heat from the area or item being cooled is transferred to the refrigerant.
- *Compressor* – A device that creates the pressure differences in the system needed to make the refrigerant flow and the refrigeration cycle work.
- *Condenser* – A heat exchanger where the heat absorbed by the refrigerant is transferred to the cooler outdoor air or another cooler substance.
- *Expansion device* – A device that provides a pressure drop that lowers the boiling point of the refrigerant just before it enters the evaporator. It is also known as the metering device.

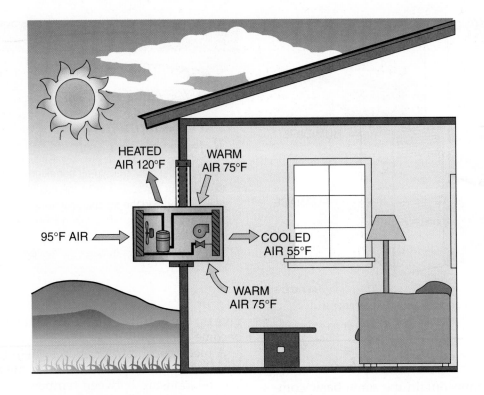

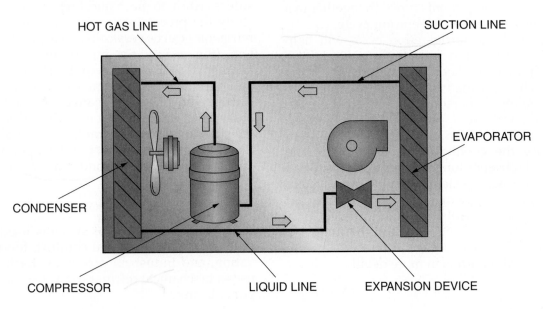

HOT GAS LINE

SUCTION LINE

EVAPORATOR

CONDENSER

COMPRESSOR

LIQUID LINE

EXPANSION DEVICE

408F04.EPS

Figure 4 ◆ Basic refrigeration cycle.

Also shown in *Figure 4* is the piping, called lines, used to connect the basic components in order to provide the path for refrigerant flow. Together, the components and lines form a closed refrigeration system. The lines are:

• *Suction line* – The tubing that carries heat-laden refrigerant gas from the evaporator to the compressor.

• *Hot gas line* (also called the discharge line) – The tubing that carries hot refrigerant gas from the compressor to the condenser.

• *Liquid line* – The tubing that carries the liquid refrigerant formed in the condenser to the expansion device.

The arrows on *Figure 4* show the direction of flow through the system. The purpose of the

Hermetic and Semi-Hermetic (Serviceable) Compressors

Hermetically sealed compressors are typically used in residential and light commercial air conditioners and heat pumps. Semi-hermetic compressors, also known as serviceable compressors, are used in large-capacity refrigeration or air conditioning chiller units. Semi-hermetic compressors can be partially disassembled for repair in the field. Hermetic compressors are sealed and cannot be repaired in the field.

HERMETIC RECIPROCATING COMPRESSOR

HERMETIC ROTARY COMPRESSOR

408SA01.EPS

SEMI-HERMETIC (SERVICEABLE) COMPRESSOR

408SA02.EPS

refrigerant is to move heat. It is the medium by which heat can be moved into or out of a space or substance. A refrigerant is any liquid or gas that picks up heat by evaporating at a low temperature and pressure and gives up heat by condensing at a higher temperature and pressure.

Remember that the operation is the same for all mechanical refrigeration systems. Only the type of refrigerant used, the size and style of the components, and the installed locations of the components and the lines will change from system to system. Other devices, called accessories, may be used in some systems to gain the desired cooling effect and to perform special functions. The events that take place within the system happen again and again in the same order. This repeating series of events is called the refrigeration cycle.

4.2.0 Refrigeration Cycle

Although you are not required to be fully knowledgeable about all aspects of HVAC operation, you need to have a basic understanding of how an HVAC system operates in order to perform your duties as an electrician. It is true that the majority of the problems in HVAC systems are electrical in nature, but it can be very difficult to pinpoint electrical failures unless you understand their corresponding impact on the mechanical (refrigeration) system. Therefore, it is important for the electrician to understand the basic refrigeration cycle in order to trace mechanical failures back to their electrical source.

The refrigeration cycle is based on two principles:

- As liquid changes to a gas or vapor, it is capable of absorbing large quantities of heat.
- The boiling point of a liquid can be changed by altering the pressure exerted on the liquid.

As shown in *Figure 4*, the refrigerant flows through the system in the direction indicated by the arrows. We will begin with the evaporator. It receives low-temperature, low-pressure liquid refrigerant from the expansion device. The evaporator is a series of tubing coils that expose the cool liquid refrigerant to the warmer air passing over the coils. Heat from the warm air is transferred through the tubing to the cooler refrigerant. This causes the refrigerant to boil and vaporize. It is important to realize that even though it has just boiled, it is still not considered "hot" because refrigerants boil at such low temperatures. So, it is a low-temperature, low-pressure refrigerant vapor that travels through the suction line to the compressor.

The compressor receives the low-temperature, low-pressure vapor and compresses it. It then becomes a high-temperature, high-pressure vapor. This travels to the condenser via the hot gas line.

The evaporator is designed so that the refrigerant is completely vaporized before it reaches the end of the evaporator. Thus the refrigerant absorbs more heat from the warm air flowing over the evaporator coils. Because the refrigerant is totally vaporized by this time, the additional heat it absorbs causes an increase in sensible (measurable) heat. This additional heat is known as **superheat.**

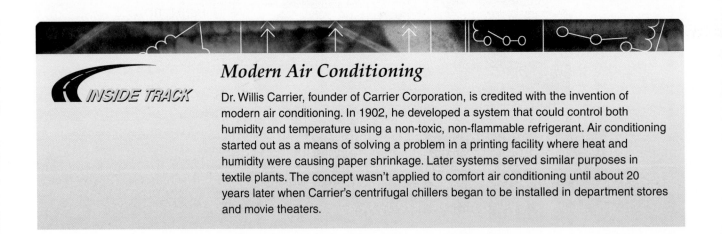

Modern Air Conditioning

Dr. Willis Carrier, founder of Carrier Corporation, is credited with the invention of modern air conditioning. In 1902, he developed a system that could control both humidity and temperature using a non-toxic, non-flammable refrigerant. Air conditioning started out as a means of solving a problem in a printing facility where heat and humidity were causing paper shrinkage. Later systems served similar purposes in textile plants. The concept wasn't applied to comfort air conditioning until about 20 years later when Carrier's centrifugal chillers began to be installed in department stores and movie theaters.

Like the evaporator, the condenser is a series of tubing coils through which the refrigerant flows. As cooler air moves across the tubing, the hot refrigerant vapor gives up heat. As it continues to give up heat to the outside air, it cools to the condensing point, where it begins to change from a vapor into a liquid. As more cooling takes place, all of the refrigerant becomes liquid. Further cooling of the saturated liquid is known as subcooling. This high-temperature, high-pressure liquid travels through the liquid line to the input of the expansion device.

The expansion device regulates the flow of refrigerant to the evaporator. It also decreases its pressure and temperature. By the use of a built-in restriction, such as a tiny hole or orifice, it converts the high-temperature, high-pressure refrigerant from the condenser into the low-temperature, low-pressure refrigerant needed to absorb heat in the evaporator.

4.3.0 Heat Pumps

A special type of air conditioner known as a heat pump is widely used in the warmer parts of the country to provide both cooling and heating. Heat pumps are extremely efficient; however, they are most effective in climates where the temperature does not generally fall below 25°F or 30°F. In colder parts of the country, heat pumps can be combined with supplemental electric, geothermal, or gas- or oil-fired, forced-air furnaces. In such arrangements, the furnace automatically takes over heating duties when the outdoor temperature falls below the efficient range of the heat pump. Like high-efficiency furnaces and boilers, heat pumps are often eligible for energy-saving incentives. Heat pumps operate by reversing the cooling cycle (*Figure 5*). For this reason, heat pumps are sometimes called reverse-cycle air conditioners. Heat pumps operate on the principle that there is always some extractable heat in the air, even though the air may be very cold. In fact, the temperature would have to be −460°F (absolute zero) for a total absence of heat to exist. In the heating mode, a special valve known as a reversing valve switches the compressor input and output so that the condenser operates as the evaporator and the evaporator becomes the condenser. Because of this role reversal, the coils in a heat pump are referred to as the outdoor coil and indoor coil instead of the condenser and evaporator.

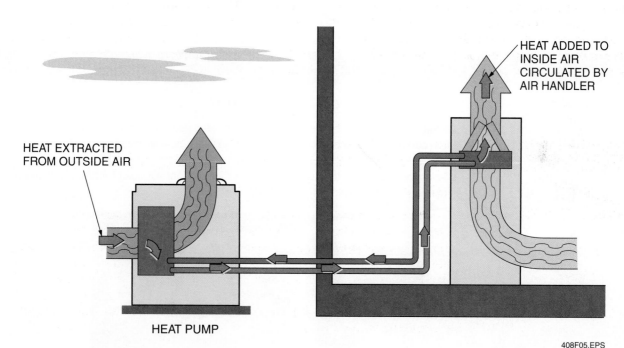

HEAT ADDED TO INSIDE AIR CIRCULATED BY AIR HANDLER

HEAT EXTRACTED FROM OUTSIDE AIR

HEAT PUMP

408F05.EPS

Figure 5 ◆ Heat pump—heating mode operation.

5.0.0 ◆ THERMOSTATS

The room thermostat (*Figure 6*) is the primary control in an HVAC system. It can be as simple as a single temperature-sensitive switch. It can also be a complex collection of sensing elements and switching devices that provide many levels of control. The term thermostat in this module generally refers to the control devices that are mounted on a wall in a conditioned space.

Most residences and many small commercial businesses such as retail stores and shops have a single thermostat. Large office buildings, shopping malls, and factories will be divided into cooling and heating zones and will have several thermostats—one for each zone. An important fact to remember about zoned heating and cooling is that each zone is independent of the others. Therefore, each thermostat must control either a separate system or the airflow from a common system.

5.1.0 Principles of Operation

Programmable electronic thermostats using electronic sensing elements are becoming increasingly popular; however, thermostats with bimetal sensing elements are still being manufactured and are still widely used. Although not quite as precise as the electronic thermostat, bimetal devices are considerably less expensive and have proven to be effective and reliable. Most bimetal thermostats will maintain the space temperature within ±2° of the setpoint. A good electronic thermostat will maintain the temperature within ±1°.

A bimetal element (*Figure 7*) is composed of two different metals bonded together; one is usually copper or brass. The other, a special metal called Invar®, contains 36% nickel. When heated, the copper or brass has a more rapid expansion rate than the Invar® and changes the shape of the element. The movement that occurs when the

408F06.EPS

Figure 6 ◆ Room thermostat.

Large Commercial Chiller Unit

As the name implies, chillers use chilled water as a heat transfer medium. Chillers are often combined with a cooling tower that serves as a heat exchanger. This is more energy efficient than other commercial air-cooled systems. Water flowing over the cooling tower absorbs the heat extracted from the indoor air. The photo below shows a large commercial chiller unit.

GOING GREEN

408SA03.EPS

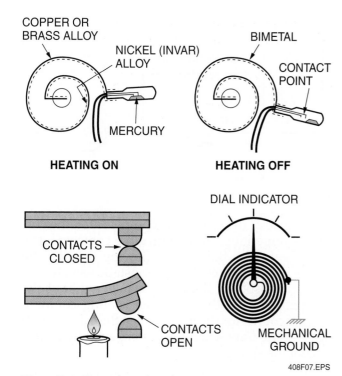

Figure 7 ◆ Bimetal sensing elements.

bimetal changes shape is used to open or close switch contacts in the thermostat.

While bimetal elements are constructed in various shapes, the spiral-wound element is the most compact in construction and the most widely used. In *Figure 7*, for example, a glass bulb containing mercury (a conductor) is attached to a coiled bimetal strip. When the bulb is tipped in one direction, the mercury makes an electrical connection between the contacts and the switch is closed. When the bulb is tipped in the opposite direction, the mercury moves to the other end of the bulb and the switch is opened. Thermostat switching action should take place rapidly to prevent arcing, which could cause damage to the switch contacts. A magnet is used to provide rapid action to help eliminate the arcing potential in some bimetal thermostats.

Most residential and small commercial thermostats are of the low voltage (24V) type. With low-voltage control circuits, there is less risk of electrical shock and less chance of fire from short circuits. Low-voltage components are also less expensive and less likely to produce arcing, coil burnout, and contact failure. Some self-generating systems use a millivolt power supply that generates about 750mV to operate the thermostat circuit. These thermostats are very similar in construction and design to low-voltage thermostats; however, they are not interchangeable with 24V thermostats.

> **WARNING!**
>
> Mercury is toxic. Even short-term exposure may result in damage to the lungs and central nervous system. Mercury is also an environmental hazard. Do not dispose of thermostats containing mercury bulbs in regular trash. Contact your local waste management or environmental authority for disposal/recycling instructions.

5.2.0 Heating-Only Thermostats

A wall-mounted heating-only thermostat typically contains a temperature-sensitive switch as shown in *Figure 8*. In this arrangement, the heating device, such as a furnace, will not come on unless the thermostatic switch is calling for heat.

When the temperature in the conditioned space reaches the thermostat setpoint, the thermostatic switch will open. Because of the residual heat in the heat exchangers and the continued rotation of the fan as it slowly comes to a stop, the temperature will overshoot the setpoint. To avoid occupant discomfort, an adjustable heat anticipator opens the thermostat before the temperature in the space reaches the setpoint. The heat anticipator is a small resistance heater in series with the switch contacts. The anticipator heats the bimetal strip, causing the contacts to open early.

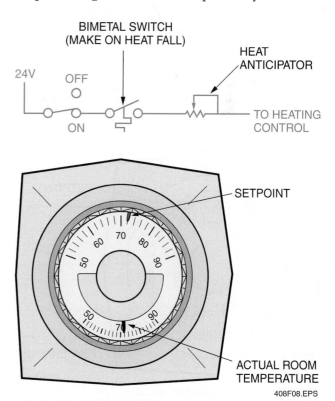

Figure 8 ◆ Heating-only thermostat.

5.3.0 Cooling-Only Thermostats

The cooling thermostat (*Figure 9*) is the opposite of a heating thermostat. When the bimetal coil heats up and unwinds, the mercury switch closes its contacts and starts the cooling system compressor. When the cooling thermostat is turned down to make the conditioned space cooler, it tips the mercury bulb so that the coil must cool more and wind tighter to turn the cooling system off.

Cooling thermostats contain a device called a **cooling compensator** to help improve indoor comfort. The cooling compensator (*Figure 10*) is a fixed resistance in parallel with the thermostatic switch. (Heating anticipators are adjustable.) No current flows through the compensator when cooling is on because it has a much higher resistance than the switch contacts. In this case, the contacts are essentially a short circuit. When the thermostat is open, however, a small current can flow through the compensator and the contactor coil.

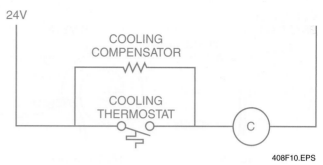

Figure 10 ◆ Cooling compensator.

Because of the size of the compensator, the current is not enough to energize the contactor. The heat created by the current flowing through the compensator makes the thermostatic switch contacts close sooner than they would without the compensator. In this way, the cooling compensator accounts for the lag between the call for cooling and the time when the system actually begins to cool the space.

5.4.0 Heating-Cooling Thermostats

When heating and cooling are combined for year-round comfort, it is impractical to use a separate thermostat for each mode. Therefore, the two are combined into one heating-cooling thermostat (*Figure 11*). When a mercury bulb design is used, a set of contacts is located at one end of the bulb for heating and the other end of the bulb for cooling (*Figure 12*). The cooling contacts close the control circuit on a rise in temperature and open the circuit on a drop in temperature. The bulb with both sets of contacts is attached to a single bimetal element.

Unless a switch is provided in the heating-cooling thermostat, the thermostat will continuously switch back and forth from heating to cooling. In effect, heating and cooling will combat each other for control. A switch provides a means to direct the control to cooling, while disconnecting the heating control circuit. Likewise, when the switch is moved to heating, the switch connects the heating components, while electrically isolating the cooling circuit. When the switch is in the center or off position, neither the heating nor cooling control circuits can be energized.

5.5.0 Heating-Cooling Automatic Changeover Thermostats

The disadvantage of a heating-cooling thermostat is that the building occupant must determine whether heating or cooling is needed at a

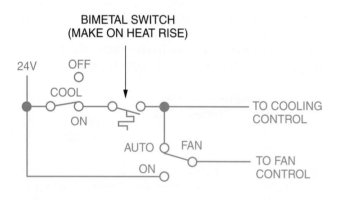

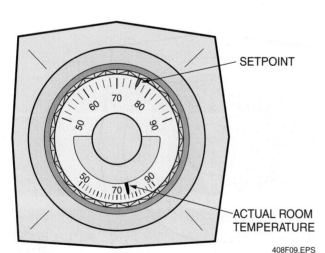

Figure 9 ◆ Cooling-only thermostat.

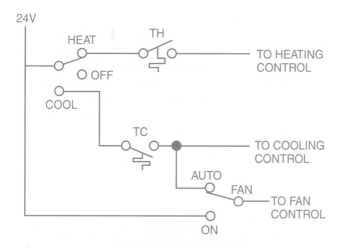

Figure 11 ◆ Heating-cooling thermostat.

408F11.EPS

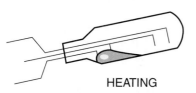

HEATING

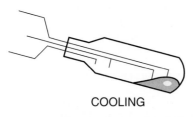

COOLING

408F12.EPS

Figure 12 ◆ Heating-cooling contacts.

particular time and set the thermostat switch accordingly. In some climates, that is very impractical; the need could change several times a day.

The **automatic changeover thermostat** automatically selects the mode, depending on the heating and cooling setpoints. The thermostat shown in *Figure 13* is essentially the same as that shown in *Figure 11*, with the exception that in *Figure 13* there is an AUTO position on the main control switch. The occupant can still select either heating or cooling. When the switch is in the AUTO position, however, the thermostat makes the selection. All that is necessary is for one of the thermostatic switches to close, indicating that the conditioned space is too warm or too cold.

A thermostat contains a built-in mechanical **differential**, which is the difference between the cut-in and cut-out points of a thermostat. The differential is normally 2°F. For example, if the heating setpoint is 70°F, the furnace will turn on at 70°F and run until the temperature is 72°F.

Automatic changeover thermostats also have a minimum interlock setting, commonly known as the **deadband**. The deadband is a built-in feature that prevents the heating and cooling setpoints

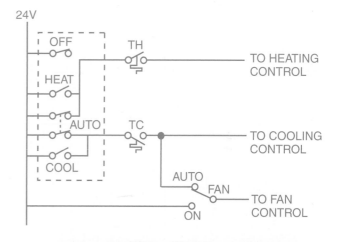

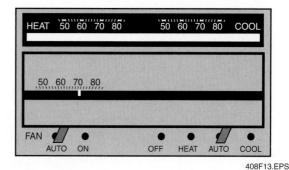

Figure 13 ◆ Automatic changeover thermostat.

408F13.EPS

from being any closer together than 3°F. The dead-band keeps the system from cycling back and forth between heating and cooling.

5.6.0 Multi-Stage Thermostats

Multi-stage thermostats (*Figure 14*) provide energy savings by cycling on stages of heating or cooling based on demand. For example, a cooling system might use a two-speed compressor. If the demand for cooling is low, the first-stage cooling circuit of the room thermostat cycles the compressor on at the lower speed and capacity until the demand is met. If the demand for cooling is high, such as on a very hot day, the second-stage cooling circuit in the thermostat energizes the higher compressor speed. Similarly, a two-stage heating thermostat could stage on burners in a gas furnace or electric elements in an electric furnace. Multi-stage room thermostats are commonly used with heat pumps. Multi-stage room thermostats come in a variety of configurations such as single-stage heat with two stages of cooling, two stages of heat with one stage of cooling, or two stages of heating and two stages of cooling.

Heat pump thermostats usually have an emergency or auxiliary heat switch. If the heat pump becomes inoperative, this switch locks out the normal heat pump operation and heats the area with supplementary electric heat until the problem can be corrected. An indicator light, usually red, is mounted on the thermostat. It will come on when the selector switch is in the emergency heat position. As soon as the unit has been repaired, return the switch to the normal operating position.

5.7.0 Programmable Thermostats

Programmable thermostats are self-contained controls with the timer, temperature sensor, and switching devices all located in the unit mounted on the wall. Early programmables looked very much like conventional thermostats (*Figure 15*). This type of thermostat contains a motor-driven time clock driving a wheel containing cams that can be set to raise and lower the temperature settings at desired intervals.

Modern electronic programmable thermostats (*Figure 16*) use microprocessors and integrated circuits to provide a wide variety of control and

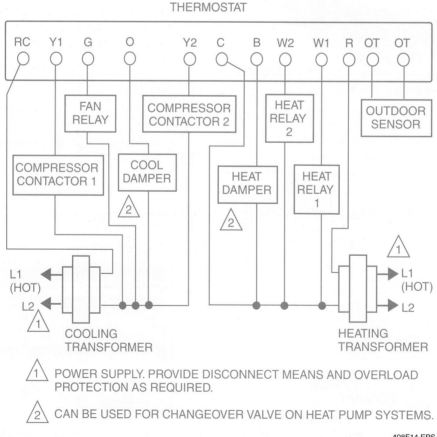

Figure 14 ◆ Multi-stage thermostat hookup.

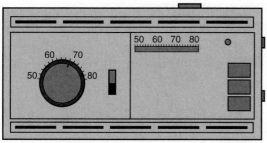

Figure 15 ◆ Programmable thermostat.

Figure 16 ◆ Electronic programmable thermostat.

energy-saving features. Their control panels use touch-screen technology, and their indicators are digital readouts rather than analog needle positioning. Different thermostats offer different features; the more sophisticated (and expensive) the thermostat, the more features it offers.

Some of the features available on electronic thermostats are:

- *Override control* – This feature allows the occupant to override the program when desired. For example, you might override the night setback on Monday night so the thermostat isn't lowered before the football game is over.
- *Multiple programs* – This feature allows the occupant to design and select different schedules for different conditions. For example, you could program a special schedule for a vacation away from home.
- *Battery backup* – This feature prevents program loss in the event of a power failure.
- *Staggered startup for multi-unit systems* – This feature avoids excessive current drain. This is an important feature in office buildings, shopping malls, and hotels.

- *Maintenance tracking* – This feature indicates when maintenance is to be performed (for example, when to replace filters).

The savings available from programmable thermostats are significant. For example, it is estimated that a setback of 10°F for both daytime and nighttime can result in a 20% energy savings. A 5°F setback will yield a 10% energy savings.

5.8.0 Line-Voltage Thermostats

Most of the thermostats you encounter will operate at low voltages (for example, 24V). There are also thermostats that operate at line voltages (for example, 240V). Line-voltage thermostats, as their name implies, control heat by directly controlling the power supply. This is in contrast to low-voltage thermostats that indirectly control heat or cooling through a relay or contactor. Line-voltage thermostats are commonly used to control baseboard heaters and floor or ceiling radiant heat grids. Conventional line-voltage thermostats use a bimetal sensing element, which does not provide precise temperature control. This is because the bimetal elements and switch contacts must be heavier to carry the larger current. Because they are heavier, they do not respond as quickly or easily as the smaller and lighter contacts in low-voltage (24V) room thermostats. Digital line-voltage thermostats provide much more precise temperature control (*Figure 17*).

5.9.0 Thermostat Installation

Even the most sensitive thermostat cannot perform correctly if it is poorly installed. Selecting the proper location for the thermostat is the first step in any installation procedure.

Figure 17 ◆ Digital line-voltage thermostat.

5.9.1 Installation Guidelines

The thermostat should be installed in the space in which it will be called upon to control the temperature and other conditioning factors. The thermostat should be installed on a solid inside wall that is free from vibration that could affect operation by making the thermostat contacts chatter. For the same reason, it should not be located on a wall near slamming doors or near stairways. The following practices should be observed when installing a thermostat:

- The installer must be a trained, experienced technician.
- The manufacturer's instructions must be carefully read prior to installing. Failure to follow them could lead to product damage or a hazardous condition.
- The rating should be checked in the instructions and on the unit to make sure the thermostat is suitable for the particular application.
- When the installation is complete, the thermostat must be operationally checked and adjusted as indicated in the installation instructions.

CAUTION

Thermostats containing solid-state devices are sensitive to static electricity when not mounted on the wall; therefore, you must discharge body static electricity before handling the instrument. This can be accomplished by touching a grounded metal pipe or conduit. Touch only the front cover when holding the device.

When unpacking the new thermostat and wallplate or **sub-base,** handle it with care. Rough handling can damage the thermostat. Save all instructional information and literature for future reference.

Locate the thermostat and wallplate or sub-base about five feet above the floor in an area with good air circulation at room temperature. Avoid locations that create the following conditions:

- Drafts or dead air spots
- Hot or cold air from ducts or diffusers
- Radiant heat from direct sunlight or hidden heat from appliances
- Heat from concealed supply ducts and chimneys
- Unheated areas behind the thermostat, such as an outside wall or garage

All thermostats must be carefully leveled (*Figure 18*). If mercury bulb thermostats are not

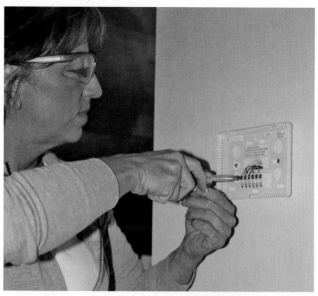

408F18.EPS

Figure 18 ◆ Thermostat mounting.

level, they will not operate properly. Electronic thermostats are leveled for appearance.

5.9.2 Thermostat Wiring

The thermostat is connected through multi-conductor thermostat wire to a terminal strip or junction box in the air conditioning unit. A standard coding method is used in the HVAC industry to designate wiring terminals and wire colors. *Figure 19* shows the coding method and illustrates how the terminals of the heating-cooling thermostat shown earlier would be designated. The terminal designation arrangement is fairly standard among thermostat and equipment manufacturers. It is not safe to assume, however, that the person who installed an existing system followed the color scheme when wiring the control circuits. There are additional codes for more complex thermostats; for example, the letter O designates orange and is connected to the reversing valve control in a heat pump. When there are multiple stages of heating or cooling, those terminals are designated with the appropriate letter plus a number.

Thermostat wiring should be done in accordance with national and local electrical codes. See *NEC Section 424.20* for limitations on using a thermostat as a disconnect device. All wiring connections should be tight. To avoid damaging the control wire conductor, always use a stripping tool designed to strip small-gauge wire. Color-coded wiring should be used where possible for easy reference should system troubleshooting be required at a later date.

THERMOSTAT WIRING CODES

TERMINAL DESIGNATION	WIRE COLOR	FUNCTION
R	RED	POWER
G	GREEN	FAN CONTROL
Y	YELLOW	COOLING CONTROL Y1 = STAGE 1 Y2 = STAGE 2
W	WHITE	HEATING CONTROL W1 = STAGE 1 W2 = STAGE 2
O	ORANGE	HEAT PUMP REVERSING VALVE CONTROL

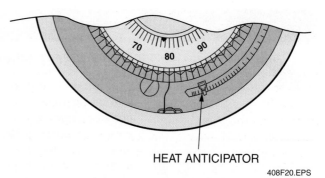

Figure 20 ◆ Thermostat heat anticipator.

408F20.EPS

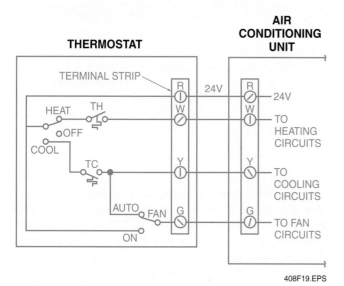

Figure 19 ◆ Thermostat wiring.

408F19.EPS

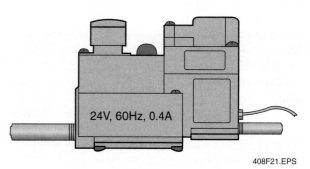

Figure 21 ◆ Gas valve electric ratings.

408F21.EPS

> **WARNING!**
>
> To prevent electrical shock or equipment damage, make sure the power is off before connecting the wiring for the current draw adjustment.

Wires should not be spliced, but if splicing is absolutely necessary, soldered splices are recommended. If wires are stapled to prevent movement, take care to ensure that the staple does not go through the wire insulation. Seal the wire opening in the wall space behind the thermostat so it is not affected by drafts within the wall stud space.

5.9.3 Checking Current Draw

The next step is to determine the current draw. This may be done by locating the current draw of the primary control in the heating unit or the heat anticipator setting on the existing thermostat (*Figure 20*). The current draw is usually printed on the furnace nameplate and/or a primary control such as the gas valve (*Figure 21*), the relay, or the oil burner control. It may also be found in the manufacturer's installation and service literature.

5.9.4 Adjusting Heat Anticipators

The heat anticipator is adjustable and should be set at the amperage indicated on the primary control. Small variations from the required setting can be made to improve performance on individual jobs.

Changing the setting of the anticipator changes the resistance of the wire resistor. This shortens or lengthens the heating cycle. Some heat anticipators have arrows to indicate the heating cycle adjustment.

Some heat anticipators have a fixed resistance; others are equipped with an adjustable dial to change the resistance. The adjustable heat anticipator has a slide wire adjustment with the pointer scale marked in tenths of an ampere. This is used to set the anticipator to match the control amperage draw of the particular furnace. Each furnace is

Thermostat Wiring

Thermostat wires come in a variety of conductor configurations. Simple two-conductor wire can be used in a heating-only installation. Thermostat wires with three, four, five, and six conductors are readily available from any HVAC parts distributor for use in more complex installations. Normally, 18-gauge thermostat wire is adequate for most installations. However, long runs of thermostat wire can produce a voltage drop that might prevent equipment operation. If you have a run of thermostat wire that seems excessively long, use 16-gauge or heavier wire to reduce voltage drop.

Heat Anticipators

Be very careful about adjusting a room thermostat heat anticipator, especially if there is no complaint of poor comfort. This is a situation where it is better to do nothing than to run the risk of creating a comfort problem.

provided with an information sticker near the burner that states the amperage drawn by the control circuit of that particular furnace. This is the amperage at which the thermostat heat anticipator should be set under ideal conditions.

For example, if the amperage draw of a control circuit is shown as 0.45A, the installer should adjust the anticipator setting to 0.45A on the scale. The heat anticipator adjustment determines the length of the thermostat call for heat cycle by artificially heating the bimetal coil. As more heat is directed at the bimetal coil, a shorter heating cycle will occur. Conversely, as less heat is directed at the bimetal coil, a longer heating cycle will occur until the thermostat satisfies the call for heat.

When the control circuit amperage is high, less of the heater wire is needed; when the control amperage is low, more of the heater wire is needed. The control circuit amperage draw should be measured for each heating system as previously described.

5.9.5 Cycle Rate

Electronic room thermostats do not contain a traditional heat anticipator. Instead, they contain a feature that allows the cycle rate (the number of times per hour the burner operates) to be adjusted, which in turn helps maintain indoor comfort. Most furnace manufacturers list the correct cycle rate for their product in the furnace installation literature. The room thermostat literature will contain directions for changing the cycle rate.

5.9.6 Final Check

The final step is to check the heating and/or cooling system when the thermostat is installed. With the thermostat in the heating mode, turn the power on, place the system switch at HEAT, and leave the fan switch in the AUTO position. Turn the setpoint dial to at least 5°F above the room temperature. The burner should come on within 15 seconds. The fan will start after a short delay. Then turn the setpoint dial to 5°F below the room temperature. The main burner should shut off within 15 seconds, but the blower may continue to run for one to two minutes. Next, set the thermostat to the cooling mode.

CAUTION

To avoid compressor damage, do not check cooling operation unless the outside temperature is at least 50°F and the crankcase heater (if so equipped) has been on for at least 24 hours.

If the outside temperature is at least 50°F, return power to the unit, set the thermostat to the COOL position, and set the fan switch to AUTO. Leave the setpoint of the thermostat at 5°F below the room temperature. The cooling system should come on either immediately or after any start delay, if the unit is so equipped. The indoor fan should come on immediately.

After the cooling system has come on, set the thermostat to at least 5°F above the room temperature. The cooling equipment should turn off within 15 seconds. Place the system switch to OFF. Move the setpoint dial to various positions. The system should not respond for heating or cooling.

5.9.7 Adjusting the Thermostat

Thermostats are calibrated or preset at the factory for accurate temperature response and will not normally need recalibration. If a mercury bulb thermostat seems out of adjustment, the first thing to do is to check for accurate leveling. If that doesn't solve the problem, check and/or adjust the calibration as follows:

Step 1 Move the temperature setting lever to the lowest setting. Set the system switch to HEAT and wait about ten minutes.

Step 2 Remove the thermostat cover and move the thermostat temperature selector lever toward a higher temperature setting until the mercury switch just makes contact. This can be done by observing the mercury droplet in the bulb.

Step 3 If the thermostat pointer and the setting lever read about the same at the instant the switch makes contact, no recalibration is needed. If recalibration is necessary, follow the manufacturer's instructions. A typical recalibration procedure is as follows:

a. With the system switch on HEAT, move the setting lever several degrees above room temperature.

b. Insert the end of a hex (Allen) wrench in the socket at the top center of the main bimetal coil. Use an open-end wrench to hold the hex nut under the coil.

c. Hold the setting lever so it will not move, and turn the Allen wrench clockwise until the mercury switch breaks contact. Remove the wrench.

d. Move the setting lever to a low setting. Wait at least five minutes until the thermostat loses any heat gained from your hands and its own operation.

e. Slowly move the setting lever up the scale until it reads the same as the thermostat pointer.

f. Reinsert the Allen wrench. Holding the setting lever so it won't move, carefully turn the Allen wrench counterclockwise until the mercury just makes contact (turn it no further than necessary).

g. Recheck the calibration. Note that the calibration process may have to be repeated if the calibration is still off.

h. When you are satisfied with the calibration, replace the cover and set the thermostat lever switches for the desired operation.

6.0.0 ◆ HVAC CONTROL SYSTEMS

Most HVAC control systems are designed to automatically maintain the desired heating, cooling, and ventilation conditions set into the system. The controls for a small system such as a window air conditioner are very simple—a couple of control switches and a thermostat. As the system gets larger and provides more features, the controls become more complicated. For example, add gas heating to a packaged cooling unit, and the size and complexity of the control circuits will more than double. Make it a heat pump instead, and the control complexity will triple.

Large commercial systems may use pneumatic and electronic controls in conjunction with conventional electrical controls. These systems may have thirty or forty control devices, whereas a window air conditioner has just a handful.

The good news is that there are only a few different kinds of control devices. Once you learn to recognize them and understand the role each plays, it won't matter how many are used to control a particular system.

All automatic control systems have the following basic characteristics in common:

- The sensing element (thermistor, thermostat, pressurestat, or humidistat) measures changes in temperature, pressure, and humidity.
- The control mechanism translates the changes into energy that can be used by devices such as motors and valves.
- The connecting wiring, pneumatic piping, and mechanical linkages transmit the energy to the motor, valve, or other devices that act at the point where the change is needed.
- The device then uses the energy to achieve some change. For example, motors operate compressors, fans, or dampers. Valves control the flow of gas to burners or cooling coils and permit the flow of air in pneumatic systems. Valves also control the flow of liquids in chilled-water systems.
- The sensing elements in the control detect the change in conditions and signal the control mechanism.
- The control stops the motor, closes the valves, or terminates the action of the component being used. As a result, the call for change is ended.

6.1.0 Motor Speed Controls

Greater efficiency can be achieved by varying the speed of compressors and fan motors to adapt to changing heating and cooling loads. In this way, the equipment consumes only as much power as is needed to meet the demand. For example,

HVAC equipment commonly uses variable-speed blowers. In addition, two-speed compressors are used in small systems, while larger systems use unloaders or multiple compressors to adapt to changing loads.

The triac (*Figure 22*) is the most common motor control device. A knob on the control device is used to adjust a potentiometer; the setting of the potentiometer determines the motor speed. This circuit is similar to that of a light dimmer.

Modern electronic motor controls use microprocessor chips to achieve continuous control. The furnace control system shown in the simplified diagram in *Figure 23* controls the combustion system as well as the blower speed. The microprocessor monitors information such as the length of the last heating cycle and how often the furnace is cycling on and off. It then optimizes the heating cycle by selecting the appropriate fan speed and adjusting the length of the low-fire and high-fire cycles to match the conditions in the space. It is also able to sense any changes within the duct system such as a dirty filter or a closed zone valve and vary the blower speed to maintain the correct airflow in the system.

6.2.0 Lockout Control Circuit

The purpose of the lockout relay in a control circuit is to prevent the automatic restart of the HVAC equipment. If the lockout relay has been activated, the system may be reset only by interrupting the power supply to the control circuit; for example, resetting the thermostat (in the case of an HVAC system), or turning the main power switch off and then on again.

In the circuit in *Figure 24*, the lockout relay coil, because of its high resistance, is not energized during normal operation. However, when any one of the safety controls opens the circuit to the compressor contactor coil, current flows through

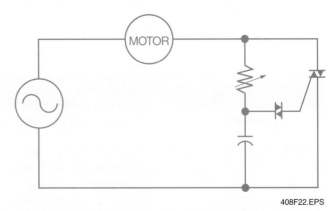

408F22.EPS

Figure 22 ◆ Motor speed control using triacs.

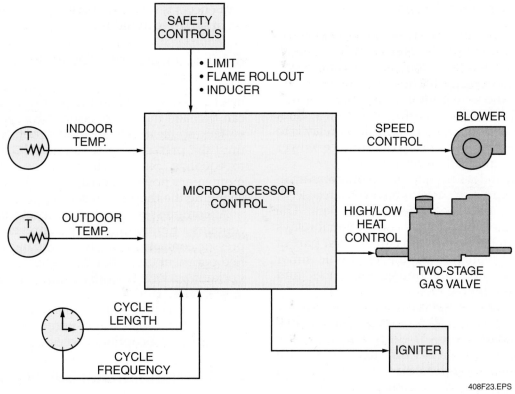

Figure 23 ◆ Electronic variable-speed furnace control.

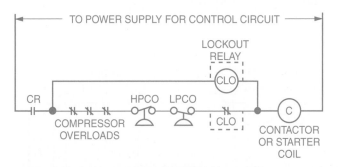

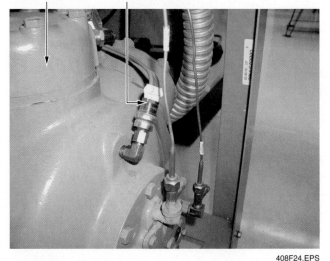

408F24.EPS

Figure 24 ◆ Lockout relay used in an HVAC control circuit.

the lockout relay coil, causing it to become ener-gized and to open its contacts. These contacts remain open, keeping the compressor contactor circuit open until the power is interrupted after the safety control has reset. Performance depends on the resistance of the lockout relay coil being much greater than the resistance of the compres-sor contactor coil.

If the lockout relay becomes defective, it should be replaced with an exact duplicate to maintain the proper resistance balance. This type of relay is sometimes called an impedance relay.

It is permissible to add a control relay coil in parallel with the contactor coil when a system demands another control. The resistance of the contactor coil and the relay coil in parallel decreases the total resistance and does not affect the operation of the lockout relay.

 WARNING!
Never put additional lockout relays, lights, or other load devices in parallel with the lockout relay coil. Doing so might defeat the lockout and create a very hazardous situation.

6.3.0 Time Delay Relays

The purpose of a time delay relay is to delay the normal operation of a compressor or motor for a predetermined length of time after the control system has been energized. The length of the delay depends on the time built into the relay coil and may vary from a fraction of a second to several minutes. A common use for a time delay relay is to delay the startup or shutdown of a furnace blower to improve heating efficiency (*Figure 25*).

Time delay relays used to stage compressors in multiple compressor systems have been replaced, for the most part, with electronic timers. The electronic timers can control conventional relays or may have relay contacts as an integral part of the timer. Advantages of the electronic timer include more accurate timing sequences and greater reliability.

A sequencer is a special type of time delay relay used to energize resistance heaters in a heat pump air handler or electric furnace. The sequencer is basically a 24V time delay relay with contacts heavy enough to carry the current of a large electric heater. It is common for heat pump air handlers to be equipped with several banks of heaters, each one controlled by its own sequencer. To avoid the current surge if all heaters were energized at once, the timing sequence of the sequencers is selected to bring the various heaters on in stages to avoid the surge.

6.4.0 Compressor Short-Cycle Timer

Attempting to start a compressor against high head pressure or when the supply voltage is low can damage the compressor. When a refrigerant system shuts down, it should not be restarted until the pressures in the system have had time to equalize. Short-cycling can be caused by a momentary power interruption or by an occupant changing the thermostat setting.

A compressor short-cycle protection circuit contains a timing function that prevents the compressor contactor from re-energizing for a specified period of time after the compressor shuts off. Lockout periods typically range from 30 seconds to several minutes.

Short-cycle timers are available as self-contained modules that can be direct-wired into a unit. They are often sold as optional accessories.

6.5.0 Control Circuit Safety Switches

Refrigerant system compressor control circuits normally include several different types of safety switches. These include pressure switches, freeze-stats, and outdoor thermostats.

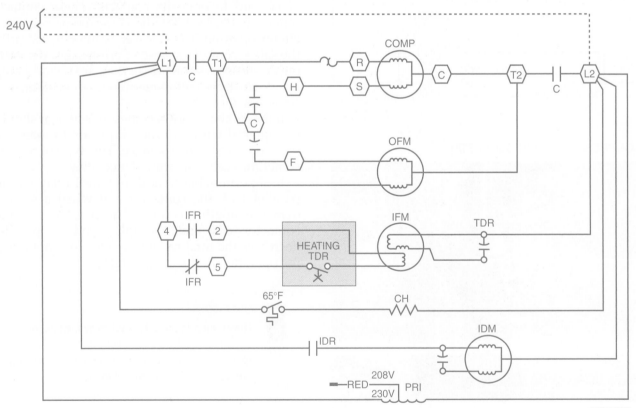

Figure 25 ◆ Time delay relay used to control a blower.

6.5.1 Pressure Switches

Many systems use one or more pressure switches in the compressor control circuit. These are safety devices designed to protect the compressor. A pressure switch is normally closed and wired in series with the compressor contactor control circuit. *Figure 24* shows an example of low-pressure cutout (LPCO) and high-pressure cutout (HPCO) switches used in such circuits.

High-pressure switches are designed to open if the compressor head pressure is too high. Low-pressure switches are designed to open if the suction pressure is too low. Pressure switches use a bellows mechanism that presses against switch contacts that have preset open and close settings. When the system pressure begins to rise above or drop below the normal operating pressure, the related high-pressure or low-pressure switch will open, causing the compressor contactor to de-energize.

Some manufacturers use a type of low-pressure switch called a loss of charge switch that removes power to the equipment if the refrigerant charge is low.

6.5.2 Freezestat

Another type of safety switch is the freeze-up protection thermostat (FPT), commonly called a freezestat. Its purpose is to prevent evaporator coil freeze-up. The freezestat switch is a normally closed bimetal switch that is usually attached to one of the endbells in the evaporator coil (*Figure 26*). It will open if the refrigerant temperature drops below a predetermined setpoint.

6.5.3 Outdoor Thermostats

Some equipment uses an outdoor thermostat to shut off the equipment when the ambient temperature drops below a predetermined setpoint (typically between 55°F and 65°F). This prevents equipment damage.

6.6.0 Furnace Controls

Common furnace controls include the fan control, limit control, thermocouple, and inducer proving switch. These controls are described in this section. Many manufacturers have replaced the bimetal fan control with an electronic timer.

6.6.1 Limit Control

A limit control (*Figure 27*) is also a heat-actuated switch with a bimetal sensing element. It is positioned near the heat exchangers. This is a safety control that is wired into the primary side of the transformer. If the temperature inside the heating chamber or plenum reaches approximately 200°F, the power will be shut off to the transformer, which also stops all power to the temperature control circuit. Many manufacturers combine the limit switch with the fan switch.

6.6.2 Thermocouple

A thermocouple (*Figure 28*) is a device that uses dissimilar metals to control electron flow. The hot junction of the thermocouple is located in the pilot flame of a gas-fired furnace. When heat is applied to the welded junction, a small voltage is produced. This small voltage is measured in millivolts (mV) and is the power used to operate the pilot safety control. The output of a thermocouple is about 30mV. This simple device can cause many problems if the connections are not kept clean and tight or the pilot flame is inadequate to generate the correct voltage.

FPT

408F26.EPS

Figure 26 ◆ Freezestat.

408F27.EPS

Figure 27 ◆ Example of a high-temperature limit switch.

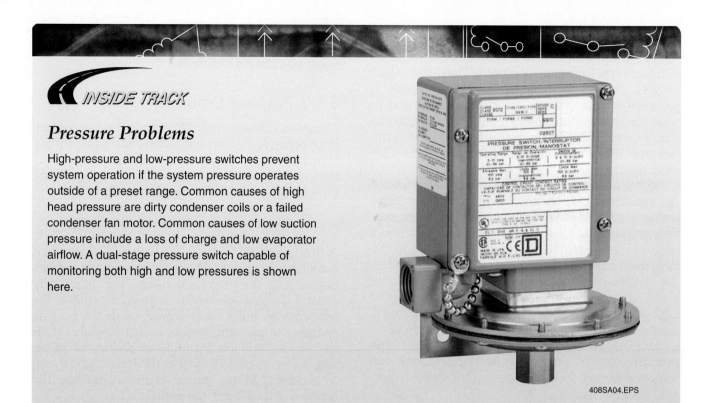
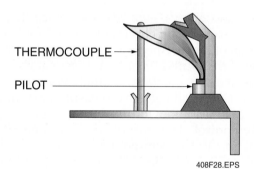

THERMOCOUPLE

PILOT

408F28.EPS

Figure 28 ◆ Thermocouple used in a safety pilot.

It takes time for the thermocouple to heat up enough to generate current. For that reason, the manual gas feed on the gas valve must usually be held down for 45 to 60 seconds when lighting the pilot.

The thermocouple can be tested by using either an adapter, as shown in *Figure 29*, or by disconnecting the cable from the gas valve and measuring the voltage between the sensing rod and the end of the cable. In this unloaded method, the meter should show at least 18mV. In a loaded test with the adapter installed, the reading should be at least 7mV.

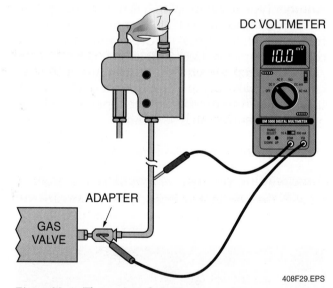

DC VOLTMETER

ADAPTER

GAS VALVE

408F29.EPS

Figure 29 ◆ Thermocouple testing.

6.6.3 Firestat

The firestat is a fire safety device mounted in the fan compartment to stop the fan if the return air temperature reaches about 160°F. It is usually a bimetal-actuated, normally closed switch that must be manually reset before the fan can operate.

6.6.4 Airflow Control

An airflow switch (*Figure 30*), also known as a *sail switch,* is often installed in ductwork as a safety device. It is used in electric heating systems to guarantee that air is circulating through the air distribution system before the heating elements are turned on and in cooling systems to verify that there is air flowing across the condenser and evaporator before the compressor is turned on. Another common use is to energize an electronic air cleaner.

6.6.5 Inducer Switch

Newer furnaces use an inducer motor to induce (produce) a draft. These furnaces have either a centrifugal switch or a pressure-operated switch in the heating control circuit to monitor the operation of the inducer motor. If the inducer motor fails to operate, the switch opens and prevents control voltage from being applied to the furnace gas valve.

The centrifugal switch has a set of contacts located near the inducer motor shaft. The centrifugal force created by the spinning shaft throws the contacts outward, closing a switch that allows control power to be applied to the furnace gas valve.

The disadvantage of a centrifugal switch is that it can stick in either the closed or open position. This can keep the unit from firing up or in some cases can allow the unit to continue firing without a demand from the thermostat.

The pressure switch (*Figure 31*) is the more common type of inducer switch. It has tubing connected to the housing and in some cases to the burner enclosure. When the inducer motor is operating, a negative pressure is created that causes the pressure switch to close, allowing control power to be applied to the furnace gas valve.

6.7.0 Heat Pump Defrost Controls

When heat is transferred from the cold outdoor air to the refrigerant in the outdoor coil, moisture condenses on the coil. Because of the temperature relationships, this moisture will freeze on the coil, even at temperatures above 40°F. Over time, the frost will build up on the coil, blocking airflow and preventing effective heat transfer. This is

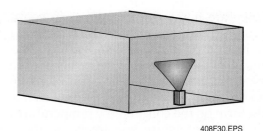

Figure 30 ◆ Sail switch.

PRESSURE SWITCH

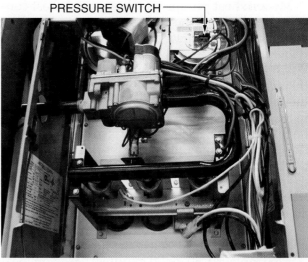

Figure 31 ◆ Combustion airflow pressure-sensing switch.

normal operation and more likely to be a problem at temperatures between 28°F and 40°F than it is at lower temperatures because the colder the air, the less moisture it contains.

To prevent ice buildup on the outdoor coil, most heat pumps have a defrost function. In air-to-air heat pumps, the heating cycle is reversed, placing the unit in the cooling mode. Hot refrigerant then flows through the outdoor coil to melt the frost buildup. This usually lasts about ten minutes. During that time, the heat pump is not providing any heat. In fact, cool air is blowing off the indoor coil into the conditioned space, just as it would on a hot day. To prevent discomfort to building occupants during the defrost cycle, an electric heater located between the indoor blower and the indoor coil is automatically turned on.

There are many different ways to control the defrost function. The most popular method uses a combination of time and temperature. The logic behind this method is that defrost is needed only if the unit has been operating for a long enough period of time (for example, 90 minutes) that frost could form on the coil, and the temperature is low enough that moisture will freeze on the coil. If both conditions do not exist, defrost is unnecessary. Other defrost methods check for airflow across the frost-blocked coil or monitor outdoor coil temperature.

Modern heat pumps use electronic defrost controls in which all the defrost circuitry, including the relays, is mounted on a printed circuit board (*Figure 32*). The timer is electronic and therefore less likely to fail. A temperature sensor, often a thermistor, sends temperature information from the outdoor coil to the board. An electronic clock in the timing logic receives a signal as long as the heating thermostat is closed. If defrost is interrupted by the opening of the heating thermostat, the timing logic will remember where it left off. If the coil temperature still indicates the presence of frost when the unit cycles on again, the unit will automatically go back into the defrost mode.

A jumper or other selection device on the board allows the installer to select the defrost frequency (for example, 30, 50, or 90 minutes) based on local conditions.

An important feature of many of these boards is a built-in defrost test cycle. It is activated by jumpering across two terminals on the board (*Figure 33*). The 90-minute defrost cycle is overridden, and the usual 10-minute running cycle can be reduced to just a few seconds, if desired. A major advantage of the electronic defrost control is that it is completely self-contained. If there is a defrost control problem, it is not necessary to isolate the failed component. Instead, the entire board is replaced.

Electronic defrost controls may be equipped with demand defrost, which combines time, temperature, and pressure information to determine if there is enough frost buildup to require defrost. It also terminates defrost when it senses that the coil is frost-free, regardless of whether the timed cycle is complete. Some electronic defrost controls contain a short-cycle timer.

7.0.0 ◆ HVAC DIGITAL CONTROL SYSTEMS

In large commercial buildings and complexes, the HVAC control systems use microprocessors to monitor and control the operation of up to several hundred elements of the heating, ventilating, and air conditioning systems. A microprocessor is a high-speed instruction executer. You give it a list of things to do (instructions) and it simply executes or carries out those instructions over and over again.

A goal of today's HVAC control designs is to build an integrated building system that is able to collect information and process that information to achieve maximum building comfort, energy usage, indoor air quality, and building safety and security.

Controlling the HVAC equipment is only one aspect of today's control systems, but it is an important one. When correctly integrated, it will prolong equipment life, increase customer satisfaction, aid in the repair of the equipment, and sometimes even communicate a problem before it can damage the equipment.

DEFROST CONTROL BOARD

408F32.EPS

Figure 32 ◆ Electronic defrost control.

408F33.EPS

Figure 33 ◆ Defrost speed-up jumper.

 ## Electrostatic Discharge

We've all experienced static electricity discharges after walking across a carpeted floor and touching a light switch or doorknob. These startling discharges are strong enough to seriously damage or destroy delicate components on electronic devices. Before touching, removing, or replacing any electronic device in HVAC equipment, always ground yourself by touching the equipment chassis, a gas pipe, or metallic electrical conduit. This will release any static charge in your body and prevent damage to the electronic device.

7.1.0 Direct Digital Control

HVAC, lighting, and other building systems controlled by traditional electromechanical devices may appear to function well. But slow response, calibration shifts, mechanical wear, and the inability to coordinate operation with other systems result in wasted energy and an inconsistent level of comfort.

A central computer can be used to integrate all building systems. It can perform complex control sequences automatically, improving the building environment and lowering costs at the same time. *Figure 34* illustrates a centralized building management system. The greatest strength of this system is its ability to communicate with the real world and to understand and sort the signals it receives. This communication helps it to reject obviously erroneous signals that older controllers were unable to detect. Great care goes into the design of the controller input/output (I/O) function because it is like a nerve system to the controller. It receives all of the signals from the system and directs them to the microprocessor to be used based on the instructions stored within its memory. *Figure 35* shows an example of a processor module used to control the interface between the building management system and the HVAC equipment.

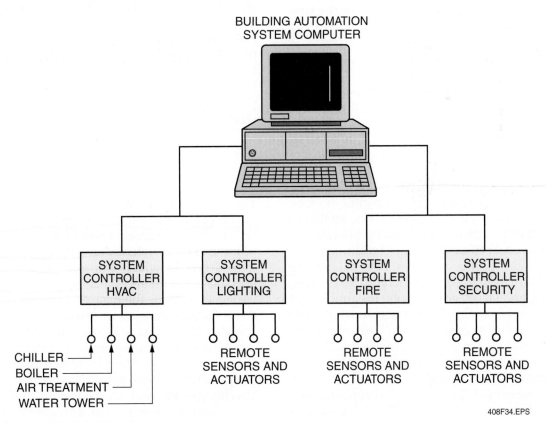

408F34.EPS

Figure 34 ◆ Centralized building management system.

Figure 35 ◆ HVAC system control module.

408F35.EPS

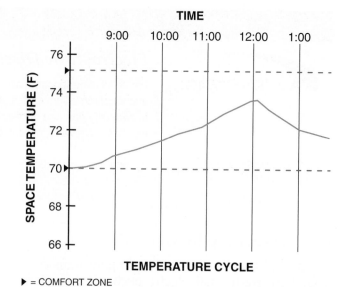

TIME

TEMPERATURE CYCLE

▶ = COMFORT ZONE

408F36.EPS

Figure 36 ◆ Changes are small and occur gradually.

The true power and flexibility of the controller is provided by the software programs that operate and control the instructions it performs. Computers are not like humans. Our brains can sort through the subtleties of human communication, interpreting the information we receive in order to decide what was intended. Computers do only what they are told, and only in the specified order. They require a rigid, mathematically precise communication system. Each symbol or group of symbols must mean exactly the same thing every time, and every statement is taken literally. This is true for all software programs and applies for the simplest to the most complex.

7.2.0 Controlling Devices

The microprocessor can execute instructions and obtain remarkable results, and it can do so at astounding speeds. If it has no way to communicate with other devices, however, it is useless. Communication can be done by dividing the signals into two categories: digital and analog. These signals are called points in the HVAC industry. The first group includes external digital devices. These might be such things as relays, switches, lights, and other devices that can be operated in either a full on or full off (binary) condition.

The second group includes external analog devices. Devices in this group include thermistors, photocells, and DC motor controls. The problem of interfacing to an analog device is somewhat more complicated. In this case, a signal with an infinite number of values (analog) is converted to a form that can be represented and manipulated by a two-state or binary device (digital).

Most real-world processes are continuously changing (*Figure 36*). Physical quantities such as pressure, temperature, liquid levels, and fluid flow tend to change value rather gradually. Changes of this type produce a large number of discrete values before ever reaching a final state. The problem of converting from analog form to digital form requires the use of a circuit called an **analog-to-digital converter.**

The analog and digital signals can be further divided into inputs and outputs. Analog in (AI) signals are from sensors for characteristics such as temperature, pressure, and humidity. Analog out (AO) signals are analog commands, such as reset of system setpoints. Digital in (DI) signals are contact closures or openings, showing status or alarm conditions in a two-position mode. Digital out (DO) signals are two-position commands like start/stop or open/close states.

7.3.0 Example of a Digital Control System

The key to a successful control system is the integration of all the sensors and unit controllers. We will look at the integration of a typical constant volume HVAC system. A constant air volume system is one in which the volume of supply air remains constant and the temperature of the air is varied to achieve the desired comfort conditions. It is a single-zone system. If there is more than one zone, each must have its own dedicated unit. The constant air volume system contrasts with a variable air volume system in which the supply air temperature is held constant and the volume of air changes to meet the changing demand. In a variable air volume system, a single unit can serve several zones.

A schematic of a constant volume system is shown in *Figure 37*. The control system is made up of several control loops: economizer control of mixed air, heating-cooling sequencing, and humidification-dehumidification sequencing.

When the unit fan is energized, as sensed by a static pressure sensor in the supply duct, the damper control system becomes activated. A mixed air sensor maintains the mixed air temperature by modulating the outdoor air, return air, and exhaust dampers. When the outdoor air temperature exceeds the setting of the outdoor air sensor, the outdoor and exhaust air dampers return to the minimum open position, as programmed, to provide ventilation. The return air damper takes the corresponding open position. In large buildings, it is often required that some outside ventilation air be provided at all times. Therefore, the damper has a minimum open position.

A space temperature sensor through the controller maintains the space temperature by modulating the heating coil valve in sequence with the chilled water coil valve. A space humidity sensor through the controller maintains the space humidity. Upon a drop in space relative humidity, the humidifier steam valve modulates toward the open position, subject to a duct-mounted high-limit humidity sensor. With a rise in space relative humidity, the humidifier steam valve modulates to the closed position, followed by the opening of the chilled water coil valve to provide dehumidification. During the dehumidification cycle, the space temperature sensor modulates the heating coil valve to maintain space temperature conditions.

A low-temperature controller, with its capillary located on the discharge side of the heating coil, will de-energize the unit fan, close the outdoor and exhaust dampers, and open the return damper if the discharge air temperature drops below its setting. Whenever the unit fan is de-energized, as sensed by the supply duct static pressure sensor, the damper control system will be de-energized, closing the outdoor damper, exhaust damper, and humidifier steam control valve.

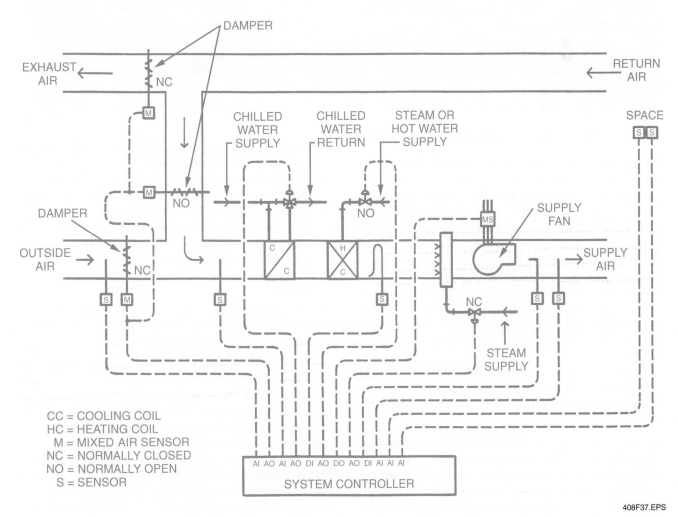

Figure 37 ◆ Typical constant volume HVAC system.

408F37.EPS

Thermistors

Thermistors are temperature-sensitive semiconductor devices. Their resistance varies in a predictable way with variations in temperature. This allows them to be used in a variety of HVAC control applications. Thermistors have either a positive or negative coefficient of resistance. If the resistance increases as the temperature rises, it has a positive coefficient. If the resistance increases as the temperature drops, it has a negative coefficient. Thermistors come in different sizes and shapes to suit a variety of applications. In HVAC systems, they are used to sense the air temperature in room thermostats, the motor winding temperature when embedded in motor windings, the outdoor coil temperature as part of a defrost circuit, and countless other applications for which an accurate measure of temperature is needed.

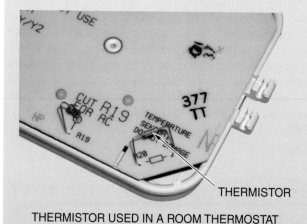

THERMISTOR USED IN A ROOM THERMOSTAT

408SA05A.EPS

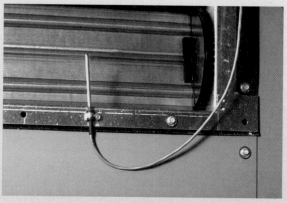

THERMISTOR USED TO CONTROL OUTDOOR
AIR DAMPER

408SA05B.EPS

8.0.0 ◆ CONTROL CIRCUIT REVIEW

An automatic air conditioner control circuit cannot get much simpler than the one for a room air conditioner shown in *Figure 38*. It has a main control switch to enable the unit and a thermostat with a sensing bulb that responds to the temperature in the conditioned space. The key word is automatic; once the unit is plugged in and turned on, it will cycle on and off by itself based on the temperature at the sensing bulb. The machine would still operate without the thermostat, but the occupant would have to turn it on and off as the temperature varied. (It would no longer be automatic.)

Figure 39 shows a control circuit that is more typical of a basic cooling system. It has some features that are common to most cooling control circuits. For example:

- The control devices (thermostat, compressor contactor, and fan relay) operate off 24V.
- The indoor fan has a separate control. By setting the FAN switch to ON, the occupant can use the fan for ventilation without operating the compressor. When the FAN switch is in the AUTO

position, the fan relay (IFR) coil will energize whenever the room thermostat (TC) closes.

- The outdoor fan motor (OFM) runs whenever the compressor is on. Follow this sequence: When the unit is on and TC closes, it completes the path to the coil of the compressor contactor (C). The normally open C contacts in the upper part of the circuit then close, completing the current path to the compressor and outdoor fan motor.
- The compressor will usually have some kind of overload protection. In this case, an automatic-reset thermal overload device (OL) is provided. It is embedded in the stator winding of the motor, or it can be externally mounted.

No matter how complex the control circuit appears to be, you will find the basic control arrangement shown in *Figure 39*—or something very much like it—at the heart of the circuit. Everything else will be special features to improve equipment safety or operating efficiency. *Figure 40* illustrates the point. This diagram is for a combined cooling and gas heating unit. The circuit looks different and there are several more components, but

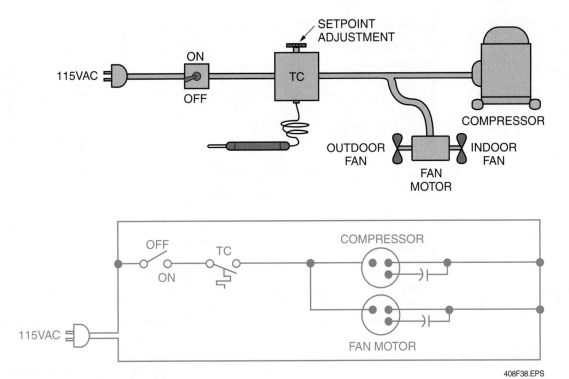

Figure 38 ◆ Room air conditioner circuit.

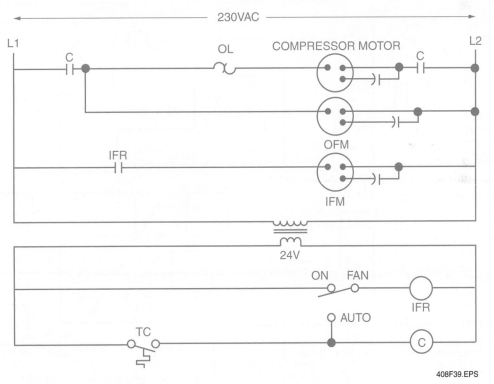

Figure 39 ◆ Typical cooling system control circuit.

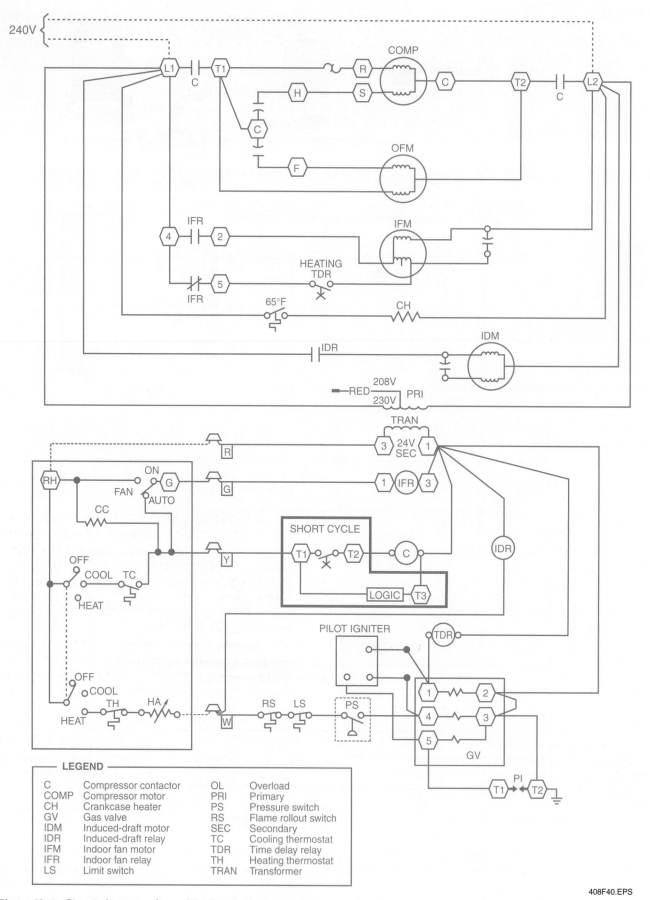

Figure 40 ◆ Circuit diagram of a cooling/gas heating system.

408F40.EPS

Crankcase Heaters

Oil circulates with the refrigerant and keeps the compressor lubricated. Crankcase heaters are used to keep the oil and refrigerant from separating, which could result in compressor failure. Crankcase heaters come in a variety of sizes (wattages) and shapes. One type, called a bellyband heater, clamps around the bottom of a welded hermetic compressor shell. Other heaters are inserted in a well in the base of the compressor shell that allows the heater to be immersed in the crankcase oil. Some equipment is designed so that power is applied to the crankcase heater at all times; others are controlled by thermostats that apply power to the heater only when the outdoor temperature drops below a certain point.

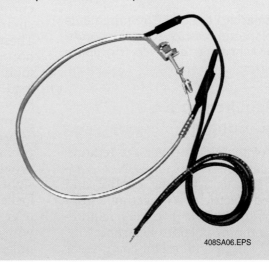

408SA06.EPS

if you trace the cooling control function, you will see that it is essentially the same. The differences are that the heating controls have been added near the bottom of the diagram, along with a heating-cooling thermostat. The cooling control has more extras (such as a compressor short-cycle protection circuit, a crankcase heater, and a current-sensitive overload device). Take away these components, and you have a circuit that is identical to the one in *Figure 39*.

Figure 40 appears different because instead of drawing L1 and L2 and the two sides of the transformer secondary as the verticals on a ladder, they are shown emanating from common terminals. This reflects the way the circuit is actually wired and is a common method used by manufacturers to draw control circuits. Ladder diagrams are nice, but they are not supplied by all manufacturers.

Some of the features of the circuit shown in *Figure 40* are:

- The unit has a two-speed indoor fan that runs on high speed for cooling and low speed for heating. In heating, the indoor fan is controlled by the time delay relay.
- Operation of the inducer fan must be proven before the gas valve is turned on. The inducer

pressure switch (PS), located in the draft hood, will close when the induced-draft motor is running at the required speed. If the induced-draft fan stops, the stack pressure will drop and the switch will open, de-energizing the gas supply. The induced-draft motor (IDM) is energized by the induced-draft relay (IDR) as soon as the heating thermostat closes.

- The heating section has two additional safety devices in series with the gas valve. A flame rollout switch (RS) will open if the burner flame escapes from the burner box. This usually indicates insufficient air for combustion or a leak in the heat exchanger. The limit switch (LS) is a heat-sensitive switch that extends into the air stream. If the heat is excessive, it will shut off the gas valve. One cause of excessive heat is insufficient air flowing over the heat exchangers. This could be caused by a blower failure or a dirty filter.
- This diagram has a pilot duty compressor overload device in addition to the line duty thermal overload included in *Figure 39*. The additional control is a current-sensitive device that will open the OL contacts in series with the compressor contactor if the motor current is excessive.

Circuit Analysis

THINK ABOUT IT

Use the diagram in *Figure 40* to determine the answers to these questions.

1. Does the induced-draft motor turn off if the rollout switch opens?
2. If you connect a voltmeter from L2 to terminal 4 of IFR while the compressor contactor is de-energized, what voltage, if any, would the meter read?
3. Is the indoor fan relay (IFR) energized or de-energized in the heating mode?

9.0.0 ◆ *NEC*® REQUIREMENTS

This section summarizes the *NEC*® requirements for HVAC controls, compressors, room air conditioners, baseboard heaters, and electric space heating cables.

9.1.0 *NEC*® Requirements for HVAC Controls

Several sections of the *NEC*® cover the requirements for HVAC controls. For example, *NEC Article 424, Part III* deals with the control and protection of fixed electric space heating equipment; *NEC Article 440, Part V* covers requirements for motor compressor controllers; and *NEC Article 440, Part VI* deals with motor compressor and branch circuit overload protection.

In general, a means must be provided to disconnect heating equipment, including motor controllers and supplementary overcurrent protective devices, from all ungrounded conductors. The disconnecting means may be a switch, circuit breaker, unit switch, or a thermostatically controlled switching device. Both the selection and use of disconnecting devices are governed by the type of overcurrent protection and the rating of any motors that are part of the equipment.

In certain heating units, supplementary overcurrent protective devices other than the branch circuit overcurrent protection are required. These supplementary overcurrent devices are normally used when heating elements rated at more than 48A are supplied as a subdivided load. In this case, the disconnecting means must be on the supply side of the supplementary overcurrent protective device and within sight of it. This disconnecting means may also serve to disconnect the heater and any motor controllers, provided the disconnecting means is within sight of the controller and heater, or it can be locked in the open position. If the motor is rated over ⅛hp, a disconnecting means must comply with the rules for motor disconnecting means unless a unit switch is used to disconnect all ungrounded conductors.

A heater without supplementary overcurrent protection must have a disconnecting means that complies with rules similar to those for permanently-connected appliances. A unit switch may be the disconnecting means in certain occupancies when other means of disconnection are provided as specified in the *NEC*®.

Figure 41 summarizes some of the *NEC*® requirements for HVAC controls, including thermostats, motor controllers, disconnects, and

Heat Pump Installations

INSIDE TRACK

Most heat pump installations require that supplemental electric heaters be installed in the air handler. They are required because the output of the heat pump declines as the outdoor temperature drops. The size of the electric heaters depends on a number of variables. Some utilities require that the electric heaters be sized to satisfy the building's full heating load if the compressor is inoperative. In some colder climates that could mean more than 20kW of electric heat. If a heat pump is being used to replace a fossil-fuel furnace in an existing structure, and a large amount of electric resistance heat is required, carefully evaluate the existing electrical service to ensure that the service can handle the additional load. In some cases, a new electrical service will be required.

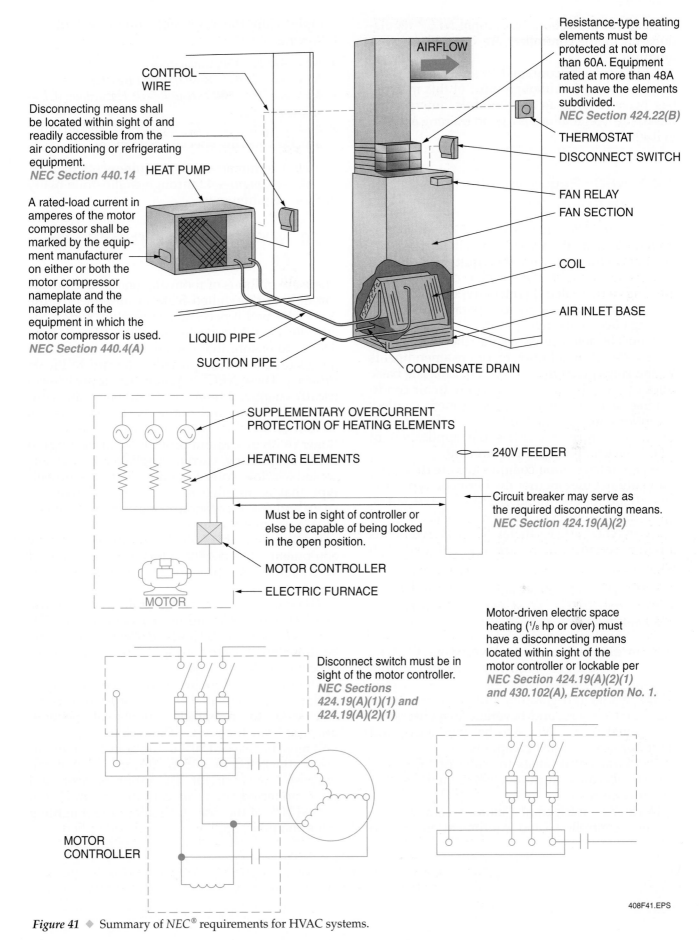

CONTROL WIRE

Disconnecting means shall be located within sight of and readily accessible from the air conditioning or refrigerating equipment.
NEC Section 440.14

HEAT PUMP

A rated-load current in amperes of the motor compressor shall be marked by the equipment manufacturer on either or both the motor compressor nameplate and the nameplate of the equipment in which the motor compressor is used.
NEC Section 440.4(A)

AIRFLOW

Resistance-type heating elements must be protected at not more than 60A. Equipment rated at more than 48A must have the elements subdivided.
NEC Section 424.22(B)

THERMOSTAT
DISCONNECT SWITCH
FAN RELAY
FAN SECTION
COIL
AIR INLET BASE

LIQUID PIPE
SUCTION PIPE
CONDENSATE DRAIN

SUPPLEMENTARY OVERCURRENT PROTECTION OF HEATING ELEMENTS

HEATING ELEMENTS

240V FEEDER

Circuit breaker may serve as the required disconnecting means.
NEC Section 424.19(A)(2)

Must be in sight of controller or else be capable of being locked in the open position.

MOTOR CONTROLLER
ELECTRIC FURNACE
MOTOR

Motor-driven electric space heating ($1/8$ hp or over) must have a disconnecting means located within sight of the motor controller or lockable per *NEC Section 424.19(A)(2)(1) and 430.102(A), Exception No. 1.*

Disconnect switch must be in sight of the motor controller.
NEC Sections 424.19(A)(1)(1) and 424.19(A)(2)(1)

MOTOR CONTROLLER

408F41.EPS

Figure 41 ◆ Summary of *NEC*® requirements for HVAC systems.

overcurrent protection. Additional *NEC®* regulations on motor controllers are covered in other modules.

You are also encouraged to study *NEC Articles 424 and 440*. Although most of this material has been covered in this module, interpreting these *NEC®* regulations is a good training exercise in itself.

9.2.0 *NEC®* Requirements for Compressors

NEC Article 440 contains provisions for motor-driven equipment, as well as for the branch circuits and controllers for the equipment. It also takes into account the special considerations involved with sealed (hermetic-type) motor compressors in which the motor operates under the cooling effect of the refrigerant.

It must be noted, however, that the rules of *NEC Article 440* are in addition to, or are amendments to, the rules given in *NEC Article 430*. The basic rules of *NEC Article 430* also apply to air conditioning and refrigerating equipment, unless exceptions are indicated in *NEC Article 440*. *NEC Article 440* further clarifies the application of *NEC®* rules to this type of equipment.

Where refrigeration compressors are driven by conventional motors (not the hermetic type), the motors and controls are subject to *NEC Article 430*. They are not subject to *NEC Article 440*.

Other *NEC®* articles that will be covered in this module (besides *NEC Articles 430 and 440*) include:

- *NEC Article 422, Appliances*
- *NEC Article 424, Fixed Electric Space Heating Equipment*

Room air conditioners are covered in *NEC Article 440, Part VII*. However, they must also comply with the rules of *NEC Article 422*.

Household refrigerators and freezers, drinking water coolers, and beverage dispensers are considered by the *NEC®* to be appliances, and their application must comply with *NEC Article 422* and must also satisfy the rules of *NEC Article 440*, because such devices contain sealed motor compressors.

Hermetic refrigerant motor compressors, circuits, controllers, and equipment must also comply with the applicable provisions of the following:

- *NEC Article 460, Capacitors*
- *NEC Article 470, Resistors and Reactors*
- *NEC Articles 500 through 503, Hazardous (Classified) Locations*
- *NEC Articles 511, 513 through 517,* and all other articles covering special occupancies

Table 1 summarizes the requirements of *NEC Article 440*. *Figures 42* through *45* illustrate many of these requirements.

9.3.0 *NEC®* Requirements for Room Air Conditioners

There are millions of room air conditioners in use throughout the United States. Consequently, the *NEC®* deemed it necessary to provide *NEC Article 440, Part VII*, beginning with *NEC Section 440.60*, to ensure that such equipment would be installed so as not to provide a hazard to life or property. These *NEC®* requirements apply to electrically energized room air conditioners that control temperature and humidity. In general, this section of the *NEC®* considers a room air conditioner (with or without provisions for heating) to be an alternating current appliance of the air-cooled window, console, or in-wall (through-wall) type that is installed in a conditioned room or space and that incorporates one or more hermetic refrigerant motor compressor(s).

Furthermore, this *NEC®* provision covers only equipment rated at 250V or less, single-phase, and such equipment is permitted to be cord- and plug-connected.

Three-phase room air conditioners, or those rated at over 250V, are not covered under *NEC Article 440, Part VII*. This type of equipment must be directly connected using a wiring method as described in *NEC Chapter 3*.

The majority of room air conditioners covered under *NEC Article 440* are cord- and plug-connected to receptacle outlets of general-purpose branch circuits. The rating of any such unit must not exceed 80% of the branch circuit rating if connected to a 15A, 20A, or 30A general-purpose branch circuit. The rating of cord- and plug-connected room air conditioners must not exceed 50% of the branch circuit rating if lighting units and other appliances are also supplied.

Table 1 Summary of *NEC*® Requirements for Hermetically Sealed Compressors

Application	*NEC*® Regulation	*NEC*® Reference
Marking on hermetic compressors	Hermetic compressors must be provided with a nameplate containing the manufacturer's name, trademark, or symbol, identifying designation, phase, voltage, frequency, rated-load current, locked-rotor current, and the words *thermally protected,* if appropriate.	*NEC Section 440.4(A)*
Marking on controllers	Controllers serving hermetically sealed compressors must be marked with the maker's name, trademark, or symbol, identifying designation, voltage, phase, and full-load and locked-rotor currents (or hp rating).	*NEC Section 440.5*
Ampacity and rating	Conductors for hermetically sealed compressors must be sized according to *NEC Tables 310.16 through 310.19* or calculated in accordance with *NEC Section 310.15,* as applicable.	*NEC Section 440.6*
Highest rated motor	The largest motor is considered to be the motor with the highest rated-load current.	*NEC Section 440.7*
Single machine	The entire HVAC system is considered to be one machine, regardless of the number of motors involved in the system.	*NEC Section 440.8*
Rating and interrupting capacity	The disconnecting means for hermetic compressors must be selected on the basis of the nameplate rated-load current or branch circuit selection current, whichever is greater, and locked-rotor current, respectively.	*NEC Section 440.12(A)*
Cord-connected equipment	For cord-connected equipment, an attachment plug and receptacle are permitted to serve as the disconnecting means.	*NEC Section 440.13*
Location	A disconnecting means must be located within sight of the equipment. The disconnecting means may be mounted on or within the HVAC equipment.	*NEC Section 440.14*
Short circuit and ground fault protection	Amendments to *NEC Article 240* are provided here for circuits supplying hermetically sealed compressors against overcurrent due to short circuits and grounds.	*NEC Section 440.21*
Rating of short circuit and ground fault protective device	The rating must not exceed 175% of the compressor rated-load current; if necessary for starting, the device may be increased to a maximum of 225%.	*NEC Section 440.22(A)*
Compressor branch circuit conductors	Branch circuit conductors supplying a single compressor must have an ampacity of not less than 125% of either the motor compressor rated-load current or the branch circuit selection current, whichever is greater.	*NEC Section 440.32*
	Conductors supplying more than one compressor must be sized for the total load plus 25% of the largest motor's full-load amps.	*NEC Section 440.33*
Combination load	Conductors must be sufficiently sized for the other loads plus the required ampacity for the compressor as required in *NEC Sections 440.32 and 440.33.*	*NEC Section 440.34*
Multi-motor load equipment	Conductors must be sized to carry the circuit ampacity marked on the equipment as specified in *NEC Section 440.4(B).*	*NEC Section 440.35*
Controller rating	Controllers must have both a continuous-duty full-load current rating and a locked-rotor current rating not less than the nameplate rated-load current or branch circuit selection current, whichever is greater, and locked-rotor current, respectively.	*NEC Section 440.41(A)*
Application and selection of controllers	Each motor compressor must be protected against overload and failure to start by one of the means specified in *NEC Sections 440.52(A)(1) through (4).*	*NEC Section 440.52(A)*
Overload relays	Overload relays and other devices for motor overload protection that are not capable of opening short circuits must be protected by a suitable fuse or inverse time circuit breaker.	*NEC Section 440.53*
Equipment on 15A or 20A branch circuit; time delay required	Short circuit and ground fault protective devices protecting 15A or 20A branch circuits must have sufficient time delay to permit the motor compressor and other motors to start and accelerate their loads.	*NEC Section 440.54(B)*

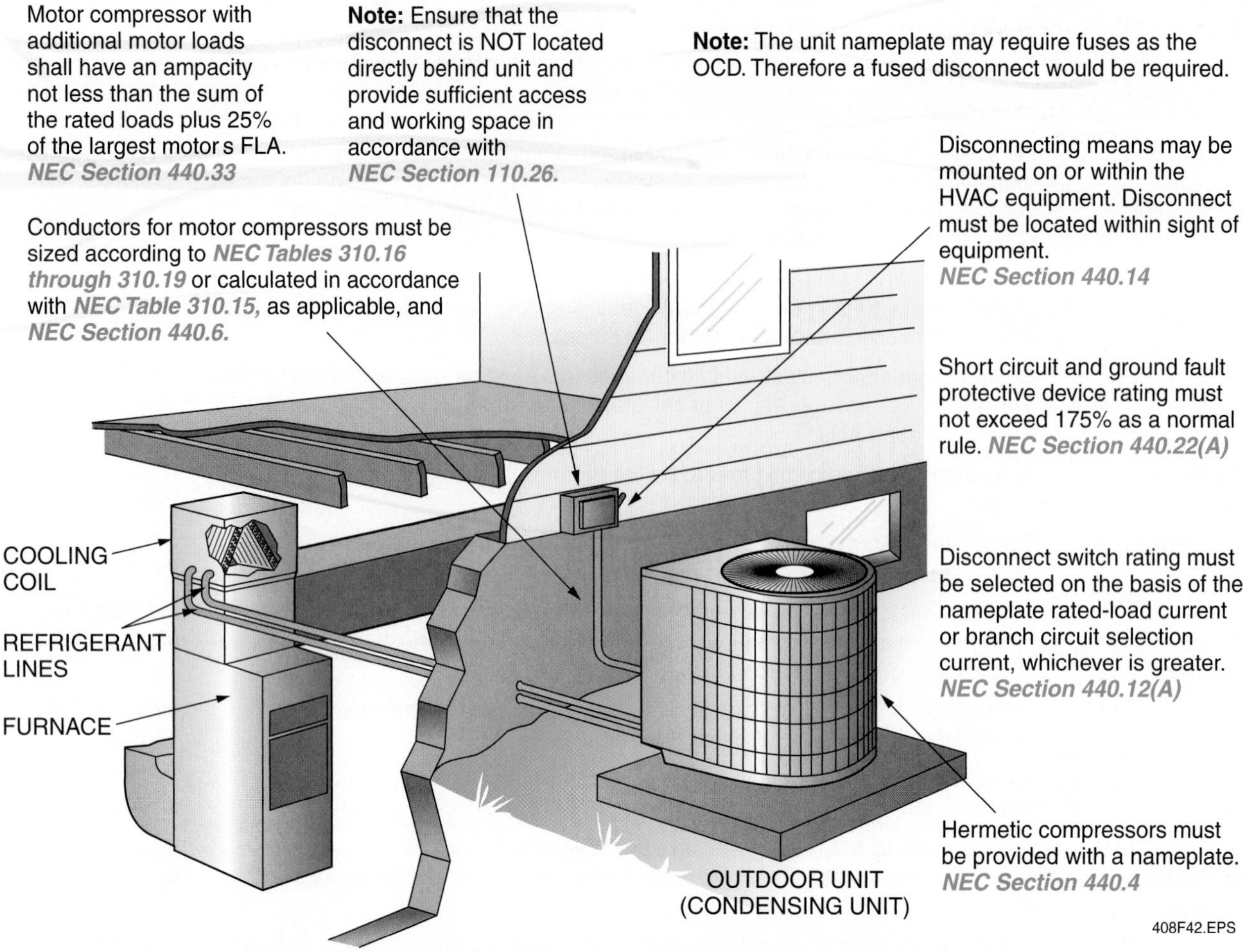

Motor compressor with additional motor loads shall have an ampacity not less than the sum of the rated loads plus 25% of the largest motor s FLA. *NEC Section 440.33*

Note: Ensure that the disconnect is NOT located directly behind unit and provide sufficient access and working space in accordance with *NEC Section 110.26.*

Note: The unit nameplate may require fuses as the OCD. Therefore a fused disconnect would be required.

Conductors for motor compressors must be sized according to *NEC Tables 310.16 through 310.19* or calculated in accordance with *NEC Table 310.15,* as applicable, and *NEC Section 440.6.*

Disconnecting means may be mounted on or within the HVAC equipment. Disconnect must be located within sight of equipment. *NEC Section 440.14*

Short circuit and ground fault protective device rating must not exceed 175% as a normal rule. *NEC Section 440.22(A)*

Disconnect switch rating must be selected on the basis of the nameplate rated-load current or branch circuit selection current, whichever is greater. *NEC Section 440.12(A)*

COOLING COIL

REFRIGERANT LINES

FURNACE

OUTDOOR UNIT (CONDENSING UNIT)

Hermetic compressors must be provided with a nameplate. *NEC Section 440.4*

408F42.EPS

Figure 42 ◆ *NEC®* regulations governing motor compressors.

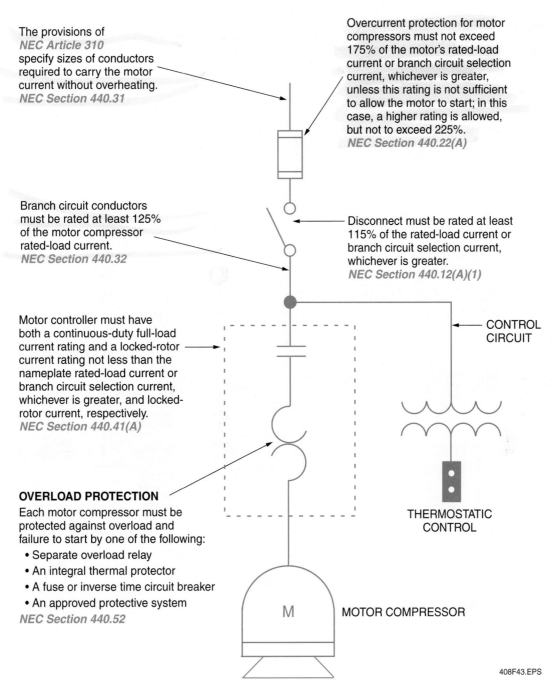

The provisions of *NEC Article 310* specify sizes of conductors required to carry the motor current without overheating. *NEC Section 440.31*

Overcurrent protection for motor compressors must not exceed 175% of the motor's rated-load current or branch circuit selection current, whichever is greater, unless this rating is not sufficient to allow the motor to start; in this case, a higher rating is allowed, but not to exceed 225%. *NEC Section 440.22(A)*

Branch circuit conductors must be rated at least 125% of the motor compressor rated-load current. *NEC Section 440.32*

Disconnect must be rated at least 115% of the rated-load current or branch circuit selection current, whichever is greater. *NEC Section 440.12(A)(1)*

CONTROL CIRCUIT

Motor controller must have both a continuous-duty full-load current rating and a locked-rotor current rating not less than the nameplate rated-load current or branch circuit selection current, whichever is greater, and locked-rotor current, respectively. *NEC Section 440.41(A)*

THERMOSTATIC CONTROL

OVERLOAD PROTECTION
Each motor compressor must be protected against overload and failure to start by one of the following:
• Separate overload relay
• An integral thermal protector
• A fuse or inverse time circuit breaker
• An approved protective system
NEC Section 440.52

M — MOTOR COMPRESSOR

408F43.EPS

Figure 43 ◆ Compressor branch and control circuits.

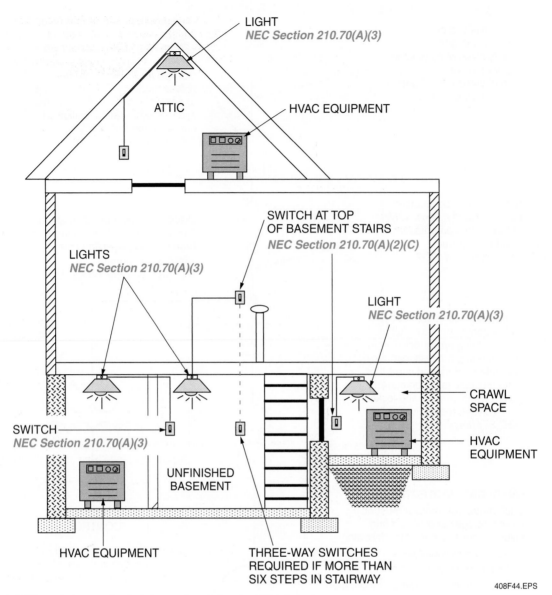

LIGHT
NEC Section 210.70(A)(3)

ATTIC

HVAC EQUIPMENT

SWITCH AT TOP
OF BASEMENT STAIRS
NEC Section 210.70(A)(2)(C)

LIGHTS
NEC Section 210.70(A)(3)

LIGHT
NEC Section 210.70(A)(3)

CRAWL
SPACE

HVAC
EQUIPMENT

SWITCH
NEC Section 210.70(A)(3)

UNFINISHED
BASEMENT

HVAC EQUIPMENT

THREE-WAY SWITCHES
REQUIRED IF MORE THAN
SIX STEPS IN STAIRWAY

408F44.EPS

Figure 44 ◆ *NEC*® requirements for lighting and switches for HVAC equipment.

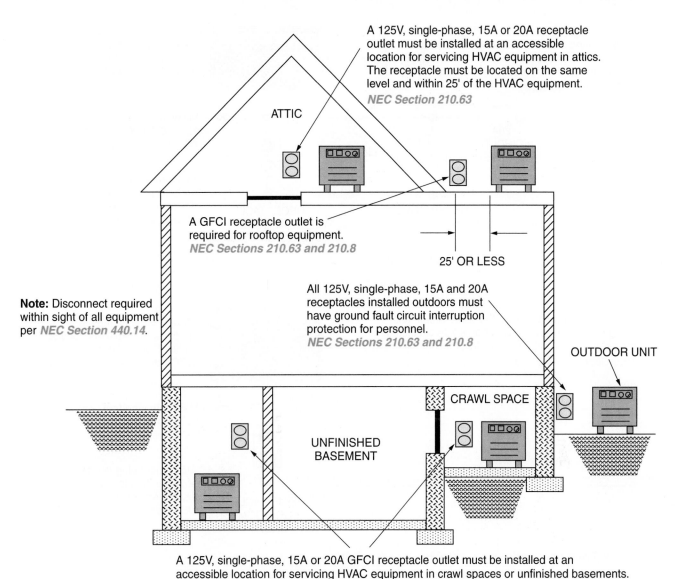

A 125V, single-phase, 15A or 20A receptacle outlet must be installed at an accessible location for servicing HVAC equipment in attics. The receptacle must be located on the same level and within 25' of the HVAC equipment.
NEC Section 210.63

ATTIC

A GFCI receptacle outlet is required for rooftop equipment.
NEC Sections 210.63 and 210.8

25' OR LESS

Note: Disconnect required within sight of all equipment per *NEC Section 440.14*.

All 125V, single-phase, 15A and 20A receptacles installed outdoors must have ground fault circuit interruption protection for personnel.
NEC Sections 210.63 and 210.8

OUTDOOR UNIT

CRAWL SPACE

UNFINISHED BASEMENT

A 125V, single-phase, 15A or 20A GFCI receptacle outlet must be installed at an accessible location for servicing HVAC equipment in crawl spaces or unfinished basements. The receptacle must be located on the same level and within 25' of the HVAC equipment.
NEC Sections 210.63 and 210.8

408F45.EPS

Figure 45 ◆ *NEC*® requirements for locating 125V receptacles at HVAC equipment.

Figures 46 and *47* depict the *NEC*® application rules for room air conditioners. Note that the attachment plug and receptacle are allowed to serve as the disconnecting means. In some cases, the attachment plug and receptacle may also serve as the controller, or the controller may be a switch that is an integral part of the unit. The required overload protective device may be supplied as an integral part of the appliance and need not be included in the branch circuit calculations.

Equipment grounding, as required by *NEC Section 440.61*, may be handled by the grounded receptacle.

9.4.0 *NEC*® Requirements for Electric Baseboard Heaters

All requirements of the *NEC*® apply for the installation of electric baseboard heaters, especially *NEC Article 424, Fixed Electric Space Heating Equipment.* In general, electric baseboard heaters must not be used where they will be exposed to severe physical damage unless they are adequately protected from such damage. Heaters and related equipment installed in damp or wet locations must be approved for such locations and must be constructed and installed so that water cannot

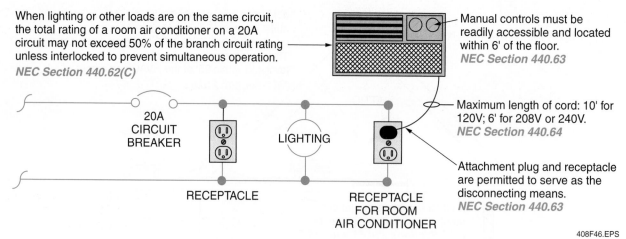

When lighting or other loads are on the same circuit, the total rating of a room air conditioner on a 20A circuit may not exceed 50% of the branch circuit rating unless interlocked to prevent simultaneous operation. *NEC Section 440.62(C)*

Manual controls must be readily accessible and located within 6' of the floor. *NEC Section 440.63*

Maximum length of cord: 10' for 120V; 6' for 208V or 240V. *NEC Section 440.64*

Attachment plug and receptacle are permitted to serve as the disconnecting means. *NEC Section 440.63*

20A CIRCUIT BREAKER

LIGHTING

RECEPTACLE

RECEPTACLE FOR ROOM AIR CONDITIONER

408F46.EPS

Figure 46 ◆ Branch circuits for room air conditioners.

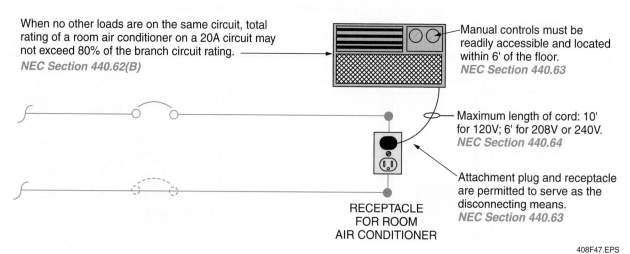

When no other loads are on the same circuit, total rating of a room air conditioner on a 20A circuit may not exceed 80% of the branch circuit rating. *NEC Section 440.62(B)*

Manual controls must be readily accessible and located within 6' of the floor. *NEC Section 440.63*

Maximum length of cord: 10' for 120V; 6' for 208V or 240V. *NEC Section 440.64*

Attachment plug and receptacle are permitted to serve as the disconnecting means. *NEC Section 440.63*

RECEPTACLE FOR ROOM AIR CONDITIONER

408F47.EPS

Figure 47 ◆ Fixed air conditioner connected to a branch circuit.

enter or accumulate in or on wired sections, electrical components, or ductwork.

Baseboard heaters must be installed to provide the required spacing between the equipment and adjacent combustible material, and each unit must be adequately grounded in accordance with *NEC Article 250*.

Figure 48 lists some of the *NEC*® regulations governing the installation of electric baseboard heaters.

9.5.0 *NEC*® Requirements for Electric Space Heating Cables

Always make certain that the heating cable is connected to the proper voltage. A 120V cable connected to a 240V circuit will melt the cable. A 240V heating cable connected to a 120V circuit will produce only 25% of the rated wattage of the cable. See *Figure 49* for a summary of *NEC*® requirements governing the installation of electric space heating cable.

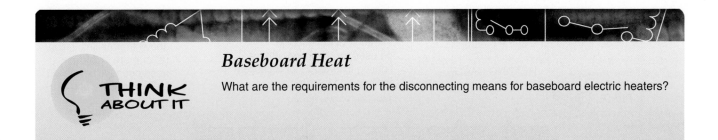

Baseboard Heat

What are the requirements for the disconnecting means for baseboard electric heaters?

THINK
ABOUT IT

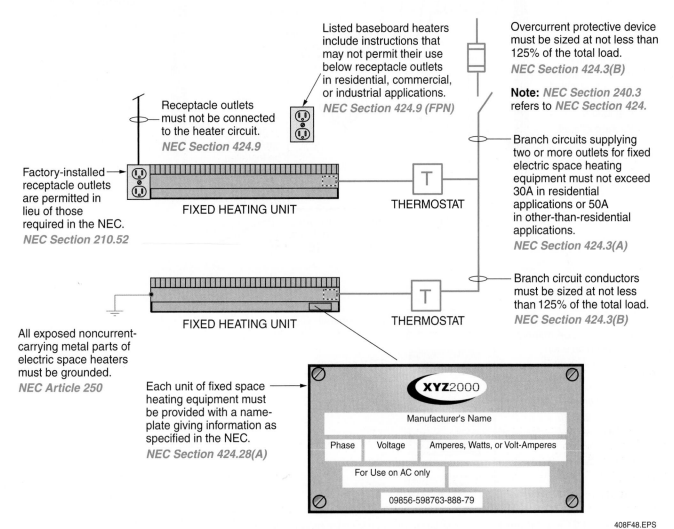

Figure 48 ◆ *NEC® installation guidelines for electric baseboard heaters.*

408F48.EPS

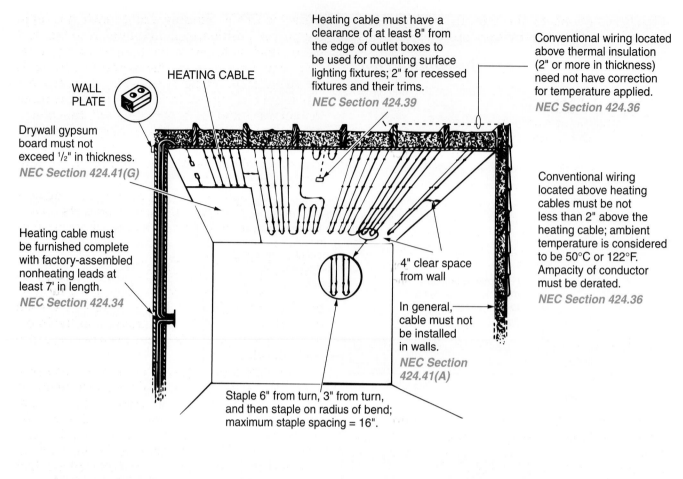

Heating cable must have a clearance of at least 8" from the edge of outlet boxes to be used for mounting surface lighting fixtures; 2" for recessed fixtures and their trims.
NEC Section 424.39

Conventional wiring located above thermal insulation (2" or more in thickness) need not have correction for temperature applied.
NEC Section 424.36

WALL PLATE

HEATING CABLE

Drywall gypsum board must not exceed ¹/₂" in thickness.
NEC Section 424.41(G)

Conventional wiring located above heating cables must be not less than 2" above the heating cable; ambient temperature is considered to be 50°C or 122°F. Ampacity of conductor must be derated.
NEC Section 424.36

Heating cable must be furnished complete with factory-assembled nonheating leads at least 7' in length.
NEC Section 424.34

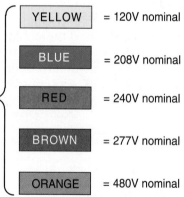

4" clear space from wall

In general, cable must not be installed in walls.
NEC Section 424.41(A)

Staple 6" from turn, 3" from turn, and then staple on radius of bend; maximum staple spacing = 16".

Each unit length of heating cable must have a permanent legible marking on each nonheating lead located within 3" of the terminal end; they must be color-coded as shown here.
NEC Section 424.35

YELLOW	= 120V nominal
BLUE	= 208V nominal
RED	= 240V nominal
BROWN	= 277V nominal
ORANGE	= 480V nominal

408F49.EPS

Figure 49 ◆ *NEC*® regulations for the installation of electric space heating cable.

10.0.0 ◆ TROUBLESHOOTING

Troubleshooting of heating and cooling systems covers a wide range of electrical and mechanical problems, from finding a short circuit in the power supply line, through adjusting a pulley on a motor shaft, to tracing loose connections in complex control circuits. However, in nearly all cases, the electrician can determine the cause of the trouble by using a systematic approach, checking one part of the system at a time in the right order.

WARNING!

HVAC systems contain high voltages, rotating components, and hot surfaces. Be extremely careful when troubleshooting this equipment.

Table 2 is arranged so that the problem is listed first. The possible causes of the problem are listed in the order in which they should be checked.

Finally, solutions to the various problems are given, including step-by-step procedures where necessary.

WARNING!

If damaged, the terminals of hermetic and semi-hermetic compressor motors in pressurized systems have been known to blow out when disturbed. To avoid injury, do not disconnect or connect wiring at the compressor terminals. When testing compressors, do not place test probes on the compressor terminals. Instead, use test points downstream from the compressor. To be safe, measurements and connecting/disconnecting wiring should only be done at the compressor terminals when the system pressure is at 0 psig.

The capacitors used in motor circuits can hold a high-voltage charge after the system power is turned off. Always discharge capacitors before touching them.

To better illustrate the use of these solutions to heating and cooling equipment problems, assume that an air conditioner fan or blower motor is operating, but the compressor motor is not. Glance down the left-hand column in the troubleshooting charts until you locate the problem titled *Compressor motor and condenser motor will not start, but the blower motor operates.* Begin with the first item under *Responses*, which tells you to check the thermostat system switch to make sure

it is set to COOL. Finding that the switch is set in the proper position, you continue on to the next item, which is to check the temperature setting. You may find that the temperature setting is above the room temperature, so the system is not calling for cooling. Set the thermostat below room temperature, and the cooling unit will function.

This example is, of course, very simple, but most of the heating and cooling problems can be just as simple if a systematic approach to troubleshooting is used.

Systems that use electric resistance heaters have special troubleshooting requirements. *Table 3* is an example of a troubleshooting chart for electric heat.

This section provided basic troubleshooting and repair procedures for common HVAC problems. For more information, also see the module entitled *Motor Controls.* Motor controls are used extensively in HVAC applications, and the information contained in this module can be very helpful to the electrician.

Manufacturers of HVAC equipment also provide troubleshooting and maintenance manuals for their equipment. These manuals can be one of the most helpful tools for troubleshooting specific HVAC equipment. When unpacking equipment, controls, and other components for the system, always save any manuals or instructions that accompany the items. File them in a safe place so that you and other maintenance personnel can readily find them. Many electricians like to secure this type of manual on the inside of a cabinet door within the equipment. This way, it will always be available when needed.

Table 2   HVAC Troubleshooting Chart (1 of 2)

Malfunction	Responses	Corrective Action
Compressor motor and condenser motor will not start, but the blower motor operates.	Check the thermostat system switch to ascertain that it is set to COOL.	Make necessary adjustments to settings.
	Check the thermostat to make sure that it is set below room temperature.	Make necessary adjustments.
	Check the thermostat to see if it is level. Mercury bulb thermostats must be mounted level; any deviation will ruin their calibration.	Remove cover plate, place a spirit level on top of the thermostat base, loosen the mounting screws, and adjust the base until it is level; then tighten the mounting screws.
	Check all low-voltage connections for tightness.	Tighten.
	Make a low-voltage check with a voltmeter on the condensate overflow switch; the condensate may not be draining.	The float switch is normally found in the condensate pump. Repair or replace.
	Make a low-voltage check of the pressure switches.	Replace if defective. Call a qualified HVAC technician to check the system charge.
Compressor, condenser, and blower motors will not start.	Check the thermostat system switch setting to ascertain that it is set to COOL.	Adjust as necessary.
	Check the thermostat setting to make sure it is below room temperature.	Adjust as necessary.
	Check the thermostat to make sure it is level.	Correct as required.
	Check all low-voltage connections for tightness.	Tighten.
	Check for a blown fuse or tripped circuit breaker.	Determine the cause of the open circuit and then replace the fuses or reset the circuit breaker.
	Make a voltage check of the low-voltage transformer.	Replace if defective.
	Check the electrical service against minimum requirements (correct voltage and amperage).	Update as necessary.
Condensing unit cycles too frequently.	Check room thermostat location.	Move if necessary.
	Check all low-voltage wiring connections for tightness.	Tighten.
	The equipment could be oversized.	Call qualified HVAC technician to check system capacity.
Inadequate cooling with condensing unit and blower running continuously.	Check for refrigerant leak or undercharge.	Call in a qualified HVAC technician to check for leaks in the refrigerant lines and/or to add charge.
	Check for undersized equipment.	Call qualified HVAC technician to check system capacity.
Condensing unit cycles, but the evaporator blower motor does not run.	Check all low-voltage connections for tightness.	Correct structural deficiencies with insulation, awnings, etc., or install properly sized equipment.
	Make a voltage check of the blower relay.	Tighten.
	Make electrical and mechanical checks on the blower motor.	Replace if necessary.
	Check for correct voltage at motor terminals. Mechanical problems could be bad bearings or a loose blower wheel. Bearing trouble can be detected by turning the blower wheel by hand (with the current off) and checking for excessive wear, roughness, or seizure.	Repair or replace defective components.
Unit shows continuous short cycling of evaporator blower and provides insufficient cooling.	Make electrical and mechanical checks.	Repair or replace motor if necessary.

Table 2 HVAC Troubleshooting Chart (2 of 2)

Malfunction	Responses	Corrective Action
Thermostat calls for heat, but the blower motor does not operate.	Check all low-voltage connections for tightness. Check all line-voltage connections for tightness. Check for blown fuses or a tripped circuit breaker in the line. Check the low-voltage transformer. Make a low-voltage check of the fan relay. Make electrical and mechanical checks on the blower motor.	Tighten. Tighten. Determine the reason for the open circuit and replace the fuses or reset the circuit breaker. Replace if defective. Repair or replace if necessary. Repair or replace the motor if defective.
Thermostat calls for heat; blower motor operates, but it delivers cold air.	Make a visual and electrical check on the heating elements. Make an electrical check of the heater limit switch; begin by disconnecting all power to the unit, then use an ohmmeter to check for continuity between the two terminals of the switch. Make an electrical check of the sequencer. Most are rated at 24V and have one set of normally open auxiliary contacts for pilot duty. Check the electric service entrance and related circuits against the minimum recommendations.	If not operating, continue to the next check. If the limit switch is open, repair or replace. If the sequencer coil is open or grounded, repair or replace. Upgrade if necessary.
Thermostat calls for heat and blower motor operates continuously; system delivers warm air, but the thermostat is not satisfied.	Check all joints in the ductwork for air leaks. Check for a dirty air filter. Make a visual and electrical check of the electric heating element. Make an electrical check of the heater limit switch as described previously. Check for undersized equipment.	Make all defective joints tight. Clean or replace filter. Repair or replace if necessary. Repair or replace. Call HVAC technician to check system capacity.
Electric heater cycles on limit switches, but the blower motor does not operate.	Make an electrical check of the fan relay. Make electrical and mechanical checks on the blower motor. Check the line connections for tightness.	Repair or replace if defective. Repair or replace if defective. Make any necessary changes.
There is excessive noise at the return air grille.	Check the return duct to make sure it has a 90° bend. Make a visual check of the blower unit to ascertain that all shipping blocks and angles have been removed. Check the blower motor assembly suspension and fasteners.	Correct if necessary. Remove if necessary. Tighten if necessary.
There is excessive vibration at the blower unit.	Visually check for vibration isolators (which isolate the blower coil from the structure). Visually check to ascertain that all shipping blocks and angles have been removed from the blower unit. Check the blower motor assembly suspension and fasteners.	If missing, install as recommended by the manufacturer. Remove if necessary. Tighten if necessary.

Table 3   Electric Heat Troubleshooting Chart

Generally the Cause Make these checks first.	Occasionally the Cause Make these checks only if the first checks failed to locate the trouble.	Rarely the Cause Make these checks only if other checks failed to locate the trouble.
Problem: No heat—unit fails to operate		
Power failure Blown fuses or tripped circuit breaker Open disconnect switch or blown fuse Thermostat switch not in proper position Element open	Control transformer Control relay or contactor Bad thermal fuse	Faulty wiring Loose terminals
Problem: Not enough heat—unit cycles too often		
Limit controls Low air volume Dirty blower wheel Dirty filters Element fuse Heat sequencer Heat relay Thermostat heat anticipator improperly set	Control relay or contactor Fan control Thermostat level Thermostat location Outdoor thermostat Blower bearings Blower motor	Loose terminals Low voltage Faulty wiring Loose blower wheel Ductwork small or restricted
Problem: Too much heat—unit cycle is too long		
Thermostat heat anticipator incorrectly set Thermostat out of calibration	Thermostat level Thermostat location	Control relay or contactor
Problem: No heat—unit runs continuously		
Thermostat faulty Control relay or contactor	Faulty wiring	Thermostat not level Thermostat location
Problem: Cost of operation too high		
Blower motor Dirty filters Outdoor thermostat setting incorrect	Low air volume Insufficient building insulation Excessive building infiltration	Low voltage Fan control Blower belt broken or slipping Dirty blower wheel Ductwork small or restricted
Problem: Mechanical noise		
Blower bearings Blower out of balance Blower belt slipping	Blower motor Cabinet	Low voltage Control transformer
Problem: Air noise		
Blower Cabinet Dirty coil	Ductwork or grilles restricted	Blower wheel Dirty filters
Problem: Odor		
Foreign substances (dirt or dust) on heating elements	Low air volume Humidifier containing stagnant water Water or moisture	Faulty wiring Loose terminals Control transformer Dirty filters Dirty or plugged heat exchanger

 INSIDE TRACK

Troubleshooting Systems with Electronic Controls

Systems that use electronic controls often have a built-in diagnostic system. Sometimes, faults are announced by a flashing LED on the control board. The troubleshooter counts the number of flashes between pauses, then interprets the fault code using a chart, as shown here. Unfortunately, there is no industry standard when it comes to fault messages or fault indications on electronic controls. Each manufacturer may have its own scheme. Even within one manufacturer, the fault message scheme may vary from product to product, with the same indication having entirely different meanings on two different products. However, most manufacturers attach labels to their products that list the various fault indications and what they mean. These fault code labels make the job of troubleshooting much easier.

LED ⟶

408SA07A.EPS

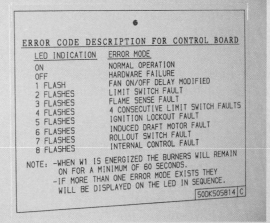

ERROR CODE DESCRIPTION FOR CONTROL BOARD

LED INDICATION	ERROR MODE
ON	NORMAL OPERATION
OFF	HARDWARE FAILURE
1 FLASH	FAN ON/OFF DELAY MODIFIED
2 FLASHES	LIMIT SWITCH FAULT
3 FLASHES	FLAME SENSE FAULT
4 FLASHES	4 CONSECUTIVE LIMIT SWITCH FAULTS
5 FLASHES	IGNITION LOCKOUT FAULT
6 FLASHES	INDUCED DRAFT MOTOR FAULT
7 FLASHES	ROLLOUT SWITCH FAULT
8 FLASHES	INTERNAL CONTROL FAULT

NOTE: –WHEN W1 IS ENERGIZED THE BURNERS WILL REMAIN ON FOR A MINIMUM OF 60 SECONDS.
     –IF MORE THAN ONE ERROR MODE EXISTS THEY WILL BE DISPLAYED ON THE LED IN SEQUENCE.

50DK505814 C

408SA07B.EPS

1. The component in an air conditioning or refrigeration system in which the heat from the area or item being cooled is transferred to the refrigerant is the _____.
   a. compressor
   b. condenser
   c. expansion device
   d. evaporator

2. Low-voltage (24V) control circuits are used in HVAC systems because _____.
   a. step-down transformers produce only 24V output
   b. 24V power is safer
   c. it is mandated by law
   d. it is the same voltage used to operate the compressor and fan motors

3. A heat anticipator is used in a heating-only thermostat to _____.
   a. turn on the heat just before the temperature reaches the thermostat setpoint
   b. keep the thermostat warm on cold days
   c. turn on a light to let the occupants know when the furnace is about to come on
   d. open the thermostat just before the heat in the conditioned space reaches the thermostat setpoint

4. A cooling compensator compensates for the lag between the thermostat call for cooling and the time when the system actually begins cooling the conditioned space.
   a. True
   b. False

5. The differential in a thermostat is _____.
   a. the difference between the cut-in and cutout points of the thermostat
   b. the difference between the cooling and heating setpoints
   c. normally at least 5°F
   d. the difference between the settings of the heat anticipator and the cooling compensator

6. Bimetal line-voltage themostats are commonly used _____.
   a. to control 24V loads
   b. for precise temperature control
   c. to control electric baseboard heaters
   d. to control gas-fired furnaces

7. A good location for a thermostat is _____.
   a. anywhere that it will receive direct sunlight
   b. about five feet above the floor on an outside wall
   c. about five feet above the floor on an inside wall
   d. in a corner where there is minimum air circulation

8. In a standard thermostat wiring scheme, the R terminal would be connected to the _____.
   a. transformer common terminal
   b. secondary of the 24V control transformer
   c. reversing valve
   d. heating thermostat

9. In a standard thermostat wiring scheme, the G terminal would be connected to the _____.
   a. primary control
   b. cooling control
   c. fan control
   d. reversing valve

10. Which statement about an electronic room thermostat is true?
    a. The heating cycle rate is determined by the furnace manufacturer.
    b. The heating cycle rate is adjustable but determined by the room thermostat manufacturer.
    c. The heating cycle rate is always greater than 6 cycles per hour but less than 12 cycles per hour.
    d. The heating cycle rate is fixed and is determined by the room thermostat manufacturer.

11. The purpose of a lockout relay is to _____.
    a. ensure no one turns on the power when you want it off
    b. ensure the compressor does not turn on until the fans are running
    c. protect the compressor motor from startup surges
    d. prevent the automatic restart of a compressor or motor in HVAC equipment

12. The purpose of a compressor short-cycle timer is to _____.
    a. prevent the compressor from restarting before system pressures can equalize
    b. ensure the compressor runs for at least five minutes once it cycles on
    c. keep track of how often the compressor turns on
    d. run the compressor every five minutes to make sure it does not freeze up

13. When a heat pump is operating in the defrost mode, what usually happens in the conditioned space?
    a. The unit continues to provide compression heat.
    b. No heat is provided to the conditioned space.
    c. The furnace takes over.
    d. An electric heater is cycled on to provide heat.

14. In the electrical control circuit of a combined heating/cooling system, the outdoor fan motor _____.
    a. is on continuously
    b. runs at low speed in the heating mode
    c. is on whenever the compressor is on
    d. is on whenever the compressor is off

15. The provisions for room air conditioners are referenced in _____.
    a. NEC Article 424, Part V
    b. NEC Article 440, Part VII
    c. NEC Article 427, Part IV
    d. NEC Article 430, Part XIII

Summary

HVAC control systems are made up of a variety of electrical, electronic, and pneumatic controls. These controls perform two basic functions: They automatically turn functions on and off, and they protect the system from damage.

The ability to analyze HVAC control systems is important because a large percentage of the problems that occur in HVAC systems are control circuit faults. No matter how complex the control system is, it consists of individual components that are combined to form control functions such as heating, cooling, fan control, and defrosting. If you know how the components work, you can figure out how they fit together and how the system functions as a whole. Once you have these skills, you can troubleshoot any system.

Notes

Trade Terms
Introduced in This Module

Analog-to-digital converter: A device designed to convert analog signals such as temperature and humidity to a digital form that can be processed by logic circuits.

Automatic changeover thermostat: A thermostat that automatically selects heating or cooling based on the space temperature.

Bimetal: A control device made of two dissimilar metals that warp when exposed to heat, creating movement that can open or close a switch.

Compressor: A component of a refrigeration system that converts low-pressure, low-temperature refrigerant gas into high-temperature, high-pressure refrigerant gas.

Condenser: A heat exchanger that transfers heat from the refrigerant flowing inside it to the air or water flowing over it.

Cooling compensator: A fixed resistor installed in a thermostat to act as a cooling anticipator.

Deadband: A temperature band, usually 3°F, that separates heating and cooling in an automatic changeover thermostat.

Differential: The difference between the cut-in and cut-out points of a thermostat.

Evaporator: A heat exchanger that transfers heat from the air flowing over it to the cooler refrigerant flowing through it.

Expansion device: A device that provides a pressure drop that converts the high-temperature, high-pressure liquid refrigerant from the condenser into the low-temperature, low-pressure liquid refrigerant entering the evaporator; also known as the liquid metering device or metering device.

Heat transfer: The transfer of heat from a warmer substance to a cooler substance.

Invar®: An alloy of steel containing 36% nickel. It is one of the two metals in a bimetal control device.

Mechanical refrigeration: The use of machinery to provide cooling.

Refrigerant: A fluid (liquid or gas) that picks up heat by evaporating at a low temperature and pressure and gives up heat by condensing at a higher temperature and pressure.

Refrigeration cycle: The process by which a circulating refrigerant absorbs heat from one location and transfers it to another location.

Sub-base: The portion of a two-part thermostat that contains the wiring terminals and control switches.

Subcooling: Cooling a liquid below its condensing temperature.

Superheat: The measurable heat added to the vapor or gas produced after a liquid has reached its boiling point and completely changed into a vapor.

Thermostat: A device that is responsive to ambient temperature conditions.

This module is intended to present thorough resources for task training. The following reference works are suggested for further study. This optional material is for continuing education rather than for task training.

Remote-Mounted Thermostats, Latest Edition. Syracuse, NY: Carrier Corporation.

NCCER makes every effort to keep these textbooks up-to-date and free of technical errors. We appreciate your help in this process. If you have an idea for improving this textbook, or if you find an error, a typographical mistake, or an inaccuracy in NCCER's Contren® textbooks, please write us, using this form or a photocopy. Be sure to include the exact module number, page number, a detailed description, and the correction, if applicable. Your input will be brought to the attention of the Technical Review Committee. Thank you for your assistance.

Instructors – If you found that additional materials were necessary in order to teach this module effectively, please let us know so that we may include them in the Equipment/Materials list in the Annotated Instructor's Guide.

Write: Product Development and Revision
National Center for Construction Education and Research
3600 NW 43rd St., Bldg. G, Gainesville, FL 32606

Fax: 352-334-0932

E-mail: curriculum@nccer.org

Craft _____   Module Name _____

Copyright Date _____   Module Number _____   Page Number(s) _____

Description _____

_____

_____

_____

(Optional) Correction _____

_____

_____

(Optional) Your Name and Address _____

_____

_____

Heat Tracing and
Freeze Protection

Santa Clara University – Purpose Drives Innovation

Design with purpose. This succinct philosophy guided the Santa Clara University team in its quest to build a sustainable solar house that is functional, elegant, intelligent, and innovative. The home's computers sense interior and exterior conditions and make automatic adjustments for thermal comfort and efficient energy usage.

26409-08

26409-08
Heat Tracing and Freeze Protection

Topics to be presented in this module include:

1.0.0	Introduction	9.2
2.0.0	Pipeline Heat Tracing Applications	9.2
3.0.0	Pipeline Electric Heat Tracing Systems	9.3
4.0.0	Equipment Selection and Installation for Pipe Heat Tracing Systems	9.14
5.0.0	Roof, Gutter, and Downspout De-Icing Systems	9.21
6.0.0	Component Selection and Installation for Roof, Gutter, and Downspout De-Icing Systems	9.23
7.0.0	Snow-Melting and Anti-Icing Systems	9.25
8.0.0	Component Selection and Installation for Snow-Melting and Anti-Icing Systems	9.27
9.0.0	Domestic Hot-Water Temperature Maintenance Systems	9.29
10.0.0	Component Selection and Installation for Domestic Hot-Water Temperature Maintenance Systems	9.30
11.0.0	Floor Heating and Warming Systems	9.30
12.0.0	Component Selection and Installation for Floor Heating Systems	9.32

Overview

Electrical heat tracing involves the warming of piping or equipment by lining the surface with insulated electrical wiring that produces a regulated amount of external or surface heat. Heat tracing is primarily used to prevent flowing or contained liquids from freezing, which can cause damage to the piping or equipment. It is also used to maintain fluidity in liquids and to prevent surfaces from freezing.

A non-electrical method frequently used to heat trace piping or equipment is steam tracing. However, there are some negative aspects associated with steam tracing, including the potential of severe burns due to steam contact and the difficulty of maintaining the system at a regulated temperature due to the dependence on a separate boiler system. These and other reasons have caused electrical tracing to be the method of choice in heat tracing and freeze protection.

Heat tracing has many applications, but regardless of the application, heat tracing works on the principle of heat transfer from one surface to another through physical contact of the two surfaces.

Objectives

When you have completed this module, you will be able to do the following:

1. Identify and describe the purpose of electric heat tracing equipment used with pipelines and vessels.
2. Select, size, and install electric heat tracing equipment on selected pipelines and vessels in accordance with the manufacturer's instructions and *National Electrical Code®* (*NEC®*) requirements.
3. Identify and describe the purpose of electric heating equipment used with roof, gutter, and downspout de-icing systems.
4. Select, size, and install selected roof, gutter, and downspout de-icing systems in accordance with the manufacturer's instructions and *NEC®* requirements.
5. Identify and describe the purpose of electric heating equipment used with snow-melting and anti-icing systems.
6. Select, size, and install selected snow-melting and anti-icing systems in accordance with the manufacturer's instructions and *NEC®* requirements.
7. Identify and describe the purpose of electric heat tracing equipment used with domestic hot-water temperature maintenance systems.
8. Select, size, and install selected electric heat traced domestic hot-water systems in accordance with the manufacturer's instructions and *NEC®* requirements.
9. Identify and describe the purpose of electric floor heating/warming systems.
10. Select, size, and install selected electric floor heating/warming systems in accordance with the manufacturer's instructions and *NEC®* requirements.

Trade Terms

Constant wattage heating cable
Dead leg
Dew point
Heat source
Heat tracing systems
Mineral-insulated heating cable
Parallel resistance heating cable
Power-limiting heating cable
Self-regulating cable
Series resistance heating cable
Startup temperature
Vessel

26413-08 Introductory Skills for the Crewleader

26412-08 Special Locations

26411-08 Medium-Voltage Terminations/Splices

26410-08 Motor Operation and Maintenance

26409-08 Heat Tracing and Freeze Protection

26408-08 HVAC Controls

26407-08 Advanced Controls

26406-08 Specialty Transformers

26405-08 Fire Alarm Systems

26404-08 Basic Electronic Theory

26403-08 Standby and Emergency Systems

26402-08 Health Care Facilities

26401-08 Load Calculations – Feeders and Services

ELECTRICAL LEVEL FOUR

ELECTRICAL LEVEL THREE

ELECTRICAL LEVEL TWO

ELECTRICAL LEVEL ONE

CORE CURRICULUM: Introductory Craft Skills

409CMAP.EPS

1. Pencil and paper
2. Appropriate personal protective equipment
3. Copy of the latest edition of the *National Electrical Code*®

Before you begin this module, it is recommended that you successfully complete *Core Curriculum; Electrical Level One; Electrical Level Two; Electrical Level Three; Electrical Level Four,* Modules 26401-08 through 26408-08.

This course map shows all of the modules in *Electrical Level Four.* The suggested training order begins at the bottom and proceeds up. Skill levels increase as you advance on the course map. The local Training Program Sponsor may adjust the training order.

1.0.0 ◆ INTRODUCTION

Applying heat to pipelines carrying process fluids is a common requirement in many industries including petroleum, plastics, chemicals, pharmaceuticals, power generation, and food processing. Pipeline electric cable heating systems, called heat tracing systems, maintain the pipeline systems at or above a specified temperature. This is done to prevent the pipelines from freezing, to maintain fluid temperature and flow in a pipeline, and/or to prevent condensation in a pipeline. Non-electrical steam and hot-liquid heat tracing systems can also be used for these purposes. However, electrical heating is often the most practical and least expensive method of pipeline heat tracing, especially when long pipe runs are involved. This module focuses on the circuitry and components used for electric pipeline heat tracing systems. It also introduces other widely used types of electric heating (heat tracing) systems. The electric heating systems covered in this module include:

- Pipeline heat tracing systems
- Roof, gutter, and downspout de-icing systems
- Snow melting and anti-icing systems
- Domestic hot-water temperature maintenance systems
- Floor heating and warming systems

2.0.0 ◆ PIPELINE HEAT TRACING APPLICATIONS

The main reasons for using electric heat tracing on pipelines are freeze protection and temperature maintenance of the process fluids. In cold climates, the thermal insulation installed around pipes is not enough to prevent the process fluids from freezing. Pipelines that carry water, fuel, or chemicals usually need to be heat traced to prevent freezing. Heat tracing is also applied to tanks and vessels to compensate for heat loss through the thermal insulation. Liquids that drop below their freezing point will form a solid obstruction in the piping system, thereby stopping fluid flow. In addition, as the liquid turns into a solid, it may also expand, causing severe damage to the piping. Pipes installed below the ground frost level are sometimes an exception. This is because the temperature in this area remains relatively constant regardless of the outdoor air temperature.

Many industrial processes require temperature maintenance of pipelines to prevent the loss of heat in process fluids flowing through them or contained in tanks or vessels. Temperature maintenance can be divided into two main categories: broad temperature control and narrow temperature control. Broad temperature control is used to maintain the viscosity of process fluids in order to keep them flowing. Typical applications of broad temperature control are on fuel lines, cooking oil lines, and grease disposal lines. Narrow temperature control is used to keep process fluids within a narrow temperature range to maintain fluid viscosity and to prevent degradation of the fluid composition. Typical applications of narrow temperature control are on sulfur, acrylic acid, food syrup, and sugar solution lines.

Temperature maintenance is also used to prevent condensation from forming in piping systems. The dew point is the temperature at which liquid first condenses when vapor is cooled. If an object's surface temperature drops below the dew point, condensation will occur. The resulting moisture can cause operating difficulties, especially in gas burners where the element may become fouled by the liquid. In addition, natural gas that contains moisture may freeze not only the system control valves, but also the entire system. Moisture is especially harmful to compressors, and in the presence of hydrogen sulfide, it can produce harmfully corrosive sulfuric acid. Hydrogen sulfide is produced in some oil refineries during the crude oil distillation process.

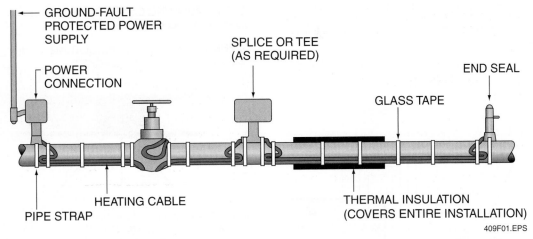

Figure 1 ◆ Basic pipeline electric heat tracing system.

3.0.0 ◆ PIPELINE ELECTRIC HEAT TRACING SYSTEMS

A basic pipeline electric heat tracing system consists of a power distribution system, electric heating cables, and interface components (*Figure 1*). Systems that are more complex may also incorporate a control and monitoring function with controllers, sensors, and alarms.

3.1.0 Heat Tracing System Power Distribution

A typical power distribution scheme used with electrical heat tracing systems is shown in *Figure 2*. As shown, three-phase utility power is applied to a dedicated step-down transformer for the heat tracing system. The transformer supplies a reduced voltage to a distribution panel that contains the main circuit breaker and branch circuit breakers. The main circuit breaker is used to disconnect power from the panel board and to protect the panel buses, the transformer, and the wiring between the transformer and the distribution panel. The branch circuit breakers are ground fault interrupting circuit breakers used to apply the operating voltages to the power connection boxes of the one or more heat tracing circuits. Typical cable supply voltages are 120/277VAC and 208/240VAC.

3.2.0 Heat Tracing System Cables

There are several types of cables used for electrical pipeline heat tracing, as well as other heating applications. These cables fall into several categories:

- Self-regulating cables
- Power-limiting cables
- Mineral insulated (MI) cables

3.2.1 Self-Regulating Cables

Self-regulating cables are used for applications that require a heat output that varies with changes in the temperature of the air or other medium surrounding the heating cable. A self-regulating cable increases its heat output with a decrease in the ambient temperature and decreases its heat output with an increase in the ambient temperature. Typically, this type of cable is used for pipeline freeze protection and temperature maintenance on metal and plastic pipes. It is also widely used for other applications such as snow melting, anti-icing, and roof and gutter de-icing.

One type of self-regulating cable (*Figure 3*) is made in the form of a heater strip consisting of two parallel bus wires embedded in a polymeric self-regulating conductive core heating element. The entire element is then surrounded by an insulating jacket, a braided tin-copper shield, and an outer jacket. Another type of self-regulating cable is similar in construction, but uses a polymeric self-regulating conductive fiber as the heating element instead of the core-type heating element. The self-regulating conductive core and conductive fiber heating elements are made of a polymer mixed with conductive carbon. This composition creates multiple electrical paths for conducting current between the parallel bus wires along the entire length of the cable. The core or fiber heating element of a self-regulating heating cable can be visualized as an array of parallel resistors connected between the bus wires, with each resistor

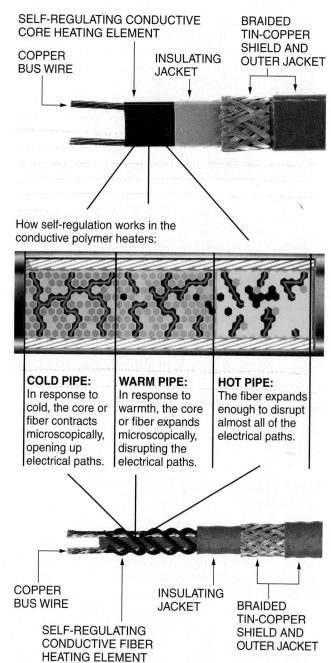

SELF-REGULATING CONDUCTIVE CORE HEATING ELEMENT

COPPER BUS WIRE

INSULATING JACKET

BRAIDED TIN-COPPER SHIELD AND OUTER JACKET

How self-regulation works in the conductive polymer heaters:

COLD PIPE:
In response to cold, the core or fiber contracts microscopically, opening up electrical paths.

WARM PIPE:
In response to warmth, the core or fiber expands microscopically, disrupting the electrical paths.

HOT PIPE:
The fiber expands enough to disrupt almost all of the electrical paths.

COPPER BUS WIRE

SELF-REGULATING CONDUCTIVE FIBER HEATING ELEMENT

INSULATING JACKET

BRAIDED TIN-COPPER SHIELD AND OUTER JACKET

409F03.EPS

Figure 3 ◆ Typical construction of self-regulating heat tracing cables.

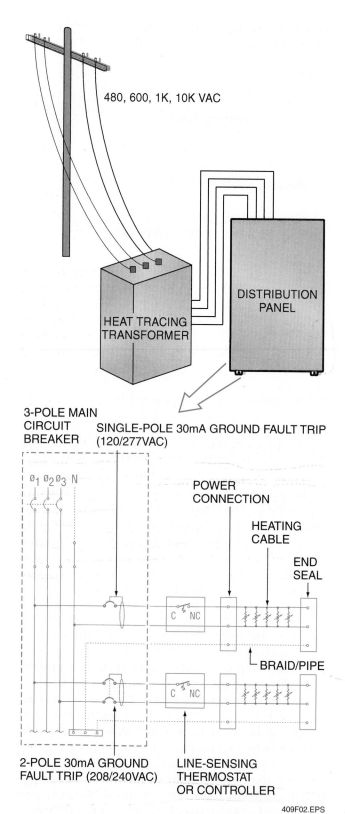

409F02.EPS

Figure 2 ◆ Heat tracing system power distribution.

having a value related to temperature. For this reason, self-regulating cables are also referred to as parallel resistance heating cables.

Self-regulating heating cables produce heat from the flow of current through the polymeric core or fibers. In both types of cables, the number of electrical paths between the bus wires changes in response to temperature variations. As the temperature surrounding the heater cable gets colder, the conductive core or fibers contract microscopically. This decreases the electrical resistance and creates numerous parallel resistance electrical paths between the bus wires. Increased current flows across these parallel paths, producing heat that warms the core or fiber along the length of the cable.

As the temperature surrounding the heater cable gets warmer, the conductive core or fibers expand microscopically. This increases the electrical resistance and decreases the number of electrical paths between the bus wires. The result is decreased current flow and a lower heat output in response to the rise in the ambient temperature. Through the process of contraction and expansion, the core or fiber heating element acts as a self-regulating thermostat protecting the cable from damage due to high temperatures. This feature allows this type of cable to be overlapped on pipes and valves during installation without causing damage to the cable from overheating. *Figure 4* shows a graph of resistance and power versus temperature for a self-regulating heat tracing cable.

Because of their parallel resistance construction, self-regulating heating cables can be cut to length and spliced in the field. Self-regulating cables with a core heating element, also called monolithic cables, are used on metal and plastic pipes to maintain temperatures up to 225°F. Those with fiber heating elements are used on metal pipes for process temperature maintenance up to 250°F.

3.2.2 Power-Limiting Cables

Power-limiting heating cables are a type of parallel-resistance, self-regulating heating cable. They are used in process temperature maintenance applications on metal pipes that require a high power output and/or high temperature exposure. The power-limiting heater cable (*Figure 5*) consists of a coiled resistor alloy-heating element wrapped around two parallel bus wires. It is connected alternately between the two bus wires at fixed intervals (usually every 2' or 3') along the cable length. The distance between the two contact

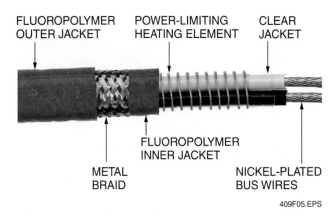

Figure 5 ◆ Construction of a power-limiting heating cable.

points forms a heating zone. For this reason, power-limiting heating cables are also referred to as zoned heating cables. Power-limiting heating cables can be cut to length at the job site. However, the portion of cable between the cut and the next connection point of the heating and bus wires will not receive electricity and therefore will not provide any heating.

The resistor alloy-heating element used in power-limiting heating cables has a positive temperature coefficient (PTC). This means the resistance of the heating element increases as its temperature increases and decreases as its temperature decreases. The result is an increase or decrease of power output as the temperature around the cable decreases or increases, respectively. This feature allows this type of cable to be overlapped on pipes and valves during installation without causing damage to the cable from overheating. This is because the temperature of the heating element is reduced at the points where the cable overlaps.

3.2.3 Mineral-Insulated Cables

Mineral-insulated (MI) cables are a type of constant-wattage, series resistance heating cables. They are widely used for tracing long pipelines in applications where the temperature or power output requirements exceed the capabilities of self-regulating or power-limiting heating cables. MI cables (*Figure 6*) consist of one or two heating conductors made of nichrome, copper, or other metals insulated with magnesium oxide and encapsulated by a metal sheath made of copper, stainless steel, nickel (Alloy 825), or another suitable metal. Unlike self-regulating and power-limiting cables, MI cables cannot be cut to length and terminated in the field. They must be ordered from the manufacturer and are cut to specified lengths and assembled with sealed connections at the factory.

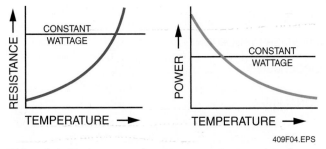

Figure 4 ◆ Resistance and power versus temperature for self-regulating heat tracing cable.

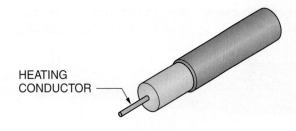

HEATING
CONDUCTOR

SINGLE-CONDUCTOR CABLE

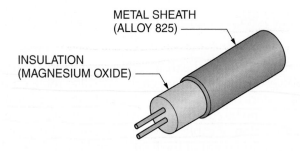

METAL SHEATH
(ALLOY 825)

INSULATION
(MAGNESIUM OXIDE)

DUAL-CONDUCTOR CABLE

409F06.EPS

Figure 6 ◆ Construction of mineral-insulated cables.

3.2.4 Long Line Cable Installations

Heat tracing of long pipelines is required in many applications, such as when transporting fluids between different processing stages in petrochemical facilities or to storage or transfer facilities, to tank farms, or to piers for ocean transport vessels. Such applications may involve cable lengths ranging between 1,250' and 5,000' that are powered from a single source. For many of these long line applications, specially designed self-regulating and mineral-insulated heating cable systems are used.

For some long pipeline heat tracing applications, specially designed series resistance heating cables are used. They are generally used when the circuit length exceeds the ratings of conventional self-regulating heating cables. Series-resistance heating cables (*Figure 7*) are constant wattage heating cables that have one or two resistance conductors designed for use in single- or three-phase systems. Typically, the resistance conductors are electrically insulated and protected by a high-temperature, heavy-wall fluoropolymer jacket, fluoropolymer insulation, a metal grounding braid, and a fluoropolymer outer jacket. Resistance heating of the conductor provides the cable heat. The wattage output depends on the total circuit length and

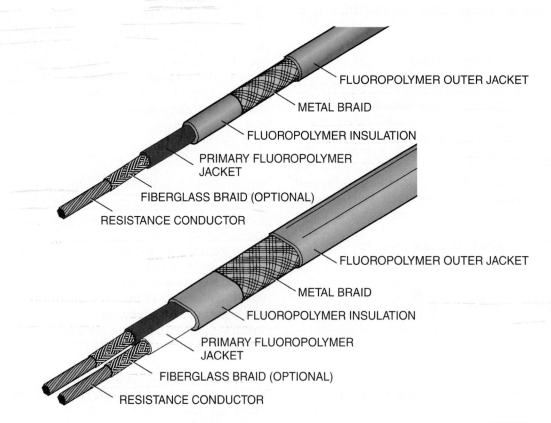

FLUOROPOLYMER OUTER JACKET

METAL BRAID

FLUOROPOLYMER INSULATION

PRIMARY FLUOROPOLYMER
JACKET

FIBERGLASS BRAID (OPTIONAL)

RESISTANCE CONDUCTOR

FLUOROPOLYMER OUTER JACKET

METAL BRAID

FLUOROPOLYMER INSULATION

PRIMARY FLUOROPOLYMER
JACKET

FIBERGLASS BRAID (OPTIONAL)

RESISTANCE CONDUCTOR

409F07.EPS

Figure 7 ◆ Construction of a series resistance heating cable.

voltage applied. Series resistance cables are commonly used in circuit lengths of 5,000' or more, with one power supply at voltages up to 600VAC.

3.2.5 Heating Cable Components and Accessories

Installation of a heat tracing system requires the use of appropriate components and accessories to interface the cable with the power source, to properly support and terminate the cable(s), and to make splices in cable runs as needed to achieve a properly functioning system. The specific components and accessories must be designed for use with the particular type of cable being installed, its application, and its environment. Always refer to the manufacturer's catalogs and installation instructions to determine the proper components and accessories to use. Except for specific model numbers, the components and accessories used with self-regulating and power-limiting cable heating systems are similar in design. *Figure 8* shows a typical self-regulating/power-limiting heating cable system. *Figure 9* shows those used with a mineral-insulated cable system. The names shown in the figures for the various components and accessories also describe the purpose for each item.

Selecting Heat Tracing Cables

Select a self-regulating heat tracing cable based on the application and the temperature requirements. The temperature produced by a specific type of cable depends on the type of polymeric core or fiber it contains and on the watts per foot it produces. When selecting heat tracing cables, refer to the manufacturer's cable data sheets and product literature and follow the recommendations pertaining to your application.

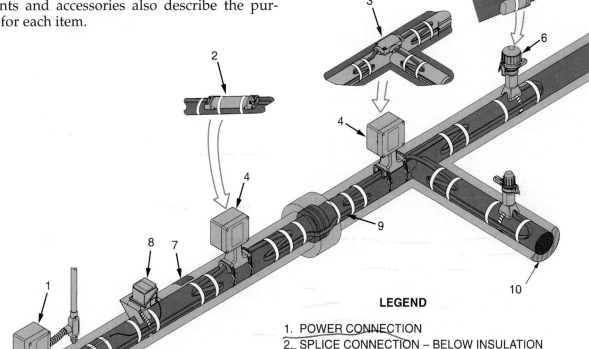

LEGEND

1. POWER CONNECTION
2. SPLICE CONNECTION – BELOW INSULATION
3. TEE CONNECTION – BELOW INSULATION
4. SPLICE/TEE CONNECTION – ABOVE INSULATION
5. END SEAL – BELOW INSULATION
6. END SEAL – ABOVE INSULATION
7. ATTACHMENT TAPE, LABELS
8. THERMOSTAT
9. SELF-REGULATING CABLE
10. INSULATION

409F08.EPS

Figure 8 ◆ Self-regulating/power-limiting heating system.

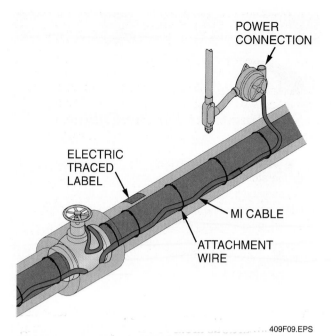

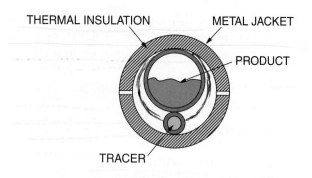

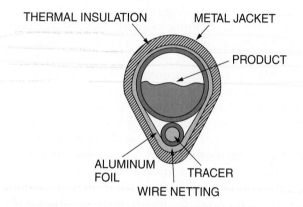

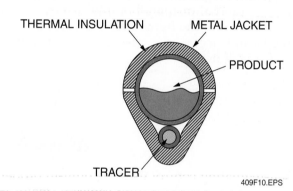

Figure 9 ◆ Mineral-insulated cable system.

3.2.6 Thermal Insulation Used with Heat Tracing Systems

Pipe systems with heat tracing cables are insulated in the same way as pipe systems without heat tracing. The only difference is that the heat tracing cables are installed first under the layer of thermal insulation. It is important to insulate the heat tracing cables and pipeline in order to reduce unnecessary heat loss to the surrounding atmosphere. In some cases, the traced line is wrapped with aluminum foil and then covered with insulation. The foil increases the radiation heat transfer. For proper transfer of heat, it is essential that the space between the pipeline and the tracer cable be kept free of particles of insulating material. The transfer of the tracer heat may be improved by putting a layer of heat conductive cement (graphite mixed with sodium silicate or other binders) between the tracer cable and the pipeline. *Figure 10* shows various tracer and insulation configurations. Straight piping may be weather-protected with metal jacketing or a polymeric or mastic system. When metal jacketing is used, it should be sealed with closure bands and a sealant applied to the outer edge where the bands overlap. Some insulating materials commonly used with heat tracing are:

- Rigid cellular urethane
- Polyisocyanurate foam
- Fiberglass
- Calcium silicate
- Cellular glass

Figure 10 ◆ Heat tracer/insulation configurations.

Some important factors to be considered when selecting an insulation material used with heat tracing are:

- Thermal characteristics
- Mechanical properties
- Chemical compatibility
- Moisture resistance
- Personal safety characteristics
- Fire resistance

3.3.0 Heat Tracing System Control

Control of heat tracing systems is used to adjust the output of the heating source to keep the pipes from freezing or to maintain process piping at correct temperatures. This section describes various heat tracing system control methods.

Skin-Effect Electric Heat Tracing Systems

Skin-effect electric heat tracing systems are used to heat trace long transfer pipelines up to 12 miles in length. These systems must be specifically designed for each application. In skin-effect heat tracing systems, heat is produced by a combination of a ferromagnetic heat tube and a temperature-resistant, electrically insulated conductor that passes through the tube. At the far end, the two components are connected together; at the near end, the two are connected to an AC supply. Current flows through the insulated conductor and into the heat tube where, by skin and proximity effects, the return current passes through and is concentrated toward the inner surface of the heat tube, causing the outer surface of the heat tube to remain at ground potential. The current flowing in the heat tube is converted to heat that is supplemented by heat generated in the insulated conductor. The heat tube may be directly welded or cemented to the pipeline, allowing the heat to be transferred by conduction through the coupling media from the heat tube to the pipeline.

3.3.1 Temperature Control

Temperature control of a heat tracing system can be accomplished in several ways. The method selected depends on the application requirements and on how critical temperature maintenance is to the process. The most widely used control methods are:

- Self-regulating control
- Ambient-sensing control
- Proportional ambient-sensing control
- Line-sensing control
- Dead-leg control

3.3.2 Self-Regulating Control

A self-regulating control uses the built-in properties of self-regulating heating cables to maintain the desired pipe temperature. It is used for basic freeze protection and broad temperature maintenance applications. Because no moving parts are involved, self-regulating control is the most economical and reliable method of control. Self-regulating control should not be used for applications that require maintaining a narrow operating temperature range, that involve temperature-sensitive fluids, or where energy consumption is a major concern.

3.3.3 Ambient-Sensing Control

An ambient-sensing control uses a mechanical thermostat or electronic controller to measure the ambient temperature and turn on the heat tracing system only when the surrounding temperature drops below a predetermined setpoint. For this reason, it is more energy efficient than a self-regulating control and therefore is widely used in freeze protection systems. *Figure 11* shows an ambient-sensing thermostat typical of those used in non-hazardous areas to directly control a single heat tracing circuit or as a pilot control of a contactor used to turn on and off multiple heat tracing circuits.

409F11.EPS

Figure 11 ◆ Ambient-sensing thermostat used to control heat tracing circuits.

3.3.4 Proportional Ambient-Sensing Control

A proportional ambient-sensing control (PASC) treats all pipes as if they contained stagnant fluid and provides power to the heat tracing system based on the ambient temperature.

A PASC uses a programmable electronic controller connected to remote temperature input modules and contactor control modules to regulate one or more heat tracing cables. The PASC system monitors the ambient temperature and operates to continuously adjust the heat tracing input to compensate for heat loss that occurs because of changing ambient temperature conditions. In order to maintain the desired temperature, a programmed algorithm calculates the cycle time during which the heating circuits are energized. The heat tracing cable is energized more often as the temperature decreases and less often as the temperature increases. On cold days, the heat tracing cable is turned on more often to maintain correct pipe temperatures. On warm days, it may not be turned on at all. A PASC is used when a self-regulating control is not precise enough. It is suitable for freeze protection, as well as all broad temperature-control and many narrow temperature-control applications. It is also one of the most energy-efficient forms of temperature control.

3.3.5 Line-Sensing Control

A line-sensing control uses a mechanical thermostat or electronic controller to measure the actual pipe temperature, not the ambient temperature surrounding the pipe. With this method of control, each heating circuit must have a separate control. Each heating circuit is turned on and off independently when signaled by its respective control device. Line-sensing controls are used in narrow temperature maintenance situations because they provide the highest degree of temperature control

using the least amount of energy. Because this method requires the use of multiple control devices, it has the highest installation cost. *Figure 12* shows a line-sensing thermostat typical of those used in hazardous areas to directly control a single heat tracing circuit or as a pilot control of a contactor used to turn on and off multiple heat tracing circuits.

3.3.6 Dead-Leg Control

The **dead-leg** method of control uses a segment of pipe (dead leg) in a piping system as a single control point for the larger system. The dead leg contains system fluid in a permanent no-flow condition. A thermostat measures the temperature of the fluid in the dead leg and uses it to represent the rest of the system. Dead-leg control provides much better control than ambient-sensing control, at about the same cost. It is more economical than line-sensing control because it uses only one thermostat. However, temperature control is imprecise, especially in high-flow areas of the system. Today, the use of dead-leg control systems is on the decline. They are being replaced by electronic control systems such as PASC that duplicate the dead leg electronically.

3.4.0 Heat Tracing System Monitoring

Several schemes can be incorporated into heat tracing systems to monitor their operation and

Figure 12 ◆ Typical line-sensing thermostat used to control heat tracing circuits.

409F12.EPS

detect any failures. The method selected depends on the application requirements and on how critical temperature maintenance is to the process. More than one monitoring method can be used in a system if the application requires it. The most widely used monitoring methods are:

- Ground fault
- Current
- Continuity
- Temperature

3.4.1 Ground Fault Monitoring

A ground fault monitoring system monitors the current leakage from the system heating cable, power wiring, and components to ground using ground fault (leakage) circuit breakers and/or current-sensing devices that measure current (*Figure 13*). *NEC Section 427.22* and other local codes require the use of ground fault equipment with heat tracing circuits. The function of ground fault circuit breakers is to protect the equipment, not personnel. The ground fault circuit breakers used in heat tracing circuits are designed to trip at 30mA, rather than at 5mA like the GFCIs used for personnel protection. Personnel protection GFCIs are not used in heat tracing applications because they often cause nuisance trips.

Ground fault trips are usually caused by mechanical damage to the heating cable or power wiring or by moisture in the junction boxes. When current leakage in the heat tracing circuit exceeds the trip setting of the ground fault circuit breaker or other protective device, it shuts off the circuit. If the protective device is a ground fault circuit breaker, it may have an auxiliary contact used to trigger a remote trip alarm.

INSIDE TRACK

Induction and Impedance Electric Heat Tracing Systems

Induction heat tracing systems are used in some high-temperature, high-power applications where very rapid heating is needed. Induction heat tracing systems use the piping as a heating element by placing it in the magnetic field of an AC source. Low-resistance wire is coiled around a conductive pipeline, and the alternating current flowing through the coiled wire generates a rapidly changing magnetic field that induces eddy currents and hysteresis losses. These eddy currents and hysteresis losses create a large amount of heat, similar to the way heat is produced in a transformer. A disadvantage of using induction heat tracing is that it does not produce uniform heating.

Impedance heat tracing systems are used to generate heat in a pipeline or vessel wall. This is done by causing current to flow through the pipeline or vessel by direct connection to an AC voltage source. A dual-winding isolation transformer isolates the distribution system from the heating system.

3.4.2 Continuity Monitoring

Continuity monitoring verifies that voltage is present at the end termination of a heat tracing cable run. This can be done by using an end-of-circuit

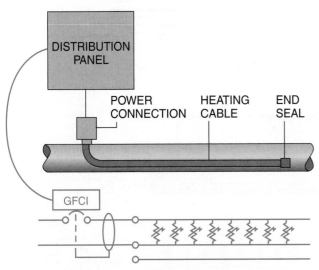

GROUND FAULT MONITORING; GFCI STATUS

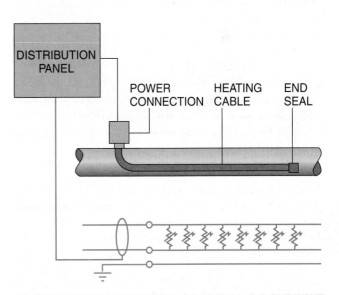

GROUND FAULT MONITORING OF ACTUAL G-F CURRENT

409F13.EPS

Figure 13 ◆ Simplified heat tracing circuit with ground fault monitoring.

LED that remains on when the circuit is operative and shuts off if there is an open in the circuit (*Figure 14*). In some systems, an end-of-circuit transmitter is used. Installed at the end of the line, it communicates the status of the circuit (continuity or open) to a centralized receiver. This type of system has the capability to send a low-voltage monitoring signal down the heat tracing line even when it is shut down.

3.4.3 Current Monitoring

Current monitoring uses a heat tracing controller or current-monitoring relay to signal an alarm when the current in the heat tracing circuit is too low or too high. This type of circuit will generate an alarm if there is a loss of power to the heating cable or if there is an open or damage to the heating cable bus wires or branch circuit wiring. *Figure 15* shows a simplified heat tracing circuit with current monitoring.

3.4.4 Temperature Monitoring

Temperature monitoring continuously measures the temperature of the pipe with a digital controller connected to a resistance temperature detector (RTD) temperature sensor attached to the pipe (*Figure 16*). If the high and/or low temperature setpoints are exceeded, the controller generates an alarm. Low-temperature alarms can be generated by low power or loss of power to the heating cable, wet or missing thermal insulation, heating cable damage, or controller failure. High-temperature alarms can be caused by fluid temperatures that exceed the preset threshold or by controller failure.

3.5.0 Typical Heat Tracing System Operation

Depending on the equipment manufacturer and the specific application, a wide variety of schemes

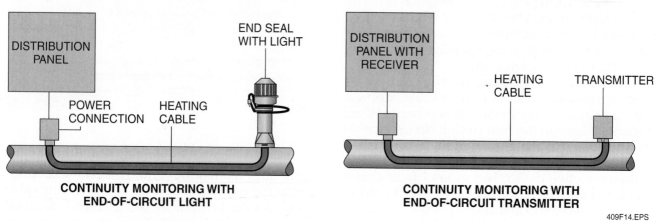

CONTINUITY MONITORING WITH
END-OF-CIRCUIT LIGHT

CONTINUITY MONITORING WITH
END-OF-CIRCUIT TRANSMITTER

409F14.EPS

Figure 14 ◆ Simplified heat tracing circuit with end-of-circuit continuity monitoring.

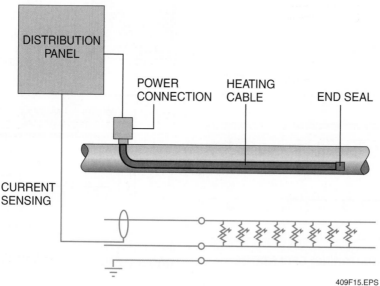

409F15.EPS

Figure 15 ◆ Simplified heat tracing circuit with current monitoring.

can be used to control and monitor electric pipe heat tracing systems. This section briefly describes a circuit configuration (*Figure 17*) used by one heat tracing equipment manufacturer to control and monitor a heat tracing system consisting of multiple circuits. The system consists of a

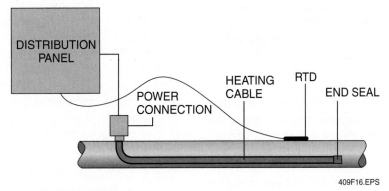

Figure 16 ◆ Simplified heat tracing circuit with temperature monitoring.

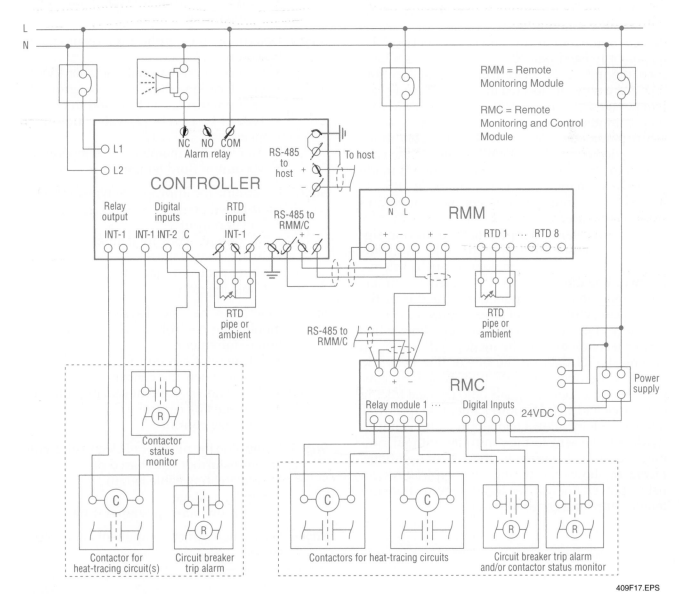

Figure 17 ◆ Typical heat tracing system circuit configuration.

microprocessor-based controller that regulates the heating circuits by selecting different modes of control, including ambient control, PASC, or line-sensing control. The controller can be operated as a standalone unit, or it can be networked to one or more remote monitoring and control modules (RMCs) and/or remote monitoring modules (RMMs). In response to signals from the controller, each RMC provides multiple relay-controlled outputs that cause the associated heating cable circuits to be switched on or off. Each RMM collects ambient temperature and/or pipe temperature inputs from RTD sensors located in the individual heater circuits. The RMM processes this information and forwards it to the controller. Based on the status inputs, the controller determines which circuits need to be turned on and which ones need to be turned off. It then signals the RMC to energize or de-energize the circuits, respectively, by means of the appropriate heating cable power contactors.

Data is communicated between the various devices in the monitoring and control network by means of twisted-pair RS-485 cables. This allows each module to be placed in the best location. RMCs are typically located in the heat tracing system power distribution panel, whereas RMMs are placed at the local sensor location. A system's setup parameters, status, and alarm conditions can be viewed by the system operator at the controller. In some systems, this information is made available at a remote location by an RS-232/RS-485 link to a PC running heat tracer monitoring and control software.

In some systems, temperature-monitoring signals are communicated to the controller by means of power line carrier technology. In such systems, a power line interface unit (PLI) receives temperature and continuity inputs transmitted over the heat tracing circuit power wires from special end-of-line heat cable circuit transmitters. Power line carrier technology uses frequency-shift keying to encode digital data on the power line network. Digital ones and zeros are transmitted by coupling high-frequency signals onto the heat tracing bus wires and the AC power line. Use of power line technology allows temperature and continuity data to be sent to the controller without running additional wiring to the sensor locations. *Figure 18* shows the wiring for a typical system using power line carrier technology to transmit temperature and continuity status.

4.0.0 ◆ EQUIPMENT SELECTION AND INSTALLATION FOR PIPE HEAT TRACING SYSTEMS

The design of a complex pipe heat tracing system and the selection of its components requires extensive knowledge of the industrial processes and process fluids used in numerous industries. For this reason, the design and component selection for a pipe heat tracing system is usually performed by a qualified master electrician, engineer, and/or manufacturer's engineering representative.

The component selection process involves the following stages and considerations:

Step 1 Determine the application:
- Freeze protection
- Temperature maintenance
- Both freeze protection and temperature maintenance

Step 2 Determine the application requirements based on:
- Pipe size and material (metal or PVC)
- Pipe location (above or below ground)
- Thermal insulation type and thickness
- Type of process fluid (water, fuel, oil, or grease)
- Ambient and maintenance temperatures
- Maximum fluid temperature

Step 3 Determine the appropriate type of heating cable based on:
- Power output
- Maximum exposure temperature
- Available voltages
- Outer protective jacket
- Cable length(s)

Step 4 Determine the type of system temperature monitoring control based on the application requirements:
- Self-regulating control
- Ambient-sensing control
- Proportional ambient-sensing control
- Line-sensing control
- Dead-leg control

Step 5 Determine the number of electrical circuits and transformer load based on:
- Total heating cable length
- Supply voltage
- Minimum **startup temperature**

Step 6 Select the correct components and accessories.

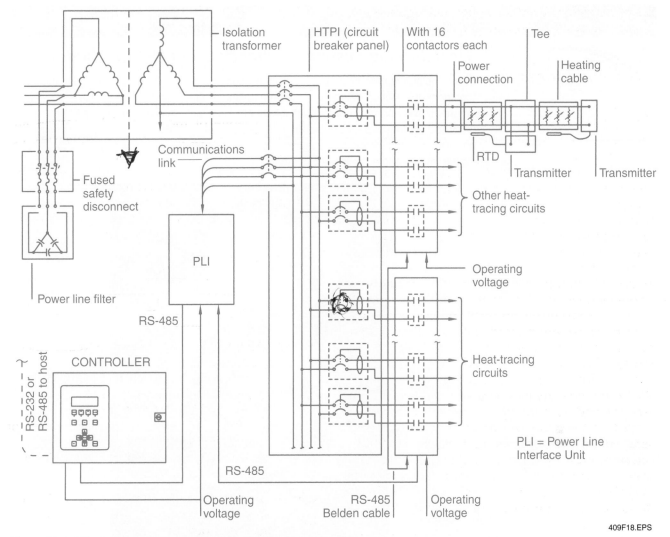

Figure 18 ◆ Heat tracing system using power line carrier technology.

The minimum number of circuits required for an installation is determined by dividing the total length of heating cable needed for the job by the maximum circuit length in feet that can be used for that particular type of cable. The maximum circuit length is determined by the type of cable, supply voltage, startup temperature, and circuit breaker size. Manufacturers provide this information in their design and/or installation literature. *Table 1* shows some examples of data from one manufacturer's design guide used to determine the maximum circuit lengths of different cables.

Table 1	Example of Data Used to Determine Maximum Circuit Length					
			Circuit Breaker Size			
			Maximum Circuit Length			
Heating Cable	**Supply (VAC)**	**Startup Temperature**	**15A**	**20A**	**30A**	**Maximum Amp/Foot**
5XL1-CR or -CT	120	40°F	165	220	250	0.072
		0°F	110	145	220	0.108
8XL1-CR or -CT	120	40°F	120	160	190	0.100
		0°F	85	115	170	0.139
5XL2-CR or -CT	208/277	40°F	285	380	450	0.042
		0°F	190	255	385	0.042

Heat Tracing System Application and Design Guides

Most equipment manufacturers have application and design information to aid in the component selection of their various heat tracing products and systems. This information is normally readily available at the manufacturer's website. It is also provided in product publications like the ones shown here. Even if you are not directly involved with the design process, reading these documents will provide you with a better understanding of these systems.

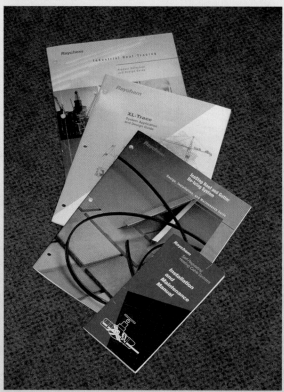

409SA01.EPS

4.1.0 Installation Guidelines

Always install the heat tracing system and its components in strict accordance with the system design specifications, the manufacturer's instructions, and all prevailing local codes. *NEC Article 427* covers the requirements for fixed electric heating equipment installed on pipelines and vessels in unclassified areas. The *NEC®* requirements are covered later in this section.

4.1.1 Installation of Heating Cable and Components

 WARNING!

To prevent injury from falling when installing heat tracing on overhead piping, use appropriate fall protection equipment and follow all safety procedures.

Some general guidelines for the installation of heat tracing cable and its components are listed here.

Prior to installation:

- Inspect all the cables, components, and accessories to make sure they are the ones specified for the job and that they are not damaged. Make sure that the rating of the heating cable is suitable for the service voltage available.
- Verify that the heating cable has not been damaged during shipping by testing it for continuity and electrical insulation resistance. Test the insulation resistance between the cable conductors and the metallic grounding braid in accordance with the cable manufacturer's instructions. Use a 2,500VDC megger. The minimum insulation resistance should meet the value specified by the cable manufacturer, typically greater than 20 megohms, regardless of length.
- Make sure there is enough heating cable available to perform the job. The minimum amount of heating cable required is equal to the total length of pipe plus the additional cable needed to make all the electrical connections and to heat trace each component such as pipeline valves, flanges, and pipe supports. Information on how much additional cable to provide for tracing each of these items can be found in the job specifications and/or in the manufacturer's installation instructions. Typically, two feet of cable is added for each electrical power connection, splice, tee, and end seal. *Table 2* lists some recommendations for the length of additional cable needed to heat trace pipeline valves, flanges, and supports.

NOTE

If the heating cable is spiraled (wrapped) around the pipe instead of being installed in a straight run, additional cable length must be allowed per foot of pipe. The amount of additional cable depends on the pipe size. For a given pipe size, the amount of cable and the distance between the spirals, called the pitch, is determined by multiplying the total pipe length by an appropriate spiral factor. Equipment manufacturers normally provide tables in their installation instructions that give the spiral factors and cable lengths to add per foot for the different sizes of pipe.

- Check and/or test each control device before installation for the correct operating temperature range, proper span, and setpoints.
- Store the heating cable and its components in a clean, dry place where they will be protected from mechanical damage.

During installation:

- Handle heating cable in a way that avoids:
  - Exposing the cable to sharp edges
  - Using excessive pulling force
  - Kinking or crushing the cable
  - Walking on or driving over the cable
- Heating cable must be installed on a clean, smooth portion of the pipeline, avoiding any sharp bends or jagged edges. It should be positioned to avoid damage due to impact, abrasion, or vibration, while still maintaining proper heat transfer. Always install heat tracing cable in accordance with the manufacturer's instructions.

Table 2 Typical Lengths of Additional Cable Required to Heat Trace Pipeline Components

Pipe Size (Inches)	Screw Valves	Flange Valves	Butterfly Valves	Pipe Supports	Pipe Elbows	Pipe Flange
1	6"	6"	6"	6"	3"	3"
2	6"	1"	6"	1'	6"	6"
3	1"	2'	1'	1'	6"	6"
4	2'	3'	1'	1'	9"	6'
6	3'	4'	1'	2'	9"	1'
8	4'	5'	2'	2'	9"	1'
10	5'	7'	2'	3'	1'	1'
12	7'	8'	2'	3'	1'	1'
14	8'	10'	3'	4'	1'	2'

Figure 19 shows how a heating cable is installed on some common pipeline components.

- Heating cable must be installed in a manner that facilitates the removal of valves, small in-line devices, and instruments without completely removing the cable or excessive thermal insulation or cutting the heating cable.
- With the exception of self-regulating cable, avoid overlapping heating cable sections. This can cause excessive temperatures at overlap points.
- Heating cable cold leads must be positioned to penetrate the thermal insulation in the lower 180° segment. This helps to minimize water entrance.
- Attach the heating cable to the pipes using fiberglass tape, aluminum tape, or plastic cable ties per the manufacturer's instructions. If using plastic cable ties, make sure they have a temperature rating that matches the system exposure rating.

CAUTION

Do not use metal attachments, vinyl electrical tape, or duct tape to fasten heating cable to the pipe.

Do not twist the bus wires together in a self-regulating cable as this will cause a short circuit.

- Protect the heating cable ends from moisture and mechanical damage if they are to be left exposed before connection.

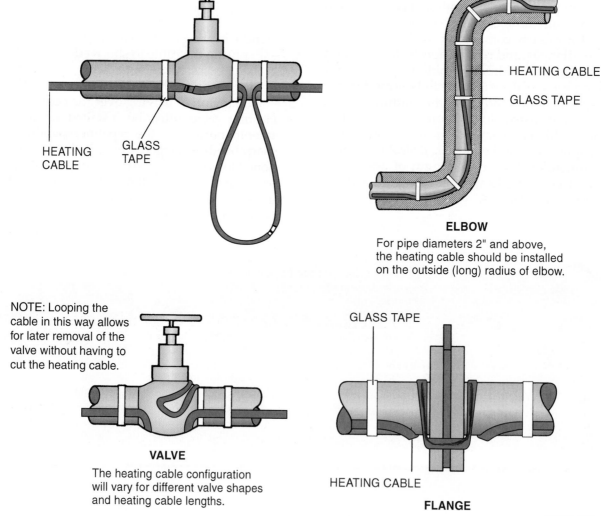

HEATING CABLE

GLASS TAPE

HEATING CABLE

GLASS TAPE

ELBOW

For pipe diameters 2" and above, the heating cable should be installed on the outside (long) radius of elbow.

NOTE: Looping the cable in this way allows for later removal of the valve without having to cut the heating cable.

VALVE

The heating cable configuration will vary for different valve shapes and heating cable lengths.

GLASS TAPE

HEATING CABLE

FLANGE

409F19A.EPS

Figure 19 ◆ Installing a heating cable on common pipeline components. (1 of 2)

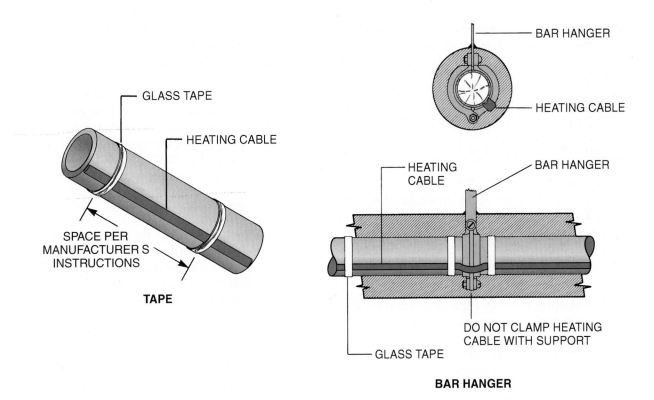

GLASS TAPE

HEATING CABLE

SPACE PER
MANUFACTURER S
INSTRUCTIONS

TAPE

BAR HANGER

HEATING CABLE

HEATING
CABLE

BAR HANGER

DO NOT CLAMP HEATING
CABLE WITH SUPPORT

GLASS TAPE

BAR HANGER

HEATING
CABLE

SUPPORT
SHOE

GLASS
TAPE

(SIDE VIEW)

GLASS
TAPE

HEATING
CABLE

(SECTION VIEW)

PIPE SUPPORT SHOE

409F19B.EPS

Figure 19 ◆ Installing a heating cable on common pipeline components. (2 of 2)

Pulling Heating Cable

INSIDE TRACK

When pulling heating cable, it helps to use a reel holder that plays the cable out
smoothly with little tension. The heating cable should be strung loosely but close to the
pipe being traced to avoid interference with supports and equipment.

The following conditions must be satisfied when installing temperature sensors:

- Sensors must be mounted to avoid the direct temperature effects of the heating cable.
- Sensors must be mounted 3' to 5' from any junction where two or more sections of cable meet or join.
- Sensors must be located 3' to 5' from any **heat source** or heat sink in the system.
- Where a pipeline runs through areas with different ambient conditions, such as inside and outside a heated building, separate sensors and associated controls are required for each area.

4.1.2 Installation of Thermal Insulation

Before installing thermal pipe insulation over the heat tracing system components, the heat tracing cable should be visually checked for signs of damage and electrically tested for continuity and electrical insulation resistance integrity. All testing should be done by properly trained personnel. Test the electrical insulation resistance between the cable conductors and the metallic grounding braid in accordance with the cable manufacturer's instructions. Use a 2,500VDC megger. Regardless of the heating cable length, the resistance reading should meet the value specified by the cable manufacturer, usually 20 megohms or above. Record the original value of each circuit for future reference.

Correct temperature maintenance of a heat traced pipe system requires that dry thermal insulation be properly installed. To lessen the chances of heating cable damage, install the thermal insulation as soon as possible after the heating cable and related devices have been attached. The insulation material and its thickness should be listed in the job specifications. If not, follow the recommendations of the heat tracing equipment manufacturer. All pipe work, including fittings, wall penetrations, and other potential heat loss areas, must be completely insulated. The outside of the pipe insulation should be marked *Electric Traced* at appropriate intervals on alternate sides to warn people of the pipe tracing. Also mark the location of all heating cable system components on the outside of the insulation.

4.1.3 NEC® Requirements

NEC Article 427 covers the requirements for fixed electric heating equipment installed on pipelines and vessels in unclassified areas. If the equipment is installed in hazardous (classified) areas, *NEC Articles 500 through 516* also apply.

Some of the *NEC®* requirements for fixed electric heating equipment installed on pipelines and vessels are summarized below. For complete *NEC®* requirements, refer to *NEC Article 427*.

- Branch circuits and overcurrent devices supplying the equipment shall be sized at not less than 125% of the total heater load per *NEC Sections 210.20(A) and 427.4*.
- Per *NEC Section 427.10*, the equipment must be identified as being suitable for:
  - The chemical, thermal, and physical environment
  - Installation in accordance with the manufacturer's drawings and instructions
- The external surfaces of any equipment that operates at temperatures above 140°F (60°C) shall be physically guarded, isolated, or thermally protected to prevent contact with personnel (*NEC Section 427.12*).

Overcurrent Protective Devices

INSIDE TRACK

The overcurrent protective devices used in a heat tracing system must be sized in accordance with the design specifications. Guidelines for correct sizing of overcurrent devices are normally given in the equipment manufacturer's design literature or computer selection programs. Ground fault circuit breakers or other ground fault protection devices with a nominal 30mA trip level are mandatory in most areas. (See *NEC Section 427.22* for an exception for industrial establishments.) Since these devices need an effective ground path, the *NEC®* requires that a conductive covering of the heating cable be connected to an electrical ground.

- The heating elements shall be attached to the pipeline or vessel surface being heated with appropriate fasteners other than the thermal insulation. Where the heating element is not in direct contact with the surface, a means shall be provided to prevent overtemperature of the heating element (*NEC Sections 427.14 and 427.15*).
- The presence of electric heating equipment shall be identified using permanent caution signs or markings at the intervals specified in *NEC Section 427.13*.
- Exposed heating equipment shall not bridge expansion joints unless provision is made for expansion/contraction (*NEC Section 427.16*).
- Ground fault protection for equipment is required except where there is an alarm indication of ground faults and only qualified personnel maintain the system and/or continuous operation is necessary for safe operation of the equipment or process (*NEC Section 427.22*).
- Equipment must be provided with a readily accessible disconnecting means such as indicating switches or circuit breakers. A factory-installed cord and plug that is accessible to the equipment may be used if the equipment is rated at 20A or less and 150V or less (*NEC Section 427.55*).

5.0.0 ◆ ROOF, GUTTER, AND DOWNSPOUT DE-ICING SYSTEMS

Roof, gutter, and downspout de-icing systems are used to prevent ice dams and icicles from forming on roofs and in gutters and downspouts (*Figure 20*). Ice dams form because melting snow and ice freeze when they reach the colder roof edge. Once ice dams are formed, water tends to collect behind them. The water backs up and seeps under the roofing materials, where it can leak into the building and cause water damage. Another danger is icicles that form when water dripping off the roof flows over ice-filled gutters. When outside temperatures rise, these heavy icicles can become a safety hazard as they break away from the gutters and roof.

Roof, gutter, and downspout de-icing systems provide a continuous heated path for melting snow and ice. The heated path keeps the water above freezing temperatures so that it can drain from the roof to a safe discharge area. *Figure 21* shows the components used with a typical roof, gutter, and downspout de-icing system. In this system, the continuous heating path is created by

409F20.EPS

Figure 20 ◆ Roof and gutter icing.

self-regulating, low-wattage heating cables specifically designed for roof, gutter, and downspout de-icing applications. As current flows through the cable's self-regulating polymeric core, the heat output is adjusted in response to the outside temperature.

The method used to control the roof de-icing system determines its capability to operate when needed as well as the amount of power consumed. Control is usually exercised in one of three ways:

- Manual control
- Ambient thermostat control
- Automatic moisture/temperature control

Manual control involves turning a switch on and off to control the system power contactor. Someone must monitor the outside temperature and manually turn on the system whenever it drops below freezing. Less prone to human error is an automatic roof de-icing system control that uses an ambient-sensing thermostat. When it senses that the outside temperature is below freezing, it automatically turns on the roof and gutter system, energizing the heating cable(s).

The most energy-efficient method of roof de-icing system control is an automatic moisture/temperature controller. It consists of a control panel, one or more aerial-mounted snow sensors, and one or more ice sensors that are located in the gutters and at other critical areas that need to be monitored. This arrangement turns on the roof, gutter, and downspout de-icing system when it begins to snow or when ice begins to form. *Figure 22* shows typical wiring for single and multiple circuit roof de-icing systems.

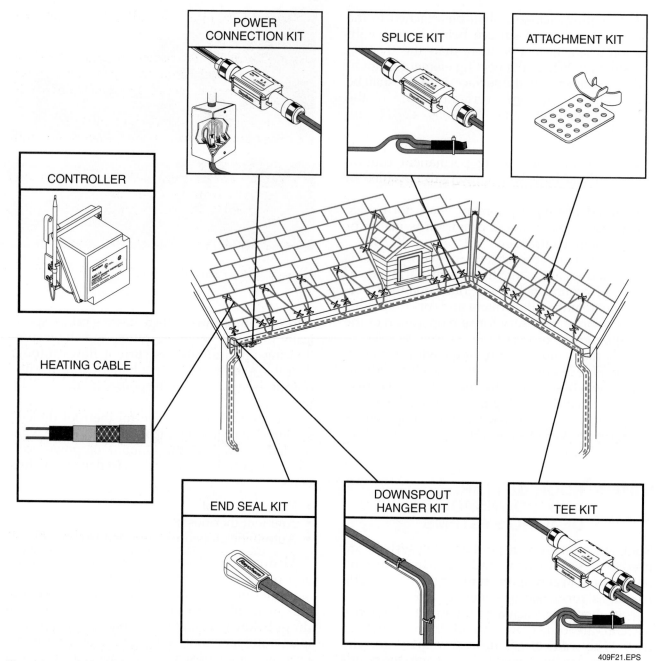

Figure 21 ◆ Components of a typical roof, gutter, and downspout de-icing system.

409F21.EPS

Electric Panel for Roof De-Icing

INSIDE TRACK

Some roof de-icing systems use mesh-type heating panels or mats instead of self-regulating heating cables to control ice and snow build-up and prevent ice dams from forming. For new roofing installations, the mats can be installed under the roofing material, making them invisible.

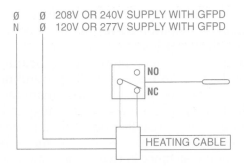

TYPICAL CONTROL WIRING – SINGLE CIRCUIT

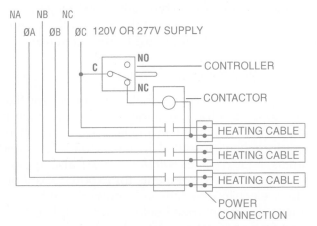

TYPICAL CONTROL WIRING – MULTIPLE CIRCUITS

409F22.EPS

Figure 22 ◆ Typical wiring for single and multiple circuit roof, gutter, and downspout de-icing systems.

6.0.0 ◆ COMPONENT SELECTION AND INSTALLATION FOR ROOF, GUTTER, AND DOWNSPOUT DE-ICING SYSTEMS

Each equipment manufacturer provides specific guidelines for the component selection and installation of its roof de-icing systems. Some general guidelines for the component selection process are described here.

Step 1 Determine the cable layout based on the roof construction:
- Type of construction (sloped roof, sloped roof with gutters, or flat roof)
- Location of roof/wall intersections
- Location of valleys, gutters, and downspouts

Step 2 Determine the method of cable attachment based on the roof construction materials and surface area:
- Roof, gutters, downspouts, and drip edges

- Mechanical or adhesive attachment clips

Step 3 Determine the method of system temperature control based on dependability, precision, and energy consumption requirements:
- Manual control
- Ambient thermostat control
- Automatic moisture/temperature control

Step 4 Determine the number of electrical circuits and transformer load based on:
- Total heating cable length
- Supply voltage
- Minimum startup temperature

Step 5 Select the correct components and accessories.

6.1.0 Installation Guidelines

Always install roof, gutter, and downspout de-icing systems in strict accordance with the system design specifications, the manufacturer's instructions, and all prevailing local codes. *NEC Article 426* covers the requirements for fixed outdoor electric de-icing and snow melting equipment. The *NEC®* requirements are covered later in this section.

6.1.1 Installation of Heating Cable and Components

Some general guidelines for the installation of roof, gutter, and downspout systems are listed here.

 WARNING!

To prevent injury from falling off roofs when installing roof, gutter, and downspout systems, use appropriate fall protection equipment and follow all safety procedures.

Prior to installation:

- Inspect all the cables, components, and accessories to make sure they are the ones specified for the job and that they are not damaged. Make sure that the rating of the heating cable is suitable for the service voltage available.
- Verify that the heating cable has not been damaged during shipping by testing it for continuity and electrical insulation resistance. Test the insulation resistance between the heating cable conductors and the metallic grounding braid in accordance with the cable manufacturer's

instructions. Use a 2,500VDC megger. The minimum insulation resistance should meet the value specified by the cable manufacturer, typically 1,000 megohms or more, regardless of length.

- Check and/or test each control device before installation for the correct operating temperature range, proper span, and setpoints.
- Store the heating cable and its components in a clean, dry place where they will be protected from mechanical damage.

During installation:

- Handle heating cable in a way that avoids:
  - Exposing the cable to sharp edges
  - Using excessive pulling force
  - Kinking or crushing the cable
  - Walking on or driving over the cable
  - Covering the heating cable with any roofing material unless it is a mat-type system designed for such use
- Install the components and attachment accessories first in the locations indicated on the project drawings.
- Install the heating cable. Start at the end seal and work backwards. Make sure the heating cable provides a continuous path for water to flow off the roof (*Figure 23*). Leave drip loops where appropriate. Also, be sure to leave a drip loop at each component so that water will not track down the heating cable and into the component.
- Loop and secure the heating cable at the bottom of the downspouts in accordance with the manufacturer's instructions to prevent any ice buildup from the draining water and to protect the heating cable from mechanical damage.
- Use UV-resistant cable ties wherever you require two heating cables to stay together.

409F23.EPS

Figure 23 ◆ Heating cable providing a continuous path for water to flow off the roof.

- Verify that the heating cable has not been damaged during installation by testing it for continuity and electrical insulation resistance. Test the insulation resistance between the heating cable conductors and the metallic grounding braid in accordance with the cable manufacturer's instructions. Use a 2,500VDC megger. If the heating cable is installed on a metal gutter, downspout, and/or metal roof, also measure the insulation resistance between the braid and metal surface. The minimum insulation resistance should meet the value specified by the cable manufacturer, typically 1,000 megohms or more, regardless of length. Record the original value for each circuit for future reference.

6.1.2 NEC® Requirements

NEC Article 426 covers the requirements for fixed outdoor electric de-icing and snow-melting equipment installed in unclassified areas. If the equipment is installed in hazardous (classified) areas, *NEC Articles 500 through 516* also apply.

The *NEC®* requirements for exposed fixed outdoor electric de-icing and snow-melting equipment are summarized below. For complete *NEC®* requirements, refer to *NEC Article 426*.

- Branch circuits and overcurrent devices supplying the equipment shall be sized at not less than 125% of the total heater load per *NEC Sections 210.20(A) and 426.4*.
- Per *NEC Section 426.10*, the equipment must be identified as being suitable for:
  - The chemical, thermal, and physical environment
  - Installation in accordance with the manufacturer's drawings and instructions
- The external surfaces of any equipment that operates at temperatures above 140°F (60°C) shall be physically guarded, isolated, or thermally protected to prevent contact with personnel (*NEC Section 426.12*).
- The presence of electric heating equipment shall be identified using permanent caution signs or markings at the intervals specified in *NEC Section 426.13*.
- Exposed heating equipment shall not bridge expansion joints unless provision is made for expansion/contraction [*NEC Section 426.21 (C)*].
- Ground fault protection for equipment is required except for mineral-insulated, metal-sheathed cable embedded in a noncombustible material [*NEC Sections 210.8(A)(3) Exception and 426.28*].

- Equipment must be provided with a readily accessible disconnecting means such as branch circuit switches or circuit breakers. A factory-installed cord and plug that is accessible to the equipment may be used if the equipment is rated at 20A or less and 150V or less (*NEC Section 426.50*).

7.0.0 ◆ SNOW-MELTING AND ANTI-ICING SYSTEMS

Snow-melting and anti-icing systems are used to prevent the buildup of snow and ice on concrete surfaces of driveways, walkways, stairways, loading ramps, building entryways, and bridge road surfaces (*Figure 24*), where they can create hazards for people and/or vehicles.

Figure 25 shows the components used with one manufacturer's snow-melting and anti-icing system. In this system, heating is provided by one or more self-regulating heating cables specifically designed for snow-melting and anti-icing applications. The cables are cut on the job. They are installed in a serpentine pattern with 12" spacing between loops, securely fastened to rebar or wire

409F24.EPS

Figure 24 ◆ Bridge road surface protected by snow-melting and anti-icing system.

mesh, then completely encased in concrete. All electrical connections and terminations are made in externally accessible junction boxes. The capability of the heating cable to melt snow is proportional to its wattage rating. As current flows through the cable's self-regulating polymeric core, the heat output is adjusted in response to the ambient temperature of the surrounding concrete.

Some snow-melting and anti-icing systems use one or more factory-made electrical snow-melting mats instead of regular heating cable. The heating mats consist of self-regulating heating cable looped in a serpentine pattern within a screened mesh framework. Like the regular heating cable, the mats are embedded in concrete and electrically connected at externally accessible junction boxes.

The method used to control the snow-melting and anti-icing system determines its ability to operate when needed as well as the amount of power consumed. Control is done in one of three ways:

- Manual control
- Ambient thermostat control
- Automatic snow control

With the exception of specific models of equipment, the manual control and ambient thermostat control methods used with snow-melting and anti-icing systems are similar to those described earlier for roof de-icing systems. Ambient thermostat control is used to prevent the formation of ice on the pavement surface during all cold weather conditions. This control method keeps all points on the surface above 32°F during freezing periods.

The most energy-efficient method of control is an automatic snow controller working in conjunction with a snow sensor. When the snow sensor detects rain or snow in the presence of low temperatures, it signals the snow controller to energize the contactor that supplies power to the snow-melting and anti-icing system. When the precipitation stops or the temperature rises above freezing, the controller de-energizes the system. An off-delay timing function incorporated into the controller can be used to extend the melting time. It keeps the melting equipment in operation for a preset time after the sensor no longer detects precipitation accompanied by below-freezing temperatures. *Figure 26* shows typical wiring for a snow-melting and anti-icing system under the control of an automatic snow controller.

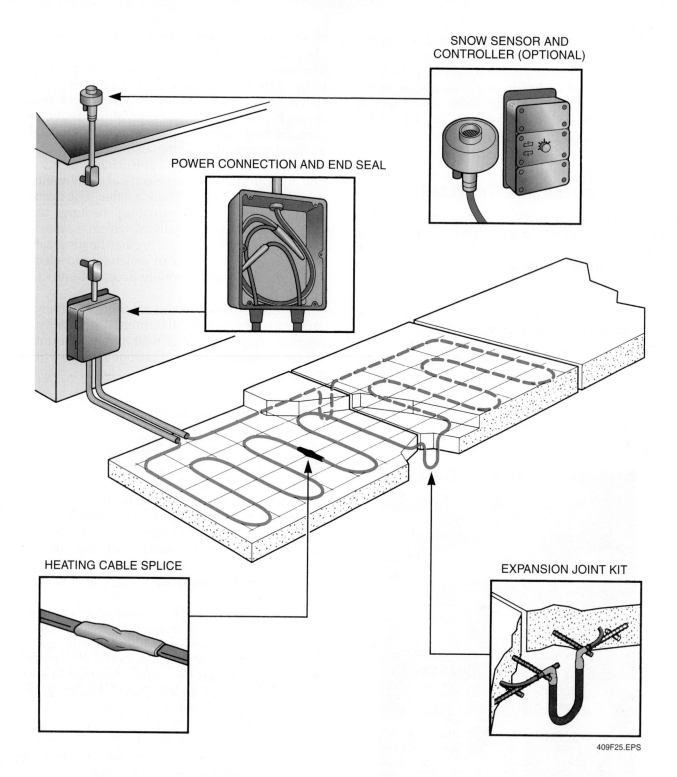

SNOW SENSOR AND
CONTROLLER (OPTIONAL)

POWER CONNECTION AND END SEAL

HEATING CABLE SPLICE

EXPANSION JOINT KIT

409F25.EPS

Figure 25 ◆ Components of a typical snow-melting and anti-icing system.

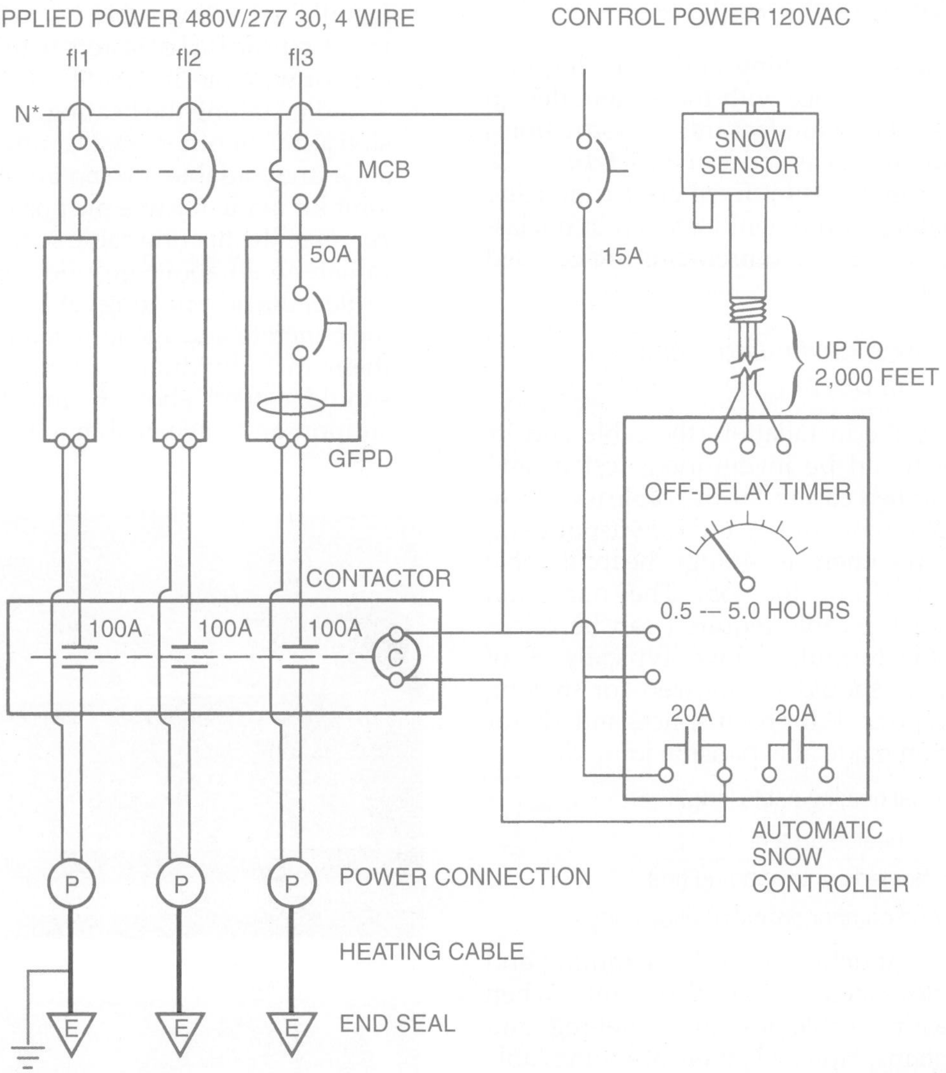

Figure 26 ◆ Typical wiring for a snow-melting and anti-icing system controlled by an automatic snow controller.

8.0.0 ◆ COMPONENT SELECTION AND INSTALLATION FOR SNOW-MELTING AND ANTI-ICING SYSTEMS

Each equipment manufacturer provides specific guidelines for the component selection and installation of its snow-melting and anti-icing systems. The component selection process for a regular heating cable (non-mat) system involves the following stages and considerations:

Step 1 Determine the spacing between the cables based on the ambient temperature and other weather characteristics of the region.

Step 2 Determine the cable layout and total length of cable based on the application and surface area to be heated.

Step 3 Determine the method of system temperature control based on dependability, precision, and energy consumption requirements:
• Manual control
• Ambient thermostat control
• Automatic snow control

Step 4 Determine the number of electrical circuits and transformer size based on:
• Total heating cable length
• Supply voltage
• Minimum startup temperature

Step 5 Select the correct components and accessories.

8.1.0 Installation Guidelines

Always install snow-melting and anti-icing systems in strict accordance with the system design specifications, the manufacturer's instructions, and all prevailing local codes. *NEC Article 426* covers the requirements for fixed electric de-icing and snow-melting equipment installed in unclassified areas. The *NEC*® requirements are covered later in this section.

8.1.1 Installation of Heating Cable and Components

Before starting the installation, the cable and its components should be inventoried, tested, and stored as described earlier for the cable and components used in roof, gutter, and downspout systems. Make sure there is enough heating cable available to perform the job. The minimum amount of heating cable required can be calculated using the formula below. Typically, 3' of additional cable should be allowed for making power connections, 1' for each splice, and 1½' for each connection made at expansion joint kits.

$$\text{Total heating cable length} = \frac{\text{heated area (ft}^2) \times 12}{\text{heated cable spacing (in.)}}$$

$$+ \text{ end and component cable allowances}$$

The general guidelines for cable handling and installation described earlier also apply when installing heating cable for snow-melting and anti-icing systems. However, there are some cable and component installation guidelines unique to snow-melting and anti-icing systems installed in concrete. They are as follows:

- Install the heating cable in a serpentine pattern covering the area to be heated. Space the cable in accordance with the job specifications, typically 12" on center. This distance should be maintained within 1". Some locations require closer spacing to provide more snow-melting or anti-icing capability.
- Do not route heating cable within 4" of the edges of the pavement, drains, anchors, or other objects in the concrete.

- Install the heating cable 1½" to 2" below the finished surface and fasten it to the reinforcing bars or wire mesh (*Figure 27*).
- Avoid installing the heating cable across expansion joints or other crack-control joints. If doing so is unavoidable, use an approved expansion joint kit made for this purpose.
- Arrange the heating cable so that both ends terminate in an above-ground weatherproof box. Protect the non-heating cables that run between the concrete and the junction box by installing them in individual conduits. These conduits should extend about 6" into the concrete for support and have bushings at both ends.

409F27.EPS

Figure 27 ◆ Heating cables in concrete sidewalk form.

- Verify that the heating cable has not been damaged during installation or during the placement of the concrete by testing it for continuity and electrical insulation resistance. Test the insulation resistance between the cable conductors and the metallic grounding braid in accordance with the cable manufacturer's instructions. Use a 2,500VDC megger. The minimum insulation resistance should meet the value specified by the cable manufacturer regardless of cable length. Record the original value for each circuit for future reference.

8.1.2 NEC® Requirements

NEC Article 426 covers the requirements for all fixed outdoor electric de-icing and snow-melting equipment. With the exception of the requirements that pertain only to exposed equipment, the *NEC Article 426* requirements described earlier for roof, gutter, and downspout de-icing systems also apply to snow-melting and anti-icing equipment. The requirements unique to equipment embedded in concrete or similar material are summarized as follows. For complete *NEC®* requirements, refer to *NEC Article 426*.

- Embedded panels or units shall not exceed $120W/ft^2$ [*NEC Section 426.20(A)*].

- Spacing between embedded adjacent cable runs shall not be less than 1" on center [*NEC Section 426.20(B)*].
- Embedded panels, units, or cables shall be placed on a masonry or asphalt base at least 2" thick, and there must be at least 1½" of masonry or asphalt applied over the finished surface [*NEC Section 426.20(C)(1)*].
- Embedded panels, units, or cables shall be fastened in place by approved means while the masonry or asphalt finish is applied [*NEC Section 426.20(D)*].

9.0.0 ◆ DOMESTIC HOT-WATER TEMPERATURE MAINTENANCE SYSTEMS

Electrical heat tracing can be used for temperature maintenance of domestic hot-water pipeline systems in places such as hospitals, schools, prisons, and commercial high-rise office buildings. For these applications, specially designed self-regulating heating cable is installed on all the domestic hot-water pipe branches and risers within the building, underneath the standard pipe insulation (*Figure 28*). The heating cable adjusts its

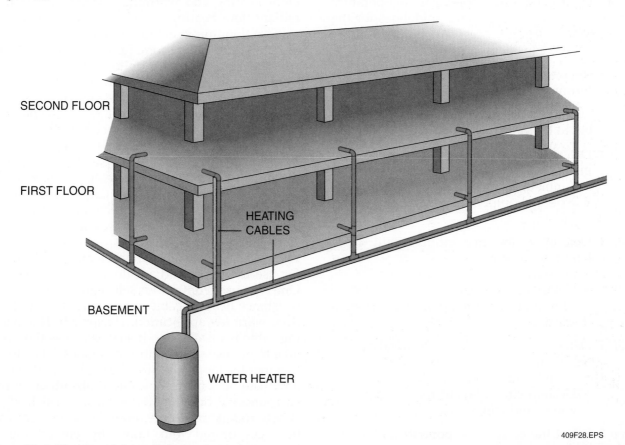

SECOND FLOOR

FIRST FLOOR

HEATING CABLES

BASEMENT

WATER HEATER

409F28.EPS

Figure 28 ◆ Heat traced domestic hot-water system.

power output to compensate for supply-pipe heat losses that result from variations in water and ambient temperature. The power output of the cable is increased or decreased only at the points where the heat losses occur. Unlike conventional domestic hot-water systems, heat traced hot-water systems do not require overheating the supply water to allow for cooling. System design is also simplified and costs are reduced because return water lines, circulating pumps, and balancing valves used in conventional domestic hot-water systems are not needed.

With the exception of the specific model, the components and accessories used in heat traced domestic hot-water systems are the same as those used with other electrical pipe heat tracing systems. These include power connections, heating cable, end seals, cable splices, and tees. The system control and monitoring methods are also similar.

10.0.0 ◆ COMPONENT SELECTION AND INSTALLATION FOR DOMESTIC HOT-WATER TEMPERATURE MAINTENANCE SYSTEMS

Each equipment manufacturer provides specific guidelines for the component selection and installation of its domestic hot-water temperature maintenance systems. General guidelines for the component selection process involve the following stages and considerations:

Step 1 Review the plumbing design plan and determine its application.

Step 2 Determine the temperature requirements based on the application.

Step 3 Determine the type of self-regulating heating cable based on the application:
- Ambient temperature
- Nominal pipe temperature requirements

Step 4 Determine the maximum circuit cable length based on:
- Cable type
- Circuit breaker size
- Minimum water temperature at system startup

Step 5 Determine the number of electrical circuits based on:
- Total heating cable length
- Desired circuit layout for floors, risers, zones, and wings

Step 6 Select the correct components and accessories.

10.1.0 Installation

Installation and testing of heat tracing cables used for domestic hot-water temperature maintenance is done in the same way as described earlier for heat tracing other types of pipe systems. The heating cables are typically installed on copper tubing beneath fiberglass insulation. At least 2' of additional cable should be allowed for making power connections and splices, and 3' should be allowed for heat tracing at pipe tees. Leave a service loop of extra heating cable at connections or at places in the piping where future service is expected.

10.2.0 *NEC*® Requirements

NEC Article 427 covers the requirements for fixed electric heating equipment installed on pipelines and vessels, including domestic hot-water heat tracing systems. Refer to the *NEC Article 427* requirements summarized earlier in the section on heat tracing pipe systems. *NEC Article 517* contains specific requirements for health care facilities.

11.0.0 ◆ FLOOR HEATING AND WARMING SYSTEMS

Floor heating and warming systems, also called radiant floor heating systems, are used to heat or warm a floor mass that in turn radiates the heat to the cooler air in the room. Rooms with floor radiant heat tend to have a uniform air temperature from the ceiling to the floor, whereas a room heated with a forced air furnace will be cold near the floor and hot near the ceiling. Floor heating systems are commonly used as either main or supplementary heating to heat or warm concrete floors, floors over unheated spaces, and tile and marble floors. Some floor heating systems are used to warm hardwood and carpeted floors. To prevent frost-heave damage, floor systems are sometimes used to warm the soil beneath cold room and freezer floors.

Different types of heating elements are made for use in electric radiant floor heating systems. One type uses continuous self-regulating or series resistance heating cable specifically designed for floor warming applications (*Figure 29*). This heating cable is installed so it snakes across the floor area to be heated in continuous loops. The closer together the loops, the more heat generated in that area. The heat from the cable is absorbed by the surrounding floor materials from which it is slowly radiated into the room. Heating cable can be incorporated into floors by embedding it in mortar; running it in conduits embedded in

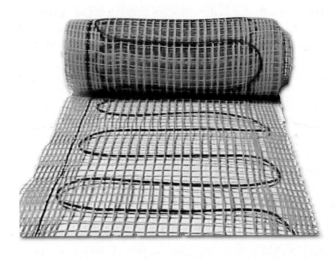

Figure 29 ◆ Basic concrete and masonry electric radiant floor heating system.

concrete, sand, or soil; or mounting it to the exposed bottom surface of concrete floors.

Many floor warming systems use one or more factory-made electrical heating mats instead of a continuous cable. These are typically installed over a wood floor joist system or concrete slab floor system. Some heating mats consist of an electric cable fastened to a flexible mesh support structure. The mat is unrolled, cut to size, taped or stapled in place, then embedded in thin-set masonry beneath the finished floor material (*Figure 30*). Some systems use mesh mats that are installed on a plywood subfloor or concrete substrate over which carpet padding and carpet is installed.

Temperature control of a floor warming system is usually managed using an appropriate thermostatic control installed in accordance with the manufacturer's instructions. For systems using continuous cable, the sensing bulb of the thermostatic control is

409F30.EPS

Figure 30 ◆ Typical electric floor heating mat installation.

installed in the floor at the same elevation as the heating cable, midway between adjacent runs of cable, and away from the room walls. The bulb should be installed inside a PVC conduit so that it can be easily removed for service, if necessary.

Radiant Floor Heating Systems

INSIDE TRACK

Radiant floor heating systems have been used for many centuries. Evidence shows that around 1300 B.C.E., a radiant floor heating system was used in a Turkish king's palace. Over a period ranging from about 80 B.C.E. to 324 C.E., the Romans used improved versions of floor radiant heating systems in upper-class houses and public baths throughout their empire. These systems consisted of chambers built under the floor or tile flues built into a stone floor. Heated combustion gases, produced by a fire enclosed in a chamber located at one end of the floor, were routed through the floor chambers or flue tiles to exhaust vents placed in the outside walls at the other end of the floor.

12.0.0 ♦ COMPONENT SELECTION AND INSTALLATION FOR FLOOR HEATING SYSTEMS

There is a wide diversity in floor heating and warming system applications, equipment, and related installation methods. Because of this diversity, it is impractical to give general guidelines for component selection and system installation. Each equipment manufacturer has readily available product literature that gives system design and component selection criteria for specific products and floor heating and warming applications. Floor heating and warming systems must be installed in strict accordance with the system design specifications, the manufacturer's instructions, and all prevailing local codes. When installing floor heating and warming systems, follow the general guidelines given earlier in this module for storing, handling, and testing the heating cable.

12.1.0 *NEC*® Requirements

NEC Article 424 covers the requirements for fixed electric space heating equipment. If the equipment is installed in hazardous (classified) areas, *NEC Articles 500 through 516* also apply. The *NEC*® requirements for heating cables installed in concrete or poured masonry floors are summarized below. For complete *NEC*® requirements, refer to *NEC Article 424*.

- Heating cable shall not extend beyond the room or area where it originates [*NEC Section 424.38(A)*].
- Embedded cable shall be spliced only where necessary and only by approved means (*NEC Section 424.40*).

- Constant wattage cables shall not exceed 16½ watts per linear foot of cable [*NEC Section 424.44(A)*].
- Spacing between adjacent cable runs shall not be less than 1" on center [*NEC Section 424.44(B)*].
- Embedded cable shall be fastened in place by approved nonmetallic means while the concrete is applied [*NEC Section 424.44(C)*].
- There shall be spacing between the heating cable and any metal embedded in the floor, unless the cable is grounded metal-clad cable [*NEC Section 424.44(D)*].
- Protect the non-heating cable leads exiting the floor by installing them in conduit [*NEC Section 424.44(E)*].
- Ground fault circuit interrupter protection for personnel shall be provided for cables installed in electrically heated floors of bathrooms and in hydromassage bathtub locations [*NEC Section 424.44(G)*].
- Cable installations shall be inspected and approved before cables are concealed (*NEC Section 424.45*).
- Radiant heating panels or panel sets installed in concrete or poured masonry shall not exceed 33W/ft² of heated area [*NEC Section 424.98(A)*].
- Heating panels or panel sets shall not be installed where they bridge expansion joints unless provision is made for expansion and contraction [*NEC Sections 424.98(C) and 424.99(C)(1)*].
- Radiant heating panels or panel sets installed under floor coverings shall not exceed 15W/ft² of heated area [*NEC Section 424.99(B)*].

Electric Floor Heating

Often, the bathroom is the only room requiring additional warmth in the morning. Using a floor warming system in the bathroom allows the homeowner to keep the remainder of the home at a lower temperature, resulting in a significant energy savings over time.

GOING
GREEN

1. As the ambient temperature surrounding a self-regulating heating cable gets colder, the _____.
   a. cable's heat output decreases
   b. cable's conductive polymeric core expands
   c. resistance of the bus wires decreases
   d. resistance of the polymeric core decreases

2. The resistor alloy-heating element in a _____ heating cable has a positive temperature coefficient.
   a. power-limiting
   b. self-regulating
   c. mineral-insulated
   d. series-resistance

3. Which of the following methods of heat tracing system control uses a programmed algorithm to calculate the cycle time during which the heating circuits are energized?
   a. Line-sensing control
   b. Ambient-sensing control
   c. Proportional ambient-sensing control
   d. Dead-leg control

4. Which type of heat tracing monitoring system measures current leakage from the system cables?
   a. Current monitoring
   b. Ground fault monitoring
   c. Continuity monitoring
   d. Temperature monitoring

5. The type of heating cable circuit control must be considered when determining the maximum circuit length for a heat tracing cable run.
   a. True
   b. False

6. When determining the minimum amount of heating cable needed to heat trace a pipeline, add about _____ of cable for each splice.
   a. 1'
   b. 2'
   c. 3'
   d. 4'

7. When installing temperature sensors in a pipe heat tracing system, the sensors should be located _____ from any heat source or heat sink.
   a. less than 1'
   b. 1' to 3'
   c. 3' to 5'
   d. 5' to 8'

8. The continuity and electrical insulation resistance of heating cables should be tested _____.
   a. before installation
   b. during installation
   c. during and after installation
   d. before and after installation

9. Which of the following methods of control for a roof de-icing system is the most energy efficient?
   a. Manual control
   b. Ambient thermostat control
   c. Automatic moisture/temperature control
   d. Automatic snow control

10. Any de-icing or snow-melting cable that operates at temperatures above _____ must be suitably guarded to prevent contact with personnel.
    a. 110 degrees F
    b. 120 degrees F
    c. 130 degrees F
    d. 140 degrees F

11. The most efficient method of snow-melting control is _____ control.
    a. manual
    b. ambient thermostat
    c. automatic snow
    d. end component

12. The requirements for the installation of electrical snow-melting and de-icing systems are defined in _____.
    a. *NEC Article 424*
    b. *NEC Article 425*
    c. *NEC Article 426*
    d. *NEC Article 427*

13. The spacing between embedded adjacent cable runs for a snow-melting and de-icing system must not be less than _____ on center.
    a. 1"
    b. 6"
    c. 10"
    d. 12"

14. Domestic hot-water temperature maintenance cable is typically _____.
    a. cut to the exact length required
    b. used with uninsulated systems
    c. installed on PVC pipe
    d. installed on copper tubing

15. Constant wattage cables used in floor heating/warming systems must not exceed _____ per linear foot of cable.
    a. 10W
    b. 12½W
    c. 16½W
    d. 120W

Summary

Electrical heat tracing systems keep pipelines at or above a given temperature for the purposes of freeze protection and temperature maintenance. In cold climates, freeze protection is needed because the thermal insulation installed around pipes is not enough to prevent the process fluids from freezing. Temperature maintenance of pipelines is required in many industrial processes to prevent heat loss in process fluids flowing through the pipelines or contained in tanks or vessels.

Roof, gutter, and downspout electrical de-icing systems prevent ice dams and icicles from forming on roofs and in gutters and downspouts. They provide a continuous heated path that keeps melting roof water above freezing temperatures so that it is free to drain from the roof to a safe discharge area.

Electrical snow-melting and anti-icing systems are used to prevent the buildup of snow and ice on concrete surfaces of driveways, walkways, stairways, loading ramps, building entryways, and bridge road surfaces, where they can create hazards for people and/or vehicles.

Electrical pipe heat tracing can be used for temperature maintenance of domestic hot-water systems in places such as hospitals, schools, prisons, and commercial high-rise office buildings. The heating cable adjusts its power output to compensate for supply-pipe heat losses resulting from variations in water temperature and ambient temperatures. Heat traced hot-water systems are economical because they do not require overheating the supply water to allow for cooling.

Floor heating systems, commonly called radiant floor heating systems, are used to heat or warm a floor mass that in turn radiates the heat to the cooler air in the room. Floor heating systems are commonly used as main or supplementary heating to heat or warm concrete floors, floors over unheated spaces, and tile and marble floors. Some floor heating systems are used to warm hardwood and carpeted floors. To prevent frost-heave damage, floor systems are sometimes used to warm the soil beneath cold room and freezer floors.

Notes

Trade Terms
Introduced in This Module

Constant wattage heating cable: A heating cable that has the same power output over a large temperature range. It is used in applications requiring a constant heat output regardless of varying outside temperatures. Zoned parallel resistance, mineral-insulated, and series-resistant heating cables are examples of constant wattage heating cables.

Dead leg: A segment of pipe designed to be in a permanent no-flow condition. This pipe section is often used as a control point for a larger system.

Dew point: The temperature at which liquid first condenses when vapor is cooled.

Heat source: Any component that adds heat to an object or system.

Heat tracing systems: Steam, hot-fluid, or electrical systems that maintain piping systems, vessels, and certain types of structures at or above a given temperature.

Mineral-insulated heating cable: A type of constant wattage, series resistance heating cable used in long line applications where high temperatures need to be maintained, high temperature exposure exists, or high power output is required.

Parallel resistance heating cable: An electric heating cable with parallel connections, either continuous or in zones. The watt density per lineal length is approximately equal along the length of the heating cable, allowing for a drop in voltage down the length of the heating cable.

Power-limiting heating cable: A type of heating cable that shows positive temperature coefficient (PTC) behavior based on the properties of a metallic heating element. The PTC behavior exhibited is much less (a smaller change in resistance in response to a change in temperature) than that shown by self-regulating heating cables.

Self-regulating cable: A type of heating cable that inversely varies its heat output in response to an increase or decrease in the ambient temperature.

Series resistance heating cable: A type of heating cable in which the ohmic heating of the conductor provides the heat. The wattage output depends on the total circuit length and on the voltage applied.

Startup temperature: The lowest temperature at which a heat tracing cable is energized.

Vessel: A container such as a barrel, drum, or tank used for holding fluids or other substances.

This module is intended to present thorough resources for task training. The following reference works are suggested for further study. These are optional materials for continuing education rather than for task training.

IEEE Standard 515, IEEE Standard for the Testing, Design, Installation, and Maintenance of Electrical Resistance Heat Tracing for Industrial Application. Piscataway, NJ: IEEE, 1997.

Industrial Heat Tracing Product Selection and Design Guide. Menlo Park, CA: Raychem HTS.

National Electrical Code® Handbook, Latest Edition. Quincy, MA: National Fire Protection Association.

Self-Regulating Heating Cable Systems. Menlo Park, CA: Raychem HTS.

NCCER makes every effort to keep these textbooks up-to-date and free of technical errors. We appreciate your help in this process. If you have an idea for improving this textbook, or if you find an error, a typographical mistake, or an inaccuracy in NCCER's Contren® textbooks, please write us, using this form or a photocopy. Be sure to include the exact module number, page number, a detailed description, and the correction, if applicable. Your input will be brought to the attention of the Technical Review Committee. Thank you for your assistance.

Instructors – If you found that additional materials were necessary in order to teach this module effectively, please let us know so that we may include them in the Equipment/Materials list in the Annotated Instructor's Guide.

Write: Product Development and Revision
National Center for Construction Education and Research
3600 NW 43rd St., Bldg. G, Gainesville, FL 32606

Fax: 352-334-0932

E-mail: curriculum@nccer.org

Craft

Module Name

Copyright Date

Module Number

Page Number(s)

Description

(Optional) Correction

(Optional) Your Name and Address

Motor Operation and Maintenance

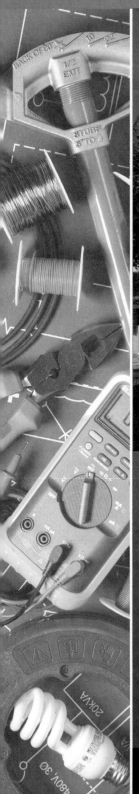

Université de Montréal and McGill University - Polar Revolution

Team Montréal is starting a Polar revolution to make solar energy popular in cold northern climates. A unique feature of the home is the use of artificial intelligence for temperature control and energy use. The "house" will search the internet for weather forecasts to predict the amount of energy it will be able to produce in the days to come balanced against how much it will need for its occupants. The system will recommend energy use choices to meet upcoming demands. The system controls heating, cooling, lighting, shading, and ventilation, all with one interface.

26410-08

26410-08
Motor Operation and Maintenance

Topics to be presented in this module include:

1.0.0   Introduction 10.2
2.0.0   Squirrel Cage Motors 10.4
3.0.0   Motor Maintenance 10.7
4.0.0   Motor Bearing Maintenance .10.10
5.0.0   Motor Insulation Testing 10.16
6.0.0   Receiving and Storing Motors 10.22
7.0.0   Troubleshooting Motors 10.23
8.0.0   Motor Installation and Commissioning Guidelines 10.25

Overview

Motors require a comprehensive maintenance program in order to provide uninterrupted service. Unfortunately, it is far too common to see motors put into operation and never touched again until the motor fails to operate. Premature motor failures can be attributed to a number of problems, including lack of cleaning, lack of lubrication, and excessive loads.

Always use the proper tools and techniques when performing maintenance and testing on motors. If a motor does not start, don't assume the problem is in the motor. Many operational motors have been changed out only to find that the supply voltage is absent. If the motor control circuitry proves to be operational but the motor still won't run, follow the manufacturer's procedures for testing the motor before replacing it.

Objectives

When you have completed this module, you will be able to do the following:

1. Recognize the factors related to motor reliability and life span.
2. Measure motor winding insulation resistance and compensate for temperature.
3. Identify motors needing replacement.

Trade Terms

Breakdown
Generator
Insulation class
Leakage

Megger®
Soft foot
Totally enclosed motor

Required Trainee Materials

1. Pencil and paper
2. Copy of the latest edition of the *National Electrical Code®*
3. Appropriate personal protective equipment

Prerequisites

Before you begin this module, it is recommended that you successfully complete *Core Curriculum; Electrical Level One; Electrical Level Two; Electrical Level Three;* and *Electrical Level Four*, Modules 26401-08 through 26409-08.

This course map shows all of the modules in *Electrical Level Four*. The suggested training order begins at the bottom and proceeds up. Skill levels increase as you advance on the course map. The local Training Program Sponsor may adjust the training order.

26413-08 Introductory Skills for the Crewleader

26412-08 Special Locations

26411-08 Medium-Voltage Terminations/Splices

26410-08 Motor Operation and Maintenance

26409-08 Heat Tracing and Freeze Protection

26408-08 HVAC Controls

26407-08 Advanced Controls

26406-08 Specialty Transformers

26405-08 Fire Alarm Systems

26404-08 Basic Electronic Theory

26403-08 Standby and Emergency Systems

26402-08 Health Care Facilities

26401-08 Load Calculations – Feeders and Services

ELECTRICAL LEVEL THREE

ELECTRICAL LEVEL TWO

ELECTRICAL LEVEL ONE

CORE CURRICULUM: Introductory Craft Skills

ELECTRICAL LEVEL FOUR

410CMAP.EPS

1.0.0 ◆ INTRODUCTION

Three-phase, squirrel cage induction motors are the most common motors in heavy commercial and industrial use. The useful life of a squirrel cage induction motor depends largely upon the condition of its insulation, which should be suitable for the operating requirements, along with any required maintenance. Electric motors are sometimes called rotating electrical machinery, electric machinery, or just machines.

When a motor malfunctions, it is due to either mechanical or electrical causes. These failures can be traced to one or more of the following causes:

- *Worn or tight bearings* – Worn or tight bearings are usually caused by failing bearing lubrication or improper lubrication. If the bearings have failed or are worn to such an extent that they allow the rotor to drag on the stator, the rotor, stator, and stator windings can be irreparably damaged. When bearings begin to fail, the condition can be detected by periodically checking the motor for excessive heat, noise, or vibration at the bearings before damage occurs. Heat and vibration can be measured using appropriate testers at the motor bearing housing. Bearing misalignment within the motor can also cause excessive wear or tightness. This is usually caused by a condition called **soft foot.** This occurs when one or two of the mounting feet of a motor are spaced away from an uneven mounting surface and are then anchored down against the mounting surface. This can cause warpage of the motor frame and misalignment of the bearings. Proper mounting surfaces and shimming must be used to prevent soft foot when a motor is installed.
- *Vibration* – Motor vibration can be caused by a bent shaft, misalignment of shafts between a motor and a directly coupled load, or loose anchorage. Vibration may result in excessive bearing wear or failure. Dial indicators can be used to check for bent shafts or misalignment between the motor and a directly coupled load.
- *Moisture* – Motor insulation must be kept reasonably dry, although many applications make this practically impossible unless a **totally enclosed motor** is used. Infrequently used or stored motors must be kept dry to prevent premature insulation failure.
- *Dust and dirt* – Dust and dirt can restrict ventilation and increase a motor's operating temperature. Overheating can cause premature bearing and winding failure. Cleaning is usually accomplished by periodically blowing out ventilation passages or fins with compressed air. The compressed air must be dry and regulated to a safe pressure for use.
- *Insulation deterioration* – Assuming no other mechanical or electrical causes, the winding insulation of all motors will deteriorate over time and will eventually break down, causing arcing and/or burnout of the windings. A megohmmeter, commonly called a **Megger®**, is used to periodically check the insulation resistance of the windings over time to chart the progress of insulation deterioration. Megohmmeters are also used as a troubleshooting tool for motors. Some megohmmeters are also capable of performing other motor performance and evaluation tests as well.
- *Overloading* – Overloading a motor can cause overheating and eventual bearing failure or premature insulation failure of the windings. Overloading usually occurs because the driven load has been changed to accommodate a higher capacity, heavier capacity, or an increased duty cycle. When driven loads are altered for these reasons, existing motors must be evaluated to be certain that they can withstand the new load.
- *Single-phasing* – The loss of a single phase of a polyphase power source for a running motor can cause the failure of the remaining phase windings of the motor if the motor is not protected by the proper equipment. This can occur in older installations where failure of one voltage phase due to a partial primary power failure or a failed fuse does not cause removal of all voltage phases from the motor. These types of installations should be updated using proper motor controllers and/or loss-of-phase protection.

1.1.0 Usual Service Conditions

The National Electrical Manufacturers Association (NEMA) has defined standards for usual service conditions for motors. When operated within the limits of the following NEMA usual service conditions, standard motors will perform in accordance with their ratings:

- *Ambient or room temperature not over 40°C* – If the ambient temperature is over 40°C (104°F) due to inadequate airflow in the motor location, the motor service factor must be reduced or a larger motor used. The larger motor will be loaded below full capacity, so the temperature rise will be lower and overheating will be reduced.
- *Altitude does not exceed 3,300' (1,000 meters)* – Motors are typically designed to operate at altitudes below 3,300'. Motors having Class A or B insulation systems and lower temperature rises as defined by NEMA will operate satisfactorily at altitudes above 3,300' in locations where the

decrease in ambient temperature compensates for the increase in temperature rise. Motors having a service factor of 1.15 or higher will operate satisfactorily at unity service factor and an ambient temperature of 40°C at altitudes above 3,300' up to 9,000'.

- *Voltage variation of not more than 10% of name-plate voltage* – Operation outside these limits or a voltage imbalance can result in overheating or loss of torque and may require the use of a larger motor.
- *Frequency variation of not more than 5% of name-plate frequency* – Operation outside of these limits results in a substantial speed variation and causes both overheating and reduced torque.
- *Plugging, jogging, or multiple starts* – Unless rated for this use, standard motors should not be used.

1.2.0 Unusual Service Conditions

Motors are often exposed to damaging atmospheres such as excessive moisture, steam, salt air, abrasive or conductive dust, lint, chemical fumes, and combustible or explosive dust or gases. To protect such motors, special enclosures or encapsulated windings and special bearing protection are required. The location of supplementary enclosures must not interfere with the ventilation of the motor.

Motors exposed to damaging mechanical or electrical loading such as unbalanced voltage conditions, abnormal shock or vibration, torsional impact loads, plugging, jogging, multiple starts, or excessive thrust or overhang loads may require special mountings or protection designed for the specific installation.

1.3.0 Effects of Overloading and Single-Phasing

Many times a motor that was originally of adequate capacity for a given load is later found to be inadequate. The following are some reasons for this problem:

- More severe duty imposed on the machine tool, such as a different die job in a punch press
- Heavier material or material of different machining characteristics
- Changed machine operation

Connecting measuring instruments in the motor circuit may disclose the reason for motor overheating, failure to start the load, or other abnormal symptoms. It is frequently desirable to connect recording instruments in the motor circuit for other purposes, such as analyzing the output of the machine or to gauge processing operations.

Control circuits for many older induction motor installations were not provided with thermal overload relays or loss-of-phase protection, and single phasing of polyphase induction motors on such circuits frequently led to motor winding failure. This results from the blowing of one fuse while the motor is up to speed and under load. Under such conditions, the portion of the winding remaining in the circuit will attempt to carry the load until it fails due to overheating.

Increasing the load on the motor beyond its rated capacity increases the operating temperature, which shortens the life of the insulation. Momentary overloads usually do no damage; consequently, there is a tendency to use thermal overload protection in modern control systems for motors rated at 600V or less. Monitoring of the actual winding temperature is the ideal method of determining if a motor is overloaded. Other methods, such as motor protection relays and electrical trip units for medium-voltage and 600V motors, are used to mimic thermal measurement and protect a motor from protracted overload conditions. These devices calculate (model) winding temperature by monitoring current and time using a microcomputer inside the device.

1.4.0 Insulation Systems

An insulation system is an assembly of insulating materials in association with conductors and the supporting structural parts of a motor. Insulation systems are divided into classes according to the thermal endurance of the system for temperature rating purposes. Four classes of insulation systems are used in motors: A, B, F, and H. Do not confuse these insulation classes with motor designs, which are also designated by letter.

- *Class A* – Class A insulation systems have a suitable thermal endurance when operated at the limiting Class A temperature of 105°C. Typical materials used include cotton, paper, cellulose acetate films, enamel-coated wire, and similar organic materials impregnated with suitable substances.
- *Class B* – Class B insulation systems have suitable thermal endurance when operated at the limiting Class B temperature of 130°C. Typical materials include mica, glass fiber, and other materials (organic or inorganic), with compatible bonding substances having suitable thermal stability.
- *Class F* – Class F insulation systems have suitable thermal endurance when operated at the limiting Class F temperature of 155°C. Typical

materials include mica, glass fiber, and other materials (organic or inorganic), with compatible bonding substances having suitable thermal stability.

- *Class H* – Class H insulation systems have suitable thermal endurance when operated at the limiting Class H temperature of 180°C. Typical materials used include mica, glass fiber, silicone elastomer, and other materials (organic or inorganic), with compatible bonding substances (for example, silicone resins) having suitable thermal stability.

2.0.0 ◆ SQUIRREL CAGE MOTORS

The following sections briefly describe starting methods and common electrical failures for polyphase squirrel cage motors. *Figure 1* shows the basic elements of a squirrel cage motor.

2.1.0 Starting Configurations

Squirrel cage motors are usually designed for across-the-line starting, which means that they can be connected directly to the power source by means of a suitable contactor. For large motors and in some other types of motors, the starting currents are very high. Usually, the motor is built to withstand starting currents, but since these currents are often six times or more the rated-load current, there may be a large voltage drop in the power system. Some method of reducing the starting current must therefore be employed to limit the system voltage drop to a tolerable value.

Reduced-voltage starting significantly reduces the torque and acceleration characteristics of any motor. Therefore, it is necessary to have the motor unloaded or nearly so when the motor is started. Reduction of starting voltage is the primary method of reducing motor starting current. Reduction of the starting current may be accomplished by the use of a variable frequency motor controller or by any one of the following reduced-voltage starting methods:

- *Primary resistor or reactance* – Primary resistor or reactance starting employs series reactance or resistance to reduce the current on the first step and then after a preset time interval, the motor is connected directly across the line. It can be used with any standard motor.
- *Autotransformer* – Autotransformer starting employs autotransformers to reduce the voltage and current on the first step and then after a preset time interval, the motor is connected directly across the line. It can be used with any standard motor.

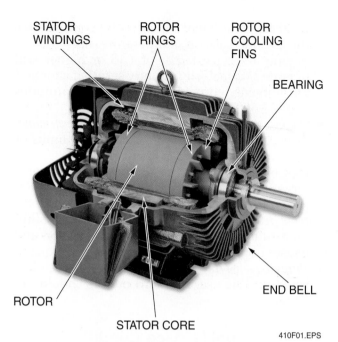

Figure 1 ◆ Squirrel cage motor construction.

410F01.EPS

- *Wye-delta* – Wye-delta starting impresses the voltage across the Y connection to reduce the current on the first step and then after a preset time interval, the motor is connected in delta, permitting full-load operation. The motor must have a winding capable of wye-delta connection.
- *Part-winding* – Part-winding starting employs motors with two separate winding circuits. Upon starting, only one winding circuit is engaged, and current is reduced. After a preset time interval, the full winding of the motor is put directly across the line. It must have a motor with two separate winding circuits. To avoid possible overheating and subsequent damage to the winding, the time between the connection of the first and second windings is limited to a four-second maximum.

All of these starting methods require special starters designed for the particular method and controlled between the start and run functions by an adjustable timer or current relay.

2.2.0 Typical Squirrel Cage Motor Winding Failures

Severe electrical, mechanical, or environmental operating conditions can damage the stator winding of a three-phase squirrel cage motor. Such winding failures and their causes are shown in *Figure 2*. Motors that have suffered severe core damage because of overheating or mechanical damage from the rotor are usually discarded. This is because severe core damage generally results in

Single-phased undamaged winding (caused by a missing voltage phase in a wye-connected motor)

Single-phased damaged winding (caused by a missing voltage phase in a delta-connected motor)

Phase-to-phase short (caused by contamination, abrasion, vibration, or voltage surge)

Turn-to-turn short (caused by contamination, abrasion, vibration, or voltage surge)

Shorted coil (caused by contamination, abrasion, vibration, or voltage surge)

Winding grounded at edge of slot (caused by contamination, abrasion, vibration, or voltage surge)

410F02A.EPS

Figure 2 ◆ Stator winding failures and causes (2 sheets).

Winding grounded in slot (caused by contamination, abrasion, vibration, or voltage surge)

Shorted connection (caused by contamination, abrasion, vibration, or voltage surge)

Unbalanced voltage damage (severe for one phase, slight for another phase, and caused by unbalanced power source loads or poor connections)

Overload damage to all windings
Note: Under-voltage and over-voltage (exceeding NEMA standards) will result in the same type of insulation deterioration.

Locked rotor overload damage to all windings (may also be caused by excessive starts or reversals)

Voltage surge damage (caused by lightning, capacitor discharges, voltage kickbacks caused by de-energizing large inductive loads, or voltage disturbances caused by solid-state devices)

410F02B.EPS

Figure 2 ◆ Continued

abnormal core losses or loss of efficiency if the motor is rewound. In an emergency, some motors with severe damage can be rewound after some or most of the damaged core laminations are restacked. However, this will result in a shortened operating life and these motors should be replaced as soon as possible. Depending on the type of facility and the number and size of the motors used within the facility, motors below some facility-established horsepower rating may not be salvaged when they fail, nor included in periodic predictive testing programs. However, the cause of failure for any motor should be investigated and corrected to prevent rapid failures of any replacement motors.

3.0.0 ◆ MOTOR MAINTENANCE

Motor failures account for a high percentage of industrial and heavy commercial electrical repair work. The care given to an electric motor while it is being stored or operated affects the life and usefulness of the motor. A motor that receives good maintenance will outlast a poorly treated motor many times over. Maintaining a motor in good operating condition requires periodic preventive maintenance actions along with periodic predictive testing. Predictive testing is required to determine if any faults exist or if faults are likely to occur in the near future. Once a motor is determined to be in the process of failing, it should be removed and replaced. If a motor has been replaced before its core has been damaged, it can usually be sent to a motor rebuild facility for salvage as a spare.

The frequency and thoroughness of the maintenance and testing depends on such factors as:

- Number of hours and days the motor operates
- Importance of the motor
- Nature of service
- Environmental conditions

Most heavy-duty commercial/industrial squirrel cage motors are totally enclosed (TE) motors. TE motors are available in the following configurations:

- *TEFC motor* – A TEFC motor is a fan-cooled (FC) motor cooled by a fan attached to the rear shaft of the motor. The fan is covered by a shroud that directs the fan air through cooling passages and/or fins external to the motor. TEFC motors are the most commonly used type of TE motor. They are also available in versions rated for severe service duty in corrosive environments, for washdown duty in food service, and for explosionproof duty in hazardous environments (*Figure 3*).

410F03.EPS

Figure 3 ◆ TEFC explosionproof motor.

- *TEBC motor* – A TEBC motor is a blower-cooled (BC) motor cooled by a separately powered blower attached to the motor. TEBC motors are typically used in heavy-duty variable speed motor applications.
- *TEAO motor* – A TEAO motor is a finned air-over (AO) motor that powers and is cooled by the slipstream from a large fan.
- *TENV motor* – A TENV motor is a non-ventilated (NV) motor used in smaller-horsepower variable speed applications.

WARNING!

TENV motors run at very hot temperatures. Do not touch these motors during or immediately after operation.

NOTE

Variable speed motors are specified in the U.S. as inverter-grade motors that can withstand heat and surges up to 2,400V.

3.1.0 Tools for Maintenance and Troubleshooting

In addition to typical equipment such as voltmeters and ammeters, motor maintenance equipment should also include an insulation resistance tester such as a multi-voltage megohmmeter. *Table 1* provides a list of practical tools and test equipment for electrical maintenance of motors and other electrical apparatus. Some of these tools have been covered in previous modules.

Table 1   Tools and Test Equipment for Electrical Maintenance

Tools or Equipment	Application
Multimeters, voltmeters, ohmmeters, clamp-on ammeters, wattmeters, clamp-on power factor meter	Measure circuit voltage, resistance, current, and power. Useful for circuit tracing and troubleshooting.
Potential and current transformers, meter shunts	Increase range of test instruments to permit the reading of high-voltage and high-current circuits.
Motor rotation and phase tester	Determines the proper three-phase connections for the desired rotation direction of a motor prior to connecting the motor. Also checks the phasing of the power source so that the motor can be connected to the desired phases.
Tachometer	Checks rotating machinery speeds.
Recording meters	Provide a permanent record of voltage, current, power, temperature, etc., on charts for analytic study.
Insulation resistance tester, thermometer, psychrometer	Test and monitor insulation resistance; use a thermometer and psychrometer for temperature and humidity correction.
Vibration testors and transistorized stethoscope	Detects faulty rotating machinery bearings and leaky valves.
Milli or microhmmeter	Precise measurement of coil resistance.
Dial indicator with mounting devices	Used to check for bent motor shafts and performing shaft alignment for directly coupled motor loads.

Transistorized Stethoscopes

A transistorized stethoscope is equipped with a transistor-type amplifier and is used to determine the condition of motor bearings. A little practice in interpreting its usage may be required, but in general, it is relatively simple to use. When the stethoscope is applied to a motor bearing housing, a smooth purring sound should be heard if the bearing is operating normally. If a thumping, grinding, or growling sound is detected, it could be an indication of a failing bearing.

3.2.0 Basic Care and Maintenance

Every motor in operation or storage should be observed regularly as recommended by the motor manufacturer to identify exposure to any adverse condition for which it is not rated. Operating motors should also be checked for excessive dust, debris, or lint on or about the motor. Ensure that airflow is not blocked by objects or room conditions and that the cooling fins are not dust-coated. Low-pressure compressed air may be used to blow motor housing ventilation paths or cooling fins clean.

WARNING!

Before performing any motor maintenance procedures, other than external visual inspection or noise monitoring, always lock out and tag the equipment according to approved procedures. If guards covering rotating components must be removed to gain access to motor bearings, remove the guards in accordance with approved facility procedures and use great care near the rotating components. When using compressed air, exercise caution by wearing appropriate personal protective equipment and using an airflow tip with air pressure limited to 30 psi or less. Excess air pressure can result in injury.

Ventilation and Temperature Control

Adequate ventilation and temperature control are essential for proper motor operation. Inspect the fan and auxiliary cooling system as part of the regular maintenance program. Many larger motors include resistance temperature detectors (RTDs) embedded in the windings and bearing housings. This provides for continuous monitoring of the winding and bearing temperatures to alert maintenance personnel to abnormal conditions.

Inspect the motor starter and verify that the motor reaches the proper speed each time it is started. Check other motor parts and accessories such as belts, gears, couplings, chains, and sprockets for excessive wear or misalignment and periodically lubricate the motor as recommended by the manufacturer (discussed in detail later in this module). Be alert to any unusual noise, which may be caused by metal-to-metal contact, such as bad bearings. Note any abnormal odor, which might indicate scorched insulation varnish.

Use an infrared thermometer (*Figure 4*) to check the bearing housings for evidence of excess heat and the motor itself for excess temperature rise as listed on the motor nameplate. Compare to previously logged data. Use a vibration tester (*Figure 5*) to determine if excess vibration is occurring as compared to previously logged data. A stethoscope may be used to detect unusual bearing noise during rundown when the motor is powered down.

Vibration testers indicate the vibration velocity in inches per second (in/sec), millimeters/second (mm/sec), or centimeters/second (cm/sec), depending on the model selected. Other models are also available that display only the total peak value of the velocity amplitude in either in/sec or mm/sec. A printer or a data logger is used with some models to capture data for further analysis.

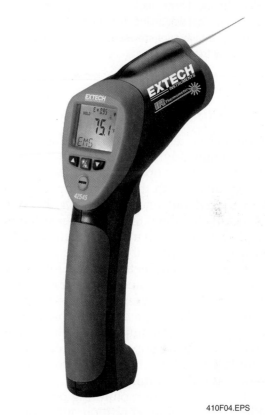

410F04.EPS

Figure 4 ◆ Infrared thermometer.

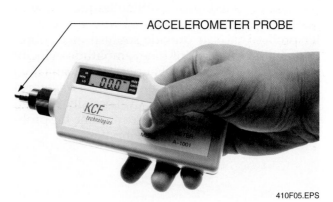

ACCELEROMETER PROBE

410F05.EPS

Figure 5 ◆ Vibration tester.

410F06.EPS

Figure 6 ◆ Online motor analysis instrument.

3.3.0 Periodic Predictive Testing

Periodic predictive testing can be conducted on two levels. Online (dynamic) tests can be conducted with the motor operating. Offline (static) tests are conducted with the motor stopped and the power locked out and tagged. Of the two types of testing, offline testing for insulation resistance and polarization index is the most common and will be described in detail in a later section. Online motor analysis testing is a relatively new technology that evaluates a number of motor parameters on a non-interference basis while a motor is running. This type of testing uses computerized portable test equipment such as that shown in *Figure 6*. It is normally done in facilities with many large motors and may be conducted more frequently than offline testing. This test equipment can be used for monitoring power circuit quality, overall motor condition, load conditions, and performance efficiency. It can provide data collection and trending information to aid in periodic maintenance evaluation in addition to the data collected during offline testing. Some of these instruments are equipped with accelerometer probes that can provide sophisticated vibration data in the form of time/amplitude velocity waveforms and/or Fast-Fourier Transform (FFT) velocity spectrum waveforms. The waveform data is normally analyzed by specially trained personnel and/or software to determine the probable causes of any abnormal vibration.

4.0.0 ◆ MOTOR BEARING MAINTENANCE

AC motors account for a large percentage of industrial and heavy commercial maintenance and repair, with many motor failures caused by faulty bearings. Consequently, most industrial facilities place a great amount of emphasis on proper care of motor bearings. Motor reliability is improved when a carefully planned lubrication schedule is followed.

If an AC motor failure does occur, the first step is to find out why the motor failed. There are various causes of motor failures, including excessive load, binding or misalignment of motor drives, wet or dirty environments, and bearing failures. Motors equipped with sealed bearings are much less prone to bearing failure.

4.1.0 Frequency of Lubrication

The frequency of motor lubrication depends not only on the type of bearing, but also on the motor application and its service environment. Small- to medium-size motors equipped with ball bearings (except sealed bearings) should be greased every three to six years if the motor duty is normal. For larger motors or for severe applications (high temperature, wet or dirty locations, or corrosive atmospheres), lubrication may be required more often. In most cases, the lubrication schedule for a motor is specified in the manufacturer's operation and service literature.

Figure 7 shows a form that can be used to record annual equipment lubrication. The form can be

YEAR _____ EQUIPMENT LUBRICATION SCHEDULE

Equipment Location and/or Number	Lubricant	Jan	Feb	Mar	Apr	May	Jun	Jul	Aug	Sept	Oct	Nov	Dec

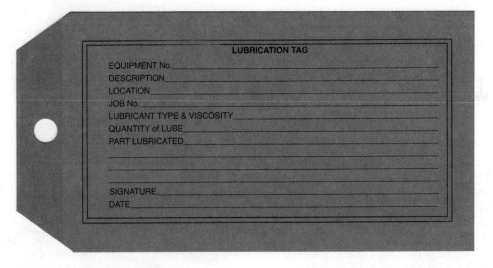

LUBRICATION TAG

EQUIPMENT No._____
DESCRIPTION_____
LOCATION_____
JOB No._____
LUBRICANT TYPE & VISCOSITY_____
QUANTITY of LUBE_____
PART LUBRICATED_____

SIGNATURE_____
DATE_____

410F07.EPS

Figure 7 ◆ Typical equipment lubrication schedule and tag.

developed based on the equipment manufacturer's recommendations and user experience. Normally, motors and their driven equipment are both included in the schedule. The specified type and quantity of lubricant can also be included. Multiple annual sheets can be set up for a series of years to accommodate equipment that requires lubrication at intervals greater than one year. Also shown is a typical lubrication tag that is replaced each time a motor is lubricated. The tag is used to indicate that the scheduled lubrication has occurred. Large facilities may use specialized software programs to schedule periodic maintenance tasks for various types of equipment. Typical tasks include lubrication, vibration testing, insulation testing, temperature measurements, and so on. These programs issue daily, weekly, monthly, and annual task reminders as necessary for each piece of equipment.

4.2.0 Lubrication Procedure

Before lubricating a ball bearing motor, clean the bearing housing, grease gun, and fittings. Exercise care to keep out dirt and debris during lubrication. If multiple lubricants are allowed in accordance with the motor manufacturer's operation and service literature, use only one of them when periodically relubricating a motor. Do not mix lubricants.

To lubricate a motor, remove the relief plug from the bottom of the bearing housing (*Figure 8*). This prevents excessive pressure from building up inside the bearing housing during lubrication. If possible, run the motor and add grease to the grease fitting until it begins to flow from the relief hole. Some manufacturers specify that only a certain amount of lubricant be injected into the

bearing housing. A grease meter can be used in conjunction with a grease gun to precisely measure the amount of grease injected (*Figure 9*).

After grease is injected, run the motor for approximately 5 to 10 minutes to expel any excess grease, then reinstall the relief plug and clean the bearing housing. Avoid over-lubrication. When excessive grease is forced into a bearing, churning of the grease may occur, resulting in high bearing temperatures and eventual bearing failure.

4.3.0 Checking Bearings

 WARNING!

In some cases, it may be necessary to remove guards over rotating motor shafts in order to accomplish bearing checks. The motor must be locked out and tagged before removing the guards, and then powered up to perform the checks. Always follow the facility safety procedures when performing checks with guards removed and the equipment operating. Stay clear of any rotating shaft components when performing these checks.

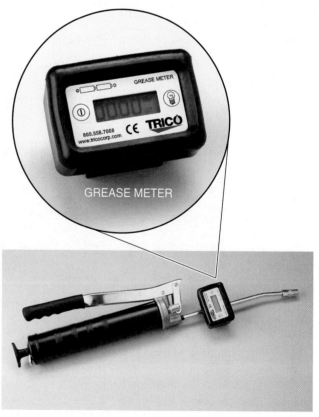

Figure 9 ◆ Grease gun and meter.

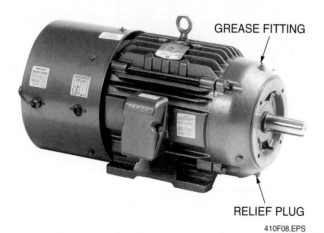

410F08.EPS

Figure 8 ◆ Grease fitting and relief plug for a TEFC motor.

Three simple methods are commonly used to check bearings during motor operation. These are listening for unusual noises, temperature measurements, and vibration testing. If the bearing housing is unusually hot, or if a growling or grinding sound is being emitted from the area of the bearings, one of the bearings is probably nearing failure. Special stethoscopes can be used to listen to the bearings while the motor is running or while it runs down after being stopped. A thunk sound that occurs just before rotation stops is usually indicative of failure.

As an absolute minimum, bearing temperature and/or vibration testing should be done in conjunction with the equipment lubrication schedule both prior to and after lubricating the equipment. Enter the temperature and/or vibration data in a log and compare it to previously logged data. Note any increase in temperature and/or vibration. In some cases, it may be necessary to decouple the load to see if a defective load is causing the problem.

NOTE

Keep in mind that some bearings may safely operate in a higher temperature range than other bearings, even in ranges exceeding 85°C. Check the manufacturer's specifications for the motor under test.

Motors should also be checked for endplay, which is the backward and forward movement in the shaft. Ball bearing motors typically will have $\frac{1}{32}$" to $\frac{1}{16}$" of end movement.

WARNING!

For bent shaft or shaft alignment checks, the motor must be locked out and tagged before any guards are removed and any checks are performed.

Other inspections may include periodically checking for misalignment or bent shafts, and for excessive belt tension. A magnetic base dial indicator can be used to check for a bent shaft on an uncoupled motor or driven load. The indicator is connected to either the motor or the load, or to a nearby solid ferrous metal surface with the indicator feeler in contact with the end of the shaft (*Figure 10*). Rotate the shaft back and forth only between the edges of any keyway in the shaft.

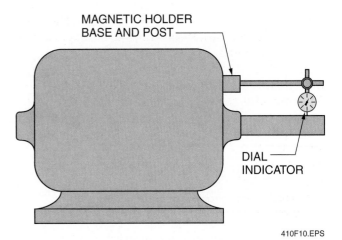

410F10.EPS

Figure 10 ◆ Checking for a bent shaft on a motor.

There should be no shaft misalignment (runout) detected for either the motor or the load; otherwise replace the motor or the load, whichever has the bent shaft.

Parallel or angular misalignment (runout) of an uncoupled motor shaft to a directly driven load shaft (*Figure 11*) can be checked using a magnetic V-base holder or clamp base holder with a dial indicator at the rear of one of the shafts (usually the motor shaft). Manually rotate the first shaft while the indicator feeler is in contact with the rear of the other shaft (usually the load shaft) (*Figure 12*). If the coupling halves are on the shafts and the outside diameter of the couplings is accurately machined in relation to the center of the coupling, they can be used in the measurements; otherwise, the one on the shaft where the dial indicator will be used must be repositioned or removed. When positioning the dial indicator

410F11.EPS

Figure 11 ◆ Typical direct-drive configuration without coupling.

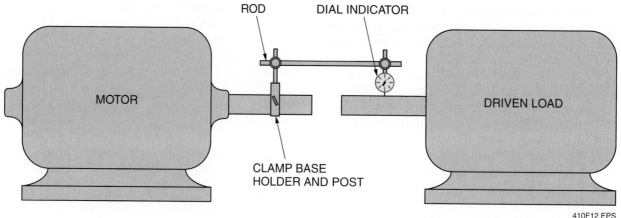

ROD   DIAL INDICATOR

MOTOR

DRIVEN LOAD

CLAMP BASE
HOLDER AND POST

410F12.EPS

Figure 12 ◆ Checking shaft alignment between a motor and directly-driven load.

feeler against that shaft, make sure that it is first positioned at the rear of the shaft and is depressed enough to cause the indicator needle to rotate 180 degrees. Then rotate the dial markings so that the zero reading is at the needle position.

To make the alignment measurements, rotate the shaft with the holder base back and forth so that the indicator feeler does not enter any keyway on the other shaft. Note the locations of the minimum and maximum readings on the other shaft. Move the indicator to the front of that shaft and repeat the measurements. If the minimum and maximum readings are the same, there is a parallel shaft runout and the load or motor will have to be moved or shimmed equally at both ends to retain the parallelism and correct the runout. If the readings are different, there is angular shaft runout and the motor or load will have to be moved or shimmed unequally at each end to remove the angular runout. Then, any resulting parallel runout will have to be corrected.

By trial and error, loosen the anchoring, as necessary, and move or shim the motor/load in the direction necessary to equalize the maximum and minimum readings at both locations to correct any angular runout. Repeat the alignment measurements for any parallel runout until the difference

between the minimum and maximum readings is essentially zero around the shaft at both positions of the dial indicator. Then anchor the motor/load and recheck for any resulting runout. If it is possible to rotate the other shaft, reverse the indicator setup and repeat the measurements as a double check. Before anchoring the motor/load, make sure that a soft foot condition was not caused by moving and/or shimming the motor/load.

Figure 13 shows a typical dial indicator with various holders along with stainless-steel slotted shims. Stainless-steel shims are preferred in order to prevent rust deterioration over time. The shim assortment includes shims of various thicknesses. The thickness is marked on the shim so that records can be kept of the total thickness used under each foot of the equipment. This enables easier reinstallation of the same equipment if it must be removed.

NOTE

Other methods of shaft alignment are sometimes used. These include laser alignment or mechanically accurate adapter frames (bell housings) for some pumps/motors.

MAGNETIC BASE HOLDER   CLAMP HOLDER WITH POST   TYPICAL DIAL INDICATOR

SLOTTED SHIMS

410F13.EPS

Figure 13 ◆ Typical dial indicator with various holders and stainless-steel slotted shims.

5.0.0 ◆ MOTOR INSULATION TESTING

An insulation resistance test indicates the condition of motor insulation and may indicate insulation breakdown or the presence of moisture, dirt, or other conductive material. Insulation resistance testing is high-energy testing and must be done in compliance with employer regulations, government regulations, and the tester manufacturer's instruction manuals.

5.1.0 Insulation Resistance Tests

Insulation resistance tests give an indication of the condition of insulation, particularly with regard to contamination by moisture and dirt. The actual value of the resistance will vary depending on the equipment type, size, voltage rating, and winding configuration.

The principal purpose of periodic insulation resistance tests is in comparing, or trending, relative values of insulation resistance of the same apparatus taken under similar conditions through its service life. Such tests usually reveal deterioration of insulation and may indicate approaching failure.

Measuring insulation resistance is straightforward. Identify any two points between which there is insulation, and make a connection across them using a megohmmeter. The measured value represents the equivalent resistance of all the insulation that exists between the two points and any component resistance present between the two points. Three-phase motors may have from three to twelve motor leads present in the motor terminal box, depending on the winding configuration and starting method. Most common are the nine leads present in the dual-voltage squirrel cage motor.

Before a motor is placed in service, test the insulation resistance of each winding segment to ground, and to all other winding segments connected to ground. A look at the connection diagrams on the motor nameplate will indicate point numbers for each test connection. The positive lead is connected to frame and ground along with all other motor leads except for the winding segment under test. The negative lead is connected to one end of the segment under test with the other end(s) isolated or insulated. The insulation resistance is tested from one segment to all other segments and to ground. The test is repeated until each winding segment has been individually tested.

This test is easily performed on a motor in storage or before connection to the electric supply. Some facilities require testing only the resistance of all windings to earth. The motor feeder conductors may be included in the test to avoid de-terminating and then re-terminating the motor connections. In this case, the test is often performed from the motor starter or local disconnect switch with one test lead connected to ground, and the other to any T lead to the motor.

WARNING!

When accessing or working within starter compartments, you may be exposed to shock hazards within the compartment and to arc flash hazards from the motor control center (MCC) bus and surrounding compartments. Follow company policies and use all appropriate personal protective equipment.

If the insulation resistance test indicates potential problems, isolate the feeders from the motor windings and test between individual phases or winding segments to identify the problem. Typical insulation resistance values used by one contractor for new equipment installations are shown in the *Appendix*. Other contractors may use different values. Specific values may also be obtained from the equipment manufacturer. In all cases where comparison is used over time for trending data, the same test connections must be used.

WARNING!

When using an insulation tester, the motor frame must be directly connected to the building/earth ground. Only qualified individuals may use this equipment.

Megohmmeters (*Figure 14*) are available in several varieties. Some are powered by a hand-cranked generator, others are battery powered, and some have rechargeable batteries or a line-powered internal power supply. Some supply test voltages as low as 50V, but the most common is 500V, with some going as high as 10,000V or more. In all cases, the test voltage is DC.

Before making resistance tests, you must first understand the concept of absorption current. The insulation between two connection points can be thought of as a dielectric, thus forming a

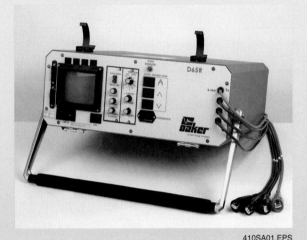
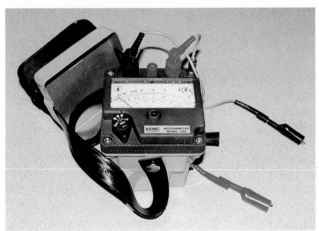

410F14.EPS

Figure 14 ◆ Typical megohmmeter.

capacitance. A phenomenon known as dielectric absorption occurs whereby the dielectric soaks up electrons and then releases them when the potential is removed. This is in addition to the current that charges the capacitance, and it occurs much more slowly. It is dependent on the nature of the dielectric. Two items where this is of concern are capacitors and wound equipment. Such current is referred to as absorption current, or I_A

in *Figure 15A*. Absorption current consists of two components: polarization current and electron drift. Polarization current is due to the reorientation of molecules in the insulation impregnating materials. It can take several minutes for this to occur and the polarization current to drop to zero. The second component is current caused by the gradual drift of electrons through most organic materials. This drift current is minor and occurs until the electrons and ions become trapped by the mica surfaces commonly found in insulation systems. For clean and dry motors, the insulation resistance between about 30 seconds and a few minutes is primarily determined by the absorption current.

This phenomenon may be demonstrated by taking a large capacitor, charging it to its rated voltage, and allowing it to remain at that voltage for some length of time. Then discharge the capacitor quickly and completely until a voltmeter placed across it reads zero. Remove the voltmeter and again allow the capacitor to sit for some length of time with its leads open circuited. Take another voltmeter reading. Any voltage present is due to the dielectric absorption phenomenon. Some types of capacitors exhibit this to a greater degree than other types. The larger the capacitor, the more apparent the effect.

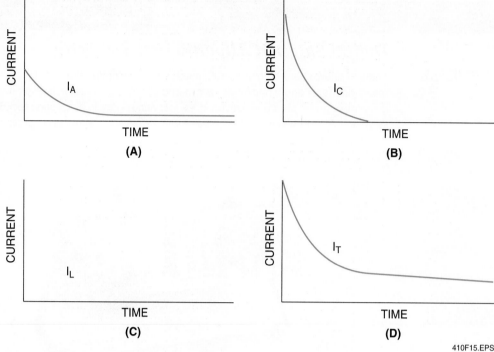

Figure 15 ◆ Current curves related to insulation testing.

The current required to charge whatever capacitance is present is known as charging current, or I_C in *Figure 15B*. Like the dielectric absorption current, it decays to zero, but more quickly. It is this current that typically determines how long it takes to make an accurate megohm measurement. When the reading appears to stabilize, it means that the charging current has decayed to a point where it is negligible with respect to the leakage current.

The current that flows through the insulation is the leakage current, or I_L in *Figure 15C*. The voltage across the insulation divided by the leakage current through it equals the insulation resistance. Thus, to accurately measure insulation resistance, the dielectric absorption current and the charging current must be allowed to decay to the point where they are truly negligible with respect to the leakage current.

The total current that flows is the sum of the three components just mentioned, or I_T in *Figure 15D*. It decays from an initial maximum and approaches a constant value, which is the leakage current alone. The megohm reading is dependent on the voltage across the insulation and the total current. It increases from an initial minimum and approaches a constant value, which is the true measured insulation resistance.

Motor manufacturers, installers, users, and repairers find megohm testing very useful in

determining the quality of the insulation in the motor. A single insulation resistance measurement, along with experience and guidelines as to what reading to expect, can indicate whether a motor is fit for use. Information of real value is obtained when a measurement is made when a motor is new and again at least every year while it is in service. Because the temperature of the windings has a great effect on the insulation resistance reading, the motor industry has standardized a winding temperature of 40°C as a reference where specific recommendations can be made about the minimum acceptable megohm value for each type of motor. When measurements are made at a winding temperature other than 40°C, they can be corrected to reflect what the reading would have been at 40°C by using a temperature coefficient, Kt, obtained from a chart such as the one shown in *Figure 16*.

Figure 17 shows the actual readings obtained on a motor over several years (the dashed line) and the same readings corrected to a standard 40°C reference (the solid line). The temperature-corrected readings show a gradual decline characteristic of the motor's normal aging process until, in 1997, a sharp drop occurred. That drop signaled a problem with the insulation that would probably lead to an insulation failure, which would be costly in terms of repair and down time. Therefore, at the earliest practical opportunity, this motor was sent

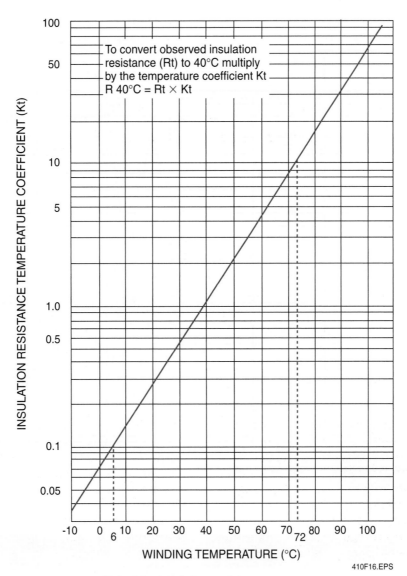

Figure 16 ◆ Approximate temperature coefficient for insulation resistance.

410F16.EPS

out for rewinding. The rewinding restored the high megohm readings, and the steady, gradual decline resumed. Today, this motor would probably just be replaced instead of rewound. *Figure 18* shows just the temperature-corrected readings that are normally kept on motors.

Another factor that affects insulation resistance readings is moisture. Motors may have excessive moisture in and around the insulation from either high humidity or having been submerged. This is a temporary effect, and by letting the motor dry out naturally or accelerating the process by baking, the readings will increase. It is important not to start the motor until a satisfactory reading is obtained, because to do so might cause an insulation failure.

The tests used to evaluate the effect of absorption current on resistance are the dielectric absorption ratio (DAR), typically applied on smaller machines, and the polarization index tests.

The condition of insulation may be inferred by evaluating the effect of the dielectric absorption current on resistance over time. The DAR, also called a 60/30 test, is determined by the following formula:

$$DAR = \frac{\text{resistance after 60 seconds}}{\text{resistance after 30 seconds}}$$

The resistance readings will gradually rise due to a decrease in dielectric absorption.

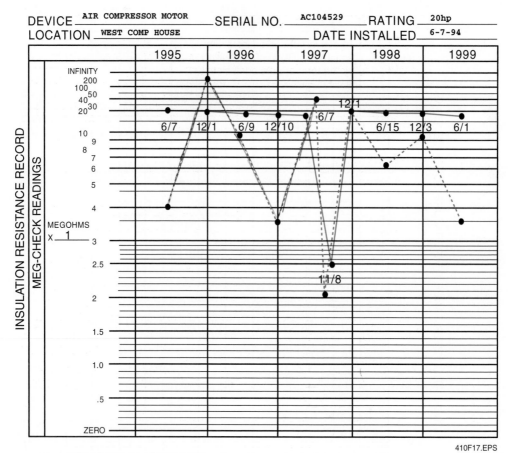

DEVICE ___AIR COMPRESSOR MOTOR___ SERIAL NO. ___AC104529___ RATING ___20hp___

LOCATION ___WEST COMP HOUSE___ DATE INSTALLED ___6-7-94___

Figure 17 ◆ Chart of insulation resistance readings.

410F17.EPS

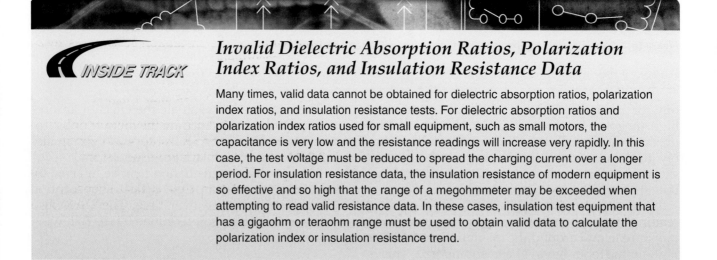

Invalid Dielectric Absorption Ratios, Polarization Index Ratios, and Insulation Resistance Data

Many times, valid data cannot be obtained for dielectric absorption ratios, polarization index ratios, and insulation resistance tests. For dielectric absorption ratios and polarization index ratios used for small equipment, such as small motors, the capacitance is very low and the resistance readings will increase very rapidly. In this case, the test voltage must be reduced to spread the charging current over a longer period. For insulation resistance data, the insulation resistance of modern equipment is so effective and so high that the range of a megohmmeter may be exceeded when attempting to read valid resistance data. In these cases, insulation test equipment that has a gigaohm or teraohm range must be used to obtain valid data to calculate the polarization index or insulation resistance trend.

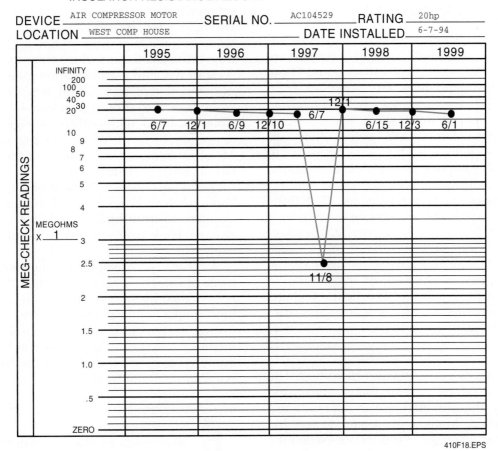

INSULATION RESISTANCE RECORD

DEVICE __AIR COMPRESSOR MOTOR__ SERIAL NO. __AC104529__ RATING __20hp__

LOCATION __WEST COMP HOUSE__ DATE INSTALLED __6-7-94__

410F18.EPS

Figure 18 ◆ Temperature compensation record for a motor.

5.2.0 Determining the Polarization Index

Knowing the polarization index of a motor or generator can be useful in appraising the fitness of the machine for service. It is commonly used for motors over 200hp. The index is calculated from measurements of the winding insulation resistance and is intended to evaluate the condition of the insulation.

Before measuring the insulation resistance, remove all external connections to the motor and completely discharge the windings to the grounded frame. Apply either 500VDC or 1,000VDC between the winding and ground using a direct-indicating, power-driven megohmmeter. (Use the higher value for motors rated 500V and above.) Apply the voltage for ten minutes and keep it constant for the duration of the test.

The polarization index (PI), sometimes called the polarization index ratio (PIR), is calculated as the ratio of the ten-minute value to the one-minute value of the insulation resistances measured consecutively:

$$PIR = \frac{\text{resistance after 10 minutes}}{\text{resistance after 1 minute}}$$

The recommended minimum value of polarization index for AC and DC motors and generators is 2.0. Machines having a PI under 2.0 are considered suspect. However, a PI over 4.0 may indicate dry, brittle insulation that is also suspect. The polarization index is useful in evaluating windings for the following:

- Buildup of dirt or moisture
- Gradual deterioration of the insulation (by comparing results of tests made earlier on the same motor)
- Fitness for overpotential tests
- Suitability for operation

5.3.0 Insulation Testing Considerations

The procedures for performing insulation resistance testing and evaluation of the results are found in the test equipment instructions. These procedures and guidelines are based on *IEEE Standard 42-2000, Recommended Practice for Testing Insulation Resistance of Rotating Machinery.*

When testing components before they are installed in a system, it is sometimes necessary to verify that the components meet their insulation resistance specifications. Wire and cable, connectors, switches, transformers, resistors, capacitors, and other components are specified to have a certain minimum insulation resistance. The megohmmeter is the proper instrument to use. Remember, the part may have a limitation on the voltage that may be applied, or the insulation resistance may be specified at a particular voltage. Consult the manufacturer for any voltage restrictions. Pay careful attention to these restrictions to avoid improper comparisons and/or damage to the part.

Use caution when selecting the test voltage. The voltage rating of the part or motor across the points where the measurement is made should not be exceeded. Many users prefer to use the highest available voltage that does not exceed the rating. In other cases, the customary test voltage is 500V, regardless of a higher rating, because higher voltage can make it harder to obtain proper data.

6.0.0 ◆ RECEIVING AND STORING MOTORS

When an electric motor is received on a job site, always follow the manufacturer's recommendations for unloading, uncrating, and installing the motor. Failure to follow these recommendations can cause injury to personnel and possible damage to the motor. Once the motor has been uncrated, check for damage that might have occurred during shipment. Check the motor shaft to verify that it turns freely. Clean the motor of any debris, dust, moisture, or any foreign matter that might have accumulated during shipment. Make sure that it is dry before putting it into storage or service.

 WARNING!
Never start a wet or damp motor.

Motors may be put into storage when the project on which they are to be used is not complete or where they will be used as backups. The first consideration when storing motors for any length of time is the location in which they are to be stored. A clean, dry, and warm location (one that does not undergo severe changes in temperature over a 24-hour period) should be selected whenever possible. Ambient temperature changes cause condensation to form on and in the stored motor. Moisture in motor insulation can cause motors to fail on startup; therefore, guarding against moisture is vital when storing motors of any type. Some motors have internal heaters to protect against moisture buildup while the motor is in storage. These heaters must be monitored while in use.

A means for transporting the motor from the place of storage is also important. Motors should not be lifted by their rotating shafts. Doing so can damage the alignment of the rotor in relationship to the stator. Eyebolts on motor frames are intended only for lifting the motor. Some factory-mounted accessories or covers also have eyebolts, but they must not be used to lift the motor.

 WARNING!
Never handle a motor by its shaft. Motor shaft keyways have sharp edges that can cause severe cuts.

The following is a list of recommendations for proper motor storage:

- Keep the motors clean and dry.
- Supply supplemental heating in the storage area, if necessary, or connect and monitor the internal heaters if so equipped.
- Store motors in an orderly fashion (i.e., grouped by horsepower, etc.).
- Rotate motor shafts periodically as specified by the motor manufacturer. Maintain a shaft rotation log (*Figure 19*) for each motor.
- Motors should not need additional lubrication while in storage.
- Protect shafts and keyways during storage and while transporting motors from one location to another.
- Test motor winding resistance upon receiving and periodically while in storage. Maintain a winding resistance test log for each motor.

Eyebolt Lifting Capacity

The eyebolt lifting capacity is based on a lifting alignment that corresponds to the eyebolt centerline. The eyebolt capacity lessens as deviation from this alignment increases.

Stored Motor Maintenance

To prevent shaft bowing, large horizontal motors in storage should have their rotors rotated 180 degrees at least once a month.

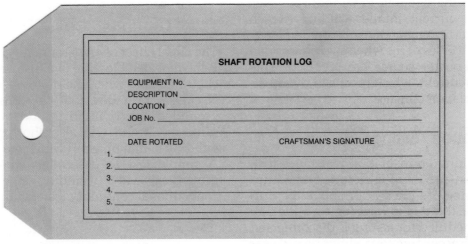

SHAFT ROTATION LOG

EQUIPMENT No. _____

DESCRIPTION _____

LOCATION _____

JOB No. _____

DATE ROTATED   CRAFTSMAN'S SIGNATURE

1. _____
2. _____
3. _____
4. _____
5. _____

410F19.EPS

Figure 19 ◆ Typical shaft rotation log tag.

7.0.0 ◆ TROUBLESHOOTING MOTORS

To detect defects in electric motors, the windings are typically tested for ground faults, opens, shorts, and reverses. The methods used in performing these tests may depend on the type of motor being tested. If a defect is found, the motor must be removed and replaced.

WARNING!

Before testing or troubleshooting a motor beyond a simple visual/auditory inspection, disconnect the power and follow the proper lockout/tagout procedures.

- *Grounded winding* – A winding becomes grounded when it makes electrical contact with the metal frame of the motor. Causes may include end plate bolts contacting conductors or windings, winding wires pressing against the laminations at the corners of slots damaged during rewinding, or conductive parts of the centrifugal switch contacting the end plate.
- *Open circuits* – Loose connections or broken wires can cause an open circuit in an electric motor.
- *Shorts* – A short occurs when two or more windings of separate phases contact each other at a point, creating a path other than the one intended. This condition may develop in a relatively new winding if the winding is wound too tight at the factory and excessive force was used to place the wires in position. Shorts may also be caused by excessive heat caused by overloads, which can degrade the insulation. Shorted turns are caused by loose or broken turns within a winding, which contact each other in a manner that effectively causes the turns within the winding to be reduced. This causes high current in the winding, which results in overheating and insulation damage.
- *Reverses* – Reverses occur when two phases of a three-phase power source are reversed. This causes high current, which results in overheating and insulation damage.

7.1.0 Insulation Testing

Before performing an insulation resistance test, first check to ensure that the circuits and equipment to be tested are rated at or above the output voltage of the megohmmeter, and that all equipment scheduled for testing is disconnected and locked out according to facility or employer procedures.

Check the megohmmeter and other testing equipment for proper operation in accordance with the manufacturer's instructions. Before conducting tests, temporarily ground all the leads of the equipment under test. Then conduct the insulation tests as prescribed in the manufacturer's instructions. For winding resistance tests, the test voltage is applied for one minute or until the reading is essentially stable. After testing, discharge the windings by placing a ground on the windings and maintaining it for four times the duration of the test.

Record the temperature and humidity immediately after the test. Also, note the condition of the environment and the running or idle time of the equipment.

After the readings have been recorded, correct them for the winding temperature using a temperature chart (supplied with most megohmmeters). For enclosed motors without internal winding temperature monitoring, the winding temperature must be estimated. The winding temperature of motors that have not been operating for eight hours or more can be considered to be the same temperature as the outside motor skin temperature as measured with a thermometer (infrared or other). For motors that have been operating, the winding temperature can be estimated by adding the skin temperature to the ambient air temperature.

For new equipment, the corrected insulation resistance readings should be similar to those shown in the *Appendix* or as specified by the manufacturer. Elaborate motor insulation tests such as time resistance and step voltage tests use various megohmmeter reading ratios and indexes (i.e., absorption ratio and polarization index) to determine the effects of temperature and humidity on insulation **breakdown** (*Figure 20*).

7.2.0 Grounded Coils

The usual effect of one grounded coil in a winding is the repeated blowing of a fuse or tripping of the circuit breaker when the motor disconnect switch is closed, provided that the machine frame is grounded. Two or more grounds will produce the same result and will also short out part of the winding in that phase in which the grounds occur. A simple test to determine whether or not a ground exists in the winding can be made using a conventional continuity tester. Before testing with

410F20.EPS

Figure 20 ◆ Bench testing motor insulation.

The One Megohm Rule of Thumb

The one megohm rule is an old minimum standard that is still used. It was originally developed for equipment manufactured 30 to 40 years ago. Newer equipment uses winding insulation that has much higher insulation resistance than earlier insulating material.

The old rule states that for equipment rated at 1,000V or less, the minimal acceptable rating is one megohm or more. For equipment rated over 1,000V, the minimal acceptable reading is one megohm minimum plus one megohm per kilovolt rating. The rule only applies to winding temperatures corrected to 40°C (104°F). To correct for the ambient temperature, the reading must be adjusted up for winding temperatures above 40°C and down for winding temperatures below 40°C. The reasoning is that equipment insulation resistance decreases at higher temperatures and increases at lower temperatures. The readings are adjusted as follows:

- For every 10° above 40°C, double the megohm reading.
- For every 10° below 40°C, divide the megohm reading in half.

For example, a 2,300V, 600hp motor has a megohm reading of 2.3 megohms. The motor has been operating and the motor winding temperature is 60°C. The minimum acceptable reading is 3.3 megohms at 40°C. The temperature-corrected reading is as follows:

$$60°C - 40°C = 20°C$$
$$20°C \div 10°C = 2$$

Therefore, the insulation resistance reading must be doubled twice, as follows:

$$2.3 \times 2 = 4.6$$
$$4.6 \times 2 = 9 \text{ (rounded off)}$$

The temperature-corrected reading is 9 megohms. It is acceptable because it is greater than the 3.3 megohm minimum requirement.

---

this instrument, first make certain that the motor disconnect switch is open, locked out, and tagged. Place one test lead on the frame of the motor and the other in turn on each of the line wires on the load side of the motor controller. If there is a grounded coil at any point in the winding, the lamp of the continuity tester will light, or in the case of a meter, the reading will show low or zero resistance.

7.3.0 Water-Damaged Motors

Motors that have been water damaged can sometimes be dried out and salvaged if rust is minimal. The NEMA guidelines for handling water-damaged equipment recommend that salvage be done only by trained personnel in consultation with the motor manufacturer. This normally requires that these motors be sent to a manufacturer-authorized outside motor repair facility where they are disassembled, cleaned, and dried using ovens, steam heat, forced air, electric heat, or infrared heat in accordance with the motor manufacturer's instructions. When dried they are subjected to extensive insulation testing and bearing replacement or lubrication, and then reassembled.

8.0.0 ◆ MOTOR INSTALLATION AND COMMISSIONING GUIDELINES

Motor must be securely installed on a flat, rigid foundation or mounting surface to minimize vibration and maintain alignment between the motor and load shafts. Failure to provide a proper mounting surface may cause vibration, misalignment, and bearing damage.

Foundation caps and sole plates are designed to act as spacers for the equipment they support. If these devices are used, be sure that they are evenly supported by the foundation or mounting surface.

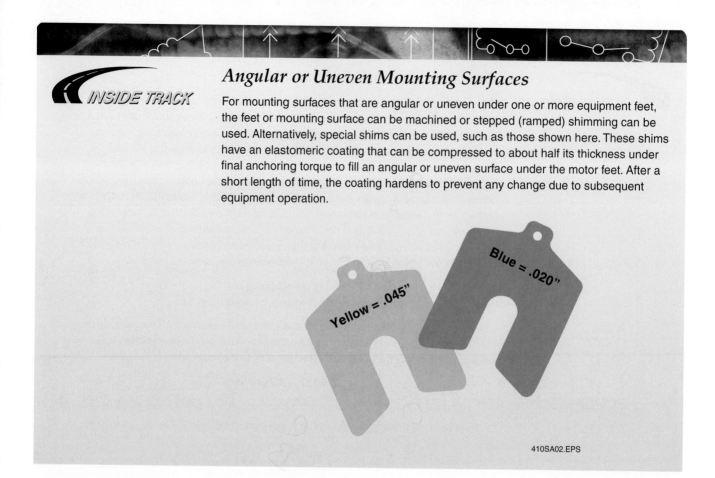
8.1.0 Alignment

Accurate alignment of the motor with the driven equipment is extremely important. Many different types of flexible couplings are available with a range of torque capabilities, bore sizes, and materials based on horsepower requirements. Two typical types that are used to reduce vibration between motors and their loads are shown in *Figure 21*. One is a jaw type that is sometimes called a Lovejoy® coupling. They are secured to the shafts with square keys and setscrews. The flexible insert between the coupling halves is available in various materials with varying degrees of flexibility depending on the application. They are also available with snap-in flexible inserts that can be externally removed and replaced without disturbing the coupling halves. The coupling halves are available as sintered iron (a magnetic alloy), cast iron, aluminum, or stainless steel. These types of joints are fail-safe because they will still work even if the insert fails. Installation of the coupling is easy and alignment is not as critical as with other kinds of couplings.

The other type of coupling shown in *Figure 21* has internal splines in the coupling halves and the flexible insert between them. The coupling halves are secured to the shafts with square keys and setscrews. The couplings are available with different materials for the coupling halves and the flexible insert between them depending on the application and torque requirements. Normally, alignment of these types of joints is critical, but depending on the flexibility of the insert material, some types can tolerate a certain amount of angular or parallel runout. Also note the balance compensator on the load shaft in this application that is used to reduce vibration caused by the load.

For direct-drive motors, use flexible couplings if possible. Consult the drive or equipment manufacturer for more information. Mechanical vibration and roughness during operation may indicate poor alignment. Use dial indicators to check alignment as discussed previously. Maintain the manufacturer's recommended spacing between coupling hubs.

For belt-drive motors, make sure the pulley ratio recommended by the manufacturer is not exceeded. Align the sheaves carefully to minimize belt wear and axial bearing loads. Belt tension should be sufficient to prevent belt slippage at rated speed and load. However, belt slippage may occur during starting.

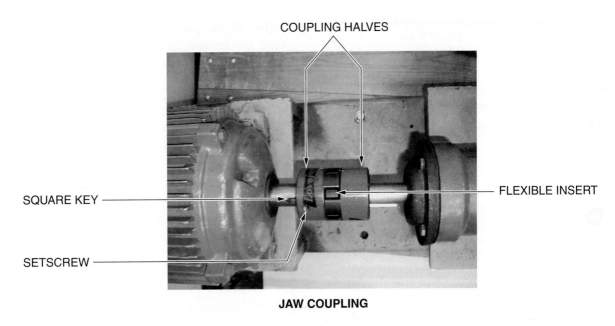

COUPLING HALVES

SQUARE KEY

SETSCREW

FLEXIBLE INSERT

JAW COUPLING

COUPLING HALVES

SQUARE KEY

SETSCREW

SPLINED FLEXIBLE INSERT

BALANCE COMPENSATOR

SPLINED COUPLING

410F21.EPS

Figure 21 ◆ Typical vibration-reducing couplers.

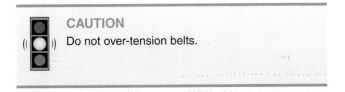

CAUTION

Do not over-tension belts.

8.2.0 Endplay Adjustment

Except for motors designed for vertical mounting that have end thrust bearings, the axial position of a horizontal motor frame with respect to its load is extremely important. The motor bearings are not designed for excessive external axial thrust loads. Improper adjustment will cause failure. To establish the electrical center (live center) of the motor shaft, loosen the bearings (if applicable), and operate the motor with no load. Mark the shaft position, in relation to the end bell, on the shaft. Shim/reposition and tighten the bearings as necessary. After coupling the motor to the load, make sure to check that the marked position is maintained when the motor is operating.

8.3.0 First-Time Startup

Before making any pre-start checks, make sure that all power to the motor and any accessories is locked out and tagged. Make sure the motor shaft will not cause mechanical rotation of the load shaft. Then perform the following checks:

- Make sure that the mechanical installation is secure.
- If the motor has been in storage or idle for some time, check the insulation integrity using an insulation resistance tester.

- Verify the desired direction of the motor and corresponding power source phases using motor rotation and phase rotation testers.
- Lubricate the motor in accordance with the manufacturer's instructions.
- Connect the wiring and inspect all electrical connections for proper termination, clearance, mechanical strength, and electrical continuity.
- Perform a final insulation resistance test, including the wiring to the motor.
- Remove any protective covering from the motor shaft.
- Replace all panels and covers that were removed during installation.
- Apply power and momentarily start (bump) the motor to verify correct rotation. If wrong, again lock out and tag the motor and switch two phases. Then bump the motor again to check.
- Start the motor and ensure that operation is smooth without excessive vibration or noise. If so, run the motor for one hour with no load connected.

- After one hour of operation, again lock out and tag the motor. Connect the load to the motor shaft as previously described. For direct-drive couplings, make sure that the coupling is aligned and is not binding.
- Verify that all coupling guards (*Figure 22*) and protective devices are installed and that the motor is properly ventilated.

8.4.0 Coupled Startup

The first coupled startup should be with no load applied to the driven equipment. Apply power and verify that the driven equipment is not transmitting excessive vibration back to the motor though the coupling or the foundation. Vibration should be at an acceptable level; if not, a balance compensator may be required on the load shaft. Run for approximately one hour with the driven equipment in an unloaded condition. The equipment can now be loaded and operated within specified limits. Do not exceed the nameplate ratings for amperes or for temperature rise for steady continuous loads.

410F22.EPS

Figure 22 ◆ Typical coupling guard over rotating components.

Installation Checklist

Some contractors or facilities may use an installation checklist similar to one shown here.

DESCRIPTION

EQUIP. #		
CIRCUIT #		
INITIAL MEGGER TEST @ _____ VOLTS	A to ↓ B to ↓ C to ↓	A to B A to C B to C
ELECT. COMPLETE	INITIAL DATE	
FINAL ALIGN COMPLETE	INITIAL DATE	
INSTRUMENTATION COMPLETE	INITIAL DATE	
FINAL MEGGER TEST INCLUDES FEED WIRES	A to ↓ B to ↓ C to ↓	A to B A to C B to C
ROTATION CHECK	INITIAL DATE	
COUPLING BELTS GUARD	INITIAL DATE	
TEST RUN	INITIAL DATE	
OPERATING AMPS	A B C	

NW GRAPHICS   FORM # EQ-18   6/91

ACCEPTED FOR SERVICE_____ DATE_____

ALIGNMENT READINGS

I.D.

O.D.

PIPING COMPLETE	INITIAL DATE
FLUSH	INITIAL DATE
LUBE	INITIAL DATE
LEVEL ALIGN GROUT	INITIAL DATE
FINAL ALIGNMENT	INITIAL DATE
MECHANICAL OK TO RUN	INITIAL DATE

ACCEPTED FOR SERVICE_____ DATE_____

410SA03.EPS

8.5.0 Doweling

After proper alignment is verified, dowel pins are inserted through the mounting feet of some motors into the foundation. This will maintain the correct motor position should motor removal be required. If doweling is desired, drill dowel holes through diagonally opposite motor feet and into the foundation or sole plate. Ream all holes. Install the proper size dowels.

1. Single-phasing means the loss of two phases in a running motor.
   a. True
   b. False

2. If the load connected to a motor is increased beyond the motor's rating, which of the following is most likely to occur?
   a. Contact between the rotor and stator
   b. Decreased operating temperature
   c. Increased operating temperature and decreased motor life
   d. Loss of bearing lubrication

3. An insulation system that, by experience or accepted test, can be shown to have a suitable thermal endurance when operated at a limiting temperature of 105°C is a Class ____ insulation system.
   a. A
   b. B
   c. F
   d. H

4. Typical materials used in a Class H insulation system include ____.
   a. cotton
   b. paper
   c. cellulose acetate films
   d. silicone elastomer

5. Typical materials used in a Class F insulation system include ____.
   a. cotton
   b. paper
   c. cellulose acetate films
   d. mica

6. An insulation system that, by experience or accepted test, can be shown to have a suitable thermal endurance when operated at a limiting temperature of 130°C is a Class ____ insulation system.
   a. A
   b. B
   c. F
   d. H

7. Motors with two separate winding circuits employ ____ starting.
   a. primary resistor
   b. autotransformer
   c. wye-delta
   d. part-winding

8. Which of the following may be eliminated as a concern in good motor maintenance?
   a. Number of hours and days the motor operates
   b. Manufacturer of the motor
   c. Environmental conditions
   d. Importance of the motor in the production scheme

9. Determine proper electrical connections before electrical connection of a motor by ____.
   a. checking with a rotation tachometer
   b. using a rotation and phase tester
   c. using a recording meter
   d. observing the rotation of the motor

10. Which of the following is usually an indication of a bad motor bearing?
    a. Hot bearing housing
    b. Low current draw
    c. Low pull-in torque
    d. Sparking at the brushes

11. When using compressed air to clean motors, which of the following precautions should be taken?
    a. Make sure the air is warmer than the ambient motor temperature.
    b. Use air pressure within safety standards.
    c. Avoid using a pulse nozzle.
    d. Use only a high-velocity nozzle.

12. A type of motor bearing with a very low rate of failure is a ____ bearing.
    a. radial
    b. pin
    c. sealed
    d. ball

13. The polarization index is the ratio of the resistance after _____ minute(s) to the resistance after _____ minute(s).

   a. 1; 5

   b. 1; 10

   c. 5; 1

   d. 10; 1

14. Which of the following statements is true concerning motors in storage?

   a. Shafts should not be moved or turned until the motor is put in use.

   b. Shafts should be turned according to manufacturer instructions.

   c. Shafts should be turned 180 degrees once a year to distribute bearing grease.

   d. Shafts should be turned once every three years to keep them from rusting.

15. The best attachment point for lifting motors is/are the _____.

   a. eyebolt

   b. shaft

   c. base

   d. end bells

Summary

This module has covered various motor maintenance techniques, including lubrication, testing,
storage and replacement. In determining
which motors are likely to fail, it is important to
remember that motor failures are generally
caused by poor maintenance as well as overloading, age, vibration, or contamination problems.

Notes

Trade Terms
Introduced in This Module

Breakdown: The failure of insulation to prevent the flow of current, sometimes evidenced by arcing. If the voltage is gradually raised, breakdown will begin suddenly at a certain voltage level. Current flow is not directly proportional to voltage. Once a breakdown current has flowed, especially for a period of time, the next gradual application of voltage will often show breakdown beginning at a lower voltage than initially.

Generator: (1) A rotating machine that is used to convert mechanical energy to electrical energy. (2) General apparatus, equipment, etc., that is used to convert or change energy from one form to another.

Insulation class: Category of insulation based on the thermal endurance of the insulation system used in a motor. The insulation system is chosen to ensure that the motor will perform at the rated horsepower and service factor load.

Leakage: AC or DC current flow through insulation and over its surfaces, and AC current flow through a capacitance. Current flow is directly proportional to voltage. The insulation and/or capacitance are thought of as a constant impedance unless breakdown occurs.

Megger®: A trade term used for a megohmeter. An instrument or meter capable of measuring resistances in excess of 200 megohms. It employs much higher test voltages than are used in ohmmeters, which measure up to 200 megohms.

Soft foot: When one or two of the mounting feet of a motor are spaced away from an uneven mounting surface and are then anchored down against the mounting surface. This condition can cause warpage of the motor frame and misalignment of the bearings.

Totally enclosed motor: A motor that is encased to prevent the free exchange of air between the inside and outside of the case.

Electrical Equipment and Wiring

Maximum Voltage Rating	Minimum DC Test Voltage	Normal DC Test Voltage for New Installations	Minimum Insulation Resistance at 20°C for New Installations
250V	250	500	25 megohms
600V	500	1,000	100 megohms
5,000V	2,500	2,500	1,000 megohms
8,000V	2,500	2,500	2,000 megohms
15,000V	2,500	5,000	5,000 megohms
25,000V	5,000	5,000	20,000 megohms

Transformer Windings

Transformer Winding Rating	Minimum DC Test Voltage	Normal DC Test Voltage for New Installations	*Minimum Insulation Resistance at 20°C for New Installations in Megohms	
			Oil-Filled	Dry-Type
0–600V	500	1,000	100	500
601–5,000V	1,000	2,500	1,000	5,000
>5,000–V	2,500	5,000	5,000	25,000

*Note: Refer to manufacturer's literature for specific acceptance criteria. Actual insulation resistance of transformer winding is dependent on transformer type, insulation, kVA rating, and winding temperature.

Motor Winding Insulation Resistance Testing

Motor Winding Rating	Minimum DC Test Voltage	Normal DC Test Voltage for New Installations	*Minimum Insulation Resistance at 20°C for New Installations
0–600V	500	1,000	15 megohms
601–5,000V	1,000	2,500	30 megohms
5,000–15,000V	2,500	5,000	50 megohms

*Note: Minimum insulation resistance for rotating equipment in megohms = equipment kV + 1

410A01.EPS

This module is intended to present thorough resources for task training. The following reference work is suggested for further study. This is optional material for continuing education rather than for task training.

National Electrical Code® Handbook, Latest Edition. Quincy, MA: National Fire Protection Association.

CONTREN® LEARNING SERIES – USER UPDATE

NCCER makes every effort to keep these textbooks up-to-date and free of technical errors. We appreciate your help in this process. If you have an idea for improving this textbook, or if you find an error, a typographical mistake, or an inaccuracy in NCCER's Contren® textbooks, please write us, using this form or a photocopy. Be sure to include the exact module number, page number, a detailed description, and the correction, if applicable. Your input will be brought to the attention of the Technical Review Committee. Thank you for your assistance.

Instructors – If you found that additional materials were necessary in order to teach this module effectively, please let us know so that we may include them in the Equipment/Materials list in the Annotated Instructor's Guide.

Write: Product Development and Revision
National Center for Construction Education and Research
3600 NW 43rd St., Bldg. G, Gainesville, FL 32606

Fax: 352-334-0932

E-mail: curriculum@nccer.org

Craft _____ Module Name _____

Copyright Date _____ Module Number _____ Page Number(s) _____

Description _____

_____

_____

_____

(Optional) Correction _____

_____

_____

(Optional) Your Name and Address _____

_____

_____

Medium-Voltage Terminations/Splices

Texas A&M University – Pieces Come Together for a Solar Home

The A&M "groHome" concept is based on interchangeable and interconnected "groWall" units, some of which will have all of a home's kitchen, bath, or entertainment utilities built into them. This makes all of the electricity and plumbing easy to get to, and gives the homeowner great flexibility.

26411-08

26411-08
Medium-Voltage Terminations/Splices

Topics to be presented in this module include:

1.0.0   Introduction 11.2
2.0.0   Medium-Voltage Power Cable 11.3
3.0.0   Splicing 11.7
4.0.0   Terminations 11.26
5.0.0   High-Potential (Hi-Pot) Testing 11.32

Overview

When terminating or splicing medium-voltage cables or conductors, only trained and qualified personnel should attempt to make these types of connections, and then only approved materials and tools may be used. Typical electrical insulating tape or putty covering a mechanical connection is not an acceptable method to use when terminating conductors carrying voltages exceeding 600 volts. Because of the high potential difference between phase conductors and ground, electrons can and will easily find breaks or weak points in unapproved terminations or splices in conductors carrying high voltage. This will cause phase-to-phase shorts or ground faults to occur.

Some cables are designed with several layers of different types of insulating materials, such as an outer jacket, metallic shield, semi-conductor layer, another shield, and finally the conductor. Termination or splicing kits specifically designed for each type of cable must be used when terminating or splicing these conductors in order to provide the same degree of insulating protection over the termination as is provided on the cable itself. In addition to these special termination and splicing kits, methods and procedures used to install these kits must follow rigid standards of cleanliness and workmanship. Contamination can be a big problem in improperly made connections.

Objectives

When you have completed this module, you will be able to do the following:

1. Select the proper materials and tools for medium-voltage terminations and splices.
2. Prepare medium-voltage cable for terminations and splices.
3. Complete cable assemblies using terminations and splices.
4. Inspect and test medium-voltage terminations and splices.

Trade Terms

Armor
Dielectric constant (K)
High-potential (hi-pot) test
Insulation shielding
Pothead
Ribbon tape shield
Shield
Shielding braid
Splice
Strand shielding
Stress

Required Trainee Materials

1. Pencil and paper
2. Appropriate personal protective equipment
3. Copy of the latest edition of the *National Electrical Code*®

Prerequisites

Before you begin this module, it is recommended that you successfully complete *Core Curriculum; Electrical Level One; Electrical Level Two; Electrical Level Three; Electrical Level Four,* Modules 26401-08 through 26410-08.

This course map shows all of the modules in *Electrical Level Four.* The suggested training order begins at the bottom and proceeds up. Skill levels increase as you advance on the course map. The local Training Program Sponsor may adjust the training order.

26413-08 Introductory Skills for the Crewleader

26412-08 Special Locations

26411-08 Medium-Voltage Terminations/Splices

26410-08 Motor Operation and Maintenance

26409-08 Heat Tracing and Freeze Protection

26408-08 HVAC Controls

26407-08 Advanced Controls

26406-08 Specialty Transformers

26405-08 Fire Alarm Systems

26404-08 Basic Electronic Theory

26403-08 Standby and Emergency Systems

26402-08 Health Care Facilities

26401-08 Load Calculations – Feeders and Services

ELECTRICAL LEVEL FOUR

ELECTRICAL LEVEL THREE

ELECTRICAL LEVEL TWO

ELECTRICAL LEVEL ONE

CORE CURRICULUM: Introductory Craft Skills

411CMAP.EPS

1.0.0 ◆ INTRODUCTION

This module provides an overview of splicing and termination methods commonly used in medium-voltage cable systems. Both taping systems and manufactured slip-on kits are used to make splices and terminations. This module discusses the materials and methods used to terminate, splice, and test shielded cable systems rated up to 35kV.

1.1.0 Straight Splices

Every splice must provide mechanical and electrical integrity at least equal to the original conductor insulation. Furthermore, all terminations and splices must be made under the best environmental conditions possible on the job site. The slightest amount of moisture, dirt, or other foreign material can render the splice useless.

Splices can be completed by tapering the ends of the cables and taping with several layers of high-dielectric insulating tape covered with several layers of environmental jacketing tape. An alternative method is to use a stress control void-filling material and heat-shrink insulating and environmental tubing to rebuild the cable to its original insulation thickness. With this method, tapering, also known as penciling, is not required.

A typical straight splice for a single, unshielded conductor (ranging from 601V to 5,000V) is shown in *Figure 1*. In general, the conductor jacket and insulation should be removed for one-half of the connector length plus ¼". Then, a protective wrap of electrical tape is applied to the jacket ends to protect the jacket and insulation while cleaning the conductor and securing the connector, as shown in *Figure 1(A)*.

After performing the preceding operations, remove the tape and also remove additional jacket and insulation in an amount equal to 25 times the thickness of the overall insulation, as shown in *Figure 1(B)*, extending in both directions from the first insulation cut. For example, if the thickness of the insulation is ¼" (0.25"), the outside edge of the splice will be:

$$25 \times 0.25" = 6.25" \text{ or } 6¼"$$

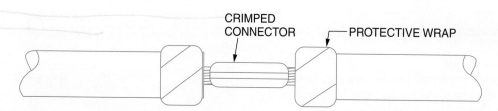

A. APPLY A PROTECTIVE WRAP OF TAPE TO THE JACKET ENDS TO PROTECT THE JACKET AND INSULATION WHILE CLEANING AND SECURING THE CONNECTOR.

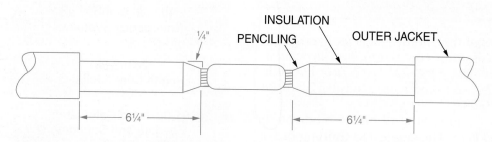

B. REMOVE ADDITIONAL JACKET AND INSULATION BY AN AMOUNT EQUAL TO 25 TIMES THE THICKNESS OF THE OVERALL INSULATION.

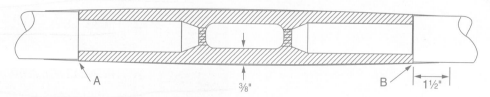

C. INSULATE AND TAPE THE SPLICE.

411F01.EPS

Figure 1 ◆ Making a basic splice.

WARNING!

Although any conductor splice or termination must be done with care, extreme caution must be exercised with higher voltages. In general, the higher the voltage, the more fine-tuning the splice or termination will require. Splicing is a skill that requires special training. Many localities and industries require certification of training and experience in order to splice, terminate, and test medium- and high-voltage cable systems.

Therefore, the outside edge of the splice will be 6 ¼" beyond the insulation cut. To continue, pencil the insulation with a sharp knife or insulation cutter, taking care not to nick the conductor.

Refer to *Figure 1(C)*. Starting from A or B, wrap the splice area with high-voltage insulating rubber splicing tape, stretching the rubber tape about 25% and, if using lined tape, unwinding the interliner as you wrap. The finished splice should be built up to one and one-half times the factory insulation, if rubber (two times, if thermoplastic), as measured from the top of the connector. For example, if the rubber insulation is ¼", the finished splice would be ⅜" thick when measured from the top of the connector.

To finish off the splice and ensure complete protection, cover the entire splice with four layers of vinyl electrical tape, extending about 1½" beyond the edge of the splicing compound.

2.0.0 ◆ MEDIUM-VOLTAGE POWER CABLE

There is a tremendous variety of medium-voltage power cables in use today. Five of the most common are shown in *Figure 2* and listed below:

- Ribbon tape shield
- Drain wire shielded
- Cable UniShield®
- Concentric neutral (CN)
- Jacketed concentric neutral (JCN)

INSIDE TRACK

Making Terminations and Splices

Correct cable preparation, proper installation of all components, and good workmanship are all required to make splices and terminations like those shown here. Manufacturers have done much in the last few years to develop products that make splicing and terminating cables easier, but the expertise, skills, and care of the installer are still necessary to make dependable splices and terminations.

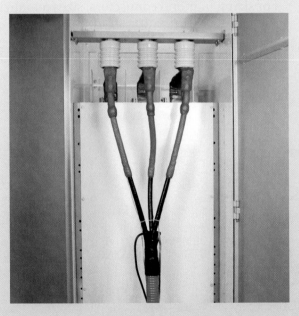

411SA01.EPS

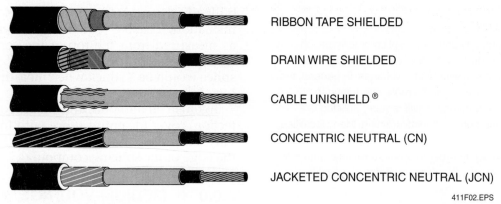

RIBBON TAPE SHIELDED

DRAIN WIRE SHIELDED

CABLE UNISHIELD ®

CONCENTRIC NEUTRAL (CN)

JACKETED CONCENTRIC NEUTRAL (JCN)

411F02.EPS

Figure 2 ◆ Basic types of high-voltage cable.

Beyond cable shield types, two common configurations are used:

- Single-conductor configurations, consisting of one conductor per cable or three cables for a three-phase system
- Three-conductor configurations, consisting of three cables sharing a common jacket

Despite these visible differences, all medium-voltage power cables are essentially the same, consisting of the following:

- Conductor
- Strand shielding
- Insulation
- Insulation shielding system (semi-conductive and metallic)
- Jacket

Each component is vital to an optimal power cable performance and must be understood in order to make a dependable splice or termination. See *Figure 3*.

2.1.0 Medium-Voltage Cable Components

Refer to *NEC Article 328*. Conductors used with dielectric cables are available in four basic configurations (*Figure 4*). Descriptions of these configurations follow.

- *Concentric stranding (Class B)* – Not commonly used in modern shielded power cables because of the penetration of the extruded strand shielding between the conductor strands, making the strand shield difficult to remove during field cable preparation.
- *Compressed stranding* – Compressed to 97% of the concentric conductor diameters. This compression of the conductor strands blocks the penetration of an extruded strand shield, thereby making it easily removable in the field. For sizing lugs and connectors, sizes remain the same as with concentric stranding.

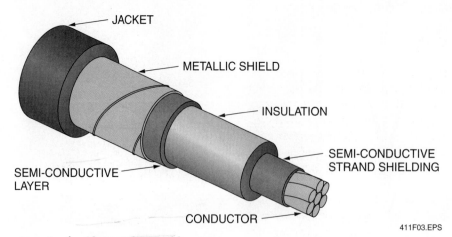

JACKET

METALLIC SHIELD

INSULATION

SEMI-CONDUCTIVE
STRAND SHIELDING

SEMI-CONDUCTIVE
LAYER

CONDUCTOR

411F03.EPS

Figure 3 ◆ Basic components of medium-voltage cable.

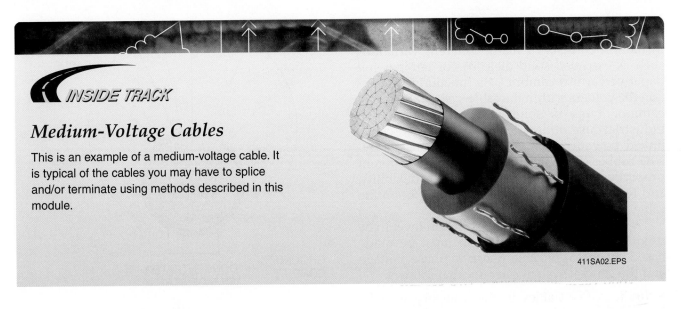

Medium-Voltage Cables

This is an example of a medium-voltage cable. It is typical of the cables you may have to splice and/or terminate using methods described in this module.

411SA02.EPS

CONCENTRIC STRANDING   COMPRESSED STRANDING   COMPACT STRANDING   SOLID

411F04.EPS

Figure 4 ◆ Basic cable configurations.

- *Compact stranding* – Compacted to 90% of the concentric conductor diameters. Although this conductor has full ampacity ratings, the general rule for sizing is to consider it one conductor size smaller than concentric or compressed cable. This reduced conductor size results in a proportional reduction of all cable layers, a consideration when sizing for molded rubber devices.
- *Solid wire* – This conductor is not commonly used in industrial shielded power cables.

2.2.0 Strand Shielding

Strand shielding is the semi-conductive layer between the conductor and insulation that compensates for air voids between the materials.

Air is a poor insulator, having a nominal dielectric strength of only 76V/mil, whereas most cable insulations have dielectric strengths over 700V/mil. Without strand shielding, an electrical potential exists that will overstress these air voids.

As air breaks down or ionizes, it produces corona discharges. This forms ozone, which chemically deteriorates the cable insulation. The semi-conductive strand shielding eliminates this potential by simply shorting out the air.

Modern cables are generally constructed with an extruded strand shield. See *Figure 5*.

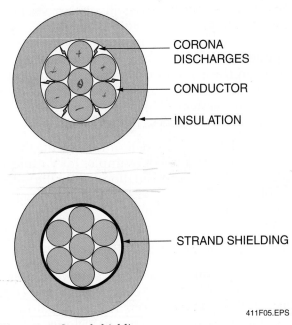

CORONA DISCHARGES

CONDUCTOR

INSULATION

STRAND SHIELDING

411F05.EPS

Figure 5 ◆ Strand shielding.

2.3.0 Insulation

Insulation consists of many different variations such as extruded solid dielectric or laminar (oil paper or varnish cambric). Its function is to contain the voltage within the cable system. The most common types of solid dielectric insulation in industrial use today are polyethylene, cross-linked polyethylene (XLP), and ethylene propylene rubber (EPR). Each is preferred for different properties such as superior strength, flexibility, and temperature resistance, depending upon the cable characteristics required.

The selection of the cable insulation level to be used in a particular installation must be made based on the applicable phase-to-phase voltage and the general system category that follows:

- *100% level* – Cables in this category may be applied where the system is provided with relay protection such that ground faults will be cleared as rapidly as possible, but in any case within one minute. Although these cables are applicable to the great majority of cable installations on grounded systems, they may also be used on other systems for which the application of cable is acceptable, provided the above clearing requirement is met in completely de-energizing the faulted section.
- *133% level* – This insulation level corresponds to that formerly designated for ungrounded systems. Cables in this category may be applied in situations where the clearing time requirements of the 100% level category cannot be met, and yet there is adequate assurance that the faulted section will be de-energized in a period not exceeding one hour. In addition, they may be used when additional insulation strength over the 100% level category is desirable.
- *173% level* – Cables in this category should be applied on systems where the time required to de-energize a grounded section is indefinite. Their use is also recommended for resonant grounded systems. Consult the manufacturer for insulation thickness requirements.

The insulation level is determined by the thickness of the insulation. For example, 15kV cable at 100% has 175 mils of insulation, while the same type of cable at 133% has 220 mils. This variance holds true up to 1,000 kcmil.

2.4.0 Insulation Shield System

The outer shielding is comprised of two conductive components: a semi-conductive (semi-con) layer under a metallic layer. The principal functions of a shield system (*Figure 6*) include:

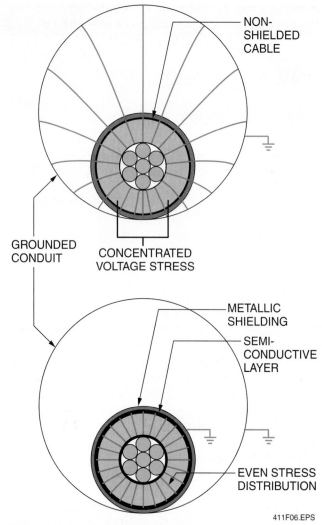

Figure 6 ◆ Insulation shield system.

- Confining the dielectric field within the cable
- Obtaining a symmetrical radial distribution of voltage stress within the dielectric
- Protecting the cable from induced potentials
- Limiting radio interference
- Reducing the hazard of shock
- Providing a ground path for leakage and fault currents

NOTE

The shield must be grounded for the cable to perform these functions.

The semi-conductive component is available either as a tape or as an extruded layer. Some cables have an additional layer painted on between the semi-con and the cable insulation. Its function is similar to strand shielding—to eliminate the problem of air voids between the insulation and the metallic component (in this case, the metallic shielding). In effect, it shorts out the air that underlies the metallic shield, preventing corona and its resultant ozone damage.

The metallic shield is the current-carrying component that allows the insulation shield system to perform the functions mentioned earlier. The various cable types differ most in this layer. Therefore, most cables are named after their metallic shield (for example, tape shielded, drain wire shielded, or UniShield®). The shield type (cable identification) thus becomes important information to know when selecting devices for splices and terminations.

2.5.0 Jacket

The jacket (*Figure 7*) is the tough outer covering that provides mechanical protection, as well as a moisture barrier. Often, the jacket serves as both an outer covering and the semi-conductive component of the insulation shield system, combining two cable layers. However, because the jacket is external, it cannot serve as the semi-conductive component of the complete system. Typical materials used for cable jackets are PVC, neoprene, and lead. Frequently, industrial three-conductor cables have additional protection in the form of an armor layer.

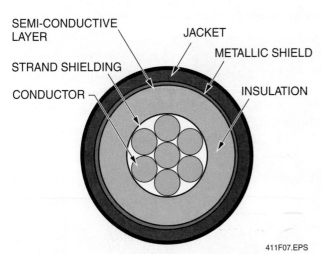

Figure 7 ◆ Jacket and insulation.

SEMI-CONDUCTIVE LAYER
JACKET
METALLIC SHIELD
STRAND SHIELDING
CONDUCTOR
INSULATION

411F07.EPS

3.0.0 ◆ SPLICING

Whenever possible, splicing should be avoided. However, splicing is often an economic necessity. There can be many reasons for making splices, including:

- Insufficient supplied length of cable due to reel limitations or cumbersome cable length
- Cable failures
- Cable damaged after installation
- Insufficient room for cable training radius
- Excessive twisting of cable would otherwise result
- A tap into an existing cable (tee or wye splices)

For damaged cable and taps, the option is to either splice the cable or replace the entire length. In many cases, the economy of modern splicing products makes splicing an optimal choice. See *Figure 8*.

Whatever the reason to splice, good practice dictates that splices have the same rating as the cable. In this way, the splice does not derate the cable and become the weak link in the system.

3.1.0 Splicing Steps

The five basic tasks in building a splice are:

- Preparing the surface
- Joining the conductors with connector(s)
- Reinsulating
- Reshielding
- Rejacketing

3.1.1 Preparing the Surface

Careful surface preparation is essential to making effective cable splices. It is necessary to begin with good cable ends. For this reason, it is a common practice to cut off a portion of the cable after pulling to ensure an undamaged end. Always use sharp, high-quality tools. When the various layers are removed, cuts should extend only partially through the layer. For example, when removing cable insulation, be careful not to cut completely

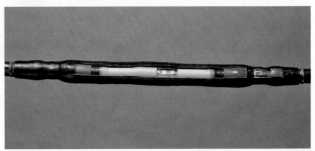

Figure 8 ◆ Cutaway view of a splice.

411F08.EPS

Extruded Semi-Conductors

Some extruded semi-conductors will peel easier if they are heated slightly. A heat-shrink blowgun works well.

Cable Preparation Tools

It is essential to use the proper tool when preparing conductors for termination or splicing. This photo shows a cable preparation tool used for removing the outer cable jacket. It can also be used for stripping the insulation.

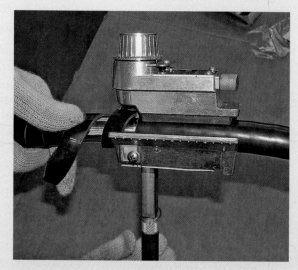 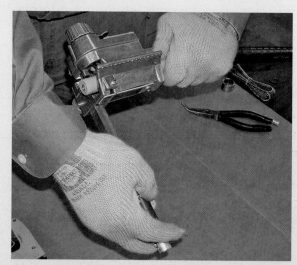

411SA03.EPS

This semi-conductor scoring tool has a blade that is adjustable for both cutting depth and angle. Once set to the proper depth, the blade is set perpendicular to the cable and a ring cut is made around the cable at the desired point. After the ring cut is made, the blade is angled so that each rotation around the cable results in a spiral cut.

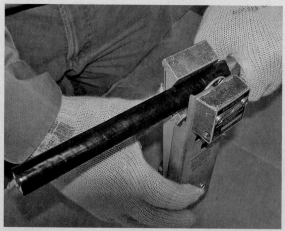

411SA04.EPS

through and damage the conductor strands. Specialized tools are available to aid in the removal of the various cable layers. Another good technique for removing polyethylene cable insulation is to use a nylon string as the cutting tool.

When penciling is required (not normally necessary for molded rubber devices), a full, smooth taper is necessary to eliminate the possibility of air voids.

It is necessary to completely remove the semiconductive layer(s) and the resulting residue. Two methods are commonly used to remove the residue:

- *Abrasives* – Research has proven that a 120-grit abrasive cloth is fine enough to protect the medium-voltage interface and yet coarse enough to remove semi-con residue without loading up the abrasive cloth. This abrasive must have nonconductive grit. DO NOT use emery cloth or any other abrasive that contains conductive particles because these could embed themselves into the cable insulation. When using an insulation diameter-dependent device (for example, molded rubber devices), take care not to abrade the insulation below the minimums specified for the device.
- *Solvents* – Use the solvent recommended by the cable manufacturer and by the splicing kit manufacturer. Any solvent that leaves a residue should be avoided. DO NOT use excessive amounts of solvent because this can saturate the semi-con layers and render them nonconductive. Always wipe the cable from the conductor back toward the jacket.

CAUTION

Always read the material safety data sheet (MSDS) before using any solvent. Avoid solvents that present health hazards, especially with long-term exposure. Choose a nontoxic solvent whenever possible.

3.1.2 Joining the Conductors with Connectors

After the cables are completely prepared, the rebuilding process begins. If a premolded splice or heat-shrink splice is being installed, the appropriate splice components must be slid onto the cable(s) before the connection is made. The first step is reconstructing the conductor with a suitable connector. Connector selection is based on the conductor material (copper or aluminum). A

suitable connector for medium-voltage cable splices is a compression or crimp type. Do not use mechanical-type connectors, such as split-bolt connectors.

Aluminum conductors should be connected with an aluminum-bodied connector pre-loaded with contact aid (anti-oxide paste) to break down the insulating aluminum oxide coating on both the connector and conductor surfaces.

When the conductor is copper, connect it with either copper or aluminum-bodied connectors.

It is recommended that a UL-listed connector be used that can be applied with any common crimping tool. This connector should be listed for use at medium voltages. In this way, the choice of the connector is at the discretion of the user and is not limited by the tools available.

3.1.3 Reinsulating

There are several methods used to reinsulate splices:

- *Taping* – One method of reinsulating a splice is the tape method. This approach is not dependent upon cable types and dimensions. Tape has a history of dependable service and is generally available. However, wrapping tape on a medium-voltage cable can be time consuming and error prone, since the careful buildup of tape requires accurate half-lapping and constant tension in order to reduce air voids.

- *Heat-shrink tubing* – Heat-shrinkable products use cross-linked polymer tubing that shrinks when heated to its transition temperature. This provides a tight, even seal around the connection area and the cable components. These products can be installed using a flame torch or an electric hot air gun. Heat-shrink splices are range taking, which means that they can be used on a wide range of cable types. The only information required to select a heat-shrink splice is the conductor size and the voltage range of the cable. By using shim tubing, heat-shrink splice kits can be used as reducer splices over a very broad range of cable sizes. These products are also available with an adhesive coating on the inside that provides a superior water seal.

- *Molded rubber splices* – These factory-made splices are designed for the convenience of the installer. In many cases, these splices are also factory tested and designed to be installed without the use of special installation tools. Most molded rubber splices use EPR as the reinsulation material. The EPR must be cured during the molding process. Either a peroxide cure or a sulfur cure can be used.

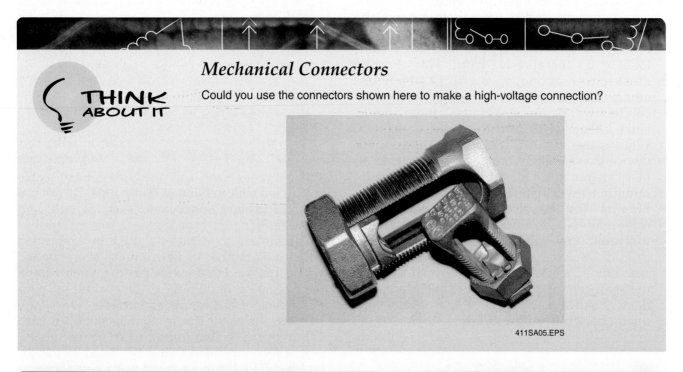

THINK ABOUT IT

Mechanical Connectors

Could you use the connectors shown here to make a high-voltage connection?

411SA05.EPS

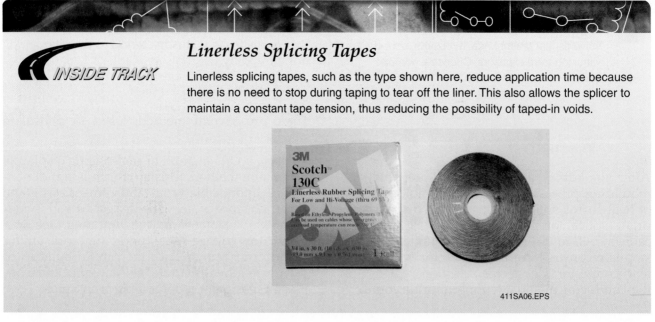

INSIDE TRACK

Linerless Splicing Tapes

Linerless splicing tapes, such as the type shown here, reduce application time because there is no need to stop during taping to tear off the liner. This also allows the splicer to maintain a constant tape tension, thus reducing the possibility of taped-in voids.

411SA06.EPS

Tape Splice Kits

Tape splice kits contain all the necessary tapes, along with detailed instructions. These versatile kits ensure that the proper materials are available at the job site and make an ideal emergency splice kit.

3.1.4 Reshielding

The cable's two shielding systems (strand shield and insulation shield) must be rebuilt when constructing a splice. The same methods are used as outlined in the reinsulation process: tape, heat-shrink, and molded rubber.

For a tape splice, the cable strand shielding is replaced by a semi-conductive tape. This tape is wrapped over the connector area to smooth the crimp indents and connector edges. The insulation shielding system is replaced by a combination of tapes. The semi-conductive layer is replaced with the same semi-conductive tape used to replace the strand shield.

The cable's metallic shield is generally replaced with a flexible woven mesh of tin-plated copper braid. This braid is for electrostatic shielding only and is not designed to carry shield currents. For conducting shield currents, a jumper braid is installed to connect the metallic shields of the two cable ends. This jumper must have an ampacity rating equal to that of the cable shields.

The insulation shielding of a heat-shrink splice contains a layer of conductive material bonded to the outside of the insulation tubing, similar to the way the cable is designed. The bonded conductive layer is continuous across the splice, so the placement of the tubing is not as critical as it is with a premolded splice. In addition, because the tube has the inner insulation and outer conductive layer bonded together in one dual-wall tube, it prevents air pockets between the insulation and semi-conductive layers of the splice. The metallic shield is reinstated with ground braids and stainless steel spring clamps to ensure proper fault carrying capacity of the splice.

For a molded rubber splice, conductive rubber is used to replace the cable strand shielding and the semi-conductive portion of the insulation shield system. Again, the metallic shield portion must be jumpered with a metallic component of equal ampacity.

A desirable design parameter of a molded rubber splice is that it be installable without special tools. To accomplish this, very short electrical interfaces are required. These interfaces are attained through proper design shapes of the conductive rubber electrodes.

Reshielding

This is an example of a partially completed splice of a shielded single-conductor power cable. The splice is being made using a heat-shrinkable splice kit for use with shielded power cables.

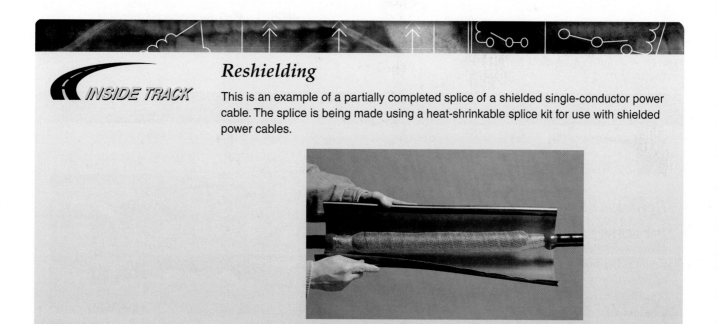

411SA07.EPS

3.1.5 Rejacketing

Rejacketing is accomplished in a tape splice by using rubber splicing tape overwrapped with vinyl tape.

In a heat-shrink splice, rejacketing is completed by using a heavy-wall, adhesive-coated tubing or, where space is limited, a wraparound sleeve.

In a molded rubber splice, rejacketing is accomplished by the proper design of the outer semi-conductive rubber, effectively resulting in a semi-conductive jacket.

When a molded rubber splice is used on internally shielded cable such as tape shield, drain wire shield, or UniShield® cables, a shield adapter is used to seal the opening that results between the splice and cable jacket.

3.2.0 Inline Tape Splices

This section provides an example of a tape splicing procedure using one manufacturer's splicing kit. The procedure is applicable to metallic and wire shield cable utilizing voltage ratings from 5kV to 25kV, with a normal maximum temperature rating of 90°C and 130°C emergency operation.

Figure 9 shows the various cable sections. Become thoroughly familiar with these sections before continuing in this module, and refer to *Figure 9* as often as needed for reference.

3.2.1 Preparing the Cable

Prepare the cable as follows:

Step 1 Train the cable in position and cut it carefully so that the cable ends will butt squarely.

Step 2 Thoroughly scrape the jacket 3" beyond dimension A in *Figure 10* to remove all dirt, wax, and cable-pulling compound so the tape will bond, making a moisture-tight seal.

Step 3 Remove the cable jacket and nonmetallic filler tape from the ends to be spliced for distance A (*Figure 10*) plus one-half the connector length.

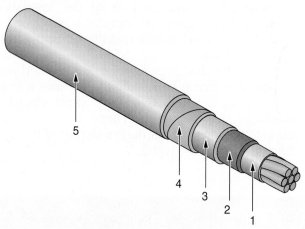

1. STRAND SHIELDING – Semi-conductive material

2. PRIMARY INSULATION – Ethylene propylene or cross-linked polyethylene

3. INSULATION SHIELD – Semi-conductive material

4. CABLE METALLIC SHIELDING – Metallic or wire

5. JACKET – PVC, neoprene, polyethylene

411F09.EPS

Figure 9 ◆ High-voltage cable construction.

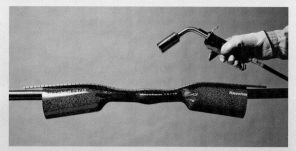

411SA08.EPS

Selecting a Splice Method

For the versatility to handle practically any splicing emergency, for situations where only a few splices need to be made, or when little detail is known about the cable, the most effective splice is made with either a tape kit or a heat-shrink kit.

For those times when cable size, insulation diameter, and shielding type are known and when numerous splices will be made, use heat-shrink or molded rubber splices for dependability and simplicity, as well as quick application.

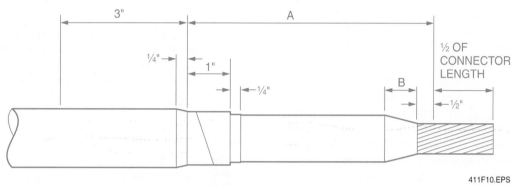

411F10.EPS

Figure 10 ◆ Recommended cable preparation procedures for an inline splice.

 CAUTION
Be careful not to cut into the metallic shielding or insulation when removing the jacket.

 CAUTION
Do not use emery cloth, because it contains metal particles.

Step 4 Remove the cable metallic shielding to expose 1" beyond the cable jacket. If shielded wire cable is used, see *Figure 11*. Be careful not to nick the semi-conductive material.

Step 5 Tack the metallic shielding in place with solder.

 CAUTION
Do not overheat the thermoplastic insulation.

Step 6 Remove the semi-conductive material, leaving ¼" exposed beyond the cable metallic shielding. Be sure to remove all traces of this material from the exposed cable insulation by cleaning it with non-conductive abrasive cloth.

Step 7 Remove the cable insulation from the end of the conductor for a distance of ½" plus one-half the length of the connector. Be careful not to nick the conductor.

Step 8 Pencil the insulation at each end, using a penciling tool or a sharp knife, for a distance equal to B in *Figure 10*. Buff the tapers with an abrasive cloth so no voids will remain after the joint is insulated.

Step 9 Taper the cable jacket smoothly at each end for a distance equal to ¼". Buff the tapers with an abrasive cloth.

Step 10 Clean the entire area by wiping it with solvent-saturated cloth from the cable preparation kit. Solvent may be used on thermoplastic insulation, but the area must be absolutely dry and free of all solvent residue, especially in the conductor strands and underneath the cable shield, before proceeding to the next step.

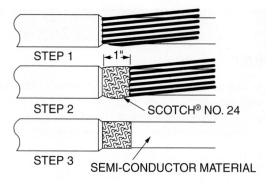

1. Remove cable jacket per previous instructions. Be careful not to cut any of the wires.

2. Wrap two unstretched layers of Scotch® No. 24 electrical shielding tape over shield wires for 1" beyond cable jacket. Tack in place with solder.

3. Cut shielding wires off flush with leading edge of tape.

411F11.EPS

Figure 11 ◆ Wire shield procedure.

3.2.2 Connecting the Conductors

Connect the conductors as follows:

Step 1 Join the conductors with an appropriate connector. Use a crimp connector to connect thermoplastic-insulated cables. Follow the connector manufacturer's directions. Use anti-oxidant paste for aluminum conductors. When using a solder-sweated connector, protect the cable insulation with temporary wraps of cotton tape. Be sure to remove all anti-oxidant paste from the connector area after crimping the aluminum connector.

 CAUTION

Do not use acid core solder or acid flux.

Step 2 Fill the connector indents with Scotch No. 13 semi-conductive tape, as shown in *Figure 12*. Smoothly tape two half-lapped layers of highly elongated tape over the exposed conductor and connector area. Cover all threads of the semi-conductive strand shielding and overlap the cable insulation with ¹⁄₁₆" semi-conductive tape. Stretching the tape increases its conductance and will not harm it in any way.

3.2.3 Applying Primary Insulation

Apply the primary insulation as follows:

Step 1 Apply Scotch No. 23 medium-voltage splicing tape (*Figure 13*) half-lapped over the hand-applied semi-conductive tape and exposed insulation up to ¼" from the cable semi-conductive material. Continue to apply the tape half-lapped in successive level, wound layers until the thickness is equal to the dimension shown in *Figure 13*. The outside surface should taper gradually along distance C, reaching its maximum diameter over the penciled insulation.

 NOTE

To eliminate voids in critical areas, highly elongate the splicing tape in the connector area and at the splice ends near the cable semi-conductive material. Stretch the tape just short of its breaking point. Doing so will not alter its physical or electrical properties. Throughout the rest of the splice area, less elongation may be used. Normally, No. 23 tape is stretched to 75% of its original width in these less-critical areas. Always attempt to exactly half-lap to produce a uniform buildup.

Step 2 Wrap one half-lapped layer of Scotch No. 13 tape over the splicing tape, extending over the cable semi-conductive material and onto the cable metallic shielding ¼" at each end. Highly elongate this tape when taping over the edge of the cable metallic shielding.

Step 3 Wrap one half-lapped layer of Scotch No. 24 electrical shielding tape over the semi-conductive tape, overlapping ¼" onto the cable metallic shielding at each end. Solder the ends to the metallic shielding.

 CAUTION

Do not lay a solder bead across the **shielding braid**. Do not use acid core solder or acid flux.

Manufacturer's Instructions

High-quality products usually include detailed installation instructions. Always follow these instructions carefully. A suggested technique is to check off each step as it is completed. Some manufacturers also offer hands-on training programs designed to teach the proper installation of their products. It is highly recommended that inexperienced splice and termination installers take advantage of such programs.

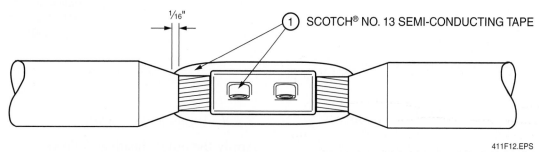

1/16"

① SCOTCH® NO. 13 SEMI-CONDUCTING TAPE

411F12.EPS

Figure 12 ◆ Connecting conductors.

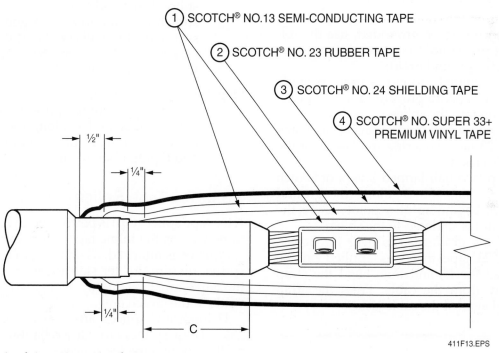

① SCOTCH® NO.13 SEMI-CONDUCTING TAPE

② SCOTCH® NO. 23 RUBBER TAPE

③ SCOTCH® NO. 24 SHIELDING TAPE

④ SCOTCH® NO. SUPER 33+ PREMIUM VINYL TAPE

1/2"

1/4"

1/4"

C

411F13.EPS

Figure 13 ◆ Applying primary insulation.

Step 4 Wrap one half-lapped layer of Scotch No. 33+ vinyl plastic electrical tape, covering the entire area of the shielding braid. Stretch it tightly to flatten and confine the shielding braid.

Step 5 Attach Scotch No. 25 ground braid as shown in *Figure 14*, soldering it to the cable metallic shielding at each end. Use a wire or strap having at least the same ampacity as the shield (No. 6 AWG is usually adequate).

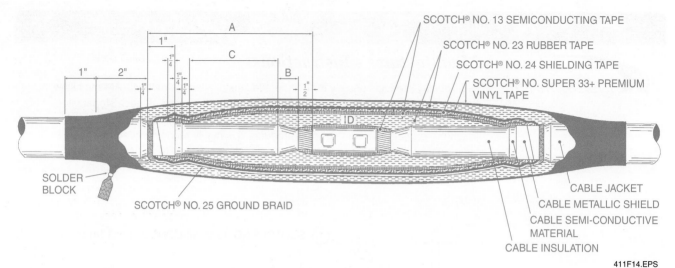

SCOTCH® NO. 13 SEMICONDUCTING TAPE
SCOTCH® NO. 23 RUBBER TAPE
SCOTCH® NO. 24 SHIELDING TAPE
SCOTCH® NO. SUPER 33+ PREMIUM VINYL TAPE

SOLDER BLOCK

SCOTCH® NO. 25 GROUND BRAID

CABLE JACKET
CABLE METALLIC SHIELD
CABLE SEMI-CONDUCTIVE MATERIAL
CABLE INSULATION

411F14.EPS

Figure 14 ◆ Summary of an inline tape splice.

CAUTION

Do not overheat the thermoplastic insulation.

Step 6 If a splice is to be grounded, use the following procedure to construct a moisture seal at the ground braid:

- If a stranded ground is used, provide solder block to prevent moisture penetration into the splice.
- Wrap two half-lapped layers of Scotch No. 23 tape, covering the last 2" of the cable jacket.
- Wrap two half-lapped layers of Scotch No. 23 tape for 3" along the ground strap, beginning at a point where the ground is soldered to the shield.
- Lay the ground strap over the Scotch No. 23 tape that was applied on the jacket for 1", and then bend the strap away from the cable. Press the strap hard against the cable jacket.

NOTE

Whenever possible, substitute Scotch No. 130C linerless splicing tape for Scotch No. 23 tape.

3.2.4 *Applying the Outer Sheath*

Apply the outer sheath as follows:

Step 1 Wrap four half-lapped layers of Scotch No. 23 tape (or equivalent) over the entire splice 2" beyond the splicing tape. Highly stretch the tape to form a good moisture seal and eliminate voids. Make sure the tape covers the ground braid to complete the moisture seal.

Step 2 Wrap two half-lapped layers of Scotch No. 33+ over the entire splice 1" beyond the splicing tape and 1" along the grounding braid. A summary of the end preparation and taping is shown in *Figure 14*.

3.3.0 Tee Tape Splice

A tee tape splice (tap) is performed in basically the same way as the inline tape splice, except a third conductor is attached, as shown in *Figure 15*. Tee splices require greater skill to construct because the tape must be highly elongated to prevent voids in the crotch area.

Heat-shrink splices are also available up to 25kV. These splices provide a reliable and efficient method of making tee splices.

3.4.0 Manufactured Termination and Splice Kits

Manufactured termination and splice kits consist of cable preparation materials, semi-conductor material, stress control means, jacketing, sealing, and grounding materials. The stress control and

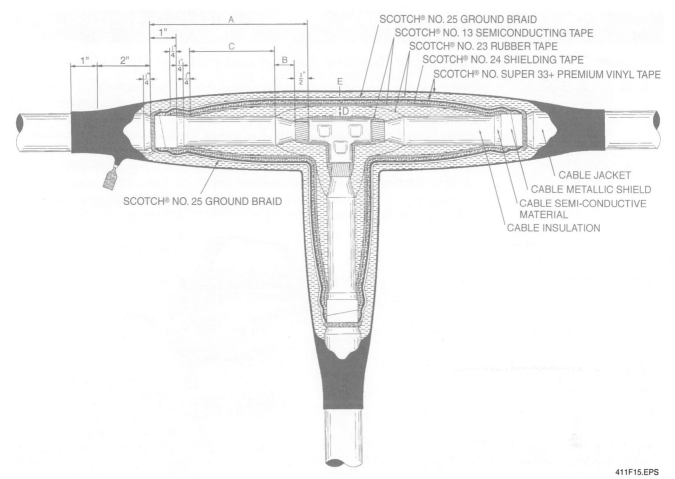

Figure 15 ◆ Summary of a tee tape splice.

jacket materials are applied using a shrink-to-fit method. Both heat shrink and cold shrink materials are available.

Heat-shrink splice kits have large use ranges and allow the connection of different conductor sizes, different insulation types (for example, paper insulation to polymeric insulation), and different cable constructions (for example, three-conductor cable spliced to single-conductor cable).

The following instructions are for a heat-shrink kit used on copper tape shielded cable. This same kit will splice UniShield®, wire shield, lead-sheathed cable, or any combination thereof.

To use a quick inline splicing kit, proceed as follows:

Step 1 Select the appropriate kit for the cable diameter using the selection charts provided by the kit manufacturer.

Step 2 Verify that the ground braid or bond wire has a cross-section equivalent to that of the cable's metallic shield.

Step 3 Cut back the cable to the dimensions specified in the kit manufacturer's instructions for the type of cable being spliced. See *Figure 16*. If necessary, abrade the insulation to remove embedded semi-con.

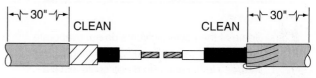

Figure 16 ◆ Preparing the cables.

The Use of Propane Torches and Hot-Air Heat Guns

Making proper splices and terminations using heat-shrink tubing and materials requires that the proper heat level be applied to the heat-shrink material. One manufacturer recommends that a clean-burning propane torch, like the one shown here, be used for heating the heat-shrink tubing materials when making splices and terminations on high-voltage cables. The torch should have an output heat rating of about 90,000 Btu/hr. Heat-shrink splices and terminations on low-voltage cables and/or smaller high-voltage cables (up to 15kV) can be made using either a propane torch or a hot-air heat gun. The propane torch should have a heat output rating of about 20,000 Btu/hr; the hot-air heat gun should have a heat output rating of about 750°F to 1,000°F.

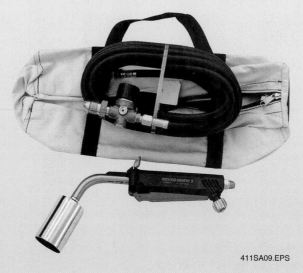

411SA09.EPS

Step 4 Clean the cable jackets for the length of the tubes.

WARNING!
Some solvents are toxic and must be handled with extreme caution. Refer to the solvent manufacturer's material safety data sheet (MSDS) for safety precautions associated with the solvent being used.

NOTE
Some of the newer solvents do not evaporate quickly and need to be removed using a clean, lint-free cloth. Failure to do so could leave a residue on the surface and interfere with the electrical integrity of the cable.

Step 5 Place the nested tubes and the rejacketing tube over the cable as shown in *Figure 17*. Protect the tubes from the sharp ends of the conductor as they are placed over the cable.

Step 6 Install the connector. Wipe the insulation down using an oil-free solvent. See *Figure 18*.

Step 7 Tightly wrap the exposed conductor with the stress relief material (SRM) provided in the splicing kit. Be sure to fill all gaps and low spots around the connector. Also install SRM over the semi-con cutback areas. See *Figure 19*.

NOTE
If the connector diameter is larger than the insulation diameter, apply two overlapped layers of SRM along the length of the entire connector.

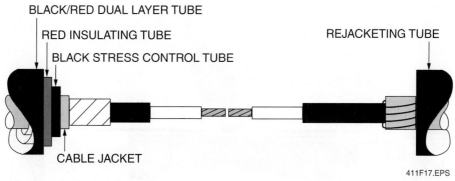

Figure 17 ◆ Placing the nested tubes on the cable.

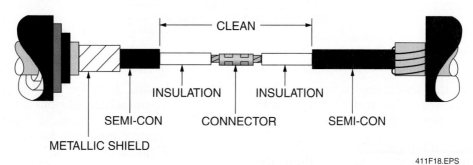

Figure 18 ◆ Installing the connector.

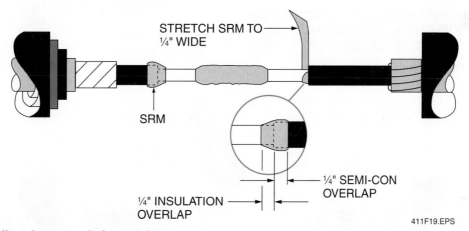

Figure 19 ◆ Installing the stress relief material.

 WARNING!

To avoid risk of accidental fire or explosion when using gas torches, always check all connections for leaks before igniting the torch and follow the manufacturer's safety instructions.

Step 8 Center the stress control tube over the splice and shrink it into place. See *Figure 20.*

When shrinking the stress control or insulating tube, apply the outer tip of the flame using a smooth brushing motion. Be sure to keep the flame moving to avoid scorching the material.

Unless otherwise instructed, start shrinking at the center of the tube, then work the flame outward on all sides to ensure uniform heat.

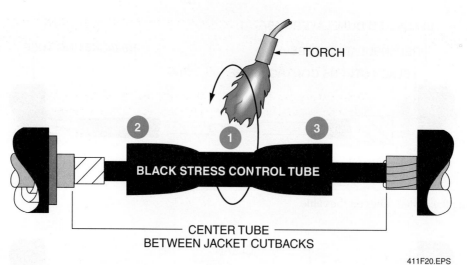

Figure 20 ◆ Shrinking the stress control tube.

Step 9 Position the insulating tube over the stress control tube and shrink it into place. See *Figure 21*.

Step 10 Apply sealant to the tube ends to provide a positive environmental seal. Build the sealant to the level of the insulating tube. See *Figure 22*.

Step 11 Position the insulating/conductive tube over the insulating tube and seal the ends before shrinking it into place. See *Figure 23*.

Step 12 Install metallic shielding mesh using manufacturer-supplied constant-tension clamps for metallic tape shielding or crimp connectors for wire shielding. See *Figure 24*.

Step 13 Position the rejacketing tube and shrink it into place. Check the ends for correct adhesive flow. See *Figure 25*.

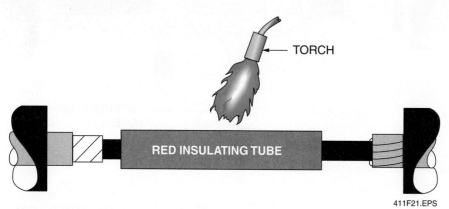

Figure 21 ◆ Shrinking the insulating tube.

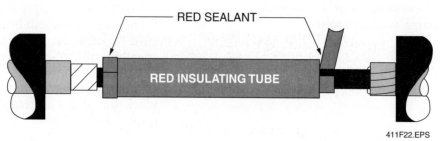

Figure 22 ◆ Applying sealant.

Installing Multiple Tubes

When installing multiple tubes, make sure that the surface of the last tube is still warm before positioning and shrinking the next tube. If the installed tube has cooled, reheat the entire surface.

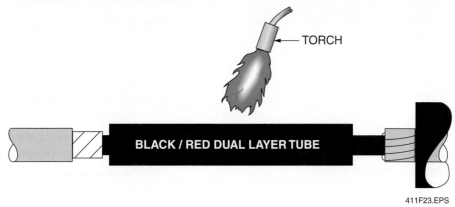

Figure 23 ◆ Shrinking the insulating/conductive tube.

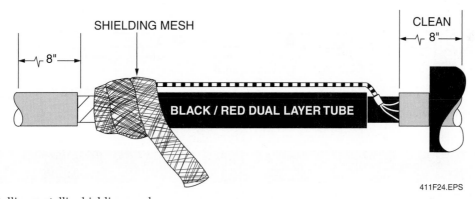

Figure 24 ◆ Installing metallic shielding mesh.

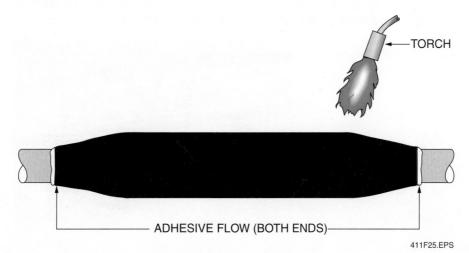

Figure 25 ◆ Shrinking the rejacketing tube and checking for adhesive flow.

3.5.0 Quick Inline Splicing Kits

There are several new types of cable splicing kits that have appeared on the market within the past few years. All greatly simplify cable splicing and should be used when the exact type and dimensions of the cable are known.

The following instructions are for a cold-shrink kit used on ribbon shielding cable. Kits are also available for wire shield and UniShield®. Complete instructions are included in each kit, and once you understand the directions in this module for ribbon shielding cable, the directions for the other types should be readily understood as well.

3.5.1 Preparing the Cables

Prepare the cables as follows:

Step 1 Clean each cable jacket by wiping it with a dry cloth about 2" from each end.

Step 2 See *Figure 26*. Remove the jacket to distances B and C for cables X and Y, respectively.

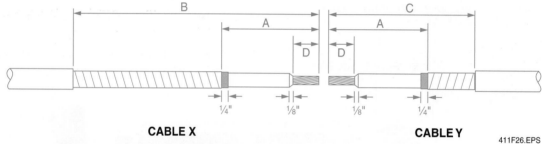

NOTE

The dimensions shown in *Figure 26* will vary with the type of kit used.

Step 3 Remove the metallic shielding for distance A in *Figure 26*.

Step 4 Remove the cable semi-conductive material (semi-con), allowing it to extend ¼" beyond the metallic shielding (dimension A in *Figure 26*).

Step 5 Remove the insulation for dimension D in *Figure 26* and taper the edges ⅛" at approximately 45°.

Figure 26 ◆ Cable preparation details.

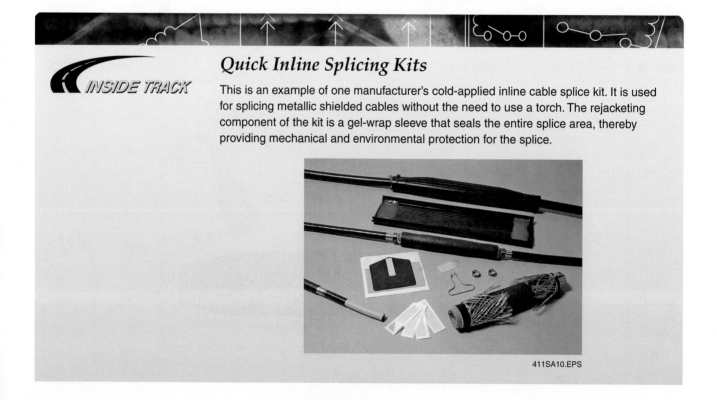

Quick Inline Splicing Kits

This is an example of one manufacturer's cold-applied inline cable splice kit. It is used for splicing metallic shielded cables without the need to use a torch. The rejacketing component of the kit is a gel-wrap sleeve that seals the entire splice area, thereby providing mechanical and environmental protection for the splice.

411SA10.EPS

Step 6 Clean the exposed insulation using the cleaning pads in the kit. Do not use solvent or abrasive on the cable semi-con layer. If abrasive must be used on the insulation, do not reduce the diameter below that specified for minimum splice application.

Step 7 Apply two highly elongated, half-lapped layers of Scotch No. 13 semi-conductive tape starting at 1" on the cable metallic shield and extending ½" onto the cable insulation. Leave a smooth leading edge and tape back to the starting position. See *Figure 27*.

Step 8 On cable X, apply one half-lapped layer (adhesive side out) of the orange vinyl tape supplied in the kit over the metallic shield, beginning approximately ¼" on the previously applied No. 13 tape and ending over the cable jacket (*Figure 27*).

3.5.2 Splicing the Installation

Splice the installation as follows:

Step 1 Slide the longer PST cold-shrink insulator onto the X jacket and the shorter PST onto the Y jacket, directing the pull tabs away from the cable ends, as shown in *Figure 28*.

Step 2 Apply a few wraps of vinyl tape over cable X.

Step 3 Lubricate the exposed insulation, No. 13 tape, and orange vinyl tape of cables X and Y with silicone grease. Do not allow the grease to extend onto the metallic shield of cable Y.

Step 4 Lubricate the splice bore with silicone grease and install it onto cable X, leaving the conductor exposed for the connector, as shown in *Figure 29*.

NOTE

Rotating the splice while pulling will ease the installation. Make certain all splicing components (PSTs and splice body) are located properly on their respective cable ends before installing the connector.

Step 5 Remove the vinyl tape from the cable X conductor.

Step 6 Install a connector crimped per the manufacturer's instructions.

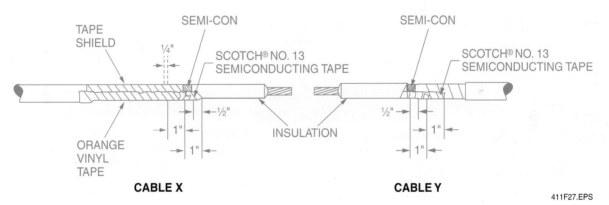

CABLE X **CABLE Y**

411F27.EPS

Figure 27 ◆ Components of a ribbon shielding cable splice.

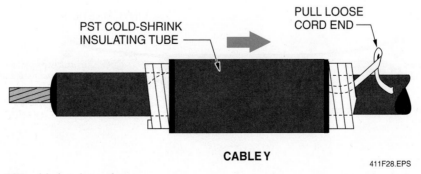

CABLE Y

411F28.EPS

Figure 28 ◆ Installing PST cold-shrink insulator.

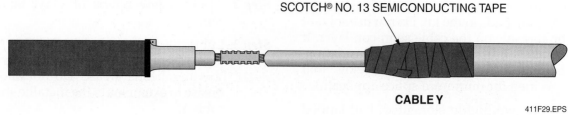

SCOTCH® NO. 13 SEMICONDUCTING TAPE

CABLE Y

411F29.EPS

Figure 29 ◆ Wrapping No. 13 tape onto the cable.

Step 7 Slide the splice body into its final position over the connector, using the bumps formed on the splice ends as guides for centering. See *Figure 30*.

Step 8 Remove the orange vinyl tape and wipe off any remaining silicone grease.

3.5.3 Installing the Shield Continuity Assembly

Install the shield continuity assembly as follows:

Step 1 Position the shield continuity assembly over the splice body and hold it in place with a strap or two of vinyl tape at each end of the splice body (*Figure 31*). Form the shield continuity strap over the splice shoulder on each side (*Figure 32*).

NOTE

The assembly coils must be facing the cable and positioned so that they will only make contact with the metallic shielding when applied. Avoid positioning the flat strap over the splice grounding eyes.

Step 2 Unwrap the coil and straighten it for 1" to 2", then hold the coil and shield strap in place with your thumb. Pull to unwrap the coil around the cable and rewrap it around the cable metallic shield and itself. Cinch (tighten) the applied coil after the final wrap. Repeat for the other end of the splice.

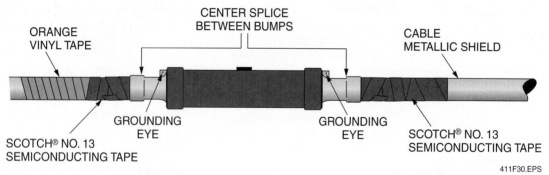

ORANGE VINYL TAPE

CENTER SPLICE BETWEEN BUMPS

CABLE METALLIC SHIELD

GROUNDING EYE

GROUNDING EYE

SCOTCH® NO. 13 SEMICONDUCTING TAPE

SCOTCH® NO. 13 SEMICONDUCTING TAPE

411F30.EPS

Figure 30 ◆ Use bumps formed on splice ends as guides for centering.

COIL

VINYL MASTIC PAD

411F31.EPS

Figure 31 ◆ Wrap vinyl tape at each end of the splice body.

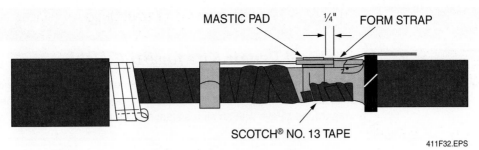

Figure 32 ◆ Form the shield continuity strap over the splice shoulder on each side.

Step 3 To seal the shield strap, cut the supplied strip of vinyl mastic into four equal lengths. After removing the liner, place one pad under the strap, with the mastic side toward the strap and located at the end of the splice body (*Figure 32*). Place a second pad over the strap and first pad and press together, mastic-to-mastic. This will be wrapped with No. 13 tape. Repeat for the other end of the splice body.

Step 4 Beginning just beyond the splice body, wrap No. 13 tape over the vinyl mastic pads extending onto the splice for approximately ¼" and returning to the starting point.

3.5.4 Installing Cold-Shrink Insulators

Install the cold-shrink insulators as follows:

Step 1 Position each PST so its leading edge will butt against the splice body grounding eye.

Step 2 Remove the core by unwinding it counterclockwise and tugging (*Figure 33*).

3.5.5 Grounding the Splice

If the installation calls for grounding the splice, fasten the alligator ground clamp (provided with most kits) to the exposed shield continuity strap approximately at the center of the splice. A ground strap conductor can then be fastened to the alligator clamp. Discard the black plastic shoe packaged with the ground clamp.

3.6.0 Paper-Insulated Cable Splices

Many older facilities may have paper-insulated, lead-covered (PILC) power cable in the ground. You may be required to splice new circuits into this existing cable. Splicing PILC power cable is a special type of splice that must not only rebuild the various layers of the power cable, but must also create an oil seal system to lock in the PILC cable. The most common methods for PILC cable splices are hand-taped, lead-wiped, compound-filled splices or heat-shrink splices.

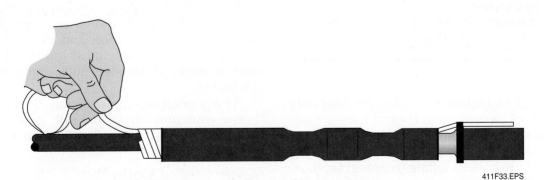

Figure 33 ◆ Remove core by unwinding counterclockwise.

Lead Shield Repair Kits for PILC Cable

This is an example of a repair kit available from one manufacturer for the repair of a damaged lead sheath in paper-insulated, lead-covered (PILC) cables. The kit consists of a wraparound sleeve with an oil-resistant sealing mastic.

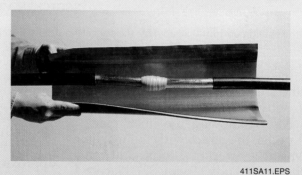

411SA11.EPS

4.0.0 ◆ TERMINATIONS

A Class 1 cable termination (or more simply, a Class 1 termination) provides the following:

- Some form of electric stress control for the cable insulation shield terminus
- Complete external leakage insulation between the conductor(s) and ground
- A seal to prevent the entrance of the external environment into the cable and to maintain the pressure, if any, within the cable system

NOTE

This classification encompasses the conventional potheads for which the original *IEEE Standard 48-1962* was written. With this new classification or designation, the term pothead is henceforth dropped from usage in favor of Class 1 termination.

A Class 2 termination is one that provides only some form of electric stress control for the cable insulation shield terminus and complete external leakage insulation, but no seal against external elements. Terminations falling into this classification would include stress cones with rain shields or special outdoor insulation added to give complete leakage insulation and the more recently introduced skip-on terminations for cables having extruded insulation when not providing a seal as in Class 1.

A Class 3 termination is one that provides only some form of electric stress control for the cable insulation shield terminus. This class of termination would be for use primarily indoors. Typically, this would include hand-wrapped stress cones (tapes or pennants), as well as slip-on stress cones.

Some Class 1 and Class 2 terminations have external leakage insulation made of polymeric material. It is recognized that there is some concern about the ability of such insulation to withstand weathering, ultraviolet radiation, contamination, and leakage currents and that a test capable of evaluating the various materials would be desirable. There are a number of test procedures available today for this purpose. However, none of them has been recognized and adopted by the industry as a standard. Consequently, this module does not include such a test.

These IEEE classes make no distinction between indoor and outdoor environments. This is because contamination and moisture can be highly prevalent inside most industrial facilities such as paper plants, steel mills, and petrochemical plants.

As a general recommendation, if there are airborne contaminants or fail-safe power requirements are critical, use a Class 1 termination.

4.1.0 Stress Control

In a continuous shielded cable, the electric field is uniform along the cable axis and there is variation in the field only in a radial section. This is illustrated in *Figure 34*, which shows the field distribution over a radial section.

Terminations

This is an example of the types of high-voltage cable terminations encountered at a power substation.

411SA12.EPS

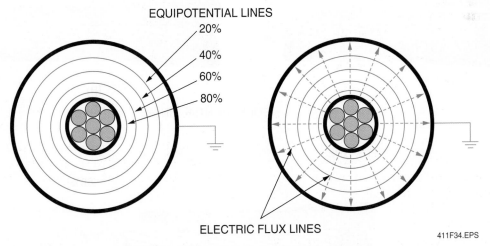

Figure 34 ◆ Field distribution over a radial cable section.

The spacing of the electric flux lines and corresponding equipotential lines is closer near the conductor than at the shield, indicating a higher electric stress on the insulation at the conductor. This stress increase or concentration is a direct result of the geometry of the conductor and shield in the cable section and is accommodated in practical cables by insulation thickness sufficient to keep the stress within acceptable values.

When terminating a shielded cable, it is necessary to remove the shield to a point some distance from the exposed conductor, as shown in *Figure 35(A)*. This is to secure a sufficient length of insulation surface to prevent breakdown along the interface between the cable insulation and the insulating material to be applied in the termination. The particular length required is determined by the operating voltage and the properties of the insulating materials. This removal of a portion of the shield results in discontinuity of the electrical field. See *Figures 35(B)* and *35(C)*. The electric flux lines originating along the conductor are seen to converge on the end of the shield, with the attendant close spacing of the equipotential lines signifying the presence of high electric stresses in this area. This stress concentration is of much greater magnitude than that occurring near the conductor in the continuous cable and as a result, steps must be taken to reduce the stresses occurring near the end of the shield if cable insulation failure is to be avoided.

All terminations must at least provide stress control. This stress control may be accomplished by two commonly used methods:

- Geometric stress control
- Capacitive stress control

4.1.1 Geometric Stress Control

This method involves an extension of the shielding (*Figure 36*), which expands the diameter at which the terminating discontinuity occurs and thereby reduces the stress at the discontinuity. It also reduces stresses by enlarging the radius of the shield end at the discontinuity.

4.1.2 Capacitive Stress Control

This method consists of a material possessing a high **dielectric constant (K)**, generally in the range of K30, and also a high dielectric strength. (The dielectric constant K is a measurement of the ability of a material to store a charge.)

The value of K is generally an order of magnitude higher than the cable insulation. Located at the end of the shield cutback, the material capacitively changes the voltage distribution in the electrical field surrounding the shield terminus. Lines of electrical flux are regulated to equalize the electrical stresses in a controlled manner along the entire area where the shielding has been removed, as shown in *Figures 37(A)* and *37(B)*.

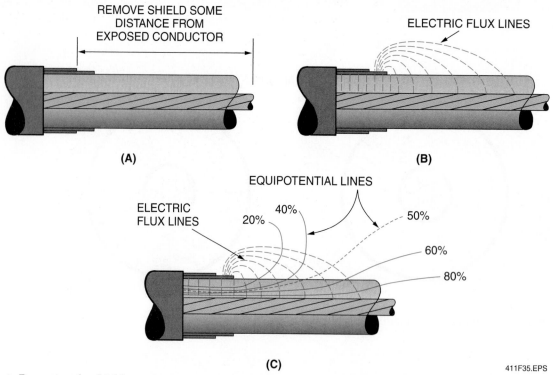

Figure 35 ◆ Removing the shield.

411F35.EPS

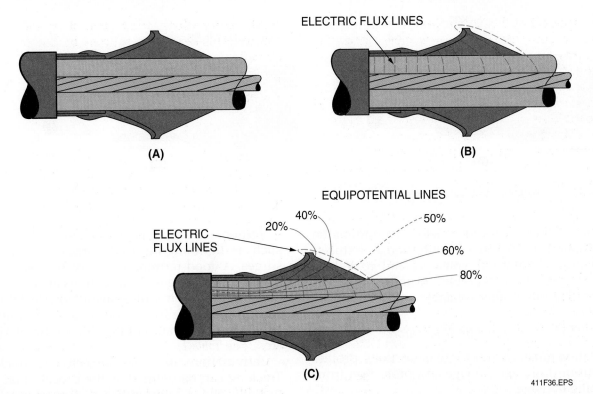

ELECTRIC FLUX LINES

(A)

(B)

EQUIPOTENTIAL LINES

40%

20%

50%

ELECTRIC
FLUX LINES

60%

80%

(C)

411F36.EPS

Figure 36 ◆ Geometric stress control.

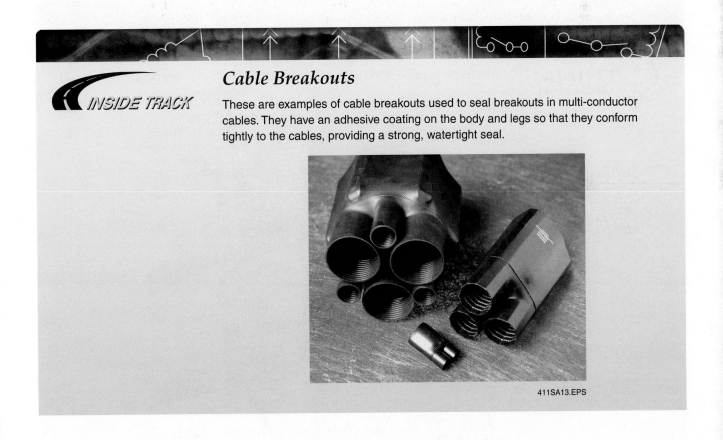

Cable Breakouts

These are examples of cable breakouts used to seal breakouts in multi-conductor cables. They have an adhesive coating on the body and legs so that they conform tightly to the cables, providing a strong, watertight seal.

411SA13.EPS

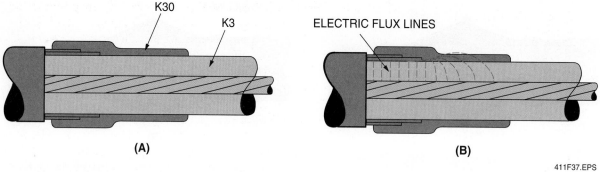

Figure 37 ◆ Equalizing electrical stresses.

411F37.EPS

By changing the electrical field surrounding the termination, the stress concentration is reduced from several hundred volts per mil to values found in continuous cable (usually less than 50V/mil at rated cable voltage).

4.1.3 External Leakage Insulation

Insulation must provide for two functions: protection from flashover damage and protection from tracking damage.

Terminations can be subjected to flashovers such as lightning-induced surges or switching surges. A good termination must be designed to survive these surges. Terminations are assigned a basic lightning impulse insulation level (BIL) according to their insulation class. For example, a 15kV class termination has a BIL rating of 110kV crest. See *Table 1*.

Terminations are also subjected to tracking. Tracking can be defined as the process that produces localized deterioration on the surface of the

Disc Assemblies

What is the purpose of the large circular assemblies shown in this picture?

THINK ABOUT IT

411SA14.EPS

Table 1 Insulation Classes and BIL Ratings

Insulation Class (kV)	BIL (kV crest)
5.0	75
6.7	95
15.0	110
25.0	150
34.5	200
46.0	250
69.0	350

insulator, resulting in the loss of insulating function by the formation of a conductive path on the surface.

A termination can be considered an insulator having a voltage drop between the conductor and the shield. As such, a leakage current develops between these points. The magnitude of this leakage is inversely proportional to the resistance on the insulation surface. See *Figure 38*.

Both contamination (dust, salt, and airborne particles) and moisture (humidity, condensation, and mist) will decrease this resistance. This results in surface discharges referred to as tracking. In an industrial environment, it is difficult to prevent these conditions. Therefore, a track-resistant insulator must be used to prevent failures.

A high-performance Class 1 termination contains insulator skirts that are sized and shaped to form breaks in the moisture/contamination path, thus reducing the probability of tracking problems. In addition, these skirts can reduce the physical length of a termination by geometrically locating the creepage distance over the

convolutions of the insulator, an important factor when space is a consideration.

Good termination design dictates that the insulators be efficiently applied under normal field conditions. This means the product should be applied by hand with a minimum of steps, parts, and pieces. Because they are inorganic materials, both porcelain and silicone rubbers provide excellent track resistance. Heat-shrink termination materials are the only non-tracking materials available and have shown superior long-term reliability.

4.2.0 Sealing to the External Environment

In order to qualify as a Class 1 termination, the termination must provide a seal to the external environment. Both the conductor/lug area and the shielding cutback area must be sealed. These seals keep moisture out of the cable to prevent degradation of the cable components. Several methods are used to make these seals, including factory-made seals, mastics used with heat-shrink kits, tape seals (silicone tape must be used for a top seal), and compound seals.

4.2.1 Grounding Shield Ends

All shield ends are normally grounded. However, individual circumstances may exist where the shield is grounded at only one end. The cable shield must be grounded somewhere in the system. When using solid dielectric cables, it is recommended that solderless ground connections be used, eliminating the danger of overheating cable insulation when soldering.

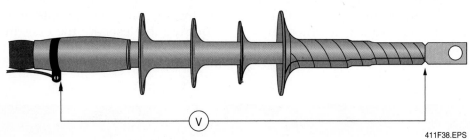

411F38.EPS

Figure 38 ◆ A terminator may be considered an insulator.

5.0.0 ◆ HIGH-POTENTIAL (HI-POT) TESTING

Suitable dielectric tests are commonly applied to determine the condition of insulation in cables, transformers, and rotating machinery. When properly conducted, high-potential (hi-pot) tests will indicate such faults as cracks, discontinuity, thin spots, or voids in the insulation, excessive moisture or dirt, faulty splices, and faulty terminations. An experienced operator may not only frequently predict the expected breakdown voltage of the item under test, but can often make a good estimate of its future operating life.

The use of direct current (DC) has several important advantages over alternating current (AC). The test equipment itself may be much smaller, lighter in weight, and lower in price. When properly used and interpreted, DC tests will give much more information than is obtainable with AC testing. There is far less chance of damage to equipment and less uncertainty in interpreting results. The high-capacitance current frequently associated with AC testing is not present to mask the true leakage current, nor is it necessary to actually break down the material being tested to obtain information regarding its condition. Although DC testing may not simulate the operating conditions as closely as AC testing, the many other advantages of using DC make it worthwhile.

The great majority of medium-voltage cable testing is done with direct current, primarily due to the smaller size of machines and the ability to easily provide quantifiable values of charging, leakage, and absorption currents. Although both AC and DC testers may damage a cable during tests, the damage in a DC test only occurs if the increasing current is not interrupted soon enough and the capacitance stored is discharged across the insulation.

Hi-pot testers for medium-voltage cable will also have a microammeter and a range dial to allow measurement of up to 5mA, at which level the circuit breaker will trip or the reactor will collapse.

WARNING!

Extreme caution must be exercised when conducting hi-pot testing of high-voltage cable. The cable will recharge itself as the absorption current migrates after the test unless grounds are constantly maintained.

Hi-pot testing may be divided into the following broad categories:

- *Design tests* – Tests usually made in the laboratory to determine proper insulation levels prior to manufacturing.
- *Factory tests* – Tests made by the manufacturer to determine compliance with the design or production requirements.
- *Acceptance tests* – Tests made immediately after installation, but prior to putting the equipment or cable into service.
- *Proof tests* – Tests made soon after the equipment has been put into service and during the guarantee period.
- *Maintenance tests* – Tests performed during normal maintenance operations or after servicing or repair of equipment or cable.
- *Fault locating tests* – Tests conducted to determine the location of a specific fault in a cable installation.

The maximum test voltage, testing techniques, and interpretation of the test results will vary somewhat, depending upon the particular type of test. Unfortunately, in many cases, the specifications do not spell out the test voltage, nor do they outline the test procedure to be followed. Therefore, it is necessary to apply a considerable amount of common sense and draw strongly upon experience when conducting these tests. There are certain generally accepted procedures, but the specific requirements of the organization requiring the tests are the ones that should govern. It is obvious that an acceptance test must be much more severe than a maintenance test, and a

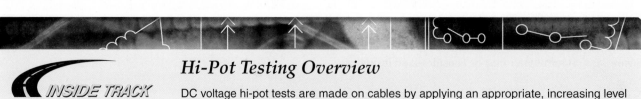

Hi-Pot Testing Overview

INSIDE TRACK

DC voltage hi-pot tests are made on cables by applying an appropriate, increasing level of DC voltage in steps and measuring the leakage current (microamperes) through the cable insulation at each step. The voltage level and measured current values are plotted on graph paper, and the shape of the resultant curve is used for determining the condition of the cable insulation being tested.

test made on equipment that is already faulted would be conducted in a manner different from a test being conducted on a piece of equipment in active service.

Special testing instruments are available for all of the tests listed here. However, testing instruments will be used mostly by electricians to perform routine production tests and acceptance tests for newly installed systems.

The general criteria for acceptance of a DC hi-pot test are a consistent leakage for each voltage step (linear) and a decrease in current over time at the final voltage. The level of charging current should also be consistent for each step.

5.1.0 Method of Application

DC hi-pot tests are performed primarily to determine the condition of the insulation. The extent of the highest breakdown voltage may or may not be of interest, but the relative condition of the insulation, in the neighborhood of and somewhat above the operating voltage, is certainly required. There are many ways of achieving these results. Two of the most popular are the leakage current-versus-voltage test and the current-versus-time test. The current-versus-time test is frequently made immediately after the current-versus-voltage test and yields the best results when performed in this manner.

Before performing a test, first determine what type of test will be performed and the appropriate test voltage. Refer to the cable and termination product supplier literature for test voltage recommendations and limitations. Next, the test configuration must be decided. When testing a circuit, the standard test setup is to test each conductor to its own shield and ground with all other conductors and their shields grounded also. The physical setup of the test equipment and circuit under test must be reviewed to ensure the safety of test participants and others. Barriers, flagging, or other means must be used to keep others away from the test area, including the other end of the cable

under test. Testing cannot be done satisfactorily when the humidity is too high or in windy conditions. Also to be determined is the method of documenting test results. *Appendix A* shows a test data sheet used by one company.

After all safety precautions have been taken, disconnect all accessory equipment, connected loads, and potential transformers from the cable or rotating machine being tested. Make sure that the shield, frame, and unused phases are grounded, and the hot side of the hi-pot tester is connected to the conductor under test.

The voltage is slowly raised in discrete steps, allowing sufficient time at each step for the leakage current to stabilize. As the voltage is raised, the leakage current will first be relatively high and then will decrease with time until a steady minimum value is reached.

The initial high value of current is known as the charging current and depends primarily on the capacitance of the item under test. The lower steady-state current consists of the actual leakage current and the dielectric absorption current (which is relatively minor as far as these tests are concerned). The value of the leakage current is noted at each step of voltage, and as the voltage is raised, a curve is plotted of leakage current versus voltage. As long as this curve is relatively flat (that is, equal increments of voltage giving equal increments of current, as shown in point A to point B in *Figure 39*), the item under test is considered to be in good condition. At some point, the current will start rising at a more rapid rate (point C). This will show up on the plot as a knee in the curve. It is very important that the voltage increments be small enough so that the starting point of this knee can be noted. If the test is carried on beyond the start of the knee, the current will increase at a much more rapid rate, and breakdown will soon occur (point D). Unless breakdown is actually desired, it is typical to halt a test as soon as the beginning of the knee is observed.

With a little experience, the operator can extrapolate the curve and estimate the point of

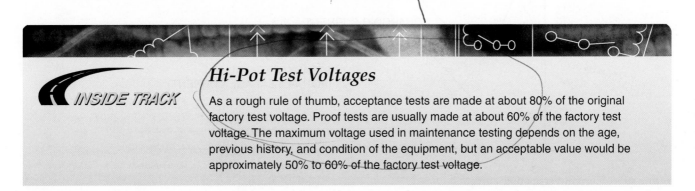

Hi-Pot Test Voltages

INSIDE TRACK

As a rough rule of thumb, acceptance tests are made at about 80% of the original factory test voltage. Proof tests are usually made at about 60% of the factory test voltage. The maximum voltage used in maintenance testing depends on the age, previous history, and condition of the equipment, but an acceptable value would be approximately 50% to 60% of the factory test voltage.

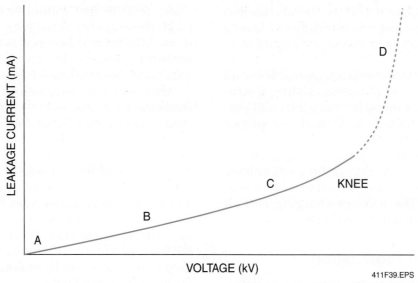

Figure 39 ◆ Current-versus-voltage curve.

breakdown voltage. This procedure enables the operator to anticipate the breakdown point without actually damaging the insulation, because if the test is stopped at the start of the knee, no harm is done.

Although the important thing to watch for is the rate of change of current as the voltage is raised, much information may be obtained from the value of the current. A comparison of the leakage current from phase to phase, a comparison of leakage current with values measured in other equipment under similar conditions, or a comparison with the leakage current values obtained on tests made of the same equipment on prior occasions will give an indication of the condition of the insulation. In general, higher leakage current is an indication of poorer insulation, but this is difficult

to evaluate from a single test. As long as the leakage current-versus-voltage curve is of the general shape shown in *Figure 39* and the knee occurs at a sufficiently high voltage (above the maximum test voltage required), the equipment being tested may be considered satisfactory.

On long cable runs or with rotating machinery having a high capacitance between windings or from winding to frame, the time for stabilization of the leakage current (decrease from E to F in *Figure 40*) may run as high as several minutes, several hours, or more. To minimize the testing time under these conditions, it is common to time the decay of current to where it definitely slows and to take a reading at each voltage step after the same fixed interval. For example, the voltage could be brought up to the first step, allowed to

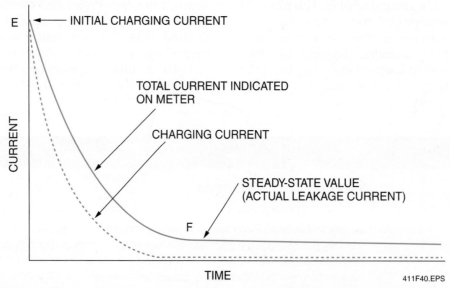

Figure 40 ◆ Current-versus-time curve.

remain at that point for four minutes, and then the current reading taken. The voltage is next raised to the second step, the same four-minute wait is allowed, and the current reading is taken again. This is continued throughout the test. Although this method will not give the true leakage current at any point, it will still produce the proper curve and permit a comparative evaluation.

When the final test voltage is reached, the tester is left on and the current-versus-time curve (*Figures 40* and *41*) is plotted by recording the current at fixed intervals as it decays from the initial high charging value to the steady-state leakage value. This curve for good cable should indicate a continuous decrease in leakage current with time or a stabilization of the current without any increase during the test.

Any increase in current during this test would indicate a bad cable or machine. The test should be stopped immediately if such an indication occurs (see *Figure 42*). After the current has stabilized and the last reading has been taken, shut off the high voltage. The kilovolt-meter on the tester will indicate the actual voltage present in the equipment under test as the tester's internal circuitry permits the charge to

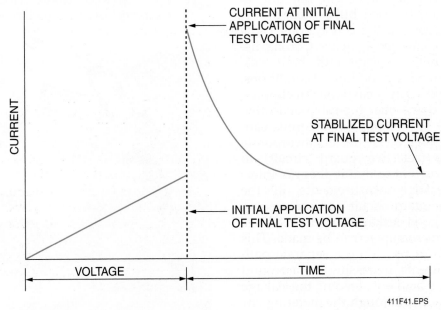

411F41.EPS

Figure 41 ◆ Current-versus-voltage and current-versus-time curves at maximum voltage.

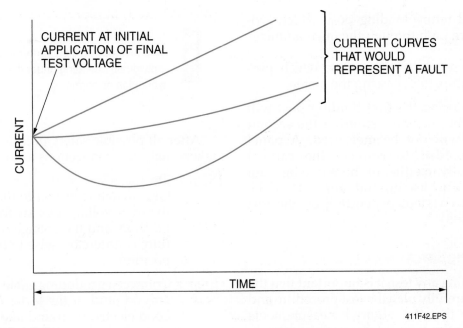

411F42.EPS

Figure 42 ◆ Faulty insulation.

gradually leak off. After the voltage reaches zero, put an external ground on the cable or machine, and disconnect it from the tester. It is important to wear rubber gloves at this point, since the effect of absorption currents may cause a buildup of voltage in the item under test until after it has been grounded for some time. The ground should be maintained until the cable is terminated and for at least five times the duration of the test.

5.2.0 Selective Guard Circuits

When making the current-versus-voltage test, a rapid climb in current may occur not due to defects in the equipment itself, but because of surface leakage or corona at the connections, cable ends, or blades of connected switches. Corona discharge is usually not significant until about 20kV DC and is worse in windy and humid conditions. Bagging the cable ends in a plastic bag or slipping an empty plastic bottle over the cable end often satisfactorily controls corona. When corona cannot be controlled, then all DC testers incorporate internal circuitry (selective guard circuit) to bypass this stray current so that it does not interfere with the leakage measurements. By the proper use of the guard circuit and corona ring when necessary, good leakage current readings even below one microampere can be made. This technique consists of providing a separate path for these stray currents, bypassing them around the metering circuit, and only feeding the leakage current to be measured through the metering circuit. In order to accomplish this, DC testers have two return paths:

- The metered return binding post, which provides a return path through the microammeter circuit
- The bypass return binding post, which provides a return path bypassing the meter

Currents entering the first binding post will be measured, but currents entering the second binding post will not be measured. A panel switch is provided to connect the cabinet ground internally to either of these two binding posts (which enables ground currents to be measured or bypassed, depending on the test being performed).

5.3.0 Connections

Before performing any test, it is important that the test operator carefully plan the test procedure and connections so that the required measurements may be taken without misinterpretation or error.

INSIDE TRACK

Hi-Pot Testing Safety

All safety precautions must be taken when performing hi-pot testing of equipment or cables. This includes de-energizing and grounding the equipment to be tested to eliminate any charge. All circuit energizing switches must be opened, locked out, and tagged. Disconnect all jumpers running from terminations to feeders, potential transformers, and lighting arresters. Current transformers might be left in the circuit if their insulation level is sufficiently high.

411SA15.EPS

 WARNING!
This test may only be performed by qualified individuals operating under the appropriate safe work plan or permit.

After all possible safety precautions have been taken, make the connections as follows:

Step 1 On the tester, make sure the main ON/OFF switch is turned to the OFF position, the high-voltage ON switch is in the OFF position, and the voltage control is turned fully counterclockwise to the zero voltage position.

Step 2 Connect a grounding cable from the safety ground stud of the tester (*Figure 43*) to a good electrical ground and make sure the connection is secure at both ends.

Step 3 Connect the return line from the item under test to the metered return or bypass return binding post of the tester, and move the grounding switch on the tester panel to the appropriate position. The return lines connected to either the bypass return or metered return binding post should be insulated for about 100V, so ordinary building grade wire may be used.

Step 4 Plug the shielded cable furnished with the tester into the receptacle at the top of the unit. If the tester has an oil-filled output receptacle, make sure that the oil level is at the mark indicated on the oil stick. If the oil level is low, add the proper amount of high-grade transformer or other insulating oil. After the cable is plugged into the receptacle, tighten the clamping nut to hold it securely in place.

Step 5 Connect the shield grounding strap from the cable to the shield ground stud on the tester adjacent to the receptacle.

Step 6 Connect the other end of the output cable to the item under test. This connection should be made mechanically secure, avoiding any sharp edges. It is usually advisable to tape the connection using a high-grade electrical tape to minimize corona at this point. If the voltages to be used are high or if for any other reason corona is anticipated, its effect may be minimized by using a corona ring or corona shield.

Step 7 Plug the line cord into a 120V, 60Hz, single-phase outlet and clip the ground wire coming from the line cord to the conduit receptacle mounting screw or other ground source.

If a bench-type tester or any other unit having an output bushing rather than a receptacle is used, the same procedure is followed except that the output cable is fastened to the bushing and should be supported along its length, preferably using nylon cord so that it stays well away from any

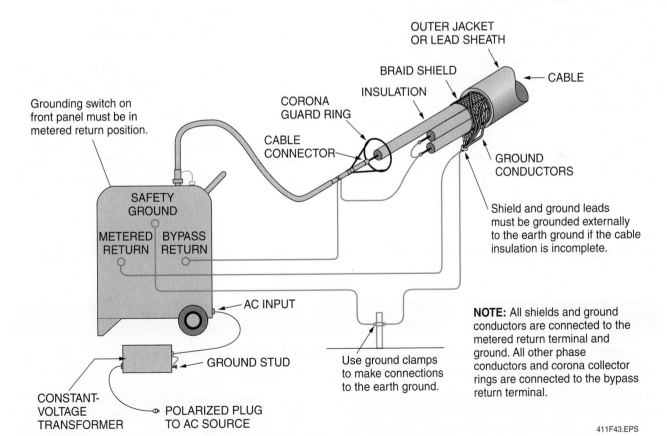

Figure 43 ◆ Hi-pot connections for testing multiple-conductor shielded cable.

grounded or other conducting surface. The high-voltage connection between the tester and the item under test should be as short and direct as possible.

When shielded cable is used, it is important that the shield be trimmed back from each end for a distance of about an inch or more for every 10,000V. The shield at the tester end should be fastened to the ground stud on the tester. The shield at the test end should be taped and no connection made to it.

If the item under test is beyond the reach of the tester cable supplied, an extension running between the end of the tester cable and the item under test must be used. If this extension is made of shielded cable, it is not only necessary to splice the central conductor to the end of the tester cable, but a jumper should also connect the shields of the two cables together at the splice, taking care to run the jumper well away from the splice itself to avoid leakage or breakdown.

The splice should be well taped. If no shielded cable is available, the single wire extension used should be supported well away from all other objects, as described previously. When operating at voltages within the corona levels, it is preferable that the tester be brought close enough to the item under test so that an extension cable is not needed. It should be noted that extension cable is a frequent cause of erratic readings, especially if the quality degrades at the high voltages used.

When line regulation is poor (because of location, heavy or intermittent equipment loads in the vicinity, or the use of portable generating equipment), it may be necessary to connect a voltage regulator (constant-voltage transformer) between the tester and the 120V supply. The use of a regulator is particularly desirable when using a tester with a sensitive vacuum tube current meter with ranges of less than 10μA full scale or when highly capacitive loads are being tested.

Since the DC output voltage of the tester is dependent on the AC line input voltage, an unstable line voltage will cause fluctuation of the tester output voltage with corresponding fluctuation of the current meter. This will be especially noticeable when measuring low current of a few microamperes or when testing loads having considerable capacitance. When the load is highly capacitive, it charges up to the test potential and tends to maintain that potential because of the high leakage resistance. A sudden drop in line voltage causes the tester output voltage to also drop, so that the load capacitance is then charged to a higher value than the tester output. Since the load is at a higher voltage than the tester, the direction of current flow will be reversed, and the current meter will be driven down the scale. A sudden increase in line voltage causes the opposite effect, with the tester voltage suddenly rising above the voltage to which the load has been charged, causing a rapid increase in the current meter reading. As a result, the current meter will fluctuate so that it might be difficult to obtain a meaningful leakage current reading.

The use of a line voltage regulator under such conditions will greatly increase the usefulness and ease of operation of the equipment. In general, the power rating of the regulator should be based on the maximum power output capability of the tester multiplied by a factor of about 1.3. For example, a tester rated at 75kV at 2.5mA can deliver an output power of:

$$75kV \times 2.5mA = 187.5W$$
$$187.5W \times 1.3 = 243.75W$$

Consequently, for this example, a 250W, harmonic-free regulator should be used. This is preferable to the standard type of unit, although either may be used with satisfactory results. If considerable testing at very light loads is done, it might also be desirable to have a smaller regulator to use under those conditions, since a small regulator operating at one-half to three-quarters of its capacity will give better results than a lightly loaded large regulator.

To use the voltage regulator, turn the microampere range switch to the highest range. On multiple voltage range units, turn the kilovolt range

Adjusting Voltage Regulator Microampere and Kilovolt Range Switches

The microampere range may be changed while the test is underway, but the kilovolt range may not. Safety interlocks within the unit automatically shut off the high voltage as the voltage range switch is turned while the test is in progress.

switch to the range that will allow testing to the maximum voltage required. The main power switch or circuit breaker may now be turned ON and the test begun.

Figures 44 through *50* illustrate typical connection diagrams. The experienced operator will soon develop the technique of making the connections most suitable to the test being performed. The diagrams are merely offered as a guide.

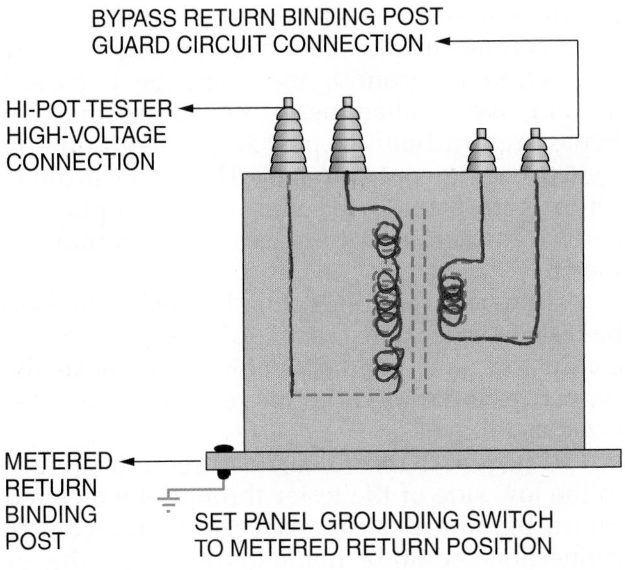

If the return path is grounded, set the panel grounding switch to the metered return position. The total load current (leakage/corona) will be read on the meter.

If the return path is not grounded, set the panel grounding switch to the bypass return position, and the effect of corona currents on the meter reading will be minimized.

> **NOTE**
>
> In the various connection diagrams given here, the guard circuit may be omitted by tying the connections shown going to the bypass return binding post to the metered return binding post instead. This is the difference shown by the connections in *Figures 48* and *49*.

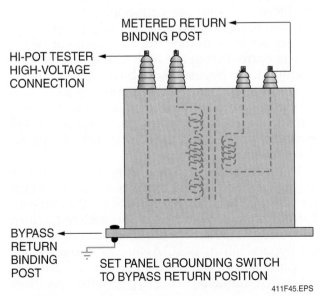

Figure 44 ◆ Testing a high-voltage winding to a grounded core or case with a secondary winding to a bypass.

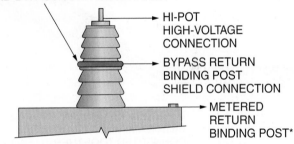

*Set panel grounding switch to metered return position

411F46.EPS

Figure 46 ◆ Measuring bushing internal leakage.

Figure 45 ◆ Testing a high-voltage winding to a low-voltage winding with the ground and case to a bypass.

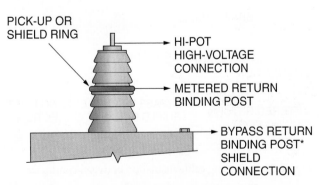

*Set panel grounding switch to bypass return position

411F47.EPS

Figure 47 ◆ Measuring bushing surface leakage.

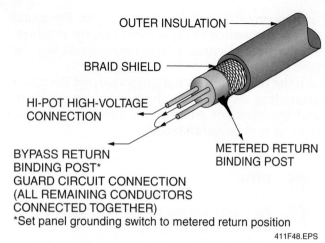

OUTER INSULATION

BRAID SHIELD

HI-POT HIGH-VOLTAGE
CONNECTION

BYPASS RETURN
BINDING POST*
GUARD CIRCUIT CONNECTION
(ALL REMAINING CONDUCTORS
CONNECTED TOGETHER)
*Set panel grounding switch to metered return position

METERED RETURN
BINDING POST

411F48.EPS

Figure 48 ◆ Testing cable insulation between one
conductor and the shield.

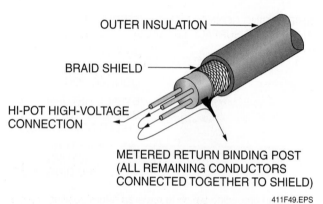

OUTER INSULATION

BRAID SHIELD

HI-POT HIGH-VOLTAGE
CONNECTION

METERED RETURN BINDING POST
(ALL REMAINING CONDUCTORS
CONNECTED TOGETHER TO SHIELD)

411F49.EPS

Figure 49 ◆ Testing cable insulation between one
conductor and all others and the shield.

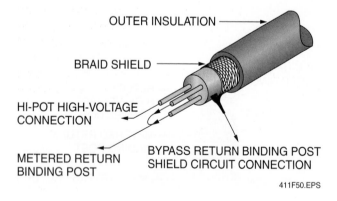

OUTER INSULATION

BRAID SHIELD

HI-POT HIGH-VOLTAGE
CONNECTION

METERED RETURN
BINDING POST

BYPASS RETURN BINDING POST
SHIELD CIRCUIT CONNECTION

411F50.EPS

Figure 50 ◆ Testing cable insulation between conductors.

5.4.0 Selective Guard Service Connections

In the connection diagrams, frequent reference is made to the bypass return or metered return binding post. The simplified sketch of the tester output circuit shown in *Figure 51* will aid in understanding the use of these terminals.

Examination of the diagram shows that any current to be measured must be returned to the tester through the metered return binding post, and any current that is not to be measured, but is to be bypassed around the meter, should be returned to the bypass return binding post.

As corona currents flow from the high-voltage connection to ground, they may be bypassed around the milliammeter by grounding the bypass return binding post with the grounding switch on the control panel. This gives a direct return path back to the high-voltage supply for corona currents and bypasses them around the meter.

Assume a three-conductor shielded cable is to be tested for leakage current between one of the conductors and the outer shield. To make the proper tester connection, the test should be analyzed as follows.

The current that is to be measured must return to the low side of the tester through the metered return binding post; therefore, the high-voltage connection should be made to the one conductor in question and the shield should be connected to the metered return binding post.

Since the measurement of current flow between the high-voltage connection and the other two conductors in the cable is not desirable, these two conductors should be connected together and returned to the bypass return binding post.

In order to minimize the measurement of the corona current from the high-voltage connection to ground, the bypass return binding post should be grounded by setting the grounding switch on the control panel to the bypass position. This will provide a return path for the corona current around the meter, and the corona current will not affect the meter readings.

To measure the leakage current between one conductor and the other two conductors (which are tied together), the return path for the two conductors should be through the metered return post, and the shield should then be connected to the bypass return post. Again, to minimize the measurement of the corona current, the grounding switch should be in the bypass return position.

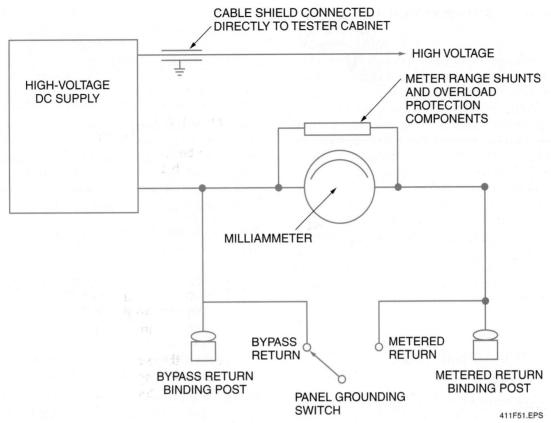

Figure 51 ◆ Simplified hi-pot tester output circuit diagram.

The corona current alone may be measured by connecting the shield and the two conductors together and connecting them to the bypass return binding post. If the grounding switch is then placed in the metered return position, the return path for the corona current will be through the microammeter, and the corona current will be the only current being measured.

The grounding switch on the control panel will either include or exclude the corona current in the meter reading, depending upon its position.

It was previously assumed that there were no restrictions placed on the cable with respect to grounding. This was done to illustrate the general case. In many cases, however, the shield is permanently grounded, and therefore certain modifications in the test setup must be made in order to achieve corona-free stability and accuracy in the leakage current readings.

If it is desired to measure the leakage current from one conductor to the shield, the shield should be connected to the metered return binding post as before. However, since the shield is now permanently grounded, the metered return binding post should also be grounded by placing the panel grounding switch in the metered return position. If the panel grounding switch were placed in the bypass return position, then the metered return post and the bypass return post would both be grounded, shorting out the microammeter. Therefore, the grounding switch must be put in the metered return position when one side of the test item is permanently grounded.

As outlined above, all unwanted currents that flow between the high-voltage point and any ungrounded elements in the cable may be returned through the bypass line and therefore be excluded from the leakage current reading. However, this does not include corona current because corona current always flows between the high-voltage point and ground. Since we cannot shunt the corona current around the microammeter by grounding the bypass return line as we did before, we must use some auxiliary method of intercepting the corona current before it reaches ground. This can be easily accomplished by the use of a corona or guard ring. A corona or guard ring is nothing more than a metal loop that encompasses the high-voltage connection and intercepts the corona current. The ring is connected to the tester through the bypass return binding post and therefore keeps the corona current out of the meter reading.

5.5.0 Corona Guard Ring and Guard Shield

When using a corona ring, the diameter should be made as small as possible, without causing an arc to jump between the high-voltage connection and the ring. This will ensure interception of as much corona current as possible. A wire should be connected to the ring and the other end connected to the bypass return binding post. This will bypass the intercepted corona current around the microammeter and allow essentially the same stability and accuracy that was obtainable previously.

Corona may also be minimized by electrically shielding or smoothing the connection. This may be done by smoothly wrapping the connection with metal foil or conductive tape or enclosing it in a round can or cover so that all jagged edges, wire points, and rough surfaces are within the metal enclosure. This corona shield (*Figure 52*) must be electrically connected to the high-voltage point and completely insulated from ground so that both the shield and high-voltage connection are at the same potential.

5.6.0 Detailed Operating Procedure

These instructions are meant to illustrate a typical operating procedure. An actual test must account for special conditions peculiar to the installation, ranges, and operating controls of the tester available, as well as the requirements of the operator. To test the system, proceed as follows:

Step 1   Assemble the necessary equipment and supplies at the test site. In addition to the tester, connecting cable, and so forth, bring a clipboard or other writing surface, pencils, colored pencils, scratch paper, sheets of 8½" by 11" graph paper (which is very convenient for actually plotting the test results while the test is underway), an electronic pocket calculator, and a watch.

Step 2   Determine the type of test to be made. For example, is the voltage going to be raised to breakdown, or is the voltage going to be raised to the specific test value only? If the item on the test is found to be defective, should the test stop before breakdown until the repairs can be made later, or should the test be continued to the preselected value regardless of breakdown so that repairs can be made immediately?

 WARNING!
This test may only be performed by qualified individuals operating under the appropriate safe work plan or permit.

Step 3   Determine the maximum test voltage to use. Since the value chosen depends not only on the type of test being made, but also on the material to be tested, it is necessary that all factors involved be considered. As a rough rule of thumb, proof testing is usually conducted at 60% of the factory test voltage. Acceptance testing would call for a maximum value of about 80% of the factory test voltage. For maintenance testing or on older work, the minimum test voltage chosen should be

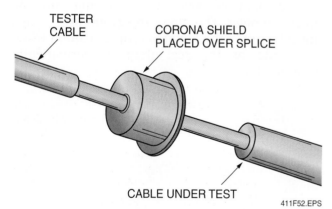

TESTER CABLE

CORONA SHIELD PLACED OVER SPLICE

CABLE UNDER TEST

411F52.EPS

Figure 52 ◆ Corona shield.

at least 1.7 times the operating voltage up to a maximum value between 50% and 60% of the factory test voltage.

In many cases, the value of the maximum test voltage to be used may be obtained from the cable manufacturer.

Step 4 Choose suitable increments so that the voltage may be raised to the final test value in about eight to ten steps (more or less to suit the individual conditions).

Step 5 Once the maximum test voltage and increments have been chosen, the graph paper may be prepared so that the test results can be plotted as the readings are taken. In this way, the condition of the cable is under constant surveillance and if testing short of breakdown is decided upon, the operator has the necessary control to stop the test at the start of the knee of the current-versus-voltage curve.

To prepare the graph paper, mark the current scale on the 8½" side and the voltage scale along the horizontal 11" side. Suitable voltage increments should be chosen to produce a convenient plot. The current scale should include values well above the expected maximum leakage current. If the magnitude of leakage current is not known, it may be advisable to either make a low-voltage test run or to hold off plotting the first few points until the magnitude is seen.

Step 6 Connect the tester to the equipment under test, as described previously.

Step 7 Make sure the adjustable voltage control is turned to the OFF position (fully counterclockwise), the high-voltage switch is in the center or OFF position, and the power switch is in the OFF position. Connect the input line cord to the 120V, 60Hz receptacle, and connect the alligator clip at the end of the line cord to ground.

Step 8 Turn the kilovoltmeter range switch and microammeter range switch to the appropriate positions.

Step 9 Turn the main ON/OFF switch or circuit breaker to the ON position.

Step 10 Operate the high-voltage ON switch. Note that the lever-type high-voltage switch has a spring return from the DOWN position but will remain in the UP position. If it is desired to keep the high voltage under strict manual control, use the DOWN position of the high-voltage switch. As soon as it is released to the OFF position, the high voltage will be turned off, and it will become necessary to return the powerstat control to zero before the high voltage can be reapplied.

If the high-voltage switch is placed in the UP position, the switch may be released and will remain in the UP position, and high voltage will be available until the switch is manually returned to the center OFF position. On units with a keyed interlock, it is necessary to turn the key to the right to enable the UP position of the high-voltage lever switch to be operated.

Step 11 Slowly rotate the variable voltage control from the extreme counterclockwise position (zero volts) to raise the voltage to the first increment, as previously determined. The voltage should be raised at a slow enough rate to avoid having the microammeter pointer go off the scale. The slower the voltage is raised, the lower the maximum charging current and microammeter reading. If the charging current is allowed to exceed the maximum current rating of the tester, the circuit breaker will trip. The circuit breaker will also trip if the unit under test fails.

Since most testers are equipped with zero return interlocks for operator safety, it is necessary to return the voltage control to the zero position before high voltage can be re-applied once it is shut off, whether due to circuit breaker tripping, kilovolt range selector switch rotation, or operation of the high-voltage ON lever switch to the OFF position.

Step 12 When the first increment of voltage is reached, watch the microammeter indicate gradually decreasing current as the cable or machine under test becomes charged. As the current drops to lower portions of the scale, switch the microammeter progressively to lower ranges until the current stabilizes. Note the length of time it takes for the current to stabilize before accurate results can be obtained. The same stabilization period is allowed at each voltage step. After the current is stabilized, whether it takes 10 seconds, a minute, or 10 minutes, record the value of the voltage and current.

Step 13 If the microampere scale for plotting the current-versus-voltage curve has been predetermined, the current reading at the first voltage step can now be plotted. If the scale has not yet been determined, record this value and after a few more increments, a scale can be determined and the plotting can take place.

Step 14 After the current is stabilized and a point is plotted, turn the microammeter range switch back to the higher range, and gradually raise the voltage to the next increment. Allow the current to stabilize for the same length of time as for the previous increment. Take a second reading and continue the test, gradually raising the voltage step-by-step and plotting the current versus voltage at each step. It is important that the voltage be raised at the same rate in each step.

Step 15 Closely watch the rate of change of current at each voltage increment. The curve should indicate a linear (even) rate of change up to the final test voltage. If this is the case, the material being tested is all right. Upon any indication of a rapid rise in current, stop the test immediately, unless, of course, you wish to test to breakdown. Even a slight knee in the curve may be an indication of imminent breakdown.

Step 16 This step-by-step raising of voltage and plotting of current should be continued until the maximum test voltage previously selected is reached. If a current-versus-time plot is also desired, the time should be noted immediately upon reaching the final voltage step. Then, as the current slowly decays, readings of current and time should be taken at suitable increments and the current-versus-time curve may be plotted. For a safe insulation system, this curve should indicate a continuous decrease in leakage current with time until the current stabilizes. At no time during this part of the test should there be any increase in current.

Step 17 At the end of the test, move the high-voltage lever switch to the OFF position. Allow the material being tested to discharge either through the internal discharge circuit of the tester or by using a hot stick and rubber gloves and grounding the output.

Step 18 The kilovoltmeter on the tester will give a direct indication of the voltage at the output terminal regardless of whether the tester is ON or OFF. The cable or equipment being tested should not be touched until the kilovoltmeter reads zero.

Step 19 After the kilovoltmeter reads zero, disconnect all test leads, and immediately reconnect all cable conductors or equipment terminals to ground and leave grounded.

 WARNING!
Rubber gloves must be worn at all times during this portion of the test. The dielectric absorption effect of cable in particular, and to some extent that of rotating equipment, will tend to restore a dangerous voltage on the disconnected item until it has been adequately grounded for a considerable length of time.

Step 20 The other phases of a multi-phase installation may be tested in a similar manner.

In many cases, once the fault occurs, its exact location may be determined without too much difficulty. However, in those cases where a fault occurs in a long run of cable, special fault-finding techniques incorporating special testing equipment could be of great assistance.

5.7.0 Go/No-Go Testing

The tests described previously provide a good indication of the condition of the equipment or cable, and with proper interpretation, may even allow prediction of the expected life of the device under test. However, to do the job properly takes time and a certain amount of experience and skill, both in conducting and evaluating the operation.

A much shorter and simpler test may be sufficient in those cases where the contractor, purchaser, or operator is only interested in knowing whether or not the installation meets a specific high-voltage breakdown requirement (that is, whether the equipment or cable will withstand X kilovolts without breakdown).

WARNING!
This test may only be performed by qualified individuals operating under the appropriate safe work plan or permit.

The procedure to be followed in conducting a go/no-go test is as follows:

Step 1 Set the microammeter range switch to the highest range.

Step 2 Gradually raise the test voltage at a rate that will keep the charging current below the microammeter's full-scale point. The voltage is raised at this steady rate until the required test value is reached.

Step 3 Maintain the voltage at this value for the length of time required by the specification, or in the absence of any specification, until after the current has stabilized for a minute or more.

Step 4 If the leakage current does not become excessive (remains below the maximum specified value or is low enough to avoid tripping the circuit breaker), the equipment has passed the test.

Step 5 Failure is indicated by a gradual or abrupt increase of current sufficient to trip the circuit breaker. This leakage current should not be confused with the charging current. The charging current may be minimized by raising the voltage at a slower rate.

Step 6 If the current seems to be rising too high, stop raising the voltage at that point and wait. If the current immediately starts decreasing, the indication is that of charging current, and the test may be continued.

Step 7 If the current holds steady or increases, it is due either to leakage current or to corona current. Methods of minimizing the effect of corona current were explained earlier.

Step 8 At the end of the test, reduce the voltage to zero, and follow the procedures recommended by the tester manufacturer.

NOTE

This type of test is not by any means a thorough analysis of cable condition. However, it is a sufficient test and is all that is necessary in a great number of cases. For example, the contractor installing the cable must immediately know if the installation is good. If it is defective, this test will reveal it in the shortest possible time. If the installation is faulty, the test cannot harm it any further, and the contractor may proceed to make the necessary repairs. The only information sought in this type of test is that the cable will not fail below a required test voltage level.

5.8.0 Insulation Resistance Measurements

The DC tester is an invaluable tool for making insulation resistance measurements at higher voltages. Using the same connections and following the same precautions given for high-potential testing, raise the voltage to the desired value and read the microammeter after the current has stabilized (no further decay). The insulation resistance may then be calculated using Ohm's law:

$$R = E \div I$$

Where:

R = insulation resistance (in megohms)

E = voltage (kV meter reading $\times$ 1,000)

I = current (in microamperes)

The insulation resistance may be calculated at each voltage step as the leakage current-versus-voltage test is made. Curves of insulation resistance versus voltage will yield much information to the experienced operator. It also helps to keep records of the insulation resistance at a specific voltage so that they can be compared from test to test or from year to year. This will give a good basis of comparison for proper insulation evaluation.

Insulation resistance may also be measured with a standard megger. These instruments will give a direct reading of insulation resistance at a 500V (or higher, depending on the model) test potential.

Insulation resistance measurements help in determining whether or not the items under test are suitable for hi-pot testing. If the insulation resistance is abnormally low, the material is defective and there is no need to make the hi-pot test. If the insulation resistance is high, then the hi-pot test may be made.

Instead of waiting for the current to stabilize before calculating the insulation resistance, readings of voltage and current can be taken at fixed time intervals during the leakage current-versus-time test, and calculations of insulation resistance as a function of time can be made. A plot of the insulation resistance versus time is known as the absorption curve and is often used when testing rotating machinery. In a good insulation system, this curve will rise rapidly at first (because of the decaying charging current) and gradually level off (an indication of stabilization of leakage current). The more pronounced the rise, the better the insulation. A relatively flat curve will indicate moist or dirty insulation.

Insulation Resistance Tests

The use of a megger for measuring insulation resistance should not be confused with a high-potential test, since insulation resistance, in general, has little or no direct relationship to dielectric or breakdown strength. For example, an actual void in the insulation may show an exceedingly high value of insulation resistance even though it would permit breakdown at a relatively low voltage.

The ratio of the insulation resistance at the end of 10 minutes to the insulation resistance at the end of one minute is called the polarization index. If the value of the polarization index is less than two, it is usually an indication of excessive moisture or contamination. Values as high as 10 or more are common when testing large motors or generators.

Caution should be exercised in evaluating the importance of insulation resistance measurements. The absolute value is not critical, but the relative order of magnitude or the change in value as the test progresses is significant. Such factors as temperature and humidity can have a very large effect on the reading.

1. Which of the following can render a high-voltage splice useless?
   a. Dirt
   b. Moisture
   c. Any foreign matter
   d. Heat

2. Which of the following cable configurations has the smallest overall diameter for the same amount of current-carrying ability?
   a. Compact stranding
   b. Compressed stranding
   c. Eccentric stranding
   d. Concentric stranding

3. Solid wire is commonly used in industrial shielded power cable.
   a. True
   b. False

4. When air breaks down or ionizes, it produces ____.
   a. moisture
   b. hydrogen sulfate
   c. photoreceptors
   d. corona discharges

5. Which of the following best describes the semi-conductive layer between the conductor and the insulation that compensates for air voids that exist between the conductor and the insulation?
   a. Metallic shield
   b. Jacket
   c. Polyethylene
   d. Strand shielding

6. Resonant grounded systems typically require an insulation level of____ percent.
   a. 100
   b. 133
   c. 153
   d. 173

7. What are the two materials normally used to construct cable shields?
   a. Semi-con and PVC plastic
   b. Metal and semi-conductive material
   c. Copper and aluminum
   d. Silver-plated aluminum (silver and aluminum)

8. What is normally the final step in a high-voltage cable splice?
   a. Rejacketing
   b. Reinsulating
   c. Reshielding
   d. Cleaning

9. When splicing aluminum conductors, which of the following should be applied to the contact areas of the connector?
   a. Antioxide paste
   b. Abrasives
   c. Solvents
   d. Vinyl insulating tape

10. One advantage of AC current testing is ____.
    a. smaller, less expensive test equipment
    b. less chance of equipment damage
    c. quantifiable test results
    d. true simulation of operating conditions

Summary

This module discussed high-voltage terminations and splices. Of the nearly limitless variety of cables in use today, five of the most common are:

- Ribbon or tape shielded
- Drain wire shielded
- Cable UniShield®
- Concentric neutral (CN)
- Jacketed concentric neutral (JCN)

All power cables are essentially the same, consisting of the following:

- Conductor
- Strand shield
- Insulation
- Insulation shield system (semi-con and metallic)
- Jacket

Each component is vital to the optimal performance of the power cable and must be understood in order to make a dependable splice or termination.

There are several methods used to terminate shielded power cables: hand taping, heat-shrink kits, molded rubber, and cold-shrink kits. Each type of splice requires special training to execute properly. Heat-shrink and molded rubber terminations are considered more reliable termination methods, especially where moisture and contamination are present.

Cable is tested using a variety of methods, including hi-pot testing, go/no-go testing, and DC insulation testing. Cable testing may only be performed by qualified personnel.

Notes

Trade Terms
Introduced in This Module

Armor: A mechanical protector for cables; usually a formed metal tube or helical winding of metal tape formed so that each convolution locks mechanically upon the previous one (interlocked armor).

Dielectric constant (K): A measurement of the ability of a material to store a charge.

High-potential (hi-pot) test: A high-potential test in which equipment insulation is subjected to a voltage level higher than that for which it is rated to find any weak spots or deficiencies in the insulation.

Insulation shielding: An electrically conductive layer that provides a smooth surface in contact with the insulation outer surface; it is used to eliminate electrostatic charges external to the shield and to provide a fixed, known path to ground.

Pothead: A terminator for high-voltage circuit conductors that keeps moisture out of the insulation and protects the cable end, along with providing a suitable stress relief cone for shielded-type conductors.

Ribbon tape shield: A helical (wound coil) strip under the cable jacket.

Shield: A conductive barrier against electromagnetic fields.

Shielding braid: A shield of small, interwoven wires.

Splice: Two or more conductors joined with a suitable connector and reinsulated, reshielded, and rejacketed with compatible materials applied over a properly prepared surface.

Strand shielding: The semi-conductive layer between the conductor and insulation that compensates for air voids that exist between the conductor and insulation.

Stress: An internal force set up within a body to resist or hold it in equilibrium.

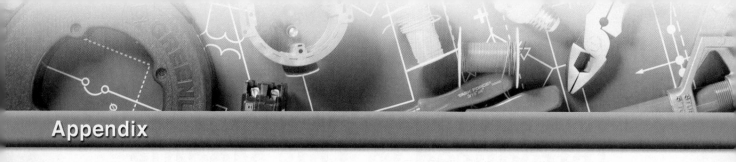

Appendix

CABLE DIELECTRIC TEST

PROJECT		DATE	24-Aug-07
JOB #			

CIRCUIT DESIGNATION	CABLE P1022-P01C		

TEST EQUIPMENT	MODEL #	SERIAL #	CAL DATE
Hypotronics	800PL	01911-08	28-Mar-07

Cable Nameplate Data: **8kV 133% 115 mil EPR, TS, PVC, CT Rated**

D-C TEST DATA				
		PHASE A	PHASE B	PHASE C
TIME STARTED		1:20	1:45	2:15
TEST VOLTAGE		CURRENT µA		
KV	7	1.5	1+	1+
KV	14	2+	1.6	2.0
KV	21	4.0	3.0	2.8
KV	28	5.0	4.0	4.3
KV	35	5.2	4.5	4.8

Cable factory test voltage	45kV
Acceptance Test DC voltage	36kV

Cable is: NEW **X** USED

Cable Size: **500KCM** Length: **750 FT**

Operating System kV: **4.16**

System is: GROUNDED **X** UNGROUNDED

Shield type: **Copper Tape**

Cable Mfr.: **General**

Temperature **82°F** Humidity **25%**

Termination manufacturer / type: **Raychem HVT-2**

Splice type and locations: **N/A**

TIME AFTER 100% TEST VOLTAGE IS APPLIED			
0 SECONDS	5.2	4.6	4.9
15 SECONDS	4.8	4.4	4.5
30 SECONDS	4.0	3.9	4.1
45 SECONDS	3.7	3.5	3.6
1 MINUTE	3.5	2.9	3.3
2 MINUTES	3.2	2.6	3.4
3 MINUTES	3.0	2.8	3.0
4 MINUTES	3.2	2.5	3.0
5 MINUTES	3.1	2.6	2.9
10 MINUTES	3.1	2.5	2.7
15 MINUTES	3.1	2.4	2.5
KVDC AFTER 1 MIN DECAY	‹1	0	‹1
STABILIZATION TIME	20 SEC.		

SAFETY CHECK:	BY:
Test set grounded	MP
Personal safety equipment	MP
Personnel clearance	MP
Lockout in place	MP / KV
Barriers in place	MP

REMARKS & FAULT LOCATION **MP - Test normal**

If leakage current rises with steady or decreasing voltage drop the voltage quickly to prevent insulation breakdown unless breakdown is desired to locate the fault.

TEST PERFORMED BY: *Megan Paye*

WITNESSED BY: *Jordan Vidler*

411A01.EPS

This module is intended to present thorough resources for task training. The following reference works are suggested for further study. These are optional materials for continuing education rather than for task training.

American Electrician's Handbook, Terrell Croft and Wilfred I. Summers. New York, NY: McGraw-Hill, 1996.

National Electrical Code® Handbook, Latest Edition. Quincy, MA: National Fire Protection Association.

NCCER makes every effort to keep these textbooks up-to-date and free of technical errors. We appreciate your help in this process. If you have an idea for improving this textbook, or if you find an error, a typographical mistake, or an inaccuracy in NCCER's Contren® textbooks, please write us, using this form or a photocopy. Be sure to include the exact module number, page number, a detailed description, and the correction, if applicable. Your input will be brought to the attention of the Technical Review Committee. Thank you for your assistance.

Instructors – If you found that additional materials were necessary in order to teach this module effectively, please let us know so that we may include them in the Equipment/Materials list in the Annotated Instructor's Guide.

Write: Product Development and Revision
National Center for Construction Education and Research
3600 NW 43rd St., Bldg. G, Gainesville, FL 32606

Fax: 352-334-0932

E-mail: curriculum@nccer.org

Craft

Module Name

Copyright Date

Module Number

Page Number(s)

Description

(Optional) Correction

(Optional) Your Name and Address

Special Locations

University of Cincinnati – A Way with Walls

The main living area of the University of Cincinnati Solar Decathlon home is a single airy space that has no walls to divide cooking, eating, and dining areas. Innovative walls are key to the home's inventive design. The south-facing wall separating the living space from the home's courtyard is glass. The glass wall lets in warming sunlight in the winter and provides great daylighting. The wall's specially produced triple-pane, low-e glass maintains excellent insulation, and louvered shades keep out unwanted summer solar heating.

26412-08

26412-08
Special Locations

Topics to be presented in this module include:

1.0.0	Introduction	12.2
2.0.0	Assembly Occupancies	12.3
3.0.0	Theaters and Similar Locations	12.5
4.0.0	Carnivals, Circuses, Fairs, and Similar Events	12.14
5.0.0	Agricultural Buildings	12.18
6.0.0	Marinas and Boatyards	12.21
7.0.0	Temporary Installations	12.26
8.0.0	Wired Partitions	12.29
9.0.0	Swimming Pools, Fountains, Hot Tubs, and Similar Installations	12.30
10.0.0	Natural and Manmade Bodies of Water	12.44

Overview

Electrical wiring requirements for most commercial and residential occupancies and common electrical equipment such as appliances and motors can be found in the first four chapters of the *National Electrical Code®*. There are, however, electrical systems in certain occupancies and structures, and electrical equipment with unique applications that require additional or special wiring methods because of their intended usage, environmental conditions, or potential exposure to electrical hazards. This module is intended to cover some of the more common electrical installations regulated by *NEC Chapter 5, Special Occupancies*, including places of assembly (*NEC Article 518*), theaters (*NEC Article 520*), carnivals and circuses (*NEC Article 525*), agricultural buildings (*NEC Article 547*), and marinas (*NEC Article 555*). It also covers *NEC Chapter 6, Special Equipment*, including wired office partitions (*NEC Article 605*), swimming pools and fountains (*NEC Article 680*), and natural or manmade bodies of water (*NEC Article 682*).

Objectives

When you have completed this module, you will be able to do the following:

1. Identify and select equipment, enclosures, devices, and wiring methods approved by the current *NEC®* for the following special occupancies or installations:
   - Places of assembly
   - Theaters
   - Carnivals, circuses, and fairs
   - Agricultural buildings
   - Marinas and boatyards
   - Temporary wiring
   - Office partitions
   - Swimming pools, fountains, hot tubs, and similar installations
   - Natural and manmade bodies of water
2. Comply with *NEC®* requirements regarding equipotential planes as they refer to bonding and grounding in water-related installations.
3. Determine electrical datum planes in water-related installations.

Trade Terms

Border light
Dead-front
Electrical datum plane
Equipotential plane
Extra-hard usage cable
Festoon lighting
Fire rating
Footlight
Isolated winding-type transformer
Landing stage
Portable switchboard
Positive locking device

Potting compound
Proscenium
Reactor-type dimmer
Resistance-type dimmer
Rheostat
Shunt trip
Site-isolating device
Solid-state phase-control dimmer
Solid-state sine-wave dimmer
Unbalanced load
Voltage gradients

Required Trainee Materials

1. Pencil and paper
2. Copy of the latest edition of the *National Electrical Code®*

26413-08 Introductory Skills for the Crewleader

26412-08 Special Locations

26411-08 Medium-Voltage Terminations/Splices

26410-08 Motor Operation and Maintenance

26409-08 Heat Tracing and Freeze Protection

26408-08 HVAC Controls

26407-08 Advanced Controls

26406-08 Specialty Transformers

26405-08 Fire Alarm Systems

26404-08 Basic Electronic Theory

26403-08 Standby and Emergency Systems

26402-08 Health Care Facilities

26401-08 Load Calculations – Feeders and Services

ELECTRICAL LEVEL FOUR

ELECTRICAL LEVEL THREE

ELECTRICAL LEVEL TWO

ELECTRICAL LEVEL ONE

CORE CURRICULUM: Introductory Craft Skills

412CMAP.EPS

Before you begin this module, it is recommended that you successfully complete *Core Curriculum*; *Electrical Level One*; *Electrical Level Two*; *Electrical Level Three*; and *Electrical Level Four*, Modules 26401-08 through 26411-08.

This course map shows all of the modules in *Electrical Level Four*. The suggested training order begins at the bottom and proceeds up. Skill levels increase as you advance on the course map. The local Training Program Sponsor may adjust the training order.

1.0.0 ◆ INTRODUCTION

Special locations such as places of assembly, theaters, carnivals, agricultural buildings, marinas, temporary installations, wired partitions, swimming pools, or bodies of water may require more stringent wiring methods than traditional occupancies. For the most part, these special wiring methods provide added protection against unique hazards that may exist due to large crowds (places of assembly or theaters), exposed cable wiring (carnivals or fairs), livestock protection (agricultural buildings), or a mixture of electricity and water (marinas, swimming pools, and other bodies of water).

Temporary wiring installations and wired partitions present their own unique electrical hazards. Temporary wiring does not afford long-term conductor or device protection, while wired partitions such as those used in office buildings are often constructed of sheet metal containing sharp edges that present a potential for ground fault conditions.

Electricians who install wiring in these locations must refer to and comply with specific articles in the *NEC*®.

INSIDE TRACK

Special Occupancies – Construction Job Site Trailers

Wiring methods for construction job site trailers such as the one shown here are regulated by *NEC Section 550.4*.

412SA01.EPS

Special Equipment – Electric Vehicle Charging Systems

As energy costs increase, hybrid and electric automobiles have become increasingly common. *NEC Article 625* covers wiring for electric vehicle charging systems. This photo shows an electrical charging system for wet-cell batteries. The code contains provisions for using a listed hybrid vehicle charging system as an emergency power backup in the event of a power failure (see *NEC Section 625.26*).

412SA02.EPS

2.0.0 ◆ ASSEMBLY OCCUPANCIES

In order to qualify as an assembly occupancy per *NEC Article 518*, a building or structure must be designed to hold 100 or more people and used for such purposes as meetings, worship, entertainment, eating and drinking, amusement, or a place to await transportation. Convention centers, churches, arenas, dance halls (*Figure 1*), auditoriums, cafeterias, and bus or train stations are some examples of places of assembly.

Often, an assembly occupancy is attached to other buildings, structures, or rooms, such as in the case of the civic center/chamber of commerce building shown in *Figure 2*. Only the portion of the building designed to assemble more than one hundred people is regulated by *NEC Article 518*. Portions of the structure such as offices and kitchens are governed by other appropriate articles in the *NEC®*.

NOTE

If the assembly area contains a projection booth, stage platform, or accommodations for the presentation of theatrical or musical performances, it is governed by *NEC Article 520* (theaters or audience areas), not *NEC Article 518*.

412F01.EPS

Figure 1 ◆ Oldest dance hall in Texas.

2.1.0 Wiring Methods in Assembly Occupancies

For the most part, the building materials used in assembly occupancies are required to have a specific **fire rating** to retard the spread of fire through adjacent spaces. Approved wiring methods allowed in fire-rated construction include metal raceways, flexible metal raceways, nonmetallic raceways encased in at least 2 inches of concrete,

Figure 2 ◆ Combination civic center/chamber of commerce under construction.

412F02.EPS

Figure 3 ◆ MC cable.

412F03.EPS

mineral-insulated, metal-sheathed (MI) cable, metal-clad (MC) cable (*Figure 3*), or armored (AC) cable. Raceways or cable assemblies must either qualify as an approved grounding conductor per *NEC Section 250.118* or contain an insulated grounding conductor sized in accordance with *NEC Table 250.122*. In portions of assembly occupancies that are not required to be of fire-rated construction, wiring methods may include nonmetallic-sheathed (NM) cable, AC cable, electrical nonmetallic tubing, and rigid nonmetallic conduit (PVC). Wiring methods for assembly occupancies are covered in *NEC Section 518.4*.

CAUTION

Local codes may prohibit certain wiring methods. Always check with the authority having jurisdiction before selecting a wiring method.

NOTE

Temporary wiring used to power display booths in trade shows and similar exhibitions commonly held in assembly occupancies is covered by *NEC Article 590, Temporary Installations*.

2.2.0 Finish Ratings

In addition to fire ratings, building codes often establish a finish rating for combustible (wood) supports when they are used in the construction of assembly buildings or structures. A finish rating for places of assembly is defined as the length of time it takes for a wood support exposed to fire to reach an average temperature rise of 121°F or an individual temperature rise of 325°F when the temperature is measured on the plane of the wood nearest the fire.

The use of electrical nonmetallic tubing and rigid nonmetallic conduit in places of assembly containing combustible supports is restricted to certain areas including club rooms, conference and meeting rooms in hotels or motels, courtrooms, dining facilities (*Figure 4*), restaurants, mortuary chapels, museums, libraries (*Figure 5*), and places of religious worship. In addition to location restrictions, the raceways must be installed and concealed in walls, floors, and ceilings that provide a thermal barrier of material with a finish rating of at least 15 minutes. See *NEC Section 518.4(C)*.

Figure 4 ◆ Dining facility in assembly occupancy.

Figure 5 ◆ Library.

3.0.0 ◆ THEATERS AND SIMILAR LOCATIONS

NEC Article 520 covers indoor or outdoor spaces used for dramatic or musical presentations, motion picture projection, or similar purposes. This article also covers specific audience seating areas within motion picture or television studios. A typical seating area is shown in *Figure 6*.

Figure 6 ◆ Theater seating area.

3.1.0 Wiring Methods in Theaters, Audience Areas, and Similar Locations

Wiring methods for theaters are covered in *NEC Section 520.5*. Fixed wiring methods include metal raceways, nonmetallic raceways encased in a minimum of 2 inches of concrete, MI cable, MC cable, or AC cable containing an insulated equipment grounding conductor sized in accordance with *NEC Table 250.122*. Areas in occupancies not required to be of fire-rated construction, as determined by applicable building codes, may contain wiring methods consisting of nonmetallic-sheathed cable, AC cable, electrical nonmetallic tubing, or rigid nonmetallic conduit.

Wiring for portable equipment such as **portable switchboards** (*Figure 7*), stage set lighting, and stage effects may be in the form of approved flexible cords and cables as long as the cords or cables are not fastened or secured by uninsulated staples or nails. See *NEC Section 520.5(B)*.

3.1.1 Number of Conductors in a Raceway

The maximum number of conductors in metallic or nonmetallic raceways as permitted by the *NEC®* for theaters or audience areas of motion picture and television studios follows the requirements of *NEC Chapter 9, Table 1*. It states that for one conductor, the raceway fill is limited to 53% of the raceway's interior cross section; for two conductors, it is 31%; and for more than two conductors, it is 40%.

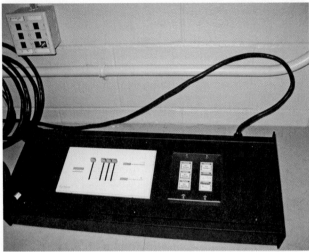

412F07.EPS

Figure 7 ◆ Portable switchboard.

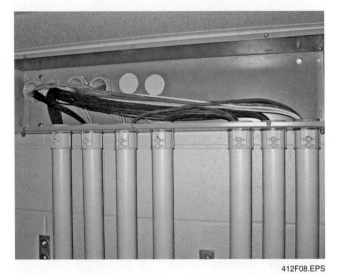

412F08.EPS

Figure 8 ◆ Gutter with more than 30 wires installed in a theater.

When conductors are installed in a gutter or wireway in areas covered by *NEC Article 520,* the total cross-sectional area of all conductors cannot exceed more than 20% of the gutter or wireway cross-section in which they are installed. This complies with the *NEC®* cross-section requirements for most conductors installed in an auxiliary gutter (*NEC Section 366.22*) or wireway (*NEC Section 376.22*). However, if the number of conductors in an auxiliary gutter or wireway exceeds 30, derating of the conductor ampacity (which is typically required in other locations) is not required in theaters or audience areas per *NEC Section 520.6.* See *Figure 8.*

3.1.2 Protective Guarding, Branch Circuits, and Portable Equipment

Electrical circuits, switches, and equipment in theaters and similar areas covered by *NEC Article 520* are often operated by people with limited or no electrical safety training. For this reason, any energized part must be enclosed or guarded to prevent electrical shock. In addition, all switches must be designed for external operation only. Dimmers, **rheostats**, and similar devices must be

installed in cabinets that enclose all energized parts (*Figure 9*).

In addition to supplying receptacles, branch circuits in theaters and similar areas may also supply stage set lighting. Receptacles and conductors in these branch circuits must have an ampere rating equal to or greater than the branch circuit overcurrent protective device to which they are connected. As with any branch circuit receptacle, the voltage rating of the receptacle must not be less than the circuit voltage. See *NEC Section 520.9.*

Portable electrical equipment such as stage or studio lighting, and power distribution equipment may be temporarily used outdoors as long as the use of such equipment is continually supervised by qualified personnel and the equipment is barricaded to prevent public access.

3.2.0 Fixed Stage Switchboards

The design and construction of switchboards installed in theaters and similar locations must be of the **dead-front** type (*Figure 10*) and comply with *NEC Article 408* unless the switchboard has

Figure 9 ◆ Switchboard panel with guard grating in place at the top.

412F10.EPS

Figure 10 ◆ Fixed stage switchboard.

been determined to be suitable for stage switchboard use based on tests performed by a qualified laboratory in which approved test standards and principles are applied. Stage switchboards are covered in *NEC Article 520, Part II*.

Stage switchboard equipment that is designed with exposed electrical parts on the back side of the board must be recessed in building walls and covered by wire mesh grilles or other approved protective means to prevent personnel contact. See *NEC Section 520.22*. Any rear access door that exposes energized parts must be of the self-closing type. If the design of the equipment is not of the dead-front type or cannot be recessed into a wall, a metal hood that extends the full length of the switchboard must be installed to protect the equipment from falling objects (*NEC Section 520.24*).

Stage lighting switchboards must include accessible overcurrent protection for stage lighting and receptacle branch circuits, including receptacles used for cord-and-plug equipment (*NEC Section 520.23*). Overcurrent protective devices for non–stage lighting controlled by dimmers may be installed in stage lighting switchboards.

3.2.1 Dimmers

Dimmers used for stage and auditorium lighting may be designed to control voltage through either the grounded or ungrounded circuit conductor supplying the controlled lighting. In installations where the ungrounded conductor is connected through the dimmer, the dimmer must be protected by an overcurrent device with a rating of no greater than 125% of the dimmer current rating. In addition, all ungrounded conductors connected to the line side of the dimmer must be opened when either the switch or circuit breaker supplying the dimmer is opened. See *NEC Section 520.25(A)*.

Older **resistance-type dimmers** and **reactor-type dimmers** may be installed in either the grounded or ungrounded lighting circuit conductor. See *NEC Section 520.25(B)*. *Figure 11* is an example of a portable resistance-type dimming panel. Newer solid-state dimmers may not be supplied with a voltage exceeding 150V between conductors unless the dimmer is rated for the voltage. *Figure 12* shows a solid-state programmable dimming system with the front cover removed. When a grounded conductor is connected to a solid-state dimmer, the conductor must be common to both the input and output circuits. The metal chassis of a solid-state dimmer must be connected to the equipment grounding conductor. See *NEC Section 520.25(D)*.

Figure 11 ◆ Portable resistance-type dimming panel.

412F11.EPS

3.2.2 Types of Switchboard Control

Stage switchboard control may include manual control, remote control, intermediate control, or any combination of these types. See *NEC Section 520.26*. In a manually controlled stage switchboard, all dimmers and switches are manually operated by handles that are mechanically linked to the control devices. *Figure 13* shows two manually controlled, portable, solid-state stage switchboard lighting controls.

In a remotely controlled switchboard, dimmers and switches are electrically controlled from a pilot-type console or panel. These pilot panels may be part of the switchboard itself or located in a separate area.

Intermediate control incorporates a design in which primary and secondary switchboards are used. The secondary switchboard functions as a patch panel or secondary panelboard, containing lighting circuit interconnections and overcurrent protection devices. It is typically in a different location than the main or primary stage switchboard.

3.2.3 Stage Switchboard Feeders

A manual or remotely operated stage switchboard may be supplied by a single feeder that is disconnected by a single disconnecting device. See *NEC Section 520.27(A)(1)*. In intermediate control, multiple feeders may supply the secondary switchboard (patch panel) as long as all feeders are part

Figure 12 ◆ Fixed, solid-state programmable dimming system.

412F12.EPS

412F13.EPS

Figure 13 ◆ Manually controlled switchboard.

of the same electrical system. When neutral conductors are included in the same raceway with multiple feeders to a secondary switchboard, the neutral conductor must be of sufficient ampacity to carry the maximum **unbalanced load** of the secondary switchboard feeders, but is not required to be of a greater ampacity than the neutral conductor supplying the primary switchboard. See *NEC Section 520.27(A)(2)*.

Installations in which individual feeders supply a single primary switchboard must provide a disconnecting means for each feeder, and a labeling system that states the number and location of each disconnecting means. See *NEC Section 520.27(A)(3)*. If the disconnecting means are located in more than one distribution panelboard, the primary stage switchboard must be designed so that barriers are installed to separate feeders from different disconnecting means locations, with labels indicating the disconnect locations.

In three-phase, four-wire, **solid-state phase-control dimmers**, the feeder neutral conductor is considered a current-carrying conductor. However, feeder neutral conductors in three-phase, four-wire, **solid-state sine-wave dimmers** are not considered as current-carrying conductors. When a combination of phase-control and sine wave control is used, the neutral conductor is considered a current-carrying conductor. See *NEC Section 520.27(B)*.

In order to calculate feeder sizes supplying stage switchboards, the total controlled load may be used as long as all feeders supplying the switchboard are (1) protected by an overcurrent device that is rated no greater than the feeder ampacity, and (2) the opening of the overcurrent device does not affect the operation of the exit or emergency lighting systems. See *NEC Section 520.27(C)*.

3.3.0 Wiring Methods for Fixed Equipment (Other Than Switchboards)

Wiring methods for fixed stage equipment including **footlights, border lights, proscenium** sidelights, portable strip lights, backstage lamps, stage receptacles, connector strips, curtain machines, and smoke ventilators are covered in *NEC Article 520, Part III*.

The maximum branch circuit rating for footlights, border lights, and proscenium sidelights is 20A unless heavy-duty lampholders are used. See *NEC Section 520.41*.

NOTE

NEC Article 210 covers rules associated with circuits supplying heavy-duty lampholders.

Conductors that supply foot, border, proscenium, portable strip lights (*Figure 14*), and connector strips must have a minimum insulation temperature rating of 125°C; however, when calculating the conductor's ampacity using *NEC Table 310.16*, the 60°C column must be used. See *NEC Section 520.42*. Cable or conductor drops from overhead connector strips must have an insulation rating of 90°C and must also have their ampacity calculated using the 60°C column of *NEC Table 310.16*. No more than 6 inches of conductor can extend into the connector strip. In any of these cases, the derating factor based on the number of conductors does not apply.

3.3.1 Footlights

When footlights are enclosed in a metal trough, the trough must be constructed of sheet metal that is at least 0.032-inch thick and treated to prevent rust. All lampholder terminals must be spaced a minimum of ½ inch from any metal trough surface, and lampholder terminations must be made by a soldered connection. See *NEC Section 520.43(A)*.

In nonmetallic troughs, individual outlet boxes for lampholders must be connected with rigid metal conduit, intermediate metal conduit, flexible metal conduit, MC cable, or MI cable. Again,

412F14.EPS

Figure 14 ◆ Strip lights.

all terminations must be soldered connections. See *NEC Section 520.43(B)*.

CAUTION

Footlights that are designed to disappear into a storage recess when not in use must be designed so that the voltage supply is automatically disconnected when the lights are retracted into the storage recess. See *NEC Section 520.43(C)*.

3.3.2 Cords and Cables for Border Lights

Because border lights produce extreme heat, certain rules apply to the type of cord or cable that can be used to supply these lights. See *NEC Section 520.44*. The cord or cable must be listed for **extra-hard usage**, must be properly supported, and can only be used in applications that require flexible conductors. The ampacity of these cords or cables is governed by *NEC Section 400.5*.

If the cord or cable is not in direct contact with heat-producing lighting or equipment, the cable or cord ampacity may be calculated using *NEC Table 520.44*, as summarized in *Table 1*.

The values in this table reflect the ampacity for multiconductor cords and cables with no more than three current-carrying copper conductors and an ambient temperature of 86°F. The derating factors in *NEC Table 520.44* must be applied when more than three current-carrying conductors are in the cable or cord. See *Table 2*.

3.3.3 Miscellaneous Fixed Equipment, Receptacles, and Outlets

Any receptacle used to supply on-stage electrical equipment must be rated in amperes and the conductors supplying the receptacle must comply with *NEC Articles 310 and 400*. See *NEC Section 520.45*. In addition, receptacles that supply portable stage-lighting equipment must be of the pendant type or installed in suitable pockets or enclosures (*NEC Section 520.46*).

Backstage lighting is usually utilitarian in design since it is out of view of the audience (*Figure 15*). When bare bulbs are used as backstage lighting, lamp guards must be installed (*Figure 16*) and a minimum space of 2 inches must be maintained between lamps and any combustible material (*NEC Section 520.47*). Electrical devices that are used to control the movement of curtains must be listed for that purpose (*NEC Section 520.48*).

Table 1 Ampacities of Listed Extra-Hard Usage Cords and Cables (Data from *NEC Table 520.44*)

Size AWG	Temperature Rating of Cords and Cables		Maximum Rating of Overcurrent Device
	75°C	90°C	
14	24	28	15
12	32	35	20
10	41	47	25
8	57	65	35
6	77	87	45
4	101	114	60
2	133	152	80

Reprinted with permission from *NFPA 70®*, the *National Electrical Code®*. Copyright © 2007, National Fire Protection Association, Quincy, MA 02269. This reprinted material is not the complete and official position of the National Fire Protection Association on the referenced subject, which is represented only by the standard in its entirety.

Table 2 Derating Factors for Extra-Hard Usage Cords and Cables (Data from *NEC Table 520.44*)

Number of Conductors	Percent of Ampacity
4–6	80%
7–24	70%
25–42	60%
43+	50%

Reprinted with permission from *NFPA 70®*, the *National Electrical Code®*. Copyright © 2007, National Fire Protection Association, Quincy, MA 02269. This reprinted material is not the complete and official position of the National Fire Protection Association on the referenced subject, which is represented only by the standard in its entirety.

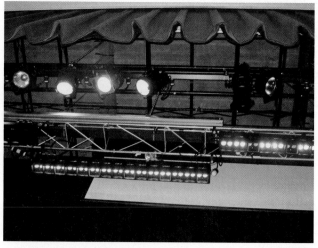

412F15.EPS

Figure 15 ◆ Backstage lighting.

Figure 16 ◆ Lamp guard.

3.3.4 Smoke Ventilators

Electrical devices controlling stage smoke ventilators must be supplied by a normally closed circuit, rated for the full voltage of the circuit to which it is connected, located above the scenery in a loft area, and mounted in a metal enclosure having a tightly sealing self-closing door. See *NEC Section 520.49*. In addition, the circuit supplying the device must be controlled by at least two externally operated switches, with one switch being located on stage at a readily accessible spot and the other switch location determined by the authority having jurisdiction for the electrical installation.

3.4.0 On-Stage Portable Switchboards

Portable stage switchboards are typically used for road show presentations and other similar portable activities (*NEC Section 520.50*). Portable on-stage switchboards may be supplied with power from road show connection panels, often referred to as patch panels, or from cord or cable power outlets. Portable on-stage switchboards are used to supply lighting and receptacles just like fixed on-stage switchboards.

3.4.1 Road Show Connection Panel (Patch Panel)

A patch panel is often used to connect portable stage switchboards to fixed lighting outlets by way of permanently installed supplementary circuits. In these installations, the panel, supplementary circuits, and outlets must follow certain rules

as outlined in *NEC Section 520.50* and summarized here:

Circuit conductors must terminate in a polarized, grounding-type inlet that is matched to the current and voltage rating of the fixed-load receptacle.

When transfer switches are used between fixed and portable switchboards, all circuit conductors must transfer at the same time.

Any exterior device that supplies current to supplementary circuits must be protected by overcurrent protective devices.

Individual supplementary circuits within the road show patch panel and theater must be protected by overcurrent protective devices installed within the panel.

NOTE

Patch panel construction is regulated by *NEC Article 408*.

3.4.2 Power Outlets and Supply Conductors

Power outlets that supply on-stage portable switchboards must be externally operated, enclosed fused switches, fused disconnects, or circuit breakers with provisions for an equipment grounding conductor. See *NEC Section 520.51*. These supply devices must be readily accessible from the stage floor, mounted on stage, or at the location of the permanent switchboard.

Extra-hard usage cords or cables must be used to connect the power outlets to the switchboard. See *NEC Section 520.53(H)(1)*. These cables or cords must terminate totally within the enclosed power outlet and switchboard enclosure, or in a connector assembly identified for that purpose.

A single-conductor current cable cannot be smaller than No. 2 AWG and the equipment grounding conductor must be at least No. 6 AWG. See *NEC Section 520.53(H)(2)*. If single conductors are installed in parallel to increase ampacity, the conductors must be the same size and length. Single-conductor supply cables may be grouped together but not bundled together.

Equipment grounding and grounded (neutral) conductors must be marked with approved marking methods in compliance with *NEC Sections 200.6, 250.119, and 310.12*. One approved method for marking neutral conductors that supply portable switchboards is a white or gray marking 6 inches from each end of the conductor (*Figure 17*). Similarly, an approved marking method for equipment grounding conductors is green or green with yellow stripes 6 inches from each end of the conductor.

CAUTION

As with any electrical system, if more than one nominal voltage exists within the same premises, each ungrounded conductor must be identified according to the system from which it originates.

Portable on-stage switchboard supply conductors may be reduced in size based on the length of conductor required between the voltage supply and switchboard, or between the voltage supply and an intermediate overcurrent protective device.

For supply conductors not longer than 10 feet, all of the conditions in *NEC Section 520.53(H)(3)* must be met in order for the conductors to be reduced in size:

- Supply conductors must have at least one-fourth the ampacity of the overcurrent device protecting them.
- Supply conductors must terminate at a single overcurrent protective device that limits the load to no greater than the supply conductor ampacity.
- Supply conductors cannot penetrate floors, walls, or ceilings or pass through doors or other traffic areas.
- Supply conductor terminations must be made using an approved method.
- Supply conductors cannot contain splices or connectors.

412F17 .EPS

Figure 17 ◆ Properly marked grounded conductor.

- Supply conductors must not be bundled together.
- Supply conductors must be supported above the floor in an approved manner.

If the supply conductors are over 10 feet in length but not longer than 20 feet, all of the conditions in *NEC Section 520.53(H)(4)* must be met in order for the conductors to be reduced in size:

- Supply conductors must have an ampacity that is at least one-half the ampacity of the overcurrent device protecting them.
- Supply conductors must terminate at a single overcurrent protective device that limits the load to no greater than the supply conductor ampacity.
- Supply conductors cannot penetrate floors, walls, or ceilings or pass through doors or other traffic areas.
- Supply conductor terminations must be made using an approved method.
- Supply conductors must be supported at least 7 feet above the floor except at termination points.
- Supply conductors must not be bundled together.
- Tap conductors must not contain spices or connectors (other than at the point of tap).

Supply conductors that are not reduced in size based on the listed conditions may pass through holes in walls per *NEC Section 520.53(H)(5)*. If the wall is fire-rated, the penetration must be in accordance with the rules of *NEC Section 300.21*.

When supply cables pass through enclosures, a protective bushing must be installed to protect the cable. See *NEC Section 520.53(I)*. In addition, the arrangement of the cables through and within the enclosure should be such that no tension is applied at the point of termination or connection.

For long lengths of supply conductors that require interconnecting connectors, *NEC Section 520.53(J)* states that no more than three mating connectors may be used where the total length of conductor from supply to switchboard does not exceed 100 feet. If the conductor exceeds 100 feet in length, one additional mating connector is permitted for each additional 100 feet of conductor.

Single-pole portable cable connectors must be of the locking type, and if installed in parallel as input devices, must be labeled with a warning that parallel connections are present. See *NEC Section 520.53(K)*. At least one of the following conditions must exist to permit the usage of single-pole separable connectors:

- Connectors can be connected or disconnected only if the supply is de-energized and the connectors are interlocked at the source.

- If sequential-interlocking line connectors are used, load connectors must be connected in the following sequence: equipment grounding conductor first, grounded conductor second, and ungrounded conductor(s) last, or a caution tag must be attached adjacent to the line connectors that lists the sequence of connection. Disconnection is in the reverse order.

The supply neutral terminal and busbar within a portable switchboard designed to operate from a three-phase, four-wire with ground supply must have a rated ampacity that is at least twice the ampacity of the largest ungrounded supply conductor terminal within the switchboard. See *NEC Section 520.53(O)(1)*.

Neutral conductors supplying portable switchboards that incorporate solid-state phase-control dimmers are considered current-carrying conductors and must be sized accordingly. See *NEC Section 520.53(O)(2)*. However, neutral conductors supplying portable switchboards that incorporate solid-state sine-wave dimmers are not considered current-carrying conductors. When single conductors not installed in raceways are used as feeders for multiphase circuits powering portable switchboards containing solid-state phase-control dimmers, the neutral conductor must have a rated ampacity that is at least 130% of the ungrounded conductors. In cases where the switchboard contains sine-wave dimmers, the neutral ampacity is only required to be at least 100% of the ungrounded conductors.

Only qualified personnel may install portable supply conductors, connect and disconnect connectors, or energize and de-energize supply circuits per *NEC Section 520.53(P)*. Portable switchboards must visibly display a label in a conspicuous place indicating this requirement. An exception to this rule states that a portable switchboard may be connected to a permanently installed supply receptacle by an unqualified person if the receptacle is protected by an overcurrent protective device that is not greater than 150A, and the switchboard complies with all of the following requirements:

- Listed multipole connectors are used for each supply interconnection.
- The general public may not have access to supply connections.
- Multiconductor cords and cables are listed for extra-hard usage, with a rated ampacity equal to or greater than the load and not less than the ampacity of the connectors used. *Figure 18* shows a small portable switchboard supplied by a multiconductor cord.

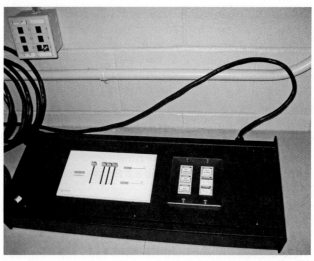

412F18.EPS

Figure 18 ◆ Portable switchboard (dimmer) supplied by a multiconductor cord.

3.4.3 Portable Switchboard Construction, Circuits, Interior Wiring, and Devices

Portable switchboards must be installed within an enclosure that offers protection to the equipment and the general public. See *NEC Sections 520.53(A) through (E)*. Wood-framed enclosures must be lined with sheet metal with a minimum thickness of 0.020-inch, galvanized, enameled, or otherwise coated to prevent rust. There cannot be any exposed energized parts within the enclosure, and all switches and circuit breakers must be enclosed and externally operated. Each ungrounded conductor in every circuit supplied by the switchboard must have overcurrent protection. Dimmer terminals must not be exposed and faceplates must be installed to prevent accidental contact. The short circuit current rating must be marked on the switchboard.

Portable on-stage switchboards must be equipped with a pilot light that is connected to the supply circuit of the switchboard. The power to the pilot light must not be interrupted by the opening of the master switch, and must be supplied by an individual branch circuit rated at no more than 15A. See *NEC Section 520.53(G)*.

Individual portable switchboard circuits that supply incandescent lighting equipment with lamps not greater than 300W must be protected by overcurrent protective devices of 20A or less (*NEC Section 520.52*). Circuits supplying incandescent lighting equipment with lamps greater than 300W must comply with the rules of *NEC Article 210*.

Conductors within the switchboard enclosure (except busbars) must be stranded per *NEC Section 520.53(F)*. Minimum insulation temperature ratings for dimmer conductors are dependent on the type of dimmer and must be at least equal to the operating temperature of the dimmer. Conductors connected to resistance-type dimmers must have a minimum insulation temperature rating of 200°C, while conductors connected to reactor-type, autotransformer, and solid-state dimmers must be rated for a minimum of 125°C.

3.5.0 Dressing Rooms

Pendant-type lampholders are not permitted in dressing rooms per *NEC Section 520.71*. Incandescent lamps installed less than 8 feet from the floor must be equipped with open-ended guards that are riveted to the outlet box cover or somehow locked in place. See *NEC Section 520.72*.

Wall switches must be installed in the dressing rooms to control lights or receptacles adjacent to mirrors and above the dressing table countertops (*NEC Section 520.73*). Switches for receptacles located adjacent to mirrors and above dressing table counters must be equipped with a pilot light located adjacent to the door outside the dressing room to indicate when the receptacles are energized (*Figure 19*). Other general-use receptacles in dressing rooms do not require switch control.

4.0.0 ◆ CARNIVALS, CIRCUSES, FAIRS, AND SIMILAR EVENTS

Carnivals, circuses, and fairs present several unique wiring and safety concerns. These include large crowds, high power demands for operating rides and concessions, outdoor installations, unpredictable animal behavior, and rapid setup and teardown of equipment. *NEC Article 525* lists the requirements for portable wiring associated with equipment and portable structures used in carnivals (*Figure 20*), circuses, fairs, and similar functions. The wiring in permanent structures such as audience areas and assembly halls located within these activities are covered by *NEC Articles 518 and 520*. The wiring in permanent amusement attractions is covered by *NEC Article 522*.

4.1.0 Overhead Conductor Clearances

Tents and concessions must maintain vertical clearances from overhead conductors in accordance with *NEC Section 225.18*. For portable structures (*Figure 21*), a clearance of 15 feet must be maintained in any direction from overhead conductors operating at 600V or less unless the conductors are supplying power to the portable structure per *NEC Section 525.5(B)(1)*. A portable structure may not be located under or within 15 feet of conductors that have an energized voltage greater than 600V per *NEC Section 525.5(B)(2)*.

412F19.EPS

Figure 19 ◆ Dressing room mirror lights.

412F20.EPS

Figure 20 ◆ Carnival.

Figure 21 ◆ Tents.

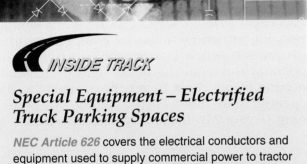

INSIDE TRACK

Special Equipment – Electrified Truck Parking Spaces

NEC Article 626 covers the electrical conductors and equipment used to supply commercial power to tractor trailers, such as those with refrigeration units. A typical electrified truck parking lot is shown here.

4.2.0 Power Sources

Areas that house service equipment must be lockable or located to prevent access to unqualified persons. The equipment must be mounted on a surface with a solid backing and protected from the weather or constructed with weatherproof enclosures and components. See *NEC Section 525.10*.

Portable structures at carnivals, circuses, or fairs often receive their power supplies from separately derived systems or different services (*Figure 22*). In these situations, if the structures are less than 12 feet apart, all equipment grounding conductors must be bonded together at portable structures per *NEC Section 525.11*. The bonding conductor must be sized in accordance with *NEC Table 250.122*, based on the largest overcurrent device supplying the portable structures, and no smaller than No. 6 AWG.

4.3.0 Wiring Methods – Cords, Cables, and Connectors

Flexible cords and cables exposed to potential physical damage must be listed for extra-hard usage per *NEC Section 525.20(A)*. Cord and cables not exposed to possible physical damage may be listed for hard usage. If the cords or cables are to be used outdoors, they must also be listed for wet locations and be sunlight resistant. Permanent wiring on portable amusement rides and other

Figure 22 ◆ Separately derived systems (generators) powering portable structures and rides.

attractions may be extra-hard usage flexible cords and cables as long as the cords or cables are not exposed to physical damage (*Figure 23*).

Single-conductor cables must be a minimum No. 2 AWG in size per *NEC Section 525.20(B)*. *NEC Section 525.20(C)* permits the use of open conductors only in **festoon lighting**, which is regulated by *NEC Article 225*. Flexible cords or cables may not contain any spices or taps between boxes or fittings. See *NEC Section 525.20(D)*.

Cord and cable connectors may not be laid on the ground unless the connector is rated for wet locations, and even then, connectors must not be located in any audience traffic area or other areas (unless guarded) that are accessible to the general public. See *NEC Section 525.20(E)*. Cords or cables may not be draped or hung on rides or similar structures unless designed for the purpose [*NEC Section 525.20(F)*]. If the cable or cord presents a tripping hazard, it may be covered with nonconductive matting as long as the matting does not present an additional tripping hazard per *NEC Section 525.20(G)*. Cables or cords may be buried to eliminate the tripping hazard.

412F23.EPS

Figure 23 ◆ Permanent wiring on amusement rides.

4.4.0 Wiring Methods – Rides, Tents, and Concessions

Per *NEC Section 525.21(A)*, all rides and similar portable structures must be equipped with a disconnect switch that is within 6 feet of the operator and within his or her sight. The disconnecting means must be readily available to the operator at all times, even when the ride is running (*Figure 24*). If the disconnecting means is

What's wrong with this picture?

412SA04.EPS

Figure 24 ◆ Disconnect switch for amusement ride.

Figure 25 ◆ Outlet box for concession.

accessible to the general public, it must be lockable. A **shunt trip** in the operator's console may be used to open the fused disconnect or circuit breaker.

Lamps that are part of portable lighting inside tents and concessions must be guarded per *NEC Section 525.21(B)*. When outlet boxes or other device boxes are installed outdoors, the boxes must be weatherproof and mounted so the bottom of the box is at least 6 inches off the ground (*Figure 25*). See *NEC Section 525.22(A)*.

NEC Section 525.23(A) states that ground fault circuit interrupter (GFCI) protection is required on all 125V, single-phase, 15A and 20A nonlocking-type receptacles that are used to assemble or disassemble rides, amusements, or concessions, or when exposed to the general public. Any equipment supplied by 125V, single-phase, 15A or 20A circuits and exposed to the general public must also be GFCI-protected. The GFCI protection may be an integral part of the attachment plug or installed directly in the power supply cord within 12 inches of the attachment plug. Locking-type receptacles and exit lighting circuits are not required to be GFCI-protected per *NEC Section 525.23(B)*.

4.5.0 Grounding and Bonding

All metal raceways, metal-sheathed cable, metal enclosures of electrical equipment, metal frames and metal parts of portable structures, trailers, trucks, or any other equipment that contains or supports electrical equipment connected to the same electrical source must be bonded together and connected to an equipment grounding conductor. See

NEC Section 525.30. The equipment grounding conductor may serve as the bonding means.

The equipment grounding conductor must be connected to the system's grounded conductor at the service disconnect. In a separately derived system such as a generator, the equipment grounding conductor must be connected at the generator or in the first disconnecting means supplied by the generator (*Figure 26*). See *NEC Section 525.31*.

 CAUTION

Never connect the grounded circuit conductor to the equipment grounding conductor anywhere on the load side of the service disconnect or generator disconnect.

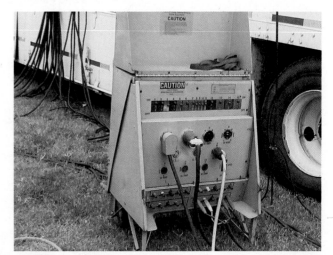

Figure 26 ◆ First disconnecting means supplied by a generator at a carnival.

5.0.0 ◆ AGRICULTURAL BUILDINGS

NEC Article 547 covers agricultural buildings in which excessive dust and dust with water may accumulate, including all areas of poultry, livestock, and fish confinement systems. It also contains provisions for agricultural buildings that contain a corrosive atmosphere that may be caused by poultry or animal excrement, corrosive particles mixed with water, and wet or damp corrosive areas caused by cleaning and sanitizing solutions.

5.1.0 Wiring Methods

Wiring systems in agricultural buildings may contain jacketed MC cable, NMC (nonmetallic sheathed, corrosion resistant) cable, copper SE (service entrance) cable, UF (underground feeder) cable, rigid nonmetallic conduit, liquidtight flexible nonmetallic conduit, and any other cable or raceway approved and suitable for the location, with approved fittings. See *NEC Section 547.5(A)*.

Per *NEC Section 547.5(B)*, cables must be secured within 8 inches of any box, cabinet, or fitting to which they connect. If nonmetallic boxes are mounted directly to an agricultural building surface, the ¼-inch airspace requirement of *NEC Section 300.6(D)* does not apply.

See *NEC Section 547.5(C)*. Enclosures, boxes, conduit bodies, and fittings installed in dusty areas of agricultural buildings must be designed to minimize the entrance of dust, with no screw or mounting holes. In damp or wet locations, these enclosures, boxes, conduit bodies, and fittings must be listed for wet locations or located to prevent the entrance of moisture or water (*Figure 27*). Wiring methods used in corrosive atmospheres must provide corrosion resistance. Aluminum and magnetic ferrous (steel) materials are not suitable for corrosive atmospheres.

> **NOTE**
> *NEC Table 110.20* lists properties for various types of electrical enclosures, including those that offer corrosion resistance.

Where flexible connections are required, such as supplying power to motors and other vibrating or moving equipment, *NEC Section 547.5(D)* requires the use of liquidtight flexible metal or liquidtight flexible nonmetallic conduit with dusttight connectors, or flexible cord listed for hard usage.

INSIDE TRACK

Fans in Poultry Buildings

This chicken farm has an elaborate system of auxiliary power supplies in the form of generators and transfer switches. Failure of a livestock ventilation system can result in death by asphyxiation (from lack of oxygen and increased carbon dioxide), heat prostration, or poisoning from the buildup of gases such as ammonia and methane. The farmer who owned this building said that he would lose most of his chickens if the ventilator fans stopped operating for as little as 20 minutes.

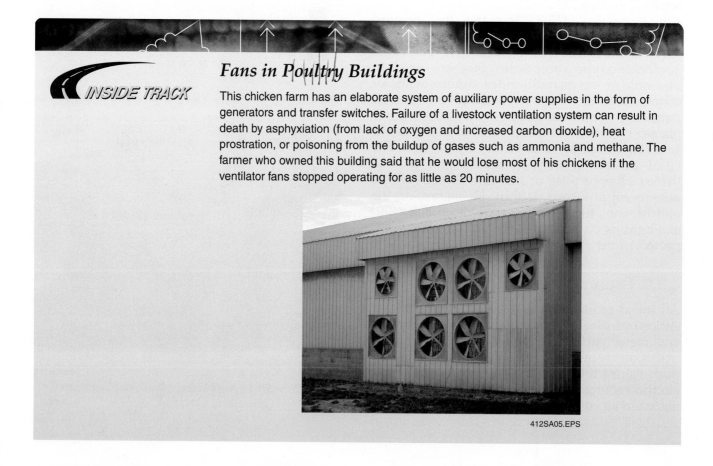

412SA05.EPS

Figure 27 ◆ Device box listed for wet location in agricultural building.

Grounding conductors installed in agricultural buildings must be copper, and if installed underground, must be insulated or covered copper. See *NEC Section 547.5(F)*.

Per *NEC Section 547.5(G)*, ground fault protection must be provided for all 125V, single-phase, 15A and 20A general-purpose receptacles installed in areas having an **equipotential plane**, outdoors, in damp or wet locations, and in dirt confinement areas for livestock. GFCI protection is not required, however, for any accessible dedicated-load receptacle if a GFCI receptacle is located within 3 feet of the dedicated load receptacle.

5.2.0 Motors and Luminaires

NEC Section 547.7 requires that all motors and rotating electrical machinery be totally enclosed or designed to reduce the entry of moisture, dust, or corrosive materials (*Figure 28*). Luminaires must also be designed to minimize the entry of dust, moisture, or corrosive particles. If luminaires are exposed to physical damage, guards must be installed. Luminaires exposed to water must be listed as watertight. See *NEC Sections 547.8(A) through (C)*.

Figure 28 ◆ Totally enclosed motor.

5.3.0 Electrical Supply from a Distribution Point

NEC Section 547.9 states that two or more agricultural buildings or structures located on the same premises may be electrically supplied from a single distribution point (*Figure 29*). If two or more buildings are supplied overhead from a distribution point, a **site-isolating device** must be installed and comply with the requirements or subject to exceptions as listed in *NEC Sections 547.9(A)(1) through (10)* and summarized as follows:

- Pole-mounted
- Simultaneously open all ungrounded service conductors from the premises wiring
- Mounted in an enclosure bonded to the grounded and grounding conductors
- System grounded conductors bonded to a grounding electrode system by means of a grounding electrode conductor at the site-isolating device

Figure 29 ◆ Agricultural building distribution point.

- Rated for the calculated load
- Not required to provide overcurrent protection
- If not readily accessible, capable of being remotely operated by a readily accessible operating handle not more than 6'–7" above grade at its highest position
- Permanently marked on or adjacent to the operating handle, identifying it as a site-isolating device

NOTE

If the electric utility provides a site-isolating device as part of their installation, no additional site-isolating device is required.

Service disconnecting means and overcurrent protection for two or more buildings supplied from a distribution point may be located at the buildings or at the distribution point. If located at the buildings, supply conductor sizes must comply with *NEC Article 220, Part V* and conductor installation must follow the requirements of *NEC Article 225, Part II*. Grounding and bonding of the supply conductors must be in accordance with *NEC Section 250.32*. If the equipment grounding conductor and the largest supply conductor are constructed from the same material, they must be the same size, or if of different materials, the equipment grounding conductor must be adjusted according to *NEC Table 250.122*.

When the service disconnecting means and overcurrent protection for two or more buildings are located at the distribution point, supply conductors to the buildings must meet the requirements of *NEC Section 250.32* and *NEC Article 225, Parts I and II*.

When an equipment grounding conductor from a distribution point to a building or structure used to house livestock is buried directly in the ground, the conductor must be insulated or covered copper [*NEC Section 547.9(D)*].

If more than one service supplies an agricultural site, and any two of the services are 500 feet or less apart from one another as measured in a straight line, each distribution point must have a permanent label or directory installed to indicate the location of each of the other distribution points and the buildings served by each [*NEC Section 547.9(E)*].

5.4.0 Equipotential Planes

NEC Section 547.10(A)(1) requires that equipotential planes be installed in indoor animal confinement areas having concrete floors with metallic equipment that may become energized and accessible to the animals. Equipotential planes are also required in outdoor areas of concrete slabs where metallic equipment may become energized and accessible to the animals [*NEC Section 547.10(A)(2)*]. See *Figure 30*. The plane must cover the area in which the animals may stand while accessing the metallic equipment.

Equipotential planes must be bonded to the grounding system with a bare or insulated copper conductor no smaller than No. 8 AWG [*NEC Section 547.10(B)*]. Bonding connections to wire mesh or other conductor elements may be made using approved pressure connectors or clamps of brass, copper, or copper alloy.

412F30.EPS

Figure 30 ◆ Concrete area with accessible metal piping and animal contact.

6.0.0 ◆ MARINAS AND BOATYARDS

NEC Article 555 covers all commercial and multi-family docking facilities, boatyards, boat basins, boathouses, and yacht clubs used for the purpose of repairing, berthing, launching, storing, or fueling boats and/or the moorage of floating buildings (*Figure 31*).

> **NOTE**
> Docking facilities for single-family dwellings are not covered by this article.

6.1.0 General Requirements for Devices, Equipment, and Enclosures

A marine power outlet is any receptacle, circuit breaker, fused switch, meter, and any other similar device approved for marine use. Electrical distribution systems within boatyards and piers (*Figure 32*) cannot exceed 600V phase-to-phase per *NEC Section 555.4*. When mounting transformers in areas covered by *NEC Article 555*, the bottoms of transformers must be above the **electrical**

datum plane as defined in *NEC Section 555.2*. Electrical service equipment cannot be installed on or in a floating structure, but must be mounted adjacent to the structure per *NEC Section 555.7*.

Electrical connections within boxes or enclosures that are not rated for submersion must be located at least 12 inches above the deck of a floating pier as stated in *NEC Section 555.9* and shown in *Figure 33*. If the junction box is rated for submersion, it may be located below the electrical datum plane for floating piers but must be above the water line. Any electrical connection on a fixed pier must be at least 12 inches above the deck of the fixed pier but never below the electrical datum plane.

Conduit may not be the sole support for electrical enclosures installed on piers above deck level. External box ears or lugs must be used to support these enclosures per *NEC Section 555.10(A)*. If internal screws are used to secure boxes or enclosures to the pier structure, the screw heads must be sealed to prevent water from entering the enclosure or box through the mounting holes. The location of electrical equipment, boxes, or enclosures must not be in the way of mooring lines per *NEC Section 555.10(B)*.

NEC Section 555.11 states that a person must be able to operate circuit breakers or switches that

Figure 31 ◆ Marina.

Figure 32 ◆ Electrical distribution system in a marina.

Floating Buildings

Wiring methods for buildings that float on water, are permanently moored, and electrically supplied by a system not located on the premises, are covered by *NEC Article 553*, not *NEC Article 555*, which only covers the moorage of these buildings.

Figure 33 ◆ Electrical connections within a box not rated for submersion must be mounted at least 12 inches above the deck.

Figure 34 ◆ Accessible circuit breakers in watertight enclosure.

have been installed in watertight enclosures without removing the enclosure cover (*Figure 34*). However, watertight enclosures housing such devices must be designed with a weep hole to discharge condensation.

6.2.0 Service and Feeder Conductor Load Calculations

All general lighting and receptacle loads for marinas and boatyards must be calculated using the standard general lighting load calculation procedures of *NEC Article 220, Part III*. When service or feeder circuit conductors are used to supply a number of receptacles for shore power to boats, the demand factors shown in *Table 3* may be applied to determine the circuit rating and conductor sizes.

Other rules may apply to shore power installations and are listed in the notes following *NEC Table 555.12*. One such rule is that if two separate receptacles having different voltages and amperage ratings are installed in a single boat slip, only the receptacle with the greater volt-ampere (kilowatt) demand needs to be included in the total load calculation. For example, if one receptacle is

Table 3   Demand Factors (Data from *NEC Table 555.12*)

Number of Receptacles	Factor Applied to the Sum of the Rating of the Receptacles
1–4	100%
5–8	90%
9–14	80%
15–30	70%
31–40	60%
41–50	50%
51–70	40%
71 and up	30%

Reprinted with permission from *NFPA 70®*, the *National Electrical Code®*. Copyright © 2007, National Fire Protection Association, Quincy, MA 02269. This reprinted material is not the complete and official position of the National Fire Protection Association on the referenced subject, which is represented only by the standard in its entirety.

rated 30A at 125V (3,750VA), and another is rated 50A at 250V (12,500VA) within the same slip, the total load calculation for these two receptacles would be figured at 12,500VA.

Consideration must be given to ambient temperatures and types of equipment when applying

Submeters

INSIDE TRACK

If individual kilowatt-hour submeters are installed to measure usage at each boat slip, a factor of 0.9 (90%) may be used instead of applying the factors of *NEC Table 555.12* to determine total load demand.

demand factors. Extremely hot or cold temperatures may result in increased demand factors for circuits with loaded heating, air conditioning, or refrigeration equipment.

6.3.0 Wiring Methods

The standard wiring methods in *NEC Chapter 3* apply to locations covered by *NEC Article 555* as long as the methods are rated for wet locations. When portable power cables are used as permanent wiring on the underside of fixed or floating piers, or where flexibility is required on the floating sections of piers (*Figure 35*), the cable must meet the following criteria per *NEC Sections 555.13(A)(2) and (B)(4):*

- Rated for extra-hard usage
- Insulation rated for a minimum of 75°C and 600V
- Listed for wet locations
- Sunlight resistant
- Manufactured with an outer jacket resistant to temperature extremes, oil, gasoline, ozone, abrasions, acids, and chemicals
- Properly supported by nonmetallic clips to pier structural members other than deck planking
- Located on the underside of piers where not subject to physical damage
- Protected by nonmetallic sleeves where penetrating structural members

412F35.EPS

Figure 35 ◆ Cable may be used on floating sections of piers.

In installations where cables are used to supply power to floating sections of a pier, an approved, corrosion-resistant junction box containing terminal blocks must be installed on each pier section for conductor terminations. Any exposed metal boxes, covers, screws, and the like must be corrosion-resistant. See *NEC Section 555.13(B)(4)(b).*

What's wrong with this picture?

412SA06.EPS

Overhead wiring in and around boatyards and marinas must be installed in a manner so as to avoid contact with masts, boat parts, and portable lifts (*Figure 36*) being moved in the yard. *NEC Section 555.13(B)(2)* requires a minimum overhead clearance of 18 feet. In addition, all electrical conductors and cables must maintain a minimum of 20 feet from the outer perimeter or any section of the boatyard that may be used to move vessels or for stepping (raising) or unstepping (lowering) boat masts per *NEC Section 555.13(B)(1)*.

Per *NEC Section 555.13(B)(5)*, rigid metal or nonmetallic conduit must be installed above decks of piers and landing stages (*Figure 37*), and below the enclosure the landing stage serves. Full-thread connections must be used to connect the conduit to the enclosure. When nonmetallic rigid conduit (PVC) is installed in these locations, nonmetallic fittings that provide a threaded connection must be used to connect the conduit to the enclosure. All installations must be approved for damp or wet locations.

6.4.0 Grounding

All metal boxes, metal cabinets, metal enclosures, metal frames of utilization equipment, and grounding terminals of grounding-type receptacles in locations covered by *NEC Article 555* must be connected to an equipment grounding conductor.

NEC Section 555.15(B) states that the equipment grounding conductor must be copper, insulated, no smaller than No. 12 AWG, and green or green with one or more yellow stripes unless the equipment grounding conductor is larger than No. 6 AWG. In these cases the equipment grounding conductor may be marked by other approved methods as described in *NEC Section 250.119*. In cases where MI cable is used, the equipment grounding conductor may be identified with an approved means at each termination.

Branch circuit equipment grounding conductors must terminate at a grounding terminal in a remote panelboard or on the grounding terminal in the main service equipment per *NEC Section 555.15(D)*. An insulated equipment grounding conductor must interconnect the grounding terminal in a remote panelboard to the grounding terminal in the service equipment. See *NEC Section 555.15(E)*.

412F36.EPS

Figure 36 ◆ Masts and lifts must clear overhead wiring.

412F37.EPS

Figure 37 ◆ Landing stage where people access or exit boats.

6.5.0 Disconnecting Means for Shore Power

NEC Section 555.17 requires that a disconnecting means be provided to disconnect each individual boat shore power receptacle from the electrical supply. The disconnecting means may be a circuit breaker or switch that opens all ungrounded supply circuit conductors and all disconnecting means must be properly marked to identify which receptacle they control. The disconnecting means must be located within 30 inches of the receptacle. Receptacles must be mounted at least 12 inches above the deck surface of the pier and not below the electrical datum plane on a fixed pier (*NEC Section 555.19*). *Figure 38* shows disconnects for receptacles in the form of circuit breakers.

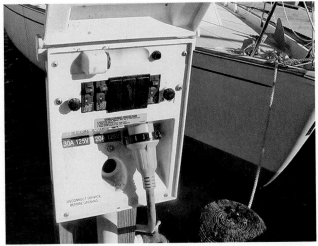

Figure 38 ◆ Watertight shore power enclosure with accessible circuit breakers.

Figure 39 ◆ GFCI-protected receptacle on non–shore power receptacle installed outdoors on pier.

6.6.0 Receptacles

The two types of receptacles in a marina are shore power receptacles and non–shore power receptacles. *NEC Section 555.19(A)(1)* states that shore power receptacles must be listed as marina power outlets, and installed in enclosures that provide protection from the weather. Strain relief must be provided for the cable when it is plugged into the receptacle. Each shore power receptacle must be protected by an individual branch circuit that corresponds to the rating of the receptacle, which can be rated at no less than 30A. Shore power receptacles rated 30A and 50A must be of the locking and grounding type, while receptacles rated 60A and 100A must be of the pin and sleeve type. See *NEC Section 555.19(A)(4)*.

GFCI protection must be provided on non–shore power receptacles, including all 15A and 20A, single-phase, 125V receptacles installed outdoors, in boathouses, and in all buildings used for storage, maintenance, or repairs where portable electrical hand tools, electrical diagnostic equipment, or portable lighting may be used (*Figure 39*). See *NEC Section 555.19(B)(1)*.

Any non–shore power receptacle installed in a marine power outlet enclosure must be marked to indicate that the receptacle is not to be used as shore power for boats. See *NEC Section 555.19(B)(2)*.

6.7.0 Hazardous Locations in and around Marinas and Boatyards

Marina or boatyard wiring that is located near fuel dispensing areas must comply with *NEC Article 514, Fuel Dispensing Facilities*. In general, all power and lighting wiring in marinas and boatyards must be located on the opposite side from fueling systems and piping on wharfs, piers, or docks. See *NEC Section 555.21(A)*.

NEC Tables 514.3(B)(1) and 514.3(B)(2) classify areas in and around fuel dispensing locations as either Class I, Division 1 or Class I, Division 2. When applying these tables to marina or boatyard fuel dispensing areas, you must first determine whether the area is considered closed construction or open construction. See *NEC Section 555.21(B)*.

Closed construction-type docks, piers, or wharfs are areas in which no space exists between the bottom of the dock, pier, or wharf and the

Related Standards

INSIDE TRACK

Additional requirements may be found in the *National Fire Protection Association Standard 303-2006®, Fire Protection Standard for Marinas and Boatyards*, and *NFPA 30A-2008®, Motor Fuel Dispensing Facilities and Repair Garages*.

waterline. The class and division designations for closed construction are as follows:

• *Class I, Division 1* – Voids, pits, or similar areas below closed-construction floating docks, piers, or wharfs that support fuel dispensers in which flammable liquid or vapor can accumulate. *Figure 40* shows a fuel dispenser mounted on a closed construction-type pier. Areas directly below the fuel dispenser are voids and are rated Class I, Division 1.

• *Class I, Division 2* – Spaces above the surface of closed-construction floating docks, piers, or wharfs that are within 18 inches horizontally in all directions extending to grade from the dispenser enclosure. Also sections of a dock, pier, or wharf that do not support fuel dispensers but directly abut and are 20 feet or more from sections that do support fuel dispensers, with air spaces between them to allow for vapor dissipation. Those sections of pier that abut the fuel dispensing pier are classified as Class I, Division 2, as illustrated in *Figure 41*.

Open construction-type docks, piers, or wharfs are built on stringers that are supported by pilings,

412F41.EPS

Figure 41 ◆ Docks or sections of pier that abut this fuel dispensing pier are classified as Class I, Division 2.

floats, pontoons, or similar devices in which space exists between the bottom of the dock, pier, or wharf and the waterline. The class and division designations for open construction are as follows:

• *Class I, Division 1* – Voids, pits, and similar areas within 20 feet of the dispenser where flammable liquids or vapors may accumulate.

• *Class I, Division 2* – Areas 18 inches above the surface of open construction-type fuel dispensing docks, piers, or wharfs, extending 20 feet horizontally in all directions from the outside edge of the dispenser, and down to the water level.

7.0.0 ◆ TEMPORARY INSTALLATIONS

Temporary wiring is covered in *NEC Article 590*. It includes wiring used to supply temporary power and lighting to construction sites (*Figure 42*), and for remodeling, maintenance, repair, demolition, holiday decorative lighting, electrical tests, emergencies, experiments, and developmental work. Temporary wiring, excluding holiday decorative lighting (*Figure 43*), must be removed immediately upon completion of the temporary need. Holiday lighting may remain up no longer than 90 days.

7.1.0 Feeder and Branch Circuit Conductors

Feeder and branch circuit conductors used in temporary wiring may be multiconductor cables that are identified for hard usage or extra-hard usage according to *NEC Table 400.4*, and must originate

412F40.EPS

Figure 40 ◆ Class I, Division 1 area below fuel dispenser in the void between the pier and the water.

Figure 42 ◆ Temporary wiring on construction site.

Figure 43 ◆ Holiday lighting.

from approved panelboards or power outlets. Per *NEC Section 590.4(C)*, Type NM cable may be used for temporary wiring in any dwelling or other structure without consideration to height limitations or concealment within walls, floors, or ceilings. Single insulated feeder and branch circuit conductors may be used to supply power for temporary testing, emergencies, experiments, and developmental work where accessible to qualified persons only. Holiday decorative lighting may also be supplied by single insulated branch circuit conductors as long as the voltage to ground does not exceed 150V, the wiring is not subjected to physical damage, and the conductors are supported on insulators no more than 10 feet apart. If the lighting is hung in festoon (swag) fashion, the conductors must be arranged so that no conductor strain is felt at the lampholder.

7.1.1 Grounding

Temporary branch circuits must include a separate equipment grounding conductor unless the branch circuit conductors are installed in a continuous metal raceway that meets the requirements of an equipment grounding conductor. See *NEC Section 590.4(D)*. Metal-clad cables that qualify as equipment grounding conductors may also be used.

7.1.2 Disconnects

Disconnecting switches or plug connectors must be installed to disconnect all ungrounded conductors supplying temporary circuits. See *NEC Section 590.4(E)*. Multiwire branch circuits must be provided with a disconnecting means that simultaneously disconnects all ungrounded conductors at the power outlet or panel from which they originate. Only identified handle ties may be used.

7.1.3 Protection and Support

All cables and conductors must be protected from pinch points, sharp corners, and other potentially damaging conditions, and proper fittings must be installed where flexible cords and cables enter boxes. See *NEC Sections 590.4(H) and (I)*. Cable assemblies must also be adequately supported with staples, cable ties, straps, or other approved means at intervals to provide protection from physical damage [*NEC Section 590.4(J)*]. Vegetation may not be used to support overhead runs of branch circuit or feeder conductors, except for holiday lighting. In these cases, the conductors or cables must be arranged and supported with strain relief or other devices to avoid conductor damage from the movement of limbs and branches.

7.2.0 Receptacles

Receptacles in temporary wiring must be of the grounding type, and must be electrically connected to the equipment grounding conductor [*NEC Section 590.4(D)*].

WARNING!

Receptacles that supply temporary power on construction sites may not be installed on branch circuits that supply temporary lighting, nor may they be connected to the same ungrounded conductor of multiwire circuits supplying temporary lighting.

7.3.0 Temporary Lighting

General illumination luminaires must be designed to protect the lamp from contact or breakage per *NEC Section 590.4(F)*. All exposed lamps must be equipped with a lamp guard (*Figure 44*). Brass lamp shells, paper-lined sockets, or other metal-cased sockets may not be installed unless the lamp shell is grounded. All holiday or other decorative lighting must be listed per *NEC Section 590.5*.

NOTE

It is not necessary to install a box for splices or connections on construction sites in multiwire cord or cable circuit conductors as long as the equipment grounding conductor's continuity is maintained. However, a box, conduit body, or terminal fitting designed with a separate hole for each conductor must be used when changing from conduit to cable [*NEC Section 590.4(G)*].

7.4.0 Wiring and Equipment Greater Than 600V

If any wiring or equipment operates at greater than 600V at any temporary location, it must be guarded by a fence, barrier, or other effective means and only accessible to authorized and qualified personnel familiar with the equipment or wiring. See *NEC Section 590.7*.

7.5.0 Ground Fault Protection

The *NEC®* applies specific rules to power derived from an electric utility company or from an on-site generator (*NEC Section 590.6*). Ground fault protection must be provided on all 125V, single-phase, 15A, 20A, or 30A temporary receptacles that are not part of the permanent wiring (*Figure 45*), unless temporary power is being supplied by the permanent wiring. If a receptacle that is part of the permanent wiring is used for temporary electric power, it too must be GFCI-protected. GFCI protection may also be incorporated into a cord set or device to comply with this rule.

The only exception to the GFCI rule is in industrial locations where a greater hazard would be introduced if power were interrupted or the system design is not compatible with GFCI protection [*NEC Section 590.6(A), Exception*]. In these locations, GFCI protection is not required as long as only qualified personnel are involved and an assured equipment grounding conductor program is in place. However, if GFCI protection becomes available and is the appropriate voltage and amperage for the application, it must be used.

7.6.0 Assured Equipment Grounding Conductor Program

If GFCI protection is not supplied, there must be a written assured equipment grounding conductor program in place at each site requiring temporary wiring. See *NEC Section 590.6(B)(2)*. The program must be enforced by at least one designated person to ensure that all cord sets, receptacles (temporary), and cord-and-plug equipment are equipped with equipment grounding conductors and maintained in accordance with applicable requirements of the *NEC®*. At a minimum, the

412F44.EPS

Figure 44 ◆ Lamp guard in place on temporary lighting at construction site.

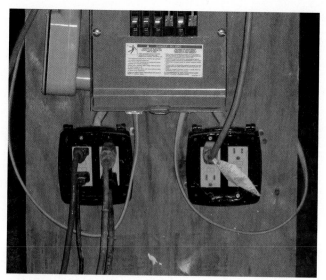

412F45.EPS

Figure 45 ◆ Receptacles on construction sites must be GFCI-protected.

assured equipment grounding conductor program must contain the following:

- The equipment grounding conductors must be electrically continuous and tested for continuity.
- Each receptacle and plug must be tested for correct grounding conductor termination. These tests must be performed before first use, when there is evidence of damage, after repairs and before next usage, and at intervals not exceeding three months.
- All test results must be recorded and available for review.

8.0.0 ◆ WIRED PARTITIONS

Wired partitions allow for convenient rearrangement of office spaces. *NEC Article 605* provides the requirements for lighting accessories and wired office partitions. When installed in interconnecting office partitions, conductors and connections must be contained within wiring channels identified for such use (*NEC Article 605.3*). When two or more wired partitions are electrically connected together, the electrical connection between the partitions must be made with a flexible assembly specifically designed for such purposes (*Figure 46*), or with a flexible cord as long as all of the following conditions are met (*NEC Article 605.4*):

- The flexible cord must be rated for extra-hard usage, No. 12 AWG or larger conductors, and contain an insulated grounding conductor.
- The partitions must be mechanically secured together (contiguous).

- The cord may be no longer than 2 feet.
- The cord must be terminated at an attachment plug-and-cord connector equipped with a strain relief device.

Per *NEC Section 605.6*, partitions that are fixed to the building surfaces must be permanently connected to the building's electrical system using the general wiring methods found in *NEC Chapter 3*. As with any multiwire branch circuit, a disconnecting means must be provided that simultaneously disconnects all ungrounded conductors.

Individual freestanding (not fixed) partitions, or groups of freestanding partitions that are mechanically secured together and do not exceed 30 feet when connected, may be connected to the building's electrical system using a single flexible cord, provided all of the conditions in *NEC Section 605.8* are met:

- The flexible cord is rated for extra-hard usage, No. 12 AWG or larger, contains an insulated equipment grounding conductor, and is no longer than 2 feet.
- The supply receptacle is on a separate circuit that serves only partitions and no other loads, and is located no farther than 12 inches from the partition connected to it.
- Individual partitions or grouped partitions cannot contain more than 13 receptacle outlets (15A, 125V).
- Individual partitions or grouped partitions cannot contain multiwire circuits.

Figure 47 shows a group of freestanding partitions that are connected to the building's electrical system using a single flexible cord.

Figure 46 ◆ Flexible assembly connecting two wired partitions.

412F46.EPS

Figure 47 ◆ Group of freestanding office partitions connected using a single flexible cord.

412F47.EPS

9.0.0 ◆ SWIMMING POOLS, FOUNTAINS, HOT TUBS, AND SIMILAR INSTALLATIONS

NEC Article 680 applies to electrical wiring and equipment associated with the construction and installation of permanently installed swimming pools (*Figure 48*), storable pools, spas and hot tubs, fountains (*Figure 49*), therapeutic tubs and tanks, and hydromassage bathtubs.

Figure 48 ◆ Permanently installed pool.

412F48.EPS

Figure 49 ◆ Fountain.

412F49.EPS

9.1.0 General Wiring Requirements

The general wiring requirements for all pools, spas, fountains, and tubs are covered by *NEC Article 680* and summarized in the following sections. These include requirements for grounding, cord-and-plug connections, clearances, electric pool heaters, underground wiring, and disconnects.

9.1.1 Grounding Requirements

Because of the proximity to water, all electrical equipment in a pool area must be grounded. The following equipment and devices must be grounded per *NEC Section 680.6*:

- Underwater and through-wall luminaires except low-voltage lighting listed for usage without a grounding conductor
- Electrical equipment located within 5 feet of the inside wall of the body of water
- All associated water-circulating equipment
- Junction boxes
- Transformer enclosures
- Ground fault circuit interrupters
- Panelboards not associated with the service equipment, but supplying electrical equipment associated with the body of water

Figure 50 is a panel located within the equipment room of a permanently installed swimming pool.

9.1.2 Cord-and-Plug Connections

Flexible cords with a maximum length of 3 feet may be used to connect fixed or stationary equipment (except underwater luminaires) to their power supply for easy removal or disconnection

412F50.EPS

Figure 50 ◆ Panel in equipment room of swimming pool.

for maintenance or repair (*NEC Section 680.7*). The cord must contain a copper equipment grounding conductor no smaller than No. 12 AWG that terminates on the supply end to a grounding-type attachment plug and is connected to a fixed metal part of the equipment on the other end. All components of the removable equipment must be bonded together.

9.1.3 Clearances

Clearances between open overhead electrical conductors and the water's edge are set forth in *NEC Table 680.8*. *Figure 51* shows overhead conductors in the voltage range of over 750V near a community swimming pool.

The values given in *Table 4* assume a maximum water level. The distances do not apply to communication, radio, and television coaxial cables. These cables must maintain a minimum height of 10 feet above swimming and wading pools, diving boards and towers, or other observation platforms. However, the values in *Table 4* do apply to network-powered broadband communications systems conductors operating at 0V to 750V to ground.

9.1.4 Electric Pool Heaters

Heater elements on electric pool water heaters must be subdivided into loads not exceeding 48A each and protected by overcurrent protective devices not greater than 60A. The ampacity of branch circuit conductors and the overcurrent device rating may not exceed 125% of the total nameplate load. See *NEC Section 680.9*.

412F51.EPS

Figure 51 ◆ Overhead conductors near a community swimming pool.

9.1.5 Underground Wiring

Wiring may not be installed directly under a pool or within 5 feet horizontally from the inside wall of the pool unless the wiring is needed to supply associated pool equipment (*NEC Section 680.10*). In these cases, the wiring must be installed in corrosion-resistant rigid metal conduit, intermediate metal conduit, or a nonmetallic raceway system (PVC). Conduit installations must conform to the minimum cover requirements as outlined in *NEC Table 680.10* and summarized in *Table 5*.

Table 4	Overhead Conductor Clearances (Data from *NEC Table 680.8*)		
	Insulated Cables Supported By a Solidly Grounded Bare Messenger Wire or Grounded Neutral Conductor	Conductors Not Insulated or Supported by Solidly Grounded Bare Messenger Wire or Grounded Neutral Conductor	
Parameters	**0–750V to Ground**	**0–15kV**	**Over 15KV–50kV**
A. Clearance in any direction to the water level, edge of water surface, base of diving platform, or permanently anchored raft	22.5 feet	25 feet	27 feet
B. Clearance in any direction to observation stands, towers, or diving platform	14.5 feet	17 feet	18 feet
C. Horizontal limit measured from the inside wall of the pool	Must extend to the outer edge of all structures described in A and B but not less than 10 feet.		

Table 5 Minimum Cover Depths (Data from *NEC Table 680.10*)

Wiring Method	Minimum Depth in Inches
Rigid Metal Conduit	6
Intermediate Metal Conduit	6
Nonmetallic Direct Burial Conduit	18
*Other Approved Raceways	18

Raceways requiring concrete encasement for burial must have a concrete envelope no less than 2 inches thick.

*Reprinted with permission from *NFPA 70®*, the *National Electrical Code®*. Copyright © 2007, National Fire Protection Association, Quincy, MA 02269. This reprinted material is not the complete and official position of the National Fire Protection Association on the referenced subject, which is represented only by the standard in its entirety.

9.1.6 Disconnecting Means

A disconnecting means that simultaneously disconnects all ungrounded conductors must be provided for all utilization equipment other than lighting (*NEC Section 680.12*). The disconnecting means must be installed in a readily accessible location at least 5 feet horizontally from the inside wall of a pool, hot tub, or spa unless a barrier that requires a reach path of 5 feet or more is permanently installed between the disconnect and pool's edge. *Figure 52* shows the meter socket and service panel for the pool, which houses the disconnecting means for all utilization equipment around the pool and in the adjacent equipment room. Notice that the service equipment is greater than 5 feet from the pool.

NOTE

The horizontal distance is the shortest path from the pool's edge to the disconnecting device.

9.2.0 Permanently Installed Pools

Permanently installed pools must follow specific wiring and equipment requirements as specified in *NEC Article 680, Parts I and II*. These include special requirements for motors and auxiliary equipments, receptacles, lighting, junction boxes, feeders, and bonding.

9.2.1 Motors

Branch circuit conductors supplying motors associated with pools must be installed in rigid or intermediate metal conduit, rigid PVC conduit, reinforced thermosetting resin conduit, or MC cable listed for the location [*NEC Section 680.21(A)(1)*]. If the motor is installed within a building, electrical metallic tubing (EMT) may be used as a raceway. All raceways or cables must contain an insulated copper equipment grounding conductor sized in compliance with *NEC Section 250.122*, but no smaller than No. 12 AWG.

When flexibility is necessary due to motor vibration, *NEC Section 680.21(A)(3)* requires that the connection from the motor be made with liquidtight flexible metal or liquidtight flexible non-metallic conduit (*Figure 53*). Cord-and-plug connections are also permitted as long as the cord

412F52.EPS

Figure 52 ◆ Service and disconnecting means for the pool equipment.

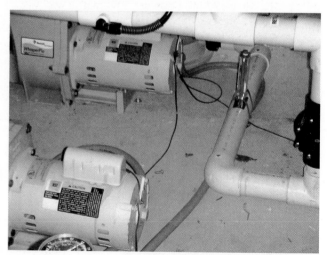

412F53.EPS

Figure 53 ◆ Pool recirculation motors connected with liquidtight flexible metal conduit.

is no longer than 3 feet and incorporates an equipment grounding conductor sized in accordance with *NEC Section 250.122* and terminated in a grounding-type attachment plug.

9.2.2 Receptacle and Area Lighting

General-purpose receptacles must be located at least 6 feet from the inside walls of a pool per *NEC Section 680.22(A)(1)*. Receptacles that are used to provide power to pump motors or other loads directly related to the circulation or sanitation systems of the pool must be located at least 10 feet from the inside wall of a pool unless the receptacle is a single grounding-type receptacle employing a locking configuration and GFCI protection. These types of receptacles must also be located at least 6 feet from the wall of the pool.

Every pool installed at a dwelling unit must be provided with at least one 125V, 15A or 20A receptacle supplied by a general-purpose branch circuit. The receptacle must be located within 20 feet of the pool's inside wall but no less than 6 feet, and at a height of no more than 6'-6" above the grade or deck level [*NEC Section 680.22(A)(3)*]. These receptacles must be GFCI-protected. All other outlets, whether by receptacle or direct connection, that supply power to pump motors from branch circuits rated 15A or 20A, 125V or 240V must also have GFCI protection for personnel [*NEC Section 680.22(B)*].

Area lighting includes all lighting around or above pool areas except those luminaires installed in or under water. *Figure 54* is an example of pole-mounted area lighting, while *Figure 55* shows low-voltage area lighting.

The installation of new luminaires, lighting outlets, or paddle fans above outdoor pools or within 5 feet horizontally from the inside walls of the pool must maintain a clearance of no less than 12 feet above maximum water level of the pool [*NEC Section 680.22(C)(1)*]. For new indoor installations, a reduced minimum clearance of 7'-6" is permitted if the branch circuit has ground fault protection and the luminaires are totally enclosed,

412F54.EPS

Figure 54 ◆ Pole-mounted area lighting around a pool.

412F55.EPS

Figure 55 ◆ Low-voltage area lighting around a pool.

or in the case of fans, the fan is identified for use beneath ceilings such as on porches or patios [*NEC Section 680.22(C)(2)*].

Existing luminaires or lighting outlets (installed before the pool was built) that are located less than 5 feet from the inside edge of the pool must also be at least 5 feet above the maximum water level, rigidly attached to the existing structure, and ground fault protected per *NEC Section 680.22(C)(3)*.

Luminaires, lighting outlets, and ceiling fans that are installed between 5 and 10 feet from the inside edge of the pool must also be GFCI-protected unless they are installed 5 feet or more above the maximum water level and rigidly attached to the structure adjacent to or enclosing the pool [*NEC Section 680.22(C)(4)*]. Luminaires located within 16 feet of the pool and powered by cord-and-plug connections must comply with the cord-and-plug requirements of *NEC Section 680.7*.

Switches must be located at least 5 feet from the pool's edge unless separated from the pool by a solid fence, wall, or other permanent barrier or the switch is listed for use within 5 feet of a pool [*NEC Section 680.22(D)*]. *Figure 56* shows an emergency shutoff switch (red switch) for an attached spa. The switch on the right is an automatic timer switch for the spa that can be set for various time periods, with an automatic shutoff after the time expires. These switches must be located a minimum of 5 feet from the pool's edge and be readily accessible.

Other outlets such as those used for remote control operations, signaling, fire alarm, or communication circuits must be located a minimum of 10 feet from the inside edge of the pool per *NEC Section 680.22(E)*.

9.2.3 Underwater Luminaires

The maximum voltage supplying underwater luminaires is limited to 150V between circuit conductors [*NEC Section 680.23(A)(4)*]. Underwater luminaires may be directly supplied by a branch circuit or by an **isolated winding-type transformer** listed as a swimming pool and spa transformer, with an ungrounded secondary and a grounded metal barrier between the primary and secondary windings [*NEC Section 680.23(A)(2)*].

In order to reduce or prevent shock during relamping, *NEC Section 680.23(A)(3)* requires any underwater luminaire that operates at more than 15V to have a ground fault circuit interrupter installed in the branch circuit supplying the luminaire. *Figure 57* shows the GFCI circuit breaker in the panel serving the pool lighting and equipment. The ground fault circuit interrupter must be installed so that any shock hazard is eliminated should contact occur between a person and any ungrounded part of the branch circuit or the luminaire to ground. Underwater luminaires supplied by a transformer and operated at less than 15V do not require ground fault protection.

Wall-mounted underwater luminaires must be installed so that the top of the lens is at least 18 inches below the water level per *NEC Section 680.23(A)(5)*. Under no conditions may an underwater luminaire be installed less than 4 inches below water level. If an underwater luminaire is installed at the bottom of the pool facing upward, it must either have the lens guarded to prevent people contact, or be listed for use without a guard [*NEC Section 680.23(A)(6)*]. Underwater luminaires that depend on water for cooling must be inherently protected against overheating when not submerged [*NEC Section 680.23(A)(7)*].

Wet-Niche Luminaires – A wet-niche luminaire in a pool or fountain is mounted in a forming shell that is completely surrounded by water, with the forming shell providing a conduit connection. All

412F56.EPS

Figure 56 ◆ Spa emergency shutoff switch and adjacent timer.

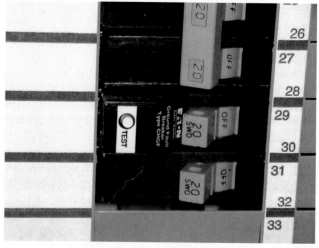

412F57.EPS

Figure 57 ◆ GFCI circuit breaker supplying the underwater luminaire.

Matching Luminaire and Niche

Underwater luminaires and forming shells must be listed and matched. To insure proper matching, listed luminaires are marked on the exterior of the luminaire to identify the compatible forming shell in which the luminaire may be installed. Likewise, forming shells are marked on the interior surfaces to identify the luminaires that may be installed inside the shell. This marking system is part of *Underwriters Laboratories Standard for Safety for Underwater Luminaires, UL 676.*

metal parts of the luminaire and forming shell that come in contact with the water must be made of brass or other approved corrosion-resistant metal [*NEC Section 680.23(B)(1)*]. *Figure 58* shows a plated copper forming shell that has been set into the concrete structure of the pool, which has not yet been plastered.

Forming shells designed with nonmetallic conduit connections (other than low-voltage systems not requiring grounding) must be equipped with a grounding lug that will accept a No. 8 AWG copper conductor. Only rigid metal conduit, intermediate

412F58.EPS

Figure 58 ◆ Plated copper forming shell.

metal conduit, liquidtight flexible nonmetallic conduit, or rigid nonmetallic (PVC) conduit may connect to a wet-niche forming shell, and must run directly from the shell to a junction box [*NEC Section 680.23(B)(2)*].

Conduit to shells may be metal or nonmetallic conduit that is approved for swimming pool installations. Metal conduit must be constructed of brass or other approved corrosion-resistant metal [*NEC Section 680.23(B)(2)(a)*]. Nonmetallic conduit must contain a No. 8 AWG insulated solid or stranded copper conductor to function as a bonding jumper between metal parts, unless a low-voltage luminaire not requiring grounding is installed. The bonding jumper must be terminated in the shell, junction box, transformer enclosure, or ground fault circuit interrupter enclosure (if installed). Once the grounding termination is made in the forming shell, it must be covered with an approved **potting compound** to protect it from deterioration. See *NEC Section 680.23(B)(2)(b)*.

If a flexible cord or cable supplies a wet-niche luminaire, the cord or cable must contain an insulated copper equipment grounding conductor no smaller than No. 16 AWG that grounds all exposed noncurrent-carrying metal parts of the luminaire and shell [*NEC Section 680.23(B)(3)*]. The grounding conductor must be terminated on a grounding terminal within the supply junction box or transformer enclosure and shell, as shown in *Figure 58*. The ends of the flexible cord jacket

Pool Electrocutions

The Consumer Product Safety Commission reported more than a dozen electrocutions and a similar number of electrical shock incidents involving circuits around swimming pools between 1997 and 2002. Electrical incidents involving underwater pool lighting were more numerous than those involving any other consumer product used in or around pools, spas, and hot tubs.
Source: *The Consumer Product Safety Commission, Safety Alert #5039*

and the termination within the luminaire must be sealed with an approved potting compound to prevent water from entering the cable [*NEC Section 680.23(B)(4)*].

The wet-niche luminaire must be secured and bonded to the forming shell by a **positive locking device** (typically an Allen-head screw) that provides a low-resistance path, requiring a wrench or other tool to remove the luminaire from the forming shell [*NEC Section 680.23(B)(5)*]. Wet-niche luminaires must be supplied with sufficient cable or cord length to allow the luminaire to be removed and placed on the deck or other dry location for servicing, without requiring the servicing technician to enter the water (*Figure 59*) per *NEC Section 680.23(B)(6)*.

Other Than Wet-Niche Luminaires – Other types of less common underwater luminaire installations include dry-niche, no-niche, and through-wall lighting assemblies. Refer to *NEC Sections 680.23(C) through (E)* when installing these devices.

Branch Circuit Wiring – Branch circuits supplying enclosures and junction boxes connected to wet-niche and no-niche luminaires, and branch circuits supplying field wiring compartments of dry-niche luminaires must be installed in rigid metal conduit, intermediate metal conduit, liquidtight flexible nonmetallic conduit, rigid PVC conduit, or reinforced thermosetting resin conduit [*NEC Section 680.23(F)(1)*]. Electrical metallic tubing (EMT) may be used if installed on or in a building. Electrical nonmetallic tubing, MC, or AC cable may only be used if installed inside a building.

Figure 59 ◆ Sufficient cord must be provided to allow the luminaire to be relamped without entering the water.

412F59.EPS

Equipment Grounding – Regardless of the type of underwater luminaire, they must all be connected to an insulated copper equipment grounding conductor containing no joints or splices, sized in accordance with *NEC Table 250.122*, and no smaller than No. 12 AWG [*NEC Section 680.23(F)(2)*]. Two exceptions to an unbroken equipment grounding conductor are as follows:

- Where more than one underwater luminaire is supplied by the same branch circuit, the equipment grounding conductor may be broken at the interconnection between the grounding terminals of junction boxes, transformer enclosures or other supply circuit enclosures to wet-niche luminaires, or between the field-wiring compartments of dry-niche luminaires.
- Underwater luminaires supplied from a transformer, GFCI, clock-operated switch, or manual snap switch installed between the panelboard and a junction box directly connected by conduit to the luminaire may have the circuit equipment grounding conductor terminated on the enclosure holding the transformer, GFCI, clock-operated switch, or manual switch.

Conductors – Per *NEC Section 680.23(F)(3)*, load-side conductors of GFCIs or transformers supplying underwater luminaires operating at no more than 15V must not be installed in raceways, boxes, or enclosures containing other conductors unless at least one of the following conditions are met:

- The other conductors are GFCI-protected.
- The other conductors are grounding conductors.
- The other conductors supply a feed-through GFCI.
- A GFCI supplying an underwater luminaire is installed in a panelboard containing other circuits protected by non-GFCI devices.

9.2.4 Transformer or GFCI Junction Boxes and Enclosures

Junction boxes (*Figure 60*) that are used as terminal points for conduit running to wet-niche luminaire niches or mounting brackets of no-niche luminaires must comply with the construction and installation requirements of *NEC Sections 680.24(A)(1) and (2)*:

- The junction box must be listed as a swimming pool junction box.
- It must be equipped with threaded hubs or a nonmetallic hub.
- The box must be made of copper, brass, plastic, or other approved corrosion-resistant material.

412F60.EPS

Figure 60 ◆ Junction box supplying wet-niche luminaire.

- The box must provide electrical continuity between all metal conduit connections and the grounding terminals by means of copper, brass, or other approved corrosion-resistant metal that is integrated into the box construction.
- If the luminaire operates at more than 15V, the junction box must be mounted at least 4 inches above ground (measured from the bottom of the box) or at least 8 inches above the maximum water level of the pool, whichever provides the greater elevation, and at least 4 feet from the inside wall of the pool unless separated by a fence or other permanent barrier.
- If the luminaire operates at 15V or less, the junction box may be mounted as a flush deck box as long as the box is filled with an approved potting compound to prevent the entry of moisture, and the flush box is located at least 4 feet from the inside wall of the pool.

Enclosures used to house transformers, GFCIs, or similar devices that directly connect (by conduit) to a niche or no-niche luminaire mounting bracket must comply with the construction and installation requirements of *NEC Sections 680.24(B)(1) and (2)*. These are the same as for wet-niche luminaires except the use of a flush deck box is not permitted. Also, *NEC Section 680.24(B)(1)* requires the use of a duct seal or another approved seal at the conduit connection to prevent air circulation between the conduit and enclosure.

Regardless of type, every enclosure must meet the following requirements:

- Enclosures must be protected from physical damage per *NEC Section 680.24(C)*.
- Enclosures must be provided with at least one grounding terminal more than the number of conduit connections provided by the enclosure per *NEC Section 680.24(D)*.
- Terminations of flexible cords into enclosures must be provided with a strain relief to protect the cord per *NEC Section 680.24(E)*.
- Grounding terminals of enclosures must be connected to the panelboard equipment grounding terminal per *NEC Section 680.24(F)*.

9.2.5 Feeders

Feeders that supply swimming pool branch circuit panelboards, feeders on the load side of the service equipment associated with pools, or the source of a separately derived system supplying pools must follow specific wiring methods and grounding rules of *NEC Section 680.25*. Some of these requirements are as follows:

- Feeders may only be installed in rigid metal conduit, intermediate metal conduit, liquidtight flexible nonmetallic conduit, rigid PVC conduit, or reinforced thermosetting resin conduit.
- Electrical metallic tubing may be used if installed in or on a building.
- Electrical nonmetallic tubing may be used if installed within a building.
- Aluminum conduit may not be installed in the pool area because of corrosion potential.

Existing Feeders from Service Panels to Subpanels

INSIDE TRACK

In existing installations, feeders from service panels to remote panelboards (subpanels) may be installed in flexible metal conduit, or in cables such as SE or other approved cables, as long as the cable contains an equipment grounding conductor within its outer sheath. This exception can be found at the end of *NEC Section 680.25(A)*.

- An equipment grounding conductor must be installed with the feeders between the pool equipment panelboard grounding terminal and the grounding terminal of the service equipment or the source of a separately derived system.
- The grounding conductor must be sized according to *NEC Section 250.122* but never smaller than No. 12 AWG.
- The grounding conductor from a separately derived system must be sized in accordance with *NEC Table 250.66* but never smaller than No. 8 AWG.
- See additional grounding requirements and exceptions for feeders to separate buildings in *NEC Section 680.25(B)(2)*.

9.2.6 Equipotential Bonding

Equipotential bonding in pool or fountain systems can be described as the interconnection of all exposed and buried, noncurrent-carrying conductive parts and surfaces (not just metal), with an approved grounded bonding conductor to reduce any **voltage gradients** in the pool or fountain area.

Per *NEC Section 680.26(B)*, the bonding conductor used must be solid copper, insulated or bare, and no smaller than No. 8 AWG. Interconnections between boxes and enclosures made with noncorrosive rigid metal conduit such as brass may also be considered as bonded. However, bonding made with a solid No. 8 AWG copper is not required to be extended to remote panelboards, electrodes, or service equipment. All bonded connections must comply with *NEC Section 250.8*.

The following equipment, surfaces, and pool components must be included in the equipotential bonding system and comply with all rules of *NEC Articles 680 and 250*:

- All conductive pool shells including concrete or concrete block shells (not vinyl liners and fiberglass pools) must be bonded. Pool shells constructed using structural reinforcing steel that is not encapsulated in a nonconductive compound must be bonded together using steel tie wires per *NEC Section 680.26(B)(1)(a)*. Pool shells constructed with encapsulated structural reinforcing steel must have a copper conductive grid installed in accordance with *NEC Section 680.26(B)(1)(b)*.
- Pool perimeter surfaces including any paved, unpaved, poured concrete or other conductive surfaces that extend outward to a maximum of 3 feet when measured horizontally from the inside walls of the pool require bonding per *NEC Section 680.26(B)(2)*. See *Figure 61*.

412F61.EPS

Figure 61 ◆ Pool perimeter conductive surfaces.

- All metal components, metal fittings, and underwater lighting niches and mounting brackets must be included in the equipotential bonding system, with the exception of low-voltage luminaires in nonmetallic niches. See *NEC Sections 680.26(B)(3) through (5)*.
- Any metal part associated with the pool water circulating system or pool covers must be bonded, including pumps and motors [*NEC Section 680.26(B)(6)*].

NOTE

Even though double-insulated water pump motors are not required to be bonded, a solid No. 8 AWG copper bonding conductor of sufficient length must be provided in the area of the water pump motor in case the motor is replaced with other than a double-insulated motor. See *NEC Section 680.26(B)(6), Exception*.

- Follow the specific manufacturer's recommendations regarding grounding and bonding of pool water heaters. *Figure 62* shows the enclosure of a pool heater bonded to the equipotential bonding system.
- Metal raceways, metal-sheathed cables, metal piping, and any fixed metal parts that are within 5 feet horizontally from the inside walls of the pool or 12 feet vertically from the maximum water level, tower, stand, platform, or diving structure must also be bonded unless separated by permanent barrier per *NEC Section 680.26(B)(7)*.

Accessories Can Complete the Bonding Requirement

This bonding requirement of *NEC Section 680.26(C)* can be fulfilled by any of the other parts that are in direct contact with the water and already required to be bonded in other sections of *NEC Article 680*, such as shells, conduit, ladders, etc., as long as the total water contact area of the bonded metal parts measures a minimum of 9 square inches.

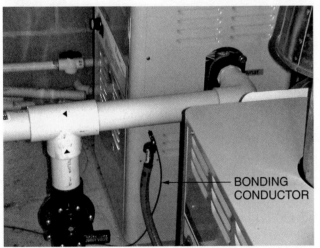

412F62.EPS

Figure 62 ◆ Pool heaters bonded to the equipotential bonding system.

- Per *NEC Section 680.26(C)*, the pool water itself must be bonded using a conductive surface with a minimum area of 9 square inches contacting the pool water.

9.2.7 Specialized Pool Equipment

Accessory equipment not directly related to the lighting and operation of the pool itself falls into the category of specialized pool equipment. Examples of specialized pool equipment include underwater audio equipment, electrically operated pool covers, and deck area heating.

Underwater Audio Equipment – Underwater speakers must be mounted in an approved metal forming shell that is recessed in the wall or floor of the pool, and covered with a metal screen that is bonded to the forming shell [*NEC Section 680.27(A)(1)*]. The screen must be held in place by a positive locking device that requires a tool to open or remove the screen. Both the forming shell and screen must be constructed of brass or other approved corrosion-resistant metal per *NEC Section 680.27(A)(3)*. The forming shell must also be

equipped with a bonding terminal that will accommodate a No. 8 AWG copper conductor.

Electrically Operated Pool Covers – All electrical equipment associated with the operation of a electric pool cover must be located at least 5 feet from the inside wall of the pool unless a permanent barrier is in place between the pool and equipment. See *NEC Section 680.27(B)(1)*. Motors that are installed below grade level must be totally enclosed, and the device controlling the pool cover motor must be located so that the operator has a full view of the pool during cover operation. Both the motor and controller must be connected to a GFCI-protected circuit [*NEC Section 680.27(B)(2)*].

Deck Area Heating – Electrically operated unit heaters or permanently wired radiant heaters that are installed within 20 feet from the inside wall of a pool must be rigidly mounted in place and guarded to prevent contact with the heating elements [*NEC Section 680.27(C)(1)*]. No type of heater may be installed over the pool or within 5 feet from the pool's edge. Permanently wired radiant heaters must be installed at least 12 feet above the pool decking per *NEC Section 680.27(C)(2)*.

 WARNING!

Electrical (not steam) heating cables that are designed to be embedded in surfaces to produce heat transfer into the decking surface are not permitted.

9.3.0 Storable Pools

Cord-connected storable pool filter pumps must be double-insulated and have any internal, nonaccessible, noncurrent-carrying metal parts grounded using an equipment grounding conductor that is part of the supply cord and terminated in a grounding-type attachment plug per *NEC Section 680.31*. The attachment plug must be designed with integral GFCI protection or the power supply cord must incorporate GFCI protection within 12 inches of the attachment plug. *Figure 63* shows a type of storable pool.

The pool filter pump is not the only equipment requiring GFCI protection. *NEC Section 680.32* requires that all electrical equipment associated with storable pools and all 125V receptacles installed within 20 feet of the inside walls of the pool have GFCI protection unless a permanent barrier is installed between the equipment or receptacle and the pool. In no case may a receptacle be installed within 6 feet of the inside walls of the pool unless a permanent barrier is installed.

Underwater luminaires in storable pools must be installed in or on the wall of the pool. See *NEC Section 680.33*. The lens, luminaire body, and transformer enclosure (if used) must be constructed of a polymeric impact-resistant material with no exposed metal parts. If the luminaire operates at 15V or less, it must be a listed cord-and-plug assembly using a transformer listed for the application, with a primary voltage rating not to exceed 150V. Underwater luminaires operating at voltages greater than 15V but less than 150V require GFCI protection.

9.4.0 Spas and Hot Tubs

Spas and hot tubs may be installed indoors or outdoors. Regardless of their location, all spas and hot tub electrical installations, except in single-family dwellings, must include a clearly labeled emergency shutoff or control switch that controls power to the recirculation and/or jet system motor (*Figure 64*). This emergency shutoff must be located at a readily accessible point for the user

412F64.EPS

Figure 64 ◆ Spa emergency shutoff switch located at least 5 feet from spa.

but no less than 5 feet away from the spa or hot tub per *NEC Section 680.41*.

Outdoor Installations – Spas and hot tubs installed outdoors must follow the same rules as permanently installed pools installed outdoors, in addition to the requirements listed in *NEC Section 680.42* and summarized as follows:

- Packaged spas or hot tubs that are listed for outdoor installations or self-contained spas or hot tubs equipped with a factory-installed control panel or panelboard may be connected using 6 feet or less of flexible metal conduit or liquidtight flexible nonmetallic conduit, or a GFCI-protected cord and plug with a maximum cord length of 15 feet.
- Metal-to-metal bonding of common frame or base components is permitted.
- Metal parts used to secure wooden stakes in place are not required to be bonded.
- If a self-contained or packaged outdoor spa/hot tub is supplied by the interior wiring of a single-family dwelling or associated building, the supply wiring method must comply with *NEC Chapter 3* and must include a copper equipment grounding conductor (No. 12 AWG or larger) that is insulated or enclosed within the outer sheath of the wiring method.

Indoor Installations – Spas and hot tubs installed indoors must also follow the same rules for permanently installed pools and comply with the wiring methods of *NEC Chapter 3*. One exception is that indoor packaged spas or hot tubs rated 20A or less may be connected by cord and plug to allow for easy removal or disconnection during maintenance or repair. In addition, the rules of *NEC Section 680.43* also apply:

412F63.EPS

Figure 63 ◆ Storable pool.

Location Measurements

Location measurements assume the shortest distance between the receptacle and the inside wall of the pool without encountering any permanent barrier such as walls, hinged doorways, etc.

- At least one 125V, 15A or 20A receptacle must be provided, and must be located no less than 6 feet and no more than 10 feet from the inside wall of the unit. No receptacle may be located closer than 6 feet from the inside walls of the spa or hot tub.
- Any 125V, 30A (or less) receptacle located within 10 feet of the inside walls of an indoor spa or hot tub must be GFCI-protected.

Wall switches must be installed at least 5 feet from the inside walls of the spa or hot tub [*NEC Section 680.43(C)*]. Luminaires, lighting outlets, and ceiling fans installed over indoor spas or hot tubs must be a minimum of 12 feet above the maximum water level in installations without GFCI protection per *NEC Section 680.43(B)(1)(a)*. For installations with GFCI protection, the minimum clearance is 7'-6" per *NEC Section 680.43(B)(1)(b)*.

The following components associated with indoor spas and hot tubs must be bonded together per *NEC Section 680.43(D)*:

- All metal fittings within or attached to the spa or hot tub, including handrails (*Figure 65*)
- Noncurrent-carrying metal parts of electrical equipment, including circulating pump motors
- Any metal raceway, piping, or metal surface within 5 feet of the inside walls of the spa or hot tub, unless separated by a permanent barrier
- Any electrical device or control not associated with the spa or hot tub, but located within 5 feet of it

Bonding may be accomplished by the interconnection of threaded metal pipe and fittings, metal-to-metal mounting on a common frame or base, or by an insulated, covered, or bare No. 8 AWG or larger solid copper bonding jumper. See *NEC Section 680.43(E)*. In addition to bonding, any electrical equipment that is located within 5 feet of the inside wall of the spa or hot tub, or any electrical equipment associated with the circulating system must also be grounded per *NEC Section 680.43(F)*. Receptacles that supply self-contained, packaged, or field-assembled spas and hot tubs must be GFCI-protected except as noted in *NEC Section 680.44*.

412F65.EPS

Figure 65 ◆ Spa handrails must be bonded.

9.5.0 Fountains

Fountains have become increasingly popular in residential and commercial landscapes. Permanently installed fountains (*Figure 66*), ornamental pools, display pools, and reflection pools must comply with *NEC Article 680, Parts I and V*. In addition, fountains that use common water from a swimming pool (*Figure 67*) must also comply with *NEC Article 680, Part II*.

412F66.EPS

Figure 66 ◆ Permanently installed fountain.

Figure 67 ◆ Fountain using pool water.

NOTE

Self-contained portable fountains and water drinking fountains are not covered by *NEC Article 680*.

9.5.1 Luminaires, Submersible Pumps, and Equipment

Luminaires, submersible pumps, and other submersible equipment installed in fountains (*Figure 68*) must have GFCI protection unless they operate at 15V or less and are supplied by a listed transformer intended for pool and fountain applications. See *NEC Section 680.51(A)*.

Luminaires may not operate at voltages over 150V between conductors, and submersible equipment may not operate at voltages over 300V

Figure 68 ◆ Luminaire and pump in fountains must have GFCI protection.

between conductors. See *NEC Section 680.51(B)*. Luminaires must also be installed so that the top of the luminaire is below the normal water level of the fountain, unless the luminaire is listed for out-of-water operation. Luminaires installed with the lens facing upward must have the lens guarded against people contact or be listed for use without a guard per *NEC Section 680.51(C)*.

NEC Section 680.51(F) requires that luminaires and other submersible equipment be installed to provide easy removal from the fountain for relamping or other maintenance without having to drain or reduce the water level. Luminaires and other submersible equipment may not be permanently embedded into the fountain's structure.

Electrical submersible equipment that must remain submersed in water for proper operation and safety must be equipped with a low-water shutoff device per *NEC Section 680.51(D)*. Submersible equipment must also be equipped with threaded conduit entries or a factory- or field-installed flexible cord no longer than 10 feet in length [*NEC Section 680.51(E)*]. If the cord extends beyond the perimeter of the fountain, the cord must be enclosed in an approved wiring enclosure. Any metal parts in contact with the water must be constructed of brass or other corrosion-resistant material.

9.5.2 Junction Boxes and Other Enclosures

Junction boxes and other enclosures associated with fountains must comply with *NEC Section 680.24*, which covers junction boxes and enclosures in permanently installed pools. In addition, underwater junction boxes and enclosures installed in fountains must also comply with the requirements listed in *NEC Section 680.52(B)* and summarized as follows:

- Designed with threaded conduit entries or compression-type sealing (grommet) fittings for cord entry
- Listed for submersible use and constructed of copper, brass, or other approved corrosion-resistant material
- Filled with an approved potting compound to prevent water entry
- Firmly supported and bonded
- Only supported by conduit if the conduit is copper, brass, stainless steel, or other approved metal conduit
- If supplied by nonmetallic conduit, provided with additional support using approved corrosion-resistant metal support devices or fasteners

9.5.3 Bonding and Grounding

Any metal piping associated with a fountain must be bonded to the branch circuit equipment grounding conductor that supplies the fountain per *NEC Section 680.53*. In addition, all electrical equipment located within 5 feet of the inside wall, any electrical equipment associated with the recirculation system, and any panelboards that are not part of the service equipment but supply electrical equipment associated with the fountain must be grounded per *NEC Section 680.54*.

NEC Section 680.55(B) requires that any electrical equipment associated with a fountain that is supplied by a flexible cord have all of its exposed noncurrent-carrying metal parts grounded using an insulated copper equipment grounding conductor that is part of the flexible cord. The other end of the equipment grounding conductor must be terminated to an equipment grounding terminal in the supply junction box or other supply enclosure. Any 15A or 20A, 125V through 250V receptacle installed within 20 feet of a fountain must be GFCI-protected per *NEC Section 680.58*.

9.5.4 Cord-and-Plug Equipment

NEC Section 680.56(A) requires that any electrical equipment associated with a fountain and supplied by a cord and plug be GFCI-protected. In addition, the flexible cord must be rated for extra-hard usage and listed for wet usage per *NEC Section 680.56(B)*. Ends and terminations of the flexible cord, and the grounding connection within the equipment, must be covered with or encapsulated in a suitable potting compound to prevent the entry of water into the cable or deterioration of the terminations [*NEC Section 680.56(C)*].

9.5.5 Signs Placed in Fountains

NEC Section 680.57(C)(1) requires that any fixed or stationary electric sign installed within a fountain be at least 5 feet from the outside edge of the fountain. This places the sign out of reach of passersby or those sitting on the fountain edge.

Fixed signs must also be equipped with a local disconnecting means in accordance with *NEC Sections 600.6 and 680.12*, and be grounded and bonded in accordance with *NEC Section 600.7*. Portable signs may not be installed in fountains or within 5 feet from the inside walls of the fountain [*NEC Section 680.57(C)(2)*].

9.6.0 Therapeutic Pools and Tubs

NEC Article 680, Part VI covers pools and tubs used for therapeutic purposes in health care facilities, gymnasiums, athletic training rooms, and similar areas. Permanently installed therapeutic pools follow the same *NEC*® requirements as permanently installed swimming pools and must also comply with *NEC Article 680, Parts I and II*.

Electrical outlets that supply self-contained, packaged, or field-assembled therapeutic tubs or hydrotherapeutic tanks must be GFCI-protected unless the listed unit includes integral GCFI protection for all electrical parts within the unit or operates from a three-phase system greater than 250V or with a heater load of more than 50A [*NEC Section 680.62(A)(1) and (2)*]. Bonding requirements for therapeutic tubs or hydrotherapeutic tanks are listed in *NEC Section 680.62(B)* and summarized as follows:

- All metal fittings within or attached to the tub or tank must be bonded together.
- All noncurrent-carrying metal parts of the water circulation system (including pump motors) must be bonded.
- Any metal-sheathed cables, raceways, metal piping, and metal surfaces within 5 feet of the inside walls of the tub or tank must be bonded unless separated by a permanent barrier.
- Any electrical device or control that is not associated with the operation of the tub or tank but is within 5 feet of the tub or tank must be bonded.

This bonding requirement may be satisfied by the interconnection of threaded metal pipe and fittings, metal-to-metal mounting on a common

Cord and Cable Types

INSIDE TRACK

NEC Table 400.4 lists cords and cables according to their trade names (e.g., portable power cable, heater cord, etc.) and type letter (SE, SEW, SJ, etc.). Cords or cables used in fountains must have a type letter that ends with W to indicate a cable or cord suitable for a wet location.

Hydromassage Bathtubs

When installing a hydromassage bathtub in an installation where the final finish will be a tiled surround, make sure the tub or surround design provides for permanent access to the electrical equipment.

frame, connections made using approved metal clamps, or by the installation of an insulated, covered, or bare solid copper bonding jumper no smaller than No. 8 AWG.

All electrical equipment located within 5 feet of the inside wall of the tub or tank, and all electrical equipment associated with the tub or tank's water circulation system, must be grounded to the equipment grounding conductor of the supply circuit per *NEC Section 680.62(D)*. In addition, *NEC Section 680.62(E)* requires that any receptacle located within 6 feet of a therapeutic tub or tank be GFCI-protected, and all luminaires installed in the area of therapeutic tubs or tanks must be totally enclosed [*NEC Section 680.62(F)*].

9.7.0 Hydromassage Bathtubs

Hydromassage bathtubs, also known as whirlpool tubs, are commonly installed in the master bathrooms of new residential construction and are often added when older bathrooms are renovated. A hydromassage bathtub is a permanently installed bathtub equipped with a recirculation pump, piping system, and jets. By design, the tub can accept, circulate, and discharge water in each usage. Hydromassage bathtubs must be installed in accordance with *NEC Article 680, Part VII*.

NEC Section 680.71 requires that hydromassage tubs be supplied by an individual branch circuit, which can be an issue when installing these tubs on the second floor in retrofit installations. In addition, a readily accessible GFCI must be provided between the supply and the tub, and all 125V receptacles located within 6 feet of the inside wall of the tub must be GFCI-protected. The tub electrical equipment must be accessible without damaging the building structure or finish per *NEC Section 680.73*, which is often accomplished using a removable panel either on the tub itself or through an adjacent wall. All metal piping systems and grounded metal parts in contact with the water must be effectively bonded using an insulated, covered, or bare solid copper bonding jumper no smaller than No. 8 AWG [*NEC Section 680.74*]. The bonding jumper must be connected to a circulating pump motor terminal intended for

this purpose. Double-insulated circulating pump motors do not require bonding. The bonding jumper is not required to be extended or attached to remote panelboards, service equipment, or any electrode.

10.0.0 ◆ NATURAL AND MANMADE BODIES OF WATER

Natural bodies of water include lakes, streams, ponds, and rivers, all of which may naturally vary in depth. *Figure 69* is an example of a natural body of water, while *Figure 70* illustrates a manmade or artificial body of water. Manmade bodies of water refer to aeration ponds, fish farm ponds, storm retention basins, water treatment ponds, and irrigation channels. Electrical installations in and around natural and manmade bodies of water are covered in *NEC Article 682*.

10.1.0 Electrical Datum Plane

The electrical datum plane is an imaginary horizontal plane based on changing water levels. Electrical datum planes are referred to extensively in *NEC Articles 555 and 682* because both articles deal with changing water levels that can affect electrical installations in and around marinas, boatyards, and natural and artificial bodies of

412F69.EPS

Figure 69 ◆ Natural body of water.

Figure 70 ◆ Manmade body of water.

412F70.EPS

water. The definition of an electrical datum plane is provided in *NEC Section 682.2* and varies based on the following conditions:

- Where land areas are affected by tidal fluctuations, the electrical datum plane is 2 feet above the highest high tide for the area.
- Where land areas are not affected by tidal fluctuations, the electrical datum plane is 2 feet above the highest water level for the area under normal conditions.
- In areas subject to flooding, the electrical datum plane is 2 feet above the prevailing high water mark determined by the authority having jurisdiction.
- For floating structures that are installed to rise and fall in response to the water level without lateral movement, the electrical datum plane is 30 inches above the point at which the floating structure touches the water and a minimum of 12 inches above the deck level.

10.2.0 Location of Equipment and Enclosures

NEC Section 682.10 requires that all electrical equipment and enclosures installed in or near natural and manmade bodies of water be listed and approved for the location. Equipment or enclosures not identified for submersible use must not be located below the applicable electrical datum plane.

Per *NEC Section 682.11*, service equipment and service disconnecting means that supply floating structures and submersible equipment must be located on land at least 5 feet from the shoreline, within sight of the structure served, identified as to which structure or equipment it serves, and

installed to allow a minimum of 12 inches between live parts and the electrical datum plane. In situations where the water level rises to the electrical datum plane, the service equipment must be designed to disconnect the electrical service when this occurs.

Electrical connections that are not identified for submersible use must be located at least 12 inches above the deck of a floating or fixed structure, but never below the established electrical datum plane (*NEC Section 682.12*).

10.3.0 GFCI Protection, Grounding, and Bonding

All 15A and 20A, single-phase, 125V through 250V receptacles that are installed outdoors, in or on floating buildings, or in storage or maintenance buildings within the electrical datum plane where portable electric hand tools, diagnostic equipment, or portable lighting may be used must be GFCI-protected per *NEC Section 682.15*. The GFCI protection device must be located at least 12 inches above the established electrical datum plane.

Equipment grounding conductors must meet the requirements listed in *NEC Sections 682.31(A) through (D)* and summarized as follows:

- Equipment grounding conductors must be insulated copper conductors no smaller than 12 AWG.
- Feeders supplying a remote panelboard must include an insulated grounding conductor that connects from the service grounding terminal to a grounding terminal and busbar in the remote panelboard.
- Branch circuit equipment grounding conductors must terminate on the grounding bus in the remote panelboard or the grounding terminal in the main service panel.
- Cord-and-plug equipment must be grounded by means of an equipment grounding conductor in the cord and connected to a grounding-type attachment plug at the supply end.

Metal piping, tanks, noncurrent-carrying metal parts that could become energized, and metal parts in contact with the water must be bonded and grounded through the equipment grounding conductor to the grounding bus in the panelboard per *NEC Section 682.32*.

10.4.0 Equipotential Planes

An equipotential plane bonds all metal parts together, including grounding grids, structural steel, conductive equipment enclosures, etc., in order to reduce or eliminate any gradient (variable) voltage potential between system components, structures, enclosures, and even the water itself. Per *NEC Section 682.33(A)*, equipotential planes must be installed in areas adjacent to all outdoor service equipment with metal enclosures or disconnecting means that are used to supply or control equipment in or on the water, and are accessible to personnel. The plane must encompass the immediate area around and directly below the equipment, and extend outward at least 36 inches in all directions from which a person would be able to stand and reach the equipment.

Equipotential planes must be bonded to the electrical grounding system using an insulated, covered, or bare solid copper conductor no smaller than No. 8 AWG [*NEC Section 682.33(C)*]. Connections of the bonding conductor to the grounding system must be made by exothermic welding or using listed stainless steel, brass, copper, or copper alloy pressure connectors or clamps.

1. In order to qualify as an assembly occupancy, a meeting hall must be designed to hold at least _____ people.
   a. 50
   b. 100
   c. 200
   d. 500

2. Wiring methods allowed in assembly occupancies include metal raceways, flexible metal raceways, nonmetallic raceways encased in _____, MI cable, MC cable, or AC cable.
   a. at least 18 inches of sand
   b. metal cable tray
   c. at least 2 inches of concrete
   d. at least 6 inches of concrete

3. The use of electrical nonmetallic tubing and rigid nonmetallic conduit in spaces of assembly occupancies containing _____ supports is restricted to certain areas.
   a. combustible
   b. conduit
   c. conductive
   d. clamping

4. Areas in occupancies covered by *NEC Article 520* that are required to be of fire-rated construction may contain wiring methods consisting of nonmetallic-sheathed cable, AC cable, electrical nonmetallic tubing, or rigid nonmetallic conduit.
   a. True
   b. False

5. In a manually controlled stage switchboard, all dimmers and switches are operated by handles that are _____ linked to the control devices.
   a. automatically
   b. mechanically
   c. electrically
   d. manually

6. Portable structures in circuses and similar areas must maintain a clearance of _____ feet in any direction from overhead conductors operating at 600V or less unless the conductors are supplying power to the portable structure.
   a. 10
   b. 15
   c. 18
   d. 20

7. The size of the bonding conductor for portable buildings at carnivals, circuses, or fairs must be accordance with *NEC Table 250.122*, based on the largest overcurrent device supplying the portable structures, but must not be smaller than No. _____ AWG.
   a. 2
   b. 6
   c. 8
   d. 12

8. When outlet boxes or other device boxes are installed outdoors at circuses, fairs, carnivals, and the like, the boxes must be weatherproof and mounted so the bottom of the box is at least _____ inches off the ground.
   a. 6
   b. 8
   c. 12
   d. 18

9. Cables in agricultural buildings must be secured within _____ inches of any box, cabinet, or fitting to which they connect.
   a. 2
   b. 6
   c. 8
   d. 12

10. If two or more agricultural buildings are supplied overhead from a distribution point, a _____ device must be installed.
    a. service-splitter
    b. current-monitoring
    c. motion-detecting
    d. site-isolating

11. If more than one service supplies an agricultural site, and any two of the services are _____ feet or less apart from one another as measured in a straight line, each distribution point must have a permanent label or directory installed to indicate the location of each of the other distribution points and the buildings served by each.
    a. 10
    b. 50
    c. 250
    d. 500

12. Electrical distribution systems within boatyards and piers cannot exceed _____ phase-to-phase.
    a. 120V
    b. 240V
    c. 480V
    d. 600V

13. Any electrical connection on a fixed pier must be at least 12 inches above the deck of the fixed pier but never below the _____.
    a. service height
    b. electrical datum plane
    c. seepage point
    d. fuel dispensing system

14. The factor applied to the sum of the rating for 35 shore power receptacles is _____.
    a. 40%
    b. 60%
    c. 70%
    d. 100%

15. Holiday lights must be removed after _____ days from the date of installation.
    a. 30
    b. 60
    c. 90
    d. 180

16. It is not necessary to install a box for splices or connections on construction sites in multiwire cord or cable circuit conductors as long as the equipment grounding conductor's _____ is maintained.
    a. termination
    b. insulation
    c. continuity
    d. location

17. When two or more wired partitions are electrically connected together, the connection may be made using a flexible cord that is no longer than _____ feet in length.
    a. 2
    b. 4
    c. 6
    d. 8

18. The overhead clearance to the water's edge of a swimming pool for insulated cables supported by a solidly grounded bare messenger wire or grounded neutral conductor in a voltage range of 0–750V to ground is _____ feet.
    a. 18
    b. 22.5
    c. 25
    d. 27.5

19. If a flexible cord or cable supplies a wet-niche luminaire, the cord or cable must contain an insulated copper equipment grounding conductor no smaller than No. _____ AWG that grounds all exposed non-current-carrying metal parts of the luminaire.
    a. 6
    b. 8
    c. 12
    d. 16

20. Where land areas are affected by tidal fluctuations, the electrical datum plane is 2 feet above the highest _____ for the area.
    a. water level
    b. water mark
    c. flood plane
    d. high tide

Summary

NEC Chapters 1 through 4 cover the minimum requirements for electrical installations in most dwelling units and commercial buildings, as well as some of the requirements associated with industrial electrical installations; however, because some areas, types of structures, or equipment by their very nature or the environment in which they are installed pose added potential electrical hazards, additional chapters in the NEC® had to be developed, approved, and written. This module presents many of the articles presented in the additional chapters covering special occupancies and special equipment. As an electrician, it is your responsibility to familiarize yourself with all of the NEC® requirements for any electrical installation you undertake, as well as any local or regional codes that may also apply.

Notes

Trade Terms
Introduced in This Module

Border light: A permanently installed overhead strip light.

Dead-front: Equipment or enclosures covered or protected in order to prevent exposure of energized parts on the operating side of the equipment.

Electrical datum plane: Horizontal plane affecting electrical installations based on the rise and fall of water levels in land areas and floating piers and landing stages.

Equipotential plane: An area in wire mesh or other conductive material embedded in or placed under concrete that may conduct current. It must be bonded to all metal structures and non-electrical, fixed equipment that could become energized, and then connected to the electrical grounding system.

Extra-hard usage cable: Cable designed with additional mechanical protection for use in areas where it may be subject to physical damage.

Festoon lighting: Outdoor lighting that is strung between supporting points.

Fire rating: Time in hours or minutes that a material or assembly can withstand direct exposure to fire without igniting and spreading the fire.

Footlight: A border light installed on or in a stage.

Isolated winding-type transformer: A transformer designed with a mechanical isolating device, such as a plate, between the primary and secondary windings.

Landing stage: A floating platform used for the loading and unloading of passengers and cargo from boats or ships.

Portable switchboard: A movable lighting switchboard used for road shows and similar presentations, usually supplied by a flexible cord or cable.

Positive locking device: A locking device such as a screw, clip, or other mechanical device that requires a tool or key to lock or unlock it.

Potting compound: An approved and listed insulating compound, such as epoxy, which is mixed and installed in and around electrical connections to encapsulate and seal the connections to prevent the entry of water or moisture.

Proscenium: The wall and arch that separate the stage from the auditorium (house).

Reactor-type dimmer: A dimming system that incorporates a number of windings or coils that are linked by a magnetic core, an inductive reactor that functions like a capacitor, and a relay. The reactor stores current from the source and reduces lamp power.

Resistance-type dimmer: A manual-type dimmer in which light intensity is controlled by increasing or decreasing the resistance in the lighting circuit.

Rheostat: A slide and contact device that increases or decreases the resistance in a circuit by changing the position of a movable contact on a coil of wire.

Shunt trip: A low-voltage device that remotely causes a higher voltage overcurrent protective device to trip or open the circuit.

Site-isolating device: A disconnecting means at the point of distribution that is used to isolate the service for the purpose of electrical maintenance, emergency disconnection, or the connection of an optional standby system.

Solid-state phase-control dimmer: A solid-state dimmer in which the wave shape of the steady-state current does not follow the wave shape of the applied voltage (i.e., the wave shape is non-linear).

Solid-state sine-wave dimmer: A solid-state dimmer in which the wave shape of the steady-state current follows the wave shape of the applied voltage (i.e., the wave shape is linear).

Unbalanced load: Three-phase loads in which the loads are not evenly distributed between the phases.

Voltage gradients: Fluctuating voltage potentials that can exist between two surfaces or components and may be caused by improper or insufficient bonding or grounding.

This module is intended to be a thorough resource for task training. The following reference work is suggested for further study. This is optional material for continued education rather than for task training.

National Electrical Code® Handbook, Latest Edition. Quincy, MA: National Fire Protection Association.

CONTREN® LEARNING SERIES – USER UPDATE

NCCER makes every effort to keep these textbooks up-to-date and free of technical errors. We appreciate your help in this process. If you have an idea for improving this textbook, or if you find an error, a typographical mistake, or an inaccuracy in NCCER's Contren® textbooks, please write us, using this form or a photocopy. Be sure to include the exact module number, page number, a detailed description, and the correction, if applicable. Your input will be brought to the attention of the Technical Review Committee. Thank you for your assistance.

Instructors – If you found that additional materials were necessary in order to teach this module effectively, please let us know so that we may include them in the Equipment/Materials list in the Annotated Instructor's Guide.

Write: Product Development and Revision
National Center for Construction Education and Research
3600 NW 43rd St., Bldg. G, Gainesville, FL 32606

Fax: 352-334-0932

E-mail: curriculum@nccer.org

Craft _____ Module Name _____

Copyright Date _____ Module Number _____ Page Number(s) _____

Description _____

_____

_____

_____

_____

(Optional) Correction _____

_____

_____

(Optional) Your Name and Address _____

_____

_____

Introductory Skills
for the Crewleader

University of Illinois at Urbana-Champaign – Assembly Line to Success

Students from University of Illinois developed their entry with the ability to mass-produce modules in a large-scale environment. Modular design allows the buyer to be able to customize the interior space, and cooling and heating is all radiant via ceiling panels with no forced air. Placement of windows and doors for daylighting was designed in parallel with an artificial lighting plan.

26413-08

ACKNOWLEDGMENTS

The NCCER wishes to acknowledge the dedication and expertise of Roger Liska, the original author and mentor for this series on leadership development.

Roger W. Liska, Ed.D., FCIOB, CPG, FAIC, PE

Clemson University

Chair & Professor

Department of Construction Science & Management

We would also like to thank the following reviewers for contributing their time and expertise to this endeavor:

J.R. Blair

Tri-City Electrical Contractors

An Encompass Company

Mike Cornelius

Tri-City Electrical Contractors

An Encompass Company

Dan Faulkner

Wolverine Building Group

David Goodloe

Clemson University

Kevin Kett

The Haskell Company

Danny Parmenter

The Haskell Company

FOREWORD

The crew leader plays a key role in the operation of a construction company. As the first line of management on the job, the crew leader is responsible for supervising workers in an effective and efficient manner. Once a construction worker is promoted to the position of crew leader, he or she becomes involved with managing people.

As a supervisor, the crew leader will use new skills. These skills include problem solving, planning, scheduling, estimating, controlling the use of job resources, such as material and equipment, and knowing how to perform the work in a safe manner. Practicing effective human relations skills, such as motivating workers and communicating with them, are also an essential part of the crew leader's job.

The purpose of this program is to provide the person who is considering be-coming a crew leader with knowledge of the basic leadership skills required for the job.

This program consists of an Instructor's Guide and a Participant's Manual. The Instructor's Guide contains a breakdown of the information provided in the Participant's Manual as well as the actual text that the participant will use. The Participant's Manual contains the material that the participant will study, along with self-check exercises and activities, to evaluate whether the participant has mastered the knowledge necessary to become an effective crew leader.

For the participant to gain the most from this program, it is recommended that the material be presented in a formal classroom setting, using a trained and experienced instructor. If the student is so motivated, he or she can study the material on a self-learning basis by using the material in both the Manual and Guide. Recognition through the National Registry is available for the participants provided the program is delivered through the Accredited Sponsor network by a Master Trainer or ICTP graduate. More details on this program can be received by calling the National Center for Construction Education and Research at 352-334-0911.

Users of this program should note that some examples provided to reinforce the material may not apply to the participant's exact work, although the process does. Furthermore, every company has its own mode of operation. Therefore, some topics may not apply to the participant's company. Topics have been included because they are important considerations for prospective crew leaders throughout the industry.

A NOTE TO NCCER INSTRUCTORS AND TRAINEES

If you are training under an Accredited NCCER Sponsor, note that you may be eligible for dual credentials for successful completion of Introductory Skills for the Crew Leader. When submitting Form 200, indicate completion of the two module numbers that apply to Introductory Skills for the Crew Leader – MT101 (from NCCER's Contren® Management Series) and 26413-08 (from NCCER's Electrical Level Four) and transcripts will be issued to you accordingly.

TABLE OF CONTENTS

CHAPTER ONE – ORIENTATION TO THE JOB

OBJECTIVES 13.1

1.0.0   GROWTH AND ECONOMICS OF THE
        CONSTRUCTION INDUSTRY 13.1

1.1.0   Changing Values of Workers 13.2

2.0.0   THE CONSTRUCTION INDUSTRY TODAY 13.4

2.1.0   Training 13.4

2.1.1   *Motivation* 13.4

2.1.2   *Understanding the New Generation* 13.4

2.1.3   *Craft Training* 13.5

2.1.4   *Supervisory Training* 13.5

2.2.0   Impact of Developing Technology 13.6

2.2.1   *Other Communication Sources* 13.6

3.0.0   GENDER AND CULTURAL ISSUES 13.6

3.1.0   Communication Styles of Men and Women 13.7

3.2.0   Language Barriers 13.7

3.3.0   Cultural Differences 13.8

3.4.0   Sexual Harassment 13.8

3.5.0   Gender and Minority Discrimination 13.9

4.0.0   THE CONSTRUCTION ORGANIZATION 13.11

4.1.0   Division of Responsibility 13.11

4.2.0   Authority and Responsibility 13.12

4.3.0   Job Descriptions 13.12

4.4.0   Policies and Procedures 13.13

CHAPTER TWO – LEADERSHIP SKILLS

OBJECTIVES 13.15

1.0.0   INTRODUCTION TO SUPERVISION 13.15

2.0.0   THE SHIFT IN WORK ACTIVITIES 13.16

3.0.0   BECOMING A LEADER 13.17

3.1.0   Characteristics of Leaders 13.17

3.1.1   *Leadership Traits* 13.18

3.1.2   *Expected Leadership Behavior* 13.18

3.2.0   Functions of a Leader 13.18

3.3.0   Leadership Styles 13.18

3.3.1   *Selecting a Leadership Style* 13.19

3.3.2   *Crew Evaluations* 13.19

3.4.0	Ethics in Leadership	13.19
4.0.0	COMMUNICATION	13.20
4.1.0	Verbal Communication	13.20
4.1.1	*The Sender*	13.21
4.1.2	*The Message*	13.21
4.1.3	*The Receiver*	13.21
4.1.4	*Feedback*	13.22
4.2.0	Non-Verbal Communication	13.22
4.3.0	Written or Visual Communication	13.23
4.4.0	Communication Issues	13.23
5.0.0	MOTIVATION	13.26
5.1.0	Employee Motivators	13.26
5.1.1	*Recognition and Praise*	13.26
5.1.2	*Accomplishment*	13.26
5.1.3	*Opportunity for Advancement*	13.26
5.1.4	*Job Importance*	13.27
5.1.5	*Change*	13.27
5.1.6	*Personal Growth*	13.27
5.1.7	*Rewards*	13.27
5.2.0	Motivating Employees	13.27
6.0.0	TEAM BUILDING	13.29
6.1.0	Successful Teams	13.29
6.2.0	Building Successful Teams	13.29
7.0.0	GETTING THE JOB DONE	13.29
7.1.0	Delegating Responsibilities	13.29
7.2.0	Implementing Policies and Procedures	13.30
8.0.0	PROBLEM SOLVING AND DECISION MAKING	13.31
8.1.0	Problem Solving vs. Decision Making	13.31
8.2.0	Types of Decisions	13.31
8.3.0	Formal Problem-Solving Techniques	13.31
8.4.0	Special Leadership Problems	13.35
8.4.1	*Poor Attitude Towards the Workplace*	13.35
8.4.2	*Inability to Work With Others*	13.35
8.4.3	*Absenteeism and Turnover*	13.35

CHAPTER THREE – SAFETY

OBJECTIVES		13.41
1.0.0	SAFETY OVERVIEW	13.41
1.1.0	Accident Statistics	13.42
2.0.0	COSTS OF ACCIDENTS	13.42
2.1.0	Insured Costs	13.42
2.2.0	Uninsured Costs	13.42
3.0.0	SAFETY REGULATIONS	13.43
3.1.0	Workplace Inspections	13.43

3.2.0   Penalties for Violations 13.44
4.0.0   SAFETY RESPONSIBILITIES 13.44
4.1.0   Safety Program 13.44
4.2.0   Safety Policies and Procedures 13.44
4.3.0   Hazard Identification and Assessment 13.45
4.4.0   Safety Information and Training 13.45
4.5.0   Safety Record Systems 13.45
4.6.0   Accident Investigation Procedures 13.46
5.0.0   SUPERVISOR INVOLVEMENT IN SAFETY 13.46
5.1.0   Safety Meetings 13.46
5.2.0   Inspections 13.46
5.3.0   First Aid 13.46
5.4.0   Fire Protection and Prevention 13.47
5.5.0   Substance Abuse 13.47
5.6.0   Accident Investigations 13.48
6.0.0   PROMOTING SAFETY 13.48
6.1.0   Meetings 13.49
6.2.0   Contests 13.49
6.3.0   Recognition and Awards 13.50
6.4.0   Publicity 13.50

CHAPTER FOUR – PROJECT CONTROL
OBJECTIVES 13.51
1.0.0   PROJECT CONTROL OVERVIEW 13.51
1.1.0   Construction Projects 13.51
1.1.1   *Development Phase* 13.51
1.1.2   *Planning Phase* 13.52
1.1.3   *Construction Phase* 13.52
2.0.0   PROJECT DELIVERY SYSTEMS 13.54
2.1.0   General Contracting 13.54
2.2.0   Design-Build 13.54
2.3.0   Construction Management 13.54
3.0.0   PLANNING 13.54
3.1.0   Why Plan? 13.55
4.0.0   STAGES OF PLANNING 13.56
4.1.0   Pre-Construction Planning 13.56
4.2.0   Construction Planning 13.56
5.0.0   THE PLANNING PROCESS 13.57
5.1.0   Establishing a Goal 13.57
5.2.0   Identifying the Work to be Done 13.57
5.3.0   Determining Tasks 13.57
5.4.0   Communicating Responsibilities 13.58

5.5.0 Follow-Up 13.58

6.0.0 PLANNING RESOURCES 13.59

6.1.0 Planning Materials 13.59

6.2.0 Planning Equipment 13.59

6.3.0 Planning Tools. 13.59

6.4.0 Planning Labor 13.59

7.0.0 WAYS TO PLAN 13.60

8.0.0 ESTIMATING 13.61

8.1.0 The Estimating Process 13.61

8.1.1 *Estimating Material Quantities* 13.65

9.0.0 SCHEDULING 13.68

9.1.0 The Scheduling Process 13.68

9.2.0 Bar Charts 13.68

9.3.0 Network Schedule 13.70

9.4.0 Short-Interval Production Scheduling 13.70

9.5.0 Updating a Schedule 13.71

10.0.0 COST AWARENESS AND CONTROL 13.73

10.1.0 Categories of Costs 13.73

10.2.0 Field Reporting System 13.73

10.3.0 Supervisor's Role in Cost Control 13.74

11.0.0 RESOURCE CONTROL 13.76

11.1.0 Control 13.76

11.2.0 Materials Control 13.76

11.2.1 *Ensuring On-Time Delivery* 13.76

11.2.2 *Preventing Waste* 13.76

11.2.3 *Verifying Material Delivery* 13.77

11.2.4 *Controlling Delivery Storage* 13.77

11.2.5 *Preventing Theft and Vandalism* 13.77

11.3.0 Equipment Control 13.77

11.4.0 Tools Control 13.78

11.5.0 Labor Control 13.78

12.0.0 PRODUCTION AND PRODUCTIVITY 13.80

LIST OF FIGURES

Figure 1 • Job Description 13.12

Figure 2 • Safety Policies and Procedures 13.13

Figure 3 • Example Project Organization Chart............... 13.16

Figure 4 • Percentage of Time Spent on
Technical and Supervisory Work.................. 13.17

Figure 5 • Communication Process......................... 13.21

Figure 6 • Tailor Your Message 13.24

Figure 7 • Project Flow.................................... 13.53

Figure 8 • Project Delivery Systems........................ 13.54

Figure 9 • Steps to Effective Planning...................... 13.59

Figure 10 • Planning Form 13.60

Figure 11 • Sample Page from a Job Diary................... 13.61

Figure 12 • Quantity Takeoff Sheet 13.63

Figure 13 • Summary Sheet................................ 13.64

Figure 14 • Footing Formwork.............................. 13.65

Figure 15 • Work Sheet with Entries........................ 13.66

Figure 16 • Bar Chart 13.69

Figure 17 • Network Schedule.............................. 13.70

CHAPTER ONE

Orientation to the Job

OBJECTIVES

Upon completion of this chapter, you should be able to:

1. Discuss the growth and economic conditions of the construction industry.

2. Describe how workers' values have changed over the years.

3. Explain the importance of training for construction industry personnel.

4. List the new technologies available, and discuss how they are helpful to the construction industry.

5. Identify the gender and minority issues associated with a changing workforce.

6. Describe what employers can do to prevent workplace discrimination.

7. Differentiate between formal and informal organizations.

8. Describe the difference between authority and responsibility.

9. Explain the purpose of job descriptions and what they should include.

10. Distinguish between company policies and procedures.

SECTION 1

1.0.0 GROWTH AND ECONOMICS OF THE CONSTRUCTION INDUSTRY

Construction is the second largest industry in the United States, employing over 5 million workers. As the population of our country increases, so will the need for new construction. Aging infrastructure and new environmental laws will contribute to this increase. Finally, more and more American companies will be needed for construction overseas.

With this increased demand for construction comes a stronger need for trained craftworkers. The U.S. Department of Labor (DOL) reported that employment in the construction industry grew 32 percent between 1997 and 2006, yet construction companies continue to have a shortage of skilled applicants.

The tight labor market and the boom in information technology, including computerization, have resulted in a decrease in available, qualified construction workers. The DOL reports that there is a shortage of approximately 240,000 workers per year in the construction industry. Because of the labor shortage, construction projects across the country are being delayed.

The construction industry is branching out in an attempt to recruit and train younger workers just out of high school and is reaching out to women and minorities.

1.1.0 Changing Values of Workers

The construction industry can no longer expect that workers will simply show up and stay throughout a project. Instead, extra efforts must be made to attract and retain a skilled workforce. To do so, construction companies must be aware of what workers value.

Because the unemployment rate is so low, it is definitely a laborer's job market. Therefore, construction companies have to try even harder to attract potential workers to learn a trade.

Workers value a number of different things. Examples include recognition and praise, a sense of accomplishment, and opportunity for advancement. What an individual values and the extent to which it is valued varies from person to person.

Research conducted by the Society for Human Resources Management (SHRM) suggests that workers are not as committed to their organizations as they were in the past. The opportunities to change jobs are plentiful, so employees rarely work at one company throughout their careers as they did in past generations. Today, workers select companies where they can learn and develop new skills. Once these skills are developed and there is nothing more to learn, the workers move on to another job.

Training and education, the use of new technologies, and opportunities for advancement and growth within the construction profession are valuable sales tools being used by the construction industry to attract job candidates. These three areas can help to retain skilled workers for longer, as well.

1. There is a shortage of skilled workers in the construction industry today because _____.

   a. there is poor long-term job security in the construction industry

   b. recruiters have recently become more selective when searching for potential workers

   c. the labor market is tight

   d. technology has replaced the need for many construction jobs, which discourages potential workers

2. One way in which workers' values have changed over the years is that _____.

   a. workers expect more time off now than they did in the past

   b. employees value learning more now than they did in the past

   c. workers want a secure job environment in which to settle down and feel comfortable

   d. employees typically value a high salary above everything

3. Workers tend to be less loyal to their companies now than they were previously because _____.

   a. they typically feel their job is too difficult

   b. they desire challenges

   c. their job involves too much change and variety

   d. they feel their job is too physically dangerous

4. Construction companies are actively recruiting _____.

   a. retired people who would like to resume working

   b. students who have just finished college

   c. junior high school students who would like to work as apprentices

   d. students who have just completed high school

5. A good way to attract and retain workers is to offer training.

   a. True

   b. False

2.0.0 THE CONSTRUCTION INDUSTRY TODAY

There have been many changes in the construction industry over the years. The continued need for training and education, advances in technology, and the addition of women and minorities to the workforce have affected the way that construction companies operate.

2.1.0 Training

The need for continuing craft training is necessary if the industry is to meet the forecasted worker demands. The Construction Labor Research Council estimates that the construction industry's replacement needs will exceed growth demands by 250 percent.

Because older workers in construction are retiring earlier and younger workers are leaving construction for jobs in other industries, the construction industry will be forced to increase its share of the new job entrants to meet future demands. The Department of Labor (DOL) concludes that the best way for the construction industry to reduce shortages of skilled workers is to create more education and training opportunities. The DOL suggests that companies and community groups form partnerships and create apprenticeship programs. Such programs could provide women, young people, and minorities an opportunity to develop the skills needed for construction work by giving them hands-on experience.

Present construction craft apprenticeship and task-training programs are graduating an average of 50,000 people per year. If the training of construction workers is not significantly increased, a severe shortage of skilled help could result. Therefore, the two major concerns of the construction industry today are:

- Developing adequate training programs to provide skilled craftworkers to meet future needs
- Providing supervisory training to ensure there are qualified leaders in the industry to supervise the craftworkers

When training workers, it is important to understand that people learn in different ways. Some people learn by doing, some people learn by watching or reading, and others need step-by-step instructions as they are shown the process. It is important to understand what kind of a learner you are because if you learn one way, you tend to teach the way you learn. Have you ever tried to teach somebody and failed, and then another person explains it in a different way and they understand it? We as mentors or trainers need to be able to ascertain what kind of learner we are addressing and teach according to their needs, not ours. Once we understand how they learn it will be much easier to teach them.

Most people learn best by a combination of styles, such as getting step-by-step instructions while they do it, or written instructions while they watch you demonstrate the process.

2.1.1 Motivation

As a supervisor or crew leader, it is important to understand what motivates your crew. We may think that money is a good motivator. Although that may be true to some extent, it is proven that at best it is only temporary once a person has achieved the ability to pay their bills. Studies show that people tend to be motivated by environment and conditions. A great workplace will provide more satisfaction than pay. If you give somebody a raise, they tend to work harder for a period of time and then the satisfaction dissipates and they want another raise. It is important to create a feeling of accomplishment for your crew. If you don't feel accomplished you tend to become frustrated, leading to dissatisfaction and lower production. A person feeling more involvement or achievement is likely to be better motivated and help to motivate others.

2.1.2 Understanding the New Generation

As the baby boomers and Generation X grew up they were taught to work hard, stay with it, and in the long run it will pay off. Expectations were to see accomplishments over a longer period of time. Our newest generation of workers, sometimes called millenials, Generation Y, or gamers, can sometimes be perceived as lazy and unmotivated, but what we must understand is that they have grown up in a fast-moving, quick reward environment (e.g., play a game, do it right, get a reward). It is better to give them small projects or break up large projects into smaller accomplishments so that they feel repetitively rewarded, thus enhancing their perception of success.

Goal Setting – Set short-term and long-term goals, including tasks to be done and expected time frames. Help the trainees to understand that things can happen to offset the short-term goals. This is one reason to set long-term goals as well. Don't set them up for failure as this leads to frustration, and frustration can lead to lack of production.

Feedback – It is important to provide feedback in a timely manner. For example, telling someone that they did a really good job last year is meaningless. Feedback can be both positive and negative. Always try to include both, when applicable.

2.1.3 Craft Training

Most of the craft training done in the construction industry is rather informal, taking place on the job site, outside of a traditional training classroom. According to the American Society for Training and Development, craft training is generally handled through on-the-job instruction by a co-worker or conducted by a supervisor.

The Society of Human Resources Management (SHRM) offers the following tips to supervisors in charge of training their employees:

1. *Establish career goals for each person in your area.* This information will give you an idea of how much time you need to devote to developing each person. Employees with the highest ambitions and those with the lowest skill levels will require the most training.

2. *Determine what kind of training to give.* Training can be on the job under the supervision of a co-worker, it can be one-on-one with the supervisor, it can involve cross training to teach a new trade or skill, or it can involve delegating new or additional responsibilities.

3. *Determine the trainee's preferred method of learning.* Some people learn best by watching, others from verbal instructions, and others from doing. When training more than one person at a time, use all three methods.

The SHRM advises that supervisors do the following when training their employees:

1. *Explain the task, why it needs to be done, and how it should be done.* Confirm that the trainees understand these three areas by asking questions. Allow them to ask questions as well.

2. *Demonstrate the task.* Break the task down into manageable chunks. Do one part at a time.

3. *Ask trainees to do the task while you observe them.* Try not to interrupt them while they are doing the task unless they are doing something that is unsafe and potentially harmful.

4. *Give the trainees feedback.* Be specific about what they did and mention any areas where they need to improve.

2.1.4 Supervisory Training

Given the increased demand for skilled craftworkers and qualified supervisory personnel, it seems logical that training would be on the rise. While some contractors have developed their own in-house training program and others participate in training offered by associations and other organizations, many others do not offer training at all.

The reasons for not developing training programs include the following:

- Lack of money to train
- Lack of time to train
- Lack of knowledge about training programs
- High rate of employee turnover
- Workforce too small
- Past involvement ineffective
- Hiring only trained workers
- Lack of interest from workers

For craftworkers to move up into supervisory and managerial positions, it will be necessary for them to continue their education and training. Those who are willing to acquire and develop new skills will have the best chance of finding stable employment. It is, therefore, critical that construction companies remove the barriers against training and devote the needed resources to training.

Your company has recognized the need for training. Participating in a leadership training program such as this will begin to fill the gap between craft and supervisory training.

2.2.0 Impact of Developing Technology

Like many industries, the construction industry has made the move to technology as a means of remaining competitive.

Benefits include increased productivity and speed, improved quality of documents, greater access to common data, and better financial controls and communication. As technology becomes a greater part of supervision, crew leaders will need to be able to use it properly. One important concern with electronic communication is to keep it as brief and factual as possible. Because the receiver has no visual or auditory clues as to the sender's intent, they can be easily misunderstood. In other words, it is a lot more difficult to tell if someone is just joking via e-mail because you can't see their face or hear the tone in their voice.

2.2.1 Other Communication Sources

Cellular telephones, voicemail, pagers, and handheld communication devices have made it even easier to stay in touch. They are particularly useful communication sources for contractors or crew leaders who are on a job site, away from their offices, or constantly on the go.

Cellular telephones allow the users to receive incoming calls as well as make outgoing calls. Unless the owner is out of the cellular provider's service area, the cell phone may be used any time to answer calls, make calls, and check voicemail or e-mail.

Handheld communication devices allow construction supervisors to plan their calendars, schedule meetings, manage projects, and access their e-mail from remote locations. These computers are small enough to fit in the palm of your hand, yet powerful enough to hold years of information from various projects. Information can be transmitted electronically to others on the project team or transferred to a desktop computer.

3.0.0 GENDER AND CULTURAL ISSUES

During the past several years, the construction industry in the United States has experienced a trend in worker expectations and diversity. These two issues are converging at a rapid pace. At no time has there been such a generational merge in the workforce ranging from The Silent Generation (1925–1945), Baby Boomers (1946–1964), Gen X (1965–1979) and the Millennials (1980–2000).

This trend, combined with diversity initiatives by the construction industry, has created a climate that is both exciting and challenging. Today, many construction companies are aggressively seeking a broad base of candidates for the construction industry. To do this effectively, they are using their own resources as well as relying on associations with the government and other construction trade organizations. All current research indicates that this industry will be more dependent on the critical skills of a diverse workforce—a workforce that is both culturally and ethnically fused. Across the United States, the construction industry is aggressively seeking to bring new workers into its ranks, specifically women and racial and ethnic minorities. Diversity is no longer solely driven by social and political issues, but by consumers who need hospitals, malls, bridges, power plants, refineries, and many other commercial and residential buildings

Some issues relating to gender and race must be addressed on the job site. These include different communication styles of men and women, language barriers associated with cultural differences, the possibility of sexual harassment, and the potential for cases of gender or racial discrimination.

3.1.0 Communication Styles of Men and Women

As more and more women move into construction, it becomes increasingly important that communication barriers between men and women are broken down and differences in behaviors are understood so that they can work together more effectively.

The Jamestown, New York Area Labor Management Committee (JALMC) offers the following explanations and tips.

1. *Women tend to ask more questions than men do.* Men are more likely to proceed with a job and figure it out as they go along, while women are more likely to ask questions first.

2. *Men tend to offer solutions before empathy; women tend to do the opposite.* Both men and women should say what they want up-front—solutions or a sympathetic ear. That way, both genders will feel understood and supported.

3. *Women are more likely to ask for help than men.* Women are generally more pragmatic when it comes to completing a task. If they need help, they will ask for it. Men are more likely to attempt to complete a task by themselves, even when it might be a good idea to seek assistance.

4. *Men tend to communicate more competitively, and women tend to communicate more cooperatively.* Both parties need to hear one another out without interruption.

3.2.0 Language Barriers

Language barriers are a real workplace challenge. The U.S. Census Bureau reported that the number of people who speak languages other than English increased to 47 million in 2000. Spanish is now the most common non-English language spoken in the United States. As the makeup of the immigration population continues to change, the number of non-English speakers will rise dramatically, and the languages being spoken will also change. Bilingual job sites are increasingly common.

Companies have the following options to overcome this challenge:

- Offer English classes either at the worksite or through school districts and community colleges
- Offer incentives for workers to learn English

As our workforce becomes more diverse, communicating with people for whom English is a second language will be even more critical. The Knowledge Center's *Manager's Tool Kit* offers the following tips for communicating across language barriers:

- Be patient. If you expect too much, too quickly, you will likely confuse and frustrate the people involved.

- Don't get angry if people don't understand at first. Give them time to process the information in a way that they can comprehend.

- If people speak English poorly but understand reasonably well, ask them to demonstrate their understanding through actions rather than words.

- Avoid humor. Your jokes will likely be misunderstood and may even be misinterpreted as a joke at the worker's expense.

- Don't assume that people are unintelligent simply because they don't understand. Many immigrants doing menial jobs in their adopted countries were doctors, lawyers, and educators in their native countries.

- Speak slowly and clearly, and avoid the tendency to yell.

- Use face-to-face communication whenever possible. Over-the-phone communication is often more difficult when a language barrier is involved.

- Use pictures or drawings to get your point across, if necessary.

3.3.0 Cultural Differences

As workers from a multitude of backgrounds and cultures are brought together, there are bound to be differences and conflicts in the workplace.

To overcome cultural conflicts, the SHRM suggests the following:

1. Define the problem from both points of view. How does each person involved view the conflict? What does each person think is wrong? This involves moving beyond traditional thought processes to consider alternate ways of thinking.

2. Uncover cultural interpretations. What assumptions are being made based on cultural programming? By doing this, the supervisor may realize what motivated an employee to act in a particular manner. For instance, an employee may call in sick for an entire day to take his or her spouse to the doctor. In some cultures, such an action is based on responsibility and respect for family.

3. Create cultural synergy. Devise a solution that works for both parties involved. The purpose is to recognize and respect other's cultural values, and work out mutually acceptable alternatives.

3.4.0 Sexual Harassment

In today's business world, men and women are working side-by-side in careers of all kinds. There are no longer male occupations or female jobs; therefore, men and women are expected to relate to each other in new ways that are uncharacteristic of the past.

As women make the transition into traditionally male industries, such as construction, the likelihood of sexual harassment increases. Sexual harassment is defined as unwelcome behavior of a sexual nature that makes someone feel uncomfortable in the workplace by focusing attention on their gender instead of on their professional qualifications. Sexual harassment can range from telling an offensive joke or hanging a poster of a swimsuit-clad man or woman in one's cubicle to making sexual comments or physical advances within a work environment.

Historically, sexual harassment was thought to be an act performed by men of power within an organization against women in subordinate positions with lesser power. However, as the number of sexual harassment cases have shown over the years, this is no longer the case.

Sexual harassment can occur in a variety of circumstances, including but not limited to the following:

- The victim as well as the harasser may be a woman or a man. The victim does not have to be of the opposite sex.

- The harasser can be the victim's supervisor, an agent of the employer, a supervisor in another area, a co-worker, or a non-employee.

- The victim does not have to be the person harassed but could be anyone affected by the offensive conduct.

- Unlawful sexual harassment may occur without economic injury to or discharge of the victim.

- The harasser's conduct must be unwelcome.

The Equal Employment Opportunity Commission (EEOC) enforces sexual harassment laws within industries. When investigating allegations of sexual harassment, the EEOC looks at the whole record including the circumstances and the context in which the alleged incidents occurred. A decision on the allegations is made from the facts on a case-by-case basis.

Prevention is the best tool to eliminate sexual harassment in the workplace. The EEOC encourages employers to take steps necessary to prevent sexual harassment from occurring. Employers should clearly communicate to employees that sexual harassment will not be tolerated. They can do so by developing a policy on sexual harassment, establishing an effective complaint or grievance process, and taking immediate and appropriate action when an employee complains.

Both swearing and off-color jokes are not only offensive to co-workers, but also tarnish a worker's image. Crew leaders need to emphasize that abrasive or crude behavior may affect opportunities for advancement.

3.5.0 Gender and Minority Discrimination

With the increase of women and minorities in the workforce, there is more room for gender and minority discrimination. Consequently, many business practices, including the way employees are treated, the organization's hiring and promotional practices, and the way people are compensated, are being analyzed for equity. More attention is being placed on fair recruitment, equal pay for equal work, and promotions for women and minorities in the workplace.

The EEOC requires that companies be equal opportunity employers. This means that organizations hire the best person for the job, without regard for race, sex, religion, age, etc. Once traditionally a male-dominated industry, construction companies are moving away from this notion and are actively recruiting and training women, younger workers, people from other cultures, and workers with disabilities to compensate for the shortage of skilled workers.

Despite the positive change in recruitment procedures, there are still inadequacies when it comes to pay and promotional activities associated with women and minorities. Individuals and federal agencies continue to win cases in which women and minorities have been discriminated against in the workplace, thus demonstrating that discrimination persists.

To prevent discrimination cases, employers should develop valid job-related criteria for hiring, compensation, and promotion. These measures should be used consistently for every applicant interview, employee performance appraisal, etc. Therefore, all employees responsible for recruitment and selection, supervision, and evaluating job performance should be trained on how to use the job-related criteria legally and effectively.

1. The construction industry should provide training for craftworkers and supervisors _____.

   a. to ensure that there are enough future workers

   b. to avoid discrimination lawsuits

   c. in order to update the skills of older workers who are retiring at a later age than they previously did

   d. even though younger workers are now less likely to seek jobs in other areas than they were 10 years ago

2. Construction companies traditionally offer craftworker training _____.

   a. that a supervisor leads in a classroom setting

   b. that a craftworker leads in a classroom setting

   c. in a hands-on setting, where craftworkers learn from a co-worker or supervisor

   d. on a self-study basis to allow craftworkers to proceed at their own pace

3. One way to provide effective training is to _____.

   a. avoid giving negative feedback until trainees are more experienced in doing the task

   b. tailor the training to the career goals of trainees

   c. choose one kind of training method and stick with it for consistency's sake

   d. encourage trainees to listen, saving their questions for the end of the session

4. One way to prevent sexual harassment in the workplace is to _____.

   a. require employee training in which the potentially offensive subject of stereotypes is carefully avoided

   b. develop a consistent policy with appropriate consequences for engaging in sexual harassment

   c. communicate to workers that the victim of sexual harassment is the one who is being directly harassed, not those affected in a more indirect way

   d. educate workers to recognize sexual harassment for what it is—unwelcome conduct by the opposite sex

5. Employers can minimize all types of workplace discrimination by hiring based on a consistent list of job-related requirements.

   a. True

   b. False

4.0.0 THE CONSTRUCTION ORGANIZATION

An organization is concerned with the relationships among the people within the company. The supervisor needs to be aware of two types of organizations. These are formal organizations and informal organizations.

A formal organization exists when the activities of the people within the work group are coordinated toward the attainment of a goal. An example of a formal organization is a work crew consisting of four carpenters and two laborers led by a supervisor all working together on a job.

A formal organization is typically documented on an organizational chart, which outlines all of the positions that make up an organization and how those positions are related. Some organizational charts even depict the people within each position and the person to whom they report as well as the people that the person supervises, which is also known as *span of control*. An efficient span of control consists of a manager who directly supervises no more than six people.

An informal organization can be described as a group of individuals having a common goal, but for different reasons. An example of this would be a plane full of passengers flying to the same city, each with a different reason for doing so, and dispersing once the city is reached.

An informal organization allows for communication among its members so they can perform as a group. It also establishes patterns of behavior that help them to work as a group. One pattern may be a specific manner of dressing, speaking alike, or developing conformity.

Both types of organizations establish the foundation for how communications flow. The formal structure is the means used to delegate authority and responsibility and to exchange information. The informal structure is used to exchange information. The "grapevine" is an example of an informal organization.

Members in an organization perform best when each member:

- Understands the job and how it will be done
- Communicates effectively with others in the group
- Knows his or her role in the group
- Knows who has the authority and responsibility

4.1.0 Division of Responsibility

The conduct of a construction business involves certain functions. In a small organization, responsibilities may be divided between one or two people. However, in a larger organization with many different and complex activities, responsibilities may be grouped into similar activity groups, and the responsibility for each group assigned to department managers. In either case, the following major departments exist in construction companies:

Executive: This office represents top management. It is responsible for the success of the company through short-range and long-range planning.

Accounting: This office is responsible for all record-keeping and financial transactions, including payroll, taxes, insurance, and audits.

Contract Administration: This office prepares and executes contractual documents with owners, sub-contractors, and suppliers.

Purchasing: This office obtains material prices and then issues purchase orders. The purchasing office also obtains rental and leasing rates on equipment and tools.

Estimating: This office is responsible for recording the quantity of material on the jobs, the takeoff, pricing labor and material, analyzing sub-contractors' bids, and bidding on projects.

Construction Operation: This office plans, controls, and supervises all construction field activities.

Other divisions of responsibility a company may create involve architectural and engineering design functions. These divisions usually become separate departments.

4.2.0 Authority and Responsibility

As the organization grows, the manager must ask others to perform many duties so that the manager can concentrate on managing. Managers typically assign activities to their subordinates or workers through a process called delegation.

When delegating activities, the supervisor assigns others the responsibility to perform the designated tasks. Delegation implies responsibility, which refers to the obligation of an employee to perform the duties.

Along with responsibility comes authority. Authority is the power to act or make decisions in carrying out an assignment. The kind and amount of authority a supervisor or worker has depends on the company for which he or she works.

Authority and responsibility must be balanced so employees can carry out their tasks. In addition, delegation of sufficient authority is needed to make an employee accountable to the supervisor for the results. Accountability is the act of holding an employee responsible for completing the assigned activities. Even though authority is delegated, the final responsibility always rests with the supervisor.

4.3.0 Job Descriptions

Many companies furnish each employee a written job description, which explains the job in detail. Job descriptions set a standard for the employee, make judging performance easier, clarify the tasks each person should handle, and simplify the training of new employees.

Each new employee should understand all the duties and responsibilities of the job after reviewing the job description. Thus, the time it takes for the employee to make the transition from being a new and uninformed employee to a more experienced member of a production team is shortened.

A job description need not be long, but it should be detailed enough to ensure there is no misunderstanding of the duties and responsibilities

of the position. The job description should contain all the information necessary to evaluate the employee's performance.

A job description should contain the following:

- Job title
- General description of the position
- Specific duties and responsibilities
- The supervisor to whom the position reports
- Other requirements, such as required tools/equipment for the job

A sample job description is shown in *Figure 1*.

Position:
Front-line Construction Supervisor – Field

General Summary:
First line of supervision on a construction crew installing underground and underwater cathodic protection systems.

Reports To:
Field Superintendent

Duties and Responsibilities:
- Oversee crew
- Provide instruction and training in construction tasks as needed
- Make sure proper materials and tools are on the site to accomplish tasks
- Keep project on schedule
- Enforce safety regulations

Knowledge, Skills, and Experience Required:
- Extensive travel throughout the Eastern United States, home base in Atlanta
- Ability to operate a backhoe and trencher
- Valid commercial driver's license with no DUI violations
- Ability to work under deadlines with the knowledge and ability to foresee problem areas and develop a plan of action to solve the situation

413F01.EPS

Figure 1 • Job Description

4.4.0 Policies and Procedures

Most construction companies have formal policies and procedures established to help the supervisor carry out his or her duties. A policy is a general statement establishing guidelines for a specific activity. Examples include policies on vacations, rest breaks, workplace safety, or checking out power hand tools.

Procedures are the ways that policies are carried out. For example, a procedure written to implement a policy on workplace safety would include guidelines for reporting accidents and general safety procedures that all employees are expected to follow.

A sample of a policies and procedures document is shown in *Figure 2*.

WORKPLACE SAFETY POLICY:

Your safety is the constant concern of this company. Every precaution has been taken to provide a safe workplace.

Common sense and personal interest in safety are still the greatest guarantees of your safety at work, on the road, and at home. We take your safety seriously and any willful or habitual violation of safety rules will be considered cause for dismissal. We are sincerely concerned for the health and well being of each member of the team.

Workplace Safety Procedure:

To ensure your safety, and that of your co-workers, please observe and obey the following rules and guidelines:

- Observe and practice the safety procedures established for the job.
- In case of sickness or injury, no matter how slight, report at once to your supervisor. In no case should an employee treat his own or someone else's injuries or attempt to remove foreign particles from the eye.
- In case of injury resulting in possible fracture to legs, back, or neck, any accident resulting in an unconscious condition, or a severe head injury, the employee is not to be moved until medical attention has been given by authorized personnel.
- Do not wear loose clothing or jewelry around machinery. It may catch on moving equipment and cause a serious injury.
- Never distract the attention of another employee, as you might cause him or her to be injured. If necessary to get the attention of another employee, wait until it can be done safely.
- Where required, you must wear protective equipment, such as goggles, safety glasses, masks, gloves, hairnets, etc.
- Safety equipment such as restraints, pull backs, and two-hand devices are designed for your protection. Be sure such equipment is adjusted for you.
- Pile materials, skids, bins, boxes, or other equipment so as not to block aisles, exits, fire-fighting equipment, electric lighting or power panel, valves, etc. FIRE DOORS AND AISLES MUST BE KEPT CLEAR.
- Keep your work area clean.
- Use compressed air only for the job for which it is intended. Do not clean your clothes with it and do not play with it.

413F02.EPS

Figure 2 • Safety Policies and Procedures

---

1. Members tend to function best within an organization when they _____.
   a. are allowed to select their own style of clothing for each project
   b. know their role
   c. do not disagree with the statements of other workers or supervisors
   d. have authority

2. A formal organization is defined as a group of individuals who have their own personal motivations but who share the same goal
   a. True
   b. False

3. A formal organization uses an organizational chart to _____.
   a. depict all companies with whom it conducts business
   b. show all customers with whom it conducts business
   c. track projects between departments
   d. show the relationships among the existing positions; may even include the actual people in the position

4. Authority is defined as _____.
   a. making decisions without ever having to discuss things with others
   b. giving an employee a particular task to perform
   c. making an employee responsible for the completion and results of a particular duty
   d. having the power to promote someone

5. A good job description should include _____.
   a. a complete organization chart
   b. any information needed to judge the job performance of an individual
   c. the company dress code
   d. the company's sexual harassment policy

6. Job descriptions are used in order to _____.
   a. specify to job candidates whether a company uses a formal or informal organizational structure
   b. give details to potential employees about how to perform specific duties
   c. give information regarding the responsibilities of the potential job candidate
   d. inform the potential candidate about the company's benefits

7. The purpose of a policy is to _____.
   a. implement company guidelines regarding a particular activity
   b. specify what tools and equipment are required for a job
   c. list all information necessary to judge an employee's performance
   d. inform employees about the future plans of the company

8. One example of a procedure would be the rules for taking time off.
   a. True
   b. False

REVIEW QUESTIONS ◆ **CHAPTER ONE, SECTIONS 4.0.0–4.4.0**

◆ **CHAPTER TWO**

Leadership Skills

OBJECTIVES

Upon completion of this chapter, you should be able to:

1. Explain the role of a crew leader.

2. List the characteristics of effective leaders.

3. Be able to discuss the importance of ethics in a supervisor's role.

4. Identify the three styles of leadership.

5. Describe the forms of communication.

6. Explain the four parts of verbal communication.

7. Demonstrate the importance of active listening.

8. Illustrate how to overcome the barriers to communication.

9. List some ways that supervisors can motivate their employees.

10. Explain the importance of delegating and implementing policies and procedures.

11. Differentiate between problem solving and decision making.

SECTION 1

1.0.0 INTRODUCTION TO SUPERVISION

For the purpose of this program, it is important to define some of the positions to which we will be referring. The term craftworker refers to a person who performs the work of his or her trade(s). The crew leader is a person who supervises one or more craftworkers on a crew. A superintendent is essentially a second-line supervisor who supervises one or more crew leaders or front-line supervisors. Finally, a project manager is responsible for managing the construction of one or more construction projects. In this training, we will concentrate primarily on the role of the crew leader.

The organizational chart in *Figure 3* on the following page depicts how each of the four positions is related in the chain of command.

Craftworkers and crew leaders differ in that the crew leader manages the activities that the skilled craftworkers on the crews actually perform. In order to manage a crew of craftworkers, a crew leader must have first-hand knowledge and experience in the activities being performed. In addition, he or she must be able to act directly in organizing and directing the activities of the various crew members.

This chapter will discuss the importance of developing effective leadership skills as a new supervisor as well as provide some tips for doing so. Effective ways to communicate with all levels of employees, build teams, motivate crew members, make decisions, and resolve problems will be covered in depth.

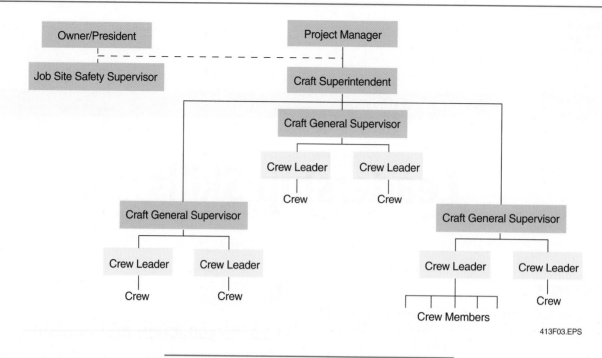

413F03.EPS

Figure 3 • Example Project Organization Chart

2.0.0 THE SHIFT IN WORK ACTIVITIES

The crew leader is generally selected and promoted from a construction work crew. The selection will frequently be based on that person's ability to accomplish tasks, to get along with others, to meet schedules, and to stay within the budget. The crew leader must lead the team to work safely and provide a quality product.

Making the transition from a craftworker to a crew leader can be difficult, especially when the new supervisor is in charge of supervising a group of peers. Crew leaders are no longer responsible for their work alone; rather, they are accountable for the work of an entire crew of people with varying skill levels and abilities, a multitude of personalities and work styles, and different cultural and educational backgrounds.

Supervisors must learn to put their personal relationships aside and work for the common goals of the team.

As an employee moves from a craftworker position to the role of a crew leader, he or she will find that more hours will be spent supervising the work of others than actually performing the technical skill for which he or she has been trained. *Figure 4* represents the percentage of time craftworkers, crew leaders, superintendents, and project managers spend on technical and supervisory work as their management responsibilities increase.

The success of the new crew leader is directly related to the ability to make the transition from crew member into a leadership role.

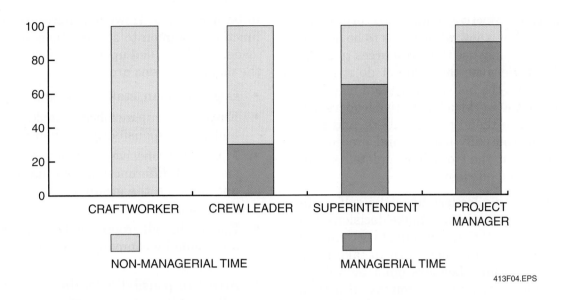

Figure 4 • Percentage of Time Spent on Technical and Supervisory Work

3.0.0 BECOMING A LEADER

It is essential that a supervisor have leadership skills to be successful. Therefore, one of the primary goals of a person who wants to become a supervisor should be to develop strong leadership skills and learn to use them effectively.

The expression "a natural-born leader" is pretty common. Although there is no overwhelming evidence for this, it is true that some people may have inherited leadership characteristics or developed traits that motivate others to follow and perform. Research shows that people who possess such talents are likely to succeed as leaders.

3.1.0 Characteristics of Leaders

Leadership traits are similar to the skills that a supervisor needs to be effective. Although the characteristics of leadership are many, there are some definite commonalities among effective leaders.

First and foremost, effective leaders lead by example. In other words, they work and live by the standards that they establish for their crew members or followers, making sure that they practice what they preach.

Next, effective leaders tend to have a high level of drive, determination, persistence, or a stick-to-it attitude. In the face of a challenging situation, leaders seek involvement and drive through adversity to achieve their goal. When faced with obstacles, effective leaders don't get discouraged; instead, they identify the potential problems, make plans to overcome them, and work toward achieving the intended goal. In the event of failure, effective leaders learn from their mistakes and apply that knowledge to their future leadership attempts.

Third, effective leaders are typically effective communicators. Accomplishing this may require that the leader overcome issues such as language barriers, gender biases, or differences in personalities to ensure that each follower or member of the crew understands the established goals of the project.

Effective leaders have the ability to motivate their followers or crew members to work to their full potential and become effective members of the team. They try to develop their crew members' skills and encourage them to improve and learn as a means to contribute more to the team effort. Effective leaders tend to demand 100% from themselves and their teams, so they work hard to provide the skills and leadership necessary to do so.

In addition, effective leaders are organized planners. They know what needs to be accomplished, and they use their resources to make it happen. Because they cannot do it alone, leaders enlist the help of their team members to share in the workload. Effective leaders delegate work to their crew members, and they implement company policies and procedures to ensure that the work is completed safely, effectively, and efficiently.

Finally, effective leaders have the self-confidence that allows them to make decisions and solve problems. If they are going to accomplish their goals, leaders must learn to take risks when appropriate. Leaders must be able to absorb information, assess courses of action, understand the risks, make decisions, and assume the responsibility for them.

3.1.1 Leadership Traits

There are many other traits of effective leaders. In fact, the list is very long. Some other major characteristics of leaders include:

- The ability to plan and organize
- Loyalty
- Fairness
- Enthusiasm
- The ability to teach and learn from others
- Initiative
- Salesmanship
- Good communication skills

3.1.2 Expected Leadership Behavior

Followers have expectations of their leaders. They look to their leaders to:

- Set the example
- Suggest and direct
- Influence their actions
- Communicate effectively
- Make decisions and assume responsibility
- Be a loyal member of the group
- Abide by company policies and procedures

3.2.0 Functions of a Leader

The functions of a leader will vary with the environment, the group being led, and the tasks to be performed. However, there are certain functions common to all situations which the leader will be called upon to perform. Some of the major functions are:

- Organize, plan, staff, direct, and control
- Empower group members to make decisions and take responsibility for their work
- Maintain a cohesive group by resolving tensions and differences among its members and between the group and those outside the group
- Ensure that all group members understand and abide by company policies and procedures
- Accept responsibility for the successes and failures of the group's performance
- Represent the group to others
- Be sensitive to the differences of a diverse workforce

3.3.0 Leadership Styles

There are three broad categories of leadership style that supervisors can adopt. At one extreme is the autocratic or dictator style of leadership, where the supervisor makes all of the decisions independently, without seeking the opinions of crew members. At the other extreme is the hands-off leadership style, where the supervisor lets the employees make all of the decisions. In the middle is the democratic style, where the supervisor seeks crew members' opinions and makes the appropriate decisions based on their input.

The following are some characteristics of each of the three leadership styles:

Autocratic leaders:

- Expect crew members to work without questioning procedures
- Seldom seek advice from crew members
- Insist on solving problems alone
- Seldom permit crew members to assist each other
- Praise and criticize on a personal basis
- Have no sincere interest in creatively improving methods of operation or production

Democratic leaders:

- Discuss problems with their crew members
- Listen
- Explain and instruct
- Give crew members a feeling of accomplishment by commending them when they do a job well
- Are friendly and available to discuss personal and job-related problems

Hands-off leaders:

- Believe no supervision is best
- Rarely give orders
- Worry about whether they are liked by their crew members

Effective leadership takes many forms. The correct style for a particular situation or operation depends on the nature of the crew as well as the work it has to accomplish. For example, if the crew does not have enough experience for the job ahead, then a direct, autocratic style may be appropriate. The autocratic style of leadership is also effective when jobs involve repetitive operations that require little decision-making.

However, if a worker's attitude is an issue, then democratic action is required. In this case, providing the missing motivational factors may increase performance and result in the improvement of the worker's attitude. The democratic style of leadership is also used when the work is of a creative nature because brainstorming and exchanging ideas with the crew members can be beneficial.

Hands-off leadership is not a very effective leadership style. In these cases, either the leadership becomes the responsibility of a crew member who takes the initiative to assume the role, or chaos results due to lack of formal leadership from the supervisor.

3.3.1 Selecting a Leadership Style

In selecting the most appropriate style of leadership, the supervisor should consider the power base from which to work. There are three elements that make up a power base:

1. Authority
2. Expertise, experience, and knowledge
3. Respect, attitude, and personality

First, the company must give a supervisor sufficient authority to do the job. This authority must be commensurate with responsibility, and it must be made known to crew members when they are hired so that they understand who is in charge.

Next, a supervisor must have an expert knowledge of the activities to be supervised in order to be effective. This is important because the crew members need to know that they have someone to turn to when they have a question or a problem, when they need some guidance, or when modifications or changes are warranted by the job.

Finally, respect is probably the most useful element of power. This usually derives from being fair to employees, by listening to their complaints and suggestions, and by utilizing incentives and rewards appropriately to motivate crew members. In addition, supervisors who have a positive attitude and a favorable personality tend to gain the respect of their crew members as well as their peers. Along with respect comes a positive attitude from the crew members.

3.3.2 Crew Evaluations

The type of leader you choose to be will also affect the way in which you perform crew evaluations. Many companies have formal evaluation forms that are used on a yearly basis to evaluate workers for pay increases. These evaluations should not come as a surprise. An effective crew leader provides constant performance feedback. It is also important to stress the importance of self-evaluation with your crew.

3.4.0 Ethics in Leadership

The construction industry demands the highest standards of ethical conduct. Every day the crew leader has to make decisions which may have ethical implications. When an unethical decision is made, it not only hurts the supervisor but also other workers, peers, and the company for which he or she works.

There are three basic types of ethics:

1. Business, or legal
2. Professional, or balanced
3. Situational

Business, or legal, ethics is adhering to all laws and regulations related to the issue. Professional, or balanced, ethics is carrying out all activities in such a manner as to be fair to everyone concerned. Situational ethics pertains to specific activities or events that may initially appear to be a gray area. For example, you may ask yourself, "How will I feel about myself if my actions are published in the newspaper or if I have to justify my actions to my family, friends, and colleagues?"

There are going to be many times when the supervisor will be put into a situation where he or she will need to assess the ethical consequences of an impending decision. For instance, should a supervisor continue to keep one of his or her crew working who has broken into a cold sweat due to overheated working conditions just because the superintendent says the activity is behind schedule? Or should a supervisor, who is the only one aware that the reinforcing steel placed by his or her crew was done incorrectly, correct the situation before the concrete is placed in the form? If a supervisor is ever asked to carry through on an unethical decision, it is up to him or her to inform the superintendent of the unethical nature of the issue, and if still requested to follow through, refuse to act.

SECTION 4

4.0.0 COMMUNICATION

Successful supervisors learn to communicate effectively with people at all levels of the organization. In doing so, they develop an understanding of human behavior and acquire communication skills that enable them to understand and influence others.

There are many definitions for communication. One is that communication is the act of accurately and effectively conveying or transmitting facts, feelings, and opinions to another person. Simply stated, communication is the method of exchanging information and ideas.

Just as there are many definitions of communication, it also comes in many forms. Verbal, non-verbal, and written are the types of communication. Each of these forms of communication will be discussed in this section.

Research shows that the typical supervisor spends about 80 percent of his or her day communicating through writing, speaking, listening, or using body language. Of that time, studies suggest that approximately 20 percent of communication is written, and 80 percent involves speaking or listening.

4.1.0 Verbal Communication

Verbal communication refers to the spoken words exchanged between two or more people when communicating. It can be done face-to-face or over the telephone.

Verbal communication consists of four distinct parts:

1. Sender
2. Message
3. Receiver
4. Feedback

The diagram in *Figure 5* depicts the relationship of these four parts within the communication process.

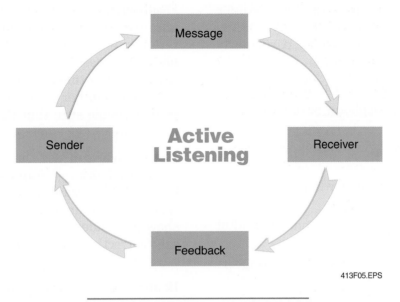

Figure 5 • Communication Process

4.1.1 The Sender

The sender is the person that creates the message to be communicated. In verbal communication, the sender actually says the message aloud to the person(s) for whom it is intended.

The sender must be sure to speak in a clear and concise manner that can be easily understood by others. This is not an easy task; it takes practice. Some basic speaking tips are:

- Avoid talking with anything in your mouth (food, gum, pencils, etc.).

- Avoid swearing.

- Don't speak too quickly or too slowly. In extreme cases, people tend to focus on your rate of speech instead of on what you are saying.

- Be aware of your tone. Remember, it's not so much what you say, but how you say it.

- Enunciate to prevent misunderstandings (many letters sound similar).

- Don't talk in a monotone; rather, put some enthusiasm and feeling in your voice.

4.1.2 The Message

The message is what the sender is attempting to communicate to the audience. A message can be a set of directions, an opinion, or a feeling. Whatever its function, a message is an idea or fact that the sender wants the audience to know.

Before speaking, the sender should determine what it is he or she wants to communicate. The sender should then organize what to say, ensuring that the message is logical and complete. Taking the time to clarify your thoughts prevents rambling, not getting the message across effectively, or confusing the audience. It also permits the sender to get to the point quickly.

In delivering the message, the sender should assess the audience. It is important not to talk down to your audience. Remember that everyone, whether in a senior or junior position, deserves respect and courtesy. Therefore, the sender should use words and phrases that the audience can understand and avoid technical language or slang. In addition, the sender should use short sentences, which give the audience time to understand and digest one point or fact at a time.

4.1.3 The Receiver

The receiver is the person to whom the message is communicated. For the communication process to be successful, it is important that the receiver understands the message as the sender intended. Therefore, the receiver must listen to what is being said.

The first step to becoming a good listener involves realizing that there are many barriers that get in the way of our listening, particularly on a busy construction job site.

Some barriers to effective listening are:

- Noise, visitors, telephone, or other distractions
- Preoccupation, being under pressure, or daydreaming
- Reacting emotionally to what is being communicated
- Thinking about how to respond instead of listening
- Giving an answer before the message is complete
- Personal biases to the sender's communication style
- Finishing the sender's sentence

Some tips for overcoming these barriers are:

- Take steps to minimize or remove distractions; learn to tune-out your surroundings
- Listen for key points
- Take notes
- Try not to take things personally
- Allow yourself time to process your thoughts before responding
- Let the sender communicate the message without interruption
- Be aware of your personal biases, and try to stay open-minded

There are many ways for a receiver to show that he or she is actively listening to what is being said. This can even be accomplished without saying a word. Examples include maintaining eye contact, nodding your head, and taking notes. It may also be accomplished through feedback.

4.1.4 Feedback

Feedback refers to the communication that occurs after the message has been sent by the sender and received by the receiver. It involves the receiver responding to the message.

Feedback is a very important part of the communication process because it allows the receiver to communicate how he or she interpreted the message, and it permits the sender to ensure that the message was understood as intended. In other words, feedback is a checkpoint to make sure that the receiver and the sender are on the same page.

The receiver can use the opportunity of providing feedback to paraphrase back what was heard. When paraphrasing what you heard, it is best to use your own words. That way, you can show the sender that you interpreted the message correctly and could explain it to others if needed.

In addition, the receiver can clarify the meaning of the message and request additional information when providing feedback. This is generally accomplished by asking questions.

4.2.0 Non-Verbal Communication

Unlike verbal or written communication, non-verbal communication does not involve the spoken word or the written word. Rather, non-verbal communication refers to things that you can actually see when communicating with others. Examples include facial expressions, body movements, hand gestures, and eye contact.

Non-verbal communication can provide an external signal of an individual's inner emotions. It occurs simultaneously with verbal communication; often, the sender of the non-verbal communication is not even aware of it.

Because it can be physically observed, non-verbal communication is just as important as the words used in conveying the message. Often, people are influenced more by non-verbal signals than by spoken words. Therefore, it is important that you be conscious of the non-verbal cues that you send because you don't want the receiver to interpret your message incorrectly based on your posture or an expression on your face. After all, these things may have nothing to do with the current communication exchange; instead, they may be carrying over from something else going on in your day.

4.3.0 Written or Visual Communication

Some communication will have to be written or visual. Written or visual communication refers to communication that is documented on paper or transmitted electronically via the computer using words or visuals.

Many messages on a job have to be communicated in text form. Examples include weekly reports, requests for changes, purchase orders, and correspondence on a specific subject. These items are written because they must be recorded for business and historical purposes. In addition, some communication on the job will have to be visual. Items that are difficult to explain verbally or by the written word can best be explained through diagrams or graphics. Examples include the plans or drawings used on a job.

When writing or creating a visual message, it is best to assess the reader or the audience before beginning. It is essential that the reader can read the message and understand the content; otherwise, the communication process will be unsuccessful. Therefore, the writer should consider the actual meaning of words or diagrams and how others might interpret them. In addition, the writer should make sure that all handwriting is legible if the message is being handwritten.

Here are some basic tips for writing:

- Avoid emotion-packed words or phrases
- Be positive whenever possible
- Avoid using technical language or jargon
- Stick to the facts
- Provide an adequate level of detail
- Present the information in a logical manner
- Avoid making judgments, unless asked to do so
- Proofread your work; check for spelling and grammatical errors
- Make sure that the document is legible
- Be prepared to provide a verbal or visual explanation, if needed

Here are some basic tips for creating visuals:

- Provide an adequate level of detail
- Ensure that the diagram is large enough to be seen
- Avoid creating complex visuals; simplicity is better
- Present the information in a logical order
- Be prepared to provide a written or verbal explanation of the visual, if needed

4.4.0 Communication Issues

It is important to note that each person communicates a little differently; that is what makes us unique as individuals. And as the diversity of the workforce changes, communication will become even more challenging. We must consider that our audience consists of individuals from different ethnic groups, cultural backgrounds, educational levels, or economic status groups. Therefore, we must learn to assess our audience to determine how to effectively communicate with each individual.

The key to effective communication is to acknowledge that people are different and be able to adjust your communication style to meet the needs of your audience or the person on the receiving end of your message. This involves relaying your message in the simplest way possible, avoiding the use of words that people may find confusing. Be aware of how you use technical language, slang, jargon, and words that have multiple meanings. Present the information in a clear, concise manner. Avoid rambling and always speak clearly, using good grammar.

In addition, you have to be able to communicate your message in multiple ways or adjust your level of detail or terminology to ensure that everyone understands your meaning as intended. For instance, you may have to draw a map for a visual person who cannot comprehend directions in a verbal or written form, or overcome language barriers on the job site by using graphics or visual aids to relay your message.

Figure 6 shows how to tailor your message to your audience.

| VERBAL INSTRUCTIONS | VERBAL INSTRUCTIONS | WRITTEN | DIAGRAM/MAP |
| Experienced Crew | Newer Crew | INSTRUCTIONS | |

| "Please drive to the supply shop to pick up our order." | "Please drive to the supply shop. Turn right here and left at Route 1. It's at 75th Street and Route 1. Tell them the company name and that you're there to pick up our order." | 1. Turn right at the exit.
2. Drive 2 miles to Route 1. Turn LEFT.
3. Drive 1 mile (pass the tire shop) to 75th Street.
4. Look for the supply store on the right… | 
7501 N. Highway 1
(800) 555-7567 |

Different people learn in different ways. Be sure to communicate so you can be understood.

413F06.EPS

Figure 6 • Tailor Your Message

PARTICIPANT EXERCISE A

1. Read the following verbal conversations, and identify any problems:

Conversation I:

Judy: Hey Roger…

Roger: What's up?

Judy: Has the site been prepared for the job trailer yet?

Roger: Job trailer?

Judy: The job trailer – it's coming in today. What time will the job site be prepared?

Roger: The trailer will be here about 1:00 PM?

Judy: The job site! What time will the job site be prepared?

Conversation II:

John: Hey, Mike. I need your help.

Mike: What is it?

John: You and Joey go over and help Al's crew finish laying out the site.

Mike: Why me? I can't work with Joey. He can't understand a word I say.

John: Al's crew needs some help, and you and Joey are most qualified to do the job.

Mike: I told you, I can't work with Joey.

PARTICIPANT EXERCISE A ◆ LEADERSHIP SKILLS

Conversation III:

Ed: Hey, Jill.

Jill: Sir?

Ed: Have you received the latest DOL, EEO requirement to be sure the OFCP administrator finds our records up to date when he reviews them in August?

Jill: DOL, EEO, and OFCP?

Ed: Oh, and don't forget the MSHA, OSHA, and EPA reports are due this afternoon.

Jill: MSHA, OSHA, and EPA?

_____

_____

_____

_____

_____

_____

_____

Conversation IV:

Susan: Hey, Bob, would you do me a favor?

Bob: Okay, Sue. What is it?

Susan: I was reading the concrete inspection report and found the concrete in Bays 4A, 3B, 6C, and 5D didn't meet the 3000 psi strength requirements. Also, the concrete inspector on the job told me the two batches that came in today had to be refused because they didn't meet the slump requirements as noted on page 16 of the spec. I need to know if any placement problems happened on those bays, how long the ready mix trucks were waiting today, and what do we plan to do to stop these problems in the future.

_____

_____

_____

_____

_____

2. Read the following written memos, and identify any problems:

Memo I:

Let's start with the transformer vault $285.00 due. For what you ask? Answer practically nothing I admit, but here is the story. Paul the superintendent decided it was not the way good ole Comm Ed wanted it, we took out the ladder and part of the grading (as Paul instructed us to do) we brought it back here to change it. When Comm Ed the architect or Doe found out that everything would still work the way it was, Paul instructed us to reinstall the work. That is the whole story there is please add the $285.00 to my next payout.

_____

_____

_____

_____

_____

Memo II:

Let's take rooms C 307-C-312 and C-313 we made the light track supports and took them to the job to erect them when we tried to put them in we found direct work in the way, my men spent all day trying to find out what do so ask your Superintendent (Frank) he will verify seven hours pay for these men as he went back and forth while my men waited. Now the Architect has changed the system of hanging and has the gall to say that he has made my work easier, I can't see how. Anyway we want an extra two (2) men seven (7) hours for April 21 at $55.00 per hour or $385.00 on April 28th Doe Reference 197 finally resolved this problem. We will have no additional charges on Doe Reference 197, please note.

_____

_____

_____

_____

PARTICIPANT EXERCISE A ◆ LEADERSHIP SKILLS

5.0.0 MOTIVATION

The ability to motivate others is a key leadership skill that effective supervisors must possess. Motivation refers to behavior set into action because of a need felt by the individual to perform. It's also the word we use to describe the amount of effort that a person is willing to put forth to accomplish something. For example, a crew member who skips breaks and lunch in an effort to complete a job on time is thought to be highly motivated, but a crew member who does the bare minimum or just enough to keep his or her job is considered unmotivated.

Employee motivation has dimension because it can be measured. Examples of how motivation can be measured include determining the level of absenteeism, the percentage of employee turnover, and the number of complaints, as well as the quality and quantity of work produced.

5.1.0 Employee Motivators

Different things motivate different people in different ways. Consequently, there is no one-size-fits-all approach to motivating crew members. It is important to recognize that what motivates one crew member may not motivate another. In addition, what works to motivate a crew member once may not motivate that same person again in the future.

Frequently, the needs that motivate individuals are the same as those that create job satisfaction. They include:

- Recognition and praise
- Accomplishment
- Opportunity for advancement
- Job importance
- Change
- Personal growth
- Rewards

A supervisor's ability to satisfy these needs increases the likelihood of high morale within a crew. Morale refers to individuals' attitudes toward the tasks that they are expected to perform. High morale, in turn, means that employees will be motivated to work hard, and they will have a positive attitude about coming to work and doing their jobs.

5.1.1 Recognition and Praise

Recognition and praise refer to the need to have good work appreciated, applauded, and acknowledged by others. This can be accomplished by simply thanking employees for helping out on a project, or it can entail more formal praise, such as an award for Employee of the Month.

Some tips for giving recognition and praise:

- Be available on the job site so that you have the opportunity to witness good work.
- Know good work, and praise it when you see it.
- Look for good work and look for ways to praise it.
- Give recognition and praise only when truly deserved; otherwise, it will lose its meaning.
- Acknowledge satisfactory performance, and encourage improvement by showing that you have confidence in the ability of your crew members to do above-average work.

5.1.2 Accomplishment

Accomplishment refers to a worker's need to set challenging goals and achieve them. There is nothing quite like the feeling of achieving a goal, particularly a goal one never expected to accomplish in the first place.

Supervisors can help their crew members attain a sense of accomplishment by encouraging them to develop performance plans, such as goals for the year. In addition, supervisors can provide the support and the tools (such as training and coaching) necessary to help their crew members achieve these goals.

5.1.3 Opportunity for Advancement

Opportunity for advancement refers to an employee's need to gain additional responsibility and develop new skills and abilities. It is important that some employees know that they are not stuck in dead-end jobs. Rather, they want a chance to grow with the company and to be promoted in recognition of excelling in their work.

Effective leaders encourage their crew members

to work to their full potentials. In addition, they share information and skills with their employees in an effort to help them to advance within the organization.

5.1.4 Job Importance

Job importance refers to an employee's need to feel that his or her skills and abilities are valued and make a difference. Employees who do not feel that they make a difference tend to have difficulty justifying why they should get out of bed in the morning and go to work.

Supervisors should attempt to make every crew member feel like an important part of the team, as if the job wouldn't be possible without their help.

5.1.5 Change

Change refers to an employee's need to have variety in work assignments. Change is what keeps things interesting or challenging. It prevents the boredom that results from doing the same task day after day with no variety.

5.1.6 Personal Growth

Personal growth refers to an employee's need to learn new skills, enhance abilities, and grow as a person. It can be very rewarding to master a new competency on the job, particularly one that you were not trained to do. Similar to change, personal growth prevents the boredom associated with doing the same thing day after day without developing any new skills.

Supervisors should encourage the personal growth of their employees as well as themselves. Learning should be a two-way street on the job site; supervisors should teach their crew members and learn from them as well. In addition, crew members should be encouraged to learn from each other.

5.1.7 Rewards

Rewards are compensation for hard work. Rewards can include a crew member's base salary or go beyond that to include bonuses, give-aways, or other incentives. They can be monetary in nature (salary raises, holiday bonuses, etc.), or they can be non-monetary, such as free merchandise or other prizes (shirts, coffee mugs, jackets, etc.).

5.2.0 Motivating Employees

To increase motivation in the workplace, supervisors must individualize how they motivate different crew members. It is important that supervisors get to know their crew members and determine what motivates them as individuals. Once again, as diversity increases in the workforce, this becomes even more challenging; therefore, effective communication skills are essential.

Here is a list of some tips for motivating employees:

- Keep jobs challenging and interesting. Boredom is a guaranteed de-motivator.

- Communicate your expectations. People need clear goals in order to feel a sense of accomplishment when they are achieved.

- Involve your employees. Feeling that their opinions are valued leads to pride in ownership and active participation.

- Provide sufficient training. Give employees the skills and abilities they need to be motivated to perform.

- Mentor your employees. Coaching and supporting your employees boosts their self-esteem, their self-confidence, and ultimately their motivation.

- Lead by example. Become the kind of leader that employees admire and respect, and they will be motivated to work for you.

- Treat employees well. Be considerate, kind, caring, and respectful; treat employees the way that you want to be treated.

- Avoid using scare tactics. Threatening employees with negative consequences can backfire, resulting in employee turnover instead of motivation.

- Reward your crew for doing their best by giving them easier tasks from time to time. It is tempting to always give your best employees the hardest or dirtiest jobs because you know they will do the jobs correctly.

- Reward employees for a job well done.

- Recognize a job well done and praise it.

PARTICIPANT EXERCISE B

You are the supervisor of a masonry crew. Sam Williams is the person whom the company holds responsible for ensuring that equipment is operable and distributed to the jobs in a timely manner.

Occasionally, disagreements with Sam have resulted in tools and equipment arriving late. Sam, who has been with the company 15 years, resents having been placed in the job and feels that he outranks all the crew supervisors.

Sam figured it was about time he talked with someone about the abuse certain tools and other items of equipment were receiving on some of the jobs. Saws were coming back with guards broken and blades chewed up, bits were being sheared in half, motor housings were bent or cracked, and a large number of tools were being returned covered with mud. Sam was out on your job when he observed a mason carrying a portable saw by the cord. As he watched, he saw the mason bump the swinging saw into a steel column. When the man arrived at his workstation, he dropped the saw into the mud.

You are the worker's supervisor. Sam approached as you were coming out of the work trailer. He described the incident. He insisted, as crew leader, you are responsible for both the work of the crew and how its members use company property. Sam concluded, "You'd better take care of this issue as soon as possible! The company is sick and tired of having your people mess up all the tools!"

You are aware that some members of your crew have been mistreating the company equipment.

1. How would you respond to Sam's accusations?

_____

_____

_____

_____

_____

_____

_____

_____

_____

_____

_____

_____

_____

2. What action would you take regarding the misuse of the tools?

_____

_____

_____

_____

_____

3. How can you motivate the crew to take better care of their tools? Explain.

_____

_____

_____

_____

_____

_____

_____

_____

_____

PARTICIPANT EXERCISE B ◆ LEADERSHIP SKILLS

6.0.0 TEAM BUILDING

Organizations are making the shift from the traditional boss-worker mentality to one that promotes teamwork. The manager becomes the team leader, and the workers become team members. They all work together to achieve the common goals of the team.

The Society for Human Resources Management (SHRM) reports that there are a number of benefits associated with teamwork. They include the ability to complete complex projects more quickly and effectively, higher employee satisfaction, and a reduction in turnover.

6.1.0 Successful Teams

Successful teams are comprised of individuals who are willing to share their time and talents in an effort to reach a common goal—the goal of the team. Members of successful teams possess an Us attitude; they consider what's best for the team and put their egos aside.

Some characteristics of successful teams include the following:

- Everyone participates, and every team member counts.
- There is a sense of mutual trust and interdependence.
- Team members are empowered.
- They communicate.
- They are creative and willing to take risks.
- The team leader has strong people skills and is committed to the team.

6.2.0 Building Successful Teams

To be successful in the team leadership role, the crew leader should contribute to a positive attitude within the team.

There are several ways in which the team leader can accomplish this. First, he or she can work with the team members to create a vision or purpose of what the team is to achieve. It is important that every team member is committed to the purpose of the team, and the team leader is instrumental in making this happen.

Team leaders within the construction industry are typically assigned a crew. However, it is beneficial for the team leader to be involved in selecting the team members. Selection should be based on a willingness of people to work on the team and the resources that they are able to bring to the team.

When beginning a new team, team leaders should do the following:

- Explain the purpose of the team. Team members need to know what they will be doing, how long they will be doing it (if they are temporary or permanent), and why they are needed.
- Help the team establish goals or targets. Teams need a purpose, and they need to know what it is they are responsible for accomplishing.
- Define team member roles and expectations. Team members need to know how they fit into the team and what is expected of them as members of the team.
- Plan to transfer responsibility. Teams should be responsible for the tasks to be accomplished. With this responsibility comes a level of autonomy and empowerment to make group decisions.

7.0.0 GETTING THE JOB DONE

A supervisor is unable to complete all of the activities within a job independently; rather, other people are relied on to get everything done. In addition, supervisors must implement policies and procedures to make sure that the work is done correctly.

7.1.0 Delegating Responsibilities

Construction jobs have crews of people with various experiences and skill levels available to assist in the work. The supervisor's job is to draw from this expertise to get the job done well and in a timely manner.

Once the various activities that make up the job have been determined, the supervisor must identify the person or persons who will be responsible for completing each activity.

This requires that the supervisor be aware of the skills and abilities of the people on the crew. Then, the supervisor must put this knowledge to work in matching the crew's skills and abilities to specific tasks that must be accomplished to complete the job.

Upon matching crew members to specific activities, the supervisor must then delegate the assignments to the responsible person(s). The delegation of responsibilities is generally communicated verbally by the supervisor talking directly to the person that has been assigned the activity. However, there may be times when work is assigned indirectly through written instructions or verbally through someone other than the supervisor.

Some tips for delegating work are:

- Delegate work to a crew member that can do the job properly. If it becomes evident that he or she does not perform to the standard desired, either teach the crew member to do the work correctly or turn it over to someone else who can.

- When delegating, make sure that the crew member understands what to do and the level of responsibility. Make clear the results desired, and specify the boundaries and deadlines for accomplishing the results as well as the resources available.

- Identify the standards and methods of measurement for progress and accomplishment, along with the consequences of achieving or not achieving the desired results. Discuss the task with the crew member, and check for understanding by asking questions. Allow the crew member to contribute feedback or make suggestions about how the task should be performed in a safe and quality manner.

- After delegating the task, give the crew member the time and freedom to get started without feeling the pressure of too much supervision. When making the work assignment, be sure to tell the crew member how much time there is to complete it, and confirm that this time is consistent with the job schedule.

- After a crew member completes a task, examine and evaluate the result. Then, give the crew member some feedback as to how well it has been done. Get the crew member's comments. The information obtained at this time is valuable and will enable the supervisor to know what kind of work to assign that crew member in the future. It will also give the supervisor a means of measuring his or her own effectiveness in delegating work.

7.2.0 Implementing Policies and Procedures

Every company establishes policies and procedures that employees are expected to follow and the supervisors are expected to implement. Company policies and procedures are essentially guides for how the organization does business. They can also reflect organizational philosophies such as putting safety first or making the customer the number-one priority. Examples of policies and procedures include safety guidelines, credit standards, and billing processes.

Here are some tips for implementing policies and procedures.

- Learn the purpose of each policy. That way, you can follow it and apply it properly and fairly.

- If you're not sure how to apply a company policy or procedure, find out.

- Apply company policies and procedures. Remember that they combine what's best for the customer and the company. In addition, they provide direction on how to handle specific situations and answer questions.

- Use common sense when applying company policies or procedures. Some situations may warrant bending a company policy; others must be implemented in order to protect the company (for example, not permitting visitors without hard hats in dangerous construction areas).

Sometimes, supervisors issue orders to their crew members. Basically, an order initiates, changes, or stops an activity. Orders may be general or specific, written or oral, and formal or informal. The decision of how an order will be issued is up to the supervisor, but it is governed by the policies and procedures established by the company.

Some tips on issuing orders are:

- Make orders as specific as possible.

- Avoid giving general or vague orders unless it is impossible to foresee all of the circumstances that could occur in carrying out the order.

- It is not necessary to write orders for simple tasks unless the company requires that all orders be written.

- Write orders for more complex tasks that will take considerable time to complete or orders that are permanent.

- Consider the order, the audience to whom it applies, and the situation under which it will be implemented to determine the appropriate level of formality for the order.

8.0.0 PROBLEM SOLVING AND DECISION MAKING

Like it or not, problem solving and decision making are a large part of every supervisor's daily work. There will always be problems to be resolved and decisions to be made, especially in fast-paced, deadline-oriented industries such as construction.

8.1.0 Problem Solving vs. Decision Making

Sometimes, the difference between problem solving and decision making is not clear. Decision making refers to the process of choosing an alternative course of action in a manner appropriate for the situation. Problem solving involves determining the difference between the way things are and the way things should be, and finding out how to bring the two together. The two activities are interrelated because in order to make a decision, you may also have to use problem-solving techniques.

8.2.0 Types of Decisions

Some decisions are routine or simple. These types of decisions can be made based on past experiences. An example would be deciding how to get to and from work. If you've worked at the same place for a long time, you are already aware of the options for traveling to and from work (take the bus, drive a car, carpool with a co-worker, take a taxi, etc.). Based on past experiences with the options identified, you can make a decision of how best to get to and from work.

On the other hand, some decisions are non-routine or more difficult. These types of decisions require more careful thinking about how to carry out an activity by using a formal problem-solving technique. An example is planning a trip to a new vacation spot. If you are not sure how to get there, where to stay, what to see, etc., one option is to research the area to determine the possible routes, hotel accommodations, and attractions. Then, you will have to make a decision about which route to take, what hotel to choose, and what sites to visit, without the benefit of direct past experience with the options.

8.3.0 Formal Problem-Solving Techniques

To make non-routine decisions, the following procedure can be used as a part of formal problem solving.

Step 1   Recognize that a problem exists.

A problem is the difference between the way things are and the way that you want them to be. Solving a problem refers to eliminating the differences between what's existing and what's desired, or making things turn out the way you want them.

Consider the following example:

It is 3:30 p.m. Friday afternoon, and you are walking back to the job-site trailer after making the afternoon rounds on your job. As you exit the building under construction, you notice a lot of usable conduit in the scrap pile. This bothers you, so you proceed to ask the electrician in charge why so much good conduit is in the scrap pile. Mary, the head electrician, states she was not aware that there was so much usable conduit being scrapped.

You both decide to take a closer look at the scrap pile to see just how much is salvageable. You both conclude that a lot of usable conduit is being thrown away. Upon further questioning of the electrician, you find that she has been sick the last two days, and Mary left her helper, Tom, to do the work.

Investigating even further, you find that Tom just graduated from high school and is enrolled in the first year of the electrical program. According to his instructor, Tom appears willing to learn and performs satisfactorily in the classroom but only knows the basics of electrical work.

Rereading the scenario, the first thing to do to solve the problem is to identify the signals (statements) that indicate that a problem exists. In this case, the main signal is that a lot of usable conduit is in the scrap pile.

There may be other areas related to the one noted above. If so, what are they? As the reader can see, the signal is the thing that will motivate the supervisor to find a solution to the problem.

Step 2 Determine what you know or assume you know about the problem, and separate facts from non-facts.

This second step involves answering questions about the problems. Specifically, the what (what is the problem? what caused the problem?), when (when did it occur?), where (where did it happen?), and who (who caused the problem? who is involved?) should be determined. Once these questions have been considered, the information must be categorized as factual or non-factual.

Referring to the example above, some responses to an analysis of the problem may be:

What?
- There is too much usable conduit in the scrap pile.
- There is too much waste of materials on this job site.
- The job has gone over budget because of this waste.

When?
- The problem was discovered on Friday afternoon.

Where?
- The problem is the scrap pile near the building entrance.

Who?
- Mary, the electrician in charge, was off sick for two days, and she left a helper in charge.

- The electrician acted incorrectly by leaving the helper in charge.
- The helper, Tom, doesn't know how to do his job.

Based on the analysis above, some of the statements are facts while the others are assumptions, judgments, expressions of frustration, and/or blame. To solve the real problem, the supervisor must eliminate the non-factual items and deal only with the facts.

Facts:
- There is too much usable conduit in the scrap pile.
- The problem was discovered on Friday afternoon.
- The problem is in the scrap pile near the building entrance.
- The electrician in charge was off sick for two days, and she left a helper in charge.

Non-facts:
- There is too much waste of materials on this job site.
- The job has gone over budget because of this waste.
- The electrician acted incorrectly by leaving the helper in charge.
- The helper doesn't know how to do his job.

Step 3 State the problem.

Using only facts, formulate a problem statement(s). One should note that a fact is a statement of the problem when it meets the following criteria:

- A fact notes the difference between what is and what should be.
- A fact indicates that the present situation has potential for change.

Using the factual statements from the example case, the problem statements are as follows:

- There is much good conduit in the scrap pile that could be used for the job.
- The electrician's helper may not know how to cut or bend conduit to minimize waste.

Step 4   Develop objectives that will eliminate the problem.

Describing the desired end result (an objective) is a turning point in the problem-solving sequence. At this point, decision making comes into play. It basically involves choosing to do one of two things: to act on only one idea that comes into mind or to act on what appears to be the most effective of several alternatives.

Written or stated objectives are the basis of the decision-making process. The more precise the objective, the easier it will be to choose the most effective and efficient plan of action.

Continuing with the example above, some of the objectives may be the following:

- By the end of business on Monday, I will have found out whether or not Tom knows how to cut conduit in a way that will reduce waste.

- By the end of business on Tuesday, I will have had a discussion with Mary to ensure that she realizes that training helpers is a part of her job.

- By the end of business on Wednesday, Mary will have prepared an on-the-job training program for Tom.

- By the end of business on Friday, Tom will have begun an on-the-job training program on how to select and cut conduit efficiently.

Step 5   Develop alternate solutions, and select one that will solve the problem.

At this point, the supervisor should list all possible actions that can be taken to accomplish the objectives (solve the problem). When developing the list, judgments should not be made about each alternative. Instead, each possible action should be listed, and then the supervisor should weigh the alternatives to make a decision.

The previous example might have the following list of alternatives:

- Talk to Tom to determine whether or not he is aware of how to cut conduit efficiently; if not, provide training to enable him to do so.

- Provide an on-the-job training program.

- Discuss the problem with Mary and determine the best solution together.

- Inform Mary that it's her responsibility as a lead person to provide training where and whenever needed. Therefore, instruct her to develop and implement a training program.

- Follow up after the program to see if, in fact, Tom can now perform satisfactorily.

Step 6   Develop a plan of action.

Once the solution to the problem has been determined, the next step is to decide who is responsible for carrying out the plan and the deadline for accomplishing it, if not already stated in the objective(s).

For our example, one definite plan of action would be:

1. Meet with Mary to discuss the problem. Get a commitment from her that by Monday she will have discussed the problem with her helper and will report back with the outcome.

2. Upon meeting with Mary the second time, obtain a commitment from her to develop a one-week on-the-job training program for Tom. Inform Mary that it will be her duty to implement the training program and follow up on it. The plan is due by Tuesday.

3. Meet with Mary to review and make any needed revisions to her plan.

4. Follow up next week to ensure that Tom has successfully completed the training program.

Once all six steps of the formal problem-solving technique are completed, the supervisor must follow up to ensure that the action plan was completed as intended. If not, corrective measures should be taken to see that the plan is carried out. In addition, the supervisor should take steps to prevent the same problem from occurring again in the future.

PARTICIPANT EXERCISE C

1. Referring back to the problem example, explain how you would handle this situation.

_____

_____

_____

_____

_____

_____

_____

_____

_____

_____

_____

_____

_____

2. Give an example of a decision that you make which does not require a formal problem-solving technique.

_____

_____

_____

_____

_____

_____

_____

_____

_____

_____

_____

_____

_____

_____

3. Using the six-step formal problem-solving technique, solve a problem that you are now having or have recently experienced on your job. Write out all the steps.

_____

_____

_____

_____

_____

_____

_____

_____

_____

_____

_____

_____

_____

_____

_____

_____

_____

_____

_____

_____

_____

_____

_____

_____

_____

PARTICIPANT EXERCISE C ◆ LEADERSHIP SKILLS

8.4.0 Special Leadership Problems

Because they are responsible for leading others, it is inevitable that supervisors will encounter problems on occasion and be forced to make decisions about how to respond. Some problems will be relatively simple to resolve, like covering for a sick crew member who has taken a day off from work. Other problems will be complex and much more difficult to handle.

There are some complex problems that are relatively common. A few of the major employee problems include:

- Poor attitude towards the workplace
- An inability to work with others
- Absenteeism and turnover

8.4.1 Poor Attitude Towards the Workplace

Sometimes, employees have poor attitudes towards the workplace because of bad relationships with their fellow employees, negative outlooks on their supervision, or a dislike of the job in general. Whatever the case, it is important that the supervisor determine the cause of the poor attitude.

The best way to determine the cause of a poor attitude is to talk with that employee one-on-one, listening to what the employee has to say and asking questions to uncover information. Once this conversation has occurred and the facts have been assessed, the supervisor can determine how to correct the situation and turn the negative attitude into a positive one.

If the supervisor discovers that the problem stems from factors in the workplace or the surrounding environment, the supervisor has several choices. First, the supervisor can move the worker from the situation to a more acceptable work environment. Next, the supervisor can change that part of the work environment found to be causing the poor attitude. Finally, the supervisor can take steps to change the employee's attitude so that the work environment is no longer a negative factor.

8.4.2 Inability to Work with Others

Sometimes supervisors will encounter situations where an employee has a difficult time working with others on the crew. This could be a result of personality differences, an inability to communicate, or some other cause. Whatever the reason, the supervisor must address the issue and get the crew working as a team.

The best way to determine the reason for why individuals don't get along or work well together is to talk to the parties involved. The supervisor should speak openly with the employee as well as the other individual(s) to find out why.

Once the reason for the conflict is found, the supervisor can determine how to respond. There may be a way to resolve the problem and get the workers communicating and working as a team again. On the other hand, there may be nothing that can be done that will lead to a harmonious situation. In this case, the supervisor would either have to transfer the employee to another crew or have the problem crew member terminated. This latter option should be used as a last measure and should be discussed with one's superiors or Human Resources Department.

8.4.3 Absenteeism and Turnover

Absenteeism and turnover are big problems on construction jobs. Without workers available to do the work, jobs are delayed, and money is lost.

Absenteeism refers to workers missing their scheduled work time on a job. Absenteeism has many causes, some of which are inevitable. For instance, people get sick, they have to take time off for family emergencies, and they have to attend family events such as funerals. However, there are some causes of absenteeism that could be prevented by the supervisor.

The most effective way to control absenteeism is to make the company's policy clear to all employees. Companies that do this find that chronic absenteeism diminishes as a problem. New employees should have the policy explained to them. This explanation should include the number of absences allowed and the reasons for which sick or personal days can be taken. In addition, all workers should know how to inform their supervisors when they miss work and understand the consequences of exceeding the number of sick or personal days allowed.

Once the policy on absenteeism is explained to employees, supervisors must be sure to implement it consistently and fairly. If the policy is administered equally, employees will likely follow it. However, if the policy is not administered equally and some employees are given exceptions, then it will not be effective. Consequently, the rate of absenteeism will increase.

Despite having a policy on absenteeism, there will always be employees who are chronically late or miss work. In cases where an employee abuses the absenteeism policy, the supervisor should discuss the situation directly with the employee. The supervisor should confirm that the employee understands the company's policy and insist that the employee comply with it. If the employee's behavior continues, disciplinary action may be in order.

Turnover refers to the loss of an employee that is initiated by that employee. In other words, the employee quits and leaves the company to work elsewhere or is fired for cause.

Like absenteeism, there are some causes of turnover that cannot be prevented and others that can. For instance, it is unlikely that a supervisor could keep an employee who finds a job elsewhere earning twice as much money. However, supervisors can prevent some employee turnover situations. They can work to ensure safe working conditions for their crew, treat their workers fairly and consistently, and help promote good working conditions. The key to doing so is communication. Supervisors need to know the problems if they are going to be able to successfully resolve them.

Some of the major causes of turnover include:

- *Uncompetitive wages and benefits* – Workers may leave one construction company to go to another that pays higher wages and/or offers better benefits. They may also leave to go to another industry that pays more.

- *Lack of job security* – They leave to find more permanent employment.

- *Unsafe project sites* – They leave for safer projects.

- *Unfair/inconsistent treatment by their immediate supervisor*

- *Poor working conditions*

Essentially, the same actions described above for absenteeism are also effective for reducing turnover. Past studies have shown that maintaining harmonious relationships on the job site will go a long way in reducing both turnover and absenteeism. This will take effective leadership on the part of the supervisor.

PARTICIPANT EXERCISE D

Case I:

On the way over to the job trailer, you look up and see a piece of falling scrap heading for one of the laborers. Before you can say anything, the scrap material hits the ground about five feet in front of the worker. You notice the scrap is a piece of conduit. You quickly pick it up, assuring the worker you will take care of this matter.

Looking up, you see your crew on the third floor in the area from which the material fell. You decide to have a talk with them. Once on the deck, you ask the crew if any of them dropped the scrap. The men look over at Bob, one of the electricians in your crew. Bob replies, "I guess it was mine. It slipped out of my hand."

It is a known fact that the Office of Occupational Safety and Health Administration (OSHA) regulations state that an enclosed chute of wood shall be used for material waste transportation from heights of 20 feet or more. It is also known that Bob and the laborer who was almost hit have been seen arguing lately.

1. Assuming Bob's action was deliberate, what action would you take?

   _____

   _____

   _____

   _____

   _____

PARTICIPANT EXERCISE D ◆ LEADERSHIP SKILLS

2. Assuming the conduit accidentally slipped from Bob's hand, how can you motivate him to be more careful?

_____

_____

_____

3. What follow-up actions, if any, should be taken relative to the laborer who was almost hit?

_____

_____

_____

4. Should you discuss the apparent OSHA violation with the crew? Why or why not?

_____

_____

_____

5. What acts of leadership would be effective in this case? To what leadership traits are they related?

_____

_____

_____

Case II:

Mike James had just been appointed supervisor of a tile-setting crew. Before his promotion into the management ranks, he had been a tile setter for five years. His work had been consistently of superior quality.

Except for a little good-natured kidding, Mike's co-workers had wished him well in his new job. During the first two weeks, most of them had been cooperative while Mike was adjusting to his supervisory role.

At the end of the second week, a disturbing incident took place. Having just completed some of his duties, Mike stopped by the job site wash station. There he saw Steve and Ron, two of his old friends who were also in his crew, washing.

"Hey Ron, Steve, you should not be cleaning up this soon. It's at least another thirty minutes until quitting time," said Mike. "Get back to your work station, and I'll forget I saw you here."

"Come off it, Mike," said Steve. "You used to slip up here early on Fridays. Just because you have a little rank now, don't think you can get tough with us." To this Mike replied, "Things are different now. Both of you get back to work, or I'll make trouble." Steve and Ron said nothing more, and they both returned to their work stations.

From that time on, Mike began to have trouble as a supervisor. Steve and Ron gave him the silent treatment. Mike's crew seemed to forget how to do the most basic activities. The amount of rework for the crew seemed to

be increasing. By the end of the month, Mike's crew was behind schedule.

1. How do you think Mike should have handled the confrontation with Ron and Steve?

_____

_____

2. What do you suggest Mike could do about the silent treatment he got from Steve and Ron?

_____

_____

_____

3. If you were Mike, what would you do to get your crew back on schedule?

_____

_____

_____

4. What acts of leadership could be used to get the crew's willing cooperation?

_____

_____

_____

5. To which leadership traits do they correspond?

_____

_____

_____

1. A supervisor differs from a craftworker in that _____.

   a. a supervisor need not have direct experience in those job duties that a craftworker typically performs

   b. a supervisor can expect to oversee one or more craftworkers in addition to performing some of the typical duties of the craftworker

   c. a supervisor is exclusively in charge of overseeing, since performing technical work is not part of this role

   d. a supervisor's responsibilities do not include being present on the job site

2. Among the many traits effective leaders should have is _____.

   a. the ability to communicate the goals of a project

   b. the drive necessary to carry the workload by themselves in order to achieve a goal

   c. a perfectionist nature, which ensures that they will not make useless mistakes

   d. the ability to make decisions without needing to listen to others' opinions

3. Of the three styles of leadership, the _____ style would be effective in dealing with a craftworker's negative attitude.

   a. hands-off, followed by autocratic

   b. autocratic, followed by hands-off

   c. democratic

   d. autocratic

4. Verbal communication involves _____.

   a. feedback, in which the receiver of a message should parrot back the content of the message in order to let the sender know that the message was understood

   b. the message, which can contain jargon and technical terms used within the profession

   c. the sender, who should remember to concentrate more on the content of the message rather that on the manner in which it was expressed

   d. the receiver, who should first focus on removing barriers to effective listening

5. One way to overcome barriers to effective communication is to _____.

   a. avoid taking notes on the content of the message, since this can be distracting

   b. react objectively, rather than subjectively, to the message

   c. anticipate the content of the message and interrupt if necessary in order to show interest

   d. think about how to respond to the message while listening

6. A good way to motivate employees is to use a one-size-fits-all approach, since employees are members of a team with a common goal.

   a. True

   b. False

7. A supervisor can effectively delegate responsibilities by _____.

   a. refraining from evaluating the employee's performance once the task is completed, since it is a new task for the employee

   b. doing the job for the employee to make sure the task is done correctly

   c. allowing the employee to give feedback and suggestions about the new task

   d. communicating information to the employee generally in written form

8. Problem solving differs from decision making in that _____.

   a. problem solving involves identifying discrepancies between the way a situation is and the way it should be

   b. decision making involves separating facts from non-facts

   c. decision making involves eliminating differences

   d. problem solving involves determining an alternative course of action for a given situation

◆ **CHAPTER THREE**

Safety

OBJECTIVES

Upon completion of this chapter, you will be able to:

1. Demonstrate an understanding of the importance of safety.

2. Give examples of direct and indirect costs of workplace accidents.

3. Identify safety hazards of the construction industry.

4. Explain the purpose of the Occupational Safety and Health Act (OSHA).

5. Discuss OSHA inspection programs.

6. Identify the key points of a safety program.

7. List the steps to train employees on how to perform new tasks safely.

8. Identify a supervisor's safety responsibilities.

9. Explain the importance of having employees trained in first aid and Cardio-Pulmonary Resuscitation (CPR) on the job site.

10. Describe the signals of substance abuse.

11. List the essential parts of an accident investigation.

12. Describe the ways to maintain employee interest in safety.

SECTION 1

1.0.0 SAFETY OVERVIEW

The construction industry loses millions of dollars every year because of job-site accidents. Work-related injuries, sickness, and deaths have caused untold human misery and suffering. Project delays and budget overruns from construction injuries and fatalities are huge, and work-site accidents erode the overall morale of the crew.

Construction work has dangers that are part of the job. Examples include working on scaffolds, using cranes in the presence of power lines, and operating heavy machinery. Despite these hazards, experts believe that applying preventive safety measures could drastically reduce the number of accidents.

A job of the crew leader is to carry out the company's safety program and make sure that all workers are performing their tasks safely. To be successful, the crew leader must:

- Be aware of the costs of accidents.

- Understand all federal, state, and local governmental safety regulations.

- Be involved in training workers in safe-work methods.

- Conduct safety meetings.

- Get involved in safety inspections, accident investigations, and fire protection and prevention.

Crew leaders have the most impact for ensuring that all jobs are performed safely by their crew members. Providing employees with a safe working environment by preventing accidents and enforcing safety standards will go a long way towards maintaining the job schedule and enabling a job's completion on time and within budget. This chapter will present the major duties of a supervisor in relation to jobsite safety.

1.1.0 Accident Statistics

The National Institute for Occupational Safety and Health (NIOSH) estimates that more than 7 million people, or 5 percent of the workforce, are currently employed in the construction industry. Each day, construction workers face the risk of falls, machinery accidents, electrocutions, and other potentially fatal occupational hazards.

NIOSH statistics show that more than 1,100 construction workers are killed on the job each year, more fatalities than in any other industry. According to the 2006 survey of the National Census of Fatal Occupational Injuries, falls were the leading cause of deaths in the construction industry, accounting for over 40 percent of the fatalities. Nearly half of the fatal falls occurred from roofs, scaffolds, or ladders. Roofers, structural metal workers, and painters experienced the greatest number of fall fatalities.

The number of fatalities in the construction industry is exceedingly high, but the number of injuries that occur in the workplace is even higher. NIOSH reports that approximately 15 percent of all workers' compensation costs are spent on injured construction workers. The causes of injuries on construction sites include falls, coming into contact with electric current, fires, and mishandling of machinery or equipment. According to NIOSH, back injuries are the leading safety problem in workplaces.

SECTION 2

2.0.0 COSTS OF ACCIDENTS

Occupational accidents are expensive. The American Medical Association reported that the costs due to work-related injuries and illnesses were estimated to be $171 billion per year. These costs affect the employee, the company, and the construction industry as a whole.

Organizations encounter both direct and indirect costs associated with workplace accidents. Examples of direct costs include workers' compensation claims and sick pay; indirect costs include increased absenteeism, loss of productivity, loss of job opportunities due to poor safety records, and negative employee morale attributed to workplace injuries. There are many other related costs involved with workplace accidents. A company can be insured against some of them, but not others. To compete and survive, companies must control these as well as all other employment-related costs.

2.1.0 Insured Costs

Insured costs related to injuries or deaths include the following:

- Compensation for lost earnings
- Medical and hospital costs
- Monetary awards for permanent disabilities
- Rehabilitation costs
- Funeral charges
- Pensions for dependents

Insurance premiums or charges related to property damages include:

- Fire
- Loss and damage
- Use and occupancy
- Public liability

2.2.0 Uninsured Costs

Uninsured costs related to injuries or deaths include the following:

- First aid expenses
- Transportation costs
- Costs of investigations
- Costs of processing reports
- Down time on the job site while investigations are in progress

Uninsured costs related to wage losses include:

- Idle time of workers whose work is interrupted
- Time spent cleaning the accident area
- Time spent repairing damaged equipment
- Time lost by workers receiving first aid
- Costs of training injured workers in a new career

Uninsured costs related to production losses include:

- Product spoiled by accident
- Loss of skill and experience
- Lowered production or worker replacement
- Idle machine time

Associated costs may include the following:

- Difference between actual losses and amount recovered
- Costs of rental equipment used to replace damaged equipment
- Costs of surplus workers used to replace injured workers
- Wages or other benefits paid to disabled workers
- Overhead costs while production is stopped
- Impact on schedule
- Loss of bonus or payment of forfeiture for delays

Uninsured costs related to off-the-job activities include:

- Time spent on injured workers' welfare
- Loss of skill and experience of injured workers
- Costs of training replacement workers

Uninsured costs related to intangibles include:

- Lowered employee morale
- Increased labor conflict
- Unfavorable public relations
- Loss of bid opportunities because of poor safety records
- Loss of client goodwill

3.0.0 SAFETY REGULATIONS

To reduce safety and health risks and the number of injuries and fatalities on the job, the federal government has enacted some laws and regulations, including the Occupational Safety and Health Act (OSHA) of 1970. Its purpose is "to assure so far as possible every working man and woman in the Nation safe and healthful working conditions and to preserve our human resources."

To promote a safe and healthy work environment, OSHA issues standards and rules for working conditions, facilities, equipment, tools, and work processes. It does extensive research into occupational accidents, illnesses, injuries, and deaths in an effort to reduce the number of occurrences and adverse effects. In addition, OSHA regulatory agencies conduct workplace inspections to ensure that companies follow the standards and rules.

3.1.0 Workplace Inspections

To enforce OSHA, the government has granted regulatory agencies the right to enter public and private properties to conduct workplace safety investigations. The agencies also have the right to take legal action if companies are not in compliance with the Act. These regulatory agencies employ OSHA Compliance Safety and Health Officers (CSHOs), who are chosen for their knowledge in the occupational safety and health field. The CSHOs are thoroughly trained in OSHA standards and in recognizing safety and health hazards.

States with their own occupational safety and health programs conduct inspections. To do so, they enlist the services of qualified state CSHOs.

Companies are inspected for a multitude of reasons. They may be randomly selected, or they may be chosen because of employee complaints, due to an imminent danger, or as a result of major accidents or fatalities.

OSHA can assess significant financial penalties for safety violations. A superintendent can be held criminally liable for repeat violations.

3.2.0 Penalties for Violations

OSHA has established monetary fines for the violation of their regulations. The penalties are as follows:

- Willful Violations: $59,000 – $70,000

- Repeated Violations: Maximum $70,000

- Serious, Other-Than-
  Serious, Other Specific
  Violations: Maximum $7,000

- OSHA Notice Violation: $1,000

- Failure to Post OSHA
  200 Summary: $1,000

- Failure to Properly Maintain
  OSHA 200 and 100 Logs: $1,000

- Failure to Promptly and Properly
  Report Fatality/Catastrophe: $5,000

- Failure to Permit Access to
  Records Under 1904 Regulations: $1,000

- Failure to Follow Advance
  Notification Requirements
  Under 1903.6 Regulations: $2,000

- Failure to Abate – for Each
  Calendar Day Beyond
  Abatement Date: Maximum $7,000

- Retaliation Against Individual
  for Filing OSHA Complaint: $10,000

In addition, there can be criminal penalties or personal liability for not complying with OSHA. States may also inflict penalties for any violations to their safety and health programs.

SECTION 4

4.0.0 SAFETY RESPONSIBILITIES

Each employer must set up a safety and health program to manage workplace safety and health and to reduce injury, illness, and fatalities. The program must be appropriate to the conditions of the workplace. It should consider the number of workers employed and the hazards to which they are exposed while at work.

To be successful, the safety and health program must have management leadership and employee participation. In addition, training and informational meetings play an important part in effective programs. Being consistent with safety policies is the key.

4.1.0 Safety Program

The supervisor plays a key role in the successful implementation of the safety program. The supervisor's attitudes towards the program set the standard for how crew members view safety. Therefore, the supervisor should follow all program guidelines and require crew members to do the same.

Safety programs should consist of the following:

- Safety policies and procedures
- Hazard identification and assessment
- Safety information and training
- Safety record system
- Accident investigation procedures
- Appropriate discipline for not following safety procedures

4.2.0 Safety Policies and Procedures

Employers are responsible for following OSHA and state safety standards. Usually, they incorporate OSHA and state regulations into a safety policies and procedures manual. Such a manual is presented to employees when they are hired.

Basic safety requirements should be presented to new employees during their orientation to the company. If the company has a safety manual, the new employee should be required to read it and sign a statement indicating that it is understood. If the employee cannot read, the employer should have someone read it to the employee and answer any questions that arise. The employee should then sign a form stating that he or she understands the information.

It is not enough to tell employees about safety policies and procedures on the day they are hired and never mention them again. Rather, supervisors should constantly emphasize and reinforce the importance of following all safety policies and procedures. In addition, employees should play an active role in determining job safety hazards and find ways that the hazards can be prevented and controlled.

4.3.0 Hazard Identification and Assessment

Safety policies and procedures should be specific to the company. They should clearly present the hazards of the job. Therefore, supervisors should also identify and assess hazards to which employees are exposed. They must also assess compliance with OSHA and state standards.

To identify and assess hazards, OSHA recommends that employers conduct inspections of the workplace, monitor safety and health information logs, and evaluate new equipment, materials, and processes for potential hazards before they are utilized.

Supervisors and employees play important roles in identifying hazards. It is the supervisor's responsibility to determine what working conditions are unsafe and inform employees of hazards and their locations. In addition, they should encourage their crew members to tell them about hazardous conditions. To accomplish this, supervisors must be present and available on the job site.

The supervisor also needs to help the employee be aware of and avoid the built-in hazards that are part of the construction industry. Examples include working on high buildings and in tunnels that are deep underground, on caissons, in huge excavations with earthen walls that have to be buttressed by heavy steel I-beams, and other naturally dangerous projects.

In addition, the supervisor can take safety measures, such as installing protective railings to prevent workers from falling from buildings, as well as scaffolds, platforms, and shoring.

4.4.0 Safety Information and Training

The employer should provide periodic information and training to new and long-term employees. This should be done as often as necessary so that all employees are adequately trained. Special training and informational sessions should be provided when safety and health information changes or workplace conditions create new hazards.

Whenever a supervisor assigns an experienced employee a new task, the supervisor must ensure that the employee is capable of doing the work in a safe manner. The supervisor can accomplish this by providing safety information or training one-on-one or in a group.

The supervisor should do the following:

- Define the task.
- Explain how to do the task safely.
- Explain what tools and equipment to use and how to use them safely.
- Identify the necessary personal protective clothing.
- Explain the nature of the hazards in the work and how to recognize them.
- Stress the importance of personal safety and the safety of others.
- Hold regular safety meetings with the crew's input.
- Review Material Safety Data Sheets that may be necessary.

4.5.0 Safety Record Systems

OSHA law requires that employers keep records of hazards identified and document the severity of the hazard. The information should include the likelihood of employee exposure to the hazard, the seriousness of the harm associated with the hazard, and the number of exposed employees.

In addition, the employer must document the actions taken or plans for action to control the hazards. While it is best to take actions immediately, it is sometimes necessary to develop a plan, which includes setting priorities and deadlines and tracking progress in controlling hazards.

Employers who are subject to the recordkeeping requirements of the Occupational Safety and Health Act of 1970 must maintain for each establishment a log of all recordable occupational injuries and illnesses.

Logs must be maintained and retained for five years following the end of the calendar year to which they relate. Logs must be available (normally at the establishment) for inspection and copying by representatives of the Department of Labor, the Department of Health and Human Services, or states accorded jurisdiction under the Act. Employees, former employees, and their representatives may also have access to these logs.

An MSDS is designed to provide both workers and emergency personnel with the proper procedures for handling or working with a substance that may be dangerous. An MSDS will include information such as physical data (melting point, boiling point, flash point, etc.), toxicity, health effects, first aid, reactivity, storage, disposal, protective equipment, and spill/leak procedures. These sheets are of particular use if a spill or other accident occurs.

4.6.0 Accident Investigation Procedures

In the event of an accident, the employer should investigate the cause of the accident and determine how to avoid it in the future. According to OSHA, the employer must investigate each work-related death, serious injury or illness, or incident having the potential to cause death or serious physical harm. The employer should document any findings from the investigation as well as the action plan to prevent future occurrences. This should be done immediately, with photos if possible.

SECTION 5

5.0.0 SUPERVISOR INVOLVEMENT IN SAFETY

To be effective leaders, supervisors must be actively involved in the safety program. Supervisory involvement includes conducting safety meetings and inspections, promoting first aid and fire protection and prevention, preventing substance abuse on the job, and investigating accidents.

5.1.0 Safety Meetings

A safety meeting may be a brief informal gathering of a few employees or a formal meeting with instructional films and talks by guest speakers. The size of the audience and the topics to be addressed determine the format of the meeting. Small, informal safety meetings are typically conducted weekly.

Safety meetings should be planned in advance, and the information should be communicated to all employees affected. In addition, the topics covered within safety meetings should be timely and practical.

5.2.0 Inspections

Supervisors must make regular and frequent inspections to prevent accidents from happening. They must also control the number of accidents that occur. They need to inspect the job sites where their workers perform their tasks. It is recommended that this be done before the start of work each day and during the day at different times.

Supervisors must correct or protect workers from existing or potential hazards in their work areas. Sometimes, supervisors may be required to work in areas controlled by other contractors. Here, the supervisors may have little control over unsafe conditions. In such cases, they should immediately bring the hazards to attention of the contractor at fault, their superior, and the person responsible for the job site.

Supervisors' inspections are only valuable if action is taken to correct what is wrong or hazardous. Consequently, supervisors must be alert for unsafe acts in their work sites. When an employee performs an unsafe action, the supervisors must explain to the employee why the act was unsafe, ask that the employee not do it again, and request cooperation in promoting a safe working environment. The supervisor may decide to document what happened and what the employee was asked to do to correct the situation. It is then very important that supervisors follow up to make certain that the employee is complying with the safety procedures. Never allow a safety violation to go uncorrected.

5.3.0 First Aid

First aid is described as "the immediate and temporary care given the victim of an accident or sudden illness until the services of a physician can be obtained." The victim of an accident or sudden illness at a construction site poses more problems than normal since he or she may be working in a remote location. Frequently, the construction site is located far from a rescue squad, fire department, or hospital, presenting a problem in the rescue and transportation of the victim to a hospital. The worker may also have been injured by falling rock or other materials, so special rescue equipment or first-aid techniques are often needed.

The primary purpose of first aid is to provide immediate and temporary medical care to employees involved in accidents, as well as employees experiencing non-work-related health emergencies, such as heart pains or heart attacks. To meet this objective, every supervisor should be aware of the location and contents of first-aid kits available on the job site. Emergency numbers should be posted in the job trailer. The supervisor should be trained in administering and teaching first aid. OSHA has specific requirements for the contents of first aid kits. In addition, OSHA requires that at least one person trained in first aid be present at the job site at all times. In addition, it is recommended that someone on site be certified in CPR.

The employer benefits by having personnel trained in first aid at each job site in the following ways:

1. The immediate and proper treatment of minor injuries may prevent them from developing into more serious conditions. As a result, medical expenses, lost work time, and sick pay will be eliminated or reduced.

2. It may be possible to determine if professional medical attention is needed.

3. Valuable time can be saved when a trained individual prepares the patient for treatment when professional medical care arrives. As a result, lives can be saved.

The American Red Cross and the United States Bureau of Mines provide basic and advanced first aid courses at nominal costs. The American Red Cross trains on both first aid and CPR. The local area offices of these organizations can provide further details regarding the training available.

5.4.0 Fire Protection and Prevention

NIOSH estimates that fires and explosions kill 175 and injure more than 5,000 workers each year.

The necessity of fire protection and prevention is increasing as new building materials are introduced. Some building materials are highly flammable. They produce great amounts of smoke and gases, which cause difficulties for fire fighters. Other materials melt when ignited and may spread over floors, preventing fire-fighting personnel from entering areas where this occurs.

OSHA has specific standards for fire safety. They require that employers provide proper exits, fire-fighting equipment, and employee training on fire prevention and safety. It is important that supervisors understand and practice these fire-prevention techniques. For more information, consult OSHA guidelines.

5.5.0 Substance Abuse

Unfortunately, drug and alcohol abuse is a reality within our country. Drug abuse means inappropriately using drugs, whether they are legal or illegal. Some people use illegal street drugs, such as cocaine or marijuana. Others use legal prescription drugs incorrectly by taking too many pills, using other people's medications, or self-medicating. Others consume alcohol to the point of intoxication.

It is essential that supervisors enforce company policies and procedures about substance abuse. Supervisors must work with management to deal with suspected drug and alcohol abuse and should not handle these situations themselves. These cases are normally handled by the human resources department or designated manager. There are legal consequences of drug and alcohol abuse and the associated safety implications. That way, the business and the employee's safety are protected. If you suspect that an employee is suffering from drug or alcohol abuse, you should immediately contact your supervisor and/or human resources department for assistance.

Construction supervisors have to deal with immediate job safety because of the dangers associated with construction work. It is the supervisor's responsibility to make sure that safety is maintained at all times. This may include removing workers from a work site where they may be endangering themselves or others.

For example, suppose several crew members go out and smoke marijuana or have a few beers during lunch. Then, they return to work to erect scaffolding for a concrete pour in the afternoon. If you can smell marijuana on the crew members' clothing or the alcohol on their breath, you must step in and take action. Otherwise, they might cause an accident that could delay the project or cause serious injury or death.

It is often difficult to detect drug and alcohol abuse. The best way is to search for identifiable effects, such as those mentioned above or sudden changes in behavior that are not typical of the employee. Some examples include:

- Unscheduled absences; failure to report to work on time
- Significant changes in the quality of work
- Unusual activity or lethargy
- Sudden and irrational temper flare-ups
- Significant changes in personal appearance, cleanliness, or health

There are other more specific signs which should arouse suspicions, especially if more than one is exhibited. Among them are:

- Slurring of speech or an inability to communicate effectively
- Shiftiness or sneaky behavior, such as an employee disappearing to wooded areas, storage areas, or other private locations
- Wearing sunglasses indoors or on overcast days to hide dilated or constricted pupils
- Wearing long-sleeved garments, particularly on hot days, to cover marks from needles used to inject drugs
- Attempting to borrow money from co-workers
- The loss of an employee's tools and company equipment

5.6.0 Accident Investigations

There are two times when a supervisor will be involved with an accident investigation. The first time is when an accident, injury, or report of work-connected illness takes place. If present on site, the supervisor should proceed immediately to the place the incident occurred to see that proper first aid is being provided. He or she will also want to make sure that other safety and operational measures are taken to prevent another incident.

After the incident, the supervisor will need to make a formal investigation and submit a report. An investigation looks for the causes of the accident by examining the situation under which it occurred and talking to the people involved. Investigations are perhaps the most useful tool in the prevention of accidents in the future.

For years, a prominent safety engineer was confused as to why sheet metal workers fractured their toes frequently. The supervisor had not made a thorough accident investigation, and the injured workers were embarrassed to admit how the accidents really occurred. It was later discovered they used the metal reinforced cap on their safety shoes as a "third hand" to hold the sheet metal vertically in place when they fastened it. The sheet metal was inclined to slip and fall behind the safety cap onto the toes and cause a fracture. With a proper investigation, further accidents could have been prevented.

There are four major parts to an accident investigation. The supervisor will be concerned with each one. They are:

- Describing the accident
- Determining the cause of the accident
- Determining the parties involved and the part played by each
- Determining how to prevent re-occurrences

SECTION 6

6.0.0 PROMOTING SAFETY

The main way for supervisors to encourage safety is through example. Supervisors should be aware that their behavior sets standards for their crew members. If a supervisor cuts corners on safety, then the crew members may think that it is okay to do so as well.

The key to effectively promote safety is good communication. It is important to plan and coordinate activities and to follow through with safety programs. The most successful safety promotions occur when employees actively participate in planning and carrying out activities.

1. A construction worker performing a job safely involves _____.
   a. conducting safety inspections after an incident occurs to determine what went wrong
   b. undergoing required training after each accident that occurs
   c. being aware of potential risks
   d. not worrying about the costs of accidents, but instead focusing on potential of injury or death

2. With regard to workplace safety, the crew member is held completely responsible for _____.
   a. reporting hazardous conditions
   b. ensuring that everyone is performing their job in a safe manner
   c. conducting safety meetings
   d. enforcing safety standards among the employees

3. In order to ensure workplace safety, the supervisor should _____.
   a. hold formal safety meetings featuring various films and guest speakers
   b. have crew members conduct on-site safety inspections
   c. notify contractors and their supervisor of hazards in a situation where a job is being performed in an unsafe area controlled by other contractors
   d. hold crew members responsible for making a formal report and investigation following an accident

4. A superintendent can be sent to jail for repeat safety violations.
   a. True
   b. False

Some activities used by organizations to help motivate employees on safety and help promote safety awareness include:

- Meetings
- Contests
- Recognition and awards
- Publicity

6.1.0 Meetings

For information on safety meetings, refer to Section 5 of this chapter.

6.2.0 Contests

Contests are a great way to promote safety in the workplace. Examples of safety-related contests include:

- Sponsoring housekeeping contests for the cleanest job site or work area
- Challenging employees to come up with a safety slogan for the company or department
- Having a poster contest that involves employees or their children creating safety-related posters
- Recording the number of accident-free workdays or person-hours
- Giving safety awards (hats, T-shirts, other promotional items or prizes)

One of the positive aspects of contests is their ability to encourage employee participation. It is important, however, to ensure that the contest has a valid purpose. For example, the posters created in a poster contest can be displayed throughout the organization as reminders.

6.3.0 Recognition and Awards

Recognition and awards serve several purposes. Among them are acknowledging and encouraging good performance, building goodwill, reminding employees of safety issues, and publicizing the importance of practicing safety standards. There are countless ways to recognize and award safety. Examples include:

- Supplying food at the job site when a certain goal is achieved
- Providing a reserved parking space to acknowledge someone for a special achievement
- Giving gift items such as T-shirts or gift certificates to reward employees
- Giving plaques to a department or an individual
- Sending a letter of thanks
- Publicly honoring a department or an individual for a job well done

Creativity can be used to determine how to recognize and award good safety on the worksite. The only precautionary measure is that the award should be meaningful and not perceived as a bribe.

6.4.0 Publicity

Publicizing safety is the best way to get the message out to employees. An important aspect of publicity is to keep the message accurate and current. Safety posters that are hung for years on end tend to lose their effectiveness. It is important to keep ideas fresh.

Examples of promotional activities include posters or banners, advertisements or information on bulletin boards, payroll mailing stuffers, and employee newsletters. In addition, merchandise can be purchased that promotes safety, including buttons, hats, T-shirts, and mugs.

PARTICIPANT EXERCISE E

1. What are the two major types of costs caused by accidents?

   _____

   _____

2. List three causes of accidents.

   _____

   _____

   _____

3. What federal government agency regulates job site safety? _____

4. True or false: Every supervisor should be trained in first-aid procedures.   T   F

5. List the six steps to informing and training employees how to do a new task.

   _____

   _____

   _____

   _____

   _____

   _____

6. Why are job site inspections important?

   _____

   _____

   _____

   _____

   _____

   _____

7. True or false: Investigations may be the most useful tool in preventing future accidents.   T   F

8. List three signs that may suggest substance abuse.

   _____

   _____

   _____

9. List three ways that supervisors can promote safety.

   _____

   _____

   _____

PARTICIPANT EXERCISE E ◆ SAFETY

◆ **CHAPTER FOUR**

Project Control

OBJECTIVES

Upon completion of this chapter, you will be able to:

1. Describe the three phases of a construction project.

2. Define the three types of project delivery systems.

3. Define planning and describe what it involves.

4. Explain why it is important to plan.

5. Describe the two major stages of planning.

6. Explain the importance of documenting one's work.

7. Describe the estimating process.

8. Explain how schedules are developed and used.

9. Identify the two most common schedules.

10. Explain short-interval production scheduling (SIPS).

11. Describe the different costs associated with building a job.

12. Explain the supervisor's role in controlling costs.

13. Illustrate how to control the main resources of a job: materials, tools, equipment, and labor.

14. Define the terms *production* and *productivity* and explain why they are important.

SECTION 1

1.0.0 PROJECT CONTROL OVERVIEW

The contractor, the project manager, and the crew leader each have management responsibilities for their assigned jobs. For example, the contractor's responsibility begins with obtaining the contract, and it does not end until the client takes ownership of the project. Next, the project manager is generally the person with overall responsibility for coordinating the project. Finally, the crew leader is responsible for coordinating the installation of the work of one or more workers, one or more crews of workers within the company and, on occasion, one or more crews of subcontractors.

This chapter provides the necessary information to effectively and efficiently control a project. It examines estimating, planning and scheduling, resources, and cost control.

1.1.0 Construction Projects

Construction projects are made up of three phases. The first is the development phase, the second is the planning phase, and the third is the construction phase.

1.1.1 Development Phase

A new building project begins when an owner has an idea or concept that requires his or her company or community to construct a new facility or add to an existing facility. The development process is the first stage of planning a new building project. This process involves land research and feasibility studies to ensure that the project has merit. Architects or engineers draw the conceptual drawings that define the project graphically. They then provide the owner with sketches of room layouts or elevations and make suggestions about what construction materials should be used.

During the development phase, an estimate of the proposed project will be established, which anticipates all costs or expenses. Once an estimate has been determined, the financing of the project will be discussed with lending institutions. The architects/engineers and/or the owner will start preliminary reviews with government agencies. These reviews include zoning, building restrictions, landscape requirements, and environmental impact studies.

Also during the development phase, the owner must analyze the project's cost and potential to ensure that its costs do not exceed its market value and that the project provides a good return on investment. If the project passes the "value does not exceed cost" criteria, the architect will proceed to the next phase.

1.1.2 Planning Phase

When the architect/engineer begins to develop the preliminary drawings and specifications, other design professionals such as structural, mechanical, and electrical engineers are brought in. They perform the calculations, make a detailed technical analysis, and check details of the project for accuracy.

The design professionals create drawings and specifications. These drawings and specifications will be used to communicate the necessary information to the contractors, subcontractors, suppliers, and workers that contribute to a project.

During the planning phase, the owners hold many meetings to refine estimates, adjust plans to conform to regulations, and secure a construction loan. If the project is a condominium, an office building, or a shopping center, then a marketing program must be developed. In addition, selling of the project must start before actual construction begins.

Next, contract documents to provide the owner with a complete set of drawings, specifications, and bid documents are produced. After the contract documents are complete, the owner will select the method to obtain contractors. The owner may choose to negotiate with several contractors or select one through competitive bidding. In either case, after the owner selects a contractor and awards the contract, construction begins.

Note that safety must also be considered as part of the planning process. A safety supervisor may walk through the site as part of the pre-bid process.

1.1.3 Construction Phase

A general contractor enlists the help of mechanical, electrical, elevator, and other specialty contractors, called subcontractors, to complete the construction phase. The general contractor will usually perform one or more parts of the construction, and rely on subcontractors for the remainder of the work. The general contractor is responsible for management of all the trades necessary to complete the project.

As construction nears completion, the architect/engineers, owner, and government agencies start their final inspections and acceptance of the project. If the project has been managed by the general contractor, the subcontractors have performed their work, and the architect/engineers have regularly inspected the project to ensure that local codes have been followed, then the inspection procedure can be completed in a timely manner. This results in a satisfied client and a profitable project for all.

On the other hand, if the inspection reveals faulty workmanship, poor design or use of materials, and violation of codes, then the inspection and acceptance will become a lengthy battle and may result in a dissatisfied client and an unprofitable project.

Figure 7 shows the flow of a typical project.

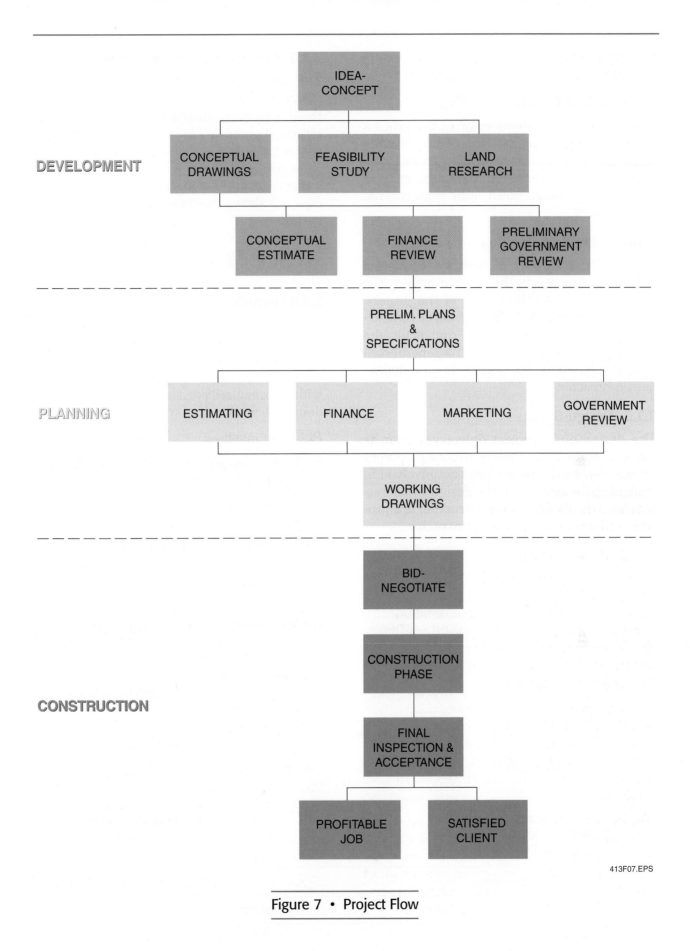

413F07.EPS

Figure 7 • Project Flow

2.0.0 PROJECT DELIVERY SYSTEMS

Project delivery systems refer to the process by which projects are delivered, from development through construction. Project delivery systems focus on three main systems: general contracting, design-build, and construction management.

2.1.0 General Contracting

The traditional project delivery system uses a general contractor. In this type of project, the owner determines the design of the project, then solicits proposals from general contractors. After selecting a general contractor, the owner contracts directly with the general contractor, who builds the project as an independent contractor.

2.2.0 Design-Build

The design-build project delivery system is different from the general contracting delivery system. General contracting is not involved in designing the project. In the design-build system both the design and construction of a project are handled by a single contractor.

2.3.0 Construction Management

The construction management project delivery system uses a construction manager to facilitate the design and construction of a project. Construction managers are very involved in project control; their main concerns are controlling time, cost, and quality of the project. There are various forms of construction management. *Figure 8* shows the three main project delivery systems.

3.0.0 PLANNING

Planning can be defined as determining the method that will be used to carry out the different tasks to complete a project. It involves deciding what needs to be done and coming up with an organized sequence of events or plan for doing the work.

Planning involves:

- Determining the best method for performing the job
- Identifying the responsibilities of each person on the work crew
- Determining the duration of each activity

	GENERAL CONTRACTING	DESIGN-BUILD	CONSTRUCTION MANAGEMENT
OWNER	Designs project (or hires architect)	Hires general contractor	Hires construction management company
GENERAL CONTRACTOR	Builds project (with owner's design)	Designs project, builds project	Builds, may design (hired by construction management company)
CONSTRUCTION MANAGEMENT COMPANY			Hires and manages general contractor and architect

413F08.EPS

Figure 8 • Project Delivery Systems

- Identifying what tools will be needed to complete a job
- Ensuring that the required materials are at the work site when needed
- Making sure that heavy construction equipment is available when required
- Working with other contractors in such a way as to avoid interruptions and delays

3.1.0 Why Plan?

With a plan, a supervisor can direct work efforts efficiently and can use resources such as personnel, materials, tools, equipment, and work methods to their full potential.

Some reasons for planning include:

- To control the job so that it is built on time and within cost
- To lower job costs
- To prepare for unknowns or unexpected occurrences, such as bad weather, and develop alternate plans
- To promote and maintain favorable employee morale

PARTICIPANT EXERCISE F

1. In your own words, define planning, and describe how a job can be done better if it is planned. Give an example.

_____

_____

_____

_____

_____

_____

2. Consider the job that you are working on now to answer the following:

   a. List the material(s) being used.

   _____

   _____

   _____

   _____

   _____

   b. List each member of the crew with whom you work and what each does.

   _____

   _____

   _____

   _____

   _____

   c. List the kinds of equipment being used.

   _____

   _____

   _____

   _____

   _____

   _____

   _____

3. List some suggestions for how your present job could be done better, and describe how you would plan for each of the suggestions.

   _____

   _____

   _____

   _____

   _____

   _____

   _____

   _____

PARTICIPANT EXERCISE F ◆ PROJECT CONTROL

4.0.0 STAGES OF PLANNING

There are various times when planning is done for a construction job. The two most important occur before a project begins, in the pre-construction phase and during the completion of the construction.

4.1.0 Pre-Construction Planning

The pre-construction stage of planning occurs before the start of construction. Except in a fairly small company or for a relatively small job, the supervisor usually does not get directly involved in the pre-construction planning process, but it is important to understand what it involves.

There are two phases of pre-construction planning. The first is when the proposal, bid, or negotiated price for the job is being developed. This is when the estimator, the project manager, and the field supervisors develop a preliminary plan for how the work will be done. This is accomplished by applying experience and knowledge from previous projects. It involves determining what methods, personnel, tools, and equipment will be used and what level of productivity they can achieve.

The second phase occurs after the contractor is awarded the contract. This phase requires a thorough knowledge of any project documents that pertain to the project. During this stage, the actual work methods and resources needed to perform the work are selected. Here, supervisors might get involved, but they must adhere to work methods, production rates, and resources that fit within the estimate prepared before the contract was awarded when planning. If the project requires a method of construction different from what is normal, the supervisor will usually be informed of what method to use.

4.2.0 Construction Planning

During construction, the supervisor is directly involved in planning on a daily basis. This planning consists of selecting methods of completing tasks before beginning them. Effective planning exposes likely difficulties. It enables the supervisor to minimize the unproductive use of personnel and equipment. Effective planning also provides a gauge to measure job progress. Effective supervisors develop what is known as look-ahead plan. These are plans that take into consideration actual circumstances as well as projections two to three weeks into the future. Developing this look-ahead plan helps to ensure that all resources will be available on the project when needed.

All construction jobs consist of several activities. One of the characteristics of an effective supervisor is the ability to reduce each job to its simpler parts and organize a plan for handling each task.

Others establish time and cost limits for the project, and the supervisor's planning must fit within those constraints. Therefore, it is important to understand the following factors that may affect the work:

- Site and local conditions, such as soil types, accessibility, or available staging areas
- Climate conditions that should be anticipated during the project
- Timing of all phases of work
- Types of materials to be installed and their availability
- Equipment and tools required and their availability
- Personnel requirements and availability
- Relationships with the other contractors and their representatives on the job

On a simple job, these items can be handled almost automatically. However, larger or more complex jobs force the supervisor to give these factors more formal consideration and study.

5.0.0 THE PLANNING PROCESS

The planning process consists of the following five steps:

1. Establishing a goal
2. Identifying the work activities that must be completed in order to achieve the goal
3. Determining what tasks must be done to accomplish those activities
4. Communicating responsibilities
5. Following up to see that the goal is achieved

5.1.0 Establishing a Goal

A goal is something that needs to be accomplished. There are two types of goals: goals that are set in our personal lives and goals that are set by others in our jobs.

Personal goals are those goals that a person sets for himself or herself. Examples of personal goals include purchasing a new car, moving to a new area, or taking a vacation with friends or family.

Professional or job-related goals are those goals that the organization or project team establishes. Examples of job-related goals include constructing column forms, placing concrete, or setting a water heater for a school building. Although supervisors can provide input on how a job goal will be accomplished, they rarely, if ever, establish the goal.

5.2.0 Identifying the Work to be Done

The second step in planning is to identify the work to be done to achieve the goal. In other words, it is a series of activities that must be done in a certain sequence. The topic of breaking down a job into activities is covered later in this chapter. At this point, the supervisor should know that, for each activity, one or more objectives must be set.

An objective is a statement of what is desired at a specific time. An objective must:

- Mean the same thing to everyone involved
- Be measurable, so that everyone knows when it has been reached
- Be achievable
- Have everyone's full support

Examples of practical objectives include the following:

- By 4:30 p.m. today my crew will have completed laying two-thirds of the 12" interior block wall.
- By closing time Friday, all electrical panels will be installed to standard.

Notice that both examples meet the first three requirements of an objective. In addition, it is assumed that everyone involved in completing the task is committed to achieving the objective. The advantage in writing objectives for each work activity is that it allows the supervisor to evaluate whether or not the plan and schedules are being followed. In addition, objectives serve as subgoals that are usually under the supervisor's control.

Some construction work activities, such as installing 12-inch-deep footing forms, are done so often that they require little planning. However, other jobs, such as placing a new type of mechanical equipment, require substantial planning. This type of job requires that the supervisor set specific objectives.

Any time supervisors are faced with a new or complex activity, they should take the time to establish objectives that will serve as guides for accomplishing the task at hand. These guides can be used in the current situation, as well as in similar situations in the future.

5.3.0 Determining Tasks

To plan effectively, the supervisor must be able to break a work activity down into the smaller tasks. Large jobs include a greater number of tasks than small ones, but all jobs can be broken down into manageable components.

When breaking down an activity into tasks, the supervisor should make each task identifiable and definable. A task is identifiable when you know the types and amounts of resources it requires. A task is definable if you can assign a specific time to it. For purposes of efficiency, the supervisor's job breakdown should not be too detailed or complex, unless the job has never been done before or must be performed with strictest efficiency.

A suitable breakdown for the work activity, *install 12-inch × 12-inch vinyl tile in a cafeteria*, might be the following:

Step 1	Lay out
Step 2	Clean floor
Step 3	Spread adhesive
Step 4	Lay tile
Step 5	Clean floor
Step 6	Wax floor

The supervisor could create even more detail by breaking down any one of the tasks, such as *lay tile*, into subtasks. In this case, however, that much detail is unnecessary and wastes the supervisor's time and the project's money. However, breaking tasks down further might be necessary in a case where the job is very complex or the analysis of the job needs to be very detailed.

Every work activity can be divided into three general parts:

- Preparing
- Performing
- Cleaning up

One of the most frequent mistakes made in the planning process is forgetting to prepare and to clean up. The supervisor must be certain that preparation and cleanup are not overlooked with every job breakdown.

After identifying the various activities that make up the job and developing an objective for each activity, the supervisor must determine what resources the job requires. Resources include labor, equipment, materials, and tools. In most jobs, these resources are identified in the job estimate, and the supervisor must only make sure that they are on the site when needed. On other jobs, however, the supervisor may have to determine, in part, what is required as well as arrange for timely delivery.

5.4.0 Communicating Responsibilities

A supervisor is unable to complete all of the activities within a job independently. Other people must be relied on to get everything done. Therefore, construction jobs have a crew of people with various experiences and skill levels to assist in the work. The supervisor's job is to draw from this expertise to get the job done well and in a timely manner.

Once the various activities that make up the job have been determined, the supervisor must identify the person or persons who will be responsible for completing each activity. This requires that the supervisor be aware of the skills and abilities of the people on the crew. Then, the supervisor must put this knowledge to work in matching the crew's skills and abilities to specific tasks that must be accomplished to complete the job.

Upon matching crew members to specific activities, the supervisor must then communicate the assignments to the crew. Communication of responsibilities is generally handled verbally; the supervisor often talks directly to the person to which the activity has been assigned. There may be times when work is assigned indirectly through written instructions or verbally through someone other than the supervisor. Either way, the crew members should know what it is they are responsible for accomplishing on the job.

5.5.0 Follow-Up

Once the activities have been delegated to the appropriate crew members, the supervisor must follow up to make sure that they are completed effectively and efficiently. Follow-up involves being present on the job site to make sure that all of the resources are available to complete the work; ensuring that the crew members are working on their assigned activities; answering any questions; or helping to resolve any problems that occur while the work is being done. In short, follow-up means that the supervisor is aware of what's going on at the job site and is doing whatever is necessary to make sure that the work is completed as scheduled.

Figure 9 reviews the planning steps.

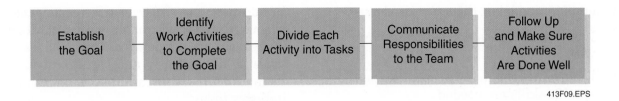

| Establish the Goal | Identify Work Activities to Complete the Goal | Divide Each Activity into Tasks | Communicate Responsibilities to the Team | Follow Up and Make Sure Activities Are Done Well |

413F09.EPS

Figure 9 • Steps to Effective Planning

6.0.0 PLANNING RESOURCES

Once a job has been broken down into its tasks or activities, the next step is to assign the various resources needed to perform them.

6.1.0 Planning Materials

The materials required for the job are identified during pre-construction planning and are listed on the job estimate. The materials are ordered from suppliers who have previously provided quality materials on schedule and within estimated cost.

The supervisor is usually not involved in the planning and selection of materials, since this is done in the pre-construction phase. However, the supervisor is involved in planning materials for tasks such as job-built formwork and scaffolding. Also, the supervisor occasionally may run out of a specific material, such as fasteners, and need to order more. In such cases, the next higher supervisor should be consulted, since most companies have specific purchasing policies and procedures.

The supervisor does, however, have a major role to play in the receipt, storage, and control of the materials after they reach the job site.

6.2.0 Planning Equipment

Planning the use of construction equipment involves identifying the types of equipment needed, the tasks each type must perform, and the time each piece is needed. Much of this is planned during the pre-construction phase; however, it is up to the supervisor to work with the home office to make certain that the equipment reaches the job site on time.

The supervisor should also coordinate equipment with other contractors on the job. Sharing equipment can save time and money and avoid duplication of effort. In addition, if the equipment breaks down, the supervisor must know whom to contact to resolve the problem. The supervisor should also designate some time for routine equipment maintenance to avoid equipment failure.

Finally, when various pieces of equipment will be working in a relatively small area, or one piece will work in conjunction with another (such as a loader and a dump truck, or a bulldozer and a scraper), the supervisor must coordinate the use of the equipment. An alternate plan must be ready in case one piece of equipment breaks down, so that the other equipment does not sit idle. This planning should be done in conjunction with the home office or the supervisor's immediate superior.

6.3.0 Planning Tools

A supervisor is responsible for planning what tools will be used on the job. This includes:

- Determining the tools required
- Informing the workers who will provide the tools (company or worker)
- Instructing the workers how to use the tools safely and effectively
- Determining what controls to establish for tools

6.4.0 Planning Labor

All tasks require some sort of labor because the supervisor cannot complete all of the work alone.

When planning labor, the supervisor must:

- Identify the skills needed to perform the work.
- Determine how many people having those specific skills are needed.
- Decide who will actually be on the crew.

In many companies, the project manager or job superintendent determines the size and makeup of the crew. Then, the supervisor is expected to accomplish the goals and objectives with the crew provided. Even though the supervisor may not be involved in staffing the crew, the supervisor will be responsible for training the crew members to ensure that they have the skills necessary to do the job.

In addition, the supervisor is responsible for keeping the crew adequately staffed at all times so that jobs are not delayed. This involves dealing with absenteeism and turnover, two common problems that affect the construction industry today.

SECTION 7

7.0.0 WAYS TO PLAN

It is recommended that supervisors carry a small note pad to be used for planning and note taking. That way, thoughts about the project can be recorded as they occur, and pertinent details will not be forgotten. The supervisor may also choose to use a planning form, such as the one illustrated in *Figure 10*.

DAILY WORK PLAN

"PLAN YOUR WORK AND WORK YOUR PLAN = EFFICIENCY"

Plan of _____   Date _____

PRIORITY	DESCRIPTION	✓ When Completed ✗ Carried Forward

413F10.EPS

Figure 10 • Planning Form

As the job is being built, the supervisor should refer back to these plans and notes to see that the tasks are being done in sequence and according to plan. This task is referred to as analyzing the job. Experience shows that jobs that are not built according to work plans usually end up costing more and taking more time; therefore, it is important that supervisors refer back to the plans periodically.

The supervisor is involved with many activities on a day-to-day basis. As a result, it is easy to forget important events if they are not documented. To help keep track of events such as job changes, interferences, and visits, the supervisor should keep a job diary.

A job diary is a small book in which the supervisor records activities or events that take place on the job site which may be important at a later date. When recording in a job diary, the supervisor should make sure that the information is accurate, factual, complete, consistent, organized, and up-to-date.

The supervisor should also be sure to follow company policy in determining which events should be recorded. However, if there is a doubt about what to include, it is better to have too much information than too little.

Figure 11 illustrates a sample page from a job diary.

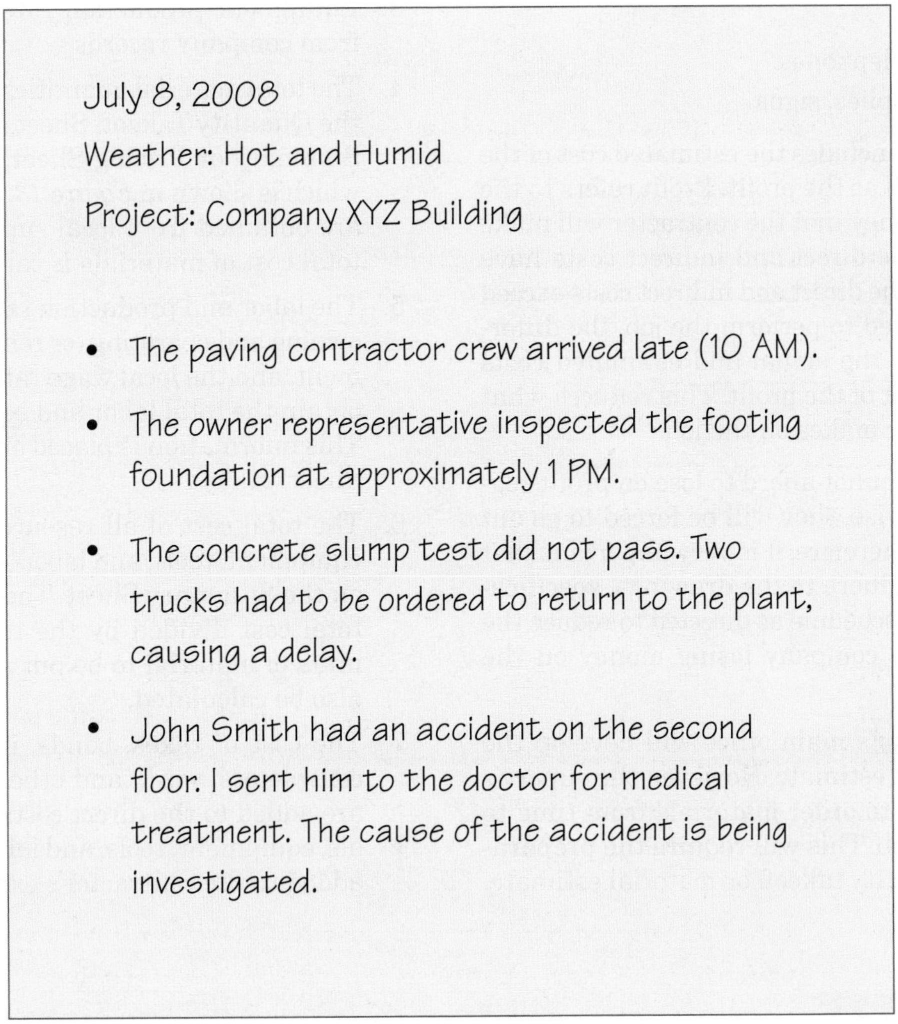

July 8, 2008

Weather: Hot and Humid

Project: Company XYZ Building

- The paving contractor crew arrived late (10 AM).

- The owner representative inspected the footing foundation at approximately 1 PM.

- The concrete slump test did not pass. Two trucks had to be ordered to return to the plant, causing a delay.

- John Smith had an accident on the second floor. I sent him to the doctor for medical treatment. The cause of the accident is being investigated.

413F11.EPS

Figure 11 • Sample Page from a Job Diary

8.0.0 ESTIMATING

Before a project is built, an estimate of the cost needs to be prepared. An estimate is the process of calculating the quantities of costs needed to build the job. There are two types of costs to consider, direct and indirect costs.

Direct costs include:

- Materials
- Labor
- Tools
- Equipment

Indirect costs refer to overhead items such as:

- Office rent
- Utilities
- Job site telephones
- Office supplies, signs

The bid price includes the estimated cost of the project as well as the profit. Profit refers to the amount of money that the contractor will make after all of the direct and indirect costs have been paid. If the direct and indirect costs exceed those estimated to perform the job, the difference between the actual and estimated costs must come out of the profit. This reduces what the contractor makes on the job.

Contractors cannot afford to lose on profit regularly; otherwise, they will be forced to go out of business. Therefore, it is very important that supervisors adhere to the drawings, specifications, and job schedule as directed to reduce the chance of the company losing money on the project.

The contractor's main office will develop the actual project estimate. However, the supervisor will have to order materials from time to time on the job. This will require the preparation of a quantity takeoff or material estimate.

8.1.0 The Estimating Process

A complete estimate is developed as follows:

1. Using the drawings and specifications, an estimator records the quantity of the materials needed to construct the job. This is called a quantity takeoff. The information is placed on a Quantity Takeoff Sheet (sometimes called a Work Sheet), such as that shown in *Figure 12*.

2. The second step is to calculate the amount of time it takes to perform an amount of work. These are known as production rates. Most companies keep records of these rates for the type and size of the jobs that they perform. The estimating department in a company has these figures.

3. Equipment production rates are obtained from company records.

4. The total material quantities are taken from the Quantity Takeoff Sheet and placed on a Summary or Pricing Sheet, an example of which is shown in *Figure 13*. Material prices are obtained from local suppliers, and the total cost of materials is calculated.

5. The labor and production rates, the costs of owning and operating (or renting) the equipment, and the local wage rates are added to obtain the total labor and equipment costs. This information is placed on the Summary Sheet.

6. The total cost of all resources—materials, equipment, tools, and labor—is then totaled on the Summary Sheet. The unit cost—the total cost divided by the total number of units of material to be put into place—can also be calculated.

7. The cost of taxes, bonds, insurance, subcontractors' work, and other indirect costs are added to the direct costs of the materials, equipment, tools, and labor. This total is added to the contractor's expected profit.

WORK SHEET

Takeoff By: _____

Checked By: _____

DATE _____

SHEET _____ of _____

PROJECT _____

ARCHITECT _____

PAGE #

| REF. | DESCRIPTION | DIMENSIONS | | | | EXTENSION | QUANTITY | UNIT | TOTAL | | REMARKS |
		NO.	LENGTH	WIDTH	HEIGHT				QUANTITY	UNIT	

413F12.EPS

Figure 12 • Quantity Takeoff Sheet

SUMMARY SHEET

PAGE #

By:

DATE _____

SHEET _____ of _____

PROJECT _____

WORK ORDER # _____

TITLE: _____

DESCRIPTION	QUANTITY		MATERIAL COST		LABOR MAN HOURS FACTORS					LABOR COST		ITEM COST	
	TOTAL	UT	PER UNIT	TOTAL	CRAFT	PR UNIT	TOTAL	RATE	COST PR	PER	TOTAL	TOTAL	PER UNIT
	MATERIAL								LABOR		TOTAL		

Figure 13 • Summary Sheet

413F13.EPS

8.1.1 Estimating Material Quantities

The crew leader may be required to estimate quantities of materials.

To estimate the amount of a certain type of material required to perform a job, a set of construction drawings and specifications will be needed. The appropriate section of the technical specifications and page(s) of drawings should be carefully reviewed to determine the types and quantities of materials required. The quantities are then placed on the Work Sheet. For example, the specification section on finished carpentry should be reviewed along with the appropriate pages of drawings before taking off the linear feet of door and window trim.

If an estimate is required because not enough materials were ordered to complete the job, the estimator must also determine how much more work is necessary. Once this is known, the supervisor can then determine specifically how many more materials are needed. The construction drawings will also be used in this process.

For example, assume you are the supervisor of a crew building footing formwork for the construction shown in *Figure 14*.

You have used all of the materials provided for the job, yet you have not completed it. You study the drawings and see that the formwork is comprised of two side forms, each 12" high. The total length of footing for the entire project is 115'-0". You have completed 88'-0" to date; therefore, you have 27'-0" remaining (115'-0" – 88'-0" = 27'-0"). Your job is to prepare an estimate of materials that you will need to complete the job. In this case, only the side forms will be estimated (the miscellaneous materials will not be considered here).

*   Footing length to complete: 27'-0"
*   Footing height: 1'-0"

Refer to the Work Sheet in *Figure 15* for a final tabulation of the side forms needed to complete the job.

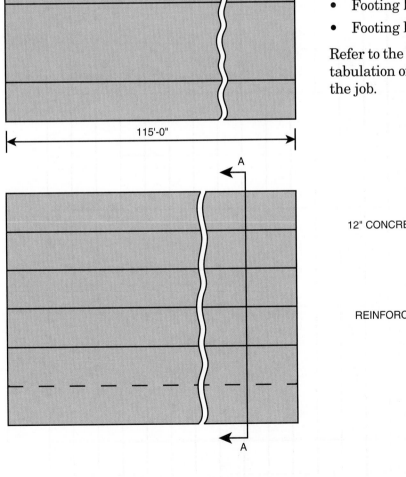

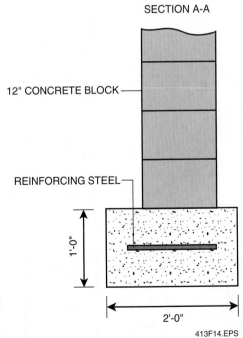

115'-0"

SECTION A-A

12" CONCRETE BLOCK

REINFORCING STEEL

1'-0"

2'-0"

413F14.EPS

Figure 14 • Footing Formwork

WORK SHEET

Takeoff By: RWH
Checked By:

DATE 2/1/15
SHEET 01 of 01

PROJECT _Sam's Diner_
ARCHITECT _654b_

REF.	DESCRIPTION	NO	DIMENSIONS			EXTENSION	QUANTITY	UNIT	TOTAL		REMARKS
			LENGTH	WIDTH	HEIGHT				QUANTITY	UNIT	
	Footing Side Forms	2	27'0"		1'0"	2x27x1	54	SF	54	SF	

Figure 15 • Work Sheet with Entries

413F15.EPS

1. Using the same footing as described in the example above, calculate the quantity (square feet) of formwork needed to finish 203 linear feet of the footing. Place this information directly on the Work Sheet on the previous page.

2. You are the supervisor of a carpentry crew whose task is to side a warehouse with plywood sheathing. The wall height is 16 feet, and there is a total of 480 linear feet of wall to side. You have done 360 linear feet of wall and have run out of materials. Calculate how many more feet of plywood you will need to complete the job. If you are using 4' × 8' plywood panels, how many will you need to order, assuming no waste? Write your estimate on the Work Sheet on the previous page.

9.0.0 SCHEDULING

Planning and scheduling are closely related and are both very important to a successful construction job. Planning involves determining which activities must be completed and how they should be accomplished. Scheduling involves establishing start and finish times or dates for each activity.

When scheduling a job, a construction schedule is generally completed. A construction schedule for a project shows:

- Operations listed in sequential order
- Units of construction
- Duration of activities
- Estimated date to start and complete each activity
- Quantity of materials to be installed

There are two major types of construction schedules used today. The first is the bar chart, and the second is the network schedule, which is sometimes called the critical path method (CPM) or precedence diagram.

9.1.0 The Scheduling Process

The following is a brief summary of the steps required to develop a schedule.

Step 1 Make a list of all of the activities that will be performed to build the job, including individual work activities and special tasks such as inspections or the delivery of materials.

At this point, the supervisor should just be concerned with generating a list, not with determining how the activities will be accomplished, who will perform them, how long they will take, or in what sequence they will be completed.

Step 2 Use the list of activities created in Step 1 to reorganize the work activities into a logical sequence.

When doing this, the supervisor should keep in mind that certain steps cannot happen until others have been completed. For example, footing must be excavated before concrete can be placed.

Step 3 Assign a duration or length of time that it will take to complete each activity and determine the start time for each.

Each activity will then be placed into a final format, known as a schedule. This step is important because it helps the supervisor to ensure that the activities are being completed on schedule.

9.2.0 Bar Charts

Bar charts can be used for both short-term and long-term jobs. However, they are especially helpful for jobs of short duration.

Bar charts provide management with the following:

- A visual concept of the overall time required to complete the job through the use of a logical method rather than a calculated guess
- A means to review each part of the job
- Coordination requirements between crafts
- Alternative methods of performing the work

A bar chart can be used as a control device to see whether or not the job is on schedule. If the job is not on schedule, immediate action can be taken in the office and the field to correct the problem and ensure that the activity is completed on schedule.

A bar chart is illustrated in *Figure 16*.

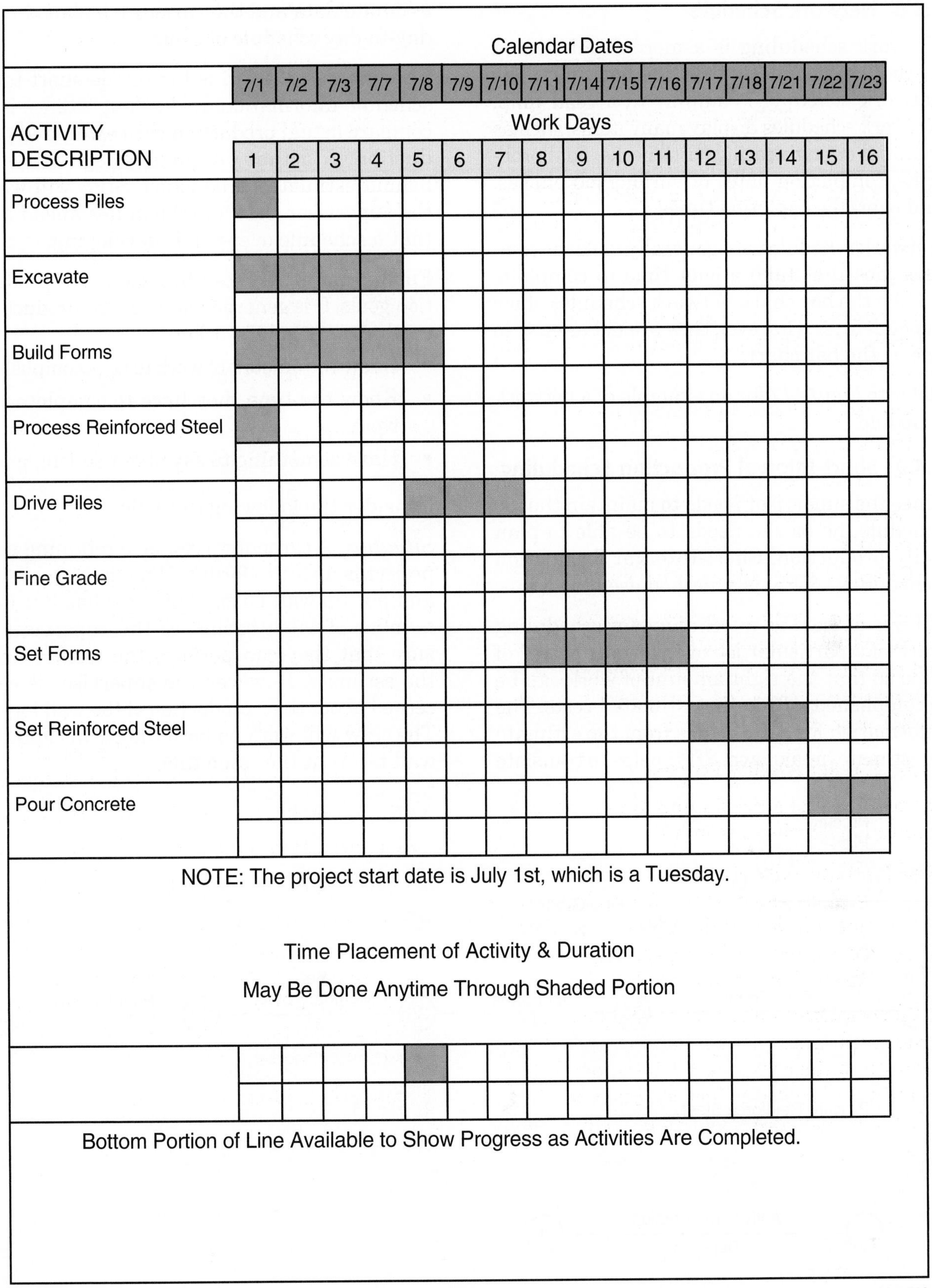

Calendar Dates

ACTIVITY DESCRIPTION	7/1	7/2	7/3	7/7	7/8	7/9	7/10	7/11	7/14	7/15	7/16	7/17	7/18	7/21	7/22	7/23
Work Days	1	2	3	4	5	6	7	8	9	10	11	12	13	14	15	16
Process Piles	▓															
Excavate	▓	▓	▓	▓												
Build Forms	▓	▓	▓	▓	▓											
Process Reinforced Steel	▓															
Drive Piles					▓	▓	▓									
Fine Grade								▓	▓							
Set Forms								▓	▓	▓	▓					
Set Reinforced Steel												▓	▓	▓		
Pour Concrete															▓	▓

NOTE: The project start date is July 1st, which is a Tuesday.

Time Placement of Activity & Duration

May Be Done Anytime Through Shaded Portion

				▓											

Bottom Portion of Line Available to Show Progress as Activities Are Completed.

413F16.EPS

Figure 16 • Bar Chart

9.3.0 Network Schedule

Network scheduling is a more sophisticated method of scheduling than the bar chart. The network schedule is similar to a road map. Network schedules display many items, such as material or engineering bottlenecks, field problems, completion dates of various job phases, and overall completion times.

Network schedules are generally used for complex jobs that take a long time to complete. Unlike the bar chart, network schedules show dependencies. However, they are not as easy to use as the bar chart.

Refer to *Figure 17* for an example of a network schedule.

9.4.0 Short-Interval Production Scheduling

Since the supervisor needs to maintain the job schedule, he or she needs to be able to plan daily production. Short-Interval Production Scheduling (SIPS) is a method used to do this.

SIPS is defined as a technique for scheduling work of a particular job over a short period of time so that the right amount of work can be accomplished within the estimated costs. The information for SIPS comes from the estimate or estimate breakdown. SIPS helps to translate estimate data and the various job plans into a day-to-day schedule of events.

The purpose of SIPS is to provide short-term schedule information. Its key importance is to compare actual production with estimated production. If actual production begins to slip behind estimated production, SIPS will warn the supervisor that a problem lies ahead and that a schedule over-run is developing.

Furthermore, SIPS can be used to set production goals. It is generally agreed that production can be improved if workers:

- Know the amount of work to be accomplished
- Know the time they have to complete the work
- Have something to say about setting goals

Consider the following example:

Situation: A carpentry crew on a retaining wall project is about to form and pour catch basins and put up wall forms. The crew has put in a number of catch basins, so the supervisor is sure that they can perform the work within the estimate. However, the supervisor is concerned about their production of the wall forms. The crew will work on both the basins and the wall forms at the same time.

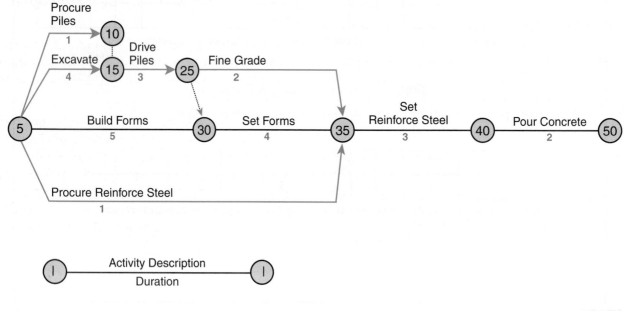

413F17.EPS

Figure 17 • Network Schedule

1. The supervisor notices the following in the estimate or estimate breakdown:

   a. Production factor for wall forms = 16 work-hours per 100 square feet

   b. Work to be done by measurement = 800 square feet

   c. Total time = 128 work-hours ($800 \times 16 \div 100$)

2. The carpenter crew consists of the following:

   a. One carpenter supervisor

   b. Four carpenters

   c. One laborer

3. The supervisor determines the goal for the job should be set at 128 work-hours (from *Step 1*).

4. If the crew remains the same (six workers), the work should be completed in about 21 crew-hours (128 work hours $\div$ 6 workers = 21.33 crew-hours).

5. The supervisor then discusses the production goal (completing 800 square feet in 21 crew-hours) with the crew and encourages them to work together to meet the goal of getting the forms erected within the estimated time.

Therefore, SIPS is needed to translate production into work-hours or crew-hours and to schedule work so that the crew can accomplish it within the estimate. Also, the establishment of a production target or goal provides the motivation to produce more than the estimate requires.

The supervisor's progress will be based on productivity, which is a measure of how well the crew maintains the schedule. The supervisor must be able to read and interpret the job schedule. On some jobs, the beginning and expected end date for each activity along with the expected crew or worker's production rate will be provided on the form. The supervisor can use this information to plan work more effectively, to realistically set goals for the crew, and to measure whether or not they were accomplished within the scheduled time.

It is important, therefore, that all job resources are on the job, in location, and ready to begin the job on time and stay on schedule. Before starting an activity, the supervisor must know or determine what materials, tools, equipment, and labor are needed to complete the job. Next he or she must know or determine when the various resources are needed. Finally, the supervisor must follow up to ensure that they are available on the job site.

The supervisor should not wait until the day of or the day before a job starts to check on the availability of needed resources. Rather, this should be done at least three to four working days before the start of the job. It should be done even earlier for larger jobs. This advance preparation will help prevent any events that could potentially delay the job or cause it to fall behind schedule.

9.5.0 Updating a Schedule

No matter what type of schedule is used, it must be kept up-to-date to be useful to the supervisor. Schedules that are inaccurate offer no value.

The person responsible for scheduling in the office handles the updates. This person uses information gathered from job field reports to update the schedule.

The supervisor is usually not directly involved in updating schedules. However, he or she may be responsible for completing field or progress reports used by the company. It is critical that the supervisor fill out any required forms or reports completely and accurately so that the schedule can be updated with the correct information.

1. Which of the following statements is true when comparing bar charts to network schedules?

   a. The bar chart is particularly good for short-term jobs, but is harder to use than the network schedule.

   b. The bar chart is good for long-term jobs, but it is harder to use than the network schedule.

   c. The network schedule is more sophisticated than the bar chart, but does not indicate dependencies like the bar chart.

   d. The network schedule is more sophisticated because it can indicate dependencies, but it is more difficult to use.

2. Short-interval scheduling can help crew leaders in the field by _____.

   a. offering a comparison of actual production to estimated production

   b. providing mainly short-term, but also long-term, schedule information

   c. determining the equipment and materials necessary to complete a task

   d. providing the information necessary for coming up with an estimate or an estimate breakdown

3. One way for supervisors to get a job back on schedule or to increase productivity if things fall behind is to _____.

   a. refrain from spending money on hiring more crew members

   b. make sure the necessary tools, materials, and equipment are on the job site a few hours in advance

   c. communicate to crew members the exact amount of time available to complete a task

   d. set goals for crew members instead of directly involving them in the process

4. Which of the following describes how schedules are typically kept up-to-date?

   a. The supervisor handles schedule updates.

   b. Office personnel update schedules using field reports.

   c. Information from job field reports or progress reports is not used for schedule updates.

   d. Information from bar charts and network schedules is used for schedule updates.

10.0.0 COST AWARENESS AND CONTROL

Being aware of costs and controlling them is the responsibility of every employee on the job, and it is the supervisor's job to ensure that employees uphold this responsibility. Control refers to the comparison of estimated performance against actual performance and following up with any needed corrective action. Supervisors who use cost-control practices are more valuable to the company than those who do not.

On a typical construction site, hundreds of activities are going on at the same time. The more going on, the harder it is to control the activities involved or constructing the building. It is therefore very important that the construction crew leader be constantly aware of the costs of a project and efficiently and effectively control the various resources used on the job.

When resources are not controlled, the cost of the job increases. For example, a plumbing crew of four people is installing soil pipe and does not have enough wyes. Three crew members wait while one crew member goes to the supply house for a part that costs only a few dollars. It takes the crew member an hour to get the part, so four hours of productive work were lost. In addition, the travel cost for retrieving the supplies must be added.

10.1.0 Categories of Costs

There are two general categories of costs associated with every job, estimated costs and actual costs. Although different, they are related in determining whether a job was built efficiently.

Estimated costs, as described in Section 8 of this chapter, are calculated before construction begins. These are the costs that are required to build the job. The estimated cost is the contract cost agreed between the contractor and the client or the general contractor and a specialty contractor to do the work.

The estimated project cost is a sum of all individual jobs and estimated costs. Estimates for each job serve as measures of what each should cost when completed. It allows the contractor to measure whether or not the job is being built for the estimated cost. It is similar to using a measuring tape to check a measurement of a piece of work in the field. If it was installed correctly, the measurement should be the same as that indicated on the drawings. If it was not, the work will have to be corrected to achieve the proper measurement.

The same is true of costs; if the job was built for the estimated cost, then the contractor will receive the estimated profit. The company will have been successful and will grow, providing job security for the supervisor. However, if the job is not built for the estimated cost, the contractor will not make the expected profit. If loss continues, the company may have to lay off employees.

Actual costs are the costs for which the contractor is responsible, such as the cost of the delivered materials and the payroll for the employees. Again, it is important that the supervisor remember that a major responsibility is to build the job so that actual costs of doing the work are below or equal to the estimated costs. This should be accomplished without sacrificing the quality required in the contract drawings or specifications.

10.2.0 Field Reporting System

The total estimated cost comes from the job estimates, but the actual cost of doing the work is obtained from an effective field reporting system. A field reporting system is made up of a series of forms, which are completed by different people, including the crew leader. Each company has its own forms and methods for obtaining information. The new supervisor needs to find out what his or her company's methods are. The information and the process of how they are used are described below.

First, the number of hours each person worked on each task must be known. This information comes from daily time cards. Once the accounting department knows how many hours each employee worked on each activity, it can calculate the total cost of the labor by multiplying the number of hours worked by the wage rate for each worker. The cost for the labor to do each task will be calculated as the job progresses, and it will be compared with the estimated cost. This comparison will also be done at the end of the job.

When material is put in place, a designated person in the company will measure the quantities from time to time and send this information to the home office and possibly the supervisor. Knowing this information, along with the actual cost of the material and the amount of hours it took the various workers to install it, a comparison will be made to determine if the actual cost of installing a specific material is below, above, or equal to the estimate. Obviously, if the cost is greater than the estimate, management and the crew leader will have to take action to reduce the cost to get it below or equal to the estimate.

A similar process is used to determine if the costs to operate equipment or the production rate are above, below, or equal to the estimated cost and production. Forms are available for obtaining this information.

If this comparison process is to be of use, the information obtained from field personnel must be correct. It is important that the supervisor be accurate in reporting. As noted above, each company does this activity differently. The supervisor is responsible for finding and carrying out his or her role in the field reporting system.

10.3.0 Supervisor's Role in Cost Control

The supervisor is the construction representative in the field, where the work takes place. Therefore, the supervisor has a great deal to do with determining job costs.

If the actual costs are at or below the estimated costs, the job is progressing as planned and scheduled, and the company will realize the expected profit. However, if the actual costs exceed the estimated costs, one or more problems may result in the company losing its expected profit and maybe more. No company can remain in business if it continually loses money.

If losses are occurring, the supervisor will be called on to get the costs back in line. How can the supervisor do this? To answer this question, one needs to know some of the major reasons why the costs are running higher than the calculated estimate. Once the reason or reasons are known, the next step is for the supervisor to correct the problem.

Noted below are some of the causes why actual costs exceed estimated costs and suggestions on what the supervisor can do to bring the costs back in line. Before starting any action, however, the supervisor should check with his or her superior to see that the action proposed is acceptable and within the company's policies and procedures.

1. *Cause:* Late delivery of materials, tools, and/or equipment

   Corrective Action: Plan ahead to ensure that job resources will be available when needed

2. *Cause:* Inclement weather

   Corrective Action: Have alternate plans ready

3. *Cause:* Unmotivated workers

   Corrective Action: Implement effective leadership skills

4. *Cause:* Accidents

   Corrective Action: Implement effective safety program

The supervisor should note that there are many other methods to get the job done on time if it gets off schedule. Examples include working overtime or increasing the size of the crew. However, these examples will increase the cost of the job, so they should not be done without the approval of the project manager.

1. The concept of cost awareness and control means _____.

   a. the supervisor must hire a person in charge of making a monthly budget

   b. the supervisor is responsible for cost reduction, whereas crew members should focus on the actual construction of the project

   c. estimated performance is compared with actual performance

   d. the supervisor must make sure the actual costs of the job are more than the estimated costs

2. It is important that the crew leader be aware of the costs of doing a project in order to prevent layoffs.

   a. True

   b. False

3. Crew leaders can help play a role in controlling job costs by _____.

   a. implementing and enforcing safety policies

   b. dismissing unmotivated employees

   c. stopping construction work during bad weather

   d. completing the project at the actual cost that the contractor and client agree upon in the beginning

4. Job costs can be controlled by _____.

   a. giving employees the full responsibility of controlling resources

   b. making sure actual costs at least match estimated costs

   c. using the field reporting system to calculate the labor costs associated with each task

   d. using the field reporting system to determine the estimated costs

11.0.0 RESOURCE CONTROL

The supervisor's job is to ensure that all jobs are built safely according to the drawings and specifications, within the estimate and schedule. To accomplish this, the supervisor must closely control how resources of materials, equipment, tools, and labor are used. Waste must be minimized whenever possible.

11.1.0 Control

Control involves measuring performance and correcting deviations from plans to accomplish objectives. Control anticipates deviation from plans and takes measures to prevent its occurrence in the future.

An effective control process can be broken down into the following steps:

Step 1 Establishing standards and dividing them into measurable units.

For example, a standard can be created using experience gained on a typical job, where 2,000 LF of 1¼" copper water pipe was installed in five days. Dividing 2,000 by 5 gives 400. Thus, the average installation rate in this case for 1¼" copper water pipe was 400 LF/day.

Step 2 Measuring performance against a standard.

On another job, the length of the same pipe placed during an average day was only 300 LF. Thus, this average production of 300 LF/day did not meet the average standard of 400 LF/day established above.

Step 3 Correcting operations to assure that the standard is met.

In *Step 2* above, if the plan called for the job to be completed in five days, the supervisor would have to take action to ensure that this happens. If 300 LF/day is the actual average daily rate, it will have to be increased by 100 LF/day to meet the standard.

11.2.0 Materials Control

The supervisor's responsibility in materials control will depend on the policies and procedures of the company. In general, the supervisor is responsible for ensuring on-time delivery, preventing waste, controlling delivery storage, and preventing theft of materials.

11.2.1 Ensuring On-Time Delivery

The supervisor must make sure that the materials for each day's work are on the job site when needed, not waiting until the morning that the materials are needed to obtain them. Rather, the supervisor should use the one-plus rule. This means that one week before needing the materials, the supervisor should know where they are, and they must be on the job site one day before they are needed.

The one-plus rule is just a guide. The exact amount of time may need to be adjusted depending upon the situation. To do this effectively, the supervisor must be familiar with the drawings and the activities to be performed. He or she can then determine how many and what types of materials are needed.

If other people are responsible for providing the materials for a job, the supervisor must follow up to make sure that the materials will be available when they are needed. Otherwise, delays will occur as crew members stand around waiting for the materials to be delivered.

11.2.2 Preventing Waste

Waste in construction can add up to loss of critical and costly materials and may result in delaying the job. Because of this, waste has to be prevented. The supervisor must ensure that every crew member knows how to properly use the materials in the most effective manner. The supervisor should monitor the crew to make certain that it is not wasting materials.

An example of materials waste is a carpenter who saws off a piece of lumber from a fresh piece, when the size needed could have been found in the scrap pile. Another example of waste involves installing a fixture or copper pipe incorrectly. The time spent installing the item incorrectly is wasted because the task will need to be redone. In addition, the materials may be lost if the fixture or copper pipe was damaged while being installed incorrectly.

11.2.3 Verifying Material Delivery

In some companies, a supervisor may be responsible for the receipt of materials delivered to the work site. When this happens, the supervisor should require a copy of the shipping ticket and check each item on the shipping ticket against the actual materials to see that the correct amounts were received.

In addition, the supervisor should check the condition of the materials to verify that nothing is defective before signing the invoice. This can be difficult and time consuming because it means that cartons must be opened and their contents must be examined. However, it is necessary because a signed invoice indicates that all of the materials were received unharmed. If the supervisor discovers damaged materials without checking prior to signing the invoice, he or she will be unable to prove that the materials came to the site in that condition.

Once the shipping ticket is checked and signed, the supervisor should give the original or a copy to the superintendent or project manager. The shipping ticket will then be filed for future reference because it serves as the only record the company has to check bills received from the supply house.

11.2.4 Controlling Delivery Storage

Another very important element of materials control is related to where the materials will be stored on the job site. There are two factors in determining the appropriate storage location. The first is convenience. If possible, the materials should be stored near where they will be used. The time and effort saved by not having to carry the materials long distances will greatly reduce the installation costs.

Next, the materials must be stored in a secure area where they will not be damaged. It is important that the storage area suit the materials being stored. For instance, materials that are sensitive to temperature such as chemicals or paints should not be stored in areas that are too hot or too cold. Otherwise, waste will occur.

11.2.5 Preventing Theft and Vandalism

Theft and vandalism of construction materials increase costs because these materials are needed to complete the job; therefore, they must be replaced in order to do so. The time lost because the needed materials are missing adds significantly to the cost. In addition, the insurance that the contractor purchases will increase as the theft and vandalism rate grows.

The best way to avoid theft and vandalism is not to expose anything so that it can be stolen. If all materials delivered to the job site have not been installed by the end of the workday, they should be stored along with any tools in a secure location such as in a construction trailer or truck on the job site. If the job site is fenced or the building can be locked to prevent unauthorized people from entering the premises, the materials can be stored within.

11.3.0 Equipment Control

The supervisor will not be responsible for long-term equipment control. However, the equipment required for a specific job may be the supervisor's responsibility. The first step in determining the kind and amount of equipment needed is to identify the work that will require the machinery or equipment to be transported from the shop or rental yard. Regardless of which worker uses the equipment, it is the supervisor who is responsible for informing the shop where it is being used and seeing that it is returned to the shop upon completion of the task.

Although this may sound like a fairly simple task, it is not. It is common for equipment to lie idle at a job site because the job has not been properly planned and the equipment arrived early. An example is when the wire-pulling equipment and materials arrive at a job site before the conduit has been installed. The result is that this equipment and materials could be used somewhere else instead of awaiting the installation of the conduit. In addition, it is possible that this unused equipment could be damaged, lost, or stolen.

To control equipment usage, the supervisor must ensure that the equipment is operated in accordance with its design and that it is being used within time and cost guidelines. In addition, it is important that equipment be maintained and repaired as indicated by the Preventive Maintenance Schedule. Delaying the maintenance and repairs will result in more costly equipment operation in the long run.

The supervisor is responsible for the proper operation of all other equipment resources, including cars and trucks. Reckless or unsafe operations of vehicles will likely result in damaged equipment and a delayed or unproductive job. This, in turn, could affect the supervisor's job security.

The supervisor should also ensure that all equipment is secured at the close of each day's work in an effort to prevent theft. If the equipment is still being used for the job, the supervisor should ensure that it is locked in a safe place; otherwise, it should be returned to the shop.

11.4.0 Tools Control

Among construction companies, various policies govern who provides tools to employees. Some companies provide all the tools, while others furnish only the larger power tools. The crew leader should find out about and enforce any company policies related to tools.

Tools control is a twofold process. First, the supervisor must control the issue, use, and maintenance of all tools provided by the company. Next, the supervisor must control how the tools are being used to do the job. This refers to tools that are issued by the company as well as tools that belong to the workers.

Using the proper tools correctly saves time and energy. In addition, proper tool usage reduces the chance of damage to the tool being used. Proper usage also reduces injury to the worker using the tool or to nearby workers.

Tools must be adequately maintained and properly stored. Making sure that tools are cleaned, dried, and oiled prevents rust and ensures that the tools are in the proper working order.

In the event that tools are damaged, it is essential that they be repaired or replaced promptly. Otherwise, an accident or injury could result.

Company-issued tools should be taken care of as if they are the property of the user. Workers should not abuse tools simply because they are not their own.

One of the major causes of time lost on a job is the time spent searching for a tool. To prevent this from occurring, a storage location for company-issued tools and equipment should be established. The supervisor should make sure that all company-issued tools and equipment are returned to this designated location after use. Similarly, workers should make sure that their personal toolboxes are organized so that they can readily find the appropriate tools. In addition, they should return their tools to their toolboxes when they are finished using them.

Studies have shown that the key to an effective control system for tools is limiting the following:

- The people allowed access to stored tools
- The people held responsible for tools
- The ways in which a tool can be returned to storage

11.5.0 Labor Control

The hardest aspect of any job is to control labor. Improving labor productivity results in greater savings and lower construction costs. Many studies have shown that construction personnel are only 35 to 40 percent efficient; that is, in one hour of a typical day, only 35 to 40 percent of the time is spent doing the assigned task. The remaining 60 to 65 percent of the time is devoted to financially unproductive or indirect tasks.

Time and motivation studies indicate that a typical worker's day consists of the following:

- Receiving instructions
- Reaching for materials, tools, and equipment
- Searching for materials, tools, and equipment
- Walking with materials, tools, and equipment
- Doing the work of his or her trade
- Working on assignments other than those within his or her trade
- Waiting for other workers to complete their tasks
- Waiting for materials, tools, and equipment
- Walking and loading
- Standing idle for no apparent reason
- Talking about matters other than work
- Delaying for personal reasons
- Planning, discussing, and studying the layout

To improve productivity, the supervisor needs to reduce the time that the worker is not being productive.

Studies also reveal that when a worker does not do a reasonable day's work, it is because there is a lack of one or more of the following:

- Information
- Work (or too many interferences)
- Tools
- Motivation
- Ability
- Desire

An effective program consisting of planning, scheduling, and supervision can correct the first three issues. An effective leadership and training program can correct the last three issues.

The kind and amount of labor will vary from job to job. The supervisor may not be expected to plan the labor needed for the entire job. However, he or she may be asked to offer recommendations based on the quantity and character of the work to be completed in the immediate future.

Frequently, the amount of work to be completed is stated in percentages. For example, 30 per-cent of the job has to be completed within two weeks. Once the percentage is established, it is necessary to determine the amount of unskilled and skilled labor required to meet the objective.

To estimate the amount of labor required, the supervisor will have to determine the following two things:

- The kind of work
- The estimated amount of time needed to complete the tasks

The supervisor should be able to identify the kind of work immediately. However, he or she may have to rely on past experiences to determine the amount of time needed to complete the task(s).

There are a few essential labor-related practices that should be followed in an effort to complete the job efficiently. They are as follows:

- Ensure that all workers have the required resources when needed
- Ensure that all personnel know where to go and what to do after each task is completed
- Make reassignments as needed
- Ensure that all workers have completed their work properly

PARTICIPANT EXERCISE H

1. How does your company minimize waste?

   _____
   _____
   _____
   _____
   _____

2. How does your company control the use of small tools on the job?

   _____
   _____
   _____
   _____
   _____

3. List five ways that you feel that your company could control labor.

   _____
   _____
   _____
   _____
   _____
   _____
   _____
   _____
   _____
   _____

PARTICIPANT EXERCISE H ◆ PROJECT CONTROL

12.0.0 PRODUCTION AND PRODUCTIVITY

Pay for time on the job is based on the work that needs to be performed. Using time productively results in more work being completed. Many savings that we have discussed are directly related to the effective usage of time.

Planning ahead is basic to a high rate of productivity. By thinking ahead, a supervisor can eliminate wasted time. The supervisor and crew will be prepared to perform the required work, and they will have the materials, tools, and equipment required to do so available.

The time on the job should be for business, not for taking care of personal problems. Anything not work-related should be handled after hours, away from the job site. Planning after-work activities, arranging social functions, or running personal errands should be handled after work or during a break.

Organizing fieldwork can save time. The key to effectively using time is to work smarter, not necessarily harder. For example, most construction projects require that the contractor submit a set of as-built drawings at the completion of the work. These drawings describe how the materials were actually installed. The best way to prepare these drawings is to mark a set of working drawings as the work is in progress. That way, pertinent details will not be forgotten and time will not be wasted trying to remember how the work was done.

Supervisors are very familiar with the terms production and productivity. Production is the amount of construction put in place. It is the quantity of materials installed on a job, such as 1,000 linear feet of waste pipe installed in a given day. On the other hand, productivity depends on the level of efficiency of the work. It is the amount of work done per hour or day, by one worker or a crew.

Production levels are set during the estimating stage. The estimator determines the total amount of materials to be put in place from the drawings. After the job is complete, the actual amount of materials installed can be assessed, and the actual production can be compared to the estimated production.

The quantity of materials used should be equal to or less than estimated. If it is not, either the estimator has made a mistake, undocumented changes have occurred, or rework has caused the need for additional materials. Whatever the case, the supervisor should use effective control techniques to ensure the efficient use of materials.

Productivity relates to the amount of materials put in place by the crew over a certain time period. The estimator uses company records during the estimating stage to determine how much time and labor it will take to place a certain quantity of materials. From this information, the estimator calculates the productivity necessary to complete the job on time.

For example, it might take a crew of two people 10 days to paint 5,000 square feet. The crew's productivity per day is obtained by dividing 5,000 square feet by 10 days. The result is 500 square feet per day. The supervisor can compare the daily production of any crew of two painters doing similar work with this average, as discussed in *Planning Resources* in this chapter.

When bidding a job, most companies use a cost per labor hour factor to determine the price of a job. For example, a 10-hour day converts to 160 labor hours (2 painters for 10 days at 8 hours per day). If the hourly labor rate the company charges is $30, the labor cost would be $4,800. The estimator would then add the cost of material, equipment, tools, and overhead costs, along with a profit factor, to determine the price of the job.

In addition, information gathered from the field through field reporting will allow the home office to calculate the actual productivity and compare it to the estimated figure. If the actual production was below that estimated, actions would have to be taken during construction to improve the productivity. In this way, the job can be built at or below the estimated cost and within the scheduled time.

1. One of the three phases of a construction project is _____.

   a. the development phase, in which design professionals produce drawings and other plans that are then shown to the contractor and to other workers

   b. the construction phase, in which architectural and governmental agencies give a final inspection of the design, adherence to codes, and materials used

   c. the planning phase, in which estimates of all expenses are made

   d. the construction phase, in which the owner selects a contractor

2. One of the three types of project delivery systems is _____.

   a. design-build, in which the owner designs how something is to be constructed

   b. general contracting, in which the contractor is in charge of the design

   c. construction management, in which the contractor builds according to the owner's design

   d. general contracting, in which the owner hires a contractor

3. One reason it is important to plan is _____.

   a. it will help in completing the job at a lower cost

   b. to prepare for the unexpected

   c. it will prevent all delays from occurring

   d. it will guarantee a successful project

4. The planning process includes _____.

   a. setting objectives that identify the work to be done, even though workers might not approve of them

   b. breaking down large jobs into subtasks

   c. visiting the job site to see how well the project is going

   d. involving the supervisors, who typically have the responsibility of establishing goals for their workers

5. A job diary should typically indicate _____.

   a. items such as job interruptions and visits

   b. only the major incidents of the day

   c. the estimated time for each job task related to a particular project

   d. the supervisor's ideas for improving employee morale

6. One example of a direct cost when building a job is _____.

   a. office rent

   b. job-site telephones

   c. equipment

   d. utilities

7. The project estimate is prepared _____.

   a. just after the project begins

   b. after the project is completed

   c. just before the project is going to end

   d. before a project begins

8. An example of overhead is _____.

   a. utilities

   b. labor

   c. material

   d. tools

9. The process of determining how much a job will cost is known as _____.

    a. the estimate

    b. the material quantity

    c. the direct cost

    d. the indirect cost

10. One of the steps of the control process typically involves _____.

    a. establishing a network schedule

    b. determining if performance was up to established standards

    c. getting an estimated cost for a project

    d. figuring out the actual cost of a project

ADDITIONAL RESOURCES ON THE WEB

American Medical Association (AMA)   www.ama-assn.org

American Society for Training and Development (ASTD)   www.astd.org

Architecture, Engineering, and
  Construction Industry (AEC)   www.aecinfo.com

CIT Group   www.citgroup.com

Equal Employment Opportunity
  Commission (EEOC)   www.eeoc.gov

National Association of Women
  in Construction (NAWIC)   www.nawic.org

National Census of Fatal
  Occupational Injuries (NCFOI)   http://www.bls.gov/iif/oshcfoi1.htm

National Center for Construction Education
  and Research   www.nccer.org

National Institute of Occupational
  Safety and Health (NIOSH)   www.cdc.gov/niosh

National Safety Council   www.nsc.org

Occupational Safety and Health Administration (OSHA)   www.osha.gov

Society for Human Resources Management (SHRM)   www.shrm.org

United States Census Bureau   www.census.gov

United States Department of Labor   www.dol.gov

USA Today   www.usatoday.com

NCCER makes every effort to keep these textbooks up-to-date and free of technical errors. We appreciate your help in this process. If you have an idea for improving this textbook, or if you find an error, a typographical mistake, or an inaccuracy in NCCER's Contren® textbooks, please write us, using this form or a photocopy. Be sure to include the exact module number, page number, a detailed description, and the correction, if applicable. Your input will be brought to the attention of the Technical Review Committee. Thank you for your assistance.

Instructors – If you found that additional materials were necessary in order to teach this module effectively, please let us know so that we may include them in the Equipment/Materials list in the Annotated Instructor's Guide.

Write: Product Development and Revision
National Center for Construction Education and Research
3600 NW 43rd St, Bldg G, Gainesville, FL 32606

Fax: 352-334-0932

E-mail: curriculum@nccer.org

Craft ___ Module Name ___

Copyright Date ___ Module Number ___ Page Number(s) ___

Description ___

(Optional) Correction ___

(Optional) Your Name and Address ___

Glossary

Addressable device: A fire alarm system component with discrete identification that can have its status individually identified or that is used to individually control other functions.

Air sampling detector: A detector consisting of piping or tubing distribution from the detector unit to the area or areas to be protected. An air pump draws air from the protected area back to the detector through the air sampling ports and piping or tubing. At the detector, the air is analyzed for fire products.

Alarm: In fire systems, a warning of fire danger.

Alarm signal: A signal indicating an emergency requiring immediate action, such as an alarm for fire from a manual station, water flow device, or automatic fire alarm system.

Alarm verification: A feature of a fire control panel that allows for a delay in the activation of alarms upon receiving an initiating signal from one of its circuits. Alarm verification must not be longer than three minutes, but can be adjustable from 0 to 3 minutes to allow supervising personnel to check the alarm. Alarm verification is commonly used in hotels, motels, hospitals, and institutions with large numbers of smoke detectors.

Americans with Disabilities Act (ADA): An act of Congress intended to ensure civil rights for physically challenged people.

Ampere turn: The product of amperes times the number of turns in a coil.

Analog-to-digital converter: A device designed to convert analog signals such as temperature and humidity to a digital form that can be processed by logic circuits.

Approved: Acceptable to the authority having jurisdiction.

Armature: The assembly of windings and metal core laminations in which the output voltage is induced. It is the rotating part in a revolving field generator.

Armor: A mechanical protector for cables; usually a formed metal tube or helical winding of metal tape formed so that each convolution locks mechanically upon the previous one (interlocked armor).

Audible signal: An audible signal is the sound made by one or more audible indicating appliances, such as bells, chimes, horns, or speakers, in response to the operation of an initiating device.

Authority having jurisdiction (AHJ): The authority having jurisdiction is the organization, office, or individual responsible for approving equipment, installations, or procedures in a particular locality.

Automatic changeover thermostat: A thermostat that automatically selects heating or cooling based on the space temperature.

Automatic fire alarm system: A system in which all or some of the circuits are actuated by automatic devices, such as fire detectors, smoke detectors, heat detectors, and flame detectors.

Autotransformer: Any transformer in which the primary and secondary connections are made to a single winding. The application of an autotransformer is a good choice where a 480Y/277V or 208Y/120V, three-phase, four-wire distribution system is used.

Bank: An installed grouping of a number of units of the same type of electrical equipment, such as a bank of transformers, a bank of capacitors, or a meter bank.

Base speed: The rating given on a motor's nameplate where the motor will develop rated horsepower at rated load and voltage.

Beam construction: Ceilings having solid structural or solid nonstructural members projecting down from the ceiling surface more than 4" (100mm) and spaced more than 3' (0.9m) center to center.

Bimetal: A control device made of two dissimilar metals that warp when exposed to heat, creating movement that can open or close a switch.

Border light: A permanently installed overhead strip light.

Breakdown: The failure of insulation to prevent the flow of current, sometimes evidenced by arcing. If the voltage is gradually raised, breakdown will begin suddenly at a certain voltage level. Current flow is not directly proportional to voltage. Once a breakdown current has flowed, especially for a period of time, the next gradual application of voltage will often show breakdown beginning at a lower voltage than initially.

Building: A structure that stands alone or that is cut off from adjoining structures by fire walls with all openings therein protected by approved fire doors.

CABO: Council of American Building Officials.

Ceiling: The upper surface of a space, regardless of height. Areas with a suspended ceiling would have two ceilings: one visible from the floor, and one above the suspended ceiling.

Ceiling height: The height from the continuous floor of a room to the continuous ceiling of a room or space.

Ceiling surface: Ceiling surfaces referred to in conjunction with the locations of initiating devices are as follows:

- *Beam construction* – Ceilings having solid structural or solid nonstructural members projecting down from the ceiling surface more than 4" (100mm) and spaced more than 3" (0.9m) center to center.

- *Girders* – Girders support beams or joists and run at right angles to the beams or joists. When the tops of girders are within 4" (100mm) of the ceiling, they are a factor in determining the number of detectors and are to be considered as beams. When the top of the girder is more than 4" (100mm) from the ceiling, it is not a factor in detector location.

Certification: A systematic program using randomly selected follow-up inspections of the certified system installed under the program, which allows the listing organization to verify that a fire alarm system complies with all the requirements of the *NFPA 72*® code. A system installed under such a program is identified by the issuance of a certificate and is designated as a certificated system.

Chimes: A single-stroke or vibrating audible signal appliance that has a xylophone-type striking bar.

Circuit: The conductors or radio channel as well as the associated equipment used to perform a definite function in connection with an alarm system.

Class A circuit: Class A refers to an arrangement of supervised initiating devices, signaling line circuits, or indicating appliance circuits (IAC) that prevents a single open or ground on the installation wiring of these circuits from causing loss of the system's intended function. It is also commonly known as a four-wire circuit.

Class B circuit: Class B refers to an arrangement of initiating devices, signaling lines, or indicating appliance circuits that does not prevent a single open or ground on the installation wiring of these circuits from causing loss of the system's intended function. It is commonly known as a two-wire circuit.

Closed circuit transition: A method of reduced-voltage starting where the motor being controlled is never removed from the source of voltage while moving from one voltage level to another.

Coded signal: A signal pulsed in a prescribed code for each round of transmission. A minimum of three rounds and a minimum of three impulses are required for an alarm signal. A coded signal is usually used in a manually operated device on which the act of pulling a lever causes the transmission of not less than three rounds of coded alarm signals. These devices are similar to the non-coded type, except that instead of a manually operated switch, a mechanism to rotate a code wheel is utilized. Rotation of the code wheel, in turn, causes an electrical circuit to be alternately opened and closed, or closed and opened, thus sounding a coded alarm that identifies the location of the box. The code wheel is cut for the individual code to be transmitted by the device and can operate by clockwork or an electric motor. Clockwork transmitters can be prewound or can be wound by the pulling of the alarm lever. Usually, the box is designed to repeat its code four times before automatically coming to rest. Prewound transmitters must sound a trouble signal when they require rewinding. Solid-state electronic coding devices are also used in conjunction with the fire alarm control panel to produce coded sounding of the system's audible signaling appliances.

Compressor: A component of a refrigeration system that converts low-pressure, low-temperature refrigerant gas into high-temperature, high-pressure refrigerant gas.

Condenser: A heat exchanger that transfers heat from the refrigerant flowing inside it to the air or water flowing over it.

Constant wattage heating cable: A heating cable that has the same power output over a large temperature range. It is used in applications requiring a constant heat output regardless of varying outside temperatures. Zoned parallel resistance, mineral-insulated, and series-resistant heating cables are examples of constant wattage heating cables.

Control unit: A device with the control circuits necessary to furnish power to a fire alarm system, receive signals from alarm initiating devices (and transmit them to audible alarm indicating appliances and accessory equipment), and electrically supervise the system installation wiring and primary (main) power. The control unit can be contained in one or more cabinets in adjacent or remote locations.

Cooling compensator: A fixed resistor installed in a thermostat to act as a cooling anticipator.

Core loss: The electric loss that occurs in the core of an armature or transformer due to conditions such as the presence of eddy currents or hysteresis.

Dashpot: A device that uses a gas or fluid to absorb energy from or retard the movement of the moving parts of a relay, circuit breaker, or other electrical or mechanical device.

Dead front: Equipment or enclosures covered or protected in order to prevent exposure of energized parts on the operating side of the equipment.

Dead leg: A segment of pipe designed to be in a permanent no-flow condition. This pipe section is often used as a control point for a larger system.

Deadband: A temperature band, usually 3°F, that separates heating and cooling in an automatic changeover thermostat.

Dew point: The temperature at which liquid first condenses when vapor is cooled.

Diac: A three-layer diode designed for use as a trigger in AC power control circuits, such as those using triacs.

Dielectric constant (K): A measurement of the ability of a material to store a charge.

Differential: The difference between the cut-in and cut-out points of a thermostat.

Digital Alarm Communicator Receiver (DACR): A system component that will accept and display signals from digital alarm communicator transmitters (DACTs) sent over public switched telephone networks.

Digital Alarm Communicator System (DACS): A system in which signals are transmitted from a digital alarm communicator transmitter (DACT) located at the protected premises through the public switched telephone network to a digital alarm communicator receiver (DACR).

Digital Alarm Communicator Transmitter (DACT): A system component at the protected premises to which initiating devices or groups of devices are connected. The DACT will seize the connected telephone line, dial a preselected number to connect to a DACR, and transmit signals indicating a status change of the initiating device.

Dwelling unit: One or more rooms for the use of one or more persons as a housekeeping unit with space for eating, living, and sleeping, as well as permanent provisions for cooking and sanitation. A one-family dwelling consists solely of one dwelling unit. A two-family dwelling consists solely of two dwelling units. A multi-family dwelling is a building containing three or more dwelling units.

Eddy currents: The circulating currents that are induced in conductive materials by varying magnetic fields; they are usually considered undesirable because they represent a loss of energy and produce excess heat.

Electrical datum plane: Horizontal plane affecting electrical installations based on the rise and fall of water levels in land areas and floating piers and landing stages.

Electrolyte: A substance in which the conduction of electricity is accompanied by chemical action (for example, the paste that forms the conducting medium between the electrodes of a dry cell or storage cell).

End-of-line (EOL) device: A device used to terminate a supervised circuit. An EOL is normally a resistor or a diode placed at the end of a two-wire circuit to maintain supervision.

Equipotential ground plane: A mass of conducting material that, when bonded together, provides a low impedance to current flow over a wide range of frequencies.

Equipotential plane: An area in wire mesh or other conductive material embedded in or placed under concrete that may conduct current. It must be bonded to all metal structures and non-electrical, fixed equipment that could become energized, and then connected to the electrical grounding system.

Evaporator: A heat exchanger that transfers heat from the air flowing over it to the cooler refrigerant flowing through it.

Exciter: A device that supplies direct current (DC) to the field coils of a synchronous generator, producing the magnetic flux required for inducing output voltage in the stator coils.

Expansion device: A device that provides a pressure drop that converts the high-temperature, high-pressure liquid refrigerant from the condenser into the low-temperature, low-pressure liquid refrigerant entering the evaporator; also known as the liquid metering device or metering device.

Extra-hard usage cable: Cable designed with additional mechanical protection for use in areas where it may be subject to physical damage.

Fault: An open, ground, or short condition on any line(s) extending from a control unit, which could prevent normal operation.

Feeder: All circuit conductors between the service equipment or the source of a separately derived system and the final branch circuit overcurrent device.

Festoon lighting: Outdoor lighting that is strung between supporting points.

Field-effect transistor (FET): A transistor that controls the flow of current through it with an electric field.

Fire: A chemical reaction between oxygen and a combustible material where rapid oxidation may cause the release of heat, light, flame, and smoke.

Fire rating: Time in hours or minutes that a material or assembly can withstand direct exposure to fire without igniting and spreading the fire.

Flame detector: A device that detects the infrared, ultraviolet, or visible radiation produced by a fire. Some devices are also capable of detecting the flicker rate (frequency) of the flame.

Footlight: A border light installed on or in a stage.

Forward bias: Forward bias exists when voltage is applied to a solid-state device in such a way as to allow the device to conduct easily.

Garage: A building or portion of a building in which one or more self-propelled vehicles carrying volatile, flammable liquid for fuel or power are kept for use, sale, storage, rental, repair, exhibition, or demonstration purposes, and all that portion of a building that is on or below the floor or floors in which such vehicles are kept and that is not separated by suitable cutoffs.

General alarm: A term usually applied to the simultaneous operation of all the audible alarm signals on a system, to indicate the need for evacuation of a building.

Generator: (1) A rotating machine that is used to convert mechanical energy to electrical energy. (2) General apparatus, equipment, etc., that is used to convert or change energy from one form to another.

Girders: Girders support beams or joists and run at right angles to the beams or joists. When the tops of girders are within 4" (100mm) of the ceiling, they are a factor in determining the number of detectors and are to be considered as beams. When the top of the girder is more than 4" (100mm) from the ceiling, it is not a factor in detector location.

Ground fault: A condition in which the resistance between a conductor and ground reaches an unacceptably low level.

Harmonic: An oscillation whose frequency is an integral multiple of the fundamental frequency.

Heat detector: A device that detects abnormally high temperature or rate-of-temperature rise.

Heat sink: A metal mounting base that dissipates the heat of solid-state components mounted on it by radiating the heat into the surrounding atmosphere.

Heat source: Any component that adds heat to an object or system.

Heat tracing systems: Steam, hot-fluid, or electrical systems that maintain piping systems, vessels, and certain types of structures at or above a given temperature.

Heat transfer: The transfer of heat from a warmer substance to a cooler substance.

High-potential (hi-pot) test: A high-potential test in which equipment insulation is subjected to a voltage level higher than that for which it is rated to find any weak spots or deficiencies in the insulation.

Horn: An audible signal appliance in which energy produces a sound by imparting motion to a flexible component that vibrates at some nominal frequency.

Hysteresis: The time lag exhibited by a body in reacting to changes in the forces affecting it; hysteresis is an internal friction.

Impedance: The opposition to current flow in an AC circuit; impedance includes resistance (R), capacitive reactance (X_C), and inductive reactance (X_L). It is measured in ohms (Ω).

Indicating device: Any audible or visible signal employed to indicate a fire, supervisory, or trouble condition. Examples of audible signal appliances are bells, horns, sirens, electronic horns, buzzers, and chimes. A visible indicator consists of an incandescent lamp, strobe lamp, mechanical target or flag, meter deflection, or the equivalent. Also called a notification device (appliance).

Initiating device: A manually or automatically operated device, the normal intended operation of which results in a fire alarm or supervisory signal indication from the control unit. Examples of alarm signal initiating devices are thermostats, manual boxes (stations), smoke detectors, and water flow devices. Examples of supervisory signal initiating devices are water level indicators, sprinkler system valve-position switches, pressure supervisory switches, and water temperature switches.

Initiating device circuit (IDC): A circuit to which automatic or manual signal-initiating devices such as fire alarm manual boxes (pull stations), heat and smoke detectors, and water flow alarm devices are connected.

Inpatient: A patient admitted into the hospital for an overnight stay.

Insulated gate bipolar transistor (IGBT): A type of transistor that has low losses and low gate drive requirements. This allows it to be operated at higher switching frequencies.

Insulation class: Category of insulation based on the thermal endurance of the insulation system used in a motor. The insulation system is chosen to ensure that the motor will perform at the rated horsepower and service factor load.

Insulation shielding: An electrically conductive layer that provides a smooth surface in contact with the insulation outer surface; it is used to eliminate electrostatic charges external to the shield and to provide a fixed, known path to ground.

Invar®: An alloy of steel containing 36% nickel. It is one of the two metals in a bimetal control device.

Isolated winding-type transformer: A transformer designed with a mechanical isolating device, such as a plate, between the primary and secondary windings.

Isolation transformer: A transformer that has no electrical metallic connection between the primary and secondary windings.

Junction field-effect transistor (JFET): A field-effect transistor formed by combining layers of semiconductor material.

Labeled: Equipment or materials to which has been attached a label, symbol, or other identifying mark of an organization acceptable to the authority having jurisdiction, which is concerned with product evaluation and whose listing states either that the equipment or material meets appropriate standards or has been tested and found suitable for use in a specified manner.

Landing stage: A floating platform used for the loading and unloading of passengers and cargo from boats or ships.

Latch: In a solid-state device, latching is the equivalent to the holding or memory circuit in relay logic. The current flowing through the relay maintains flow until the load is removed.

Leakage: AC or DC current flow through insulation and over its surfaces, and AC current flow through a capacitance. Current flow is directly proportional to voltage. The insulation and/or capacitance are thought of as a constant impedance unless breakdown occurs.

Light scattering: The action of light being reflected or refracted off particles of combustion, for detection in a modern-day photoelectric smoke detector. This is called the Tyndall effect.

Listed: Equipment or materials included in a list published by an organization acceptable to the authority having jurisdiction that is concerned with product evaluation and whose listing states either that the equipment or materials meets appropriate standards or has been tested and found suitable for use in a specified manner.

Maintenance: Repair service, including periodic inspections and tests, required to keep the protective signaling system and its component parts in an operative condition at all times. This is used in conjunction with replacement of the system and its components when for any reason they become undependable or inoperative.

Mechanical refrigeration: The use of machinery to provide cooling.

Megger®: A trade term used for a megohmmeter. An instrument or meter capable of measuring resistances in excess of 200 megohms. It employs much higher test voltages than are used in ohmmeters, which measure up to 200 megohms.

Mineral-insulated heating cable: A type of constant wattage, series resistance heating cable used in long line applications where high temperatures need to be maintained, high temperature exposure exists, or high power output is required.

Multiplexing: A signaling method that uses wire path, cable carrier, radio, fiber optics, or a combination of these techniques, and characterized by the simultaneous or sequential (or both simultaneous and sequential) transmission and reception of multiple signals in a communication channel including means of positively identifying each signal.

National Fire Alarm Code®: This is the update of the NFPA standards book that contains the former *NFPA 71*, *NFPA 72®*, and *NFPA 74* standards, as well as the *NFPA 1221* standard. The NFAC was adopted and became effective May 1993.

National Fire Protection Association (NFPA): The NFPA administers the development and publishing of codes, standards, and other materials concerning all phases of fire safety.

Noise: A term used in electronics to cover all types of unwanted electrical signals. Noise signals originate from numerous sources, such as fluorescent lamps, walkie-talkies, amateur and CB radios, machines being switched on and off, and power surges. Today, equipment must tolerate increasing amounts of electrical interference, and the quality of equipment depends on how much noise it can ignore and withstand.

Non-coded signal: A signal from any indicating appliance that is continuously energized.

Notification device (appliance): See indicating device (appliance).

N-type material: A material created by doping a region of a crystal with atoms from an element that has more electrons in its outer shell than the crystal.

Obscuration: A reduction in the atmospheric transparency caused by smoke, usually expressed in percent per foot.

Open circuit transition: A method of reduced-voltage starting where the motor being controlled may be temporarily disconnected from the source of voltage while moving from one voltage level to another.

Outpatient: A patient who receives services at a hospital but is not admitted for an overnight stay.

Overhead system service entrance: The service conductors between the terminals of the service equipment and a point usually outside the building, clear of building walls, where it is joined by a tap or splice to the service drop.

Parallel resistance heating cable: An electric heating cable with parallel connections, either continuous or in zones. The watt density per lineal length is approximately equal along the length of the heating cable, allowing for a drop in voltage down the length of the heating cable.

Path (pathway): Any conductor, optic fiber, radio carrier, or other means for transmitting fire alarm system information between two or more locations.

Photoelectric smoke detector: A detector employing the photoelectric principle of operation using either the obscuration effect or the light-scattering effect for detecting smoke in its chamber.

Portable switchboard: A movable lighting switchboard used for road shows and similar presentations, usually supplied by a flexible cord or cable.

Positive alarm sequence: An automatic sequence that results in an alarm signal, even when manually delayed for investigation, unless the system is reset.

Positive locking device: A locking device such as a screw, clip, or other mechanical device that requires a tool or key to lock or unlock it.

Pothead: A terminator for high-voltage circuit conductors that keeps moisture out of the insulation and protects the cable end, along with providing a suitable stress relief cone for shielded-type conductors.

Potting compound: An approved and listed insulating compound, such as epoxy, which is mixed and installed in and around electrical connections to encapsulate and seal the connections to prevent the entry of water or moisture.

Power supply: A source of electrical operating power, including the circuits and terminations connecting it to the dependent system components.

Power-limiting heating cable: A type of heating cable that shows positive temperature coefficient (PTC) behavior based on the properties of a metallic heating element. The PTC behavior exhibited is much less (a smaller change in resistance in response to a change in temperature) than that shown by self-regulating heating cables.

Projected beam smoke detector: A type of photoelectric light-obscuration smoke detector in which the beam spans the protected area.

Proscenium: The wall and arch that separate the stage from the auditorium (house).

Protected premises: The physical location protected by a fire alarm system.

Protected premises (local) fire alarm system: A protected premises system that sounds an alarm at the protected premises as the result of the manual operation of a fire alarm box, or as a result of the operation of protection equipment or systems, such as water flowing in a sprinkler system, the discharge of carbon dioxide, the detection of smoke, or the detection of heat.

P-type material: A material created when a crystal is doped with atoms from an element that has fewer electrons in its outer shell than the natural crystal. This combination creates empty spaces in the crystalline structure. The missing electrons in the crystal structure are called holes and are represented as positive charges.

Public Switched Telephone Network: An assembly of communications facilities and central office equipment, operated jointly by authorized common carriers, that provides the general public with the ability to establish communications channels via discrete dialing codes.

Rate compensation detector: A device that responds when the temperature of the air surrounding the device reaches a predetermined level, regardless of the rate-of-temperature rise.

Rate-of-rise detector: A device that responds when the temperature rises at a rate exceeding a predetermined value.

Reactance: The imaginary part of impedance; also, the opposition to alternating current due to capacitance (X_C) and/or inductance (X_L).

Reactor-type dimmer: A dimming system that incorporates a number of windings or coils that are linked by a magnetic core, an inductive reactor that functions like a capacitor, and a relay. The reactor stores current from the source and reduces lamp power.

Rectifier: A device or circuit used to change AC voltage into DC voltage.

Refrigerant: A fluid (liquid or gas) that picks up heat by evaporating at a low temperature and pressure and gives up heat by condensing at a higher temperature and pressure.

Refrigeration cycle: The process by which a circulating refrigerant absorbs heat from one location and transfers it to another location.

Remote supervising station fire alarm system: A system installed in accordance with the applicable code to transmit alarm, supervisory, and trouble signals from one or more protected premises to a remote location where appropriate action is taken.

Reset: A control function that attempts to return a system or device to its normal, non-alarm state.

Resistance-type dimmer: A manual-type dimmer in which light intensity is controlled by increasing or decreasing the resistance in the lighting circuit.

Reverse bias: A condition that exists when voltage is applied to a device in such a way that it causes the device to act as an insulator.

Rheostat: A slide and contact device that increases or decreases the resistance in a circuit by changing the position of a movable contact on a coil of wire.

Ribbon tape shield: A helical (wound coil) strip under the cable jacket.

Self-regulating cable: A type of heating cable that inversely varies its heat output in response to an increase or decrease in the ambient temperature.

Semiconductor: A material that is neither a good insulator nor a good conductor. Such materials contain four valence electrons and are used in the production of solid-state devices.

Separately derived system: A premises wiring system whose power is derived from a battery, a solar photovoltaic system, or from a generator, a transformer, or converter windings, and that has no direct electrical connection, including a solidly-connected grounded circuit conductor, to supply conductors originating in another system.

Series resistance heating cable: A type of heating cable in which the ohmic heating of the conductor provides the heat. The wattage output depends on the total circuit length and on the voltage applied.

Service: The conductors and equipment for delivering electric energy from the serving utility to the wiring system of the premises served.

Service conductors: The conductors from the service point to the service disconnecting means.

Service drop: The overhead service conductors from the last pole or other aerial support up to and including splices, if any, connecting to the service-entrance conductors to the building or other structure.

Service equipment: The necessary equipment, usually consisting of a circuit breaker or switch and fuses, and their accessories, located near the point of entrance of supply conductors to a building or other structure, or an otherwise defined area, and intended to constitute the main control and means of cutoff of the supply.

Service lateral: The underground service conductors between the street main, including any risers at a pole or other structure or from transformers, and the first point of connection to the service-entrance conductors in a terminal box or meter or other enclosure with adequate space, inside or outside the building wall. Where there is no terminal box, meter, or other enclosure with adequate space, the point of connection shall be considered to be the point of entrance of the service conductors into the building.

Service point: The point of connection between the facilities of the serving utility and the premises wiring.

Shield: A conductive barrier against electromagnetic fields.

Shielding braid: A shield of small, interwoven wires.

Shunt trip: A low-voltage device that remotely causes a higher voltage overcurrent protective device to trip or open the circuit.

Signal: A status indication communicated by electrical or other means.

Signaling line circuits (SLCs): A circuit or path between any combination of circuit interfaces, control units, or transmitters over which multiple system input signals or output signals (or both input signals and output signals) are carried.

Silicon-controlled rectifier (SCR): A device that is used mainly to convert AC voltage into DC voltage. To do so, however, the gate of the SCR must be triggered before the device will conduct current.

Site-isolating device: A disconnecting means at the point of distribution that is used to isolate the service for the purpose of electrical maintenance, emergency disconnection, or the connection of an optional standby system.

Smoke detector: A device that detects visible or invisible particles of combustion.

Soft foot: When one or two of the mounting feet of a motor are spaced away from an uneven mounting surface and are then anchored down against the mounting surface. This condition can cause warpage of the motor frame and misalignment of the bearings.

Solid-state device: An electronic component constructed from semiconductor material. Such devices have all but replaced the now-obsolete vacuum tube in electronic circuits.

Solid-state overload relay (SSOLR): An electronic device that provides overload protection for electrical equipment.

Solid-state phase-control dimmer: A solid-state dimmer in which the wave shape of the steady-state current does not follow the wave shape of the applied voltage (i.e., the wave shape is nonlinear).

Solid-state relay (SSR): A switching device that has no contacts and switches entirely by electronic means.

Solid-state sine-wave dimmer: A solid-state dimmer in which the wave shape of the steady-state current follows the wave shape of the applied voltage (i.e., the wave shape is linear).

Spacing: A horizontally measured dimension related to the allowable coverage of fire detectors.

Splice: Two or more conductors joined with a suitable connector and reinsulated, reshielded, and rejacketed with compatible materials applied over a properly prepared surface.

Spot-type detector: A device in which the detecting element is concentrated at a particular location. Typical examples are bimetallic detectors, fusible alloy detectors, certain pneumatic rate-of-rise detectors, certain smoke detectors, and thermoelectric detectors.

Startup temperature: The lowest temperature at which a heat tracing cable is energized.

Strand shielding: The semi-conductive layer between the conductor and insulation that compensates for air voids that exist between the conductor and insulation.

Stratification: The phenomenon in which the upward movement of smoke and gases ceases due to a loss of buoyancy.

Stress: An internal force set up within a body to resist or hold it in equilibrium.

Sub-base: The portion of a two-part thermostat that contains the wiring terminals and control switches.

Subcooling: Cooling a liquid below its condensing temperature.

Superheat: The measurable heat added to the vapor or gas produced after a liquid has reached its boiling point and completely changed into a vapor.

Supervising station: A facility that receives signals and at which personnel are in attendance at all times to respond to these signals.

Supervisory signal: A signal indicating the need for action in connection with the supervision of guard tours, the fire suppression systems or equipment, or the maintenance features of related systems.

Switchboard: A large single panel, frame, or assembly of panels on which are mounted, on the face, back, or both, switches, overcurrent and other protective devices, buses, and instruments. Switchboards are generally accessible from the rear as well as from the front and are not intended to be installed in cabinets.

Synchronous generator: An AC generator having a DC exciter. Synchronous generators are used as stand-alone generators for emergency and standby power systems and can also be paralleled with other synchronous generators and the utility system.

System unit: The active subassemblies at the central station used for signal receiving, processing, display, or recording of status change signals. The failure of one of these subassemblies causes the loss of a number of alarm signals by that unit.

Tap conductors: As defined in *NEC Article 240*, a tap conductor is a conductor, other than a service conductor, that has overcurrent protection ahead of its point of supply that exceeds the value permitted for similar conductors that are protected as described in *NEC Section 240.4*.

Thermal resistance: The resistance of a material or substance to the conductivity of heat.

Thermostat: A device that is responsive to ambient temperature conditions.

Throwover: A transfer switch used for supplying temporary power.

Thyristor: A bi-stable semiconductor device that can be switched from the off to the on state or vice versa. All they require to turn on is a quick pulse of control current. They turn off only when the working current is interrupted elsewhere in the circuit. Note that SCRs and triacs are forms of thyristors.

Totally enclosed motor: A motor that is encased to prevent the free exchange of air between the inside and outside of the case.

Transmitter: A system component that provides an interface between the transmission channel and signaling line circuits, initiating device circuits, or control units.

Triac: A bidirectional triode thyristor that functions as an electrically controlled switch for AC loads.

Trouble signal: A signal initiated by the fire alarm system or device that indicates a fault in a monitored circuit or component.

Unbalanced load: Three-phase loads in which the loads are not evenly distributed between the phases.

Underground service entrance: The service conductors between the terminals of the service equipment and the point of connection to the service lateral.

Vessel: A container such as a barrel, drum, or tank used for holding fluids or other substances.

Visible notification appliance: A notification appliance that alerts by the sense of sight.

Voltage gradients: Fluctuating voltage potentials that can exist between two surfaces or components and may be caused by improper or insufficient bonding or grounding.

Wavelength: The distance between peaks of a sinusoidal wave. All radiant energy can be described as a wave having a wavelength. Wavelength serves as the unit of measure for distinguishing between different parts of the spectrum. Wavelengths are measured in microns, nanometers, or angstroms.

Wide Area Telephone Service (WATS): Telephone company service that provides reduced costs for certain telephone call arrangements. In-WATS or 800-number service calls can be placed from anywhere in the continental United States to the called party at no cost to the calling party. Out-WATS is a service whereby, for a flat-rate charge, dependent on the total duration of all such calls, a subscriber can make an unlimited number of calls within a prescribed area from a particular telephone terminal without the registration of individual call charges.

Zone: A defined area within the protected premises. A zone can define an area from which a signal can be received, an area to which a signal can be sent, or an area in which a form of control can be executed.

Figure Credits

Module 26401-08

John Traister, 401F02, 401F03, 401F05–401F07, 401F09, 401F10

Module 26402-08

Dukane Communication Systems, 402SA01, 402F10
Mike Powers, 402F04, 402F06, 402SA03, 402F08
Hitec Power Solutions, Stafford, TX 77477, www.hitecusa.com, 402SA02
Pass & Seymour/Legrand, 402F07
Ed Cockrell, 402SA04
Cornell Communications, Inc., 402SA05
Jeron® Electronic Systems, Inc., 402SA06
Isotrol Systems, 402F11

Module 26403-08

Reprinted Courtesy of Caterpiller Inc., 403SA01–403SA03, 403SA10
Topaz Publications, Inc., 403SA04
Trojan Battery Company, 403SA05
SEC Industrial Battery Co., Ltd., 403SA06
Tim Ely, 403SA07, 403SA09
MGE UPS Systems, Inc., 403SA08
Hubbell Incorporated, 403F16

Module 26404-08

International Rectifier Corp., 404SA03
Topaz Publications, Inc., 404SA04–404SA07, 404SA09
Powerex, Inc., 404SA08
Nellis Airforce Base, 404SA10

Module 26405-08

Reproduced with permission from the National Fire Protection Association, copyright © 2005 & 2007 National Fire Protection Association. All rights reserved. 405SA01
System Sensor, 405SA02, 405F11, 405F12, 405SA04, 405SA06, 405F31, 405F36A, 405F67, 405SA11, 405SA13
Brooks Equipment Co. Inc., 405F24

Notifier, 405SA03, 405SA05, 405F28, 405F30, 405F34, 405SA08, 405SA10, 405SA09
Wheelock Inc., 405F32, 405F36B, 405F36C
W.R. Grace & Co. Conn., 405SA07
Jack Viola, 405SA12

Module 26406-08

Copyrighted material reproduced with the permission of General Electric Company, 406F02–406F04, 406SA02, 406SA03 (Pad Mounted), 406SA05, 406SA06, 406SA09
Phenix Technologies, 406SA01
Topaz Publications, Inc., 406SA03 (Pole Mounted), 406SA04, 406SA07
Jim Mitchem, 406SA08
Federal Pacific Transformer Co., 406SA10
Fluke Corporation, reproduced with permission, 406SA11

Module 26407-08

Eurotherm Controls, Inc., 407F01
Square D/Schneider Electric, 407F05, 407F06, 407SA02
Topaz Publications, Inc., 407F08, 407F10
Benshaw/Power Generation Systems International Inc., 407SA01
Eaton/Cutler-Hammer, 407F26
Extech, 407SA03
Greenlee Textron, Inc., a subsidiary of Textron, Inc. 407SA04, 407SA05

Module 26408-08

Topaz Publications, Inc., 408F02, 408SA01–408SA03, 408F06, 408F11, 408F16, 408F24, 408F27, 408F31, 408F33, 408F35, 408SA05A, 408SA07B
Courtesy of Honeywell International Inc., 408F14, 408F17, 408F18
Tim Dean, 408F18
Carrier Corporation, 408F26, 408F32, 408SA05B, 408SA07A
Square D/Schneider Electric, 408SA04

Index

Note: Figures are indicated with *f*.

A

Abrasives, surface preparation, 11.9
Absenteeism, leadership problems, 13.35–13.36
Absorption currents
    hi-pot testing and, 11.36
    insulation testing and, 10.17, 10.18*f*
Access to equipment, fire protective signaling systems, 5.35
Accidents
    costs of, 13.42–13.43
    investigation procedures, 13.48, 13.56
Accomplishments, as motivation, 13.26
Accounting departments, 13.11
Actual costs, 13.73–13.74
Addressable detectors
    conventional versus addressable commercial
        detectors, 5.8, 5.8*f*
    fire alarm initiating devices, 5.5–5.6, 5.9, 5.10
Addressable devices, 5.6, 5.9
    defined, 5.69
Addressable systems
    fire alarm control panels and, 5.6, 5.9
    fire alarm systems and, 5.5–5.6, 5.6*f*
    signaling line circuits, 5.6–5.7
    troubleshooting, 5.63
Adjustable frequency drives (AFD)
    base frequency and, 7.27
    classifications and nameplate markings, 7.29–7.30
    dynamic braking and, 7.28, 7.34–7.36, 7.35*f*, 7.36*f*
    motor controls and, 7.25–7.31, 7.26*f*, 7.28*f*, 7.30*f*, 7.31*f*,
        7.32*f*–7.33*f*, 7.33, 7.34
    NEC requirements, 7.25
    operating frequency range, 7.27, 7.28*f*
    operation, 7.25–7.28, 7.26*f*, 7.28*f*
    programmable parameters, 7.28–7.29
    pulsewidth-modulated drives, 7.25–7.26, 7.26*f*
    selection checklist, 7.32–7.33*f*
    selection criteria, 7.30
    types of adjustable speed loads, 7.30, 7.31*f*
Adjustable speed loads, 7.30, 7.31*f*
Advancement opportunities, as motivation, 13.26–13.27
Adverse locations. *See* Hazardous locations
Agricultural buildings
    electrical supplies from a distribution point, 12.19–12.20,
        12.19*f*
    equipotential planes and, 12.20, 12.20*f*
    motors and luminaires, 12.19, 12.19*f*
    wiring methods, 12.18–12.19, 12.19*f*
Air conditioning
    basic principles, 8.5–8.9, 8.6*f*, 8.9*f*
    branch circuits and, 8.36, 8.37, 8.38, 8.38*f*, 8.39*f*, 8.40*f*, 8.41,
        8.41*f*, 8.42, 8.42*f*
    chiller units, 8.10
    compressors, 8.5, 8.6*f*, 8.8
    condensers, 8.5, 8.6*f*, 8.9
    evaporators, 8.5, 8.6*f*, 8.8
    expansion devices, 8.5, 8.6*f*, 8.8, 8.9
    heat pumps, 8.9, 8.9*f*
    hot gas lines, 8.6, 8.6*f*, 8.8
    liquid lines, 8.6, 8.6*f*, 8.9
    refrigeration cycle and, 8.5, 8.6*f*, 8.8–8.9
    room air conditioner circuits, 8.36, 8.41, 8.42*f*
    suction lines, 8.6, 8.6*f*
Airflow controls, furnace controls, 8.25, 8.25*f*
Air gaps, induction motors, 7.43
Air handling units (AHU)
    conversion approximations, 5.56
    duct smoke detectors and, 5.56–5.57, 5.57*f*
Air sampling detectors, 5.9
    defined, 5.69
Air voids, 11.5, 11.7, 11.9
Alarms, 5.2. *See also* Fire alarm systems
    defined, 5.69
Alarm signals, 5.5–5.6. *See also* Fire alarm signals
    defined, 5.69
Alarm sound levels, 5.30–5.31
Alarm verification
    defined, 5.69
    initiating circuits, 5.25
Alcohol abuse. *See* Substance abuse
Alignment, motors and, 10.26, 10.27*f*
Alternate power sources, electrical distribution systems, 2.8,
    2.8*f*, 2.9*f*
Altitude, adjustable frequency drives, 7.34
Aluminum, as conductor, 4.4
Aluminum oxide (ALO), 11.9
Ambient-sensing controls, heat tracing system control, 9.9,
    9.9*f*
Ambient sound levels, 5.30–5.31
Ambient temperatures
    cooling operation checks and, 8.18
    diodes and, 4.8
    feeder conductor size adjustment and, 1.2, 1.3
    NEC requirements, 1.2, 1.3, 1.4
    solid-state devices and, 7.38
Ambient thermostat control, snow-melting and anti-icing
    systems, 9.25
Ambulatory health care occupancies, 2.2
American Medical Association, 13.42
American Red Cross, safety training, 13.47
American Society for Training and Development, 13.5
Americans with Disabilities Act (ADA)
    defined, 5.69
    fire notification appliances and, 5.26, 5.27
    manual (pull station) fire detection devices and, 5.41

Americans with Disabilities Act (ADA), *continued*
  notification device installation and, 5.53
  visual fire notification devices and, 5.27, 5.31
Ampere turns, 6.15
  defined, 6.31
Amusement rides
  permanent wiring of, 12.15–12.16, 12.16*f*
  wiring methods, 12.16–12.17, 12.17*f*
Analog addressable detectors, fire alarm initiating
    devices, 5.6, 5.9, 5.10, 5.12
Analog addressable systems, 5.5, 5.6–5.7, 5.6*f*, 5.9
Analog switching, solid-state relays, 7.3
Analog-to-digital converters
  defined, 8.53
  digital control systems and, 8.28
Ancillary control relays, 5.55, 5.56*f*
Angled-beam smoke detectors with mirrors, 5.50
Angular runout, 10.13–10.14, 10.13*f*, 10.14*f*
Anodes
  marking in diodes, 4.8–4.9, 4.9*f*
  as P-type material, 4.5, 4.5*f*
ANSI/IEEE standards, derating
    transformers, 6.26
ANSI standards, temporal three signals, 5.29
Antimony, semiconductors and, 4.4
Appliance loads, *NEC* requirements, 1.17–1.18, 1.19, 1.20,
    1.23, 1.24
Approved, 5.26
  defined, 5.69
Arc-lamp transformers, 6.12
Area lighting, permanently installed pools, 12.33–12.34,
    12.33*f*, 12.34*f*
Armatures
  AC generators, 3.2, 3.4, 3.4*f*
  defined, 3.30
Armor, 2.15, 2.17, 11.7
  defined, 11.49
Arsenic, 4.4
Assembly occupancies, 12.3–12.5, 12.3*f*, 12.4*f*, 12.5*f*
Assured equipment grounding program, temporary
    installations, 12.28–12.29
Atomic structure
  conductors and, 4.3–4.4, 4.4*f*
  electrical current and, 4.3
  electronic theory and, 4.2–4.3, 4.2*f*
  insulators and, 4.4, 4.4*f*
  valence shells and, 4.3–4.4, 4.4*f*
Attendant signals, 5.30
Attitude
  leadership problems and, 13.35, 13.37
  leadership skills and, 13.19
Audible fire notification devices
  fire notification appliances, 5.27–5.28, 5.28*f*
  NFPA standards, 5.27
  voice evacuation systems and, 5.28, 5.28*f*
Audible signals
  defined, 5.69
  as fire notification devices, 5.27–5.28, 5.28*f*
Authority
  delegating, 13.12
  leadership styles and, 13.19
Authority having jurisdiction (AHJ)
  defined, 5.69
  detector installation and, 5.47
  fire alarm control panels and, 5.26
  local codes and, 12.4
  signal considerations and, 5.30, 5.31

Autocratic leadership, 13.18–13.19
Automatic changeover thermostats, 8.12–8.14, 8.13*f*
  defined, 8.53
Automatic detectors, fire alarm initiating devices, 5.9
Automatic fire alarm systems, 5.2
  defined, 5.69
Automatic paralleling and transfer switchgear, 3.9
Automatic sensors (detectors), 5.7, 5.8*f*
Automatic sequential paralleling systems, 3.9–3.10, 3.10*f*
Automatic snow controllers, 9.26, 9.26*f*
Auto-mechanical fire detection equipment, 5.20, 5.21*f*, 5.22
Autotransformers
  buck-and-boost transformers and, 6.18–6.19, 6.19*f*
  defined, 6.31
  motor starting configurations, 10.4
  NEC requirements, 6.9, 6.10*f*, 7.16
  reduced-voltage starting motor control and, 7.16–7.18,
    7.16*f*, 7.17*f*, 7.24
  as types of transformers, 6.2, 6.7–6.9, 6.8*f*
Auxiliary power troubleshooting, 5.63, 5.65*f*
Avalanche breakover, diodes, 4.8
Awards, safety promotion, 13.50
Axial-lead silicon diodes, 4.9

B

Backstage lighting, 12.9, 12.10
Balanced ethics, 13.19–13.20
Banks
  defined, 6.31
  of transformers, 6.2
Bar charts, 13.68, 13.69*f*
Baseboard heaters, *NEC* requirements, 8.41–8.42, 8.43*f*
Base frequency, 7.27
Base speed, 7.28
  defined, 7.57
Basic lightning impulse insulation level (BIL), 11.30, 11.31
Battery and battery charger operation, 3.16–3.17, 3.16*f*
Battery back-up power
  fire alarm control panels, 5.26
  Type 3 EES and, 2.7
Battery-charging rectifiers, 6.12
Battery maintenance, 3.13–3.16
Battery-powered emergency lighting, *NEC* requirements, 3.25
Beamed ceilings, spot detector installation, 5.52
Bearing maintenance, 10.10, 10.11*f*, 10.12–10.14, 10.12*f*,
    10.13*f*, 10.14*f*
Bells, 5.4, 5.27
Belt-drive motors, alignment, 10.26
Benefits, employee turnover, 13.36
Bent shafts, 10.13–10.14, 10.13*f*, 10.14*f*, 10.15*f*
Bilingual job sites, 13.7
Bimetallic detectors, fixed-temperature heat detectors, 5.9,
    5.10, 5.10*f*, 5.11*f*
Bimetals, defined, 8.53
Bimetal sensors
  furnace limit controls, 8.23
  thermostats, 8.10–8.11, 8.11*f*
Boatyards. *See* Marinas and boatyards
Bodies of water (natural and manmade), as special
    occupancies, 12.44–12.46, 12.44*f*, 12.45*f*
Bonding requirements
  fountains, 12.43
  health care facilities, 2.9, 2.11
  pool accessories and, 12.39
Boost transformer connections, 6.19, 6.19*f*
Border lights, 12.9, 12.10
  defined, 12.50

Box shielding, 2.19
Branch circuits
    air conditioning and, 8.36, 8.37, 8.38, 8.38f, 8.39f, 8.40f, 8.41, 8.41f, 8.42, 8.42f
    critical care areas, 2.13–2.14
    general care areas, 2.12–2.13
    luminaires, 12.36
    small appliances and, 1.16, 1.17, 1.19, 1.24
Breakdown
    defined, 10.33
    hi-pot testing, 11.33–11.34
    insulation breakdown, 10.24, 10.24f
Bridge rectifiers, 4.6, 4.7f, 4.8, 4.10
Buck-and-boost transformers, 6.17–6.19, 6.18f, 6.19f
Building control circuits, 5.37, 5.37f
Buildings, 1.9
    defined, 1.41
Built-in siren drivers, FACP alarm outputs, 5.26
Busbars and connecting lugs, harmonics, 6.22
Business ethics, 13.19–13.20
Buzzers, 5.27, 5.28
Bypass isolation switches, 3.21, 3.22f

C

Cable breakouts, terminations, 11.29
Cable components, medium-voltage power cable, 11.4–11.5, 11.5f
Cable dielectric test forms, 11.51
Cable jackets, medium-voltage power cables, 11.4, 11.4f, 11.7, 11.7f
Cable preparation kits, splicing, 11.9
Cable preparation tools, 11.8
Cable protection and support, temporary installations, 12.27
Cable shield types, high-voltage power cable, 11.3, 11.4f
Cables in raceways, fire protective signaling systems, 5.37
Cable spacing, fire protective signaling systems, 5.37
Cables running between floors, 5.37
Cable Unishield, 11.3, 11.4f
CABO, 5.31
    defined, 5.69
CABO/ANSI standards, visual fire notification devices, 5.31
Calibrating thermostats, 8.19
Capacitance, dielectric absorption, 10.16–10.18
Capacitive stress control, 11.28, 11.30, 11.30f
Cardi-pulmonary resuscitation (CPR), 13.47
Carnivals, circuses, fairs, and similar events
    cords, cables, and connectors, 12.15–12.16, 12.16f
    grounding and bonding, 12.17, 12.17f
    overhead conductor clearances, 12.14, 12.15f
    power sources, 12.15, 12.15f
    rides, tents, and concessions, 12.16–12.17, 12.17f
    wiring for, 12.15–12.17, 12.16f, 12.17f
Carrier, Willis, 8.8
Cast-coil transformers, 6.6
Cathodes
    marking in diodes, 4.8–4.9, 4.9f
    as N-type material, 4.5, 4.5f
Ceiling heights
    defined, 5.69
    smoke detectors and, 5.48, 5.51, 5.52, 5.54
Ceilings
    defined, 5.69
    heat detector temperature ratings and, 5.11, 5.12
    photoelectric smoke detector installation on flat smooth ceilings, 5.49–5.50, 5.49f, 5.50f

spot detector installations on flat smooth ceilings, 5.48–5.49, 5.49f
    spot detector installations on irregular ceilings, 5.50–5.52, 5.51f
Ceiling surfaces, 5.52
    defined, 5.70
Cellular backups, 5.33, 5.34f
Cellular telephones, 13.6
Centralized building management systems, 8.27–8.28, 8.27f, 8.28f
Central station monitoring, 5.31
Centrifugal force, atomic theory, 4.2
Certification
    defined, 5.70
    splicing and, 11.3
Certified central station monitoring, 5.31
Changes, as motivation, 13.27
Charging currents, 10.18, 10.18f, 11.33
Chiller units, 8.10
Chimes
    as audible notification devices, 5.27
    defined, 5.70
Circuit boards, troubleshooting, 7.52–7.53
Circuit breakers
    GFCI circuit breakers, 12.34, 12.34f, 12.36
    harmonics and, 6.19, 6.22
    troubleshooting, 7.47, 7.47f, 7.49, 7.49f
Circuits. See also Branch circuits; Class A circuits; Class B circuits; Control circuits; Emergency system circuits for light and power; Motors and motor circuits
    defined, 5.70
    identification of, 5.4
Circuses. See Carnivals, circuses, fairs, and similar events
Class A circuits
    defined, 5.70
    initiation device circuits and, 5.38–5.39, 5.40f
    notification appliance circuits and, 5.40–5.41, 5.40f
    signaling line circuits, 5.7
Class B circuits
    defined, 5.70
    end-of-line devices and, 5.38, 5.39f
    fire alarm control panels and, 5.38
    initiation device circuits and, 5.38
    notification appliance circuits and, 5.40, 5.40f
    notification appliance installation and, 5.53
    signaling line circuits, 5.7
Class C fire extinguishers, 3.15
Classified locations, 2.16–2.19
Cleaning insulation, 11.13
Closed circuit transitions, 7.16–7.17, 7.18–7.19
    defined, 7.57
Closed-construction, docks, piers, and wharves, 12.25–12.26
Clothes dryers, NEC requirements, 1.10, 1.23, 1.24
Cloud chamber smoke detectors, 5.15, 5.16f
Code Blue, 2.5
Coded signals, 5.31
    defined, 5.70
Cogeneration, engine-driven generator sets, 3.6
Coil resistance tests, 7.50, 7.52, 7.52f
Cold-shrink insulators, 11.25, 11.25f
Color-coding
    emergency power systems and, 2.5, 2.11, 2.11f
    isolated power systems, 2.23
Combination heat detectors, fire alarm initiating devices, 5.10, 5.11f
Commercial automatic sensors, 5.8, 5.8f

Commercial Building Standard for Telecommunications Pathways and Spaces, ANSI/EIA/TIA 569, 5.34
Commercial Building Telecommunications Wiring Standard, ANSI/EIA/TIA 568, 5.34
Commercial occupancy load calculations
    lighting loads, 1.25–1.26, 1.27
    office buildings, 1.28–1.31, 1.29*f*
    retail stores with show windows, 1.26, 1.27–1.28, 1.27*f*
    types of loads, 1.24, 1.31
Commissioning guidelines, motors, 10.25–10.29, 10.27*f*, 10.28*f*
Communication
    e-mails, 13.6
    language barriers, 13.7
    men's versus women's communication styles, 13.7
    non-verbal communication, 13.22
    planning assignments and, 13.58, 13.59*f*
    styles of, 13.23, 13.24*f*
    verbal communication, 13.20–13.22, 13.21*f*, 13.24–13.25, 13.24*f*
    written or visual communication, 13.23, 13.24*f*, 13.25
Communications and monitoring
    cellular backups, 5.33, 5.34*f*
    digital communicators, 5.31–5.33, 5.32*f*
    fire alarm signals, 5.5–5.6
    monitoring options, 5.31
Communication systems
    handheld communication devices, 13.6
    health care facilities, 2.21–2.22, 2.21*f*
Compact stranding, cable components, 11.5, 11.5*f*
Compliance Safety and Health Officers (CSHOs), 13.43
Compressed stranding, cable components, 11.4, 11.5*f*
Compressors
    air conditioning, 8.5, 8.6*f*, 8.8
    branch and control circuits, 8.36, 8.39*f*
    compressor short-cycle timers, 8.19, 8.22
    defined, 8.53
    hermetic and semi-hermetic (serviceable) compressors, 8.7, 8.36, 8.37, 8.45
    NEC requirements, 8.36, 8.37, 8.38*f*, 8.39*f*
Computer and Business Equipment Manufacturers' Association (CBEMA), 6.26
Computer control. *See also* Personal computers
    fire alarm control panels, 5.24
    harmonics and, 6.19
    heat tracing system control and, 9.13–9.14
    HVAC direct digital control, 8.27–8.28, 8.27*f*, 8.28*f*
Concentric neutral (CN), 11.3, 11.4*f*
Concentric stranding, medium-voltage power cable, 11.4, 11.5*f*
Concessions, wiring methods, 12.16–12.17, 12.17*f*
Condensation, 9.2
Condensers
    air conditioning, 8.5, 8.6*f*, 8.9
    defined, 8.53
Conductor configurations, 11.4
Conductors
    atomic structure, 4.3–4.4, 4.4*f*
    splicing, 11.14, 11.15*f*
Connector strips, 12.9
Constant-current transformers, 6.9–6.10
Constant horsepower loads, 7.30, 7.31*f*
Constant torque loads, 7.30, 7.31*f*
Constant volume HVAC systems, 8.28–8.29, 8.29*f*
Constant wattage heating cables, defined, 9.36
Constant-wattage series resistance heating cables, 9.5, 9.6–9.7, 9.6*f*. *See also* Mineral-insulated (MI) heating cables
Construction equipment
    equipment control, 13.77–13.78
    planning and, 13.59
Construction industry
    changes in company operations and, 13.4–13.6
    construction company departments, 13.11
    construction organizations, 13.11–13.13
    gender and cultural issues, 13.6–13.9
    growth and economics, 13.1–13.2
Construction job site trailers, wiring methods for, 12.2
Construction management, as project delivery system, 13.54, 13.54*f*
Construction materials
    materials control, 13.76–13.77
    planning and, 13.59, 13.65, 13.66*f*, 13.67
Construction operation departments, 13.11
Construction phase of projects, 13.52, 13.53*f*
Construction planning, 13.56
Construction projects, phases of, 13.51–13.52, 13.53*f*
Contact closure, troubleshooting, 7.52, 7.52*f*, 7.53
Contactor coil resistance tests, 7.50, 7.52, 7.52*f*
Contact replacement, 7.39
Contact types, timing relays, 7.10, 7.10*f*
Contamination
    insulation resistance measurements and, 11.46
    terminations and, 11.31
Contests, safety-related, 13.49
Continuity, troubleshooting contacts, 7.52, 7.52*f*
Continuity monitoring, heat tracing system monitoring, 9.11–9.12, 9.12*f*
Continuous duty loads
    feeder conductor size adjustment, 1.2, 1.3–1.4
    lighting loads and, 1.26, 1.27
Contract administration departments, 13.11
Control circuits
    building control circuits, 5.37, 5.37*f*
    examples of heating, ventilating, and air conditioning (HVAC) controls, 8.30, 8.31*f*, 8.32*f*, 8.33
    input control circuits, 7.2–7.3, 7.3*f*
    lockout control circuits, 8.20–8.21, 8.21*f*
    NEC requirements, 8.34, 8.35*f*, 8.36
    safety, 7.44
    safety switches, 8.21*f*, 8.22–8.23, 8.23*f*
Control power transformers, 6.12
Control process, project control, 13.76
Control systems. *See also* Digital control systems
    cooling system control circuit, 8.30, 8.31*f*, 8.32*f*, 8.33
    heating, ventilating, and air conditioning (HVAC) controls, 8.20–8.26, 8.20*f*, 8.21*f*, 8.22*f*, 8.23*f*, 8.24*f*, 8.25*f*, 8.26*f*
    motor speed controls, 8.20, 8.20*f*, 8.21
Control transformers
    control power transformers, 6.12
    maximum inrush VA, 6.10–6.11, 6.11*f*
    troubleshooting, 7.50, 7.52*f*
Control units, 5.59
    defined, 5.70
Conventional initiation device circuits, 5.38–5.39, 5.39*f*, 5.40*f*
Cooling compensators
    cooling-only thermostats and, 8.12, 8.12*f*
    defined, 8.53
Cooling/gas heating system circuits, 8.30, 8.32*f*, 8.33
Cooling-only thermostats, 8.12, 8.12*f*
Cooling system control circuits, 8.30, 8.31*f*, 8.32*f*, 8.33
Cooling systems, engine-driven generator sets, 3.5
Copper, as conductor, 4.4, 4.4*f*
Cord-and-plug connections
    fountains, 12.43
    swimming pools, fountains, hot tubs, and similar installations, 12.30–12.31

Core loss
    defined, 6.31
    induction motors and, 10.4, 10.7
    transformers, 6.23
Corona currents, high potential (hi-pot) tests, 11.36,
    11.39, 11.42
Corona guard rings and guard shields, 11.42, 11.42f
Costs
    of accidents, 13.42–13.43
    direct and indirect, 13.62
    project control and, 13.73–13.74
Craft training, 13.4, 13.5
Craftworkers, 13.15–13.16, 13.17f
Crankcase heaters, 8.18, 8.33
Crest factor, harmonics, 6.25
Crew evaluations, 13.19
Crew leaders, 13.15, 13.16, 13.17f
Critical branches, essential electrical systems for health care
    facilities, 2.3–2.5, 2.4f, 2.6–2.7
Critical care areas
    hospital wiring and devices, 2.13–2.15, 2.14f
    as patient care area, 2.3
Critical path method (CPM), scheduling, 13.68
Cross-linked polyethylene insulation (XLP), 11.6
Cultural differences. See Gender and cultural issues
Current draw, thermostats and, 8.17, 8.17f
Current monitoring, heat tracing system monitoring, 9.11,
    9.12, 9.12f
Current transformers
    connections of, 6.14–6.15, 6.15f
    polarity marks, 6.14, 6.14f, 7.9
    primary and secondary windings of, 6.13–6.14
    three-phase, 6.16
    types of, 6.2–6.3, 6.3f
Current-versus-time test, high potential (hi-pot) tests,
    11.33–11.35, 11.34f, 11.35f
Curtain machines, 12.9
Customer interfaces, troubleshooting, 7.39–7.40
Cycle rates, electronic thermostats, 8.18

D

Dance halls, as assembly occupancies, 12.3, 12.3f
Dashpots, 7.9
    defined, 7.57
Dashpot timing relays, 7.12
Data systems, health care facilities, 2.21–2.22, 2.21f
DC (direct current), 6.4, 11.32
Deadbands
    defined, 8.53
    thermostats and, 8.12–8.13
Dead front, 12.6
    defined, 12.50
Dead-leg, 9.10
Dead-leg controls, heat tracing system control, 9.10
Dead legs, defined, 9.36
Decision making, leadership skills, 13.31–13.33, 13.34,
    13.35–13.37
Deck area heating, 12.39
Defense Intelligence Agency (DIAM-50-3) standards, for fire
    alarm systems, 5.3
Defrost test cycles, 8.26, 8.26f
De-icing systems
    component selection and installation, 9.23–9.25, 9.24f
    installation guidelines, 9.23–9.24, 9.24f
    installation of heating cable and components, 9.23–9.24,
    9.24f
    NEC requirements, 9.23, 9.24–9.25

for roofs, gutters, and downspouts, 9.21, 9.22, 9.22f, 9.23f
    wiring for single and multiple circuit, 9.21, 9.22f
Delegating authority, 13.12
Delegating responsibilities, 13.29–13.30
Delta connections, transformers, 6.2, 6.3–6.7, 6.5f
Demand factors
    commercial and industrial load calculations, 1.25
    farm load calculations, 1.34, 1.35f
    hotels and motels, 1.33
    load calculation procedures, 1.10–1.11, 12.22–12.23
    marinas and boatyards, 1.34
    multi-family load calculations, 1.21, 1.23
    restaurants, 1.31
Democratic leadership, 13.18–13.19
Derating transformers, harmonics, 6.26–6.27
Design-build, as project delivery system, 13.54, 13.54f
Detector mountings, fire protective signaling systems, 5.35
Determining degree of slope, spot detector installation, 5.50,
    5.51–5.52, 5.51f
Development phase of projects, 13.51–13.52, 13.53f
Dew points
    condensation in pipelines and, 9.2
    defined, 9.36
Diacs
    defined, 4.20
    electronic theory and, 4.16, 4.16f
    as thyristors, 4.14–4.15
Dial indicators, 10.13, 10.13f, 10.14, 10.14f, 10.15f
Dielectric, 10.16–10.17
Dielectric absorption
    hi-pot testing and, 11.44
    insulation testing, 10.16–10.18, 10.18f, 10.19, 10.20
Dielectric absorption ratios (DAR), 10.19, 10.20
Dielectric constant (K)
    capacitive stress control and, 11.28
    defined, 11.49
Diesel engine generators, 2.8, 2.8f, 3.2, 3.5
Differentials
    defined, 8.53
    thermostats and, 8.13
Digital Alarm Communicator and Transmitters (DACT),
    defined, 5.70–5.71
Digital Alarm Communicator Receivers (DACR)
    defined, 5.70
    digital communicators, 5.33
Digital Alarm Communicator Systems (DACS)
    defined, 5.70
    digital communicators, 5.33
Digital Alarm Communicator Transmitters (DACT), 5.33
Digital communicators
    National Fire Protection Association standards, 5.33
    RJ31-X modular jack, 5.32, 5.32f, 5.33
    Wide Area Telephone Service (WATS) and, 5.33
Digital control systems
    analog-to-digital converters and, 8.28
    controlling devices, 8.28, 8.28f
    direct digital controls, 8.27–8.28, 8.27f, 8.28f
    heating, ventilating, and air conditioning (HVAC)
    controls, 8.26–8.30, 8.27f, 8.28f, 8.29f
    microprocessors and, 8.28
Digital surge/DC hipot/resistance testers, 10.17
Digital voice evacuation messages, 5.29
Dimmers, fixed stage switchboards and, 12.7, 12.8f
Diodes
    axial-lead silicon, 4.9
    electronic theory and, 4.5–4.6, 4.5f, 4.7f, 4.8–4.9, 4.8f,
    4.9f, 4.10f

Diodes, *continued*
  forward current values and, 4.8
  germanium, 4.8
  identification, 4.8–4.9, 4.9*f*
  light-emitting diodes, 4.11–4.12, 4.11*f*
  marking of cathode and anode, 4.8–4.9, 4.9*f*
  peak current values and, 4.8
  peak inverse voltage and, 4.8, 4.9
  PN junction and, 4.5, 4.5*f*
  in rectifiers, 4.5
  silicon and, 4.8
  testing with ohmmeters, 4.9, 4.10*f*
  zener diodes, 7.7
Direct costs, 13.62
Direct current (DC), 6.4, 11.32
Direct current (DC) braking of AC motors, 7.34–7.35, 7.35*f*
Direct digital controls, 8.27–8.28, 8.27*f*, 8.28*f*
Dirt build-up, analog addressable detectors, 5.6
Disc assemblies, 11.30
Discharge lines. *See* Hot gas lines
Disconnecting means
  agricultural buildings, 12.20
  generators at carnivals, 12.17, 12.17*f*
  marinas and boatyards, 12.24, 12.25*f*
  temporary installations, 12.27
  underground conductors, 12.32, 12.32*f*
  X-ray installations, 2.20
Discrimination, gender and cultural issues, 13.9
Distribution transformers, 6.11
Docks. *See* Piers
Domestic hot-water temperature maintenance systems,
  9.29–9.30, 9.29*f*
Doorbell transformers, 6.12
Door locking arrangements, *NFPA* standards, 5.60
Doping, semiconductors, 4.4
Double-action pull stations, 5.19, 5.19*f*
Double-conversion UPS systems, static uninterruptible
  power supplies, 3.17, 3.18, 3.19*f*
Double-ended substations, electrical distribution
  systems, 2.7–2.8, 2.8*f*
Double-insulated water pumps, 12.38
Doweling, 10.29
Downspout de-icing. *See* De-icing systems
Drain wire shielding, 11.3, 11.4*f*
Dressing rooms, 12.14, 12.14*f*
Drug abuse. *See* Substance abuse
Dry chemical extinguishing systems, *NFPA* standards, 5.60
Dry contacts, 5.26
Drying water-damaged motors, 10.25
Dry-niche luminaires, 12.36
Dry sprinkler systems, 5.20, 5.21*f*, 5.22
Duct smoke detectors
  air handling units (AHU) and, 5.56–5.57, 5.57*f*
  fire alarm-related systems, 5.55–5.58, 5.57*f*, 5.58*f*
  functions of, 5.14, 5.16
  installation and connections, 5.57–5.58, 5.58*f*
  locating, 5.56
  NFPA standards, 5.15, 5.55, 5.56
  as projected beam smoke detectors, 5.14–5.15, 5.14*f*
  remote indicators, 5.58, 5.58*f*
  Underwriters Laboratories (UL) standards, 5.55
Duty cycle, adjustable frequency drives, 7.34
Dwelling units
  defined, 1.41
  lighting loads and, 1.11
Dynamic braking, adjustable frequency drives, 7.28,
  7.34–7.36, 7.35*f*, 7.36*f*

E

Eddy currents
  defined, 6.31
  harmonics and, 6.22, 6.23
Electrical datum planes
  bodies of water and, 12.44
  defined, 12.50
  marinas and boatyards and, 12.21
Electrical distribution systems, health care facilities, 2.7–2.9,
  2.8*f*, 2.9*f*, 2.10, 2.11
Electrical noise. *See* Noise
Electrical panels
  harmonics and, 6.22
  roof de-icing and, 9.22
Electrical supplies from distribution point, agricultural
  buildings and, 12.19–12.20, 12.19*f*
Electric fields, stress control, 11.26, 11.27*f*, 11.28
Electric flux lines, stress control, 11.26, 11.27*f*
Electric heat troubleshooting chart, 8.48
Electric pool heaters, 12.31, 12.38
Electric range loads, *NEC* requirements, 1.17, 1.18, 1.42
Electric space heating cables
  installation of heating cable and components, 8.42, 8.44*f*
  NEC requirements, 8.42, 8.44*f*
Electric vehicle charging systems, 12.3
Electrified truck parking spaces, 12.15
Electrocutions, underwater lighting, 12.35
Electroluminescence, 4.11, 4.11*f*
Electrolytes
  defined, 3.30
  storage batteries and, 3.11–3.12
Electromagnetic interference (EMI)
  terminal and busbar connections and, 7.42
  troubleshooting alarm systems and, 5.63
  uninterruptible power supplies, 3.17
Electromechanical braking, motor braking methods, 7.36, 7.37*f*
Electromechanical relays
  motor controls, 7.2, 7.3–7.5
  solid-state relays compared to, 7.2, 7.3–7.5
Electron flow, transistors, 4.12, 4.13
Electronic air cleaners, 8.2, 8.5*f*
Electronic controls, troubleshooting, 8.49
Electronic defrost controls, heat pumps, 8.25–8.26, 8.26*f*
Electronic theory
  atomic structure and, 4.2–4.3, 4.2*f*
  diacs and, 4.16, 4.16*f*
  diodes and, 4.5–4.6, 4.5*f*, 4.7*f*, 4.8–4.9, 4.8*f*, 4.9*f*, 4.10*f*
  light-emitting diodes and, 4.11–4.12, 4.11*f*
  rectifiers and, 4.6, 4.7*f*
  semiconductors and, 4.2, 4.3–4.4, 4.4*f*, 4.5*f*
  silicon-controlled rectifiers and, 4.14–4.15, 4.16*f*
  solid-state devices and, 4.2
  transistors and, 4.12–4.14, 4.12*f*, 4.13*f*, 4.14*f*
  triacs and, 4.16, 4.16*f*, 4.17
Electronic thermostats, 8.10, 8.14–8.15, 8.15*f*
Electrons, atomic structure, 4.2–4.3, 4.2*f*, 4.4
Electrostatic discharge. *See* Static electricity
Elevator lobby detectors, 5.59, 5.60
Elevator recall, 5.55, 5.59, 5.59*f*, 5.60*f*
Elevator shafts, fire protective signaling systems, 5.37
E-mail, 13.6
Embedded heating cables, permanently installed
  pools, 12.39
Emergency and standby power system components
  automatic sequential paralleling systems, 3.9–3.10, 3.10*f*
  engine-driven generator sets, 3.2, 3.3, 3.4–3.6, 3.4*f*
  one-line diagram of, 3.3

storage batteries, 3.11–3.17, 3.11*f*, 3.12*f*, 3.16*f*
transfer switches, 2.4, 3.7–3.9, 3.7*f*, 3.24, 3.24*f*
Emergency circuit identification, *NEC* requirements, 2.5, 3.25, 3.25*f*, 3.26
Emergency lighting
  battery-powered, 3.25
  emergency lighting units, 3.26, 3.26*f*
  lighting fixtures, 2.6
  NEC requirements, 2.18, 3.25
Emergency power systems
  batteries and, 2.3
  components, 3.2, 3.3, 3.3*f*, 3.4–3.10, 3.4*f*, 3.7*f*, 3.10*f*
  emergency system circuits for light and power, 2.3–2.7, 2.3*f*, 2.4*f*, 3.23–3.26, 3.24*f*, 3.25*f*, 3.26*f*
  health care facilities and, 2.3–2.7, 2.3*f*, 2.4*f*
  introduction, 2.2–2.3, 3.2
  legally required standby systems, 3.2, 3.21–3.22, 3.23*f*
  static uninterruptible power supplies, 3.17–3.19, 3.19*f*, 3.20
Emergency shutoff switches, permanently installed pools, 12.34, 12.34*f*
Emergency system circuits for light and power
  battery-powered emergency lighting and, 3.25
  health care facilities, 2.3–2.7, 2.3*f*, 2.4*f*, 3.23–3.25
  NEC requirements, 2.4, 2.6, 2.7, 3.23, 3.25
  for places of assembly, 3.26
Employees. *See also* Labor; Workers
  absenteeism and, 13.35–13.36
  attitudes of, 13.35, 13.37
  motivation and, 13.4, 13.26–13.27, 13.28
  safety responsibilities, 13.44–13.46
  training, 11.15, 13.2, 13.4, 13.5–13.6, 13.45
End-of-line (EOL) devices
  class B circuits and, 5.38, 5.39*f*
  defined, 5.71
  fire notification appliance circuits and, 5.41
  fire notification device installation and, 5.53
  hardwired systems and, 5.4, 5.5*f*
Endplay
  adjustment, 10.27, 10.28
  motor maintenance and, 10.13
Engine-driven generator sets
  cogeneration, 3.6
  cooling systems, 3.5
  emergency and standby power system components, 3.2, 3.3, 3.4–3.6, 3.4*f*
  fuels, 3.5–3.6
  generator engines, 3.2, 3.4
  maintenance and service, 3.6
  oversizing, 3.6
  principles of operation, 3.2, 3.4, 3.4*f*
  sizing and selection considerations, 3.4–3.6
  starting systems, 3.5
Environment, adjustable frequency drives, 7.34
Equal Employment Opportunity Commission (EEOC), 13.8
Equalizing charge, battery maintenance, 3.13
Equipment control, 13.77–13.78
Equipment grounding
  assured equipment grounding program, 12.28–12.29
  hospital receptacles and fixed equipment, 2.15–2.16
  underwater luminaires and, 12.36
Equipment systems, 2.3, 2.4
Equipotential bonding, 12.38–12.39, 12.38*f*, 12.39*f*
Equipotential ground plane, 2.16
  defined, 2.27
Equipotential lines, stress control, 11.26, 11.27*f*, 11.28

Equipotential planes
  agricultural buildings and, 12.20, 12.20*f*
  defined, 12.50
  installations in or near bodies of water, 12.45–12.46
Essential electrical systems (EES)
  NEC requirements, 2.4, 2.5, 2.6, 2.7, 3.23, 3.25
  types of, 2.3–2.7, 2.3*f*, 2.4*f*
Essential loads, 2.3, 2.3*f*
Estimated costs, 13.73–13.74
Estimating departments, 13.11
Estimations
  direct and indirect costs, 13.62
  estimating process, 13.62, 13.63*f*, 13.64*f*
  material quantities and, 13.65, 13.65*f*, 13.66*f*, 13.67
Ethics, in leadership, 13.19–13.20
Ethylene propylene rubber (EPR) insulation, 11.6
Evaporators
  air conditioning, 8.5, 8.6*f*, 8.8
  defined, 8.53
Exciters
  defined, 3.30
  engine-driven generator sets and, 3.2, 3.4, 3.4*f*
Executive departments, 13.11
Exit lighting fixtures, 3.25, 3.26, 3.26*f*
Expansion devices
  air conditioning, 8.5, 8.6*f*, 8.8, 8.9
  defined, 8.53
Experience, leadership styles, 13.19
Expertise, leadership styles, 13.19
Extension cords, marinas and boatyards, 12.23, 12.24
Exterior signs and lighting, *NEC* requirements, 1.26
External analog devices, 8.28
External digital devices, 8.28
External leakage insulation, terminations, 11.30–11.31, 11.31*f*
Extra-hard usage cable, defined, 12.50
Extruded semi-conductors, peeling, 11.8
Eyebolt lifting capacity, 10.23

F

Factory Mutual (FM) standards, fire alarm systems, 5.3
Fairs. *See* Carnivals, circuses, fairs, and similar events
Fall protection, safety, 9.16, 9.23
Falls, accident statistics, 13.42
False alarms
  fire alarm initiating devices, 5.9
  fire protective signaling systems, 5.47
Farm load calculations, 1.34–1.36, 1.35*f*
Fastening heating cable, 9.18
Fatalities, construction workers, 13.42
Fault current studies, 1.15
Faults, defined, 5.71
Faulty insulation, hi-pot testing, 11.35–11.36, 11.35*f*
Federal Bank Protection Act standards, fire alarm systems, 5.3
Feedback
  fuse checking, 7.47
  verbal communication and, 13.20, 13.21*f*, 13.22
Feeder ampacity
  feeder conductor size adjustment and, 1.2, 1.3–1.4, 1.7–1.8
  load calculation procedures and, 1.7–1.8
  NEC requirements, 1.8
Feeder circuit load calculations
  basic steps for performing, 1.11, 1.12*f*
  calculation procedures, 1.2–1.12, 1.5*f*, 1.12*f*
  commercial occupancy calculations, 1.24–1.25
  farm load calculations, 1.34–1.36, 1.35*f*
  hotels and motels, 1.33

Feeder circuit load calculations, *continued*
  introduction to, 1.2
  motors and motor circuits, 1.36
  office buildings, 1.30–1.31
  schools, 1.33–1.34
  shore power circuits for marinas and boatyards, 1.34
Feeder circuit overcurrent protection devices, 1.2
Feeder conductor size adjustments
  ambient temperatures and, 1.2, 1.3
  continuous duty loads and, 1.2, 1.3–1.4
  examples, 1.3–1.4, 1.5–1.7, 1.8
  feeder ampacity and, 1.2, 1.3–1.4, 1.7–1.8
  load calculation procedures, 1.2–1.7
  NEC requirements, 1.2–1.7
  temperature limitations for terminating conductors, 1.7
  voltage drop and, 1.2, 1.4–1.7
Feeders
  defined, 1.41
  permanently installed pools, 12.37–12.38
  stage switchboard feeders, 12.8–12.9
Festoon lighting, defined, 12.50
Field-effect transistors (FETs)
  defined, 4.20
  junction field-effect transistor (JFET), 4.14, 4.14f
Field reporting, actual costs, 13.73–13.74
Finish ratings, assembly occupancies, 12.4, 12.5, 12.5f
Fire alarm circuit identification, fire protective signaling
    systems, 5.35
Fire alarm control panels (FACP)
  addressable systems and, 5.6, 5.9
  alarm outputs, 5.26
  analog addressable systems and, 5.6, 5.9
  authority having jurisdiction and, 5.26
  class B circuits and, 5.38
  initiating circuits and, 5.25
  inputs and outputs, 5.23, 5.23f
  installation guidelines, 5.54–5.55, 5.55f
  intelligent systems, 5.23, 5.23f
  introduction, 5.23–5.24
  labeling, 5.25
  listings, 5.26
  NFPA standards, 5.26, 5.54, 5.55, 5.55f
  personal computers and, 5.24
  primary power requirements, 5.26, 5.41
  protection of, 5.54, 5.55f
  secondary power and, 5.26
  smoke detectors and, 5.6
  troubleshooting, 5.65f
  Underwriters Laboratories (UL) standards, 5.26
  user control points, 5.23f, 5.24, 5.24f
  voice evacuation systems and, 5.28, 5.29f
  water flow alarms and, 5.22
Fire alarm initiating devices. *See also* Heat detectors; Smoke
    detectors
  addressable detectors, 5.5–5.6, 5.9, 5.10
  analog addressable detectors, 5.6, 5.9, 5.10, 5.12
  automatic detectors, 5.9
  auto-mechanical fire detection equipment, 5.20, 5.21f, 5.22
  commercial automatic sensors, 5.8, 5.8f
  detection versus stages of fires, 5.7, 5.8f
  detector installation, 5.35
  detector mounting, 5.35
  false alarms and, 5.9
  fire suppression systems, 5.7
  infrared (IR) flame detectors, 5.8f, 5.15, 5.18, 5.18f,
    5.43, 5.43f
  manual (pull station) fire detection devices, 5.4, 5.7,
    5.18–5.19, 5.19f, 5.20

UL standards, 5.7
  ultraviolet (UV) flame detectors, 5.8f, 5.15, 5.18, 5.18f,
    5.43, 5.43f
Fire alarm-related systems
  ancillary control relays, 5.55, 5.56f
  door locking arrangements, 5.60
  duct smoke detectors, 5.55–5.58, 5.57f, 5.58f
  elevator recall, 5.55, 5.59, 5.59f, 5.60f
  fire suppression systems, 5.7
  installation guidelines, 5.55–5.63, 5.56f, 5.57f, 5.58f,
    5.59f, 5.60f
Fire alarm signals. *See also* Fire protective signaling systems
  communications and monitoring, 5.5–5.6
  fire notification appliances, 5.28–5.31
  NFPA standards, 5.29, 5.30
Fire alarm systems. *See also* Fire alarm control panels
    (FACP); Fire alarm initiating devices; Total premises
    fire alarm system installation guidelines
  addressable and analog addressable systems, 5.5–5.7,
    5.6f, 5.9
  codes and standards, 5.2–5.4
  communications and monitoring, 5.31–5.33, 5.32f, 5.34f
  conventional hardwired systems, 5.4, 5.5f
  Defense Intelligence Agency (DIAM-50-3) standards, 5.3
  Factory Mutual (FM) standards, 5.3
  Federal Bank Protection Act standards, 5.3
  fire notification appliances, 5.4, 5.26–5.31, 5.27f, 5.28f,
    5.29f, 5.40–5.41, 5.52–5.53, 5.54
  health care facilities, 2.21–2.22, 2.21f
  installation guidelines, 5.2, 5.34–5.35, 5.36, 5.37–5.41, 5.37f,
    5.39f, 5.40f, 5.46
  multiplex systems, 5.4–5.5, 5.5f
  National Electrical Manufacturers' Association, 5.3
  National Institute for Certification in Engineering
    Technologies standards, 5.3
  NEC requirements, 5.34
  overview of, 5.2, 5.4–5.7, 5.5f, 5.6f
  power supplies, 5.5
  system components, 5.7
  troubleshooting, 5.48, 5.63, 5.64f, 5.65f
  UL standards, 5.3
  wiring requirements, 5.34
  zones and, 5.4
Fire notification appliances
  Americans with Disabilities Act and, 5.26, 5.27
  audible notification devices, 5.27–5.28, 5.28f
  circuits, 5.40–5.41
  fire alarm signal considerations, 5.28–5.31
  fire alarm systems, 5.4
  installation of, 5.52–5.53, 5.54
  NFPA installation standards, 5.52, 5.53, 5.54
  visual fire notification devices, 5.27, 5.27f
  voice evacuation systems, 5.28, 5.29f
Fire prevention, 13.47
Fire protection, safety, 13.47
Fire protective signaling systems
  access to equipment, 5.35
  cable requirements, 5.37
  detector mountings, 5.35
  elevator shafts and, 5.37
  false alarms, 5.47
  fire alarm circuit identification, 5.35
  fire seals, 5.35
  initiating device circuits, 5.59
  NEC requirements, 5.3–5.4, 5.34
  network command centers, 5.38
  notification appliance circuits, 5.40–5.41, 5.40f
  outdoor wiring of fire alarm circuits, 5.35

power-limited circuits in raceways, 5.35
primary power requirements, 5.41
remote control signaling circuits, 5.37, 5.37f
secondary power requirements, 5.41
terminal wiring methods and, 5.37, 5.39f
wiring in air handling spaces, 5.35, 5.37
wiring in hazardous locations, 5.37
workmanship requirements, 5.34
Fire pumps, 5.61
Fire ratings, 12.3, 12.4, 12.5
defined, 12.50
Fires, 5.2
defined, 5.71
Fire seals, fire protective signaling systems, 5.35
Firestats, furnace controls, 8.24
Firestopping materials, 5.36
Fire suppression systems
dry chemical extinguishing systems, 5.60
supervision, 5.60–5.63
wet chemical extinguishing systems, 5.60
First aid, 13.46–13.47
First coupled startup, 10.28
First-time startup procedures, 10.28
Fixed electric space heating equipment, 8.34, 8.41–8.42, 8.43f
Fixed heat detectors, 5.8f
Fixed stage switchboards
design and construction, 12.6–12.7, 12.7f
dimmers and, 12.7, 12.8f
stage switchboard feeders, 12.8–12.9
switchboard control types, 12.8, 12.8f
Fixed-temperature heat detectors
bimetallic detectors, 5.9, 5.10, 5.10f, 5.11f
detection versus stages of fires, 5.8f
fusible link, 5.9, 5.9f, 5.10, 5.11f
quick metal detector, 5.9, 5.10f
Flame detectors
defined, 5.71
infrared (IR) flame detectors, 5.8f, 5.15, 5.18, 5.18f, 5.43, 5.43f
installation, 5.43, 5.43f
ultraviolet (UV) flame detectors, 5.8f, 5.15, 5.18, 5.18f, 5.43, 5.43f
Flashovers, terminations, 11.30
Flexible couplings, 10.26
Floating buildings, 12.21
Float voltage
battery charging and, 3.5, 3.16, 3.16f
battery maintenance, 3.13, 3.15, 3.16
Floor heating and warming systems
component selection and installation, 9.32
heat tracing and freeze protection, 9.30–9.31, 9.31f
NEC requirements, 9.32
radiant floor heating systems, 9.30–9.31, 9.31f
Follow-up, planning, 13.58, 13.58f
Footcandles (fc), infrared (IR) flame detectors, 5.43
Footlights, 12.9–12.10
defined, 12.50
Forced-air heating, 8.2, 8.3f
Formal organizations, 13.11
Form FA converters, 7.29
Form FB converters, 7.29
Form FC converters, 7.29
Forming shells, 12.34–12.35, 12.35f
Formulas
adjustable frequency drive speed, 7.25
derating transformers, 6.26
dielectric absorption ratios (DAR), 10.19
for feeder ampacity, 1.7
polarization index ratios (PIR), 10.21

sizing buck-and-boost transformers, 6.18–6.19
for voltage drop, 1.4, 1.6
Forward bias
defined, 4.20
diodes and, 4.5, 4.5f
diode testing and, 4.9, 4.10f
Forward current values, diodes, 4.8
Fountains, 12.41–12.43, 12.41f, 12.42f
Four-pole AC generators, 3.4, 3.4f
Freeze protection. See Heat tracing and freeze protection
Freezestats, 8.23, 8.23f
Freeze-up protection thermostats (FPT), 8.23, 8.23f
Friction braking, 7.36, 7.37f
Fuel dispensing facilities, marinas and boatyards, 12.25–12.26, 12.26f
Fuels
engine-driven generator sets, 3.5–3.6
heating, 8.2
Full load amperage (FLA), 7.7
Full-wave rectifiers, 4.6, 4.7f
Furnace controls
airflow controls, 8.25, 8.25f
firestats, 8.24
inducer switches, 8.25, 8.25f
limit controls, 8.23, 8.23f
thermocouples, 8.23–8.24, 8.24f
Fuse checks, 7.47, 7.47f
Fuse/circuit breaker checks, troubleshooting, 7.47, 7.47f, 7.49, 7.49f
Fusible line-type heat detectors, 5.15, 5.17–5.18, 5.17f
Fusible link detectors, fixed-temperature heat detectors, 5.9, 5.9f, 5.10, 5.11f

G
Gallium, semiconductors, 4.4
Gamers. See Generation Y
Garages
defined, 1.41
lighting loads and, 1.11
Gas torches, safety, 11.19
Gender and cultural issues
cultural differences, 13.7–13.8
discrimination, 13.9
language barriers, 13.7
men's versus women's communication styles, 13.7
sexual harassment, 13.8
workplace diversity, 13.6
General alarms, 5.30
defined, 5.71
General care areas
hospital wiring and devices, 2.12–2.13, 2.13f
as patient care area, 2.3
General contracting, as project delivery system, 13.54, 13.54f
General lighting loads, 1.25–1.26
General-purpose receptacles, load calculations, 1.31
General signals, 5.29
Generational differences, company operations, 13.4–13.5, 13.6
Generation Y, perception of success, 13.4–13.5
Generator engines, 3.2, 3.4
Generators
AC generators, 3.2, 3.4, 3.4f
defined, 10.33
harmonics, 6.23
megohmmeters and, 10.16
synchronous generators, 3.2, 3.30
Generator sets
emergency power systems, 2.3
engine-driven generator sets, 3.2, 3.3, 3.4–3.6, 3.4f

Generator sets, *continued*
   legally required standby systems, 3.22, 3.23*f*
   NEC requirements, 2.23, 3.23
Geometric stress control, 11.28, 11.29*f*
Germanium
   in diodes, 4.8
   in semiconductors, 4.4
   in transistors, 4.12, 4.13
GFCI attachment plugs, storable pools, 12.39–12.40
GFCI circuit breakers, underwater luminaires, 12.34,
   12.34*f*, 12.36
GFCI receptacles
   bodies of water and, 12.45
   fountains and, 12.42
   marinas and boatyards and, 12.25, 12.25*f*
   temporary installations, 12.28, 12.28*f*
Glass-break pull stations, 5.19, 5.19*f*
Goals, planning, 13.57, 13.59*f*
Going green
   adjustable frequency drives, 7.27
   electric floor heating, 9.32
   electric vehicle charging systems, 12.3
Gold, as conductor, 4.3
Go/no go testers and testing, 11.44–11.45
Grease guns, 10.12, 10.12*f*
Grease meters, 10.12, 10.12*f*
Grounded coils, troubleshooting motors, 10.24–10.25
Grounded neutral conductors
   marking for marinas and boatyards, 12.24
   marking for portable switchboard power circuits,
     12.11–12.12, 12.12*f*
Grounded windings, 10.24
Ground fault monitoring, heat tracing system
     monitoring, 9.11, 9.11*f*
Ground fault protection
   health care facilities and, 2.8
   temporary installations, 12.28, 12.28*f*
Ground faults, 5.38
   defined, 5.71
Grounding
   in anesthetizing locations, 2.18–2.19
   at carnivals, circuses, fairs, and similar events,
     12.17, 12.17*f*
   equipment grounding, 2.15–2.16, 12.28–12.29, 12.36
   fountains, 12.43
   health care facilities, 2.9, 2.11
   hospital receptacles and fixed equipment, 2.15–2.16
   installations in or near bodies of water, 12.45
   for marinas and boatyards, 12.24
   quick inline splicing kits, 11.25
   swimming pools, fountains, hot tubs, and similar
     installations, 12.30
   temporary branch circuits, 12.27
   underwater luminaires, 12.36
   in wet locations, 2.16
Grounding shield ends, terminations, 11.31
Guards
   installing before first startup, 10.28, 10.28*f*
   luminaires in agricultural buildings, 12.19
   removing for maintenance, 10.12
   switches and equipment in theaters, 12.6, 12.7*f*
Gutter de-icing. *See* De-icing systems

H

Half-wave rectifiers, 4.6, 4.7*f*
Handheld communication devices, 13.6
Hands-off leadership, 13.18–13.19
Hardwired systems, 5.4, 5.5*f*

Harmonics
   busbars and connecting lugs and, 6.22
   circuit breakers, 6.19, 6.22
   classification of, 6.21, 6.22
   defined, 6.31
   eddy currents and, 6.22, 6.23
   electrical panels and, 6.22
   harmonics analyzers, 6.25
   meters and, 6.24–6.25
   neutral conductors and, 6.19, 6.22, 6.26
   nonlinear loads and, 6.21, 6.22
   nonlinear waveforms and, 6.19
   as non-sinusoidal waveforms, 6.19, 6.20*f*
   office buildings and, 6.22
   personal computers and, 6.21, 6.24, 6.26
   problem definition and, 6.21, 6.22
   problem solutions, 6.25–6.27
   problem survey procedure, 6.24–6.25
   rates and effects of, 6.21, 6.22
   in transformers, 6.19, 6.23, 6.23*f*, 6.26–6.27
   triplen harmonics, 6.22, 6.23, 6.27
   versus 60-cycle hum in power lines, 6.20
   voltage harmonics, 6.21
Harmonics analyzers, 6.25
Harmonic voltages, 6.21
Hazardous locations
   classifying, 2.16–2.19, 12.26
   fire protective signaling systems and, 5.37
   NEC requirements, 2.16, 2.17, 5.37, 9.20, 9.32, 12.25
Hazards, identification and assessment, 13.45
Health care facilities
   communication, signaling, data, and fire alarm
     systems, 2.21–2.22, 2.21*f*
   electrical distribution systems, 2.7–2.9, 2.8*f*, 2.9*f*,
     2.10, 2.11
   emergency system circuits for light and power, 2.3–2.7,
     2.3*f*, 2.4*f*, 3.23–3.25, 3.24*f*
   essential electrical system (EES) types, 2.3–2.7, 2.3*f*, 2.4*f*
   introduction to, 2.2–2.3
   isolated power systems, 2.22–2.23, 2.22*f*
   lighting applications, 3.23
   wiring and devices, 2.11–2.20, 2.11*f*, 2.12*f*, 2.13*f*, 2.14*f*
Heat anticipators
   adjusting, 8.17–8.18
   heating-only thermostats and, 8.11, 8.11*f*
Heat detector ratings, 5.11, 5.12
Heat detectors
   combination heat detectors, 5.10, 5.11*f*
   defined, 5.71
   fixed-temperature heat detectors, 5.8*f*, 5.9, 5.9*f*, 5.10,
     5.10*f*, 5.11*f*
   fusible line-type, 5.15, 5.17–5.18, 5.17*f*
   hardwired fire alarm systems, 5.4
   heat detector ratings, 5.11, 5.12
   installation on irregular ceilings, 5.50–5.52
   rate compensation detectors, 5.15, 5.16–5.17, 5.17*f*, 5.72
   rate-of-rise heat detectors, 5.8*f*, 5.9, 5.10, 5.11*f*, 5.72
   semiconductor line-type, 5.15, 5.17, 5.17*f*
Heat exchangers, 8.2, 8.3*f*, 8.4*f*
Heating
   basic principles, 8.2
   crankcase heaters, 8.18, 8.33
Heating, ventilating, and air conditioning (HVAC)
     controls
   air conditioning principles, 8.5–8.9, 8.6*f*, 8.9*f*
   baseboard heaters, 8.41–8.42, 8.43*f*
   compressor short-cycle timer, 8.19, 8.22
   control circuit examples, 8.30, 8.31*f*, 8.32*f*, 8.33

control circuit safety switches, 8.21f, 8.22–8.23, 8.23f

control systems, 8.20–8.26, 8.20f, 8.21f, 8.22f, 8.23f, 8.24f, 8.25f, 8.26f

cooling/gas heating system circuits, 8.30, 8.32f, 8.33

digital control systems, 8.26–8.30, 8.27f, 8.28f, 8.29f

duct smoke detectors and, 5.14, 5.15, 5.16, 5.55–5.58

electric space heating cables, 8.42, 8.44f

electronic controls, 8.49

freezestats, 8.23, 8.23f

furnace controls, 8.23–8.25, 8.23f, 8.24f, 8.25f

heating principles, 8.2

heat pump defrost controls, 8.25–8.26, 8.26f

lighting requirements, 8.40f

lockout control circuits, 8.20–8.21, 8.21f

mechanical refrigeration, 8.5, 8.6f

motor speed controls, 8.20, 8.20f, 8.21

NEC requirements, 8.34, 8.35f, 8.36

outdoor thermostats, 8.23

receptacle outlets and, 8.36, 8.41f

smoke stratification phenomenon and, 5.45

system control modules, 8.27, 8.28f

thermistors, 8.30

thermostats, 8.10–8.19, 8.10f, 8.11f, 8.12f, 8.13f, 8.14f, 8.15f, 8.16f, 8.17f

time delay relays, 8.22, 8.22f

troubleshooting, 8.44–8.47, 8.48, 8.49

troubleshooting charts, 8.46–8.47, 8.48

ventilation principles, 8.2, 8.5f

Heating-cooling thermostats, 8.12–8.14, 8.13f

Heating mats, 9.31, 9.31f

Heating-only thermostats, 8.11, 8.11f

Heat pumps

air conditioning, 8.9, 8.9f

defrost controls, 8.25–8.26, 8.26f

supplemental heaters and, 8.9, 8.34

thermostats and, 8.14, 8.14f, 8.16

Heat rejection, 7.34

Heat sensors, spacing, 5.11

Heat-shrink splicing kits, 11.17–11.20, 11.17f, 11.19f, 11.20f, 11.21, 11.21f

Heat-shrink tubing, 11.10, 11.18

Heat sink compounds, bridge rectifiers, 4.10

Heat sinks

defined, 7.57

solid-state relays and, 7.2f, 7.6–7.7

thermal resistance and, 7.6–7.7

Heat sources, 9.20

defined, 9.36

Heat tracing and freeze protection. See also Heat tracing system cables

component selection, 9.7, 9.7f, 9.8f

condensation and, 9.2

de-icing systems, 9.21–9.25, 9.22f, 9.23f, 9.24f

domestic hot-water temperature maintenance systems, 9.29–9.30, 9.29f

equipment selection and installation for pipe heat tracing systems, 9.14–9.18, 9.15f, 9.18f, 9.19, 9.19f, 9.20–9.21

floor heating and warming systems, 9.30–9.31, 9.31f

heat tracing system control, 9.8–9.10, 9.9f, 9.10f

heat tracing system monitoring, 9.10–9.12, 9.11f, 9.12f

heat tracing systems, 9.3–9.14, 9.3f, 9.4f, 9.5f, 9.6f, 9.7f, 9.8f, 9.9f, 9.10f, 9.11f, 9.12f, 9.13f

installation guidelines, 9.16–9.18, 9.18f–9.19f, 9.19, 9.20–9.21

introduction, 9.2

mineral-insulated cable system, 9.3, 9.5, 9.6f, 9.7, 9.8f

NEC requirements, 9.11, 9.16, 9.20–9.21

overcurrent protection devices, 9.20

pipelines, 9.3–9.14, 9.3f, 9.4f, 9.5f, 9.6f, 9.7f, 9.8f, 9.9f, 9.10f, 9.11f, 9.12f, 9.13f

power distribution, 9.3, 9.4f

power line carrier technology, 9.14, 9.15f

process fluid freezing prevention and, 9.2, 9.14

skin-effect electric heat tracing systems, 9.9

snow-melting and anti-icing systems, 9.25, 9.26f, 9.27f

system operation, 9.12–9.14, 9.13f, 9.15f

temperature control and, 9.9

thermal insulation and, 9.8, 9.8f, 9.10, 9.20

transformers and, 9.3, 9.11

vessels and, 9.2

Heat tracing system application and design guides, 9.16

Heat tracing system applications, 9.2

Heat tracing system cables

components and accessories, 9.7, 9.7f, 9.8f

installing, 9.16–9.18, 9.18f–9.19f, 9.19, 9.20–9.21

long line cable installations, 9.6–9.7, 9.6f

mineral-insulated (MI) cables, 9.3, 9.5, 9.6f, 9.7, 9.8f

NEC requirements, 9.16, 9.20–9.21

power-limiting cables, 9.3, 9.5, 9.5f, 9.7, 9.7f

self-regulating cables, 9.3–9.5, 9.4f, 9.5f, 9.7, 9.7f

Heat tracing system control

ambient-sensing control, 9.9, 9.9f

dead-leg control, 9.10

line-sensing control, 9.10, 9.10f

personal computers and, 9.13–9.14

proportional ambient-sensing control, 9.10

self-regulating control, 9.9

temperature control, 9.9

Heat tracing system monitoring

continuity monitoring, 9.11–9.12, 9.12f

current monitoring, 9.11, 9.12, 9.12f

ground fault monitoring, 9.11, 9.11f

introduction, 9.10–9.11

NEC requirements, 9.11

temperature monitoring, 9.11, 9.12, 9.13f

Heat tracing systems. See also Heat tracing and freeze protection; Heat tracing system cables; Heat tracing system control; Heat tracing system monitoring

defined, 9.36

induction and impedance electric systems, 9.11

installation of heating cable and components, 9.16–9.18, 9.18f–9.19f, 9.19, 9.20–9.21

pipelines, 9.2, 9.3–9.14, 9.3f, 9.4f, 9.5f, 9.6f, 9.7f, 9.8f, 9.9f, 9.10f, 9.11f, 9.12f, 9.13f

typical circuit configuration, 9.13–9.14, 9.13f

Heat transfer, 8.2, 8.25

defined, 8.53

Heavy-duty lampholders, 12.9

Hermetic and semi-hermetic (serviceable) compressors

NEC requirements, 8.36, 8.37

safety, 8.45

uses of, 8.7

High-efficiency gas furnaces, 8.2, 8.4f

High potential (hi-pot) tests

adjusting voltage regulator microampere and kilovolt range switches, 11.38

connections, 11.39, 11.39f

corona currents and, 11.36, 11.39, 11.42

corona guard ring and guard shield, 11.42, 11.42f

current-versus-time test, 11.33–11.35, 11.34f, 11.35f

defined, 11.49

direct current and, 6.4, 11.32

faulty insulation and, 11.35–11.36, 11.35f

go/no-go testing, 11.44–11.45

high-voltage cable and, 11.32

insulation resistance measurements, 11.45–11.46

High potential (hi-pot) tests, *continued*
    leakage current-versus-voltage test, 11.33–11.35, 11.35*f*
    operating procedure, 11.42–11.44
    safety and, 11.32, 11.33, 11.36, 11.37, 11.42, 11.44
    selective guard circuits, 11.36–11.39, 11.37*f*, 11.39*f*, 11.40*f*
    selective guard service connections, 11.40–11.42, 11.41*f*
    tester output circuit diagrams, 11.40, 11.41*f*
    test voltages, 11.33
    types of tests, 11.33
High-pressure cutout (HPCO) switches, 8.21*f*, 8.23
High-rise buildings
    NFPA standards, 5.62–5.63
    stack effect principle and, 5.46, 5.46*f*
    suppression systems and, 5.63
High-voltage cable, high potential (hi-pot) tests, 11.32
Horns, 5.4, 5.28, 5.28*f*
    defined, 5.71
Hospital-grade AC cable, 2.16
Hospital-grade receptacles, 2.11–2.12, 2.12*f*
Hospitals, defined, 2.2
Hospital wiring and devices
    above hazardous anesthetizing locations, 2.17
    critical care areas, 2.13–2.15, 2.14*f*
    general care areas, 2.12–2.13, 2.13*f*
    grounding fixed equipment, 2.15–2.16
    grounding in anesthetizing locations, 2.18–2.19
    grounding receptacles, 2.15–2.16
    within hazardous anesthetizing locations, 2.16–2.17
    health care facilities, 2.11–2.20, 2.11*f*, 2.12*f*, 2.13*f*, 2.14*f*
    hospital-grade receptacles, 2.11–2.12, 2.12*f*
    low-voltage equipment and instruments, 2.19
    other-than-hazardous anesthetizing locations, 2.17–2.18
    X-ray installations, 2.19–2.20
Hotels and motels, service and feeder load calculations, 1.33
Hot gas lines, 8.6, 8.6*f*, 8.8
Humidifiers, 8.2, 8.5*f*
Hurricane Katrina, 2.2
Hybrid circuits, signaling line circuit, 5.7
Hydromassage bathtubs, 12.44
Hydrometers, 3.13
Hysteresis
    defined, 6.31
    transformers and, 6.23

I

IEEE Standard 48-1962, termination classes, 11.26
IEEE Standard 241, Recommended Practice for Electrical Systems in Commercial Buildings, 2.2
IEEE Standard 446, battery maintenance and, 3.15
Illness logs, 13.45
Impedance, 6.15
    defined, 6.31
Impedance heat tracing systems, 9.11
Improper charging, lead-acid batteries, 3.13
Indicating devices (appliances), 5.31
    defined, 5.71
Indirect costs, 13.62
Indium, semiconductors, 4.4
Indoor air quality (IAQ), 8.2
Indoor spa and hot tub installations, 12.40–12.41
Inducer switches, furnace controls, 8.25, 8.25*f*
Induction heat tracing systems, 9.11
Induction motors
    air gaps, 7.43
    basic care and maintenance, 10.8–10.9, 10.9*f*, 10.10
    bearing maintenance, 10.10, 10.11*f*, 10.12–10.14, 10.12*f*, 10.13*f*, 10.14*f*

    core loss and, 10.4, 10.7
    installation and commissioning guidelines, 10.25–10.29, 10.27*f*, 10.28*f*
    insulation breakdown and, 10.24, 10.24*f*
    insulation classes, 10.3–10.4
    malfunctions, 10.2
    overloading and, 10.3
    polyphase induction motors, 10.3, 10.4, 10.4*f*
    single-phase operation and, 10.2, 10.3
    squirrel cage motors, 10.2, 10.4, 10.4*f*, 10.5*f*–10.6*f*, 10.7, 10.24
Inductive interference, harmonics, 6.22
Informal organizations, 13.11
Infrared (IR) flame detectors, 5.8*f*, 5.15, 5.18, 5.18*f*, 5.43, 5.43*f*
Infrared thermometers, 10.9, 10.9*f*
Initiating circuits, fire alarm control panels, 5.25
Initiating circuit zones, 5.25
Initiating device circuits (IDC)
    defined, 5.71
    elevator recall and, 5.59
    fire protective signaling systems, 5.59
    NFPA standards, 5.59
Initiating devices. *See also* Fire alarm initiating devices
    defined, 5.71
    fire alarm systems and, 5.4
Injury logs, 13.45
Inline tape splices, 11.12–11.16, 11.12*f*, 11.13*f*, 11.14*f*, 11.15*f*, 11.16*f*
Inpatients, 2.2
    defined, 2.27
Input control circuits, solid-state relays, 7.2–7.3, 7.3*f*
Input voltage/voltage unbalance measurements, troubleshooting, 7.46–7.47, 7.46*f*
Inspections, workplace safety, 13.43, 13.46
Instantaneous peak values, meters, 6.24
Instant on, solid-state relays, 7.3
Instrument transformers, 6.2–6.3, 6.3*f*, 6.13–6.15, 6.14*f*, 6.15*f*, 6.16, 6.17
Insulated gate bipolar transistors
    defined, 7.57
    pulsewidth-modulated drives and, 7.27
Insulated gate field effect transistors (IGFET), 4.14
Insulation
    harmonics and, 6.19
    medium-voltage power cables, 11.4, 11.4*f*, 11.6
    thermal insulation, 9.8, 9.8*f*, 9.10, 9.20
Insulation breakdown, induction motors, 10.24, 10.24*f*
Insulation classes
    defined, 10.33
    induction motors, 10.3–10.4
Insulation resistance measurements
    high potential (hi-pot) tests, 11.45–11.46
    minimum ratings, 10.34
Insulation resistance temperature coefficient, 10.18–10.19, 10.19*f*, 10.20*f*, 10.21*f*
Insulation shielding, defined, 11.49
Insulation shield systems, medium-voltage power cables, 11.4, 11.4*f*, 11.6–11.7, 11.6*f*
Insulation systems, motors and motor circuits, 10.3–10.4
Insulation testing
    dielectric absorption, 10.16–10.18, 10.18*f*, 10.19, 10.20
    high potential (hi-pot) tests, 11.45–11.46
    insulation resistance tests, 10.16–10.19, 10.17*f*, 10.18*f*, 10.19*f*, 10.20, 10.20*f*, 10.21*f*
    megohm testing, 10.16–10.19, 10.17*f*, 10.18*f*, 10.19*f*, 10.20*f*, 10.21*f*

polarization index, 10.21, 11.46
safety and, 10.16
test procedures, 10.16, 10.22
troubleshooting motors, 10.24, 10.24*f*
Insulators
atomic structure and, 4.4, 4.4*f*
as terminations, 11.31, 11.31*f*
Insured costs, accidents, 13.42
Integrated circuits. *See also* Microprocessors
semiconductors and, 4.3
Intelligent systems, fire alarm control panels, 5.23, 5.23*f*
Internal splined couplings, 10.26, 10.27*f*
Invalid dielectric absorption ratios, 10.20
Invalid insulation resistance data, 10.20
Invalid polarization index ratios, 10.20
Invar
defined, 8.53
thermostats and, 8.10
Investigation procedures for accidents, 13.48, 13.56
Ionization smoke detectors, 5.8*f*, 5.11–5.12, 5.13, 5.13*f*
Isolated ground receptacles, 2.12
Isolated power systems, health care facilities, 2.22–2.23, 2.22*f*
Isolated winding-type transformers
defined, 12.50
underwater luminaires and, 12.34
Isolation transformers
buck-and-boost transformers and, 6.18, 6.18*f*
defined, 6.31
health care facilities and, 2.23
NEC requirements, 6.9, 6.10*f*

J

Jacketed concentric neutral (JCN), 11.3, 11.4*f*
Jamestown Area Labor Management Committee
(JALMC), 13.7
Job descriptions, 13.12, 13.12*f*
Job diaries, planning techniques, 13.60–13.61, 13.61*f*
Job importance, as motivation, 13.27
Job security, turnover, 13.36
Junction boxes and enclosures
fountains, 12.42
permanently installed pools, 12.36–12.37, 12.37*f*
Junction field-effect transistors (JFET)
defined, 4.20
N-channel, 4.14, 4.14*f*
P-channel, 4.14, 4.14*f*
Junction transistors, 4.12, 4.13*f*

K

Key-operated pull stations, 5.19, 5.19*f*
Keypads, 5.24, 5.24*f*
K factor, 1.5
Kiss-off tones, 5.33
Knowledge, leadership styles, 13.19
K-rated transformers, 6.23

L

Labeled, 5.25
defined, 5.71
Labor. *See also* Employees; Workers
labor control, 13.78–13.79
planning and, 13.59–13.60
Landing stage, defined, 12.50
Language barriers, 13.7
Latched, 7.7
defined, 7.57
Laundry receptacles, *NEC* requirements, 1.17

Lead-acid batteries
improper charging of, 3.13
sealed valve-regulated lead-acid batteries, 3.13, 3.18
storage batteries, 3.11–3.12, 3.11*f*
Leadership behaviors, 13.18
Leadership characteristics, 13.17–13.18
Leadership functions, 13.18
Leadership skills
communication and, 13.20–13.23, 13.21*f*, 13.24–13.25, 13.24*f*
craftworker to crew leader transitions, 13.15–13.16, 13.17*f*
delegating responsibilities, 13.29–13.30
development of leaders, 13.17–13.20
employee motivation, 13.26–13.27, 13.28
ethics in, 13.19–13.20
implementing policies and procedures, 13.30–13.31
introduction to, 13.15, 13.16*f*
participant exercises, 13.24–13.25, 13.28, 13.36–13.37
problem solving and decision making, 13.31–13.33, 13.34,
13.35–13.37
team building, 13.29
Leadership styles, 13.18–13.19
Leadership traits, 13.18
Lead identification
light-emitting diodes (LED) and, 4.11
transistors, 4.13, 4.13*f*
Lead shield repair kits, 11.26
Leakage, 7.37, 7.38
defined, 10.33
Leakage current
hi-pot testing, 11.32, 11.33–11.36
insulation testing, 10.18, 10.18*f*
safety, 7.37, 7.38
Leakage current-versus-voltage test, high potential (hi-pot)
tests, 11.33–11.35, 11.34*f*, 11.35*f*
Learning styles, training, 13.4
Legal ethics, 13.19–13.20
Legally required standby systems, 3.2, 3.21–3.22, 3.23*f*
Leveling, thermostats, 8.16
Life safety branch, 2.3–2.5, 2.4*f*, 2.6
Life Safety Code (NFPA 101), 5.3, 5.4
Life safety fire protection, 5.2, 5.4
Light-activated silicon-controlled rectifiers (LASCR), 4.15
Light-emitting diodes (LED), 4.11–4.12, 4.11*f*
Lighting applications
health care facilities, 3.23
temporary lighting, 12.28, 12.28*f*
Lighting loads
commercial occupancy load calculations, 1.25–1.26, 1.27
NEC requirements, 1.11, 1.25–1.26
service load calculations, 1.25–1.26, 1.27
Lighting requirements, heating, ventilating, and air
conditioning (HVAC) controls, 8.40*f*
Lights, fire warning, 5.4
Light scattering, defined, 5.71
Light-scattering detectors, photoelectric smoke detectors,
5.12, 5.13*f*, 5.14
Limit controls, furnace controls, 8.23, 8.23*f*
Limited care facilities, 2.3
Line detectors, 5.9
Line isolation monitors, 2.23
Linerless splicing tapes, 11.10
Line seizure, digital communicators, 5.32, 5.32*f*, 5.33
Line-sensing controls, heat tracing system controls,
9.10, 9.10*f*
Line-voltage thermostats, 8.15, 8.15*f*
Liquid dielectric test sets, 6.4
Liquid lines, air conditioning, 8.6, 8.6*f*, 8.9

Listed, 5.7
  defined, 5.71
Load calculation procedures
  basic considerations, 1.2
  basic steps, 1.11, 1.12f
  demand factors, 1.10–1.11, 12.22–12.23
  feeder ampacity, 1.7–1.8
  feeder conductor size adjustments, 1.2–1.7
  lighting loads, 1.11, 1.25–1.26, 1.27
  load center length, 1.5f
  tap rules, 1.8–1.10
Load calculations. *See also* Feeder circuit load calculations;
    Optional load calculations; Service load calculations
  commercial occupancies and, 1.24–1.31, 1.27f, 1.29f
  forms for, 1.43
  minimum size service and, 1.13–1.24, 1.14f, 1.15f, 1.16f,
    1.17f, 1.22f
  restaurants and, 1.31, 1.32f, 1.33
Load center length calculations, 1.5, 1.5f
Local codes, authority having jurisdiction, 12.4
Location measurements, 12.41
Lockout control circuits, 8.20–8.21, 8.21f
Lockout relays, safety, 8.21
Long cable runs, adjustable frequency drives, 7.29
Long line cable installations, heat tracing system cables,
    9.6–9.7, 9.6f
Loss of charge switches, 8.23, 8.24
Lovejoy couplings, 10.26
Low-pressure cutout (LPCO) switches, 8.21f, 8.23
Low-voltage bridge rectifiers, 4.10
Low-voltage equipment and instruments, hospital wiring
    and devices, 2.19
Lubrication, motor bearing maintenance, 10.10, 10.11f,
    10.12, 10.12f
Luminaires
  agricultural buildings and, 12.19, 12.19f
  fountains and, 12.42
  underwater luminaires, 12.34–12.37, 12.34f, 12.35f, 12.36f

M

Magnetic fields
  AC generators, 3.4, 3.4f
  harmonics and, 6.21, 6.22
Main service calculations, for office buildings, 1.31
Maintenance
  batteries, 3.13–3.16
  bearing maintenance, 10.10, 10.11f, 10.12–10.14, 10.12f,
    10.13f, 10.14f
  defined, 5.71–5.72
  domestic hot-water temperature maintenance systems,
    9.29–9.30, 9.29f
  engine-driven generator sets, 3.6
  fire alarm systems and, 5.4
  motor controls, 7.38–7.39
  motor maintenance, 10.7–10.10, 10.7f, 10.9f, 10.10f
  transformer maintenance testing, 6.4
Malfunctions, induction motors, 10.2
Management systems, transformers, 6.3
Manmade bodies of water, 12.44–12.46, 12.45f
Manual fire alarm station resets, 5.20
Manual (pull station) fire detection devices
  Americans with Disabilities Act and, 5.41
  double-action pull stations, 5.19, 5.19f
  fire alarm initiating devices, 5.4, 5.7, 5.18–5.19, 5.19f, 5.20
  glass-break pull stations, 5.19, 5.19f
  installation, 5.19
  key-operated pull stations, 5.19, 5.19f
  manual fire alarm station resets, 5.20

NFPA standards, 5.26, 5.42f, 5.61
  single-action pull stations, 5.19, 5.19f
  sprinkler systems and, 5.61
  total premises fire alarm system installation guidelines,
    5.41, 5.42f
  UL standards, 5.19
Manufacturer's aids, troubleshooting, 7.40–7.41, 7.41f
Manufacturer's training programs, splicing, 11.15
Marinas and boatyards
  disconnecting means, 12.24, 12.25f
  general requirements for devices, equipment, and
    enclosures, 12.21–12.22, 12.21f, 12.22f
  grounding, 12.24
  as hazardous locations, 12.25–12.26, 12.26f
  receptacles, 12.25, 12.25f
  service and feeder load calculations, 1.34, 12.22–12.23
  wiring methods, 12.23–12.24, 12.23f, 12.24f
Material safety data sheets (MSDS)
  safety and, 13.46
  solvents, 11.9, 11.18
Materials control, 13.76–13.77
Maximum inrush VA, control transformers, 6.10–6.11, 6.11f
Mechanical connectors, 11.10
Mechanical refrigeration
  defined, 8.53
  heating, ventilating, and air conditioning (HVAC)
    controls, 8.5, 8.6f
Medium-voltage power cables
  cable components, 11.4–11.5, 11.5f
  cable jackets, 11.4, 11.4f, 11.7, 11.7f
  compact stranding, 11.5, 11.5f
  compressed stranding, 11.4, 11.5f
  concentric stranding, 11.4, 11.5f
  conductor configurations, 11.4
  insulation, 11.4, 11.4f, 11.6
  insulation shield systems, 11.4, 11.4f, 11.6–11.7, 11.6f
  solid wire, 11.5, 11.5f
  strand shielding, 11.4, 11.5, 11.5f
Medium-voltage terminations and splices. *See also* High
    potential (hi-pot) tests; Splicing
  medium-voltage power cables, 11.3–11.7, 11.4f, 11.5f,
    11.6f, 11.7f
  straight splices, 11.2–11.3, 11.2f
  terminations, 11.26, 11.27, 11.27f, 11.28, 11.28f, 11.29,
    11.29f, 11.30–11.31, 11.30f, 11.31f
Meggers, 10.2
  defined, 10.33
  insulation resistance measurements, 11.45, 11.46
Megohmmeters, 10.16–10.19, 10.17f
Megohm testing, motor insulation testing, 10.16–10.19,
    10.17f, 10.18f, 10.19f, 10.20f, 10.21f, 10.22
Men's versus women's communication styles, 13.7
Mercury, safety, 8.11
Messages, verbal communication, 13.20–13.21, 13.21f
Metallic layers, insulation shield system, 11.6–11.7, 11.6f
Metal-oxide semiconductor field-effect transistors
    (MOSFET), 4.14
Meters
  crest factor and, 6.25
  go/no go testers, 11.45
  grease meters, 10.12, 10.12f
  harmonics and, 6.24–6.25
  hydrometers, 3.13
  instantaneous peak values and, 6.24
  instrument transformers and, 6.14
  megohmmeters, 10.16–10.19, 10.17f
  ohmmeters, 4.9, 4.10f
  power quality analyzers, 6.25

troubleshooting and, 7.45
true rms, 6.24–6.25
Microprocessor controls
heat tracing, 9.13–9.14
transfer switches, 3.8
Microprocessors
digital control systems and, 8.28
motor speed controls and, 8.20
vulnerability to static electricity and, 7.37, 7.37f
Millenials. *See* Generation Y
Mineral-insulated (MI) heating cables
defined, 9.36
heat tracing system cables, 9.3, 9.5, 9.6f, 9.7, 9.8f
Minimum service ratings
NEC requirements, 1.13, 1.14, 1.14f, 1.15, 1.16, 1.17,
1.17f, 1.18
service load calculations, 1.13–1.18, 1.14f, 1.15f,
1.16f, 1.17f
Misaligned shafts, 10.13–10.14, 10.13f, 10.14f, 10.15f
Mission protection fire alarm systems, 5.2
Moisture
insulation resistance testing and, 10.19, 11.45–11.46
terminations and, 11.31
Molded rubber splices, reinsulating, 11.10
Monitoring. *See also* Communications and monitoring; Heat
tracing system monitoring
online motor monitoring, 10.10
power monitoring, 7.48
Motivation
leadership skills and, 13.26–13.27, 13.28
sense of achievement and, 13.4
Motor analysis instruments, 10.10, 10.10f
Motor braking methods
DC electric braking of AC motors, 7.34–7.35, 7.35f
dynamic braking of AC drives, 7.28, 7.35–7.36, 7.36f
electromechanical braking, 7.36, 7.37f
Motor control circuits and components
electrical troubleshooting for, 7.44, 7.45, 7.46–7.47, 7.46f,
7.47f, 7.48f, 7.49–7.50, 7.51f, 7.52–7.53, 7.52f
mechanical troubleshooting, 7.41–7.42, 7.42f, 7.43–7.44
Motor controls. *See also* Reduced-voltage starting motor
controls
adjustable frequency drives, 7.25–7.31, 7.26f, 7.28f, 7.30f,
7.31f, 7.32f–7.33f, 7.33, 7.34
electromechanical relays, 7.2, 7.3–7.5
maintenance, 7.38–7.39
motor braking methods, 7.28, 7.34–7.36, 7.35f, 7.36f, 7.37f
solid-state control precautions, 7.37–7.38, 7.37f
solid-state protective relays, 7.7–7.9, 7.8f, 7.9f
solid-state relays (SSRs), 7.2–7.7, 7.2f, 7.3f, 7.5f, 7.6f
timing relays, 7.9–7.13, 7.10f, 7.11f, 7.12f, 7.13f, 7.14f,
7.15–7.16, 7.15f
troubleshooting, 7.39–7.42, 7.41f, 7.42f, 7.43, 7.44, 7.45,
7.46–7.47, 7.46f, 7.47f, 7.48, 7.49–7.50, 7.49f, 7.51f,
7.52–7.53, 7.52f
Motor control troubleshooting charts, 7.43
Motor installation checklist, 10.29
Motor loads, *NEC* requirements, 1.10, 1.13, 1.24, 1.30
Motors. *See also* Induction motors
agricultural buildings and, 12.19, 12.19f
permanently installed pools and, 12.32–12.33, 12.32f
troubleshooting, 10.23–10.25, 10.24f
Motors and motor circuits
bearing maintenance, 10.10, 10.11f, 10.12–10.14, 10.12f,
10.13f, 10.14f
feeder and service load calculations, 1.36
installation and commissioning guidelines, 10.25–10.29,
10.27f, 10.28f

insulation systems, 10.3–10.4
insulation testing, 10.16–10.19, 10.17f, 10.18f, 10.19f, 10.20,
10.20f, 10.21–10.22, 10.21f
maintenance, 10.7–10.10, 10.7f, 10.9f, 10.10f
motor starting configurations, 10.4
NEMA standards for usual and unusual service
conditions, 10.2–10.3
online motor monitoring, 10.10
polarization index and, 10.21
receiving and storing motors, 10.22, 10.23, 10.23f
squirrel cage motors, 10.2, 10.4, 10.4f, 10.5f–10.6f, 10.7, 10.24
totally enclosed motors, 10.2, 10.7
troubleshooting charts, 7.43
troubleshooting motors, 10.23–10.25, 10.24f
unusual service conditions, 10.3
usual service conditions, 10.2–10.3
winding failures, 10.4, 10.5f–10.6f, 10.7, 10.24
Motor speed controls
heating, ventilating, and air conditioning (HVAC)
controls, 8.20, 8.20f, 8.21
microprocessors and, 8.20
triacs and, 8.20, 8.20f
Motor starting configurations, programmable starting
characteristics, 7.22–7.23, 7.23f
Multiconductor cords, portable switchboards, 12.13, 12.13f
Multi-family dwellings, service load calculations, 1.16–1.18,
1.21–1.24, 1.22f
Multiple outputs, solid-state relays, 7.6, 7.6f
Multiple secondaries, in transformers, 6.7, 6.7f, 6.8
Multiple tubes, heat-shrink splicing kits, 11.21
Multiplexing, 5.4–5.5
defined, 5.72
Multiplex systems, fire alarm systems, 5.4–5.5, 5.5f
Multi-stage thermostats, 8.14, 8.14f
Music system shutoffs, 5.55, 5.56f

N

National Census of Fatal Occupational Injuries, 13.42
National Electrical Code (NEC) classifications, fuel dispensing
locations, 12.25–12.26, 12.26f
National Electrical Code (NEC) division designations, fuel
dispensing locations, 12.25–12.26, 12.26f
National Electrical Code (NEC) listing of cord and cable
types, 12.43
National Electrical Code (NEC) requirements
adjustable frequency drives, 7.25
agricultural buildings, 12.18–12.19, 12.20
air conditioner branch circuits, 8.36, 8.37, 8.38f, 8.39f,
8.40f, 8.41, 8.41f, 8.42, 8.42f
ambient temperatures, 1.2, 1.3, 1.4
amusement rides, 12.16–12.17
ancillary control relays, 5.55
appliance loads, 1.17–1.18, 1.19, 1.20, 1.23, 1.24
assured equipment grounding programs, 12.28
autotransformers, 6.9, 6.10f, 7.16
baseboard heaters, 8.41–8.42, 8.43f
basic load calculations, 1.12f
border lights, 12.10
carnivals, circuses, fairs, and similar events, 12.14, 12.15,
12.16, 12.17
clothes dryers, 1.10, 1.23, 1.24
commercial and industrial load calculations, 1.24, 1.25, 1.26
communication, signaling, data, and fire alarm systems,
2.21–2.22
compressors, 8.36, 8.37, 8.38f, 8.39f
construction job site trailers, 12.2
continuous duty loads, 1.2
cord-and-plug connections, 12.30–12.31, 12.43

National Electrical Code (NEC) requirements, continued
  critical care areas, 2.13, 2.14, 2.15
  deck area heating, 12.39
  de-icing systems, 9.23, 9.24–9.25
  demand factors, 1.10–1.11
  dimmers, 12.7
  disconnecting means, 12.24, 12.32
  domestic hot-water temperature maintenance systems, 9.30
  dressing rooms, 12.14
  electrical datum planes, 12.44
  electrical enclosures, 12.18
  electrical installations in or near bodies of water, 12.44, 12.45, 12.46
  electric pool heaters, 12.31
  electric range loads, 1.17, 1.18, 1.42
  electric space heating cables, 8.42, 8.44f
  electric vehicle charging systems, 12.3
  electrified truck parking spaces, 12.15
  emergency circuit identification, 2.5, 2.23, 3.25, 3.25f, 3.26
  emergency lighting, 2.18, 3.25
  emergency power systems, 3.2, 3.21–3.22
  emergency system circuits for light and power, 2.4, 2.6, 2.7, 3.23, 3.25
  equipotential bonding, 12.38–12.39
  essential electrical systems (EES), 2.4, 2.5, 2.6, 2.7, 3.23, 3.25
  exterior signs and lighting, 1.26
  extra-hard usage cords and cables, 12.15–12.16
  farm load calculations, 1.34
  feeder ampacity, 1.8
  feeder and service loads for schools, 1.33–1.34
  feeder circuits, 1.2
  feeder conductor size adjustments, 1.2–1.7
  finish ratings, 12.4
  fire alarm systems, 5.34
  fire protective signaling systems, 5.3–5.4, 5.34
  fixed stage switchboards, 12.6, 12.7, 12.8, 12.9
  floating buildings, 12.21
  floor heating and warming systems, 9.32
  footlights, 12.9–12.10
  fountains, 12.30, 12.41, 12.42, 12.43
  generator sets, 2.23, 3.23
  grounded neutral conductors and, 12.11–12.12
  ground fault protection, 2.8, 12.28
  grounding in anesthetizing locations, 2.18–2.19
  grounding in wet locations, 2.16
  grounding receptacles and fixed equipment, 2.15, 2.16
  grounding swimming pools, fountains, hot tubs, and similar installations, 12.30
  grounding temporary branch circuits, 12.27
  ground loop currents, 7.38
  health care facilities emergency power systems, 2.2
  heating, ventilating, and air conditioning (HVAC) controls, 8.34, 8.35f, 8.36
  heat tracing and freeze protection, 9.11, 9.16, 9.20–9.21
  heat tracing system cables, 9.16, 9.20–9.21
  heat tracing system monitoring, 9.11
  heavy-duty lampholders, 12.9
  for hospital general care areas, 2.12–2.13
  hospital-grade receptacles, 2.11
  hotels and motels, 1.33
  hydromassage bathtubs, 12.44
  inhalation anesthetizing locations, 2.16–2.19
  isolated power systems, 2.23
  isolation transformers, 6.9, 6.10f
  laundry receptacles, 1.17
  lighting and switches for HVAC equipment, 8.36, 8.40f
  lighting loads, 1.11, 1.25–1.26
  low-voltage equipment, 2.19
  luminaires and, 12.19, 12.36–12.37, 12.40, 12.42
  marinas and boatyards, 1.34, 12.21, 12.22, 12.23, 12.24, 12.25
  medium-voltage power cables, 11.4
  minimum service ratings, 1.13, 1.14, 1.14f, 1.15, 1.16, 1.17, 1.17f, 1.18
  minimum size service, 1.13, 1.14, 1.15, 1.16, 1.17, 1.17f, 1.18, 1.19, 1.20, 1.21, 1.22, 1.22f, 1.23, 1.24
  motor loads, 1.10, 1.13, 1.24, 1.30
  motors and motor circuits, 1.36
  multi-family dwelling service load calculations, 1.16–1.18, 1.17f, 1.21, 1.22f, 1.23, 1.24
  multiple building electrical supply from a single distribution point, 12.19, 12.20
  neutral conductors serving nonlinear loads, 6.26
  nonpower-limited fire alarm (NPLF) circuit cables, 5.34, 5.35
  number of conductors in raceways, 12.5–12.6
  office buildings, 1.28, 1.29, 1.30, 1.31
  on-stage electrical equipment, 12.10
  on-stage portable switchboards, 12.11–12.14, 12.11f, 12.12f, 12.13f
  outdoor wiring of fire alarm circuits, 5.35
  overcurrent protection devices, 9.20
  overhead conductor clearances, 12.14, 12.31
  patch panels, 12.11
  permanently installed pools, 12.32, 12.33, 12.34, 12.35, 12.36, 12.37, 12.38, 12.39
  phase loss/phase unbalance circuits, 7.7, 7.8
  places of assembly, 3.26, 12.3, 12.4
  pool covers, 12.39
  for raceway fill, 1.2, 1.3, 1.4
  receptacle outlets at HVAC equipment, 8.36, 8.41f
  remote control signaling circuits, 5.37
  restaurants, 1.31, 1.32f, 1.33
  retail stores with show windows, 1.26, 1.28
  room air conditioners, 8.36, 8.37, 8.39f, 8.41, 8.42f
  single-family dwellings and, 1.18–1.21
  sizing neutral conductors, 1.18, 1.19, 1.20, 1.21
  small appliance branch circuits, 1.16, 1.17, 1.19, 1.24
  smoke ventilators, 12.11
  snow-melting and anti-icing systems, 9.28, 9.29
  spas and hot tubs, 12.40, 12.41
  stage switchboard feeders, 12.8, 12.9
  standby power systems, 3.2, 3.21–3.22, 3.23f
  storable pools, 12.39–12.40
  storage batteries, 3.11, 3.11f, 3.21
  swimming pool branch circuit feeders, 12.37–12.38
  tap conductors, 1.8–1.10
  temperature limitations for terminating conductors, 1.7
  temporary installations, 12.26, 12.27, 12.28
  temporary wiring and equipment greater than 600V, 12.28
  theater or audience areas, 12.3, 12.5, 12.6
  therapeutic pools and tubs, 12.43–12.44
  thermostat wiring, 8.16
  trade show display booths, 12.4
  transformer secondary conductors, 1.8, 1.9, 1.10
  underground wiring, 12.31, 12.32
  underwater audio equipment, 12.39
  voltage drop, 1.2, 1.4–1.7
  wired partitions, 12.29
  wiring in hazardous locations, 2.16, 2.17, 5.37, 9.20, 9.32, 12.25
  X-ray installations, 2.19–2.20
National Electrical Manufacturers' Association (NEMA) classifications, for AFDs, 7.29
National Electrical Manufacturers' Association (NEMA) standards
  fire alarm systems, 5.3

for handling water-damaged equipment, 10.25
usual and unusual service conditions for motors, 10.2–10.3
National Fire Alarm Code, 5.3, 5.4
defined, 5.72
National Fire Protection Association (NFPA)
abiding by codes, 5.2
defined, 5.72
Life Safety Code (NFPA 101), 5.3, 5.4
National Fire Alarm Code (NFPA 72), 5.3, 5.4
NFPA 70, 2.2
Standard for Health Care Facilities NFPA 99, 2.2, 2.3, 2.4
Uniform Fire Code, 5.3, 5.4
National Fire Protection Association (NFPA) standards
acceptability of installations and, 5.69
audible fire notification devices, 5.27
digital communicators, 5.33
door locking arrangements, 5.60
dry chemical extinguishing systems, 5.60
duct smoke detectors, 5.15, 5.55, 5.56
elevator recall, 5.59
FACP primary and secondary power, 5.26
fire alarm control panels, 5.26, 5.54, 5.55, 5.55f
fire alarm signals, 5.29, 5.30
fire alarm systems, 5.4
fire notification appliance installation, 5.52, 5.53, 5.54
fire notification device installation, 5.53
fire protection for boatyards and marinas, 12.25
fire retardant-treated wood and building materials, 12.6
high-rise fire suppression, 5.62–5.63
initiation device circuits, 5.59
manual (pull station) fire detection devices, 5.26, 5.42f, 5.61
motor fuel dispensing facilities and repair garages, 12.25
photoelectric smoke detectors, 5.49
signaling line circuits, 5.7
smoke detectors on irregular ceilings, 5.50, 5.51, 5.52
supervised automatic sprinklers, 5.60–5.61
visual fire notification devices, 5.31
wet chemical extinguishing systems, 5.60
National Institute for Certification in Engineering Technologies (NICET) standards, 5.3
National Institute for Occupational Safety and Health (NIOSH), 13.42, 13.47
Natural bodies of water, 12.44–12.46, 12.44f
N-channel, junction field-effect transistors, 4.14, 4.14f
NEC requirements. *See National Electrical Code (NEC) requirements*
Negative ions, atomic structure, 4.2–4.3
Neon sign transformers, 6.12
Network command centers, fire protective signaling systems, 5.38
Network scheduling, 13.70, 13.70f
Neutral conductors
demand factors, 1.11
grounded neutral conductors, 12.11–12.12, 12.12f, 12.24
harmonics and, 6.19, 6.22, 6.26
sizing, 1.18–1.21
Neutrons, atomic structure, 4.2
NFPA standards. See National Fire Protection Association (NFPA) standards
Niches, matching with luminaires, 12.35
Nickel cadmium batteries, 3.12, 3.12f
Noise
audible notification devices and, 5.27
bearing maintenance and, 10.9, 10.13
defined, 5.72
solid-state devices and, 7.38
Nonautomatic/remote operation, transfer switches, 3.9

Non-coded signals, 5.31
defined, 5.72
Nonessential loads, 2.3, 2.3f
No-niche luminaires, 12.36
Nonlinear loads, harmonics, 6.21, 6.22
Nonlinear waveforms, harmonics, 6.19
Nonpower-limited fire alarm (NPLF) circuit cables, 5.34, 5.35
Nonpower-limited fire alarm plenum (NPLFP) circuit cables, 5.34
Nonpower-limited fire alarm riser (NPLFR) circuit cables, 5.34
Non-programmable overload relays, 7.7–7.8, 7.8f
Non-shore power receptacles, 12.25, 12.25f
Non-sinusoidal waveforms, harmonics as, 6.19, 6.20f
Non-verbal communication, 13.22
Notification devices (appliances). *See also* Fire notification appliances
defined, 5.72
NPN transistors, 4.12, 4.12f
N-type material
cathodes as, 4.5, 4.5f
defined, 4.20
in diodes, 4.5, 4.5f
semiconductors and, 4.4
in transistors, 4.12–4.13, 4.12f, 4.13f
Nuclei, atomic structure, 4.2
Nursing homes, 2.3

O
Objectives, planning, 13.57
Obscuration
defined, 5.72
photoelectric smoke detectors, 5.12, 5.14, 5.14f
Occupational Safety and Health Act (OSHA) of 1970
fire safety standards, 13.47
first aid requirements, 13.47
hazard identification and assessment, 13.45
penalties for violations, 13.44
policies and procedures manuals, 13.44
safety record keeping systems, 13.45
workplace inspections, 13.43
Office buildings
commercial occupancy load calculations, 1.28–1.31, 1.29f
harmonics and, 6.22
main service calculations, 1.31
NEC requirements, 1.28, 1.29, 1.30, 1.31
one-line diagrams, 1.28, 1.29f
primary feeder calculations, 1.30–1.31
Ohmmeters, testing diodes, 4.9, 4.10f
Ohm's law
insulation resistance measurements and, 11.45
LED circuits and, 4.12
On-delay and off-delay relays, 7.11
One hundred foot tap rule, 1.9
One megohm rule of thumb, 10.25
Online motor monitoring, 10.10
On-stage electrical equipment, 12.10
On-stage portable switchboards
construction of, 12.13–12.14
marking grounded conductors, 12.12f
multiconductor cord, 12.13f
power outlets and supply connections for, 12.11–12.13, 12.11f
road show connection panels, 12.11
On-time delivery, materials control, 13.76
Open area protection (OAP), 5.55
Open circuits, troubleshooting motors, 10.24
Open circuit transitions, 7.17, 7.18–7.19
defined, 7.57
Open collector outputs, 5.26

Open construction, docks, piers, and wharves, 12.25–12.26
Open grids, smoke detection, 5.44
Operating frequency range, adjustable frequency drives, 7.27, 7.28*f*
Optically isolated solid-state relays (SSRs), 7.3, 7.3*f*
Optional load calculations
    for multi-family dwellings, 1.16–1.18, 1.22*f*, 1.23–1.24
    for new restaurants, 1.33
    for schools, 1.33–1.34
    for single-family dwellings, 1.16–1.18, 1.20–1.21
Optional standby power systems, 3.2
Opto-isolation circuits, 4.12
Orbitals, atomic structure, 4.3
Outdoor spa and hot tub installations, 12.40
Outdoor thermostats, 8.23
Outdoor wiring of fire alarm circuits, 5.35
Outpatients, 2.2
    defined, 2.27
Output circuits
    solid-state relays, 7.2–7.3, 7.3*f*
    switching modes, 7.3
Outside lighting loads, 1.26
Outside screw and yoke control valves, 5.62
Outside taps of unlimited length, 1.9–1.10
Overcurrent protection devices
    for feeder circuits, 1.2
    heat tracing and freeze protection and, 9.20
    NEC requirements, 9.20
    X-ray installations, 2.20
Overhead conductor clearances
    carnivals, circuses, fairs, and similar events, 12.14, 12.15*f*
    marinas and boatyards, 12.24, 12.24*f*
    swimming pools, fountains, hot tubs, and similar installations, 12.31
Overhead piping, safety, 9.16
Overhead system service entrances
    defined, 1.41
    service conductors and, 1.11
Overheated neutrals, harmonics, 6.19, 6.22, 6.26
Overloading, induction motors, 10.3
Overvoltage and overcurrent protection, solid-state relays, 7.7

P

Paper-insulated cable splices, 11.25, 11.26
Paper-insulated lead-covered (PILC) power cables, 11.25, 11.26
Parallel resistance heating cables
    defined, 9.36
    self-regulating cables and, 9.3–9.5
Parallel runout, 10.13–10.14, 10.13*f*, 10.14*f*
Part-winding starting
    motor starting configurations, 10.4
    reduced-voltage starting motor controls, 7.18, 7.18*f*, 7.24
Patch panels, 12.11
Paths, 5.6
    defined, 5.72
Patient bathrooms, as wet locations, 2.16
Patient care areas, 2.3
Patient corridor lighting, 2.4, 2.6
Patient headwall columns, 2.13
Patients' electrical risks, 2.12, 2.15
P-channel, junction field-effect transistors, 4.14, 4.14*f*
Peak current values, diodes, 4.8
Peaked ceilings, spot detector installation, 5.51–5.52
Peak function, meters, 6.25
Peak inverse voltage (PIV), 4.8, 4.9
Peak switching, solid-state relays, 7.3
Peeling, extruded semi-conductors, 11.8
Penciling, splicing, 11.2, 11.2*f*

Periodic predictive testing, 10.7, 10.10, 10.10*f*
Periodic table of the elements, 4.6
Permanently installed pools
    equipotential bonding, 12.38–12.39, 12.38*f*, 12.39*f*
    feeders, 12.37–12.38
    junction boxes and enclosures, 12.36–12.37, 12.37*f*
    motors, 12.32–12.33, 12.32*f*
    NEC requirements, 12.32, 12.33, 12.34, 12.35, 12.36, 12.37, 12.38, 12.39
    receptacle and area lighting, 12.33–12.34, 12.33*f*, 12.34*f*
    specialized pool equipment, 12.39
    underwater luminaires, 12.34–12.36, 12.34*f*, 12.35*f*, 12.36*f*
Personal computers. *See also* Computer control
    fire alarm control panels and, 5.24
    harmonics and, 6.21, 6.24, 6.26
    heat tracing system control and, 9.13–9.14
Personal growth, as motivation, 13.27
Personality, leadership styles, 13.19
Phase 1 recalls, 5.59, 5.59*f*
Phase 2 recalls, 5.59, 5.59*f*
Photo diodes, 4.11, 4.11*f*
Photoelectric smoke detectors
    defined, 5.72
    detection versus stages of fires, 5.8*f*
    installation on flat smooth ceilings, 5.49–5.50, 5.49*f*, 5.50*f*
    light-scattering detectors, 5.12, 5.13*f*, 5.14
    NFPA standards, 5.49
    obscuration principle and, 5.12, 5.14, 5.14*f*
    projected beam smoke detectors, 5.9, 5.12, 5.14, 5.14*f*, 5.50, 5.72
    as type of smoke detector, 5.11–5.12, 5.12*f*
    types of, 5.12, 5.13*f*, 5.14–5.15, 5.14*f*
Piers, 12.21, 12.23, 12.23*f*, 12.24, 12.25, 12.26, 12.26*f*
Pipelines
    equipment selection and installation for pipe heat tracing systems, 9.14–9.18, 9.15*f*, 9.18*f*, 9.19, 9.19*f*, 9.20–9.21
    heat tracing systems, 9.2, 9.3–9.14, 9.3*f*, 9.4*f*, 9.5*f*, 9.6*f*, 9.7*f*, 9.8*f*, 9.9*f*, 9.10*f*, 9.11*f*, 9.12*f*, 9.13*f*
Pitch, spiral factors, 9.17
Places of assembly, emergency system circuits for light and power, 3.26
Planning
    assigning resources, 13.59–13.60
    construction materials and, 13.59, 13.65, 13.66*f*, 13.67
    follow-up and, 13.58, 13.58*f*
    process of, 13.57–13.58, 13.59*f*
    project components of, 13.54–13.55
    as project phase, 13.52, 13.53*f*
    stages of, 13.56
    techniques for, 13.60–13.61, 13.60*f*, 13.61*f*
Planning forms, 13.60–13.61, 13.60*f*
Pneumatic timing relays, 7.11–7.12, 7.11*f*, 7.12*f*
PN junction
    diodes and, 4.5, 4.5*f*
    transistors and, 4.12
PNP transistors, 4.12–4.13, 4.13*f*
Point-contact transistors, 4.13, 4.13*f*
Polarity, diodes, 4.5, 4.9, 4.10*f*
Polarization index, insulation testing, 10.21, 11.46
Polarization index ratios (PIR), 10.21
Policies and procedures
    accident investigation and, 13.48, 13.56
    implementing, 13.30–13.31
    manuals, 13.44
    supervisor's duties and, 13.13, 13.13*f*
Polyethylene insulation, 11.6
Polyphase induction motors, 10.3, 10.4, 10.4*f*
Pool accessories, bonding requirements, 12.39

Pool covers, 12.39
Pool equipment, 12.39
Portable structures, 12.14–12.15, 12.15f
Portable switchboards. *See also* On-stage portable
  switchboards
  defined, 12.50
Positive alarm sequences, 5.25, 5.30
  defined, 5.72
Positive ions, atomic theory, 4.2
Positive locking devices, 12.36
  defined, 12.50
Positive temperature coefficient (PTC), power-limiting
  cables, 9.5
Potential (voltage) transformers, 6.3, 6.4f, 6.13, 6.15f, 6.17
Potheads
  defined, 11.49
  terminations as, 11.26
Potting compounds, defined, 12.50
Power correction capacitors, voltage harmonics, 6.21
Power distribution. *See also* Electrical distribution systems
  heat tracing and freeze protection, 9.3, 9.4f
Power-limited circuits in raceways, fire protective signaling
  systems, 5.35
Power-limited fire alarm (FPL) cables, 5.34
Power-limited fire alarm plenum (FPLP) cables, 5.34
Power-limited fire alarm riser (FPLR) cables, 5.34, 5.37
Power-limiting heating cables
  defined, 9.36
  heat tracing system cables, 9.3, 9.5, 9.5f, 9.7, 9.7f
  positive temperature coefficient and, 9.5
Power line carrier technology, 9.14, 9.15f
Power line interface units (PLI), 9.14, 9.15f
Power monitoring, 7.48
Power quality analyzers, 6.25
Power sources, carnivals, circuses, fairs, and similar events,
  12.15, 12.15f
Power supplies
  defined, 5.72
  fire alarm systems, 5.5
  static uninterruptible power supplies, 3.17–3.18, 3.19,
    3.19f, 3.20
Power transformers, 6.2, 6.2f
Power transistors, solid-state relays, 7.3
Praise, as motivation, 13.26
Pre-construction planning, 13.56
Predictive testing, 10.7, 10.10, 10.10f
Pre-signals, 5.30
Pressure switches, 8.21f, 8.23, 8.24
Preventive maintenance, motor controls, 7.38–7.39
Primary feeder calculations, for office buildings, 1.30–1.31
Primary insulation application, splicing, 11.14–11.16,
  11.15f, 11.16f
Primary power requirements, fire alarm control panels,
  5.26, 5.41
Primary resistor or reactance starting, motor starting
  configurations, 10.4
Primary windings, transformers, 6.2, 6.2f, 6.13–6.14
Problem solving, leadership skills, 13.31–13.33, 13.34,
  13.35–13.37
Process fluid freezing prevention, heat tracing and freeze
  protection, 9.2, 9.14
Production versus productivity, 13.80
Professional ethics, 13.19–13.20
Programmable overload relays
  solid-state protective relays, 7.8–7.9, 7.9f
  trip thresholds and functions, 7.9
Programmable parameters, adjustable frequency drives,
  7.28–7.29

Programmable starting characteristics, motor starting
  configurations, 7.22–7.23, 7.23f
Programmable thermostats, 8.14–8.15, 8.15f
Project control
  components of planning, 13.54–13.55
  cost awareness and control, 13.73–13.74
  estimations and, 13.62, 13.63f, 13.64f, 13.65, 13.65f,
    13.66f, 13.67
  overview, 13.51–13.52, 13.53f
  participant exercises, 13.55, 13.67, 13.79
  planning process, 13.57–13.58, 13.59f
  planning resources, 13.59–13.60
  planning techniques, 13.60–13.61, 13.60f, 13.61f
  production and productivity, 13.80
  project delivery systems, 13.54, 13.54f
  resource control, 13.76–13.79
  scheduling, 13.68, 13.69f, 13.70–13.71, 13.70f
  stages of planning, 13.56
Projected beam smoke detectors
  as automatic detectors, 5.9
  defined, 5.72
  installation of, 5.50
  obscuration principle and, 5.12, 5.14, 5.14f
Project managers, 13.15
Propane torches and heat guns, heat-shrink tubing, 11.18
Property protection fire alarm systems, 5.2
Proportional ambient-sensing controls, heat tracing system
  controls, 9.10
Proprietary monitoring, 5.31
Prosceniums
  defined, 12.50
  sidelights, 12.9
Protected premises, 5.33
  defined, 5.72
Protected premises (local) fire alarm systems, defined, 5.72
Protons, atomic structure, 4.2
Psychiatric security rooms, 2.13
P-type material
  anodes as, 4.5, 4.5f
  defined, 4.20
  diodes and, 4.5
  semiconductors and, 4.4
  in transistors, 4.12–4.13, 4.12f, 4.13f
Publicity, safety promotion, 13.50
Public Switched Telephone Networks, 5.33
  defined, 5.72
Pull cord stations, 2.21
Pulsewidth-modulated drives, 7.25–7.26, 7.26f
Pump house
  floor plans, 1.13, 1.14f
  sample service load calculations, 1.14–1.15
Purchasing department, 13.11

Q
Quantity takeoff sheets, 13.62, 13.63f
Quick inline splicing kits
  grounding, 11.25
  installing cold-shrink insulators, 11.25, 11.25f
  installing shield continuity assembly, 11.24–11.25,
    11.24f, 11.25f
  splicing installations, 11.23–11.24, 11.23f, 11.24f
  surface preparation and, 11.22–11.23, 11.22f
Quick metal detectors, fixed-temperature heat detectors,
  5.9, 5.10f

R
Raceway fill, feeder conductor size adjustment, 1.2, 1.3, 1.4
Radiant floor heating systems, 9.30–9.31, 9.31f

Radio frequency interference (RFI), ionization smoke detectors, 5.12

Rate compensation detectors, 5.15, 5.16–5.17, 5.17*f*
 defined, 5.72

Rate-of-rise heat detectors, 5.8*f*, 5.9, 5.10, 5.11*f*
 defined, 5.72

Reactance
 defined, 6.31
 transformers and, 6.10

Reactor-type dimmers, 12.7
 defined, 12.50

Receivers, verbal communication, 13.20, 13.21–13.22, 13.21*f*

Receiving motors, 10.22, 10.23, 10.23*f*

Receptacle outlets at HVAC equipment, *NEC* requirements, 8.36, 8.41*f*

Receptacles
 critical care areas, 2.13–2.15, 2.14*f*
 general care areas and, 2.12–2.13, 2.13*f*
 general-purpose receptacles, 1.31
 GFCI receptacles, 12.25, 12.25*f*, 12.28, 12.28*f*, 12.42, 12.45
 hospital-grade, 2.11–2.12, 2.12*f*
 laundry receptacles, 1.17
 marinas and boatyards, 12.25, 12.25*f*
 permanently installed pools, 12.33–12.34, 12.33*f*, 12.34*f*
 stage receptacles, 12.9
 temporary installations, 12.27

Reciprocals, conductor adjustments, 1.4

Recognition
 as motivation, 13.26
 safety promotion and, 13.50

Record keeping systems, safety, 13.45–13.46

Rectifiers
 battery-charging rectifiers, 6.12
 bridge rectifiers, 4.6, 4.7*f*, 4.8
 defined, 4.20
 diodes and, 4.5
 electronic theory and, 4.6, 4.7*f*
 full-wave rectifiers, 4.6, 4.7*f*
 half-wave rectifiers, 4.6, 4.7*f*
 silicon-controlled rectifiers, 4.12, 4.14–4.15, 4.16*f*, 4.20
 three-phase rectifiers, 4.8, 4.8*f*

Reduced-voltage starting motor controls
 autotransformers, 7.16–7.18, 7.16*f*, 7.17*f*, 7.24
 part-winding, 7.18, 7.18*f*, 7.24
 selecting controllers, 7.23, 7.24–7.25
 solid-state, 7.20–7.23, 7.20*f*, 7.22*f*, 7.23*f*, 7.25
 wye-delta, 7.18, 7.19*f*, 7.20, 7.24

Reel holders, 9.19

Refrigerant releases, 8.5

Refrigerants, 8.5, 8.6, 8.8
 defined, 8.53

Refrigeration cycle
 air conditioning and, 8.5, 8.6*f*, 8.8–8.9
 defined, 8.53

Reinsulating, splicing, 11.9–11.10

Rejacketing
 extruded semi-conductors and, 11.8
 splicing, 11.12

Relay coil resistance tests, 7.50, 7.52, 7.52*f*

Relay contacts, 5.26

Remote control signaling circuits, *NEC* requirements, 5.37

Remote duct smoke indicators, 5.58, 5.58*f*

Remote supervising station fire alarm systems, 5.33
 defined, 5.72

Rescues. *See* First aid

Resets, 5.9
 defined, 5.72

Reshielding, splicing, 11.11–11.12

Residential and Light Commercial Telecommunications Wiring Standard, ANSI/EIA/TIA 570, 5.34

Resistance temperature detectors (RTD)
 motor maintenance and, 10.9
 temperature monitoring, 9.12

Resistance-type dimmers, 12.7, 12.8*f*
 defined, 12.50

Resource control, 13.76–13.79

Respect, leadership styles, 13.19

Responsibility versus authority, 13.12

Retail stores
 commercial occupancy load calculations, 1.26, 1.27–1.28, 1.27*f*
 floor plans, 1.27
 NEC requirements, 1.26, 1.28

Retrofitting adjustable frequency drives, 7.33

Reverse bias
 defined, 4.20
 diodes and, 4.5, 4.5*f*
 diode testing and, 4.9, 4.10*f*
 peak inverse voltage, 4.8

Reverse-cycle air conditioners. *See* Heat pumps

Reverses, troubleshooting motors, 10.24

Rewards, as motivation, 13.27

Rheostats, 12.6
 defined, 12.50

Ribbon shielding cable splices, 11.3, 11.4*f*

Ribbon tape shielding
 defined, 11.49
 splicing, 11.3, 11.4*f*

Ride-through power systems, 2.10

RJ31-X modular jack, digital communicators, 5.32, 5.32*f*, 5.33

Road show connection panels, 12.11

Roof de-icing. *See* De-icing systems

Room air conditioners
 circuits, 8.36, 8.39*f*, 8.41, 8.42*f*
 control circuits, 8.30, 8.31*f*
 NEC requirements, 8.36, 8.37, 8.39*f*, 8.41, 8.42*f*

Room thermostats, 8.10, 8.10*f*

Root-mean square (rms) values, harmonics and, 6.22, 6.24

Rotors
 AC generators, 3.2, 3.4, 3.4*f*
 squirrel cage rotors, 10.2, 10.4, 10.4*f*, 10.22

Runout, 10.13, 10.13*f*, 10.14, 10.14*f*

S

Safety
 accident costs, 13.42–13.43
 anesthetic gases and, 2.19
 auto-mechanical fire detection equipment and, 5.20
 capacitors and, 8.45
 control circuits and, 7.44
 current draw adjustment and, 8.17
 current transformers and, 6.15
 dielectric absorption effect and, 11.44
 employer and employee safety responsibilities, 13.44–13.46
 fall protection, 9.16, 9.23
 first aid, 13.46–13.47
 gas torches, 11.19
 go/no-go testing and, 11.44
 guards and, 10.12
 health programs and, 13.44
 hermetic and semi-hermetic (serviceable) compressors, 8.45
 high potential (hi-pot) tests and, 11.32, 11.33, 11.36, 11.37, 11.42, 11.44
 information and training, 13.45, 13.47

insulation testing and, 10.16
leakage current, 7.37, 7.38
lockout relays and, 8.21
management programs, 13.44
material safety data sheets, 13.46
medium-voltage termination and splices, 11.3
mercury and, 8.11
motor control maintenance and, 7.38
motor maintenance and, 10.8
motor shafts and, 10.22
overhead piping and, 9.16
overview, 13.41–13.42
penalties for violations, 13.44
policies and procedures manuals, 13.44
promotion of, 13.48–13.50
record keeping systems, 13.45–13.46
safety meetings, 13.46, 13.49
safety regulations, 13.43–13.44
shielding braids and, 11.14
solid-state devices and, 7.37
storage battery safety, 3.15
substance abuse, 13.47–13.48
supervisor involvement and, 13.44, 13.45, 13.46–13.48
TENV motors and, 10.7
transfer switches and, 3.7
troubleshooting and, 7.44, 7.50, 8.44, 10.23
troubleshooting motors, 10.23
wet motors and, 10.22
workplace inspections, 13.43, 13.46
workplace safety policy form, 13.13, 13.13f
X-rays and, 2.19
Sail switches, 8.25, 8.25f
Scheduling
bar charts and, 13.68, 13.69f
network scheduling, 13.70, 13.70f
process of, 13.68
short-interval production scheduling, 13.70–13.71
updating, 13.71
Schematic symbols
diacs, 4.16f
light-emitting diodes, 4.11, 4.11f
silicon-controlled rectifiers (SCR), 4.16f
timed contacts, 7.10, 7.10f
transistors, 4.12, 4.12f, 4.13, 4.13f, 4.14f
triacs, 4.16f
voltage (potential) transformers, 6.4
Schools, service and feeder load calculations, 1.33–1.34
Sealants, heat-shrink splicing kits, 11.20, 11.20f
Sealed valve-regulated lead-acid (VLRA) batteries, 3.13, 3.18
Secondary conductors, 1.8, 1.9, 1.10
Secondary power requirements, fire protective signaling
systems, 5.41
Secondary windings, transformers, 6.2, 6.2f, 6.13–6.14
Selective guard circuits, high potential (hi-pot) tests,
11.36–11.39, 11.37f, 11.39f, 11.40f
Selective guard service connections, high potential (hi-pot)
tests, 11.40–11.42, 11.41f
Self-exciter generators, 3.4, 3.4f
Self-regulating cables
defined, 9.36
heat tracing system cables, 9.3–9.5, 9.4f, 9.5f, 9.7, 9.7f
Self-regulating controls, heat tracing system control, 9.9
Semi-conductive layers, insulation shield systems,
11.6–11.7, 11.6f
Semiconductor line-type heat detectors, 5.15, 5.17, 5.17f
Semiconductors
defined, 4.20
electronic theory and, 4.2, 4.3–4.4, 4.4f, 4.5f

integrated circuits and, 4.3
N-type material and, 4.4
P-type material and, 4.4
silicon and, 4.4, 4.5f
solid-state electronics and, 4.2, 4.12
transistors, 4.12, 4.12f
Senders, verbal communication, 13.20–13.21, 13.21f
Sensor-controlled timer relays, 7.13
Separately derived systems
defined, 1.41
generators, 12.17
load calculations and, 1.30
power sources, 12.15, 12.15f
Series resistance heating cables, 9.5, 9.6–9.7, 9.6f
defined, 9.36
Series transformers, 6.11–6.12
Service, 1.2. See also Service load calculations
defined, 1.41
Service conductors
defined, 1.41
demand factors and, 1.11
load calculations, 1.2
service laterals and, 1.11
Service drops
defined, 1.41
service conductors and, 1.11
Service entrances, 1.11
Service equipment
defined, 1.41
selecting, 1.14f, 1.22f, 1.24, 1.25, 1.28
Service laterals
defined, 1.41
service conductors and, 1.11
Service load calculations
commercial occupancy, 1.24–1.31, 1.27f, 1.29f
farms, 1.34–1.36, 1.35f
hotels and motels, 1.33
lighting loads, 1.25–1.26, 1.27
minimum service ratings, 1.13–1.18, 1.14f, 1.15f,
1.16f, 1.17f
minimum size service, 1.13–1.24, 1.14f, 1.15f, 1.16f,
1.17f, 1.22f
motors and motor circuits, 1.36
multi-family dwellings, 1.16–1.18, 1.21–1.24, 1.22f
restaurants, 1.31, 1.32f, 1.33
schools, 1.33–1.34
shore power for marinas and boatyards, 1.34, 12.22–12.23
single-family dwellings, 1.16–1.18
sizing neutral conductors, 1.18–1.21
standard calculations, 1.18–1.19, 1.22f, 1.25
Service points
defined, 1.41
service conductors and, 1.11
Sexual harassment, 13.8
Shaft rotation logs, 10.22, 10.23, 10.23f
Shed ceilings, spot detector installation, 5.50–5.51, 5.51f
Shield continuity assemblies, 11.24–11.25, 11.24f, 11.25f
Shielded cable, removing shields, 11.28, 11.28f
Shielding braids
defined, 11.49
safety and, 11.14
Shields, 11.4
defined, 11.49
Shims, motor mounting, 10.26
Shore power circuits, 1.34
Shore power for marinas and boatyards, service load
calculations, 1.34, 12.22–12.23
Shore power receptacles, 12.25

Short-interval production scheduling (SIPS), 13.70–13.71
Shorts, troubleshooting motors, 10.24
Show window loads, 1.26, 1.28
Shunt trips, defined, 12.50
Signaling line circuits (SLC)
    addressable systems, 5.6, 5.7
    defined, 5.73
Signaling systems, health care facilities, 2.21–2.22, 2.21f
Signals. *See also* Fire alarm signals
    defined, 5.72
    fire alarm systems and, 5.4
Sign and outline lighting loads, 1.26
Sign lighting transformers, 6.12
Signs in fountains, 12.43
Silicon
    in diodes, 4.8
    semiconductors and, 4.4, 4.5f
Silicon-controlled rectifiers (SCR)
    defined, 4.20
    light-activated, 4.15
    solid-state relays and, 4.12
    as thyristors, 4.14–4.15
Silicon power rectifier diodes, 4.9
Silver, as conductor, 4.3
Single-action pull stations, manual (pull station) fire
    detection devices, 5.19, 5.19f
Single-beam detector installations, 5.49, 5.49f
Single-conductor cable configurations, 11.4
Single-conversion UPS systems, static uninterruptible
    power supplies, 3.19, 3.19f
Single-family dwellings
    with electric heat, 1.19–1.20
    floor plans, 1.16, 1.16f
    minimum size service optional calculations, 1.16–1.18,
        1.20–1.21
    minimum size service standard calculations, 1.18–1.19
    sample service load calculations, 1.16–1.18
Single-phase operation, induction motors, 10.2, 10.3
Single-phase transformers, 6.2, 6.16
Sirens, 5.28
Site-isolating devices, 12.19–12.20
    defined, 12.50
Situational ethics, 13.19–13.20
Sizing adjustable frequency drives (AFD), 7.31
Sizing and selection considerations, engine-driven
    generator sets, 3.4–3.6
Sizing buck-and-boost transformers, 6.17–6.19, 6.18f, 6.19f
Sizing neutral conductors, 1.18–1.21
Skin-effect electric heat tracing systems, 9.9
Sloped ceilings, spot detector installation, 5.50–5.52, 5.51f
Small appliance branch circuits, *NEC* requirements, 1.16,
    1.17, 1.19, 1.24
Smoke chamber definition, 5.43–5.44, 5.44f, 5.45f
Smoke detectors. *See also* Photoelectric smoke detectors
    addressable and analog addressable, 5.10, 5.12
    barrier spacing, 5.43–5.44, 5.44f
    ceiling height and, 5.48, 5.51, 5.52, 5.54
    cloud chamber smoke detectors, 5.15, 5.16f
    defined, 5.73
    duct detectors, 5.14–5.15, 5.15f, 5.16, 5.55–5.58, 5.57f, 5.58f
    fire alarm control panels and, 5.6
    fire alarm initiating devices, 5.7–5.8, 5.8f, 5.9
    hardwired fire alarm systems, 5.4
    installation precautions for, 5.47
    ionization detectors, 5.8f, 5.11–5.12, 5.13, 5.13f
    spot detector installations on flat smooth ceilings,
        5.48–5.49, 5.49f
    spot detector installations on irregular ceilings, 5.50–5.52,
        5.51f
    types of, 5.11–5.12, 5.12f
    UL standards, 5.55
    wireless smoke detection systems, 5.42
Smoke doors, 5.44
Smoke spread phenomena, 5.44, 5.45f
Smoke stratification phenomena, 5.44–5.46, 5.45f, 5.46f
Smoke ventilators, 12.9, 12.11
Snow-melting and anti-icing systems. *See also* De-icing
        systems
    automatic snow controllers and, 9.26, 9.26f
    component selection and installation, 9.27–9.29, 9.28f
    concrete installation, 9.28–9.29, 9.28f
    elements of, 9.25, 9.26f
    installation guidelines, 9.28–9.29, 9.29f
    NEC requirements, 9.28, 9.29
    wiring for, 9.25, 9.27f
Snow-melting mats, 9.25
Snow sensors, 9.25, 9.26f
Snubber networks, overvoltage and overcurrent
        protection, 7.7
Society for Human Resources Management (SHRM)
    supervising employee training and, 13.5
    team building and, 13.29
    worker commitment and, 13.2
Soft foot
    defined, 10.33
    motors and, 10.2, 10.14
Soft-start controllers, 7.20, 7.20f
Solarvoltaic systems, 4.17
Solid joist ceilings, spot detector installation, 5.52
Solid-state devices
    ambient temperature and, 7.38
    damage from reverse polarity or incorrect phase sequence
        and, 7.37
    defined, 4.20
    electrical noise and, 7.38
    electronic theory and, 4.2
    safety and, 7.37
    semiconductors and, 4.2, 4.12
Solid-state electronics, semiconductors, 4.2, 4.12
Solid-state overload relays (SSOLR), 7.7
    defined, 7.57
Solid-state phase-control dimmers
    defined, 12.50
    stage switchboard feeders and, 12.9
Solid-state programmable dimmers, 12.7, 12.8f
Solid-state protective relays
    motor controls, 7.7–7.9, 7.8f, 7.9f
    non-programmable overload relays, 7.7–7.8, 7.8f
    programmable overload relays, 7.8–7.9, 7.9f
Solid-state reduced-voltage starters, 7.20–7.23, 7.20f, 7.22f,
    7.23f, 7.25
Solid-state relays (SSRs)
    analog switching, 7.3
    defined, 7.57
    electromechanical relays compared to, 7.2, 7.3–7.5
    heat sinks and, 7.2f, 7.6–7.7
    input control circuits and, 7.2–7.3, 7.3f
    instant on, 7.3
    motor controls, 7.2–7.7, 7.2f, 7.3f, 7.5f, 7.6f
    multiple outputs and, 7.6, 7.6f
    operation, 7.2–7.3, 7.3f
    optically isolated solid-state relays (SSRs), 7.3, 7.3f
    output circuits, 7.2–7.3, 7.3f
    overvoltage and overcurrent protection and, 7.7

peak switching, 7.3
power transistors and, 7.3
precautions for working with, 7.37–7.38, 7.37*f*
reduced-voltage starters, 7.20–7.23, 7.20*f*, 7.22*f*, 7.23*f*, 7.25
silicon-controlled rectifiers and, 4.12
switching modes, 7.3
temperature considerations and, 7.2*f*, 7.6–7.7
thyristors and, 7.3, 7.4
timing relays, 7.12–7.13, 7.13*f*, 7.14*f*, 7.15
two-wire and three-wire control, 7.5–7.6, 7.5*f*
zero switching, 7.3
Solid-state sine-wave dimmers
defined, 12.50
stage switchboard feeders and, 12.9
Solid-state thermostats, static electricity, 8.16
Solid wire, 11.5, 11.5*f*
Solvents
material safety data sheet and, 11.9, 11.18
surface preparation and, 11.9
Sound loss, 5.31
Spacing
defined, 5.73
heat sensors and, 5.11
Span of control, 13.11
Spas and hot tubs, 12.40–12.41, 12.40*f*, 12.41*f*
Special occupancies
agricultural buildings, 12.18–12.20, 12.19*f*, 12.20*f*
assembly occupancies, 12.3–12.5, 12.3*f*, 12.4*f*, 12.5*f*
bodies of water (natural and manmade), 12.44–12.46, 12.44*f*, 12.45*f*
carnivals, circuses, fairs, and similar events, 12.14–12.17, 12.14*f*, 12.15*f*, 12.16*f*, 12.17*f*
introduction, 12.2
marinas and boatyards, 12.21–12.26, 12.21*f*, 12.22*f*, 12.23*f*, 12.24*f*, 12.26*f*
swimming pools, fountains, hot tubs, and similar installations, 12.30–12.44, 12.30*f*, 12.31*f*, 12.32*f*, 12.33*f*, 12.34*f*, 12.35*f*, 12.36*f*, 12.37*f*, 12.38*f*, 12.39*f*, 12.40*f*, 12.41*f*, 12.42*f*
temporary installations, 12.26–12.29, 12.27*f*, 12.28*f*
theaters and similar locations, 12.5–12.14, 12.5*f*, 12.6*f*, 12.7*f*, 12.8*f*, 12.9*f*, 12.10*f*, 12.11*f*, 12.12*f*, 12.13*f*, 12.14*f*
wired partitions, 12.29, 12.29*f*
Specialty transformers, 6.7–6.13, 6.7*f*, 6.8*f*, 6.10*f*, 6.11*f*
Speed control versus reduced-voltage starting, 7.16
Spiral factors, 9.17
pitch and, 9.17
Splices, 11.2, 11.7*f*
defined, 11.49
Splicing
applying outer sheath, 11.16, 11.16*f*
cable preparation kits, 11.9
connecting conductors, 11.14, 11.15*f*
heat-shrink splicing kits, 11.17–11.20, 11.17*f*, 11.19*f*, 11.20*f*, 11.21, 11.21*f*
inline tape splices, 11.12–11.16, 11.12*f*, 11.13*f*, 11.14*f*, 11.15*f*, 11.16*f*
joining conductors with connectors, 11.9
linerless splicing tapes, 11.10
manufacturer's training programs and, 11.15
paper-insulated cable splices, 11.25, 11.26
penciling and, 11.2, 11.2*f*
primary insulation application, 11.14–11.16, 11.15*f*, 11.16*f*
quick inline splicing kits, 11.22–11.25, 11.23*f*, 11.24*f*, 11.25*f*
reasons for, 11.7
reinsulating, 11.9–11.10
rejacketing, 11.12

reshielding, 11.11–11.12
ribbon shielding cable splices, 11.3, 11.4*f*
selecting a splice method, 11.13
straight splices, 11.2–11.3, 11.2*f*
stress relief material (SRM), 11.18, 11.19*f*
surface preparation as step in, 11.7, 11.8, 11.9
tape splice kits, 11.11
tee tape splices, 11.16, 11.17*f*
Spot detectors
as automatic detectors, 5.9
defined, 5.73
installations on flat smooth ceilings, 5.48–5.49, 5.49*f*
installations on irregular ceilings, 5.50–5.52, 5.51*f*
light-scattering detectors as, 5.12, 5.13*f*
Spot networks, 2.7
Sprinkler systems, 5.60–5.63
Squirrel cage motors
motor failures, 10.2
rotors, 10.2, 10.4, 10.4*f*, 10.22
starting configurations, 10.4
winding failures, 10.4, 10.5*f*–10.6*f*, 10.7, 10.24
Stack effect principle, smoke stratification phenomena, 5.46, 5.46*f*
Stage lighting switchboards, 12.7
Stage receptacles, 12.9
Stage switchboard feeders, 12.8–12.9
Stair enclosure doors, door locking arrangements for fire egress, 5.60
Standard UPS batteries, 3.14
Standby power systems. *See also* Emergency and standby power system components
NEC requirements, 3.2, 3.21–3.22, 3.23*f*
optional, 3.2
Starter batteries, EES diesel engine generators, 2.8
Starting configurations of squirrel cage motors, voltage drop, 10.4
Starting systems, engine-driven generator sets, 3.5
Startup temperatures, 9.14
defined, 9.36
Static electricity
electronic devices and, 7.37, 7.37*f*, 8.27
microprocessors and, 7.37, 7.37*f*
solid-state thermostats and, 8.16
Static uninterruptible power supplies
configurations and, 3.17–3.18
double-conversion UPS systems, 3.17, 3.18, 3.19*f*
single-conversion UPS systems, 3.19, 3.19*f*
stored-energy flywheel UPS systems, 3.20
Stators
AC generators, 3.2, 3.4, 3.4*f*
winding failures, 10.4, 10.5*f*–10.6*f*, 10.7, 10.9
Step-voltage regulators, 6.12, 6.13
Storable pools, 12.39–12.40, 12.40*f*
Storage, materials control, 13.77
Storage batteries
battery and battery charger operation, 3.16–3.17, 3.16*f*
battery maintenance, 3.13–3.16
electrolytes and, 3.11–3.12
emergency and standby power system components and, 3.11–3.17, 3.11*f*, 3.12*f*, 3.16*f*
lead-acid batteries, 3.11–3.12, 3.11*f*
legally required standby systems, 3.11, 3.11*f*, 3.21
NEC requirements, 3.11, 3.11*f*, 3.21
nickel cadmium batteries, 3.12, 3.12*f*
sealed VRLA battery UPS systems, 3.13, 3.18
standard UPS batteries, 3.14
Stored-energy flywheel UPS systems, 3.20

Storing motors, 10.22, 10.23, 10.23*f*
Straight-line beam detectors, 5.49, 5.49*f*
Straight splices, 11.2–11.3, 11.2*f*
Strand shielding
    defined, 11.49
    medium-voltage power cables, 11.4, 11.5, 11.5*f*
Stratification
    defined, 5.73
    smoke stratification phenomena, 5.44–5.46, 5.45*f*, 5.46*f*
Street lighting circuits, 6.9
Stress, 11.6
    defined, 11.49
Stress cones, 11.26
Stress control
    capacitive stress control, 11.28, 11.30, 11.30*f*
    electric flux lines and, 11.26, 11.27*f*
    equipotential lines and, 11.26, 11.27*f*, 11.28
    geometric stress control, 11.28, 11.29*f*
    terminations and, 11.26, 11.27, 11.27*f*, 11.28, 11.28*f*,
        11.29–11.31, 11.29*f*, 11.30*f*, 11.31*f*
Stress relief material (SRM), splicing, 11.18, 11.19*f*
Strip lights, 12.9, 12.9*f*
Strobe devices, 5.27, 5.27*f*
Sub-bases
    defined, 8.53
    of thermostats, 8.16
Subcooling, 8.9
    defined, 8.53
Submersible equipment and enclosures, locating in or near
        bodies of water, 12.45
Submersible pumps, fountains, 12.42, 12.42*f*
Submeters, boat slips, 12.22
Substance abuse, safety, 13.47–13.48
Suction lines, air conditioning, 8.6, 8.6*f*
Superheat, 8.8
    defined, 8.53
Superintendents, 13.15
Supervised automatic sprinklers, 5.60–5.61
Supervising stations, 5.33
Supervision
    cost control and, 13.74
    fire suppression systems, 5.60–5.63
    leadership skills and, 13.15
    safety programs and, 13.44, 13.45, 13.46–13.48
    technology and, 13.6
Supervisor unfairness, turnover and, 13.36
Supervisory signals, 5.22
    defined, 5.73
Supervisory signals versus trouble signals, 5.8, 5.62
Supervisory switches, 5.62
Supervisory training, 13.5–13.6
Supplemental heaters, heat pumps, 8.9, 8.34
Surface preparation
    abrasives and, 11.9
    heat-shrink splicing kits and, 11.17–11.18
    preparing cable for splicing, 11.12–11.13, 11.13*f*
    quick inline splicing kits and, 11.22–11.23, 11.22*f*
    solvents and, 11.9
    as step in splicing, 11.7, 11.8, 11.9
Swimming pools, fountains, hot tubs, and similar
        installations
    fountains, 12.41–12.43, 12.41*f*, 12.42*f*
    hydromassage bathtubs, 12.44
    spas and hot tubs, 12.40–12.41, 12.40*f*, 12.41*f*
    storable pools, 12.39–12.40, 12.40*f*
    therapeutic pools and tubs, 12.43–12.44
    wiring requirements, 12.30–12.33, 12.30*f*, 12.31*f*, 12.32*f*
Switchboard control types, 12.8, 12.8*f*

Switchboards, 1.8. *See also* On-stage portable switchboards
    defined, 1.41
Switching modes, output circuits, 7.3
Synchronous generators, 3.2
    defined, 3.30
System units, 5.26
    defined, 5.73

T
Tamper-resistant receptacles, 2.12
Tamper switches, 5.62
Tap conductors
    defined, 1.41
    load calculation procedures, 1.8–1.10
    NEC requirements, 1.8–1.10
Tape splice kits, 11.11
Taping, reinsulating, 11.9
Tap rules, 1.8–1.10
Tasks, planning, 13.57–13.58, 13.59*f*
Team building, leadership skills, 13.29
TEAO motors, 10.7
TEBC motors, 10.7
Technology, supervision, 13.6
Tee tape splices, splicing, 11.16, 11.17*f*
TEFC motors, 10.7, 10.7*f*
Telecommunications, harmonics, 6.22
Telephone/computer control, fire alarm control panels, 5.24
Telephone lines, digital communicators, 5.32, 5.33
Temperature coefficient, insulation testing, 10.18–10.19,
        10.19*f*, 10.20*f*, 10.21*f*
Temperature considerations, solid-state relays, 7.2*f*, 7.6–7.7
Temperature control
    heat tracing and freeze protection and, 9.9
    heat tracing system controls, 9.9
    motor maintenance and, 10.9
Temperature limitations, for terminating conductors, 1.7
Temperature maintenance systems, for domestic hot-water,
        9.29–9.30, 9.29*f*
Temperature monitoring
    heat tracing system monitoring, 9.11, 9.12, 9.13*f*
    resistance temperature detectors and, 9.12
Temporal 3, 5.29
Temporary installations
    assured equipment grounding program, 12.28–12.29
    feeder and branch circuit conductors, 12.26–12.27
    ground fault protection, 12.28, 12.28*f*
    receptacles, 12.27
    temporary lighting, 12.28, 12.28*f*
    wiring and equipment greater than 600V, 12.28
Ten foot tap rule, 1.8
Ten second power restoration requirement, 2.4, 2.6
Tents, 12.14, 12.15*f*, 12.16–12.17
TENV motors, 10.7
Terminal wiring methods, fire protective signaling systems,
        5.37, 5.39*f*
Termination and splice kits
    contents of, 11.16–11.17
    heat-shrink splicing kits, 11.17–11.20, 11.17*f*, 11.19*f*, 11.20*f*,
        11.21, 11.21*f*
    lead shield repair kits, 11.26
    quick inline splicing kits, 11.22–11.25, 11.23*f*, 11.24*f*, 11.25*f*
Terminations
    cable breakouts and, 11.29
    classifying, 11.26
    external leakage insulation, 11.30–11.31, 11.31*f*
    flashovers and, 11.30
    grounding shield ends, 11.31
    as insulators, 11.31, 11.31*f*